T0211184

This book provides the fundamental statistical theory of atomic transport in crystalline solids; that is the means by which processes occurring at the atomic level are related to macroscopic transport coefficients and other observable quantities.

The cornerstones of this treatment are (i) the physical concepts of lattice defects, (ii) the phenomenological description provided by non-equilibrium thermodynamics and (iii) the various methods of statistical mechanics used to link these (kinetic theory, random-walk theory, linear response theory etc.). The book brings these component parts together into a unified and coherent whole and shows how the results relate to a variety of experimental measurements. It thus provides the theoretical apparatus necessary for the interpretation of experimental results and for the computer modelling of atomic transport in solid systems, as well as new insights into the theory itself.

The book is primarily concerned with transport in the body of crystal lattices and not with transport on surfaces, within grain boundaries or along dislocations, although much of the theory here presented can be applied to these low-dimensional structures when they are atomically well ordered and regular.

*Atomic Transport in Solids* will be of interest to research workers and graduate students in metallurgy, materials science, solid state physics and chemistry.

# Atomic transport in solids

# *Atomic transport in solids*

A. R. ALLNATT
Department of Chemistry,
University of Western Ontario

and

A. B. LIDIARD
AEA Technology, Harwell and Department of Theoretical Chemistry,
University of Oxford

CAMBRIDGE
UNIVERSITY PRESS

PUBLISHED BY THE PRESS SYNDICATE OF THE UNIVERSITY OF CAMBRIDGE
The Pitt Building, Trumpington Street, Cambridge, United Kingdom

CAMBRIDGE UNIVERSITY PRESS
The Edinburgh Building, Cambridge CB2 2RU, UK
40 West 20th Street, New York NY 10011–4211, USA
477 Williamstown Road, Port Melbourne, VIC 3207, Australia
Ruiz de Alarcón 13, 28014 Madrid, Spain
Dock House, The Waterfront, Cape Town 8001, South Africa

http://www.cambridge.org

© Cambridge University Press 1993

This book is in copyright. Subject to statutory exception
and to the provisions of relevant collective licensing agreements,
no reproduction of any part may take place without
the written permission of Cambridge University Press.

First published 1993
First paperback edition 2003

*A catalogue record for this book is available from the British Library*

*Library of Congress cataloguing in publication data*

Allnatt, A. R.
Atomic transport in solids / A. R. Allnatt and A. B. Lidiard.
p. cm.
Includes bibliographical references and index.
ISBN 0 521 37514 2 hardback
1. Energy-band theory of solids. 2. Transport theory. 3. Solid state physics. 4. Crystal lattices. I. Lidiard, A. B. II. Title.
QC176.8.E4A37 1993
530.4′16—dc20 92-45679 CIP

ISBN 0 521 37514 2 hardback
ISBN 0 521 54342 8 paperback

**To Betty and to Ann**

# Contents

# *Preface*

In any area of physical science there are stages in its historical development at which it becomes possible and desirable to pull its theory together into a more coherent and unified whole. Grand examples from physics spring immediately to mind – Maxwell's electro-magnetic theory, Dirac's synthesis of wave-mechanics and matrix mechanics, the Salam–Weinberg electro-weak theory. Yet this process of unification takes place on every scale. One example of immediate concern to us in this book here is afforded by the emergence of the thermodynamics of irreversible processes, where the coming together of separate strands of theory is made plainly visible in the short book by Denbigh (1951). In connection with imperfections in crystal lattices the Pocono Manor Symposium held in 1950 (Shockley *et al.* 1952) might be said to mark another such stage of synthesis. At any rate a great growth in the understanding of the properties of crystal imperfections occurred in the years following and this carried along a corresponding achievement in understanding atomic transport and diffusion phenomena in atomic terms. Around 1980, in the course of a small Workshop held at the International Centre for Theoretical Physics, Trieste we concluded from several observations that such a stage of synthesis could soon be reached in the study of atomic transport processes in solids. First of all, despite the great growth in detailed knowledge in the area, the theory, or rather theories, which were used to relate point defects to measurable quantities (e.g. transport coefficients of various sorts) were recognizably still largely in the moulds established considerably earlier in the 1950s and 1960s, notwithstanding the advances in purely numerical techniques such as molecular dynamics and Monte Carlo simulations. There were clear limits to the quantities which could be confidently expressed in terms of the properties of point defects. Thus, despite the existence of useful and accurate theories of tracer diffusion coefficients in dilute solid solutions, other coefficients, such as intrinsic (chemical) diffusion coefficients, cross-coefficients between the flows of different chemical species, ionic mobilities, etc. were less well understood, especially in concentrated solid solutions. At the same time the Kubo–Mori theory of linear transport processes had become firmly established in other areas of the theory of condensed matter; but, although some first steps had been taken to translate this scheme to the field of atomic transport (i.e. to a stochastic, hopping type of process), it had not been shown that this provided a practical way to evaluate the transport coefficients of particular defect models. Indeed, some formulations of this Kubo–Mori theory made the prospects look decidedly unpromising!

This being the situation at that time we therefore directed part of our research towards the objective of providing a more fundamentally based and unified theory of the subject in terms of the concepts of point defects. A review which we published in 1987 (Allnatt and Lidiard, 1987b) sketched out our overall view of the theory at that later time. We felt that this indicated that a more extensive and self-contained work was necessary; the present book is the consequence.

The emphasis of the work is thus on the fundamental theory of the topic. Our aim is to establish this in a more unified and powerful form than is currently available elsewhere. The corner-stones of our treatment are threefold: (i) the physical concepts of point defects, (ii) the phenomenology of non-equilibrium thermodynamics and (iii) the various methods of statistical mechanics (statistical thermodynamics, linear response theory, kinetic theory, random-walk theory, etc.), which enable us to express measurable quantitites in terms of the properties of point defects. Inevitably there are boundaries to our narrative. Since the focus of attention is the fundamentals of atomic migration in 'bulk' crystal we are not concerned, except in passing, with surface diffusion nor with diffusion in grain boundaries or other components of the 'microstructure'. From the point of view of experimental research the distinction between processes taking place in the bulk and at interfaces often presents little difficulty. However, to the computer modeller of complex systems in which a number of distinct processes are going on, we admit that our work can help with only one part of the task: our contribution is to point to the correct phenomenology to use for those processes involving migration through the body of the crystal lattice and to show how the corresponding transport coefficients depend on circumstances (structure, thermodynamic variables, forces acting, etc.). We should have liked to do more, but space and time did not allow.

Let us now turn to the structure and content of the book in more detail. Firstly the text is inevitably fairly mathematical. Nevertheless, it has been our intention that it should be of interest to all concerned with this subject, for we believe that there are scientific gains to be made by viewing the subject in the unified way we present here. We therefore offer some suggestions to readers about how they may wish to approach the book. These must obviously be different for different classes of reader.

Chapters 1–5 are essentially introductory. Chapter 1 describes the quantities of concern – diffusion coefficients, ionic mobilities, loss factors, relaxation times, dynamical structure factors, etc. Chapter 2 summarizes the physical basis of the theories of atomic migration; this may be studied more casually by those who have a basic knowledge of structural imperfections in crystals.

The rest of this book describes the theory of the quantities introduced in Chapter 1 as it is based upon the ideas and models of point defects described in Chapter 2. Chapter 3 describes several particular calculations of the thermodynamic equilibrium concentrations of point defects as well as certain more general results used later. The particular calculations should present no difficulty to those familiar with the basic ideas of statistical thermodynamics. This chapter also touches on the mobility of point defects within the same framework. The more experienced reader may need to concentrate only on the later more general parts of this chapter.

Chapters 4 and 5 provide discussions of the phenomenology of non-equilibrium thermodynamics and of the form it takes for crystalline solids containing point defects. The first of these two chapters is more formal; the second is more physical and contains basic examples and certain 'physical' extensions. Both probably merit study, for they provide the natural framework for the results of the microscopic statistical theory as we develop it.

Chapter 6 is the heart of the work, whence it follows that all readers need to understand the way the ideas are expressed mathematically. In principle all that follows this chapter flows from the master equation (6.2.1). At a first reading §§6.2 and 6.5 (as well as the introduction) would suffice, with a more complete reading later in time for Chapter 9. Theoreticians will probably want to tackle this chapter in its entirety.

The remainder of the book illustrates and applies the concepts and methods of Chapter 6 in a variety of ways. Chapters 7–12 are principally concerned with solids in which the concentrations of defects are small but Chapter 13 is specifically devoted to highly disordered solids. Chapter 7 deals with dielectric and mechanical relaxation processes associated with complex point defects and is mathematically similar to the more abstract formalism of Chapter 6. It can therefore be viewed as an illustration of this formalism. The beginner could well study this as a way of establishing confidence with the formal structure of Chapter 6. Chapter 8 extends the approach of Chapter 7 to diffusion processes. A first reading should include §§8.2 and 8.3 and a review of the results in §8.5

Chapters 9–11 present the random-walk approach to the calculation of diffusion coefficients and of the $L$-coefficients more generally. Theoreticians may want to read all three coherently but experimentalists may find it more profitable at a first reading to turn to the second part of Chapter 11 where expressions for the *L*-coefficients for various models are gathered together and discussed in relation to experimental results. This shows how models may be confirmed and their characteristic parameters inferred. Chapter 12 emphasizes the corresponding calculation of relaxation times arising in nuclear magnetic relaxation. In particular, it compares different theories using some of the results obtained in Chapter 9. It will be necessary then to refer back to this chapter if the details of the comparison are to be fully understood. On the other hand the conclusions can be employed without that understanding.

The last chapter (13) addresses the more difficult task of calculating transport coefficients of concentrated and highly defective systems. It is in essentially two parts: (i) analytic, limited to concentrated random alloys and solid solutions and (ii) Monte Carlo, which describes the use of the technique and the results obtained on models of order–disorder alloys and like systems.

Next we set down one or two caveats. Firstly, in connection with notation, nomenclature and units we have tried to follow the recommendations of I.U.P.A.P. and I.U.P.A.C. as recorded by Cohen and Giacomo (1987) and by Mills *et al.* (1988). Among other things this implies that we use SI units; the only exception is the electron-volt (1 eV equals 96.49 kJ/mol), which is so well matched to the magnitude of the defect energies we shall be concerned with that not to follow current use of it would seem perverse. Lists of our mathematical notation follow this Preface. To simplify the mathematical arguments we have limited some of them to cubic solids, but, the fact that, for the transport coefficients of most solids of practical concern, one can define principal axes (Nye, 1985) means that the bulk of our results apply directly to the component coefficients along individual principal axes. We assume that the reader is familiar with certain areas of mathematics (e.g. Cartesian vectors and tensors, matrix manipulation, Fourier and Laplace transforms) although in no case are more than the basic ideas required. We have tried to ensure that all the mathematical arguments presented can be followed either as they stand or with the aid of a very little pencil and paper work. The longer derivations are given in appendices.

Lastly, we should like to acknowledge our indebtedness to various institutions and individuals. Our own research on the topics described here has been supported in the main by the Underlying Research Programme (now the Corporate Research Programme) of the U.K.A.E.A. and by the Natural Science and Engineering Research Council of Canada, but we have also received assistance from the U.K. Science and Engineering Research Council and the Royal Society of London. This work has been aided also by the hospitality of the Department of Theoretical Chemistry of the University of Oxford, the Department of Physics of the University of Reading, the Interdisciplinary Centre for Chemical Physics at the University of Western Ontario, the Institut für Physikalische Chemie und Elektrochemie of the University of Hanover and the former Theoretical Physics Division of the U.K.A.E.A. at Harwell. One of us (A.B.L.) would particularly like to acknowledge the support of Dr R. Bullough and Dr A. W. Penn of AEA Technology. In connection with the book itself we have benefited enormously from the encouragement, advice and comments of many friends to whom we wish to record our very warm thanks. They include: Professor D. Becker (Hanover), Dr A. H. Harker (Harwell), Professor P. Heitjans (Hanover), Dr M. J. Hutchings (Harwell), Professor P. W. M. Jacobs (London, Ontario), Professor C. W. McCombie (Reading), Professor J. Mahanty (Canberra), Professor N. H. March (Oxford), Dr G. Martin (Saclay), Professor H. Schmalzried (Hanover), Professor C. A. Sholl (Armidale) and Professor J. Strange (Canterbury). The generosity of Professor March and Professor Schmalzried in time, ideas and in other ways is to be especially recorded. We are also grateful to authors and the following publishers for permission to reproduce diagrams from works as cited: Academic Press, Akademie Verlag, American Institute of Physics, American Physical Society, ASM International, Elsevier Science Publishers, Elsevier Sequoia, the Institute of Physics, the Minerals, Metals & Materials Society, Pergamon Press, Plenum Publishing Corporation, Taylor & Francis and VCH Verlag.

Finally we wish to express our warmest thanks to our wives, Betty and Ann, for their patience, understanding and direct help in innumerable ways during the writing of this book.

# *Principal symbols*

**Capital Roman**

| | |
|---|---|
| $\mathbf{A}$ | matrix of relative jump probabilities |
| A, B | particular species of atom |
| $\mathbf{B}$ | magnetic induction (vector) |
| D | diffusion coefficient (tensor) |
| $\mathbf{D}$ | electric displacement (vector) |
| $\mathbf{E}$ | electric field (vector) |
| $F$ | Helmholtz free energy |
| $F_j$ | force on species $j$ |
| $G$ | Gibbs free energy |
| $G(\mathbf{r}, t)$ | position correlation functions |
| $G_{\beta\alpha}$ | propagator matrix derived from $P_{\beta\alpha}$ |
| $H$ | enthalpy |
| I | interstitial species (goes with A, B) |
| $I$ | nuclear spin |
| $I(\mathbf{q}, t)$ | intermediate scattering function |
| $\mathbf{J}_i$ | flux (vector) of component, $i$ |
| $\mathbf{J}_q$ | flux (vector) of heat |
| $L_{ij}$ | transport coefficient (species $i$ and $j$) |
| $\mathbf{M}$ | nuclear magnetisation (vector) |
| $N_i$ | number of species $i$ in the volume of the system |
| $N_m$ | number of stoichiometric cells ('lattice molecules') in the volume of the system |
| $N_{\beta\alpha}$ | diagonal matrix derived from the $p_\alpha^{(0)}$ |
| $P$ | pressure |
| $\mathbf{P}$ | electric polarization (vector) |
| $\mathbf{P}_{\beta\alpha}$ | matrix from the master equation |

| | |
|---|---|
| $Q$ | Arrhenius activation energy |
| $Q_i^*$ | heat of transport of species, $i$ |
| $Q_{pq}$ | matrix derived from $P_{pq}$ |
| $S$ | Entropy |
| $S_i$ | Soret coefficient |
| $S(\mathbf{q}, \omega)$ | neutron scattering dynamical structure factor |
| $S_{\beta\alpha}$ | symmetrized $P_{\beta\alpha}$ matrix |
| $T$ | thermodynamic (absolute) temperature |
| $T_1$, $T_2$, etc. | *N.M.R.* relaxation times |
| $U$ | internal energy |
| $U_{\beta\alpha}$ | matrix derived from the propagator $G_{\beta\alpha}$ |
| $U(\mathbf{l}, y)$ | generating function for random walks |
| $V$ | volume of the system |
| V | designation of vacancy (like A, B and I) |
| $W(\mathbf{q})$ | Debye–Waller factor |
| $\mathbf{X}_i$ | thermodynamic force on species $i$ |
| $\mathbf{X}_q$ | thermodynamic force due to a temperature gradient |
| $X$, $Y$ | kinetic quantities (Chap. 6) |
| $Z$ | partition function |
| $Z_i^*$ | effective charge number for species $i$ (electrotransport) |

**Lower case Roman**

| | |
|---|---|
| $a$ | lattice parameter |
| $\mathbf{a}^{(\nu)}$ | $\nu$th eigenvector of $\mathbf{S}$ |
| $b$ | neutron scattering length |
| $c_i$ | relative concentration of species $i$ $(=N_i/N_m)$ |
| $e$ | proton charge |
| $f$ | Helmholtz free energy per atom, defect etc. |
| $f$ | also correlation factors and functions |
| $g$ | Gibbs free energy per atom, defect, etc. |
| $h$ | enthalpy per atom, defect, etc. |
| i | $\sqrt{-1}$ |
| $i, j, k \ldots$ | species in the system (atoms, vacancies, etc.) |
| $k$ | Boltzmann's constant |
| $\mathbf{k}$ | wave-vector in Fourier transforms |
| $\mathbf{l}$, $\mathbf{m}$, $\mathbf{n}$ | sites in a lattice |
| $n$ | number of 'lattice molecules' (sites) per unit volume $(=1/v)$ |
| $n_i$ | number density of species $i$ |
| $p$, $q$, $r$ | configurations of complex defects (pairs) |

| | |
|---|---|
| $p_\alpha$ | probablity that the system is in state $\alpha$ |
| $\mathbf{q}$ | scattering vector |
| $q_i$ | electric charge on unit of $i$ |
| $r, s, t$ | types of defect in an assembly of defects (Chap. 3) |
| $\mathbf{r}$ | position vector $(x, y, z)$ |
| $s$ | entropy per atom, defect, etc. |
| $\mathbf{s}$ | elastic compliance tensor |
| $s, t$ | time variables |
| $u$ | internal energy per atom, defect, etc. |
| $u$ | electrical mobility |
| $v$ | volume per lattice molecule |
| $v_K$ | Kirkendall velocity |
| $w_{\beta\alpha}$ | rate of transition $\alpha \to \beta$ |
| $x, y, z$ | Cartesian co-ordinates |
| $x_i$ | mole fraction of species $i$ |
| $z$ | number of nearest neighbours |
| $z_i^*$ | electron wind coupling constant for species $i$ |

**Lower case Greek**

| | |
|---|---|
| $\alpha, \beta, \gamma \ldots$ | states of the whole system |
| $\alpha^{(\nu)}$ | $\nu$th eigenvalue of $\mathbf{S}$ |
| $\gamma$ | nuclear gyromagnetic ratio |
| $\gamma$ | natural line-width (Mössbauer) |
| $\gamma_i$ | activity coefficient of $i$ |
| $\delta$ | loss angle |
| $\delta$ | with argument, Kronecker and Dirac functions |
| $\delta$ | deficiency in chemical composition |
| $\boldsymbol{\varepsilon}$ | electric permittivity tensor |
| $\boldsymbol{\varepsilon}$ | strain tensor |
| $\varepsilon_F$ | Fermi energy |
| $\theta$ | polar angle |
| $\theta$ | thermoelectric power |
| $\theta$ | Heaviside and other step functions |
| $\kappa, \lambda, \mu$ | Cartesian indices with vectors and tensors |
| $\boldsymbol{\kappa}$ | conductivity tensor |
| $\kappa_D$ | Debye–Hückel screening length |
| $\lambda$ | Laplace transform variable |
| $\lambda_{\alpha\beta}$ | elastic dipole strain tensor |
| $\mu_i$ | chemical potential of species $i$ |

| | |
|---|---|
| $\nu$ | index for eigenvalues of **S** |
| $\nu$ | mean number (exchanges, visits, etc.) |
| $\nu$ | pre-exponential factor in transition rate |
| $\pi$, $\varpi$ | probabilities |
| $\rho$ | occupancy variable |
| $\boldsymbol{\sigma}$ | stress tensor |
| $\sigma$ | rate of entropy production |
| $\sigma$ | scattering cross-section |
| $\tau$ | relaxation time |
| $\tau$ | column matrix (to effect summation) |
| $\phi$ | azimuthal angle |
| $\phi$ | electric potential |
| $\psi$ | probability function (Chap. 13) |
| $\omega$ | angular frequency |

**Capital Greek**

| | |
|---|---|
| $\Gamma(\mathbf{q})$ | half-width at half-maximum |
| $\Gamma$ | mean number of jumps per unit time |
| $\Delta$ | difference operator |
| $\Pi$ | product operator |
| $\Sigma$ | summation operator |
| $\boldsymbol{\Phi}$ | response function (tensor) |
| $\Omega$ | number of distinct configurations of a given energy |

# 1

# *Atomic movements in solids – phenomenological equations*

## 1.1 Introduction

The processes of the migration of atoms through solids enter into a great range of other phenomena of concern to solid state physics and chemistry, metallurgy and materials science. The nature of the concern varies from field to field, but in all cases the mobility of atoms manifests itself in many ways and contributes to many other phenomena. The study of this mobility and its physical and chemical manifestations is thus a fundamental part of solid state science, and one which now has a substantial history. Like much else in this science it is for the most part concerned with crystalline solids, although it also includes much which pertains also to glasses and polymers. Unfortunately, it is not possible to treat crystalline and non-crystalline solids together in any depth. The present work, which is concerned with the fundamentals of these atomic transport processes, is thus limited to crystalline solids, although occasionally we can 'look over the fence' and see implications for non-crystalline substances. Even with the study of crystalline solids there is a further important sub-division to be made, and that is between atomic movements in *good crystal*, where the arrangement of atoms is essentially as one expects from the crystal-lattice structure, and in regions where this arrangement is disturbed, as at surfaces, at the boundaries between adjoining grains or crystallites, along dislocation lines, etc. The atomic arrangements in these regions are not by any means without structure, but the structures in question are diverse and far from easy to determine experimentally. Since the basis of much of what we have to say is the assumption that there is a known and regular arrangement of atoms in the solid, it follows that we have less to say about atomic migration in or near grain boundaries and dislocations. External surfaces, however, are often represented by two-dimensional structures of the same regular kind as the three-dimensional lattice structures we shall mostly be concerned with. This difference in dimensionality mostly makes little difference and much of what we have to say is thus directly applicable to such model surfaces.

In this chapter we begin by reviewing the representation of the phenomena which result from the movement of atoms through crystalline solids. These representations we describe as *phenomenological*, because they make essentially no specific assumptions about the way these movements are accomplished. The object is to review some of the principal ways in which these movements manifest themselves and by which they may be investigated. We begin with the classic areas of solid state diffusion (§1.2) and the electrolytic conductivity of ionic solids (§1.3). These each have their phenomenology based on Fick's law and Ohm's law respectively. These and other similar linear laws are brought together into a more coherent structure by the thermodynamics of irreversible processes and that representation is thus introduced in §1.4. Within that formalism phenomenological coefficients such as diffusion coefficients, $D$, and ionic mobilities, $u$, become related to a wider set of atomic transport coefficients, $L$. All these coefficients share one very important characteristic, namely they increase rapidly as the temperature is raised. This temperature dependence may follow an Arrhenius law, e.g.

$$D = D_0 \exp(-Q/kT), \tag{1.1.1}$$

where $Q$ is called the activation energy, $k$ is Boltzmann's constant and $T$ is the absolute temperature. Sometimes this law is followed with constant values of $D_0$ and $Q$ over many orders of magnitude in $D$: in other cases there may be two or more regions of Arrhenius behaviour each with its own parameters $D_0$ and $Q$.

The response of solids to rapidly varying electric and stress fields is the subject of §1.5 (dielectric relaxation) and §1.6 (mechanical or anelastic relaxation). These linear response laws in the case of sinusoidally varying fields lead to the Debye equations, with corresponding relaxation time parameters, $\tau$. Although there are differences in detail between dielectric and mechanical relaxation the basic extension of the steady state phenomenology (§§1.2 and 1.3) required for these two topics is the same. Correspondingly, the formalism of §1.4 can be extended in the same way to cover these and other time-dependent forces within the same linear approximation.

Then in the next three sections we discuss ways of investigating diffusive atomic movements which depend upon the nuclear properties of the atoms involved, viz. nuclear magnetic relaxation (§1.7) quasi-elastic scattering of thermal neutrons (§1.8) and Mössbauer spectrometry (§1.9). These introduce other phenomenological coefficients (e.g. the nuclear magnetic relaxation times, the incoherent neutron scattering function, etc.) which can be valuable in research into the microscopic mechanisms of diffusion and atomic transport. However, since they depend upon specific nuclear properties (neutron cross-sections, nuclear spins and moments, nuclear $\gamma$-ray transitions) which vary rather erratically through the periodic table, these techniques are useful for certain elements only. Broadly speaking, the range of application of nuclear magnetic relaxation is greatest and that of Mössbauer spectroscopy the least.

In experiments and applications for which the above phenomenologies may be used the systems are near to thermodynamic equilibrium. Other systems of interest are much further removed from it, e.g. solids rapidly cooled from higher temperatures to much lower temperatures where atomic movements are extremely slow or solids which have been irradiated at similarly low temperatures and held there. On warming such solids to higher temperatures the defects and the damage so introduced may anneal out. The equations used to represent the annealing processes are the subject of the last principal section, §1.10. Experiments of this kind draw us closer to the consideration of the underlying mechanisms of diffusion and atomic transport. This is the subject then of the next chapter.

## 1.2 Diffusion

As already mentioned the processes of the migration of atoms through solids enter into a wide range of other phenomena of concern to solid state physics and chemistry, metallurgy and materials science. Basic to these interests is the flow of atoms which follows upon the existence of gradients in chemical or isotopic composition, i.e. solid state diffusion. This is the subject of this section. We deal first with the representation of these processes by Fick's laws and then consider certain extensions.

### *1.2.1 Fick's laws*

The usual representation of a diffusion process is provided by Fick's first law which, like the law of heat conduction, is a local relation expressing the flux of atoms* of a particular identifiable kind ($i$, say) in terms of the gradient of the concentration of atoms of this kind at the same position, viz.

$$\mathbf{J}_i = -D_i \nabla n_i. \tag{1.2.1}$$

The vector $\mathbf{J}_i$ gives the number of atoms of kind $i$ crossing a unit area perpendicular to the direction of $\mathbf{J}_i$ in unit time. The concentration $n_i$ is the number of atoms of kind $i$ in unit volume. The quantity $D_i$ appearing in this relation is called the *diffusion coefficient* $D_i$. To obtain a complete description of the process it is

* We use the word atom for definiteness, but the statements made here apply equally well to the diffusion of other particles (molecules, ions, etc.) as appropriate.

necessary to combine (1.2.1) with the equation which describes the conservation of atoms of type $i$,

$$\frac{\partial n_i}{\partial t} + \nabla \cdot \mathbf{J}_i = 0. \tag{1.2.2}$$

Elimination of $\mathbf{J}_i$ between (1.2.1) and (1.2.2) gives Fick's second law

$$\frac{\partial n_i}{\partial t} = \nabla \cdot (D_i \nabla n_i). \tag{1.2.3}$$

Solutions of this partial differential equation giving $n_i$ as a function of time and position can be obtained by various means once the boundary conditions have been specified. In simple cases (e.g. where $D_i$ is a constant and the conditions are geometrically highly symmetric) it is possible to obtain explicit analytic solutions of (1.2.3); see e.g. Carslaw and Jaeger (1959). Experiments can often be designed to satisfy these conditions (see e.g. Philibert, 1985), but in many other practical situations numerical methods must be used (see e.g. Crank, 1975; Fox, 1974). Additional modern methods include: (i) the FACSIMILE code (Curtis and Sweetenham, 1985), which is based on finite difference techniques and can deal with coupled sets of diffusion and reaction equations; and (ii) the use of Laplace transforms with numerical inversion (e.g. Talbot, 1979). However these methods of solution are outside the scope of this book.

Before proceeding further we should comment on the continuum nature of the equations (1.2.1) and (1.2.2) and thus of their solutions. Although there is nothing in what we have said to prevent $D_i$ being a function of position, for the most part the use of (1.2.1) and (1.2.2) will imply a degree of structural uniformity over the region where diffusion is taking place. This requirement will be satisfied if the scale of any structural irregularities (e.g. grain size, dislocation spacing) is small compared to the size of the diffusion zone. Thus even if diffusion proceeds more rapidly within grain boundaries or along dislocation lines an overall description via (1.2.1) and (1.2.2) will still be possible as long as the scale of this microstructure is suitably small, but, of course, the effective diffusion coefficient will be an average over those which apply separately within the distinct regions. In cases where the diffusion zone is comparable in size to the scale set by the inhomogeneities it will be necessary to use (1.2.1) with a different coefficient in each distinct region. Reviews of this kind of diffusion problem and its solutions can be found in the books by Philibert (1985 especially Chap. VII) and by Kaur and Gust (1989) and in the review by Le Claire and Rabinovitch (1984). For the most part we shall be concerned with transport through the regular crystal lattice rather than within grain boundaries and along dislocations. The following remarks in particular apply to such regular lattice diffusion.

The quantity $D_i$ appearing in (1.2.3) is in general a second-rank tensor. It is nevertheless symmetric and Cartesian axes can therefore always be chosen to make it diagonal; these are called *principal axes* (Nye, 1985). There are therefore never more than three components, and in a cubic crystal all three are equal. For simplicity we shall therefore mostly assume that we are concerned with one-dimensional problems in which diffusion takes place along just one of the principal axes,* and refer simply to the diffusion coefficient, and write it as a scalar quantity, $D_i$.

So far we have said nothing about the dependence of $D_i$ upon thermodynamic variables, i.e. upon temperature, $T$, pressure, $P$, and concentration. It is well known that diffusion coefficients in solids generally depend rather strongly on temperature, being very low at low temperatures but appreciable at high temperatures. Empirically, this dependence may often, but by no means always, be expressed by the Arrhenius formula

$$D_i = D_{io} \exp(-Q/kT), \tag{1.2.4}$$

in which $D_{io}$ and $Q$ are independent of $T$. ($D_{io}$ is commonly referred to as the *pre-exponential factor* and $Q$ as the *activation energy* for diffusion.) The wide range of magnitudes which can arise is illustrated in Fig. 1.1, which shows $D_i$ for a variety of elements at extreme dilution in Si. Such simple Arrhenius behaviour should not, however, be assumed to be universal. Departures from it may arise for reasons which range from fundamental aspects of the mechanisms of atomic migration to effects associated with impurities or other microstructural features. Nevertheless the expression (1.2.4) provides a very useful standard.

The variation of $D_i$ with pressure is far less striking than that with temperature. $D_i$ decreases as the pressure is increased, typically about ten times for a pressure of 1 GPa ($10^4$ bar).

Thirdly, there are variations of $D_i$ with the concentration of the diffusing species to be considered. These range from the very slight to the striking. An example of the latter is given in Fig. 1.2, which shows that the diffusion coefficient of Pb ions at high dilution in KCl is increased about seven times as the concentration of Pb ions increases over the narrow range from $10^{-5}$ to $1.5 \times 10^{-4}$ mole fraction.

* In practice, if not in principle, this excludes low-symmetry crystals belonging to the monoclinic and triclinic classes, since for them the directions of the principal axes relative to the crystal axes depend upon physical details of the diffusion mechanism. By contrast, in crystals belonging to the higher symmetry classes (orthorhombic, trigonal, tetragonal, hexagonal and cubic classes) the principal axes are defined by symmetry alone (or in cases of 'degeneracy' by a combination of symmetry and convenience). In these crystals they are 'physically obvious'. The same remarks apply to any physical property described by a symmetrical second-rank tensor. For further details see Nye (1985).

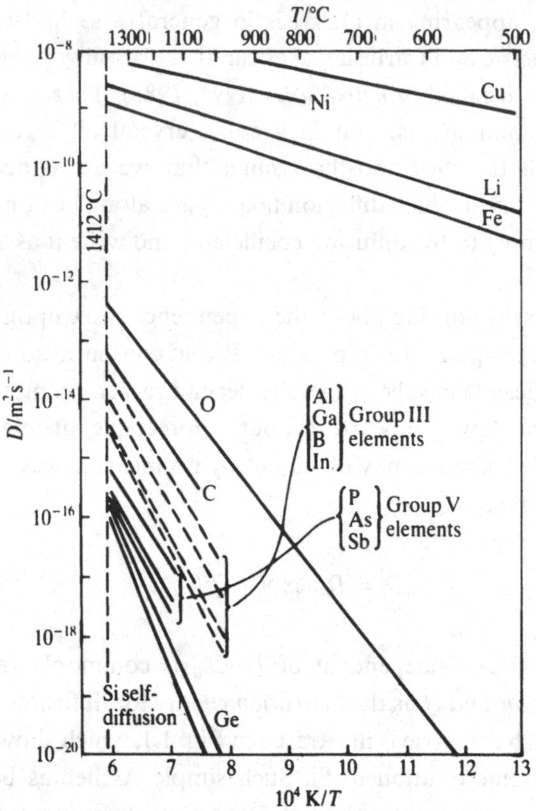


Fig. 1.1. An Arrhenius plot ($\log_{10} D$ v. $10^4/T$) of the tracer diffusion coefficients of a range of solute elements in solid crystalline silicon. (After Frank, Gösele, Mehrer and Seeger, 1984.)

### 1.2.2 *The Kirkendall effect*

This brings us to an important point. So far we have not explicitly specified the frame of reference for the diffusion equations (1.2.1)–(1.2.3). In cases where the diffusing species (solutes) are present at extreme dilution this frame can naturally be assumed to be one fixed with respect to the atoms of the solvent species. Experiments carried out with radioactive tracer atoms are mostly carried out in these conditions and the corresponding diffusion coefficients are referred to as *tracer diffusion coefficients.* Evidently we can study both solute and solvent tracer diffusion, the second being often called *self-diffusion.*

However, as the concentration of solute is increased it would seem natural to suppose that there would be at least a partial counterflow of solvent atoms so that a representation of this flow is also needed. In these circumstances some appeal to microscopic ideas, particularly to knowledge that the atoms are held in

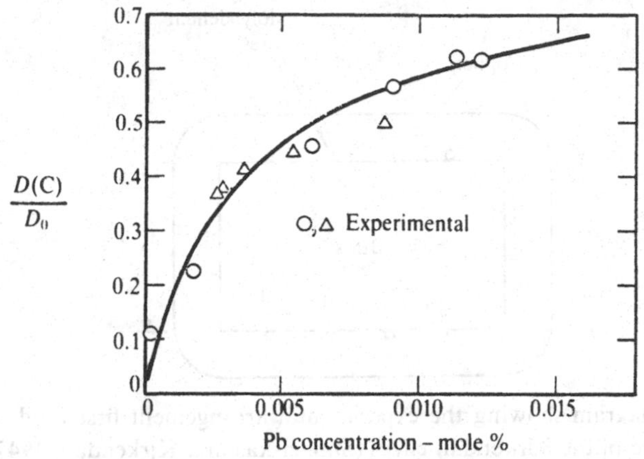


Fig. 1.2. The intrinsic diffusion coefficient of $Pb^{2+}$ ions in KCl as a function of Pb concentration at $T = 474°C$. The line represents a theoretical prediction with parameters chosen to give the best fit to the experimental points. (After Keneshea and Fredericks, 1963.)

a lattice structure, seems unavoidable if we are to adhere to the form (1.2.1). We shall therefore retain this form as the proper representation of diffusion flows relative to a frame fixed in the local crystal lattice. To express them in any other reference frame we would have to add a term $\mathbf{v}n_i$ on the right side, $\mathbf{v}$ being the velocity of the lattice in this new frame.

If we return now to our consideration of a concentrated solid solution ($A_{1-c}B_c$ say) it is clear that we shall have two equations of the form (1.2.1), one for A and one for B. There will thus be two coefficients $D_A$ and $D_B$. But evidently there is only one diffusion process, namely the intermixing of A and B. The reconciliation of these two apparently contradictory facts is found in the experimental observation that in a diffusion couple the diffusion zone (i.e. the region where $J_A$ and $J_B$ are significant) moves relative to that part of the couple where no diffusion is taking place. This is called the Kirkendall effect; see Figs. 1.3 and 1.4. It was first observed by Smigelskas and Kirkendall (1947) in a Cu–brass diffusion couple. This effect has been subsequently observed to occur with interdiffusion in compounds as well as in alloys. The rate of movement of the diffusion zone in the simplest case is just proportional to $J_A + J_B$, i.e. to the net flux of atoms relative to the local lattice. There are therefore two measurable quantities after all, namely the *chemical interdiffusion coefficient* $\tilde{D}$, which specifies the intermixing alone, and the Kirkendall movement. These are expressible in terms of $D_A$ and $D_B$. In this way both $D_A$ and $D_B$ have been determined experimentally as functions of composition for a number of alloys and solid solutions. These quantities are often referred to as *intrinsic diffusion coefficients* and we shall follow that nomenclature here too.

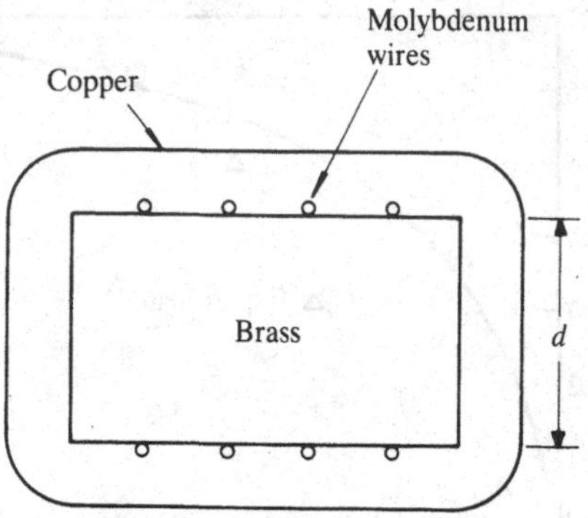


Fig. 1.3. Diagram showing the experimental arrangement first used to show the existence of the Kirkendall effect (Smigelskas and Kirkendall, 1947). Inert markers (fine wires of Mo in the original experiment) are placed at the interface between the two metals making up the diffusion couple (Cu and brass). Relative movement of the markers (disclosed by changes in the spacing $d$) shows that there is a net flux of atoms across the boundary as diffusion proceeds.

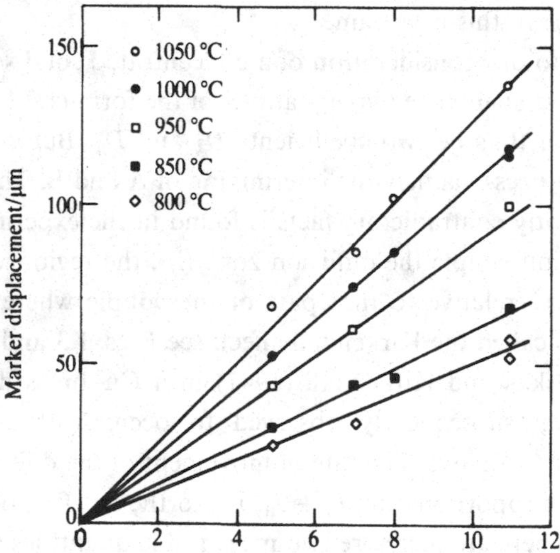


Fig. 1.4. Measurements of the Kirkendall effect arising during the interdiffusion of U and a MoU alloy. The observed displacement of the markers at the interface is shown as a function of the square root of the time, $\sqrt{t}$. The linear dependence on $\sqrt{t}$ is as expected from macroscopic diffusion theory. (After Adda and Philibert, 1966.)

### 1.2.3 *Coupled flows*

In the preceding section we have treated the flows of the various kinds of atom independently of one another, even though in the general case each $D_i$ may be viewed as a function of composition (as well as $T$ and $P$). For very many purposes that is quite adequate. Experimental situations, in particular, are often arranged to be as straightforward as possible. However, there are circumstances where we need to allow for the possibility that a gradient in the concentration of one species $\nabla n_j$ could contribute to the flux, $J_i$, of another. The obvious generalization of Fick's law (1.2.1) is then

$$J_i = \sum_{j=1}^{n} D_{ij} \nabla n_j. \tag{1.2.5}$$

For an $n$-component system there are now (for each inequivalent principal axis) $n \times n$ coefficients $D_{ij}$ in place of the $n$ coefficients $D_i$ which we had previously. Although the additional complexity of (1.2.5) presents no difficulty for the numerical methods of solution already referred to in §1.2.1, the task of determining the $n^2 D_{ij}$ coefficients experimentally is very much harder. Fortunately, there is an alternative formulation for which a theorem due to Onsager allows the number of independent coefficients to be reduced to $n(n+1)/2$. We turn to this in §1.4. Even so, the experimental difficulties associated with a determination of all the coefficients even of a binary system (3 coefficients) or a ternary system (6 coefficients) remain very substantial. It is for this reason that for circumstances where the additional effects do arise there is a substantial reliance on the theory of these coefficients.

### 1.2.4 *Soret effect and thermal diffusion*

In addition to fluxes of atoms which result from gradients of composition in an isothermal (and isobaric) system such fluxes can also result from gradients in temperature alone. This phenomenon is called *thermal diffusion*. The analogue of Fick's first law in this case would be

$$J_i = -S_i \nabla T, \tag{1.2.6}$$

but the coefficient $S_i$ may be of either sign, whereas $D_i$ is always positive. However, like $D_i$ it is a symmetrical second-rank tensor and for crystals of sufficiently high symmetry the directions of the principal axes of $D_i$ and of $S_i$ either are or may be chosen to be the same. We can therefore employ the same one-dimensional notation that we chose for isothermal diffusion.

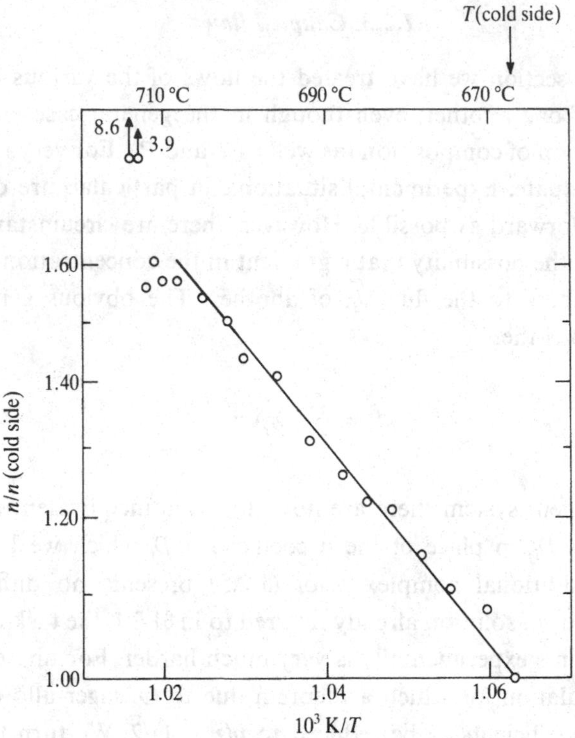


Fig. 1.5. Experimental results showing the steady-state distribution of C in a specimen of $\alpha$-Fe (i.e. body-centred cubic Fe) subjected to a (uniform) temperature gradient. The abscissa shows the C concentration relative to its value at the cold side (logarithmic scale) as a function of the inverse of the temperature at the corresponding position. The temperature gradient across the specimen was 22 K mm$^{-1}$. (After Shewmon, 1960.)

In circumstances where there is also a gradient of concentration we would combine (1.2.1) and (1.2.6) to give (in one dimension),

$$J_i = -D_i \frac{\partial n_i}{\partial x} - S_i \frac{\partial T}{\partial x}. \tag{1.2.7}$$

It is clear from this that there is the possibility of establishing a steady state in a closed system, i.e. one in which $J_i = 0$, for which

$$(\partial n_i/\partial x) = -(S_i/D_i)(\partial T/\partial x). \tag{1.2.8}$$

A temperature gradient thus may lead to the establishment of a concentration gradient in a closed system given by (1.2.8). This effect is called the *Ludwig–Soret effect*, often just the Soret effect. Results illustrating the effect are shown in Fig. 1.5.

## 1.3 Electrolytic (ionic) conductivity

There are many non-metallic ionic compounds whose electrical conductivity results from the transport of ions rather than electrons, as is usual in semiconductors and metals. Naturally, this results in the electrolytic decomposition of the solid – just as with an ionic solution or molten salt. Indeed, the classic way of verifying the dominance of electrolytic conductivity was to measure the weights of the products of this electrolytic decomposition deposited at the electrodes and to compare them with the quantity of electricity required to effect the decomposition. Purely ionic conduction will result in the liberation of the component $i$ of the compound at the rate of $|z_i|^{-1}$ moles for every Faraday of electricity passed (96 485 C), where $z_i$ is the *charge number* of $i$ in its ionic form. More modern research by contrast relies to a great extent (i) on measures of the self-diffusion coefficients, $D_i^*$, of the component species by radio-tracers or other means (§1.7) and (ii) on appropriate comparison of these with the measured direct-current, or low-frequency, conductivity $\kappa$, defined by Ohm's law as the coefficient in the relation between the electrical current density $\mathbf{J}_e$ resulting from the applied electric field, $\mathbf{E}$, viz.

$$\mathbf{J}_e = \kappa \mathbf{E}. \tag{1.3.1}$$

Like Fick's law (1.2.1), this is a *local* relation. Furthermore, $\kappa$, like $D_i$, is a symmetrical second-rank tensor, and for a crystal structure of sufficient symmetry (see p. 5) the principal axes of $\kappa$ are in the same directions as those of $D_i$. This enables us to simplify the notation again by using the same one-dimensional representation.

If the ions in the compound are labelled $i$ (=1, 2, etc.) then this conductivity, $\kappa$, is usually written in terms of mobilities as

$$\kappa = \sum_i n_i |q_i| u_i \tag{1.3.2}$$

in which $n_i$ is the number of ions of kind $i$ per unit volume, $q_i$ $(=z_i e)$ is the charge they carry and $u_i$ is the *electrical mobility* of ions of kind $i$. The appropriate comparison referred to above is obtained by means of the Nernst–Einstein relation between $u_i$ and $D_i^*$

$$\frac{u_i}{D_i^*} = \frac{|q_i|}{kT}, \tag{1.3.3}$$

or some relevant modification of it. Since we expect $D_i^*$ to follow (1.2.4) this relation tell us that the ionic mobility will also be thermally activated and thus sensitive to temperature. An example which illustrates this conclusion is shown in

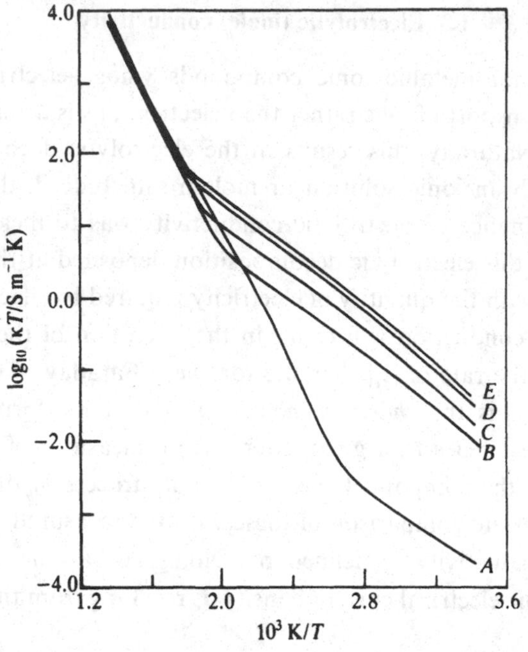


Fig. 1.6. Experimental measurements of the electrical (ionic) conductivity of single crystals of AgCl as a function of temperature. The ordinate gives the product $\kappa T$ on a logarithmic scale, while the abscissa gives the corresponding inverse temperature. Curve *A* was for a pure sample while the other crystals contained various amounts of $Cd^{2+}$ ions in substitutional solution (*B*, $60 \times 10^{-6}$; *C*, $150 \times 10^{-6}$; *D*, $370 \times 10^{-6}$; *E*, $480 \times 10^{-6}$ mole fraction). (After Corish and Mulcahy, 1980.)

Fig. 1.6, which incidentally also shows the influence of solute ions (in this case cadmium). We shall go into the derivation of this relation and the circumstances in which it is modified later. For practical aspects of the determination of ionic conductivities see Jacobs (1983).

By one or other of these means therefore a large number of substances have been identified as ionic conductors. These are almost entirely substances of closely stoichiometric composition. Departures from stoichiometry inevitably imply the addition or subtraction of electrons as well as the addition or subtraction of ions, and since electron (and hole) mobilities are very much higher than ion mobilities it needs relatively few electrons to make a major contribution to the conductivity. These ionic conductors fall roughly into two classes (i) the classic ionic conductors and (ii) so-called fast ion conductors (sometimes superionic conductors). The distinguishing characteristics are collected together in Table 1.1. Compounds with the fluorite ($CaF_2$) structure form an interesting homogeneous class by themselves, behaving like normal ionic conductors at low temperatures but undergoing a

Table 1.1. *Comparison of the characteristics of normal and fast ionic conductors*

| Characteristic | Normal | Fast |
|---|---|---|
| Structure | Simple lattice structures (e.g. cubic halides of alkali metals) | Complex and open lattice structures |
| Chemical composition | Stoichiometric | May be variable (e.g. mixed oxides) |
| Electrical conductivity | Sensitive to temperature rising to $\sim 10^{-2}$ Sm$^{-1}$ at the melting point. Arrhenius behaviour with $Q \sim$ 1–2 eV. Sensitive to purity and crystal micro-structure at lower temperatures thus giving more than one range of Arrhenius behaviour | Less sensitive to temperature, but higher in absolute value rising to $\sim 10^2$ Sm$^{-1}$ at high temperatures. Arrhenius behaviour with $Q \sim$ few $\times$ 0.1 eV. Less sensitive to purity, but dependent on composition |

broad transition around $T_c \sim 0.9$ of the melting temperature $T_m$ to a fast-ion conducting state.

Electric fields can, of course, result in the transport of ions in substances which are not predominantly ionic conductors. Transport of atoms in an electric field even occurs in metals, although here a major effect is due to the scattering of the conduction electrons off the atoms in question. This gives them an effective charge for this electrotransport which is really a kinetic quantity and is a non-integral multiple of $e$ (Huntington, 1975).

Before concluding this section we again consider non-isothermal systems. It is observed that a temperature gradient gives rise to transport processes in ionic conductors just as in metals and other solids. As in these other substances the Soret effect has been observed with solute ions. There are, however, additional effects to be observed in these materials. In so far as the different component ions $i$ have different mobilities and diffusion coefficients we might also expect the effects of a temperature gradient to differ too. This can lead to an incipient separation of oppositely charged ions in the temperature gradient, which will actually be inhibited by the establishment of an internal electric field just sufficient to ensure that the internal electric current is zero on open circuit. There is therefore an electric potential difference across the specimen, in other words a contribution $\theta_{hom}$ to the thermoelectric power, Fig. 1.7. A further contribution $\theta_{het}$ can come from the temperature dependence of the electrode-crystal contact potential giving

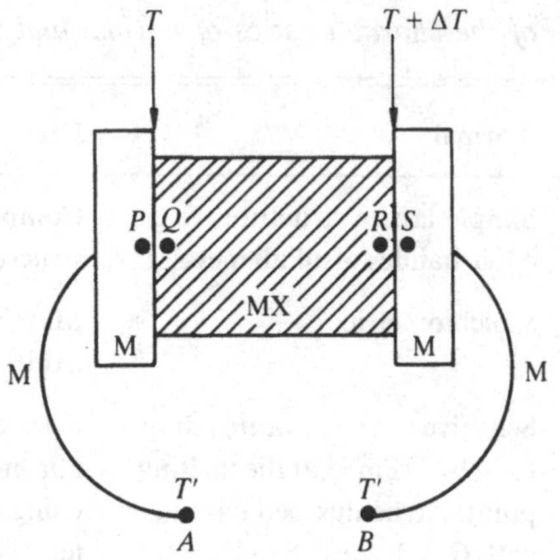


Fig. 1.7. Schematic diagram to illustrate the generation of thermoelectric power in a solid MX fitted with electrodes of the metal M. In terms of the electric potential $V$ at the various points indicated, the total thermoelectric power $\theta = (V_B - V_A)/\Delta T$ is made up of (i) a homogeneous part $\theta_{\text{hom}} = (V_R - V_Q)/\Delta T$, (ii) a heterogeneous part $\theta_{\text{het}} = [(V_S - V_R) - (V_Q - V_P)]/\Delta T = \partial\phi/\partial T$ where $\phi$ is the contact potential difference between the electrode metal M and the solid MX and (iii) a further homogeneous part arising from the temperature gradient in the leads. This last term is generally very small (μV/K) compared with the other two (mV/K). It should be recognized that this breakdown of $\theta$ into $\theta_{\text{hom}}$ and $\theta_{\text{het}}$ is different from the breakdown into Thompson and Peltier terms usual with electronic conductors, although the two can be related.

a total thermopower

$$\theta = \theta_{\text{hom}} + \theta_{\text{het}}. \tag{1.3.4}$$

Like the conductivity this thermopower can depend sensitively on temperature and purity. An illustration is given in Fig. 1.8.

Finally, we should note that the above discussion is about intrinsic properties of the material, which in turn we have implicitly supposed to be homogeneous. In practice, there may be significant qualifications to both these aspects. Firstly, the measurement of the d.c. conductivity of the material may be complicated by time-dependent processes occurring at the electrodes attached to the sample. These processes may derive from the deposition of products of electrolytic decomposition, from the existence of electrical double layers, from poor physical contact with rough surfaces, and so on. Ways of avoiding or allowing for these effects are described in a review by Franklin (1975). Related considerations arise with the thermopower, where the definition of $\theta_{\text{het}}$ demands a thermodynamically reversible electrode.

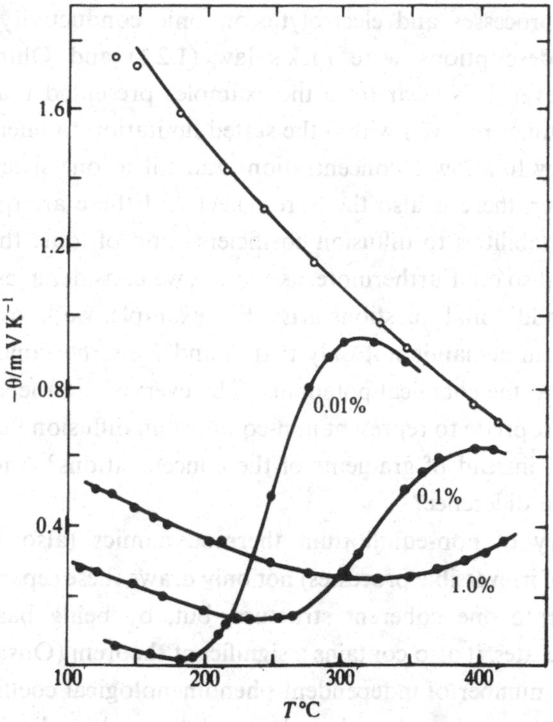


Fig. 1.8. The total thermo-electric power $\theta$ of AgCl fitted with Ag electrodes as a function of temperature, $T$: the open circles ○ are for pure AgCl and the filled circles ● are for AgCl containing various concentrations of $CdCl_2$ as indicated (mole %). (After Christy, 1961.)

Secondly, it should be emphasized that, although in this book we are interested primarily in properties which are independent of microstructure, there are physically important influences upon the ionic conductivity which do stem from microstructural regions of greater disorder, e.g. grain boundaries, dislocations and precipitates. By providing regions of locally enhanced conductivity such defects can contribute significantly to the overall measured conductivity and associated transport properties, expecially at low temperatures where the intrinsic, thermally-created disorder is generally very small. A notable example is provided by the large enhancement ($\sim 100\times$) of the ionic conductivity of Li-halides by $Al_2O_3$ grains dispersed throughout the crystal (see e.g. Maier, 1989).

## 1.4 Phenomenological equations of non-equilibrium thermodynamics

In the two preceding sections we have summarized the usual phenomenological descriptions of two important manifestations of the mobility of atoms in solids, namely diffusion processes and electrolytic or ionic conductivity. The starting points for these descriptions were Fick's law (1.2.1) and Ohm's law (1.3.1) respectively. However, it is clear from the examples presented that these simple laws may not be sufficient, even within the stated limitation to linear effects. Thus it may be necessary to allow a concentration gradient in one species to give rise to a flux of another; there is also the Soret effect and there are questions of the relation of ionic mobilities to diffusion coefficients and of ionic thermopower to the Soret effect; and so on. Furthermore, as soon as we consider a less phenomenological approach additional questions arise. For example, we know that thermodynamic equilibrium demands not only that $T$ and $P$ be the same throughout a system but also that the chemical potentials $\mu_i$ be everywhere the same; so would it not be more appropriate to represent non-equilibrium diffusion fluxes in terms of gradients of the $\mu_i$ instead of gradients of the concentrations? And if so, does it make any practical difference?

Now the theory of non-equilibrium thermodynamics (also known as the thermodynamics of irreversible processes) not only draws these separate phenomenologies together into one coherent structure, but, by being based on certain fundamental postulates, it also contains a significant theorem (Onsager's theorem) which reduces the number of independent phenomenological coefficients. Since it is a theoretical structure based on broad postulates and applicable to all three states of matter we shall defer the main description of it until Chapter 4 where we shall be in a better position to appreciate some of the detailed implications of its application to solid state processes. Here we merely introduce the phenomenological transport equations which it offers in place of those we have already presented.

The essential feature here is that we can set down one set of equations which embraces all the different circumstances leading to fluxes of atomic species (atoms, ions, electrons, etc.) in the system. It is a linear theory so that these equations, like Fick's law, Ohm's law, etc., are linear homogeneous equations which express the fluxes of these species, $J_i$ ($i = 1, 2, \ldots r$) in terms of suitably defined *forces* acting on these species, symbolized as $X_j$ ($j = 1, 2, \ldots r$). These thermodynamic forces require some explanation. When only mechanical forces are acting then they are identical with these mechanical forces, $F_j$ say. For example, for an ionic system subject to an electric field $E$ each ion of type $j$ is subject to a force $F_j = q_j E$ and thus $X_j = q_j E$. However, in the presence of a gradient of composition the appropriate force is related to the gradient of chemical potential; in an isothermal, isobaric system, in fact, $X_j = -\nabla\mu_j$, and if an electric field or other source of mechanical force were also present then we should have

$$X_j = -\nabla\mu_j + F_j. \qquad (1.4.1)$$

In an isobaric but non-isothermal system the full set of equations must include the heat flow $J_q$ and a corresponding thermal force $X_q = -\nabla T/T$ as well as the material fluxes $J_i$ and the corresponding forces $X_i$.

The set of linear phenomenological equations as expressed in terms of these quantities is

$$J_i = \sum_j L_{ij} X_j, \qquad (i, j = 1, 2, \ldots), \tag{1.4.2}$$

in which the running indices $(i, j)$ may include the terms in $q$ (for heat) if we are representing a system where $\nabla T \neq 0$. Since the $J_i$ and the $X_i$ are vectors each of the coefficients $L_{ij}$ is again a second-rank tensor. There are now several points to be made.

(1) This relation (1.4.2) like Fick's law (1.2.1), Ohm's law (1.3.1), etc. is a *local* relation.

(2) Each of the components of the set of $L$-coefficients is a function of the thermodynamic variables, $T, P$ (or more generally stress) and chemical composition at the corresponding position, but they do not depend explicitly upon the gradients of these quantities. In other words the $L_{ij}$ are independent of the forces acting at that point whether they be diffusional, electrical, mechanical or whatever. As we shall see explicitly later, this allows us to relate different transport coefficients (e.g. diffusion coefficients, electrical and mechanical mobilities, etc.); in particular it provides one way to derive relations of the Nernst–Einstein type (cf. 1.3.3).

(3) If we write the Cartesian tensor components of $L_{ij}$ as $L_{ij,\kappa\lambda}$ then Onsager's theorem tells us that

$$L_{ij,\kappa\lambda} = L_{ji,\lambda\kappa} \tag{1.4.3}$$

This relation derives from the underlying atomic dynamics of the system and ultimately from the principle of detailed balance in statistical mechanics. For the diagonal coefficients $L_{ii}$ we obviously have

$$L_{ii,\kappa\lambda} = L_{ii,\lambda\kappa} \tag{1.4.4}$$

and each is therefore a symmetrical tensor for which principal axes can be defined. The off-diagonal coefficients $L_{ij}$, by contrast, are not necessarily symmetrical, although for most of the crystal systems of interest they are. Exceptions are provided by triclinic, monoclinic and certain trigonal, tetragonal and hexagonal classes (Nye, 1985). Since we shall rarely have occasion to refer to these particular systems we shall therefore take all the $L_{ij}$ to be symmetrical second-rank tensors. We refer them to their principal axes so that we can again simplify the transport equations by expressing them in a scalar or one-dimensional form.

(4) Although we shall defer until later questions of the expression of observable quantities in terms of the $L_{ij}$ we may note here their relation to the $D_{ij}$ diffusion coefficients introduced in §1.2.3. Thus we suppose here that we are dealing with

a system at constant $(T, P)$. Furthermore, for formal simplicity let us also suppose (i) that we are here only interested in the migration of $r-1$ solute species and (ii) that forces $X_r$ on the solvent atoms have a negligible effect on the solute fluxes. (More general circumstances are considered later in Chapters 4 and 5.) Then by (1.4.2)

$$J_i = -\sum_{j=1}^{r-1} L_{ij}\frac{\partial \mu_j}{\partial x}, \tag{1.4.5}$$

for solute diffusion along the $x$-direction. Now the chemical potential $\mu_j$, as an intensive thermodynamic quantity, will be a function of the mole fractions $x_1, x_2, \ldots$ of the various species (rather than of their concentrations $n_i$). For the mole fractions we necessarily have

$$\sum_{j=1}^{r} x_j = 1, \tag{1.4.6}$$

so that each of the $\mu_j$ can be taken to be $\mu_j\,(x_1, x_2, \ldots x_{r-1})$, the solvent having again been removed from consideration. By writing

$$\frac{\partial \mu_j}{\partial x} = \sum_{k=1}^{r-1}\frac{\partial \mu_j}{\partial x_k}\frac{\partial x_k}{\partial x}, \tag{1.4.7}$$

we see that (1.4.5) gives

$$J_i = \sum_{k=1}^{r-1} D_{ik}\frac{\partial n_k}{\partial x}, \tag{1.4.8}$$

with

$$D_{ik} = \frac{1}{n}\sum_{j}^{r-1} L_{ij}\frac{\partial \mu_j}{\partial x_k}, \tag{1.4.9}$$

in which $n$ is the number of molecules of the host crystal per unit volume. Equation (1.4.8) is just the form (1.2.5) presented earlier as a generalization of the usual form of Fick's law. We observe, however, that, although by Onsager's theorem $L_{ij} = L_{ji}$, it does not follow that $D_{ik} = D_{ki}$. We would further note that by eliminating the solvent species from explicit consideration we have avoided certain formal subtleties which have little physical content, although it can be important to get them right. We shall deal with these in the course of the further discussion of non-equilibrium thermodynamics in Chapter 4.

## 1.5 Dielectric relaxation

In §§1.2–1.4 we discussed the response of solids to certain static or slowly varying forces, in particular gradients of chemical and electrical potential. Because the forces were so slowly varying, the local flux equations could be assumed also to be *local* in time, i.e. they relate the fluxes at time $t$ to the forces at the same time. In this section and the next we turn to the representation of the response to more rapidly varying electrical and mechanical forces. The range of frequencies involved is, in practice, from zero up to roughly $10^9$ Hz in both cases. From the quantum mechanical point of view even the highest of these are still very low frequencies (corresponding to $\hbar\omega \sim 10^{-7}$ eV) so that we are not concerned with *resonance* absorption processes in which there is any direct excitation of the atoms to higher electronic, vibrational or rotational states. (However, in the presence of strong static magnetic fields both electronic and nuclear spin splittings may fall in or near this range of energies; see §1.7). Nevertheless in this frequency range there is found to be a variety of *relaxation* processes, electrical, mechanical and magnetic all of which result from atomic movements of one sort or another. In this section we deal with these *electrical* relaxation processes in ionic conductors and insulators; in metals and semiconductors the free electrons dominate the electrical response and any contributions from atomic movements are unobservable. They may nevertheless be observed in mechanical relaxation (see §1.6).

In a static electric field we know that the local relation between the electric displacement **D** and the field **E** is

$$\mathbf{D} = \varepsilon \mathbf{E}. \tag{1.5.1}$$

In general, **D** and **E** are vectors and thus the electric permittivity $\varepsilon$ is a second-rank tensor. Like the others already introduced it is symmetrical (although the proof in this case derives from the energy principle rather than Onsager's theorem, Nye, 1985). For crystals of sufficient symmetry (see p. 5) the directions of the principal axes of $\varepsilon$ either are or may be chosen to be coincident with those for $D$, $\kappa$, etc. We shall therefore again use a one-dimensional notation.

Now in general it is found that when $E$ is suddenly changed the electric displacement $D$ (and equivalently the polarization $P$) is composed of a part which follows the field instantaneously and a part which follows the field only slowly (Fig. 1.9). This second part represents the phenomenon of *dielectric relaxation.* More generally the displacement at time $t$ resulting from a time-dependent field applied at earlier times $t'$ is written as

$$D(t) = \varepsilon_\infty E(t) + \int_{-\infty}^{t} \Phi(t - t')E(t')\,\mathrm{d}t'. \tag{1.5.2a}$$

In this equation the term in $\varepsilon_\infty$ specifies the instantaneous response, while the function $\Phi$ specifies the relaxation response. The first arises from the electronic polarization of the atoms of the solid in the electric field. The second represents

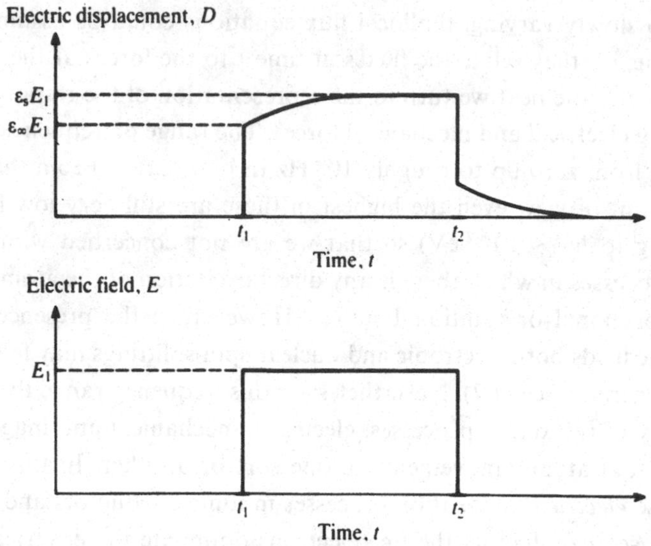


Fig. 1.9. Schematic diagram illustrating the way the dielectric displacement, $D$, may lag behind changes in the electric field, $E$.

the slower adjustment of the distribution of permanent electric dipoles to the electric field. The electric polarization $P$ follows in the usual way from

$$P(t) = D(t) - \varepsilon_0 E(t) \tag{1.5.2b}$$

in which $\varepsilon_0$ is the permittivity of free space.

If, in addition, the solid contains free charge carriers (either electrons or ions) then there will also be an equation for the current density $J_e$ like (1.3.1) which we can generally take to be an instantaneous relation. In the representation of the response to sinusoidally varying fields the two contributions – polarization and conduction – can be combined together into one complex quantity (complex permittivity or complex conductivity).

Returning to (1.5.2) it may be noted that no spatial dependence is included, since at the frequencies of concern the wavelength of electromagnetic waves is always long compared to the size of experimental specimens, so that $E$, $D$, etc. can be taken to be spatially uniform.

There are now three types of experiment to be considered: (i) the sudden removal of a constant field, (ii) the application of a sinusoidally varying field and (iii) the method of ionic thermocurrents. In the first of these a constant field $E_0$ is established until the crystal is in a steady condition, the field is then suddenly removed and the decay of the polarization is measured. From (1.5.2) it follows that $P$ at time $t$ after the instant at which the field is removed is

$$P(t) = E_0 \int_t^\infty \Phi(s)\,\mathrm{d}s. \tag{1.5.3}$$

From such experiments we should thus obtain rather direct information about the response function $\Phi$. A few successful experiments of this kind have been made (e.g. Dreyfus, 1961), but the method has been rather little used overall. By contrast, the second and third methods have been quite extensively used. We therefore deal with them in greater detail.

### *1.5.1 Response to a sinusoidally varying field: dielectric loss*

Let us represent the field by the complex quantity

$$E = E_0 \exp(\mathrm{i}\omega t). \tag{1.5.4}$$

(We can always take the real part of $E$ and of all subsequent equations to obtain relations between real physical quantities.) Insertion of (1.5.4) into (1.5.2) and a change in the variable of integration from $t'$ to $s = t - t'$ gives for the corresponding complex displacement

$$D(t) = \left[\varepsilon_\infty + \int_0^\infty \Phi(s)\,\mathrm{e}^{-\mathrm{i}\omega s}\,\mathrm{d}s\right] E_0\,\mathrm{e}^{\mathrm{i}\omega t},$$

$$\equiv (\varepsilon_1 - \mathrm{i}\varepsilon_2) E_0\,\mathrm{e}^{\mathrm{i}\omega t}, \tag{1.5.5}$$

which defines the complex permittivity.* This definition, rather than the formally more natural $\varepsilon_1 + \mathrm{i}\varepsilon_2$, is chosen because a positive $\varepsilon_2$ then corresponds to the observation that the response, $D$, always lags behind the applied force, $E$. In turn, if we write

$$\bar{\Phi}(\mathrm{i}\omega) = \int_0^\infty \Phi(s)\,\mathrm{e}^{-\mathrm{i}\omega s}\,\mathrm{d}s \tag{1.5.6}$$

this gives

$$\varepsilon_1 = \varepsilon_\infty + \mathrm{Re}\,\bar{\Phi}(\mathrm{i}\omega) \tag{1.5.7a}$$

and

$$\varepsilon_2 = -\mathrm{Im}\,\bar{\Phi}(\mathrm{i}\omega). \tag{1.5.7b}$$

The corresponding representation in terms of a complex conductivity $\kappa_1 + \mathrm{i}\kappa_2$ (connecting the complex electric current density $J_e(t) \equiv \partial D/\partial t$ to the field given by

* It should be noted that the same symbols are used by some authors to denote the complex dielectric constant, which is the ratio of the complex permittivity to $\varepsilon_0$.

(1.5.4)) is obtained from the relation

$$\kappa_1 + i\kappa_2 = i\omega(\varepsilon_1 - i\varepsilon_2). \tag{1.5.7c}$$

### *Debye equations*

So far we have said nothing about the form of $\Phi(s)$. However, it is evident from its definition (a) that it is zero for $s < 0$ because the cause ($E(t)$) must precede the effect ($D(t)$) and (b) that it is positive and falls monotonically with increasing $s$, until finally it tends to zero as $s \to \infty$. A simple form, which has these characteristics, is

$$\Phi(s) = (\Delta\varepsilon/\tau)\exp(-s/\tau), \qquad s \geq 0. \tag{1.5.8}$$

Then the transform (1.5.6) of (1.5.8) is

$$\bar{\Phi}(i\omega) = \frac{\Delta\varepsilon}{1 + i\omega\tau} \tag{1.5.9}$$

and thus

$$\varepsilon_1 = \varepsilon_\infty + \frac{\Delta\varepsilon}{1 + \omega^2\tau^2} \tag{1.5.10a}$$

and

$$\varepsilon_2 = \frac{(\omega\tau)\,\Delta\varepsilon}{1 + \omega^2\tau^2}. \tag{1.5.10b}$$

These are known as the Debye equations. The relaxation time, $\tau$, is a characteristic parameter of the relaxation process. For the most part such relaxation times in solids are thermally activated, i.e. they depend exponentially on temperature as

$$\tau^{-1} = \tau_0^{-1}\exp(-Q/kT). \tag{1.5.11}$$

Hence $\varepsilon_1$ and $\varepsilon_2$ in (1.5.10) can be regarded as functions of $\omega$ at fixed $T$ or as functions of $T$ at fixed $\omega$. Both are experimentally convenient.

It will be observed that $\varepsilon_2$ is positive (because $\Delta\varepsilon > 0$) corresponding to the fact that $D(t)$ lags in phase behind $E(t)$ by an angle $\delta$, the *loss angle*. This lag gives rise to a dielectric loss (i.e. energy absorption from the field) which is measured by

$$\tan\delta = \varepsilon_2/\varepsilon_1. \tag{1.5.12}$$

By (1.5.10) we obtain

$$\tan\delta = \frac{(\varepsilon_s - \varepsilon_\infty)\omega\tau}{\varepsilon_s + \varepsilon_\infty\omega^2\tau^2}, \tag{1.5.13a}$$

in which we have introduced the static, i.e. $\omega \to 0$, permittivity $\varepsilon_s = \varepsilon_\infty + \Delta\varepsilon$. This function takes its maximum value at $(\omega\tau)_{max} = (\varepsilon_s/\varepsilon_\infty)^{1/2}$ so that it can be written

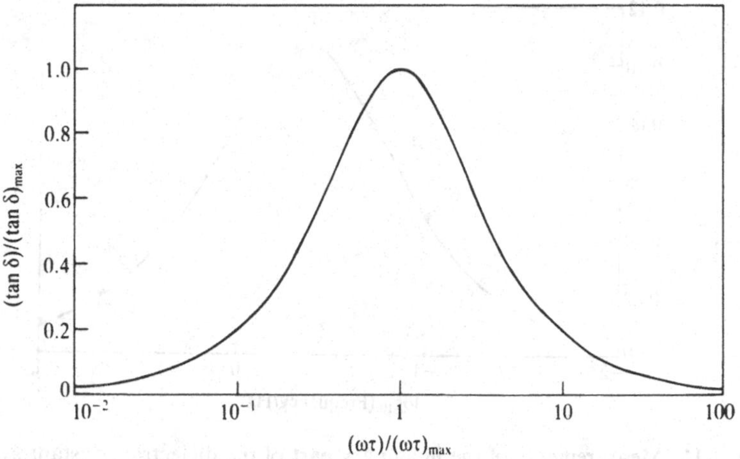


Fig. 1.10. The form of the Debye function for the tangent of the dielectric loss angle, $\tan\delta$, as a function of angular frequency $\omega$. The ordinate gives $\tan\delta$ relative to its maximum value while the abscissa gives $\omega\tau$ relative to the value for which $\tan\delta$ is a maximum (logarithmic scale).

in a normalized form as

$$\frac{\tan\delta}{(\tan\delta)_{\max}} = \frac{2(\omega\tau)/(\omega\tau)_{\max}}{1 + (\omega\tau)^2/(\omega\tau)^2_{\max}}. \tag{1.5.13b}$$

The form of this function is shown in Fig. 1.10. An experimental example of $\varepsilon_2$ is shown in Fig. 1.11.

Another feature of the Debye equations is shown in Fig. 1.12 where $\varepsilon_2(\omega)$ is presented as a function of $\varepsilon_1(\omega)$, the so-called Cole–Cole plot. It is easily verified from (1.5.10) that such a plot should be a semicircle whose centre lies on the $\varepsilon_1$-axis at a point midway between $\varepsilon_\infty$ and $\varepsilon_s$ and whose radius is $\Delta\varepsilon/2$.

We shall see later in Chapters 6 and 7 that the intrinsic response of many insulators can be represented as a sum of Debye-like relaxation modes, i.e. $\Phi(s)$ is a sum of terms like (1.5.8). However, the broad nature of the functions $\varepsilon_1(\omega)$ and $\varepsilon_2(\omega)$ (see Fig. 1.10) means that it may not be easy to separate these modes experimentally. In particular, a Cole–Cole plot (Fig. 1.12) may still appear semi-circular, although the presence of several modes may be indicated by the fact that its centre lies beneath the $\varepsilon_1$-axis. It should also be mentioned that, in practice with ionic conducting solids, it is not only relaxation modes intrinsic to the material which may be detected in these experiments but also effects associated with the electrodes or other inhomogeneities. Techniques of analysis (impedence spectroscopy) which enable one to separate out these effects are described in the book by MacDonald (1987). Even when such complexities are absent, the mere

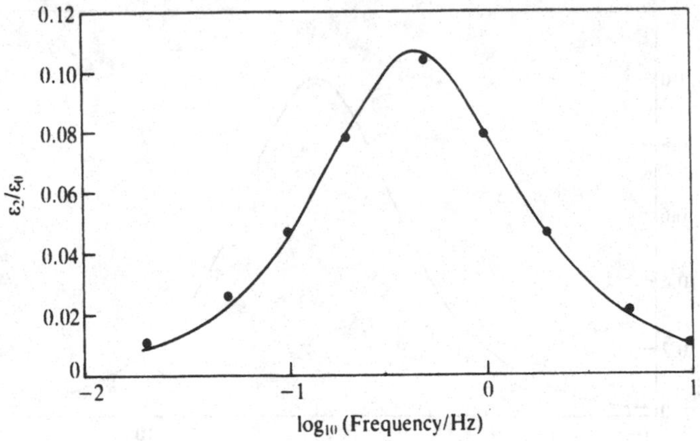


Fig. 1.11. Measurements of the imaginary part of the dielectric constant $\varepsilon_2/\varepsilon_0$ as a function of frequency for NaCl containing substitutional $Ca^{2+}$ ions at $-16.5\,°C$. The line is a Debye curve fitted to the measured points. The estimated concentration of dipoles is $2.6 \times 10^{-4}$ mole fraction. At higher temperatures an additional contribution to the dielectric loss arises from the ionic conductivity; this depends on frequency in a different way. (After Dryden and Heydon, 1978.)

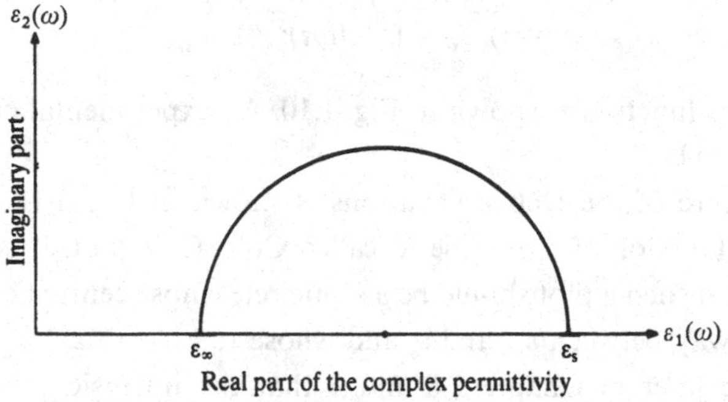


Fig. 1.12. A Cole–Cole plot of the Debye function for tan $\delta$, i.e. $\varepsilon_2(\omega)$ plotted against $\varepsilon_1(\omega)$. The centre of the semi-circle lies on the $\varepsilon_1$ axis at the mean of the high-frequency and low-frequency permittivities.

presence of a background conductivity makes the determination of relaxation modes more difficult – simply because it contributes directly to the observed dielectric loss. Thus with the field (1.5.4) the conduction current $J_e = \kappa E$ in effect contributes an out-of-phase polarization current, i.e. a term $\kappa/\omega$ to $\varepsilon_2$.

Lastly, it should be noted that the conductivity ($\kappa_1$) of some fast ion conductors (e.g. $RbAg_4I_5$ and Na $\beta$-alumina) varies with frequency in a way quantitatively similar to that associated with Debye relaxation processes – although detailed analysis shows that the relaxation function in these cases decays more slowly than (1.5.8) at long times (see e.g. Funke, 1989).

*Kramers–Kronig relations*

For completeness we conclude this section with a remark on the relation between $\varepsilon_1$ and $\varepsilon_2$. Since by (1.5.7) both are expressed in terms of the single response function $\Phi$ it follows that they are not completely independent functions, as has already been exemplified in the particular case of the Debye relaxation equations. The general relations between $\varepsilon_1$ and $\varepsilon_2$ in fact are

$$\varepsilon_1(\omega) - \varepsilon_\infty = \frac{2}{\pi}\mathscr{P}\int_0^\infty \frac{\omega'\varepsilon_2(\omega')\,d\omega'}{\omega'^2 - \omega^2} \tag{1.5.14a}$$

and

$$\varepsilon_2(\omega) = \frac{2}{\pi}\mathscr{P}\int_0^\infty \frac{(\varepsilon_1(\omega') - \varepsilon_\infty)\omega}{\omega^2 - \omega'^2}\,d\omega', \tag{1.5.14b}$$

in which $\mathscr{P}$, as usual, indicates that the principal part of the integrals is to be taken. These relations, which depend on the principle of causality, are derived in Fröhlich (1958) and in Scaife (1989). That the Debye equations (1.5.10) satisfy (1.5.14) is easily verified.

### 1.5.2 *Method of ionic thermocurrents (I.T.C.)*

An ingenious and sensitive way of studying the decay of electric polarization following the removal of a static electric field is the method of *ionic thermocurrents* due originally to Bucci *et al.* (1966) and extensively developed since by Capelletti (see e.g. Capelletti, 1986). It relies on the fact that the relaxation times $\tau$ depend sensitively upon temperature (eqns. 1.5.11) and become very long at low temperatures. The technique is as follows:

(a) Polarize the specimen in a static field at a temperature where the relaxation times are sufficiently short so that the distribution of dipoles comes to equilibrium in the field.

(b) Cool the specimen (in the field) sufficiently rapidly to a low temperature (where the relaxation times are very long) so that the polarization induced in step (a) is 'frozen-in'.

(c) With the specimen held at the low temperature reached in (b) remove the electric field. The polarization, although metastable, remains as in (a) and (b) because the relaxation times are very long at this temperature.

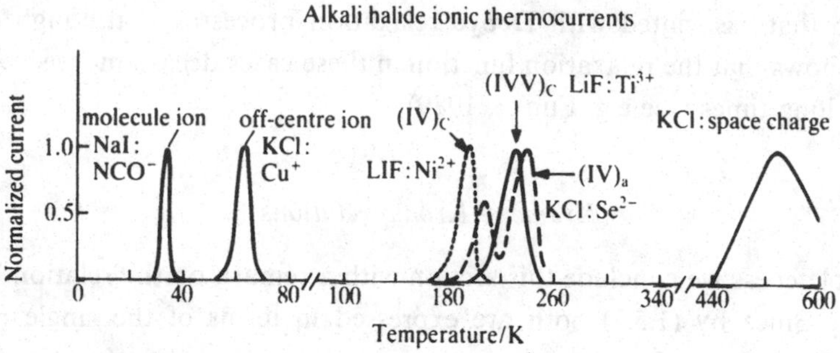


Fig. 1.13. A 'montage' of some of the I.T.C. peaks which have been identified in alkali halide crystals. Those at very low temperatures are associated with solute ions which give rise to electric dipoles that require very little thermal activation to reorient themselves. Those at intermediate temperatures are also associated with solute ions, but the activation energies associated with dipolar reorientation are greater and of roughly the same magnitude as that for the ionic conductivity in conditions where this is dominated by the presence of impurities. Lastly, broad I.T.C. peaks at high temperatures commonly arise from the decay of the electrical space-charge built up at the electrodes when the polarization field is on (stage (a)). A great many detailed studies have been made with this technique. (After Capelletti, 1986.)

(d) Finally, warm the specimen at a controlled rate (generally so that $\mathrm{d}T/\mathrm{d}t$ is constant although a constant $\mathrm{d}T^{-1}/\mathrm{d}t$ technique has also been used) and measure the depolarization current $J = \mathrm{d}P/\mathrm{d}t$ as a function of the instantaneous temperature as the frozen-in polarization melts away.

The depolarization current measured in these experiments will generally appear as a series of peaks (Fig. 1.13). One evident advantage of such I.T.C. measurements over a.c. dielectric loss measurements is the absence of anything corresponding to the background conductivity loss which can make the analysis of tan $\delta$ into separate Debye terms difficult in the a.c. experiments. Quantitative analysis of these I.T.C. measurements may make use of the following features of a well-resolved peak to which only one mode of relaxation contributes.

(1) By (1.5.3) and (1.5.8) the decay of $P$ is given by

$$J_e = \frac{\mathrm{d}P}{\mathrm{d}t} = -E_0\Phi(t)$$

$$= -\frac{E_0\,\Delta\varepsilon}{\tau}\exp(-t/\tau), \qquad (1.5.15a)$$

or, equivalently,

$$J_e = \frac{\mathrm{d}P}{\mathrm{d}t} = -\frac{P}{\tau}. \qquad (1.5.15b)$$

Thus for the initial decay of $P$ ($t$ small), i.e. for the low-temperature side of the current peak

$$J_e = \frac{dP}{dt} = -\frac{E_0\,\Delta\varepsilon}{\tau}. \tag{1.5.16}$$

With $\tau$ given by (1.5.11) and from the known rate of warming, the activation energy $Q$ governing $\tau^{-1}$ may thus be found.

(2) With a constant rate of warming $dT/dt \equiv b$ we can easily determine an equation for the temperature $T_m$ at which the current peak is a maximum by finding $(dJ_e/dT)$ from (1.5.15b) and (1.5.11). This equation is

$$T_m = \left(\frac{bQ\tau(T_m)}{k}\right)^{1/2}. \tag{1.5.17}$$

Hence, knowing $Q$, one may determine also $\tau_0$ from (1.5.17) with (1.5.11).

(3) The total area under the I.T.C. peak gives the dielectric strength $\Delta\varepsilon$ of the corresponding relaxation mode. Thus

$$\int J_e\,dT = b\int J_e\,dt$$
$$= b\,\Delta P = bE_0\,\Delta\varepsilon. \tag{1.5.18}$$

By means of these convenient features the dielectric relaxation modes of many systems have been determined (Capelletti, 1986). However, it should be noted that these determinations depend on the assumption of the Arrhenius behaviour of the relaxation times. We shall see later (Chapter 7) that it is often the case that a given $\tau$ may be determined by more than one distinct type of atomic movement. In these circumstances a simple Arrhenius dependence of $\tau$ on $T$ may be only an approximation, albeit a good one in practice.

## 1.6 Mechanical (anelastic) relaxation

Systems which display dielectric relaxation of the Debye type may often also display an analogous mechanical or anelastic relaxation. At the same time this type of relaxation can also be shown by systems in which dielectric relaxation cannot arise, most notably metals and alloys.

We are here concerned with the relation of elastic strain to applied stress instead of with the relation of electric displacement to electric field, but the two relations are both linear and can be closely analogous. As in §1.5 we begin by reviewing the static relationships and then go on to consider those that apply to time-dependent or anelastic phenomena. (A comprehensive account of this topic has been given by Nowick and Berry, 1972.)

### 1.6.1 Static elasticity

In place of (1.5.1) for static dielectric phenomena we here have as the equivalent local relation

$$\varepsilon = s\sigma, \quad \text{(tensor)}, \tag{1.6.1}$$

in which $\sigma$ denotes the stress tensor (with Cartesian components $\sigma_{\kappa\lambda}$, $\kappa$ and $\lambda = 1, 2, 3$) and $\varepsilon$ the associated strain tensor (with Cartesian components $\varepsilon_{\kappa\lambda}$). The strain tensor is defined in the usual way in terms of derivatives of the displacement field, $u$, of particles in the solid, i.e.

$$\varepsilon_{\kappa\lambda} = \frac{1}{2}\left(\frac{\partial u_\kappa}{\partial x_\lambda} + \frac{\partial u_\lambda}{\partial x_\kappa}\right). \tag{1.6.2}$$

Both tensors are of second rank and symmetric, i.e. $\sigma_{\kappa\lambda} = \sigma_{\lambda\kappa}$ and $\varepsilon_{\kappa\lambda} = \varepsilon_{\lambda\kappa}$ (Nye, 1985). It follows that the elastic compliance $s$ is a fourth-rank tensor (with Cartesian components $s_{\kappa\lambda\mu\nu}$). In general, there are thus six independent stress- and six independent strain components, and hence 36 compliances, although thermodynamic arguments show that, in fact, only 21 of these are independent (Nye, 1985).

An alternative to the tensor notation, made possible by the symmetric nature of the stress and strain tensors and widely used in practice, is the matrix notation of Voigt. Having chosen the Cartesian axes ($x_1, x_2, x_3$), we then introduce a six-component column matrix for the stresses

$$\{\sigma_1\,\sigma_2\,\sigma_3\,\sigma_4\,\sigma_5\,\sigma_6\} \equiv \{\sigma_{11}\,\sigma_{22}\,\sigma_{33}\,\sigma_{23}\,\sigma_{31}\,\sigma_{12}\}, \tag{1.6.3a}$$

and for the strains somewhat similarly

$$\{\varepsilon_1\,\varepsilon_2\,\varepsilon_3\,\tfrac{1}{2}\varepsilon_4\,\tfrac{1}{2}\varepsilon_5\,\tfrac{1}{2}\varepsilon_6\} \equiv \{\varepsilon_{11}\,\varepsilon_{22}\,\varepsilon_{33}\,\varepsilon_{23}\,\varepsilon_{31}\,\varepsilon_{12}\}. \tag{1.6.3b}$$

The strains $\varepsilon_1 \ldots \varepsilon_6$ introduced by (1.6.3b) are called *engineering strains.* The linear relations between these $\varepsilon_i$ and the stresses $\sigma_j$ corresponding to (1.6.1) are then

$$\varepsilon_i = \sum_{j=1}^{6} s_{ij}\sigma_{ij}, \tag{1.6.4}$$

in which the matrix of compliance coefficients $s_{ij}$ is derived from the compliance tensor as follows. With index $i$ corresponding to the pair $(\kappa, \lambda)$ and $j$ to the pair $(\mu, \nu)$,

$$\begin{aligned} s_{ij} &= s_{\kappa\lambda\mu\nu} && \text{for } i, j = 1, 2 \text{ or } 3 \\ s_{ij} &= 2s_{\kappa\lambda\mu\nu} && \text{for either } i \text{ or } j = 4, 5 \text{ or } 6 \\ s_{ij} &= 4s_{\kappa\lambda\mu\nu} && \text{for both } i \text{ and } j = 4, 5 \text{ or } 6. \end{aligned} \tag{1.6.5}$$

We shall use (1.6.4) as the basis of what follows. We emphasize, however, that (1.6.4) is a matrix relation and not a tensor relation. The numerical factors in (1.6.3b) and (1.6.5) are needed to ensure that (1.6.4) is equivalent to the tensor equation (1.6.1). For compactness we shall write it in matrix notation as

$$\boldsymbol{\varepsilon} = \mathbf{s}\boldsymbol{\sigma}, \quad \text{(matrix)}. \tag{1.6.6}$$

By thermodynamic arguments relating to the work done in a deformation one may deduce that $s_{ij} = s_{ji}$, i.e. the matrix $\mathbf{s}$ is symmetric. The inverse relationship to (1.6.4), giving stresses in terms of the strains, introduces the elastic stiffness constants, $c_{ij}$,

$$\sigma_i = \sum_{j=1}^{6} c_{ij}\varepsilon_j. \tag{1.6.7}$$

Evidently the matrix $\mathbf{c}$ is the inverse of the matrix $\mathbf{s}$, i.e. $\mathbf{c} = \mathbf{s}^{-1}$, whence it follows from the symmetric nature of $\mathbf{s}$ that $\mathbf{c}$ is also symmetric.

It is clear that the numerical values of the coefficients $s_{ij}$ and $c_{ij}$ depend upon the choice of the co-ordinate system $(x_1, x_2, x_3)$. In the case of the more highly symmetric crystals the best choice is fairly obvious. For example, for cubic crystals (which we shall be mostly concerned with) it is natural to choose the axes of $x_1$, $x_2$ and $x_3$ to be parallel to the three cube axes; and similarly for tetragonal and orthorhombic crystals. In less symmetrical crystals there are certain conventions in use. Tabulations of $s_{ij}$ and $c_{ij}$ coefficients for particular solids are nowadays usually made on the basis of the above definitions and conventions, when they may be called *characteristic* elastic coefficients. These bring not only the benefits of standardization, but also of a reduction in the number of distinct, non-zero coefficients. For example, for cubic crystals and with axes parallel to the cube axes, $\mathbf{s}$ has the form

$$\mathbf{s} = \begin{bmatrix} s_{11} & s_{12} & s_{12} & 0 & 0 & 0 \\ s_{12} & s_{11} & s_{12} & 0 & 0 & 0 \\ s_{12} & s_{12} & s_{11} & 0 & 0 & 0 \\ 0 & 0 & 0 & s_{44} & 0 & 0 \\ 0 & 0 & 0 & 0 & s_{44} & 0 \\ 0 & 0 & 0 & 0 & 0 & s_{44} \end{bmatrix}, \tag{1.6.8}$$

Table 1.2. *Symmetrized stresses, strains and compliances for cubic crystals*

| Symmetry Designation | Stress | Compliance | Strain |
|---|---|---|---|
| $A_g$ | $\sigma_1 + \sigma_2 + \sigma_3$ | $s_{11} + 2s_{12}$ | $\varepsilon_1 + \varepsilon_2 + \varepsilon_3$ |
| $E_g$ | $2\sigma_1 - \sigma_2 - \sigma_3$ | $s_{11} - s_{12}$ | $2\varepsilon_1 - \varepsilon_2 - \varepsilon_3$ |
| | $\sigma_2 - \sigma_3$ | $s_{11} - s_{12}$ | $\varepsilon_2 - \varepsilon_3$ |
| $T_g$ | $\sigma_4$ | $s_{44}$ | $\varepsilon_4$ |
| | $\sigma_5$ | $s_{44}$ | $\varepsilon_5$ |
| | $\sigma_6$ | $s_{44}$ | $\varepsilon_6$ |

i.e. there are only three distinct compliances $s_{11}$, $s_{12}$ and $s_{44}$. The form of **c** is similar. The forms for other crystal classes are given by Nye (1985, Chap. VIII).

Since the matrix of compliances $s_{ij}$ is symmetric it follows that one may find a similarity transformation which brings it into diagonal form, i.e. there is a transformation matrix **Q** that $\mathbf{Q}^{-1}\mathbf{s}\mathbf{Q}$ is diagonal. Under this transformation (1.6.6) becomes

$$(\mathbf{Q}^{-1}\varepsilon) = (\mathbf{Q}^{-1}\mathbf{s}\mathbf{Q})(\mathbf{Q}^{-1}\sigma), \tag{1.6.9}$$

which defines the corresponding transformed strains $(Q^{-1}\varepsilon)_i \equiv \varepsilon_i'$ and transformed stresses $(Q^{-1}\sigma)_j \equiv \sigma_j'$. By the diagonal nature of $(\mathbf{Q}^{-1}\mathbf{s}\mathbf{Q}) \equiv \mathbf{s}'$ each transformed strain element $\varepsilon_i'$ is related to the corresponding transformed stress element $\sigma_i'$ by a simple scalar, Hooke's law relation, viz.

$$\varepsilon_i' = s_{ii}'\sigma_i'. \tag{1.6.10}$$

The determination of **Q** and the diagonalization of **s** is usually achieved by using group theory on the basis of the symmetry group of the crystal class in question. The transformed stresses $\boldsymbol{\sigma}'$ and strains $\boldsymbol{\varepsilon}'$ are thus referred to as *symmetrized* stresses and strains. Table 1.2 shows these transformed quantities and their corresponding symmetry designations for cubic crystals. The form of **s** for this case is sufficiently simple (cf. 1.6.8) that the correctness of the entries can easily be verified directly. However, the symmetry designation is useful when it comes to the analysis of particular sources of anelastic relaxation (see Chapter 7).

### *1.6.2 Time-dependent responses*

After this review of the general characteristics of static elasticity theory we can proceed to the generalization needed for time-dependent elastic responses. Despite the complications introduced by the higher tensor rank of the quantitites involved here compared to those arising in dielectric theory, the linearity of both theories ensures that the essential nature of the generalization to time-dependent behaviour is the same. Indeed the formalism represented by the dielectric equations (1.5.2) *et seq.* can be taken over with only trivial changes of notation as long as they are interpreted now as matrix equations. Thus, there will be an unrelaxed matrix of compliances $\mathbf{s}_\infty$ in place of the electric permittivity tensor $\boldsymbol{\varepsilon}_\infty$ and a matrix of response coefficients in place of the tensor quantity $\Phi(t-t')$ appearing in (1.5.2a). Furthermore, for the important case of sinusoidally varying stresses the structure presented in §1.5.1 has its direct analogue here. We shall not therefore re-express all these equations again, but merely note the following points.

(i) The same symmetry considerations apply to the matrix of response functions and to the matrix of frequency-dependent compliances (analogues of $\varepsilon_1 - i\varepsilon_2$ in §1.5.1) as apply to the matrix of static compliances. Hence the analogues of eqns. (1.5.10) define increments $\Delta s'$ in the symmetrized compliances associated with the anelastic behaviour of the solid.

(ii) The real and imaginary parts of the complex compliances satisfy Kramers–Kronig relations analogous to (1.5.14).

(iii) Each type of symmetrized stress or strain (e.g. $A_g$, $E_g$ or $T_g$ for cubic solids) specifies a characteristic relaxation time (cf. 1.5.10) or set of relaxation times. These anelastic relaxation times differ from the dielectric relaxation times even when the same underlying mechanism is responsible. Examples are given later in Chapter 7. For the most part anelastic relaxation times are also thermally activated (cf. 1.5.11).

We conclude this section by noting that relaxation times closely related (if not identical) to those introduced here and in §1.5 can sometimes be obtained by spectroscopic means, especially optical and paramagnetic resonance methods. When these can be used they often allow the identification of the electronic and atomic structure of the responsible centres and the study of rates of re-orientation of these centres, either directly (e.g. Symmons, 1970; Watkins, 1975) or through measurements of lifetime broadening (e.g. Watkins, 1959; Franklin, Crissman and Young, 1975). The application of stress, for example, can be used to establish a particular distribution of these centres among the various available orientations; removal of the stress will lead to the relaxation of this distribution which can be followed by these spectroscopic means. Such specific information is of very considerable value to the interpretation of more macroscopic dielectric and anelastic relaxation processes. We return to the matter again in Chapter 7.

## 1.7 Nuclear magnetic resonance and relaxation times

The technique of nuclear magnetic resonance has been widely used for many years to give detailed information about condensed matter, especially its atomic and electronic structure. It was recognized early on that in favourable cases such measurements could also provide information on atomic movements through the influence which these had on the width of nuclear magnetic resonance lines. There are several basic causes of this width. Leaving aside inhomogeneities in the steady magnetic field, we may list (i) local internal magnetic field variations resulting from dipole–dipole interactions among the nuclear magnetic moments (ii) the interaction of nuclear electric quadrupole moments (for spins $I > 1/2$) with internal electric field gradients (iii) the interactions of nuclear magnetic moments with internal fields created by paramagnetic impurities and (iv) the coupling of nuclear magnetic moments in metals with conduction electrons through the hyperfine interaction. When the atomic nuclei are in sufficiently rapid motion the contributions from the first two are greatly reduced because local fields are averaged out and the line narrows. Hence in diamagnetic insulators, where couplings (iii) and (iv) are absent, the resonance line narrows as increases in atomic movements occur at high temperatures. In a few metals (e.g. the alkalis) it is still possible to see this effect even though in most the hyperfine interaction with the conduction electrons largely determines the width and thus prevents the observation of atomic migration by these means.

The analysis of nuclear magnetic resonance line shapes and widths proceeds via a consideration of the detailed interactions among nuclear moments and between them and other components of the solid (electrons, paramagnetic impurities, defects, etc.). This theory has been extensively developed over the past 40 years (see. e.g. Abragam, 1961; Slichter, 1978). Although this necessarily demands the use of quantum mechanics much can be represented by the semi-classical equations proposed originally by Bloch. These introduce two characteristic relaxation times, the spin–lattice relaxation time $T_1$ and the spin–spin relaxation time $T_2$. These parameters serve to characterize continuous-wave nuclear magnetic resonance line shapes and other features under many conditions. Furthermore, pulse techniques allow the direct measurement of these and other relaxation times (see e.g. Gerstein and Dybowski, 1985). We shall therefore regard them here as the basic phenomenological quantities. We introduce them next via the Bloch equations and then present the expressions for these quantities in atomic terms. The evaluation of these expressions for appropriate models of atomic movements is then the task of Chapter 12.

The phenomenological equations of Bloch provide the equations of motion of the total nuclear magnetization, **M**, in a magnetic field, **B** (with Cartesian components $B_x$, $B_y$ and $B_z$). This field in general is made up of a strong, spatially homogeneous static field $B_0$ in the $z$-direction and a much smaller applied radio-frequency field $B_1$, which is used to excite transitions among the magnetic sub-levels in the static field $B_0$ (as in continuous-wave resonance experiments) or to probe the system in other ways. The Bloch equations are

$$\frac{\mathrm{d}M_x}{\mathrm{d}t} = \gamma(M_y B_z - M_z B_y) - \frac{M_x}{T_2}, \tag{1.7.1a}$$

$$\frac{\mathrm{d}M_y}{\mathrm{d}t} = \gamma(M_z B_x - M_x B_z) - \frac{M_y}{T_2}, \tag{1.7.1b}$$

$$\frac{\mathrm{d}M_z}{\mathrm{d}t} = \gamma(M_x B_y - M_y B_x) - \frac{(M_z - M_0)}{T_1}. \tag{1.7.1c}$$

The first term in all three equations in classical terms describes the free precession of the spins about the field $\mathbf{B}$; $\gamma$ is the gyromagnetic ratio between the nuclear magnetization $\mathbf{M}$ and the nuclear angular momentum. The second terms give the rate of relaxation of the magnetization and define the relaxation times $T_2$ for the transverse components and $T_1$ for the longitudinal component. The quantity $\mathbf{M}_0$ is the equilibrium nuclear magnetization in the field $\mathbf{B}_0$ and as such is in the $z$-direction. In the absence of any transverse field, $T_1$ clearly determines the rate at which $M_z$ returns to its thermodynamic equilibrium value, $M_0$, and is referred to as the spin–lattice relaxation time. However, it is possible for the nuclear spins to be brought to a state of quasi-thermal equilibrium among themselves without being in thermal equilibrium with the lattice. The relaxation time to such a state is $T_2$ and is called the spin–spin relaxation time. It follows that $T_2 \leq T_1$.

These Bloch equations allow one to describe the shapes and saturation characteristics of resonance lines in continuous-wave experiments (Abragam, 1961, especially Chap. 3). In this way the relaxation times can be inferred and their dependence on $\omega_0 = \gamma B_0$, temperature and other parameters studied. However, elegant techniques of pulsed radio-frequency spectroscopy have also been devised and these permit the direct determination of $T_1$ and $T_2$ (see e.g. Gerstein and Dybowski, 1985). In addition, other relaxation times not contained in the Bloch equations can be operationally defined. The best known of these is $T_{1\rho}$ referred to as the *spin-lattice relaxation time in the rotating frame.* This relaxation time characterizes the decay of the magnetization when it is 'locked' parallel to $\mathbf{B}_1$ in a frame of reference rotating about $\mathbf{B}_0$ with the (precessional) frequency $\omega_0 = \gamma B_0$. In such an experiment $\mathbf{M}$ starts from $\mathbf{M}_0$ and decays effectively to zero (actually to $B_1 M_0/B_0$). $T_{1\rho}$ is shorter than $T_1$. A further relaxation time which also lies outside the scheme of Bloch's equations and which characterizes the re-ordering

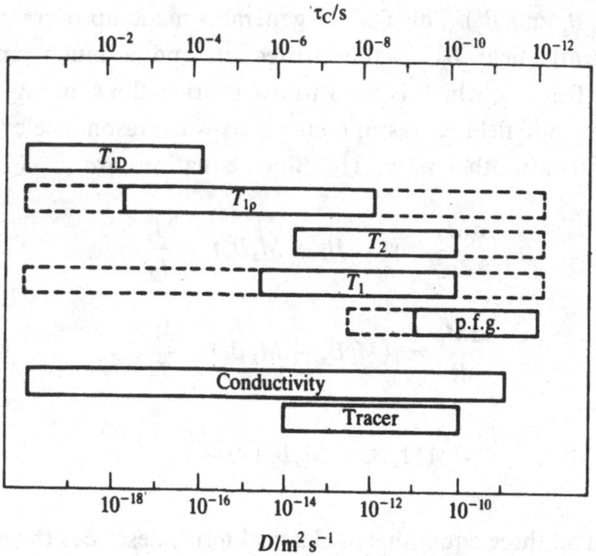


Fig. 1.14. The range of diffusion coefficients that can be studied by nuclear magnetic relaxation methods. The ranges indicated by the full lines are those which have been explored in studies of $F^-$ ion migration in crystals with the fluorite structure. The dashed lines indicate the actual and potential ranges in other systems. (After Chadwick, 1988).

of spins in the local dipolar fields can be defined. This is denoted by $T_{1D}$. This time in turn is shorter than $T_{1\rho}$. In so far as measures of all these relaxation times can give information about atomic diffusion, the ranges of diffusion coefficients which can be conveniently determined from them are shown in Fig. 1.14. All four relaxation times are determined in spatially uniform fields (both $B_0$ and $B_1$).

Another technique, known as the pulsed field-gradient method, by deliberately using inhomogeneous fields follows the diffusion of spins directly and is thus comparable to a tracer diffusion experiment. The range of values of $D$ which can be determined this way is also indicated in Fig. 1.14.

We turn now to the theory by which these relaxation times are related quantitatively to atomic movements. Apart from coupling to the spins of conduction electrons (as in metals) or of paramagnetic impurities (in non-metals) there are two basic mechanisms of relaxation to be considered in relation to atomic movements. The first is due to dipole–dipole coupling among the nuclear magnetic moments. The second arises from the interaction of nuclear electric quadrupole moments (present when the nuclear spin $I > 1/2$) with internal electric field gradients. Both these may be influenced strongly by atomic movements.

Firstly, we suppose that the system consists of just one type of nuclear spin $I$ and that we can neglect any quadrupole effects. In a quantum theory therefore we shall first specify the Hamiltonian of this system as the Zeeman term for the energy of these spins in the applied magnetic field ($\mathbf{B}_0 + \mathbf{B}_1(t)$) plus the dipolar interactions among the spins. The dipolar interaction Hamiltonian $\mathscr{H}_{ij}$ for any one pair of spins $(i, j)$ can be written in spin operator form as

$$\hbar\mathscr{H}_{ij} = \sum_{p=-2}^{2} F_{ij}^{(p)} A_{ij}^{(p)}, \tag{1.7.2}$$

in which the $A_{ij}^{(p)}$ are spin operators connecting spin states of the $(i, j)$ pair which differ in combined angular momentum $(m_i + m_j)$ by integral multiples of $\hbar$ ($\hbar p$). The summation runs over $p$ from $-2$ to 2. The form of the spin operators is given by Abragam (1961, especially p. 289). The quantities $F_{ij}^{(p)}$ come from the spatial and angular dependence of the dipole–dipole interaction and are

$$F_{ij}^{(0)} = r_{ij}^{-3}(1 - 3\cos^2\theta_{ij}), \tag{1.7.3a}$$

$$F_{ij}^{(1)} = r_{ij}^{-3}\sin\theta_{ij}\cos\theta_{ij}\exp(-i\phi_{ij}), \tag{1.7.3b}$$

$$F_{ij}^{(2)} = r_{ij}^{-3}\sin^2\theta_{ij}\exp(-2i\phi_{ij}), \tag{1.7.3c}$$

$$F_{ij}^{(-p)} \equiv F_{ij}^{(p)*}. \tag{1.7.3d}$$

Here $r_{ij}$, $\theta_{ij}$, $\phi_{ij}$ are the polar co-ordinates of the vector $\mathbf{r}_{ij}$ connecting spins $i$ and $j$ with the polar axis defined by the direction of the applied magnetic field, $\mathbf{B}_0$, but with the azimuthal plane arbitrary.

In a rigid lattice these dipolar interactions provide a relatively efficient mode of spin–spin relaxation giving values of $T_2$ typically $\sim 10^{-4}$ s. However, through the factors $F_{ij}^{(p)}$ there is a strong dependence on atomic movements and, while lattice vibrations are not very significant in this respect, diffusive jumps of the atoms from one lattice site to another average out the dipolar interactions and thus reduce the rate of spin–spin relaxation. In other words $T_2$ increases rapidly with temperature once the mean time of stay, $\tau$, of an atom on any one lattice site has become shorter than $T_2$ in the rigid lattice. In the presence of quadrupolar interactions, however, lattice vibrations are of greater significance and contribute to faster relaxation rates.

The theory of spin relaxation times then shows that these are expressible in terms of the spectral functions of the time-correlation functions of the $F^{(p)}$. These time correlations are defined by

$$F^{(p)}(s) = \sum_j \langle F_{ij}^{(p)}(t) F_{ij}^{(p)*}(t + s)\rangle, \tag{1.7.4}$$

in which $F_{ij}^{(p)}(t)$ is the value of $F_{ij}^{(p)}$ at time $t$. The angular brackets denote an equilibrium ensemble average, the result of which will be independent of $t$. From this it follows immediately that $F(-s) = F^*(s)$. Furthermore, since the system is uniform this correlation must be the same for all pairs of spins (as long as they are of the same kind, i.e. structurally the same and the same chemical and isotopic species as here assumed). The spectral densities of the $F^{(p)}(s)$ are then simply the Fourier coefficients

$$J^{(p)}(\omega) = \int_{-\infty}^{\infty} F^{(p)}(s)\,\mathrm{e}^{-i\omega s}\,\mathrm{d}s$$

$$= 2\,\mathrm{Re}\int_{0}^{\infty} F^{(p)}(s)\,\mathrm{e}^{-i\omega s}\,\mathrm{d}s. \tag{1.7.5}$$

By analyses such as those given by Abragam (1961) the following expressions for the relaxation times are obtained

$$\frac{1}{T_1} = \tfrac{3}{2}\gamma_I{}^4\hbar^2 I(I+1)(J^{(1)}(\omega_0) + J^{(2)}(2\omega_0)), \tag{1.7.6a}$$

$$\frac{1}{T_2} = \gamma_I{}^4\hbar^2 I(I+1)(\tfrac{3}{8}J^{(0)}(0) + \tfrac{15}{4}J^{(1)}(\omega_0) + \tfrac{3}{8}J^{(2)}(2\omega_0)), \tag{1.7.6b}$$

$$\frac{1}{T_{1\rho}} = \tfrac{3}{8}\gamma_I{}^4\hbar^2 I(I+1)(J^{(0)}(2\omega_1) + 10J^{(1)}(\omega_0) + J^{(2)}(2\omega_0)), \tag{1.7.6c}$$

in which $\gamma_I$ is the gyromagnetic ratio in Gaussian units or the gyromagnetic ratio in SI units multiplied by $(\mu_0/4\pi)^{1/2}$ and $\omega_0$ and $\omega_1$ are the Larmor frequencies in the static magnetic field $B_0$ and in the radio-frequency magnetic field $B_1$ respectively.

It will be noted that there is an implied dependence of the $J^{(p)}(\omega)$ and hence of $T_1$, $T_2$, etc. upon the orientation of the static magnetic field relative to the crystal axes which comes from the spatial transformation properties of the $F_{ij}^{(p)}$ functions. For a cubic crystal and a static magnetic field direction specified by polar and azimuthal angles $\theta$ and $\phi$ respectively with respect to the cubic axes the orientation dependence of the spectral functions and thus of the relaxation times is of the form

$$A_1^{(p)} + A_2^{(p)}(\sin^2 2\theta + \sin^4\theta\,\sin^2 2\phi)$$

in each case, as was shown by Wolf (1979). Furthermore at any particular $\omega$ only two of the six coefficients $A_1^{(p)}$ and $A_2^{(p)}$ $(p = 0, 1, 2)$ are independent. If these are chosen to be $A_1^{(0)}$ and $A_2^{(0)}$ the others are given by

$$A_1^{(1)} = \tfrac{1}{6}A_1^{(0)} + \tfrac{2}{9}A_2^{(0)}, \tag{1.7.7a}$$

$$A_2^{(1)} = -\tfrac{1}{9}A_2^{(0)}, \tag{1.7.7b}$$

$$A_1^{(2)} = \tfrac{2}{3}A_1^{(0)} + \tfrac{4}{9}A_2^{(0)}, \tag{1.7.7c}$$

$$A_2^{(2)} = \tfrac{1}{9}A_2^{(0)}. \tag{1.7.7d}$$

(Wolf, 1979). If $J^{(p)}(\omega)$ is averaged over all magnetic field directions (as for a powder or polycrystal) then $\langle J^{(p)}(\omega)\rangle$ is $A_1^{(p)} + 0.8A_2^{(p)}$ and $\langle J^{(0)}(\omega)\rangle : \langle J^{(1)}(\omega)\rangle : \langle J^{(2)}(\omega)\rangle$ are as 6:1:4. Some corresponding results for other crystal classes are listed by Sholl (1986). Naturally, the complexity of the expression increases as the symmetry of the crystal becomes lower.

If, as in the original paper of Bloembergen, Purcell and Pound (1948), we assume that the correlation functions $F^{(q)}(t)$ decay exponentially with a single time constant $\tau$, i.e. as

$$F^{(q)}(t) = F^{(q)}(0)\,\mathrm{e}^{-t/\tau}, \tag{1.7.8}$$

then

$$J^{(q)}(\omega) = \frac{F^{(q)}(0)2\tau}{1+\omega^2\tau^2}. \tag{1.7.9}$$

Since the movement of either atom of the pair will change $F_{ij}$ we identify $\tau$ in such a model with just one-half of the mean time of stay of an atom on any lattice site, $\tau_s$. When the thermally activated Arrhenius form is substituted for $\tau_s$ then it is found that (1.7.6) with (1.7.9) often give a good description of the observed dependence of $T_1$, $T_2$, etc. upon the temperature $T$. An illustration is given in Fig. 1.15.

As this example shows, it is possible to measure these relaxation times precisely, and thus to infer accurate values for the mean time of stay of any atom on a lattice site. However, without further analysis (1.7.8) remains an assumption. Such further analysis is the subject of Chapter 12. In principle this not only provides a test of the validity of (1.7.8), but also allows us to make predictions sufficiently accurately to distinguish the action of one specific mechanism of atomic movement from another. In practice though the circumstances where this is possible are rather few. For a start there are relatively few systems which conform to the assumptions detailed before (1.7.2), namely a single spin species (apart from those with $I = 0$) and magnetic dipole–dipole interactions as the sole mechanism of nuclear spin relaxation. When more than one isotopic or nuclear species with $I > 0$ is present then each will have its own set of relaxation times and the analogues of eqns. (1.7.6) will contains additional spectral functions deriving from the correlations between different species. This obviously makes the analysis of experimental measurements more complicated, although on the other hand more information is obtainable.

An additional important effect derives from the coupling of the nuclear electric quadrupole moment, $Q$, with an internal electric field gradient acting on the nucleus, although such interactions are only possible when $I > 1/2$. In perfect cubic crystals there can be no internal electric field gradients at the lattice sites, but in imperfect crystals such fields may indeed exist and give rise to measurable effects on the nuclear magnetic resonance line widths and line shapes. In magnitude these

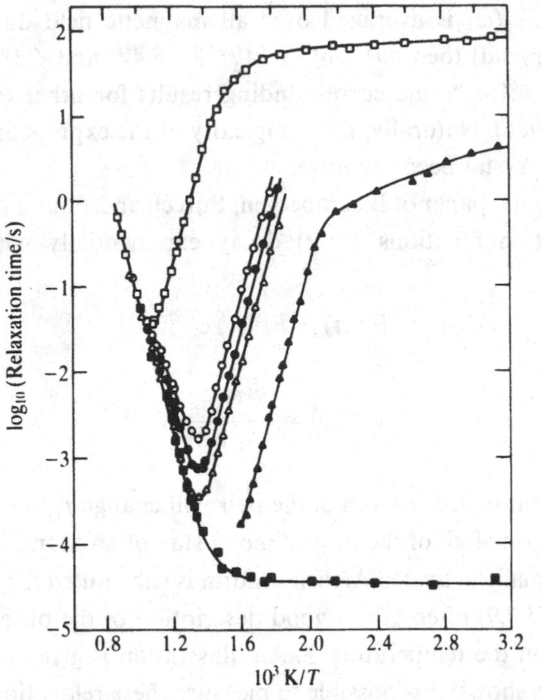


Fig. 1.15. Temperature dependence of the $^{19}F$ nuclear magnetic relaxation times for various orientations of a pure $BaF_2$ single crystal. The symbols have the following significance: □, $T_1$ for $\mathbf{B}_0$ parallel to the ⟨110⟩ crystal direction; ○, $T_{1\rho}$ at $B_1 = 12 \times 10^{-4}$ T and $\mathbf{B}_0$//⟨111⟩, ● ditto $\mathbf{B}_0$//⟨110⟩ and △ ditto $\mathbf{B}_0$//⟨100⟩; ▲, $T_{1D}$ for $\mathbf{B}_0$//⟨110⟩; ■, $T_2$ for $\mathbf{B}_0$//⟨100⟩. The V-shaped portions of the curves for $T_1$ and $T_{1\rho}$ and the shape of the curve for $T_2$ are all as expected from eqns. (1.7.6) and (1.7.9) when the Arrhenius form for $\tau$ is inserted. The spin–lattice relaxation at the lowest temperatures shown here is probably associated with the presence of paramagnetic impurity ions. (After Wolf, Figueroa and Strange, 1977.)

effects may be comparable with those arising from the dipole–dipole interaction, but since, in cubic crystals, they derive solely from the existence of imperfections they do provide specific information about these, rather than about all atoms in the crystal. In particular, this is true of motional effects.

## 1.8 Quasi-elastic neutron scattering

From measurements of the scattering of beams of slow neutrons obtained from nuclear reactors or other high intensity sources one may study a great range of structural and dynamical properties of condensed matter generally and of solids in particular. Although this has been recognised for many years now, modern high

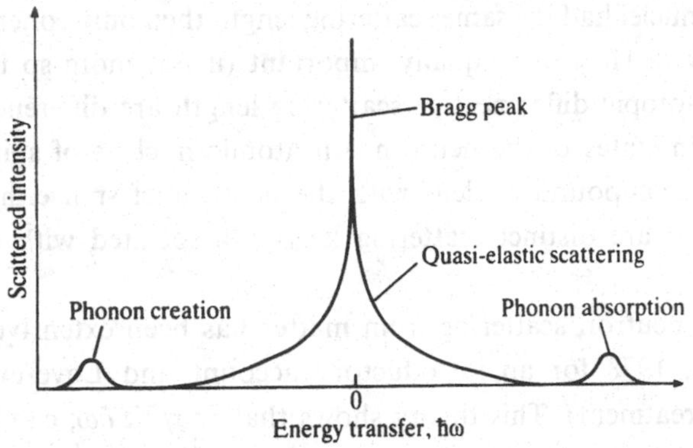


Fig. 1.16. Schematic diagram showing the intensity of thermal-neutron scattering from a crystalline solid at a particular scattering vector **q** as a function of energy transfer $\hbar\omega$. The central sharp line is a Bragg diffraction peak ($\omega = 0$) while the small peaks represent scattering by the excitation ($\omega_{\mathbf{q}} < 0$) or the absorption ($\omega_{\mathbf{q}} > 0$) of lattice phonons. The broad scattering centred around the elastic Bragg peak is the quasi-elastic scattering associated with the diffusive movements of atoms among the available lattice sites. Theory shows that this total scattering may be divided into coherent and incoherent parts (eqn. 1.8.1), although the Bragg peak comes wholly from the coherent part.

intensity sources and high resolution spectrometers have allowed the study of diffusive movements of atoms in solids to a far greater extent than formerly.

From a theoretical point of view the basic quantity obtained by such measurements is the partial differential scattering cross-section, $\partial^2\sigma/\partial\Omega\,\partial\omega$, for the scattering of a collimated mono-energetic beam of neutrons into a solid angle lying between $\Omega$ and $\Omega + d\Omega$ and with an energy loss by the neutrons lying between $\hbar\omega$ and $\hbar(\omega + d\omega)$. This quantity can be measured as a function of both scattering vector, **q**, and $\omega$ by means of triple-axis spectrometers. At the Bragg diffraction conditions elastic scattering without energy loss occurs ($\hbar\omega = 0$). Dynamical wave-like excitations such as phonons and spin waves give rise to inelastic scattering peaks among the Bragg diffraction peaks. One may also detect the diffusive motions of atoms under favourable circumstances. In this case since the movements are stochastic and relatively slow they are observable as a diffuse scattering around the elastic Bragg peaks called quasi-elastic scattering (Fig. 1.16).

For simplicity let us suppose for the moment that the solid is made up of just one kind of atom, although there will generally be various isotopes, i, present and each of these will be characterized by a distinct *scattering length*, $b_i$ which gives the (isotropic) scattering cross-section for neutrons from a single nucleus of that type, namely $4\pi b_i^2$. The presence of different isotopes distributed randomly in the solid means that the total scattering is made up of two parts, called *coherent* and *incoherent*. If all nuclei had the same scattering length then only coherent scattering would be observed. However, equally important (if not more so in the present connection) as isotopic differences in scattering length are differences associated with the two spin states of the neutron. An atomic nucleus of spin $I$ may thus form a transient compound nucleus with the neutron of spin either $I + 1/2$ or $I - 1/2$, and there are distinct scattering lengths associated with each of these possibilities.

The theory of neutron scattering from matter has been extensively developed (see e.g. Squires, 1978 for an introductory account and Lovesey, 1986 for a comprehensive treatment). This theory shows that $\partial^2\sigma/\partial\Omega\,\partial\omega$, can be written as the sum of a coherent and an incoherent part

$$\frac{\partial^2\sigma}{\partial\Omega\,\partial\omega} = \left(\frac{\partial^2\sigma}{\partial\Omega\,\partial\omega}\right)_{\text{coh}} + \left(\frac{\partial^2\sigma}{\partial\Omega\,\partial\omega}\right)_{\text{inc}}$$

$$\equiv n\frac{k_1}{k_0}[b_{\text{coh}}{}^2 S(\mathbf{q}, \omega) + b_{\text{inc}}{}^2 S_s(\mathbf{q}, \omega)], \tag{1.8.1}$$

where $n$ is the number of scattering nuclei per unit volume,

$$b_{\text{coh}} = \langle b \rangle, \tag{1.8.2}$$

and

$$b_{\text{inc}} = (\langle b^2 \rangle - \langle b \rangle^2)^{1/2}, \tag{1.8.3}$$

in which the averages indicated by the angular brackets are simple averages over the various isotopes present and their possible nuclear spin states. In (1.8.1) $\mathbf{k}_0$ is the wave-vector of the incident neutrons and $\mathbf{k}_1$ is that of the scattered neutrons, the *scattering vector* $\mathbf{q}$ being the difference $\mathbf{k}_0 - \mathbf{k}_1$. The first term in (1.8.1) gives the scattered intensity one would obtain if all the nuclei had the same average scattering length, $b_{\text{coh}}$. In such a situation the waves scattered from different atoms of the crystal would be capable of interference, which is why this term is called *coherent*. Among other things, this term gives the Bragg diffraction peaks ($\omega = 0$). The second term gives the remainder of the cross-section and is called *incoherent*. It is a sum of contributions from the individual atoms, each such contribution representing the summation (interference) of waves scattered by a single atom at various times. When more than one kind of atom is present (e.g. compounds) there is a corresponding generalization of (1.8.1) for which the interpretation is similar. (It may be noted that the differential cross-section 1.8.1 is sometimes written on the left side as $\partial^2\sigma/\partial\Omega\,\partial E$ with $E = \hbar\omega$ but the additional factor $\hbar$ which should then appear on the right side absorbed into the scattering functions.)

It should be emphasized that it is the *theory* of neutron scattering which leads to the separation into coherent and incoherent terms in (1.8.1). The direct experimental determination of the two separate functions $S(\mathbf{q}, \omega)$ and $S_s(\mathbf{q}, \omega)$ is not usually straightforward unless specimens of different isotopic composition are available, for then the coefficients $b_{\text{coh}}^2$ and $b_{\text{inc}}^2$ will vary with the composition while the scattering functions are essentially the same (apart from generally negligible differences coming from the effects of the different masses on the dynamical behaviour). However, by using particular information experimentalists are often able to separate these two contributions without the luxury of major changes in isotopic composition. For example, the incoherent cross-section for hydrogen atoms is some 40 times the coherent cross-section, while in the case of oxygen atoms there is no incoherent scattering because there is only one isotope $O^{16}$ and this has $I = 0$. With solids containing just a single isotope one has the additional possibility of separating the two terms by using spin-polarized neutron beams.

Neutron scattering theory then allows $S(\mathbf{q}, \omega)$ and $S_s(\mathbf{q}, \omega)$ to be written as the Fourier transforms of certain functions – the Van Hove pair correlation functions, $G(\mathbf{r}, t)$ and $G_s(\mathbf{r}, t)$ – which are defined in terms of the atomic dynamics of the system. Thus

$$S(\mathbf{q}, \omega) = \frac{1}{2\pi}\int_{-\infty}^{\infty} \mathrm{e}^{-\mathrm{i}\omega t} \int \mathrm{e}^{\mathrm{i}\mathbf{q}\cdot\mathbf{r}} G(\mathbf{r}, t)\, \mathrm{d}\mathbf{r}\, \mathrm{d}t,$$

$$= \frac{1}{\pi}\,\mathrm{Re}\int_{-\infty}^{\infty} \mathrm{e}^{-\mathrm{i}\omega t} \int \mathrm{e}^{\mathrm{i}\mathbf{q}\cdot\mathbf{r}} G(\mathbf{r}, t)\, \mathrm{d}\mathbf{r}\, \mathrm{d}t, \qquad (1.8.4)$$

with a similar transformation connecting $S_s(\mathbf{q}, \omega)$ to $G_s(\mathbf{r}, t)$. When the atomic movements may be described by classical mechanics the physical meaning of $G(\mathbf{r}, t)$ and $G_s(\mathbf{r}, t)$ is easily stated. Thus $G(\mathbf{r}, t)\,\mathrm{d}\mathbf{r}$ is then the conditional probability that at time $t$ there is an atom within the volume element $\mathrm{d}\mathbf{r}$ at $\mathbf{r}$ given that there was an atom at the origin ($\mathbf{r} = 0$) at $t = 0$. The two atoms referred to could be the same or different. The quantity $G_s(\mathbf{r}, t)\,\mathrm{d}\mathbf{r}$ is the conditional probability that an atom is within the volume element $\mathrm{d}\mathbf{r}$ at $\mathbf{r}$ given that it, i.e. the same atom, was at the origin ($\mathbf{r} = 0$) at $t = 0$. As Fig. 1.16 implies, the separate contributions to $S(\mathbf{q}, \omega)$ and $S_s(\mathbf{q}, \omega)$ – elastic, phonon-inelastic and quasi-elastic – are readily distinguished and can be considered separately. Henceforth we shall only be interested in the quasi-elastic component in both $S(\mathbf{q}, \omega)$ and $S_s(\mathbf{q}, \omega)$ since it is this component which derives from the diffusive movements of atoms in the solid. The corresponding functions $G(\mathbf{r}, t)$ and $G_s(\mathbf{r}, t)$ likewise will relate to the diffusive movements of atoms among the available lattice sites.

It is clear then from the interpretation of (1.8.4) that the incoherent scattering function $S_s(\mathbf{q}, \omega)$, by being directly related to the self-correlated function, is a particularly useful measure of diffusive motion. In fact, diffusion of atoms is normally studied through the *incoherent* quasi-elastic scattering (because this is related to the self-correlation function), although it is also possible to obtain information about such motions from the *coherent* quasi-elastic scattering. Thus were this motion to approximate to continuous translational diffusion we should expect

$$G_s(\mathbf{r}, t) = (4\pi Dt)^{-3/2} \exp(-r^2/4Dt), \tag{1.8.5}$$

with $D$ the self-diffusion coefficient of the atoms. We expect the probability to obey Fick's second equation with the boundary condition $G_s(\mathbf{r}, 0) = \delta(\mathbf{r})$ and (1.8.5) is then the appropriate solution. By substitution of (1.8.5) into (1.8.4) we obtain,

$$S_s(\mathbf{q}, \omega) = \frac{1}{\pi} \frac{Dq^2}{\omega^2 + (Dq^2)^2}. \tag{1.8.6}$$

The self-diffusion coefficient would then be easily obtainable from measurements of $S_s(\mathbf{q}, \omega)$. In contrast to continuous translation, the case of atoms hopping between discrete sites on a Bravais lattice requires the replacement of (1.8.6) by

$$S_s(\mathbf{q}, \omega) = \frac{1}{\pi} \frac{\Gamma(\mathbf{q})}{\omega^2 + \Gamma^2(\mathbf{q})}, \tag{1.8.7}$$

in which the function $\Gamma(\mathbf{q})$ is determined by the lattice structure, the jumps which are possible (e.g. to nearest-neighbour positions only, to nearest and next nearest, etc.) and the frequency with which they occur. An example is shown in Fig. 1.17. This shows clearly that $\Gamma(\mathbf{q})$ depends upon temperature according to the Arrhenius expression, as might be expected from the relation to diffusion. However, we should also remember that atoms in a solid, in contrast to those in a gas will be vibrating thermally about their lattice sites in the intervals between jumps. These thermal vibrations require the addition of a Debye–Waller factor – usually written as $\exp(-2W(\mathbf{q}))$ – into (1.8.7). Then

$$S_s(\mathbf{q}, \omega) = \frac{1}{\pi} \frac{\Gamma(\mathbf{q})}{\omega^2 + \Gamma^2(\mathbf{q})} e^{-2W(\mathbf{q})}. \tag{1.8.8}$$

Nevertheless, in all these cases we shall have

$$D = \pi \lim_{\omega \to 0} \left( \omega^2 \lim_{\mathbf{q} \to 0} S_s(\mathbf{q}, \omega)/q^2 \right) \tag{1.8.9}$$

for the self-diffusion coefficient. This result is derived in Appendix 1.1.

In favourable circumstances the coherent scattering function $S(\mathbf{q}, \omega)$ can also provide information about atomic migration, but the theory is less developed than for $S_s(\mathbf{q}, \omega)$ – essentially because it must deal with correlations between the

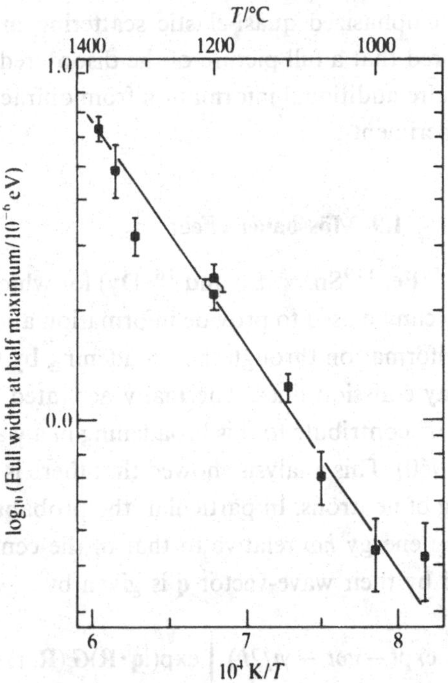


Fig. 1.17. Temperature dependence of the full-width at half maximum height (i.e. $2\hbar\Gamma(\mathbf{q})$) of the quasi-elastic peak associated with the diffusion of Co in $\beta$-Zr. These measurements were made at a scattering vector $\mathbf{q} = 15\ \mathrm{nm}^{-1}$. The straight line represents a fit to the Arrhenius expression, i.e. $\hbar\Gamma(\mathbf{q}) = 1.68 \times 10^{-2} \exp(-1.22\ \mathrm{eV}/kT)\ \mathrm{eV}$. (After Petry, Vogl, Heidemann and Steinmetz, 1987).

movements of different atoms, whereas with $S_s(\mathbf{q}, \omega)$ one is concerned with the movement of individual atoms. This contrast may be seen as analogous to that between chemical diffusion, which leads to changes of composition, and isotopic or self-diffusion in the absence of such changes. It is perhaps not surprising therefore that in special cases there may be close relations between $S(\mathbf{q}, \omega)$ and $S_s(\mathbf{q}, \omega)$, although by the same token we should generally expect any such relation to depend upon details of the system (e.g. composition, defect concentration, etc.). For this reason it is not easy to make valid general statements (however, Hutchings, 1989 gives some particular examples).

In practice only relatively fast diffusing atoms can be studied with neutrons. For, clearly, the instrumental resolution attainable in $\omega$ must be less than $Dq^2$ if quasi-elastic scattering is to be observable. At the present day this imposes a lower limit to $D$ which is $\sim 10^{-11}\ \mathrm{m}^2\ \mathrm{s}^{-1}$ or more. The method is thus applicable to relatively few systems. Interstitial solutions of hydrogen in metals and fast ion conductors are among those which have been extensively studied in this way.

Lastly, although we have emphasized quasi-elastic scattering in this section, it should always be remembered that a full picture of the disordered state of a solid material may well also require additional information from diffraction (i.e. $\omega = 0$) and inelastic scattering experiments.

## 1.9 Mössbauer effect

There are a few nuclei (e.g. $^{57}Fe$, $^{119}Sn$, $^{151}Eu$ and $^{161}Dy$) for which the technique of Mössbauer spectroscopy can be used to provide information about their atomic movements. It yields this information through the broadening by thermal motion of the otherwise sharp $\gamma$-ray emission lines. Thermally activated diffusive movements of the Mössbauer atom contribute to this broadening in a way first analysed by Singwi and Sjölander (1960). This analysis showed that there is a close analogy to the incoherent scattering of neutrons. In particular, the probability of emission of Mössbauer $\gamma$-rays having energy $\hbar\omega$ relative to that of the central Mössbauer line in a direction specified by their wave-vector $\mathbf{q}$ is given by

$$\sigma(\mathbf{q}, \omega) = 2\,\mathrm{Re}\left(\frac{1}{2\pi}\int_0^\infty \exp(-i\omega t - \gamma t/2\hbar) \int \exp(i\mathbf{q}\cdot\mathbf{R})G_s(\mathbf{R}, t)\,d\mathbf{R}\,dt\right) \quad (1.9.1)$$

in which $\gamma$ is the natural line-width. If we can assume that atoms move from one lattice site to another only at intervals long compared to the period of lattice vibrations then (1.9.1) can be written

$$\sigma(\mathbf{q}, \omega) = 2\,e^{-2W(\mathbf{q})}\,\mathrm{Re}\left(\frac{1}{2\pi}\int_0^\infty \exp(-i\omega t - \gamma t/2\hbar)\left(v \sum_n \exp(i\mathbf{q}\cdot\mathbf{R}_n)G_s(\mathbf{R}_n, t)\right)dt\right), \quad (1.9.2)$$

in which $v$ is the atomic volume, the summation is taken over all lattice sites $n$ available to the Mössbauer atoms and $G_s(\mathbf{R}_n, t)$ is now the probability that a given Mössbauer atom is at site $n$ at time $t$ given that it was at the origin at time zero. The factor $\exp -2W(\mathbf{q})$ is the Debye–Waller factor for the Mössbauer atoms. The quantity

$$I(\mathbf{q}, t) \equiv v \sum_n \exp(i\mathbf{q}\cdot\mathbf{R}_n)G_s(\mathbf{R}_n, t) \quad (1.9.3)$$

appearing in (1.9.2) is called the *intermediate scattering function.* The similarity between (1.9.1) and (1.9.2) to the incoherent scattering function for neutrons (cf. 1.8.4) is plain. Here, as there, the uncertainty over the precise value of the Debye–Waller factor means that measurements of $\sigma(\mathbf{q}, \omega)$ as a function of $\omega$ rather than $\mathbf{q}$ yield the more precise information about $G_s(\mathbf{R}_n, t)$. For further details of the techniques see Mullen (1982, 1984).

Although our later attention will mainly focus on these purely diffusional effects, it should also be mentioned that it is possible for the Mössbauer atom in some circumstances to find itself in internal electrical field gradients, which by interacting with the nuclear quadrupole moment cause the occurrence of satellite emission lines. Thermal motion of the *sources* of the field gradients can smear out these satellite lines and provide an additional source of broadening of the central Mössbauer line. Yet faster motion, however, averages the field to zero and the effect disappears. In practice the effect appears not yet to be have been used.

## 1.10 Systems farther from thermodynamic equilibrium

The phenomena we have described in the preceding sections all relate to systems which, from the point of view of the arrangement and the motions of the atoms, are all close to thermodynamic equilibrium. However, there are other circumstances where the departures from thermodynamic equilibrium are much greater, viz. solids rapidly cooled from higher temperatures (*quenched*), metals which have been heavily cold-worked and solids which have been irradiated by beams of energetic particles (e.g. fast neutrons). These treatments *damage* the solids in various ways and lead to the introduction of non-equilibrium concentrations of *defects* of various kinds. These we describe further in the next chapter. Here we merely note that they can be defined in terms of departures from the proper arrangement of atoms as required by the crystal structure and that their presence can be detected by measurements of a variety of physical properties. Correspondingly, the return of a solid to the proper equilibrium atomic arrangement, i.e. the annealing of the damage, can be followed by making such measurements as a function of time. Such annealing processes are often described by rate equations of the form

$$\frac{\mathrm{d}P}{\mathrm{d}t} = -KP^n, \tag{1.10.1}$$

where $P$ stands for the physical property measured (proportional to the concentration of defects). The quantity $K$ is called the *rate constant* while $n$ is the *order of reaction*. Examples are shown in Figs. 1.18 and 1.19. Like other phenomenological constants introduced in previous sections, the rate constants $K$ are found to be thermally activated, i.e. they depend on temperature according to the Arrhenius relationship

$$K = K_0 \exp(-Q/kT). \tag{1.10.2}$$

This can be a very useful characteristic for separating different processes with

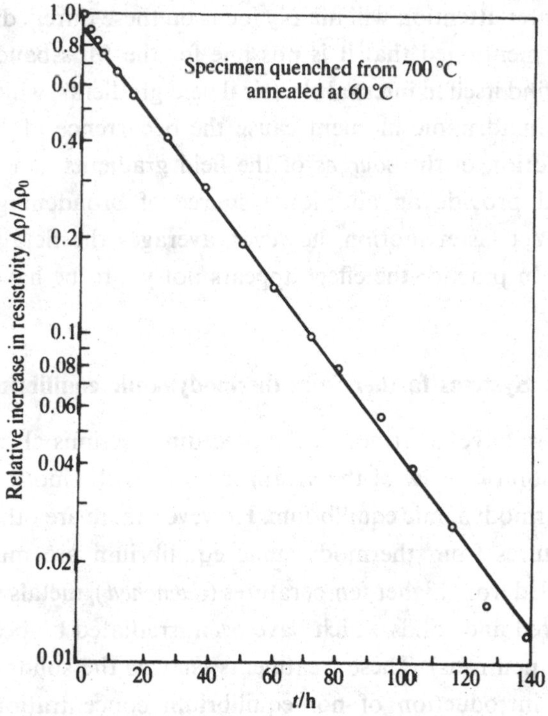


Fig. 1.18. Isothermal decay (at 60 °C) of the excess electrical resistivity, $\Delta\rho$, quenched into a Au specimen by cooling it rapidly from a high temperature. The linearity of this semi-logarthmic plot, $\log_{10}(\Delta\rho/\Delta\rho_0)$ v. time $t$, shows that eqn. (1.10.1) is obeyed with $n = 1$, i.e. the annealing follows first-order kinetics. (After Bauerle and Koehler, 1957.)

different $Q$ values in one and the same system, for small relative differences in $Q$ can mean large differences in $K$ at the same temperature.

## 1.11 Summary

In this chapter we have surveyed the macroscopic equations used to represent a number of phenomena, all of which demonstrate that atoms in solids are not permanently fixed at sites in the crystal lattice but may migrate through that lattice. Almost always these movements are found to be thermally activated and to increase rapidly with increasing temperature according to an Arrhenius relation. The corresponding activation energies are important parameters characterizing these processes.

We have not considered the mechanisms by which these atomic movements and migrations occur; that is the subject of the next chapter. Thereafter we shall develop

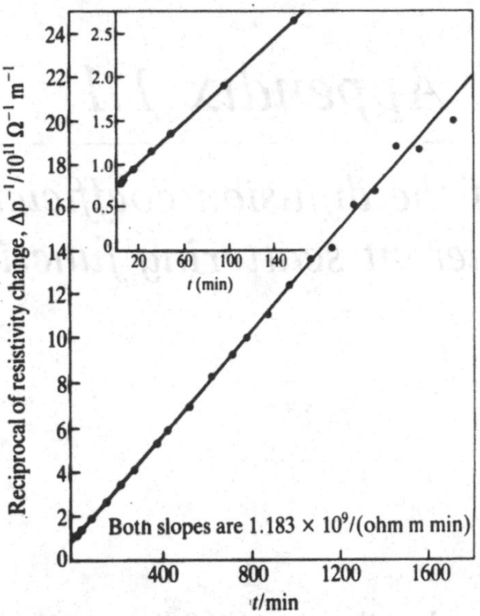


Fig. 1.19. Isothermal annealing of the excess electrical resistivity of a Cu specimen which has been irradiated with energetic protons at a very low temperature. These results were obtained at 263 K. The linearity of the plot of $1/\Delta\rho$ v. time $t$ shows that eqn. (1.10.1) is obeyed with $n = 2$, i.e. the annealing follows second-order kinetics. (After Dworschak and Koehler, 1965.)

the theory of the phenomena described here by using the models and mechanisms suggested in Chapter 2. This later development will draw together the separate phenomenologies introduced here into a coherent and unified theoretical structure, as well as providing particular theories of the macroscopic quantities such as $D$, $\kappa$, $\Delta\varepsilon$, $\Delta s_{ij}$, $\tau$, $T_1$, $T_2$, etc. in terms of microscopic, i.e. atomic, concepts and parameters.

# Appendix 1.1

## Relation of the diffusion coefficient to the incoherent scattering function

In this appendix we show how the relation (1.8.9) comes about. By definition we have

$$S_s(\mathbf{q}, \omega) = \frac{1}{2\pi} \int_{-\infty}^{\infty} e^{-i\omega t} \, dt \int e^{i\mathbf{q}\cdot\mathbf{r}} G_s(\mathbf{r}, t) \, d\mathbf{r}. \tag{A1.1.1}$$

Since the atomic motions in the systems we are concerned with are all fundamentally described by Hamiltonian mechanics (either classical or quantum), we can assume time-reversal symmetry, i.e. that $G_s(\mathbf{r}, -t) = G_s(\mathbf{r}, t)$. Whence (A1.1.1) can be written as

$$S_s(\mathbf{q}, \omega) = \frac{1}{\pi} \operatorname{Re} \int_{0}^{\infty} e^{-i\omega t} \, dt \int e^{i\mathbf{q}\cdot\mathbf{r}} G_s(\mathbf{r}, t) \, d\mathbf{r}. \tag{A1.1.2}$$

Now since $G_s(\mathbf{r}, t)$ must decrease rapidly with $\mathbf{r}$ at large $\mathbf{r}$, the intermediate scattering function (1.9.3 written in continuum form)

$$I_s(\mathbf{q}, t) = \int e^{i\mathbf{q}\cdot\mathbf{r}} G_s(\mathbf{r}, t) \, d\mathbf{r}, \tag{A1.1.3}$$

at small $q$ can be obtained by expanding the exponential, so that

$$I(\mathbf{q}, t) = 1 - \frac{q^2}{6} \langle r^2 \rangle + \cdots, \tag{A1.1.4}$$

where we have assumed that $G_s(\mathbf{r}, t)$ has at least cubic symmetry and have written

$$\langle r^2 \rangle = \int r^2 G_s(\mathbf{r}, t) \, d\mathbf{r}. \tag{A1.1.5}$$

We note that $I_s(\mathbf{q}, 0) = 1$ and that $\partial I_s/\partial t = 0$ at $t = 0$. By integrating the time integral in (A1.1.2) by parts we then obtain

$$S_s(\mathbf{q}, \omega) = -\frac{1}{\pi}\int_0^\infty \frac{1}{\omega^2}\, e^{-i\omega t}\frac{\partial^2}{\partial t^2} I_s(\mathbf{q}, t)\, dt. \tag{A1.1.6}$$

Hence by (A1.1.6) and (A.1.1.4) we obtain

$$\pi \lim_{\omega\to 0} \omega^2 \lim_{\mathbf{q}\to 0} (S_s(\mathbf{q}, \omega)/q^2) = \int_0^\infty dt\, \frac{\partial^2}{\partial t^2}\left(\frac{\langle r^2\rangle}{6}\right)$$

$$= \frac{\partial}{\partial t}\left(\frac{\langle r^2\rangle}{6}\right)_{t=\infty} - \frac{\partial}{\partial t}\left(\frac{\langle r^2\rangle}{6}\right)_{t=0}. \tag{A1.1.7}$$

Now since the atomic movements are made up of rapid vibrations and less frequent diffusive jumps, we see that at short times $\langle r^2\rangle$ will be determined merely by the vibrations and will thus be proportional to $t^2$. Hence the second term of the right side of (A1.1.7) is zero. On the other hand, at very long times it is the diffusive motions which determine $\langle r^2\rangle$ since they allow movements over long distances whereas the vibrational displacements are confined to the vicinity of lattice sites. Hence in the first term on the right side of (A1.1.7) we can put $\langle r^2\rangle = 6Dt$ (cf. §§6.5 and 10.2), whence finally

$$D = \pi \lim_{\omega\to 0} \omega^2 \lim_{\mathbf{q}\to 0} (S_s(\mathbf{q}, \omega)/q^2), \tag{A1.1.8}$$

as required.

# 2

# *Imperfections in solids*

## 2.1 Introduction

In the preceding chapter we reviewed a number of properties of crystalline solids which demonstrate the movement and migration of atoms through these solids. The atomic theory of these properties is the subject of this book, and in this chapter we review the relevant basic mechanisms. Fundamental to these are questions of structure, both the ideal lattice structure of the perfect crystal, as one would determine it by X-ray or neutron diffraction, and those local modifications of this crystalline arrangement of atoms, called imperfections or defects, which facilitate this movement of atoms through the body of the crystal. Mostly we shall be dealing with systems in which the fraction of atoms contained in imperfect regions of the crystal is small, i.e. we are dealing with *nearly perfect* crystals. Notable exceptions, which we shall consider to varying degrees in this and later chapters, include order–disorder alloys, fast ion conductors and concentrated solutions of hydrogen in metals.

There are two important and interesting general facts about these crystal imperfections. This first is that there is not an arbitrary number of distinct types, but just a few. This is true for topological reasons even if we represent the solid as a continuum, when we would classify the elementary imperfections as point defects, dislocations (linear) and surfaces (two-dimensional). In a crystal lattice there are others, such as stacking faults and grain boundaries, for which there are no analogues in a continuum, but the total number of distinct types remains small. The second general fact is that the same classification of imperfections is useful irrespective of the type of bonding between the atoms of the solid. As a first approximation one may say that where specific effects enter they are more likely to be characteristic of the structure of the perfect lattice, in which the imperfections are embedded, than of the nature of the interatomic forces. Details of these forces are, however, important quantitatively, e.g. in determining the energies and other characteristic properties of the imperfections.

The nature of these elementary imperfections is the subject of the next section, while in §2.3 we give more attention to those defects (vacant lattice sites and interstitial atoms) which are predominant in the subject of atomic migration and diffusion. In order to 'close' the system of elementary imperfections in non-metals it is necessary to include electronic defects (electrons and holes) and we consider these in §2.4. Important properties of both the structural point defects and the electronic defects are their concentrations and their mobilities. Almost equally important are their interactions and the interplay among them and other imperfections. We review these aspects in §2.5. The object of the chapter is to provide an introduction to the model systems to be considered and analysed in later chapters. The introduction in this chapter is therefore intuitive and qualitative. The quantitative analysis of point defect concentrations and mobilities is given in Chapter 3, while the quantitative theory of atomic transport as founded on these point defect models is the subject of later chapters, beginning with Chapter 5.

## 2.2 Imperfections in the crystal structure

The way atoms are arranged in crystalline structures is fundamental to solid state physics and chemistry – elementary books in the field frequently begin with the topic and more advanced ones rely on a knowledge of it. Such structures are determined by means of X-ray and neutron diffraction experiments. The geometrical arrangement of the atoms in the simpler structures and the occurrence of these structures is well known; Table 2.1 is intended just as a basic reminder. On the basis of this knowledge of lattice structure solid state physics aims to understand a wide range of intrinsic physical properties, electrical, thermal, magnetic, optical, and so on. However, these structures represent ideal or perfect crystals, whereas in practice real crystals always contain *imperfections* – which not only influence the above physical properties but which are indeed essential to the understanding of others, in particular the various atomic transport properties described in Chapter 1, as well as mechanical properties (other than the purely elastic), radiation damage and colour centres.

These imperfections range in size from atomic to macroscopic. Knowledge of their properties derives from a great number of sophisticated experiments coupled with careful calculations using the best theoretical and computational methods. Forty years ago Seitz (1952) proposed and popularized a classification based on the recognition of a few elementary imperfections which either separately or in

Table 2.1. *Crystal structures*

| Solid | Structure of common phases |
|---|---|
| Alkali metals | Body-centred cubic (b.c.c.) |
| Noble metals and Ni, Pd, Pt | Face-centred cubic (f.c.c.) |
| Transition metals (e.g. V, Cr, α-Fe, Nb, Mo, Ta, W) | Body-centred cubic (b.c.c.) |
| Transition metals (e.g. Ti, Co, Zr) | Hexagonal closed packed (h.c.p.) |
| Noble gas solids | Face-centred cubic (f.c.c.) |
| Group IV semiconductors | Diamond |
| Alkali halides, alkaline earth oxides | Rocksalt |
| III–V semiconductors | Zincblende |
| II–VI semiconductors | Wurtzite and Zincblende |
| 3d-transition metal oxides MO | Rocksalt |
| Alkaline earth fluorides, $SrCl_2$ and β-$PbF_2$; also some oxides ($CeO_2$, $ThO_2$, $UO_2$) | Fluorite |

combination and/or in interaction would allow the description of all solid state properties and processes determined by the presence of imperfections. It is one of the successes of this scheme that it has allowed the rationalization of a wide and diverse range of solid state phenomena of interest to solid state physics and chemistry, metallurgy and materials science. Atomic transport in solids is one of these areas and the theories we describe in this book depend upon a knowledge of the characteristics of the relevant imperfections. The elementary imperfections divide generally into two classes, structural and electronic – although, owing to the free-electron character of metals, this second group is only sensibly distinguished in the case of non-metals. We defer the consideration of electronic defects until §2.4 and begin with the structural imperfections because their basic characteristics are rather easily understood from the known crystal lattice structure. These imperfections are classified by their dimensionality as point defects (i.e. those of atomic size or effectively of zero dimension), linear defects or dislocations (of dimension one), surfaces and interfaces (of dimension two) and voids, precipitates, inclusions etc. (of dimension three).

The basic features of these four classes of imperfection can often be understood in terms of rather broad structural and energetic considerations, although we emphasize that their more detailed physical properties can generally only be obtained by a combination of much painstaking experimental and theoretical work. For this reason we shall define the various lattice imperfections by emphasizing atomic arrangement and structure largely without reference to thermodynamic conditions. We can legitimately do so because, for the most part, the effects of temperature ($T$) and pressure ($P$) are to change the extent of the thermal vibrations of the atoms about their lattice sites without altering the overall lattice structure, although, of course, the lattice parameters may depend to a degree upon $T$ and $P$. Naturally, when there is a change of lattice structure in a phase transition (e.g. the b.c.c. → f.c.c. transition in Fe) we must expect the structure of the imperfections to change too. Lastly, we note that these thermal vibrations mean that the characteristic energies of these imperfections will actually be thermodynamic average energies, specifically Gibbs free energies when, as usual, the thermodynamic conditions are those of fixed $T$, $P$ and chemical potential. In general, therefore they will depend upon these variables, although not necessarily strongly. We now turn to the definitions and properties of these four classes of imperfection.

### *2.2.1 Point defects**

The principal elements in this class are four in number – vacant lattice sites (vacancies), interstitial atoms, foreign atoms (variously called solute atoms, impurities, dopant atoms, etc., according to context) and, in compounds, wrong atoms, i.e. atoms present on one sub-lattice which properly belong on another. Vacancies are simply unoccupied lattice sites, Fig. 2.1. Although the atoms around the vacant site may relax their positions, in most cases the point symmetry of the lattice is retained, as shown here. Interstitial atoms are extra atoms which have been forced into the interstices of the crystal lattice, Fig. 2.2. Here a greater variety of structures may occur and the point symmetry of the interstitial site is often not retained, as exemplified in Fig. 2.2(b). When the atoms in question are of the same chemical species as those of the host lattice they are termed self-interstitials. Foreign atoms are atoms of a different chemical species from the host crystal; they

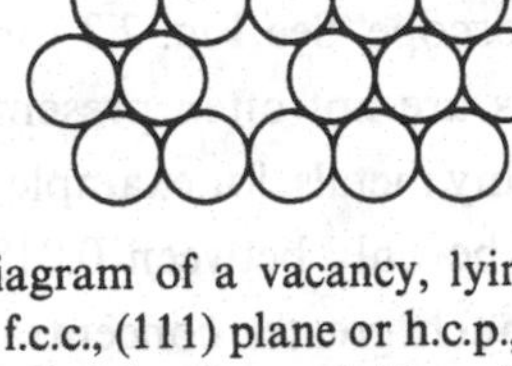

Fig. 2.1. A schematic diagram of a vacancy, lying in a plane of atoms in a close-packed lattice (e.g. f.c.c., (111) plane or h.c.p., basal plane). The usual idea of atomic size is here implied because neighbouring atoms 'touch'.

* For elementary accounts see Kittel (1986), Cottrell (1975) or Philibert (1985). For comprehensive treatments see Flynn (1972), Kröger (1974), Stoneham (1975) or Henderson and Hughes (1976).

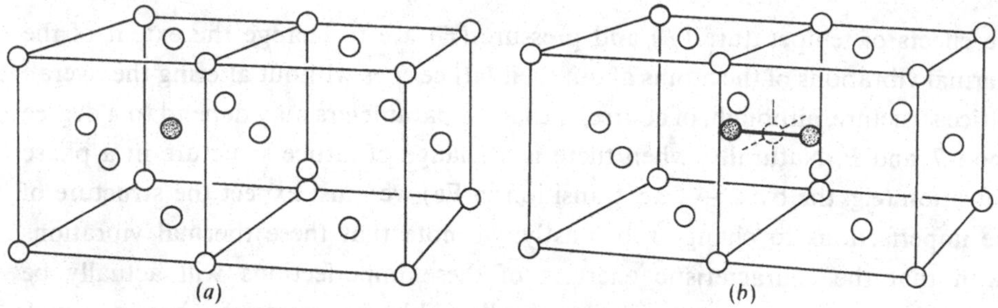


Fig. 2.2. Schematic diagram showing two types of self-interstitial (shaded) in a f.c.c. lattice: (*a*) simple interstitial at the centre of an octahedron of lattice atoms, (*b*) a dumb-bell interstitial, i.e. a pair of atoms at a lattice site normally occupied by one. (The atoms are here drawn smaller than actual size to permit this view into the lattice.)

may be substitutional when they replace a host atom or interstitial when they occupy the otherwise empty interstices of the host lattice. Of course, in concentrated substitutional solid solutions and alloys, it can make little sense to distinguish host and solute atoms, but in dilute systems the distinction is useful. In a few systems foreign atoms may distribute themselves over both interstitial sites and substitutional sites (e.g. $Cu^+$ ions in AgCl and AgBr). In intermetallic compounds atoms may be present on the wrong sub-lattice, e.g. V atoms on the Ga sub-lattice in $V_2Ga_5$ or As atoms on the Ga sub-lattice in GaAs. In some solids there may be other types of point defects; most notably, in so-called hydrogen-bonded solids (e.g. ice) it is necessary to add defects associated with movements of protons along and between the bonds (in the case of ice giving $OH^-$ and $H_3O^+$ molecules in place of $H_2O$ molecules, as well as orientational defects corresponding to no proton in a bond and two protons in a bond; see Fletcher, 1970). The atomic size of point defects means that their direct observation is generally out of the question and that their properties must be obtained from theory and inferred from measurements of physical properties. In special situations, however, direct observations at surfaces have been possible in the field-ion microscope and, more recently, in the scanning tunnelling microscope. See Fig. 2.3 for an illustration of this.

Vacancies and self-interstitials are not often present in concentrations greater than small fractions of 1%. In many metals, for example, the vacancy concentration even at the melting point may be only between 0.01% and 0.1%. Solute atoms may, of course, be present in much greater concentrations than this, depending upon their solubility. The concentration of wrong atoms in intermetallic compounds will be sensitive to departures from the stoichiometric composition, i.e. by the extra number of atoms of one type to be accommodated: it may therefore also be much larger than the concentration of vacancies.

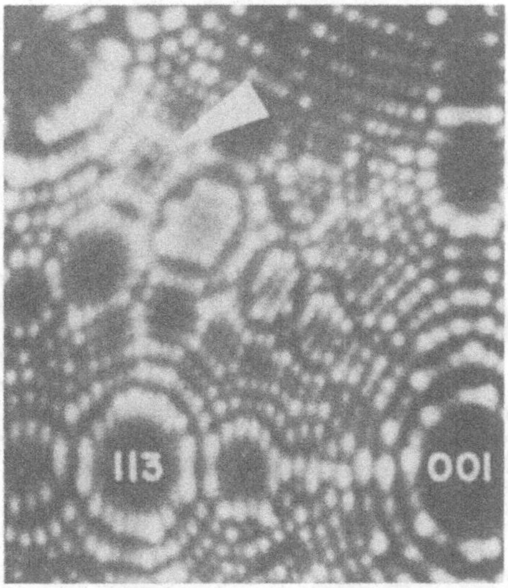


Fig. 2.3. A vacant position (arrowed) in a (203) surface of Pt as observed in the ion microscope (after Müller, 1963). For later studies of this type see Seidman (1973).

### *2.2.2 Dislocations**

These are linear defects. They may contain straight sections, but are generally curved. They do not necessarily lie in a plane; indeed helices and other non-planar dislocations are often seen. The arrangement of atoms in a dislocation line is quite different from that in the regular crystal lattice. As a result, the usual crystal lattice diffraction conditions are violated locally and dislocations thereby become visible in the electron microscope (Fig. 2.4). On the other hand, since they may be of any shape, dislocations in general are not conveniently defined in terms of this atomic arrangement in the way that point defects are. Their general definition thus introduces the notion of the *slip* of one part of the crystal lattice relative to the rest. They are then defined as the line boundary dividing a *slipped* from an *unslipped* region of the crystal (Fig. 2.5). This succinct definition nevertheless requires some elaboration. For, although physical slip on particular planes is observed when crystals are plastically deformed, what is intended here is really a thought experiment. Specifically the definition requires us (i) to switch off the forces between the atoms on opposite sides of the surface which will define the slipped region, (ii) to displace the upper part relative to the lower part by the slip vector

* For basic accounts see Kittel (1986), Cottrell (1975) or Hull and Bacon (1984). For comprehensive treatments of the subject see Nabarro (1967, 1979–) and Hirth and Lothe (1982).

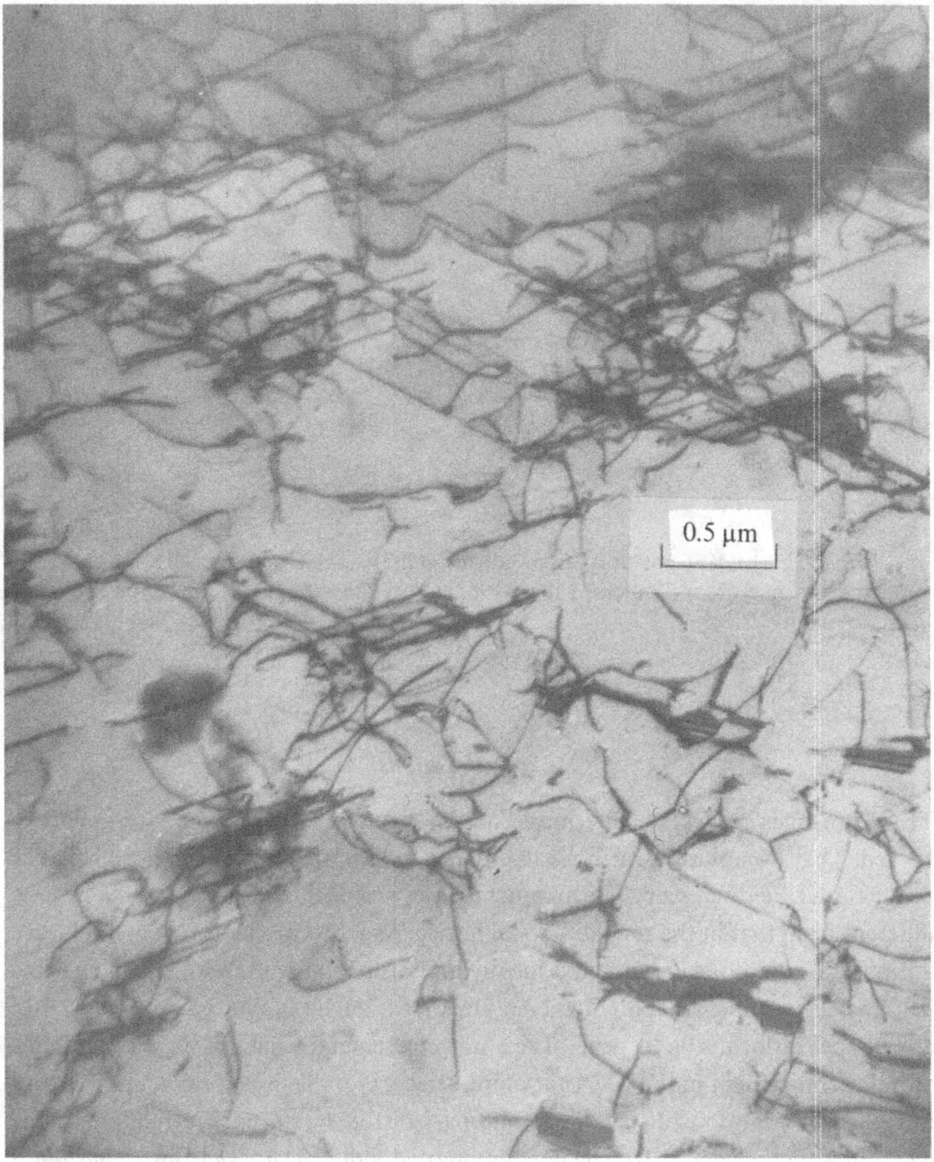


Fig. 2.4. Dislocation lines in deformed stainless steel as observed in the electron microscope. (After Whelan, 1975.)

(which must be a vector of the Bravais lattice lying in the slip surface) and (iii) to switch the atomic forces on again across the slip surface.

From this definition it follows that a dislocation is characterized by a slip plane, or more generally a slip surface, and a slip vector (more usually known as Burgers'

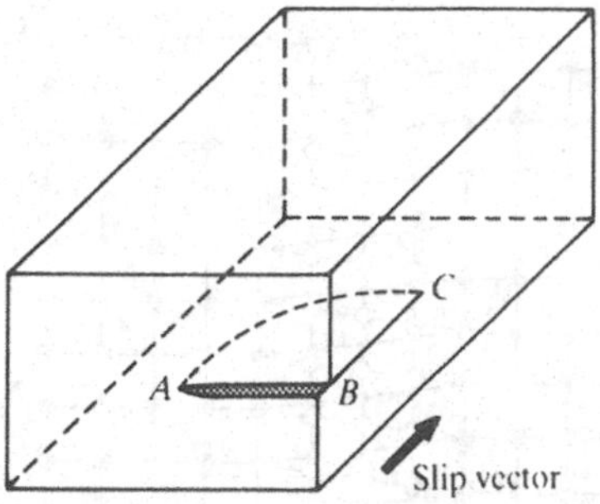


Fig. 2.5. Diagram illustrating the definition of a dislocation as the boundary between slipped and unslipped regions of the crystal. Slip over the area *ABC* by a crystal-lattice vector in the direction shown produces a dislocation (*AC*) lying in the plane *ABC* (the slip plane).

vector). Both the line of the dislocation and the slip vector lie in the slip plane. When they are perpendicular to each other the dislocation is an *edge* dislocation and when they are parallel it is a *screw* dislocation (Fig. 2.6). Obviously, when a dislocation is not straight its character will change with position along it.

A schematic diagram showing the arrangement of atoms in a pure screw dislocation in a simple cubic lattice is given in Fig. 2.7. The planes of atoms in this case are formed into a continuous helical surface, but the usual lattice coordination within the body of the solid is only disturbed in the central core region. However, where the dislocation emerges at the external surface there is necessarily a step running from the point of emergence on account of this helical arrangement; such steps facilitate the growth of crystals from solution.

Correspondingly, Fig. 2.8 shows the arrangement of atoms in and around an

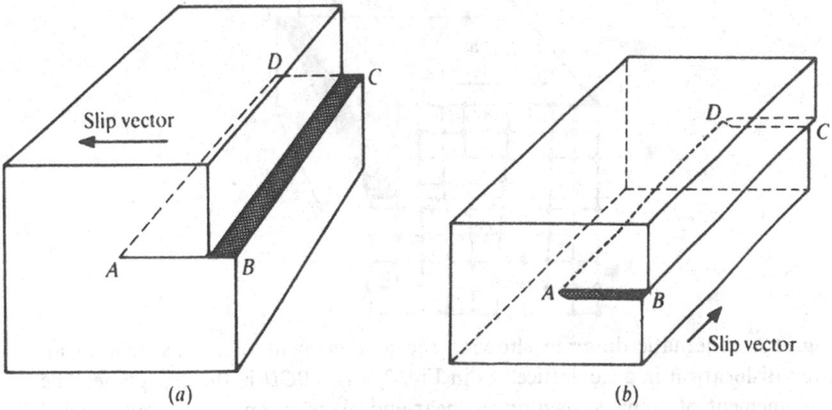


Fig. 2.6. (*a*) Slip vector and slip plane (*ABCD*) defining an edge dislocation *AD*: (b) ditto defining a screw dislocation.

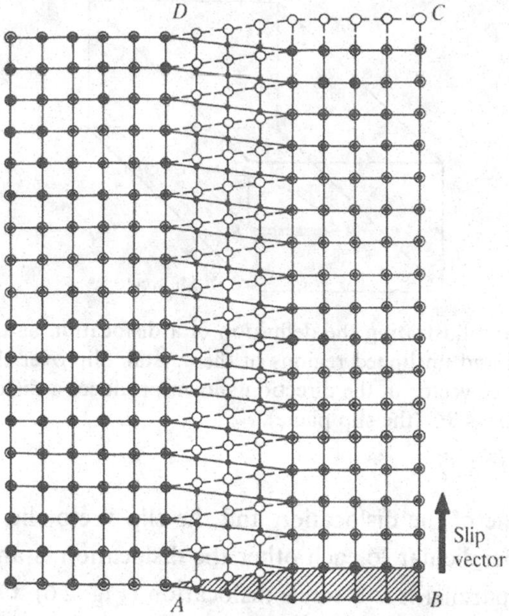


Fig. 2.7. Schematic diagram showing the arrangement of atoms around a screw dislocation in a simple cubic (s.c.) lattice. As in Fig. 2.6(*b*) *ABCD* is the slip plane. The open circles represent atoms in the plane just above the slip plane and the solid circles those in the plane just below it (size of atoms much reduced).

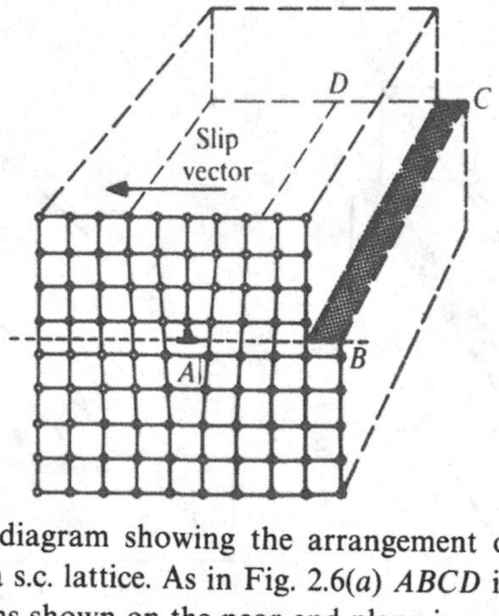


Fig. 2.8. Schematic diagram showing the arrangement of atoms around an edge dislocation in a s.c. lattice. As in Fig. 2.6(*a*) *ABCD* is the slip plane. The arrangement of atoms shown on the near-end plane is repeated on all parallel planes along the line of the dislocation *AD* (size of atoms much reduced relative to their separation).

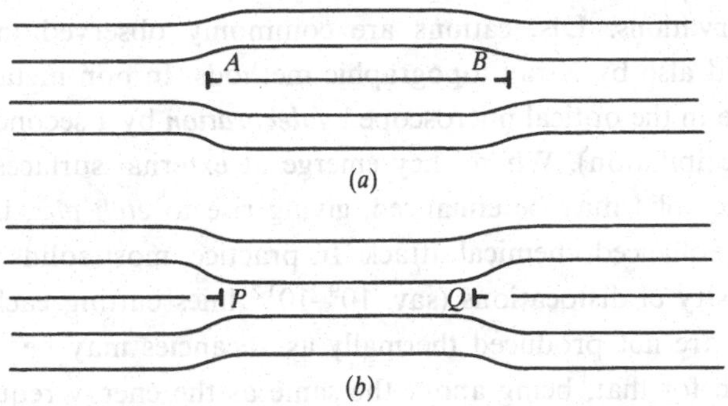


Fig. 2.9. Schematic diagram of the cross-section through planar clusters of (*a*) interstitial atoms (*b*) vacancies. In (*a*) *AB* is the extra layer formed from the interstitial atoms while in (*b*) *PQ* is the layer of vacancies, equivalent to the removal of a layer of atoms. The peripheries of the clusters define edge dislocation loops as indicated by the usual symbol for edge dislocations (cf. Fig. 2.8).

edge dislocation. An additional half-plane of atoms is visible above the slip plane; the lower edge of this defines the centre of the dislocation. The arrival of vacancies at this centre, or core, of the edge dislocation will cause the extra half-plane to be 'eaten away' so that the dislocation line *climbs* in an upwards direction normal to the original slip plane. Likewise the arrival of self-interstitial atoms will cause the dislocation line to climb in a downwards direction. In both cases the point defects have disappeared. Equally, they can be created, to the accompaniment of climb in the opposite sense to that associated with their disappearance. Thus edge dislocations may act as internal sources and sinks of vacancies and interstitials. On the other hand, a straight pure screw dislocation cannot act as a source or sink of point defects. In practice, however, dislocations are rarely if ever perfectly straight: they will therefore possess some edge character and will be able to climb in the way needed to function as a defect source or sink. It should also be recognized that dislocations can be created by the (two-dimensional) aggregation of vacancies and interstitials as illustrated in Fig. 2.9. The perimeter of the cluster defines an edge dislocation loop. It follows that dislocations play an important role in bringing vacancy and interstitial populations to equilibrium whenever the conditions affecting those populations change.

As a footnote to this introduction to dislocations we emphasize that the core structures of actual dislocations may be quite complex compared with the simple schematic arrangements shown in Figs. 2.7 and 2.8; see e.g. Vitek (1985). These complexities can affect their properties although not the overall picture which we have described.

The form, distribution and density of dislocations can be obtained by a variety of direct observations. Dislocations are commonly observed in the electron microscope, and also by X-ray topographic methods. In non-metals they can be rendered visible in the optical microscope by *decoration* by a second phase (i.e. by preferential precipitation). Where they emerge at external surfaces the chemical reactivity of the solid may be enhanced, giving rise to *etch pits*, i.e. depressions formed by the enhanced chemical attack. In practice, most solids will contain a fairly high density of dislocations (say, $10^9$–$10^{12}$ lines cutting each $m^2$ of area). However, they are not produced thermally as vacancies may be. Their energies are far too high for that; being about the same as the energy required to create one vacancy *for every lattice plane passed through.* In fact, they result from the way the specimens have been obtained and handled. In particular, mechanical deformation beyond the yield point leads to a rapid multiplication of dislocations by well-understood mechanisms (described in the texts cited at the beginning of this section). A noteworthy example, however, is provided by so-called *dislocation-free* silicon, a material whose production has been stimulated by the needs of the semiconductor industry. Even here, however, dislocations can still be created by the aggregation of point defects and in other ways. In practice therefore, specimens of dislocation-free silicon may actually contain up to $10^6$ dislocation lines per $m^2$.

### 2.2.3 *Surfaces and interfaces*

External surfaces are perhaps the most obvious type of two-dimensional imperfection, for they interrupt the lattice periodicity of the otherwise perfect infinite crystal. As a result of this interruption there may be localized electronic states at surfaces, while the arrangement of atoms may also change from that expected from the crystal lattice structure. At a coarser level there may be steps on the surface and kinks or corners along those steps (Fig. 2.10). Such sites clearly can act as sources or sinks of point defects in the same way that jog sites on edge dislocations do.

Apart from external surfaces the most visible interfaces in most solids will be the boundaries between differently oriented grains or crystallites. Even carefully grown single crystals will generally still contain low-angle boundaries separating regions of slight misorientation. Low-angle boundaries can be regarded as walls or arrays of dislocations (Fig. 2.11). High-angle boundaries present a less regular appearance, but both high- and low-angle boundaries can function as sources and sinks of vacancies and interstitials. Increasing attention, both experimental and theoretical, is being given to the arrangement of atoms in grain boundaries.

Another important type of interface is the *stacking fault*. This is an area where the correct relationship between successive lattice planes (the stacking sequence) is upset. An example for a face-centred cubic lattice is shown in Fig. 2.12. We can

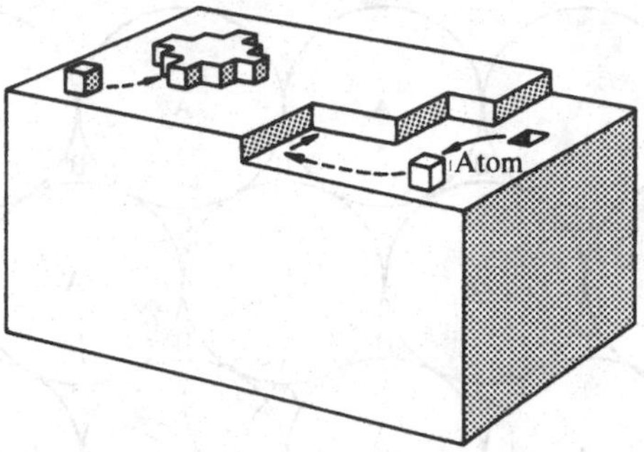


Fig. 2.10. Schematic diagram of an external (100) surface of a s.c. lattice (in which the atoms are conveniently represented by cubes). The boundary of an incomplete layer is a *step* which will generally contain re-entrant corners or *kinks*. Individual atoms may move relatively rapidly across these surface planes to become firmly attached at kinks. Vacancies can be formed in the way indicated. Clearly there are corresponding reactions for their disappearance and for the formation and disappearance of interstitials.

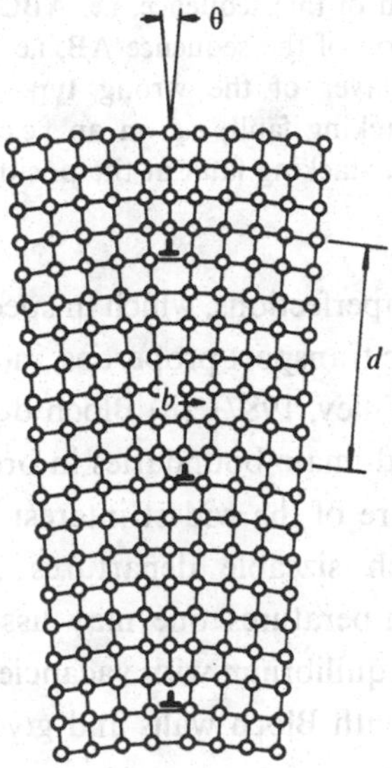


Fig. 2.11. Schematic diagram showing how a low-angle grain boundary ($\theta \ll 1$) can be represented as an array of dislocations.

see from this that, in f.c.c. metals, the aggregation of vacancies on a single (111) plane will necessarily create a stacking fault across the area as well as a dislocation around the perimeter. In the electron microscope such imperfections are visible by the interference pattern associated with the faulted area as well as by the loss of intensity from the dislocation perimeter. An example is shown in Fig. 2.13.

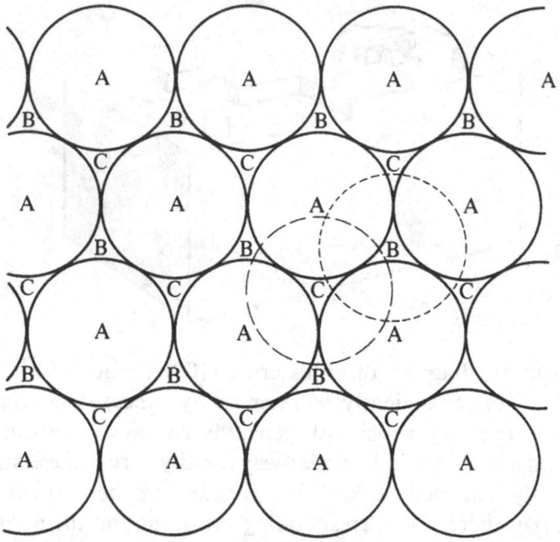


Fig. 2.12. On top of a close-packed layer of atoms A one may lay down another layer whose centres are over the points B and then another whose centres are over points C. Repetition of this sequence, i.e. ABCABC... generates a f.c.c. lattice. Likewise, repetition of the sequence AB, i.e. ABAB, generates a h.c.p. lattice. Inclusion of a layer of the wrong type in an otherwise proper sequence generates a stacking fault, e.g. in an f.c.c. lattice the sequence ... ABC AB | ABC... has a stacking fault at the position indicated.

Other two-dimensional imperfections, which in special circumstances are significant for diffusion and atomic transport properties, include (i) so-called shear-plane structures in some oxides (Tilley, 1987), (ii) Bloch domain walls in ferromagnetic crystals and (iii) anti-phase domain boundaries in order–disorder alloys. Of these the shear plane structures are of the widest interest since they provide a regular and ordered way in which sizeable departures from stoichiometry can be accommodated; at high temperatures one may assume that there is a sort of evaporation–condensation equilibrium with vacancies on the deficient sub-lattice. Point defects may interact with Bloch walls and give rise to magnetic relaxation processes.

### *2.2.4 Three-dimensional imperfections*

Several types of three-dimensional imperfections can arise in solids, depending on circumstances. They include voids (small cavities), colloids (small metallic particles in a compound host), bubbles (small cavities filled with gas atoms, e.g. He in metals), other precipitate phases and inclusions (which can occur in various shapes). All of these can be studied in the electron microscope and all are important

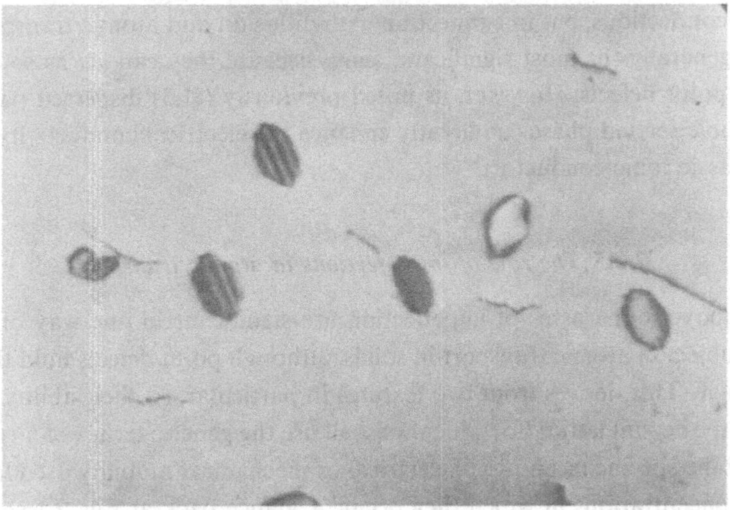

(*a*)

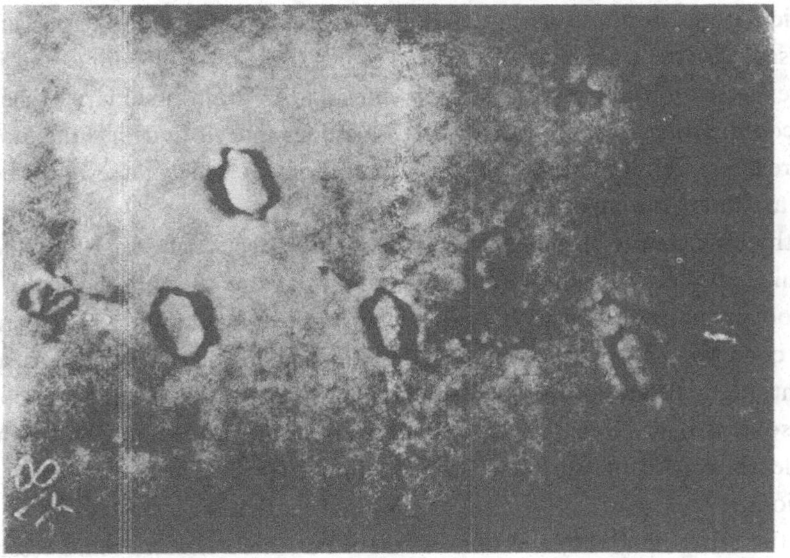

(*b*)

Fig. 2.13. Electron micrographs of hexagonal shaped dislocation loops (diameter ~100 nm) formed on {111} planes in quenched Al. The pattern in (*a*) indicates the presence of enclosed stacking fault, as would be anticipated from Fig. 2.9 if the vacancies had condensed on a single atomic plane. In (*b*) the stacking fault has been removed by annealing. (After Thompson, 1969.)

in other connections, but in connection with diffusion and atomic transport voids may be generally the most significant, again because they can act as sources and sinks of point defects. However, as noted previously (§1.3) dispersed particles of an insoluble second phase can greatly enhance the electrical conductivity of some of the classic ionic conductors.

### *2.2.5 The role of imperfections in atomic transport*

All the above four classes of imperfection are significant in one way or another for the subject of atomic transport in solids, although point defects hold the centre of the stage. This derives from two features in particular, (i) their ability to move through the crystal lattice (for which we shall use the generic term *mobility* without implying the specific meanings of electrical or mechanical mobility used later) and (ii) the concentrations in which they occur. A glance back at Fig. 2.1 shows that the existence of the vacancy allows the possibility that neighbouring atoms may move by hopping into the vacant site. Although such a move is hindered by the cohesive forces holding the atoms to their lattice positions, the energy required is generally such that at moderate and higher temperatures it will be supplied frequently by the local fluctuations in energy which occur in all systems. The rate at which atoms hop into vacancies thus increases rapidly as the temperature is increased. The same applies to interstitial atoms (Fig. 2.14). Of course, this is also to be expected for atoms in dislocation cores and in grain boundaries where the local coordination is severely disturbed. One does indeed find that the atoms in these regions are more mobile. The concentrations in which these imperfections occur are thus also important.

Of the four classes of imperfection only point defects may exist in a thermodynamic equilibrium state (although they can also be produced by non-thermodynamic processes such as energetic particle irradiation). That is to say, under conditions of thermodynamic equilibrium, point defects will be present in concentrations determined by the temperature, the pressure and the composition of the solid. Broadly speaking, these concentrations increase rapidly with temperature, decrease with pressure and, in compounds, increase with departures from stoichiometry.

On the other hand, imperfections of the other three classes, being essentially macroscopic in size, are of high energy but such low configurational entropy that any thermodynamic equilibrium concentration would be infinitesimally small. They are introduced into crystals by macroscopic processes such as mechanical deformation, conditions of solidification, entry into two-phase regions of thermodynamic stability, etc. But they are difficult to remove, so that crystals containing them may be effectively brought to thermodynamic equilibrium in all respects

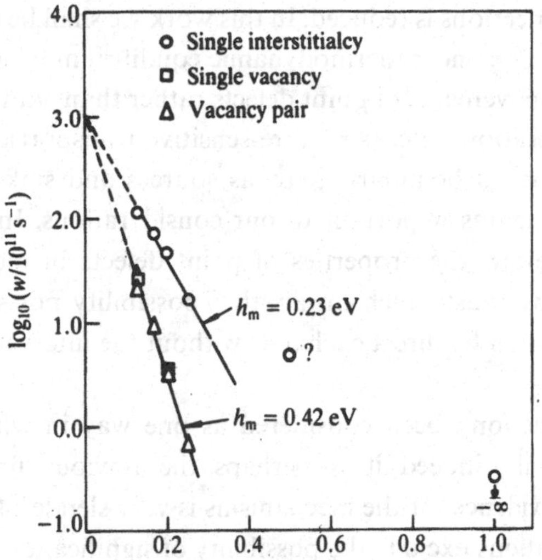


Fig. 2.14. The jump frequency $w$ as a function of reduced inverse temperature $\theta/T$ for an interstitial, a vacancy and a vacancy pair as obtained from a computer simulation of α-Fe (after Tsai, Bullough and Perrin, 1970).

apart from the presence of such imperfections. (Indeed their presence may speed the adjustment of the point-defect concentrations to changes in thermodynamic conditions.)

The upshot of this is that at moderate and high temperatures the concentrations of point defects are such that their contributions to atomic transport are predominant, whereas at lower temperatures atomic transport may be primarily confined to dislocation cores, grain boundaries and other disturbed regions. A review of diffusion along dislocations was given by Le Claire and Rabinovitch (1984) while diffusion in grain broundaries is the subject of a book by Kaur and Gust (1989).

Although dislocations, voids, etc. are not thermodynamic defects in the sense that point defects are, their high energies and their function as sources and sinks of point defects and the changes which this function implies (e.g. dislocation climb, void shrinkage, etc.) do, however, imply that a given population of such imperfections can be modified and the total degree of imperfection reduced, by *annealing* the crystal at high temperatures for long times. For example, a crystal which has been heavily deformed mechanically will contain a high concentration of dislocations all tangled together. Annealing at a high temprature will reduce the overall dislocation density and cause the dislocations to order themselves into boundaries. Likewise voids and dislocation loops in heavily irradiated metals can also be annealed by subsequently holding the specimens at a high temperature. Again, the density of these imperfections is reduced. In this work we shall be largely concerned with those situations in or near thermodynamic equilibrium which are dominated by the presence and movements of point defects rather than with atomic transport associated with dislocations, etc. (structure-sensitive transport). Nevertheless the role of dislocations, grain boundaries, etc. as sources and sinks of point defects and in other ways remains important to our considerations. In the next section we shall therefore review the properties of point defects in more detail. Before doing so, however, we must touch on another possibility not so far mentioned, namely atomic migration by direct exchange without the intervention of imperfections of any kind.

This possibility has long been considered as one way in which atoms might move through crystals; indeed it is perhaps the obvious first consideration. Nevertheless direct evidence for the mechanisms is very slender. Most calculations and computer simulations exclude the possibility of significant contributions from it, the alkali metals providing a possible exception. Were such a mechanism to operate it would, of course, imply that atomic transport was an intrinsic property of the solid essentially independent of crystal perfection and purity.

### 2.3 Point defects in more detail

The survey of lattice imperfections presented in the last section and based essentially on structural considerations enabled us to see which imperfections are significant for atomic transport processes; and it also provided us with ideas of the interplay between the different types. In particular, it enabled us to see the likely importance of point defects for the processes of atomic transport through the body of crystal lattices. In this section we shall therefore describe the physical characteristics of these point defects in more detail, partly with a view to relating their more specific characteristics to the type of solid in which they occur. Nowadays there is an enormous body of knowledge of point defects obtained from numerous experimental and theoretical investigations and we can provide only a brief survey of the principal features. Except in a few aspects we shall not be able to demonstrate the logical and inferential sequences which have led to the results we describe. We have, however, tried to avoid presenting any result which in our judgement is not likely to stand the test of time. For comprehensive accounts of the subject see Flynn (1972), Stoneham (1975) and Agullo-Lopez, Catlow and Townsend (1988).

We shall present this review as far as possible in terms of simple intuitive ideas, but refer as necessary to the results of more sophisticated analyses. Our aim remains the limited one of providing sufficient information and ideas to justify the models analysed in later chapters. The important considerations are those of defect concentrations, mobilities and interactions. In this section we consider the first two, firstly in rather broad terms and secondly more specifically for compounds. We return to the matter of interactions after we have introduced electronic defects in §2.4.

### *2.3.1 Concentrations and mobilities: generalities*

To begin, certain characteristics are physically fairly obviously common to vacancies and interstitials, whatever the substance in which they occur. Firstly, to form them requires energy, i.e. their formation is an endothermic process; for otherwise the crystal structure would not be stable. They occur in non-zero concentrations in solids in thermodynamic equilibrium because their presence increases the (configurational) entropy of the solid and so lowers the free energy below what it would otherwise be. From classical statistical mechanics we therefore anticipate that their thermodynamic equilibrium concentration will be determined, at least in part, by a Boltzmann exponential dependence on temperature, as $\exp(-g_f/kT)$. As we shall see in the next chapter this is indeed the case, although other factors may also be important (e.g. purity, stoichiometry and the presence of other phases). The quantity $g_f$ is the Gibbs free energy of formation of a single defect, i.e. the work required to form a single defect (at a particular but arbitrary position in the lattice) under conditions of constant temperature and pressure. As such it can be decomposed into enthalpy and entropy components ($h_f$ and $s_f$ respectively) as

$$g_f = h_f - Ts_f. \tag{2.3.1}$$

In the analysis of experiments it is frequently assumed that to a first approximation $h_f$ and $s_f$ are independent of temperature; this is often, though not always, justified and it would be wrong to take this as a hard and fast rule. However, in all cases we can take it that the concentration of thermally produced defects (intrinsic defects) will increase rapidly with increasing temperature.

The determination of defect concentrations experimentally is not easy, despite the fact that, in principle, any physical property of the crystal which is changed by their presence can be used. The difficulties stem from uncertainty about the magnitude of the specific property change per defect and from uncertainty about the way that property in the absence of defects would vary with changes in the parameters (i.e. $T$ and $P$) which also determine the defect concentration. There is, however, one method which has been shown to give absolute values of defect concentratons; this is a differential method based on the simultaneous measurement of two properties, namely macroscopic volume expansion and expansion of the lattice parameter.

We observe first of all that the formation of vacancies will increase the volume of the crystal. This volume change is made up of two components. The first arises because the atoms removed from lattice positions (to create the vacancies) are replaced on either external or internal surfaces. This component amounts to one atomic volume, $v$, for each vacancy created. The second component arises from the change in average lattice parameter which is caused by the relaxation of the atoms around the vacant sites. This contribution is often negative, but the magnitude depends upon details of the interatomic forces. Two conclusions follow from this. The first is that the net macroscopic volume change per vacancy created ($v_f$) will differ from $v$ (being mostly $<v$). The second, and more important, is that the difference between the fractional macroscopic volume change ($\Delta V/V$) and three times the fractional change in lattice parameter ($\Delta a/a$) is just the site fraction of vacancies, i.e.

$$\frac{\Delta V}{V} - 3\frac{\Delta a}{a} = c_v. \tag{2.3.2a}$$

Similar arguments apply if the defects are interstitials, except that the right side is now $-c_I$, i.e.

$$\frac{\Delta V}{V} - 3\frac{\Delta a}{a} = -c_I. \tag{2.3.2b}$$

Simultaneous measurements of crystal volume and lattice parameter, if of adequate precision, can thus directly measure changes in the defect concentration. This technique was pioneered by Simmons and Balluffi and has been subsequently used to demonstrate that the dominant, thermally created defects in a number of metals and rare-gas solids are vacancies rather than interstitials. Furthermore, these measurements can provide absolute measures of the defect concentration as a function of temperature. The expected exponential dependence on temperature is confirmed. An example is given in Fig. 2.15. Despite the elegance of such direct determinations of defect concentrations, it should be recognized that information on defect concentrations is mostly obtained from a host of ingenious experiments and analyses of a less direct kind, but among which atomic transport measurements feature prominantly.

As the result of this work $h_f$ and $s_f$ have been determined experimentally for many substances. Some specific examples are given in the next chapter where we consider the thermal formation of point defects more quantitatively. For the time being it will suffice to note that there is a broad correlation between $h_f$ and other properties dependent on the cohesive forces in the solid. One rough, but convenient measure of these is the melting temperature, $T_m$. Thus the noble metals (Cu, Ag and Au), with melting temperatures around 1300 K (Emsley, 1989), have enthalpies

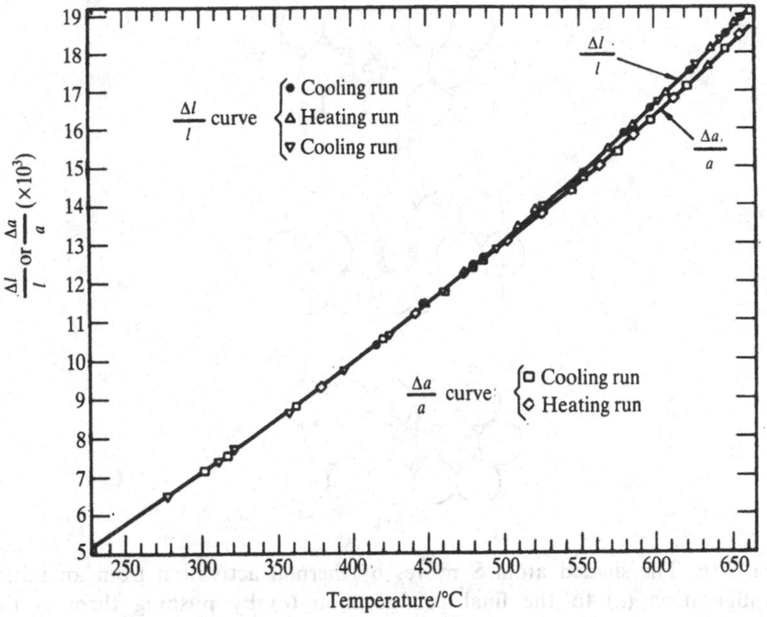


Fig. 2.15(*a*). The macroscopic thermal expansion ($\Delta l/l$) and the lattice parameter expansion ($\Delta a/a$) of Al as a function of temperature.

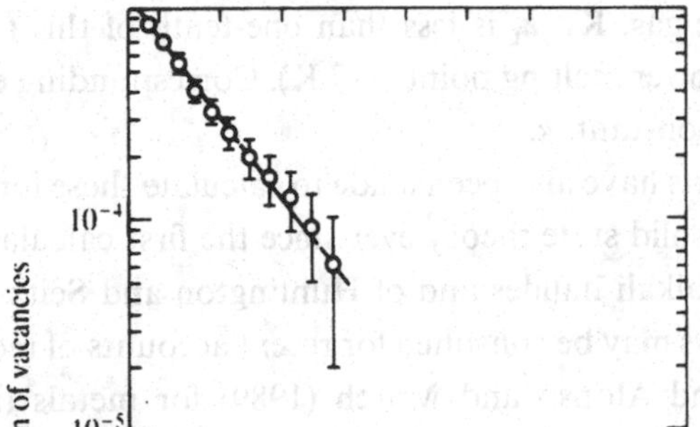

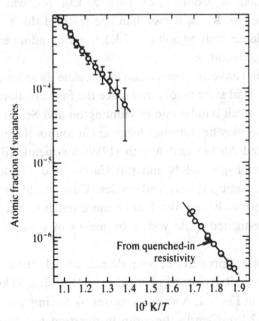

Fig. 2.15(*b*). The atomic fraction of vacancies inferred from (*a*) compared with the corresponding inferences from the additional electrical resistivity of quenched specimens. (After Simmons and Balluffi, 1960.)

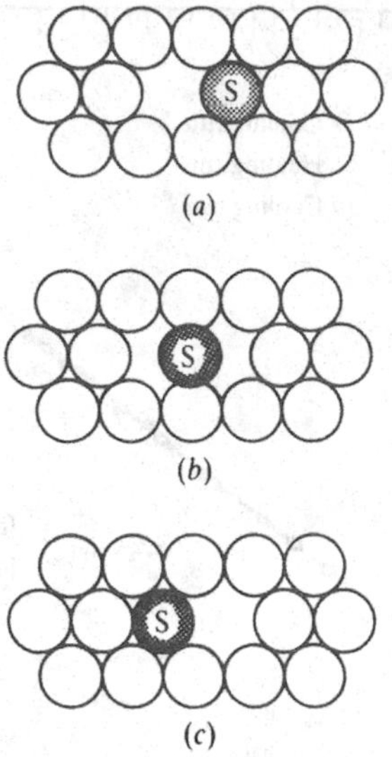

Fig. 2.16. The shaded atom S moves by thermal activation from an initial configuration (*a*) to the final configuration (*c*) by pushing through the intervening atoms (configuration (*b*)).

of vacancy formation of around 1 eV (=96.85 kJ/mol); while in the weakly cohesive solid noble gas, Kr, $h_f$ is less than one-tenth of this ($7.7 \times 10^{-2}$ eV), in line with the much lower melting point (117 K). Corresponding entropies are a few times Boltzmann's constant, $k$.

Considerable efforts have also been made to calculate these formation enthalpies and entropies from solid state theory ever since the first calculations of Mott and Littleton (1938) on alkali halides and of Huntington and Seitz (1942) on copper. The following reviews may be consulted for recent accounts of methods and results: (i) March (1978) and Alonso and March (1989) for metals (ii) Chadwick and Glyde (1977) for noble-gas solids and (iii) Catlow and Mackrodt (1982) and Harding (1990) for ionic crystals and oxides. Corresponding calculations for covalent semiconductors have proved to be more difficult, but these substances are now being investigated quite widely by means of sophisticated theoretical techniques.

The second important property of point defects, namely their mobility can also be understood from similarly broad physical considerations. Take first the case of the vacancy shown in Fig. 2.1. A vacancy moves by having a neighbouring atom move into it (Fig. 2.16). Clearly the atom in question has to squeeze between intervening lattice atoms to do so and this requires energy, as shown schematically in Fig. 2.17. The situation with interstitials is similar. These energies which must

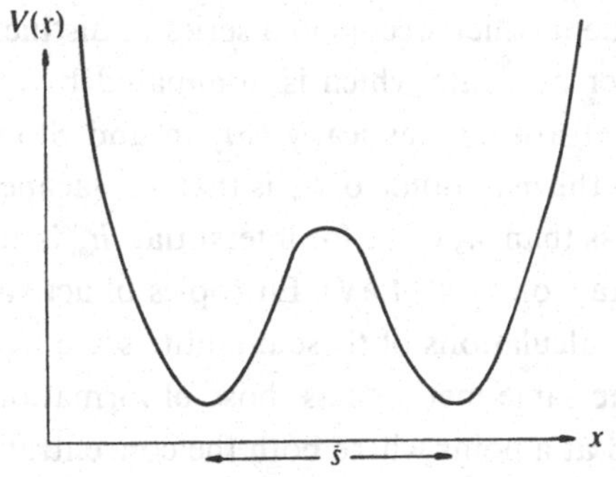


Fig. 2.17. Simple potential energy diagram to illustrate the idea of thermally activated movements.

be supplied to effect these displacements are large relative to $kT$. They are therefore provided infrequently (relative to the frequency of lattice vibrations) by local thermodynamic fluctuations. Conversely, once an atom or defect has moved as the result of such *activation* we expect the local concentration of energy to flow away from the site relatively quickly. The atom or defect thus becomes *de-activated* and waits on average for many lattice vibrations to occur before it is again activated. In other words, we expect such thermally activated atomic migration to occur in a series of discrete hops, or *jumps*, each consisting of a movement from one lattice site to the next and each jump taking place without memory of those which have gone before.* From classical statistical theory we expect the necessary local fluctuations in energy to occur with a frequency which is dominated by a Boltzmann exponential factor, say $\exp(-g_m/kT)$, where $g_m$ is a measure of the energy required (under conditions of given $T$ and $P$) to effect the jump. One may make rather similar remarks about the thermodynamic nature of $g_m$ as we made above about $g_f$, although the formal justification differs. In particular one often decomposes $g_m$ as

$$g_m = h_m - Ts_m, \tag{2.3.3}$$

and assumes $h_m$ and $s_m$ to be independent of temperature. On the whole these expectations are borne out by direct molecular dynamics simulations of model systems containing many atoms and by a very large number of experimental results. There are, however, some cases – most notably hydrogen atoms in metals at relatively low temperatures – where classical considerations are not an adequate guide (Völkl and Alefeld, 1975). Even here, however, the end result of different

* However, in some substances (fast-ion conductors) there is evidence of a memory of the transition which persists for long enough to give distinct effects upon observable properties (see e.g. Funke, 1989).

theories is still a movement which occurs in a series of distinct hops or transitions from one site to another at a rate which is dominated by an exponential factor of a similar form to that arising classically (Flynn and Stoneham, 1970; Gillan, 1988). A rough guide to the magnitude of $h_m$ is that for vacancies $h_m$ is comparable to although generally less than $h_f$, while for interstitials $h_m$ is considerably less than $h_f$ (e.g. in the noble metals only $\sim 0.1$ eV). Entropies of activation $s_m$, like $s_f$, may be only a few times $k$. Calculations of these quantities are frequently made at the same time and using the same methods as those of formation parameters.

Thus we have arrived at a point where both the concentration and the mobility of vacancies and interstitials are expected to depend exponentially on temperature. From this conclusion we may therefore confidently expect to explain one important characteristic common to most atomic transport processes in solids, namely their Arrhenius dependence on temperature (Chapter 1).

We should admit, however, that this expectation does not in itself dispose of the possibility of direct exchanges of position by pairs of neighbouring atoms (or collective movements of larger groups) for these too we should expect to be thermally activated and thus also to explain Arrhenius behaviour. In the end, these alternative 'defectless' mechanisms are mostly ruled out by various considerations, of which some of the most important are the following. Firstly, there may be experiments showing the occurrence of a Kirkendall effect, i.e. a net movement of atoms through the lattice in the course of inter-diffusion (see §1.2.2). Such an effect cannot arise with exchange or ring mechanisms. Secondly, there may be agreement between an observed Arrhenius energy (or strictly the transport coefficient itself) and that expected for a point-defect mechanism on the basis of independently determined activation energies (strictly the relevant concentrations, jump rates and other properties). Thirdly, there may be calculations showing that the activation energies for exchange mechanisms are all higher than the lowest among those for defect mechanisms.

### *2.3.2 Compound solids*

Much of what we have described so far applies generally, although it has been presented in terms that imply that we were considering monatomic substances. For instance we did not distinguish atoms of different chemical species. With compounds, including the intermetallic compounds, additional points arise which we must now consider. For example, should we not differentiate between vacancies on one sub-lattice and those on another, and do the preceding generalities apply equally to both? Obviously, because the sub-lattices occupied by the elements of the compound are distinct we must regard vacancies on the different sub-lattices as distinct defects by the nature of their surroundings. However, when we consider their formation we find that there are constraints. One is that vacancies on the different sub-lattices can be formed *only* in the correct stoichiometric proportions *unless* the formation is accompanied by some other process which involves the different types of vacancy unequally. Take, for example, a simple compound MX in which only M atoms occupy the M-sub-lattice and only X atoms occupy the X-sub-lattice. Then, in the absence of any other processes, M vacancies and X vacancies must be formed in equal numbers. At fixed overall composition anything else must be accompanied by an accumulation of atoms of one kind either at the external surface or at dislocations or other internal regions; but such an accumulation would represent another phase and thus an additional process of exchange of atoms between the MX phase and another phase has been introduced. Hence M vacancies and X vacancies must be formed in equal numbers. Vacancies formed in stoichiometric proportions in this way are called *Schottky defects.* They are the dominant thermal defects in the alkali halides, in which the enthalpies of their formation range from about 1.5 to 2.5 eV in a rough correlation with the melting temperature.

An alternative way of forming point defects in compounds while preserving their stoichiometry is to create equal numbers of interstitial atoms and vacancies separately for each component. Defects formed in this manner are called *Frenkel defects.* They are the dominant thermal defects in AgCl and AgBr (Ag vacancies and interstitials) and in compounds with the fluorite ($CaF_2$) structure (anion vacancies and interstitials).

This need to consider the creation of defects in fixed relative proportions which preserve the stoichiometry of the compound means that their characterisic formation energies are composite quantities. Thus in the above example of an MX compound the relevant Gibbs energy for Schottky defects is the energy to create one M vacancy and one X vacancy, placing both the atoms so removed at the surface or other sink sites. Such a constraint on the formation of Schottky and Frenkel defects does not, however, mean that the component defects are in any sense tied to one another in a way that would restrict their individual mobility. On the contrary, they move independently of one another with their individual jump rates – to a good, first approximation at least. These jump rates, just like those of defects in monatomic substances, will depend exponentially on temperature, each with a characteristic activation energy.

Because the process of thermal formation of Schottky and Frenkel defects demands that the components be formed in fixed proportions, it does not, however, follow that the *total* numbers of component defects will be present in these same proportions, unless there are no other defect creation processes to be considered. Take the well-studied example of the strongly ionic alkali halides. These crystals, as normally prepared and handled, are perfectly stoichiometric, but will often contain cations of a higher valency (either as impurities or as deliberate additions). Clearly the incorporation of, for example, $CaCl_2$ in the lattice of KCl requires the creation either of one potassium vacancy or of one chlorine interstitial for every calcium ion entering the potassium sub-lattice. In fact, the energies are such that in this case there is an excess of potassium vacancies. In all cases, however, there is a coupling of the concentrations of the complementary defects via a 'solubility-product' type of relation which directly reflects the composition formula of the solid. (Such relations are established in Chapter 3.)

A different example is provided by the 3*d*-transition metal monoxides, e.g. NiO, CoO, etc. In an oxygen atmosphere these crystals relatively easily take up an excess of oxygen, to an extent determined by the temperature and oxygen partial pressure. In fact, this is accomplished by the addition of oxygen to the oxygen sub-lattice with the associated creation of vacancies in the metal sub-lattice and the *creation of electron holes*, i.e. effectively by the oxidation of an equal number of metal atoms to a higher charge state. Clearly then, any equality in the numbers of metal and oxygen vacancies will be upset by the presence of this additional process. There are many other well-studied examples of this sort which show that there can be a strong interplay between the thermal formation of defects and other *reactions* involving the dissolution of solutes and changes in stoichiometry. This interplay makes the field of defects in compounds much more varied than in monatomic substances. Furthermore, the insensitivity of the defect jump rates to these changes often allows the rather direct inference of the component processes underlying observed behaviour.

### *2.3.3 Correlations among atomic movements*

The above sections have been concerned with the factors influencing the concentrations and mobilities (jump rates) of point defects. Many atomic transport coefficients (e.g. $D$, $\kappa$, $L$) depend directly on the product of the two – because, for a substitutional atom of the solid to move, it is clearly necessary for there to be a defect (vacancy or self-interstitial) next to it and for that defect to jump in such a way that it displaces the atom in question. However, when we analyse the sequence of movements of the atoms we find that this is not quite as simple as it might initially appear: in the sense that these defect mechanisms of atomic transport may imply the existence of correlations between successive moves of an atom. Consider the situation portrayed in Fig. 2.16, and suppose that this configuration has just come about by a jump of the B atom to the right from the site now vacant. Then the *next* jump of this particular atom is more likely to be back to the position it occupied beforehand than in some other direction. Such a correlation between the directions of successive moves exists even if the atom B is chemically indistinguishable from the host atoms (i.e. is an isotopic tracer atom of the host species): in that case the probability that the next jump of the atom B is just the reverse of the preceding one is just $2/z$, where $z$ is the coordination number of the lattice (Kelly and Sholl, 1987). This characteristic of defect-induced atomic migration leads to distinct factors ($\leq 1$ in magnitude) in the various transport coefficients which are additional to those representing defect concentrations and jump rates.

### *2.3.4 More highly disordered systems*

It is implicit in the above that point defects are present in low concentrations. There are, however, at least three, actively studied classes of solid in which the degree of disorder may be much larger; namely, (i) order–disorder alloys, (ii) fast ion conductors and (iii) concentrated interstitial solutions of hydrogen in metals.

Order–disorder alloys are characterized by an ordered arrangement of atoms at low temperatures which becomes progressively disordered as the temperature is raised until the long-range order disappears altogether at a critical temperature, $T_c$ (e.g. Sato, 1970). By order we here mean the arrangement of atoms on to distinct sub-lattices according to chemical type (e.g. in an alloy AB, A atoms on one, B atoms on another) – as in an intermetallic compound. Disordering occurs by the exchange of atoms of different kinds between these sub-lattices, i.e. by the creation of *wrong* atoms. The proportion of these wrong atoms rises increasingly rapidly as $T$ approaches $T_c$ until at $T_c$ all distinction between the sub-lattices disappears and there are as many wrong atoms as right ones. This order–disorder transition can be studied by the techniques of X-ray and neutron diffraction. An example of the change of order with temperature determined in this way is shown in Fig. 2.18. These changes are accompanied by striking anomalies in thermodynamic properties (e.g. specific heat).

These exchanges of atomic position are accomplished with the aid of vacancies. The thermal equilibrium population of vacancies is small, here as in most metals and alloys, but it is affected by these changes in order as $T_c$ is approached. The consequence is the presence of a term in $g_f$ directly dependent upon the degree of order, with the activation energy $g_m$ probably similarly affected. Since the degree of long-term range order varies very rapidly with $T$ as $T_c$ is approached (Fig. 2.18) the abrupt change in the activation energy of tracer diffusion at $T_c$ is immediately understandable (Fig. 2.19). Similar effects are observed in ferromagnetic metals and alloys at their Curie temperatures, where they become paramagnetic. Indeed, they are to be expected in principle at all such 'second-order' phase transitions, although, in practice, many of the transition temperatures are too low for atomic migration to be easily studied.

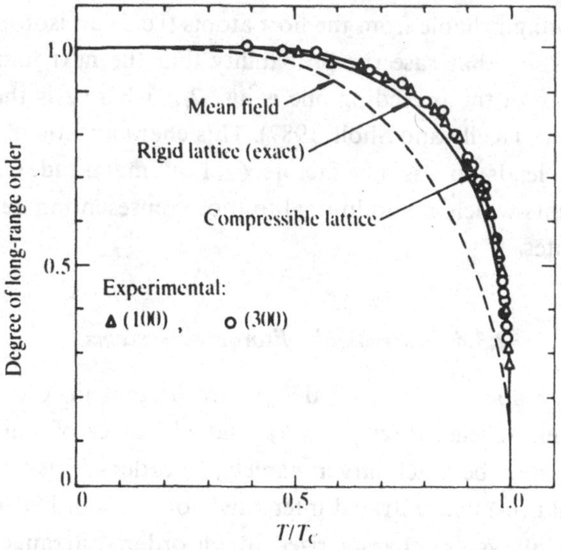


Fig. 2.18. The degree of long-range order in $\beta$-brass as a function of temperature, determined by neutron diffraction by Norvell and Als-Nielson (1970). The lines are different theoretical predictions as indicated.

The second class of highly disordered solids comprises the fast ion conductors (e.g. Laskar and Chandra, 1989). Structurally this is a very miscellaneous class (Table 1.1), but in most of its members one can distinguish a high concentration of ions which move relatively easily over a larger number of sites, in effect a high constant number of interstitial ions. The activation energy of the electrical conductivity, $\kappa$, is thus generally low, measured in units of only 0.1 eV. The basic model of such materials is one of many ions moving on a (sub-) lattice of a greater number of sites. The same basic model is often used also for solid solutions of hydrogen in metals. Sometimes more than one sub-lattice of available sites will be defined: individual ions or atoms will have different energies on the different sub-lattices. Interactions among the ions or atoms are often represented by pairwise interactions between nearest neighbours, although long-range Coulomb interactions may also be important (Funke, 1989).

The theory of atomic transport in these highly disordered solids is the subject of Chapter 13.

## 2.4 Electronic defects

The earlier reference to the effects of non-stoichiometry in ionic compounds brings us to the point where it is appropriate to consider *electronic defects*. As already mentioned in §2.2 these are of significance generally in non-metals. The ideal,

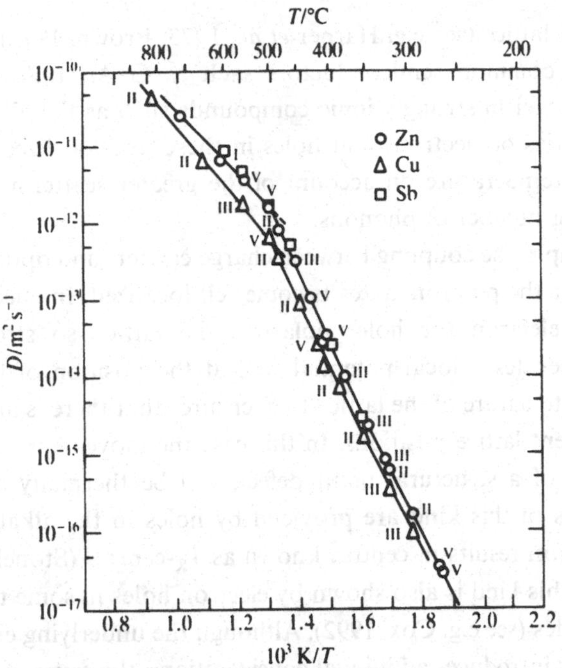


Fig. 2.19. An Arrhenius plot of the tracer diffusion coefficients of Cu, Zn and Sb in $\beta$-brass. The Roman numerals correspond to alloys of slightly different composition (in atom % Zn) as follows: I 45.7, II 47.2, III 48.0 and V 46.5. The order–disorder transition temperature changes only slightly within this range of composition. (After Kuper *et al.*, 1956).

perfect reference crystal will thus be one in which all the valence electronic states are filled and all the conduction states are empty. This perfect reference crystal is thus an electrical insulator. Additional electrons or a deficiency of electrons define electronic defects. Additional electrons may be present in the conduction band while a deficiency of electrons will lead to empty states (holes) in the valance band. Such a condition may be attained by the thermal excitation of electrons across the forbidden gap, as in an intrinsic semiconductor. Electrons may also be released into these conduction states from donor atoms (e.g. substitutional P in Si) and defects, while holes may be created in the otherwise filled valence band by the presence of acceptors (e.g. substitutional B in Si).

These conduction and valence states may be non-localized Bloch states, as in the Group IV, the III–V and the II–VI semiconductors. In the more ionic of these compound semiconductors there will be a coupling between the electron states and the optical lattice vibrations. The states of the coupled electron–lattice system (polaron states) may still be non-localized, but the effect of the coupling is to increase the effective mass of the electrons (or holes) beyond the value it would have in the rigid lattice (see e.g. Harper *et al.*, 1973; Brown, 1972). The effects are rather small in common semiconductors such as GaAs, InSb and CdTe, but become much larger in strongly ionic compounds such as the alkali halides. The electrical mobilities of electrons and holes in these non-localized states decrease with increasing temperature on account of the greater scattering of the carriers by the increasing number of phonons.

Now, in principle, the coupling between charge carriers and optical phonons can be so strong that the polaron states become self-localized. In effect, the electrical charge on the electron (or hole) polarizes the lattice so strongly that this polarization generates a local potential well at the position of the electron (or hole). The discrete nature of the lattice then ensures that there is an energy barrier between equivalent lattice positions. In this case the movements of the electronic defect, like that of a structural point defect, will be thermally activated. Well-studied examples of this kind are provided by holes in the alkali halides where the self-localization results in centres known as $V_K$-centres (Stoneham, 1975).

Behaviour of this kind is also shown by electron holes in some transition-metal and actinide oxides (see e.g. Cox, 1992). Although the underlying electronic theory of these materials introduces additional considerations, the cause of the localization is similar. For here, the filled states of highest energy are not, as in the alkali halides, centred mainly on the anions but on the transition-metal (or actinide) cations instead. The unfilled *d*-(or *f*-)shells of these ions in the appropriate charge state would, by the normal arguments of Bloch theory, lead us to expect these oxides to show a metallic conductivity. That they do not is a consequence of correlations among the movements of these d-electrons which are especially strong and which lead effectively to a localized or ionic picture. NiO, for example, thus conducts electricity only when some of the $Ni^{2+}$ ions are converted to $Ni^{+}$ (excess electron) or $Ni^{3+}$ (hole) states. NiO is thus a semiconductor; electrons and holes can be introduced thermally, by doping (e.g. Li donors) and, no less importantly, by changing the stoichiometry by equilibrium with oxygen. This simple ionic picture leads us also to expect that there will be rather strong attractions (because they are Coulombic) between holes and cation vacancies. When considering the concentration of cation vacancies (e.g. in $Ni_{1-\delta}O$) we must therefore allow for the presence not only of Ni vacancies carrying an effective charge of $-2e$, but also of those carrying a charge $-e$ or even zero.

Transition metal monoxides, like NiO, may be oxidized with varying ease, but will generally only give up oxygen reluctantly. A contrasting system is $UO_2$ which can be reduced as well as oxidized, depending on the oxygen partial pressure in the surrounding phase. Reduction of $UO_2$ is effected by the formation of oxygen vacancies and uranium ions in a lower charge state ($U^{3+}$), while oxidation of $UO_2$ results in the formation of oxygen interstitials and uranium ions in a higher charge state ($U^{5+}$). The charge states of the oxygen vacancies and interstitials may be determined by calculations (e.g. Jackson *et al.*, 1986) and by inference from measurements of the degree of non-stoichiometry as a function of oxygen partial pressure (e.g. Kröger, 1974).

These oxides of the transition metal and actinide oxides thus contain electronic charge carriers whose mobilities may be thermally activated and which therefore rise with increasing temperature. The magnitudes of these mobilities, however, remain greater than those of the structural defects.

## 2.5 Interplay and interactions of defects

In the previous sections we have described the primary defects of importance in the theory of atomic transport (and of many other physical and chemical properties of solids). It is only to be expected, however, that there will be interactions among these primary defects and that these will give rise to a variety of discernible effects. Through specific studies theory and experiment together provide guidance on what to expect. In setting out to provide a first overall view of the matter we may start from our knowledge of the nature of the disturbance which the point defect represents. Clearly the effects of the defect are many: it may perturb the lattice vibrations and scatter the phonons, it may distort the lattice elastically, in insulators it may polarize it electrically, in metals it will scatter the free electrons, in covalent solids it may disturb the bonding locally, and so on. However, it is convenient to classify the interactions roughly into three types, viz. (i) electrical (ii) elastic and (iii) electronic or atomistic. By the last terms we mean interactions either resulting from a redistribution of the electrons around the defects as they come together or from such a redistribution described in terms of atomic force laws. Of course, this last term really embraces all cases, so the first two indicate merely the nature of the dominant part of the interaction when this is adequately described in a continuum approximation. (For a review of the continuum description of point defects and associated effects see Flynn (1972, Chap. 3)).

Let us begin with electrostatic interactions. It is evident that in insulators and semiconductors there will be Coulombic interactions among all defects bearing net electrical charges, e.g. between anion vacancies and cation vacancies in an ionic crystal. Likewise there will be lesser interactions between such defects and any carrying an electric dipole moment, e.g. substitutional molecular ions such as $OH^-$. These interactions will depend upon distance in a way given by electrostatics; in the case of Coulomb interactions they are therefore necessarily of long range. In metals, of course, any net electrical charges (e.g. the missing core charge of a vacancy) are screened out by the valence electrons so that there can be no Coulombic interactions at large separations. There will nevertheless still be short-range screened Coulomb interactions, which can often be described in a simplified way (see e.g. March, 1978).

In addition to electrostatic interactions we also have elastic interactions. In the main these result from the overlap of the strain fields of the two defects, although there can also be a contribution coming from the local changes in elastic properties which are associated with the defect. An important example is provided by the interaction between the elastic strain field of a dislocation (which falls off only slowly as we go away from the dislocation) and that of a point defect. There will, of course, also be such elastic interactions between point defects themselves, but these are rather small in magnitude and fall off rapidly with separation. For detailed accounts of the elastic theory of structural imperfections see Eshelby (1956).

The limitations to dielectric and elastic continuum theories are, of course, associated with the discrete lattice arrangement of atoms, which means that one can only expect these descriptions to be valid at separations much larger than the lattice spacing. Hence only when the interactions at large separations are qualitatively important would we expect to rely only on continuum theories. An example is provided by the incorporation of the Coulomb interactions among cation and anion vacancies in, say, the alkali halides into statistical theories of the Debye–Hückel type by assuming that the interactions of a pair of defects are individually of the form $q_1q_2/4\pi\varepsilon_s r_{12}$ (where $q_1$ and $q_2$ are the net charges on the defects, $r_{12}$ is their separation and $\varepsilon_s$ is the static value of the electric permittivity of the crystal). Another is provided by the description of space-charge layers arising at surfaces and dislocations when oppositely charged point defects have different energies of formation at such sources.

Theories of defect interactions which aim to go beyond the continuum description necessarily reflect not only the lattice structure but also the electronic structure of the solid. For solids composed of atoms with closed electronic shells it will usually be possible to find a suitable semi-empirical description of the interatomic forces. This is the case for the noble-gas solids (Chadwick and Glyde, 1977) and very many ionic solids (Catlow and Mackrodt, 1982). It even forms a good starting point for many ionic solids containing open-shell ions, notable examples being some transition-metal oxides (Tomlinson, Catlow and Harding, 1990). Defects are introduced into these models in a suitable atomic or ionic form and the changes which these cause to the lattice are calculated. In metals, the main emphasis is on the redistribution of the electrons around the defect, although considerable effort has been put into the search for appropriate ways of describing interactions in terms of interatomic potential functions. Likewise, in covalent semiconductors the re-bonding together of the atoms in the defects demands a proper electronic theory. In these cases, however, there can be a strong coupling between the electronic re-distribution and the lattice relaxation on which simplified theories have foundered (Lannoo, 1986).

In the following sections we survey some of the information on defect interactions which is pertinent to our main task in this volume.

### *2.5.1 Interactions among point defects*

Since vacancies and interstitials are complementary defects (i.e. they recombine to restore perfect lattice) they may be expected to attract one another at small separations. In metals, this is certainly true. In non-metals it is true as long as the two defects do not carry net electrical charges of the same sign, for in that case the Coulomb repulsion between them will dominate the other terms in the energy. In practice, both vacancies and interstitials will be present in most solids under irradiation, but for systems close to or in thermodynamic equilibrium this will only be the case in systems displaying Frenkel disorder, e.g. ionic crystals such as AgCl and AgBr ($Ag^+$ ion interstitials and $Ag^+$ vacancies) and the alkaline earth fluorides ($F^-$ ion interstitials and $F^-$ vacancies). In these examples the interstitial and the vacancy carry equal and opposite net charges so that there is a long-range Coulomb attraction as well. The result is that Frenkel defects equilibrate very rapidly when the thermodynamic conditions change.

With the same proviso about net charges of the same sign, it is also to be expected that vacancies themselves will attract one another and lead to the formation of nearest-neighbour or other close pairs. In general this is perhaps less obvious than the existence of an attraction between vacancies and interstitials since in that case the end result is mutual annihilation and release of the whole of the energy of formation. Nevertheless, many experiments and detailed calculations show it to be so in both metals and non-metals. The fact can be rationalized in terms of a simple nearest-neighbour bond model of cohesion. (Whereas the formation of two single vacancies requires the breaking of $2z$ nearest-neighbour bonds, the formation of a nearest-neighbour vacancy pair requires the breaking of only $2z - 1$.) Also, in ionic crystals where anion and cation vacancies may carry net electrical charges of opposite sign, it is plain that there will be strong electrical attractions between them. Thus we can expect a proportion of vacancy pairs to be present in any equillibrium population of vacancy defects. With some certainty they are present in the alkali halides as pairs of anion and cation vacancies. They are also believed to play a part in diffusion in various metals.

Calculations of the interactions of interstitials in b.c.c. and f.c.c. metals again show that there is a small (free) energy of binding into pairs (and larger clusters), but these results are mainly of interest in the theory of radiation damage. The mobilities of these interstitial pairs are generally predicted to be much less than those of the corresponding single interstitials. In ionic solids showing Frenkel disorder only interstitials of one net charge will be present and pairing will bc prevented by their mutual Coulomb repulsion.

### *2.5.2 Interactions of point defects with dislocations*

We have already mentioned the important role of dislocations as sources and sinks for vacancies and interstitials. At the same time there is a strong elastic interaction (through the overlap of the stress field due to each imperfection with the strain field due to the other) such that interstitial atoms are attracted to the dilated region around the (edge) dislocation (i.e. below the slip plane in Fig. 2.8), while vacancies are attracted into the compression region (above the slip plane in Fig. 2.8). These attractions mean that any Fick's law description of the diffusion flux of point defects must be supplemented by a drift term proportional to the force on the defects in question, as envisaged in (1.4.1) and (1.4.2). In magnitude the attraction of (edge) dislocations for interstitials is somewhat greater than that for vacancies as a result of the generally greater lattice strain which interstitials cause. This has important consequences in irradiated metals and alloys. In particular, under conditions of continuous irradiation it leads to a ratio between the steady-state concentration of (self-) interstitials and the steady-state concentration of vacancies smaller than it would otherwise be. In turn, this relative excess of vacancies causes voids to form and grow, with the result that the metal swells as irradiation proceeds. A similar phenomenon appears to occur in the alkali halides, although in this case only anion vacancies and anion interstitials are involved and in place of voids one observes particles of alkali metal (colloids), i.e. regions devoid of halogen (Hughes, 1986). The growth of both voids and colloids is sensitive to various factors, but ultimately depends on the stronger attraction of (edge) dislocations for self-interstitials than for vacancies.

This attraction for self-interstitials, of course, exists equally for interstitial solutes, e.g. carbon, nitrogen and other small atoms in $\alpha$-Fe and other b.c.c. metals. When these systems are in thermal equilibrium the attraction leads to an 'atmosphere' of solute atoms surrounding the dislocation core; the drift of solute atoms under the influence of the elastic force field towards the dislocation core is just balanced by their diffusion away from it. This atmosphere – referred to as a Cottrell atmosphere – is the reason for the sharp yield point at the onset of plastic deformation in $\alpha$-Fe. Of course, when the thermodynamic conditions are changed time-dependent effects may be observed (Bullough and Newman, 1970). Related effects may be observed with loaded cracks where the region ahead of the crack tip is also subject to a dilatational stress.

### *2.5.3 Interactions of point defects with solute atoms*

The discussion and examples given so far indicate that interactions among defects occur rather widely in both metals and non-metals and that their physical origins are rather diverse. It is not surprising therefore that there are also important interactions between solute atoms and vacancies and between solute atoms and self-interstitials. These are especially important for the migration of solute atoms, and the effects are divided into several types.

#### *Substitutional solute atoms and vacancies*

The commonest situation is provided by substitutional solute atoms in interaction with vacancies. The interactions may be attractive or repulsive. In insulators, the sign of the interaction will, in first approximation, be dictated by the Coulomb interaction between any net charges borne by the solute atoms and the vacancies. For example, divalent cations dissolved in alkali halides (net charge $+e$) attract cation vacancies (net charge $-e$) but repel anion vacancies (net charge $+e$). In metals there may be analogous effects, even though the effects of any net charges are effectively screened out beyond nearest-neighbour separations by the valence electrons. Thus Cd atoms dissolved in Ag (net ion–core charge $+e$) attract Ag vacancies (net charge $-e$ since the $Ag^+$ ion core is missing). In cases where there are no net charges or where these simple arguments are invalid other terms decide the nature and extent of the interaction (e.g. covalency, solute size, etc.). On the whole, theory provides a useful guide for strongly ionic solids, the noble-gas solids and metals, but has greater difficulty in dealing with covalently bonded materials.

The magnitudes of these interactions depend very much on circumstances. Thus in metals they are relatively small; 0.1 eV may be regarded as typical at nearest-neighbour separations, but the wave-like nature of the screening electrons means that the sign of this interaction is not necessarily that indicated by the core charges. In ionic crystals electrostatic effects are dominant; solutes bearing no net charge interact weakly with vacancies, typically 0.1 eV again. But those bearing a net charge interact more strongly; typically 1/2 eV in the alkali halides, and as much as 1–2 eV in the alkaline earth oxides at the nearest-neighbour separation.

In relation to the diffusion of solute atoms, we may take it as a rough guide that, in systems where vacancies are the intrinsic defects, those solutes which attract vacancies will diffuse more rapidly than the host atoms – mainly because there is more likely to be a vacancy next to them. Conversely, those solutes which repel them will diffuse more slowly. Examples which illustrate this are provided (i) by various solutes in Al whose diffusion coefficients are shown in Fig. 2.20,

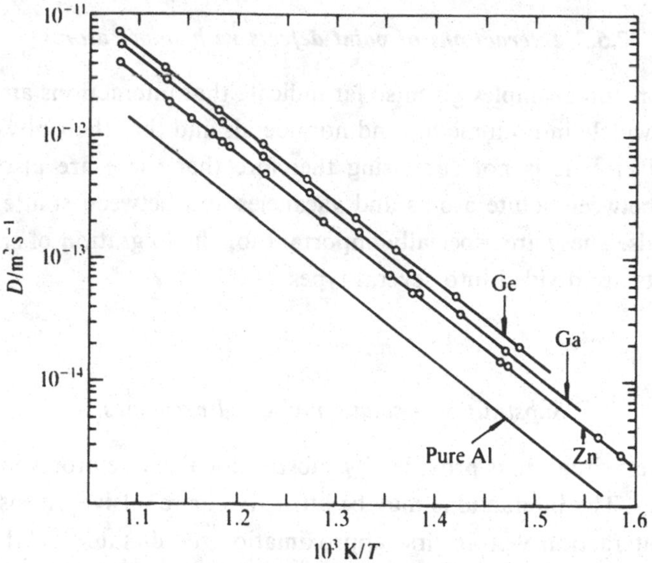


Fig. 2.20. The diffusion coefficients of Zn, Ge and Ga solutes in pure Al, together with the self-diffusion coefficient of Al. (After Peterson and Rothman, 1970a.)

and (ii) by the diffusion of multivalent cations in alkali halide crystals (Fig. 2.21). However, both qualitatively and quantitatively there are other factors which must also be considered. Firstly, there is the question of the rate at which the solute atom will be activated to jump into the nearby vacancy. One certainly expects that to differ from the corresponding rate for a host atom. Secondly, there are the rates of other vacancy movements in the vicinity of the solute atom which can, in principle, limit the overall diffusion rate. For example, it is obvious from Figs. 2.16 and 2.22 that, unless the vacancy can move around the solute atom, this atom will merely jump back and forth without making any net displacement. The dependence of these other jump rates upon the nature of the solute and the solvent is less easy to see intuitively than is the solute–vacancy interaction. For, although there will certainly be electrostatic effects, other terms such as repulsions between the ion cores (size effects), polarization effects, changes in covalent bonding, etc. also enter, depending on the circumstances. As before, theory has been useful in making predictions for ionic compounds, solid noble gases and metals, but has been less successful with covalent solids. One general use of theory has been to guide the construction of parameterized models. A widely used example for f.c.c. solids is the *five-frequency model* illustrated in Fig. 2.22 (Lidiard, 1955). Analogous models have been proposed for other lattices and will be described and analysed in later chapters. The characteristic feature of all these is that the perturbation of the vacancy movements by the solute atom is limited to its immediate vicinity.

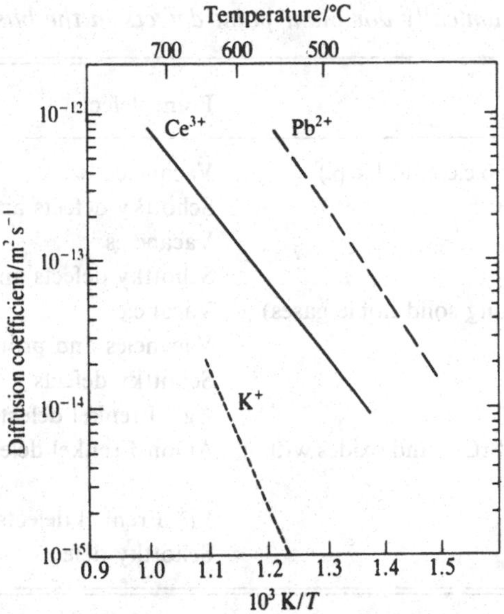


Fig. 2.21. The diffusion coefficients of the aliovalent solute ions $Pb^{2+}$ and $Ce^{3+}$ together with the self-diffusion coefficient of $K^+$ in KCl. (After Keneshea and Fredericks, 1965.)

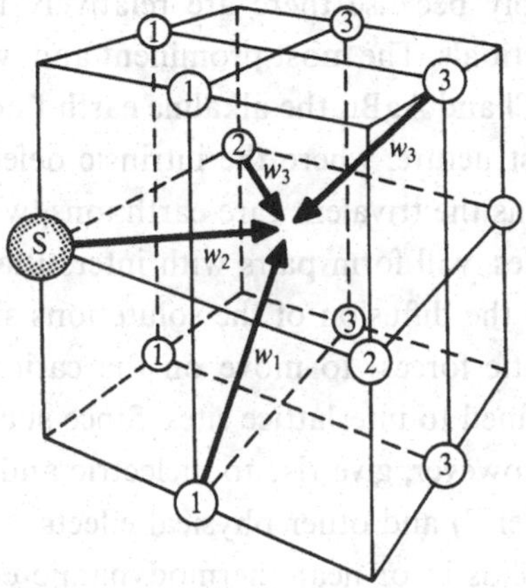


Fig. 2.22. Illustrating the five-frequency model of a solute–vacancy pair in a f.c.c. lattice. The solute atom is denoted by S and the first, second and third neighbours are as indicated. The central site (a first neighbour of S) is vacant. Jump rates of the atoms into the vacancy are as indicated. The model is completed by specifying the jump rate $w_4$ for the inverse of $w_3$ jumps and the jump rate $w_0$ for vacancy jumps among sites more remote than third neighbours.

Table 2.2. *Thermodynamically dominant point defects in the basic solid types*

| Solid | Point defects |
|---|---|
| Metals and alloys (f.c.c., b.c.c. and h.c.p.) | Vacancies |
| Intermetallic compounds | Schottky defects and exchange disorder |
| C, Si and Ge | Vacancies |
| III–V semiconductors | Schottky defects and exchange disorder |
| Molecular solids (including solid noble gases) | Vacancies |
| Ice | Vacancies and proton defects |
| Alkali halides | Schottky defects |
| AgBr and AgCl | $Ag^+$ Frenkel defects |
| Alkaline earth fluorides, $SrCl_2$, and oxides with the fluorite structure | Anion Frenkel defects |
| $Li_2O$ (anti-fluorite) | $Li^+$ Frenkel defects |
| Alkaline earth oxides | Schottky defects |

*Substitutional solute atoms and interstitials*

Some of the same general ideas can also be used to rationalize the interactions between substitutional solutes and interstitials. However, in relation to atomic movements in systems close to thermodynamic equilibrium, this is a much smaller field of application, simply because there are relatively few systems where the intrinsic defects are interstitials. The most prominent and widely studied examples are the silver halides, AgCl and AgBr, the alkaline earth fluorides and some oxides also having the fluorite structure, where the intrinsic defects are Frenkel defects (Table 2.2). Solutes such as the trivalent rare-earth ions, which dissolve readily in the alkaline earth fluorides, will form pairs with interstitial $F^-$ ions. In this case such pairs will not speed the diffusion of the solute ions since these are confined (by the strong electrostatic forces) to move on the cation sub-lattice while the interstitial $F^-$ ion is confined to interlattice sites. Since such pairs are electric and elastic dipoles, they do, however, give rise to dielectric and mechanical relaxation (§§1.5 and 1.6 and Chapter 7) and other physical effects.

In contrast to most solids in or near thermodynamic equilibrium, metals and alloys under intense irradiation by energetic particles will contain significant concentrations of self-interstitials (as well as vacancies). These interstitials may interact with substitutional solute atoms and result in their migration through the lattice. Indeed, it is necessary to take account of this interstitial-induced solute migration (as well as that induced by the vacancies) when describing the phenomenon of radiation-induced segregation in alloys (Nolfi, 1983). The rather

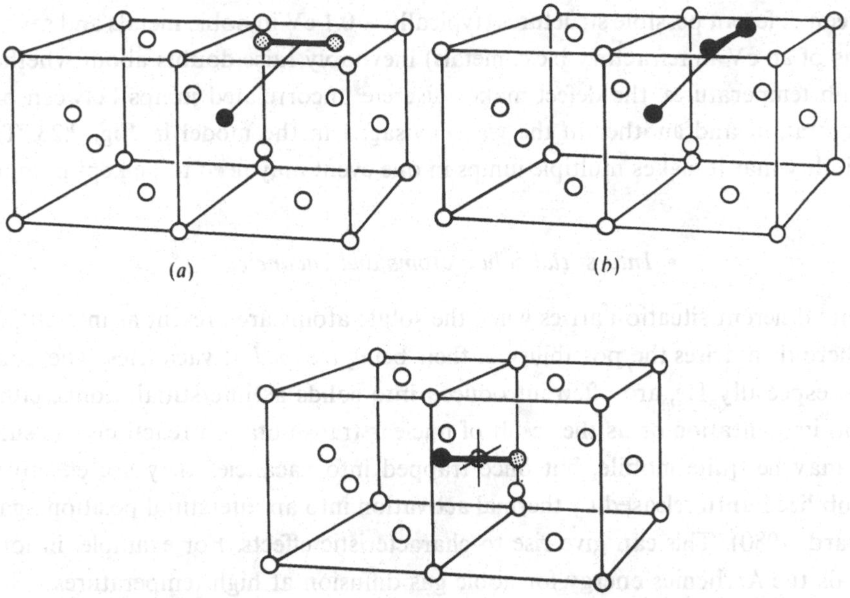


Fig. 2.23. Models of solute–interstitial complexes in an f.c.c. lattice: (*a*) an *a*-type pair (*b*) a *b*-type pair and (*c*) a mixed dumb-bell. In each case the solute atom is denoted by the filled circle while interstitial solvent atoms are shown as shaded circles. The dumb-bells are supposed to move by jumps in which one of the two atoms moves to form a new dumb-bell at a neighbouring site, while the other re-occupies the central lattice site previously straddled by the dumb-bell. In the process the axis of the new dumb-bell lines up at 90° to the previous one. In addition, the dumb-bells may rotate, without translation, about the central lattice site from one (100) orientation to another. Diffusion of the solute atoms thus proceeds via a series of movements in which they take part as mixed dumb-bells and *a*-type pairs.

few calculations of solute–interstitial interactions and migration rates indicate a variety of possible structures and modes of movement, although a general characteristic of these modes is their low activation energy. One model proposed for f.c.c. lattices on the basis of the atomistic calculations of Dederichs *et al.* (1978) is illustrated in Fig. 2.23. It will be seen that the dumb-bell self-interstitials of the host may form pairs with substitutional solute atoms in two distinct ways (*a* and *b*) and, furthermore, that by a process of internal conversion one of these forms can lead to a *mixed dumb-bell* defect. Diffusion of the solute thus proceeds through a sequence of movements of this sort. As with the vacancy mechanism, successive moves of the solute atom are correlated with one another. In conclusion though, we would remark that such models of solute migration via interstitials are less certain than those for solute migration via vacancies. The small energy differences between different possible structures (typically $\sim 0.1$ eV in noble metals and several tenths of an eV in refractory b.c.c. metals) inevitably raise doubts about whether, at high temperatures, the defect makes discrete uncorrelated jumps between one configuration and another in the way envisaged in the model in Fig. 2.23. The possibility that it makes multiple jumps in one event may need to be kept in mind.

### *Interstitial solute atoms and vacancies*

A quite different situation arises when the solute atoms are present as interstitials, for there then arises the possibility of their being *trapped* at vacancies. The noble gases, especially He, are often introduced into solids as interstitial atoms, either by ion implantation or as the result of nuclear transmutation reactions. As such, they may be quite mobile, but once trapped into vacancies they are effectively immobilized until released by thermal activation into an interstitial position again (Lidiard, 1980). This can give rise to characteristic effects. For example, in ionic crystals the Arrhenius energy for noble gas diffusion at high temperatures is low (free interstitial migration) while at low temperatures it is high (trapping regime) – just the reverse of the usual behaviour of substitutional solute ions. Similar effects can arise with C and N interstitials in $\alpha$-Fe and analogous systems. The trapping energies in these and other cases are quite large, reflecting the relatively high absolute energy associated with interstitial configurations.

In contrast to the above picture in which the solute atoms are either interstitial or trapped at vacant sites, i.e. substitutional, some dilute alloys may require a model of interstitial–vacancy pairs, i.e. the solute remains as an interstitial even when next to a vacant site. The solutes in these systems diffuse very rapidly compared to the host atoms, too rapidly in fact for the usual models of substitutional solute–vacancy pairs to hold (Chapter 11). They are referred to as *fast diffusers*. Examples are noble metal solutes in lead. In ionic crystals there are some analogous systems in which the stable position of the solute ions is 'off-site', again leading to a sort of interstitial–vacancy pair. Examples are $Li^+$ and $Cu^+$ in KCl. In these cases the energetics are well understood and the motion of the $Li^+$ and $Cu^+$ ions between off-site positions has been fully studied (Flynn, 1972).

## 2.6 Application to atomic transport

The models and examples which we have described above show that in many situations and for many properties it will be the *product* of defect concentration with the defect jump frequency which is the significant quantity. For example, we would expect the rate of diffusion of atoms in a solid in which vacancies are the dominant mobile defects to be proportional to the product of the vacancy fraction with the vacancy jump frequency. We have already touched upon the fact that since both factors depend on temperature via a Boltzmann exponential function this allows a ready explanation of properties which depend on temperature in the manner of (1.1.1). However, there are many situations where the matter is more complicated. For example, the rates of defect movement can be changed, locally at least, by interactions among defects of the same or different kinds and by interactions between defects and solute atoms. When these interactions are attractive they can also lead to the formation of compound, or complex defects, the lifetimes of which may be sufficiently long that they are conveniently regarded as distinct defects, and indeed in many cases studied as such physically. For atomic transport, however, the question is the representation of the measured properties (e.g. diffusion coefficients, ionic conductivities, n.m.r. relaxation times, etc.) in terms of the contributions from the different types of defect. The obvious, and most widely used, assumption is that the contribution of each type of defect is obtained as though defects of other types were not present and that these contributions may be simply added together. For example if we are concerned with a system containing solute atoms, vacancies and solute–vacancy pairs (say, nearest-neighbour pairs) we simply express the physical property in question ($P$) as the sum of three terms each of which is evaluated in isolation. This additivity assumption, expressed as

$$P = \sum_i P_i, \tag{2.6.1}$$

has been used extensively in the analysis of defect and atomic transport properties. Even so, although it is exact for thermodynamic quantities, for kinetic quantitites it may be only an approximation depending on the rates of interçhange and interplay of defects among the various sub-populations. For example, when the total population consists of unpaired solute atoms, unpaired vacancies and solute–vacancy pairs, the accuracy of the assumption may depend upon the rate at which vacancies are exchanged between the sub-populations of unpaired and paired vacancies. The theory which we shall develop in later chapters, in addition to providing us with the theory of those physical properties specified in Chapter 1, will also enable us to assess this approximation in particular cases.

## 2.7 Summary

In Chapter 1 we described many of the physical manifestations of diffusive atomic movements in solids and in this chapter we have introduced the atomic models which rationalize these observations. The basic concept is that of a set of elementary imperfections in an otherwise perfect crystalline solid. This idea has been successful in explaining and correlating a wide range of physical properties in all the principal classes of crystalline solids – metals, ionic crystals, oxides, molecular and covalent solids. Central to the description of diffusive atomic movements and the effects described in Chapter 1 are the ideas of point defects – vacancies, interstitial atoms, solute atoms, etc. These defects are thermodynamic defects in the sense that they will be present in solids in thermodynamic equilibrium in concentrations fixed by the temperature, pressure and chemical potentials of the constituents. We shall explore these relations more quantitatively in the next chapter by using the methods of statistical thermodynamics. These demonstrate the general tendency of the defect concentrations to follow the Arrhenius dependence on temperature which we have here presented on intuitive grounds.

Point defects can also be introduced in other ways, notably by energetic particle irradiation and plastic deformation, but in these circumstances the system will not be in thermodynamic equilibrium and the defect concentrations will always tend towards the equilibrium values. The rate at which the system moves towards equilibrium depends, of course, on the mobility of the point defects, but it also depends on the presence and density of other imperfections which can act as sources and sinks of the point defects (e.g. external surfaces, grain boundaries, dislocations). The defect jump rate itself can also be described by using the methods of statistical mechanics (together with the assumption of a transition state) and again in the next chapter we shall substantiate the intuitive argument used here to justify an Arrhenius dependence of jump rate upon temperature. This Arrhenius dependence of both defect concentrations and mobilities is important but not conclusive evidence for their role in the processes described in Chapter 1. The conclusive evidence for point defect mechanisms of these processes (as distinct from intrinsic mechanisms such as the direct exchange of atomic positions) comes from a variety of experiments, notably the Kirkendall effect, Simmons–Balluffi experiments and a range of other measurements which test the predicted dependence of defect concentration and mobility upon thermodynamic parameters.

Lastly, it should be remarked that, although in this book we are primarily concerned with non-equilibrium processes (as listed in Chapter 1), the emphasis here on the concentrations and mobilities of defects in thermodynamic equilibrium is quite proper for the circumstances we envisage. Following the pioneering work in non-equilibrium statistical mechanics by Green, Kubo, Mori and others in the 1950s, it is now well established that the coefficients of the linear response of systems subject to perturbing forces are given by the averages of appropriate kinetic quantities taken in conditions of thermodynamic equilibrium, i.e. in the absence of the perturbation (see e.g. McQuarrie, 1976; Kreuzer, 1981). In particular, it is the concentrations of defects and their rates of movement under conditions of thermodynamic equilibrium which enter into coefficients like $D$, $\kappa$, $\varepsilon$, etc. introduced in Chapter 1. We shall demonstrate this explicitly in Chapter 6. The result is foreshadowed in Chapter 4, where we describe the extended phenomenology provided by the thermodynamics of non-equilibrium processes.

# 3

# *Statistical thermodynamics of crystals containing point defects*

## 3.1 Introduction

In the previous chapter we surveyed the ideas of point defects which form the physical basis for the description of diffusion and other atomic transport phenomena. In the present chapter we shall describe the application of equilibrium statistical mechanics to these physical models under conditions of *thermodynamic equilibrium.* This will allow us to obtain the contributions of thermally created point defects to thermodynamic quantities, e.g. lattice expansions, specific heat, etc. But there is a wider interest in such calculations, as will be fully apparent in later chapters. For, although the processes of atomic transport involve systems not in overall thermodynamic equilibrium, the theory of these processes nevertheless assumes the existence of *local* thermodynamic equilibrium; i.e. of equilibrium in regions small compared to the size of the entire macroscopic specimen but large enough to allow a thermodynamic description locally. There are two aspects to this, one concerns the chemical potentials $\mu_i$ and the other the transport coefficients, $L_{ij}$ (cf. §1.4). Statistical thermodynamics allows us to calculate the chemical potentials of the various components, $i$, of any particular model system as functions of the intensive thermodynamic variables. Gradients of these chemical potentials then give the thermodynamic forces $X_i$ which generate the fluxes of atoms, defects, etc. under non-equilibrium conditions. For the transport coefficients we shall need the equilibrium concentrations of point defects and certain related quantities. Statistical thermodynamics again will provide these for particular models as a function of thermodynamic variables (e.g. $P$, $T$) without regard to the mechanism by which this equilibrium is attained (although, of course, it is implicit that the density of defect sinks and sources is adequate). No less important than the defect concentrations are the manner and rates at which defects jump from one lattice position to another. Although this matter, strictly speaking, lies outside the province of equilibrium statistical mechanics, the assumption that passage through a transition state is sufficient to lead to a transition from one defect configuration to another does allow us to use equilibrium statistical mechanics to obtain the rate at which transitions occur. It is our purpose to supply the theory of these various quantities in this chapter.

In pursuing this aim we immediately come up against one difficulty and that is the broad range of systems which are of interest (e.g. metals, alloys, elemental and compound semiconductors, strongly ionic crystals, fast ion conductors, oxides of various sorts, and so on) not to mention the factors which determine defect concentrations (e.g. temperature, pressure, composition and stoichiometry). In this respect we must therefore be very selective. Our emphasis therefore is on the methods of making such calculations rather than on the detailed discussion of the results for particular systems. To this end, and for our later purposes, we shall present a systematic framework established around the Gibbs free energy function, or thermodynamic potential. However, for the benefit of those coming new to the subject we do include a certain amount of typical detail. Many other basic examples can be found in the books by Crawford and Slifkin (1972, 1975) Flynn (1972), Kröger (1974), Schmalzried (1981) and Philibert (1985).

The structure of this chapter is therefore as follows. In the next two sections we study four examples where it is convenient to use the well-known combinatorial methods of statistical thermodynamics (Wilson, 1957; Hill, 1960) to arrive at the Gibbs free energy function. These examples are all ones where we can assume (i) the absence of any interactions among the defects and (ii) that the defect concentrations are small. Two are of monatomic substances and two are of compounds. In all four examples we can determine equations for the equilibrium concentrations of defects and we can obtain expressions for the chemical potentials of the defects and the other components. However, in the case of compounds there is an important structural requirement (imposed by the existence of the crystallographic *basis*), which means that chemical potentials for individual types of defect (e.g. cation vacancies alone) cannot be uniquely defined. It is nevertheless possible to obtain sets of defect chemical potentials which have some of the properties of ordinary chemical potentials and which can be similarly employed under appropriate constraints. These have been called *virtual* chemical potentials and we give careful attention to them in §§3.3 and 3.4.

In the second half of the chapter we turn to the description of systems in which defect interactions are significant. First (§3.5) we deal with those dilute systems where interactions manifest themselves through the formation of simple defect clusters (e.g. solute–defect pairs). Conventional combinatorial approaches to these systems provide useful approximate results, but in general they do not yield mathematically systematic approximations. An alternative systematic method based on the Mayer theory of imperfect gases, which is especially convenient at the level of approximation needed for certain theories later (Chapter 8), is given in §3.5.

This is followed by §§3.6 and 3.7 which set out the theory of internal and overall defect equilibria respectively in a form appropriate to later applications in transport theory. Section 3.8 provides an example to illustrate the use of these methods described in more general terms in §§3.5–3.7.

Next, §3.9 describes the modification to the results previously presented for ionic crystals in §3.3.1 which follows from a consideration of the long-range Coulomb interactions among the defects in such systems.

Section 3.10 turns to the more difficult problems presented by concentrated solid solutions. So-called order–disorder alloys and regular solutions are examples which have received an enormous amount of sophisticated attention – although mostly not for their defect properties, but rather for their phase transitions. Here we shall present certain qualitative indications which follow from the elementary treatments.

The last main section, §3.11, then turns to the matter of defect jump rates. This subject has also received a great deal of theoretical attention, not only in the present context but also in a wider realm of chemical reaction rates. Our treatment is therefore confined to the provision of an introduction and a summary of the principal conclusions needed for subsequent purposes.

## 3.2 Pure monatomic solids containing point defects

In this section and the next we shall provide the elementary theory of point defects in several different systems, viz. metals, noble-gas solids (more generally molecular solids), covalent elemental semiconductors, ionic crystals and intermetallic compounds. In each case all defect concentrations are assumed to be low and interactions among the component structural defects are neglected, with the result that the straightforward, elementary combinatorial methods of statistical thermodynamics may be used to obtain the thermodynamic properties of these systems. These topics are, of course, extensively discussed in many books and review articles and it is not our aim to cover all the same ground again here. Although we do aim to provide a brief introduction to the theory of defect concentrations in systems in thermal equilibrium for those who are unfamiliar with it, our more general purpose is to present the theory in such a way as to introduce the idea of the (virtual) chemical potentials of defects and to show its usefulness, as well as those limitations to the idea which arise from the existence of physical constraints on the systems (especially the conservation of matter, overall electroneutrality and, in the case of compounds, the necessity to preserve the crystal structure).

Throughout we shall work in terms of the Gibbs free-energy function $G$ as the thermodynamic potential appropriate to the usual experimental conditions of defined temperature, $T$, and pressure, $P$. (We shall continue to refer to $G$ as the Gibbs free energy, although, increasingly, the term *Gibbs energy* is being used instead, while in the German literature the term *free enthalpy* is often used for the same quantity.) We shall write down expressions for $G$ in terms of the numbers $N_i$ of the various defects and structural elements in the system and then derive appropriate equilibrium conditions as those which minimize $G$. In writing down these expressions we shall use the methods of statistical thermodynamics (see, e.g. Wilson, 1957; McQuarrie, 1976). The more familiar of these are the methods of the microcanonical ensemble or the canonical ensemble, which directly yield the entropy function, $S$, and the Helmholtz free energy function, $F$, respectively. In the applications here we shall be primarily concerned with evaluation of the configurational entropy of a system of non-interacting defects, so that the steps from $S$ and $F$ to $G$ are quite simple. Alternatively, we could use the method of the constant pressure ensemble (Hill, 1956). The results, of course, will be the same whichever method is used.

### *3.2.1 Pure monatomic solids containing vacancies in a single charge state*

To begin we consider the statistical thermodynamics of a pure monatomic solid containing vacancies. We restrict the discussion to a metal or a noble-gas solid in which the vacancies are in a single electronic state and we suppose that the vacancy concentration is so low that interactions among them can be neglected. This simple system provides an introductory example of the combinatorial approach to the configurational entropy of defect solids.

We first determine an expression for the Gibbs function $G(T, P)$ of the crystal at the temperature $T$ and uniform pressure $P$ of interest. It can be written as the sum of $G^0(N_m, T, P)$, the Gibbs function of a perfect crystal having the same total number of lattice sites, $N_m$, plus the change in Gibbs function on forming the actual crystal (which contains $N_v$ vacancies) from the perfect one, i.e.

$$G(N, N_v, T, P) = G^0(N_m, T, P) + N_v g_v^\infty - TS_{con}, \qquad (3.2.1)$$

with

$$N_m = N + N_v, \qquad (3.2.2)$$

where $N$ is the number of atoms in the actual, defective crystal. The quantity $g_v^\infty$ is the work required under conditions of constant temperature and pressure to remove one particular, but arbitrary, atom from the perfect reference solid and take it to *a state of rest at infinity*, i.e. outside and away from the solid. In this process the crystal lattice relaxes around the vacant site so created and the thermal vibrations are also altered. The energy $g_v^\infty$ is thus a thermodynamic quantity with the character of a Gibbs energy. The assumption that there are no interactions among the vacancies is reflected in the simple proportionality of the second term on the right of (3.2.1) to $N_v$. This absence of interactions also means that all $\Omega$ distinct configurations of the $N_v$ vacancies on the $N_m$ lattice sites have the same energy, so that the configurational entropy $S_{con}$ can be written from Boltzmann's equation as

$$\begin{aligned} S_{con} &= k \ln \Omega \\ &= k \ln(N_m!/N!N_v!). \end{aligned} \tag{3.2.3}$$

The expression for the Gibbs free energy of the crystal is therefore

$$G(N, N_v, T, P) = G^0(N_m, T, P) + N_v g_v^\infty - kT \ln(N_m!/N!N_v!). \tag{3.2.4}$$

Before we use this expression we note that we could have chosen the reference state to be that of a perfect solid containing the same number of *atoms* as the defective crystal. In the present example this makes no essential difference (although in the equation corresponding to (3.2.4) $g_v^\infty$ would then be replaced by $g_v$, the work required to create a vacancy by replacing the removed atom on the surface of the crystal or by incorporating it at an edge dislocation within the crystal). However, when we come to compound solids, which may be non-stoichiometric, it is better to choose the reference state in the way we have done here.

For many purposes a more convenient form for $G$, than the explicit expression (3.2.4), is obtained by expressing it in terms of the chemical potentials of the component species, i.e., in this case, of the atoms, $\mu$, and of the vacancies, $\mu_v$, defined by

$$\mu = \left(\frac{\partial G}{\partial N}\right)_{T,P,N_v}, \tag{3.2.5a}$$

and

$$\mu_v = \left(\frac{\partial G}{\partial N_v}\right)_{T,P,N}. \tag{3.2.5b}$$

From (3.2.4) these chemical potentials are therefore

$$\begin{aligned} \mu &= \left(\frac{\partial G^0}{\partial N_m}\right)_{T,P} + kT \ln(1 - c_v) \\ &\equiv \mu^0 + kT \ln(1 - c_v), \end{aligned} \tag{3.2.6a}$$

and

$$\mu_v = \mu^0 + g_v^\infty + kT \ln c_v$$
$$\equiv g_v + kT \ln c_v, \tag{3.2.6b}$$

in which

$$c_v \equiv N_v/(N + N_v)$$

is the site fraction of vacancies (here equal to the mole fraction $x_v$). The quantity $g_v$ introduced in (3.2.6b) is the work required to create a vacancy by removing an atom from one particular, but arbitrary, lattice site and incorporating it at an external surface or dislocation or other sink site. Since these chemical potentials are *intensive* quantities, while the Gibbs function is *extensive*, it follows that $G$ can be re-expressed as

$$G = N\mu + N_v\mu_v. \tag{3.2.7}$$

Now, in general, once an expression for $G$ has been established for a particular model we may proceed to determine the concentrations of defects when overall thermodynamic equilibrium is attained and to obtain expressions for other thermodynamic functions. In the present case the transport of vacancies to and from internal sources and sinks and to and from the crystal surfaces will establish an overall thermodynamic equilibrium in which the number of vacancies will be such as to minimize the Gibbs function at the given $T$ and $P$, i.e.

$$\mu_v \equiv \left(\frac{\partial G}{\partial N_v}\right)_{T,P,N} = 0. \tag{3.2.8}$$

The equilibrium fraction of vacancies is then

$$c_v = \exp(-g_v/kT). \tag{3.2.9}$$

The Gibbs energy of the crystal still contains a vacancy contribution since from (3.2.7) we find

$$G = N\mu = N[\mu^0 + kT \ln(1 - c_v)], \tag{3.2.10}$$

with $c_v$ given by (3.2.9). The presence of this vacancy term can be detected in the characteristics of the vapour pressure of relatively volatile solids (e.g. Ar and Kr, Salter, 1963).

Other thermodynamic functions for the crystal in equlibrium can be found from $G$ by standard thermodynamical manipulations. For example, the enthalpy and the entropy from (3.2.4) and (3.2.8) are

$$H = \left(\frac{\partial(G/T)}{\partial(1/T)}\right)_{P,N} = H^0 + N_v h_v \tag{3.2.11a}$$

and

$$S = -\left(\frac{\partial G}{\partial T}\right)_{P,N} = S^0 + N_v s_v - kN \ln(1 - c_v) - kN_v \ln c_v. \tag{3.2.11b}$$

$H^0$ and $S^0$, the enthalpy and entropy of the hypothetical perfect crystal, are related to $G^0$ in the standard way, i.e. as $H$ and $S$ are related to $G$ in (3.2.11a) and (3.2.11b). Similarly, the enthalpy, $h_v$, and entropy, $s_v$, of formation of a vacancy are related to the corresponding Gibbs energy $g_v$ by

$$h_v = \left(\frac{\partial(g_v/T)}{\partial(1/T)}\right)_{P,N}, \qquad s_v = -\left(\frac{\partial g_v}{\partial T}\right)_{P,N}$$

and

$$g_v = h_v - Ts_v. \tag{3.2.12}$$

We may note that the vacancy formation enthalpy is also given by (3.2.9) as

$$h_v = -\left[\frac{\partial(\ln c_v)}{\partial(1/kT)}\right]_{P,N}. \tag{3.2.13}$$

This quantity can often be directly determined from experiments since it involves only relative and not absolute vacancy concentrations. An example is provided by the technique of freezing vacancies into metals by a very rapid cooling of thin wires or foils from a high temperature $T$ down to a low one ('quenching'), and then obtaining a relative measure of $c_v(T)$ from the excess electrical resistivity resulting from this quench. (The kinetics of the disappearance of $c_v$ on annealing the specimen at intermediate temperatures has already been illustrated in Fig. 1.18.) However, absolute determinations of $c_v$ can also be made by precise simultaneous measurements of the overall linear macroscopic thermal expansion of the crystal and of the corresponding expansion of the lattice parameter by X-rays. For,

$$c_v = 3\left(\frac{\Delta l}{l} - \frac{\Delta a}{a}\right), \tag{3.2.14}$$

where $l$ is the length of the specimen and $a$ its lattice parameter. Vacancy concentrations in various metals and noble-gas solids have been determined in this way. An example has already been provided in Fig. 2.15. The enthalpy and entropy of vacancy formation so obtained are given for several metals in Table 3.1. As these values imply, it may often be assumed that $h_v$ and $s_v$ are independent of $T$. However, this may not always be an adequate approximation. For example,

Table 3.1. *Enthalpy and entropy of vacancy formation in some f.c.c. metals*

| Metal | $h_v/(\text{eV})$ | $s_v/(k)$ | Ref. |
|---|---|---|---|
| Al | 0.75 | 2.4 | a, b |
| Cu | 1.18 | 1.6–3.0 | c, d, e |
| Ag | 1.09 | ~1.5 | b |
| Au | 0.94 | 1 | f |

(a) Feder and Nowick (1958), (b) Simmons and Balluffi (1960), (c) Simmons and Balluffi (1963), (d) Differt, Seeger and Trost (1987), (e) Kluin (1992), (f) Simmons and Balluffi (1962).

For a comprehensive compilation of data on point defects in metals see Ullmaier (1991).

there appear to be rather large temperature variations in metals which exist in more than one crystal structure (see, e.g. Herzig and Köhler, 1987). But in all cases one would expect some such variation to come from the normal lattice expansion with increasing $T$ and the associated changes in interatomic forces.

Lastly, in concluding this section, we would mention that the theory for monatomic solids having interstitials as the intrinsic defects would be similar in all respects to what we have set down for vacancies. However, in reality there are few such systems.

### *3.2.2 Vacancies in a covalent semiconductor*

In this section we have in mind solids such as diamond, Si and Ge, although this group of materials is not as homogeneous from the defect point of view as one might at first expect from their general chemical classification as Group IV elements. Thus, diamond is an electrical insulator whose vacancies are only mobile at high temperatures (Collins, 1980; Evans and Qi, 1982), Si is a semiconductor which may have both vacancies and interstitials as intrinsic defects, while Ge is a semiconductor with vacancies alone as the intrinsic defects (Frank *et al.*, 1984). In both Si and Ge, however, the thermal equilibrium concentration of intrinsic defects appears to be relatively much lower than is characteristic of metals and ionic crystals. It must also be said that there is a wealth of detailed information about the electronic states and structure of vacancies and related point defects in these materials which has been obtained by a variety of spectroscopic means (see e.g. Schulz, 1989). However, we shall not call upon that for the time being.

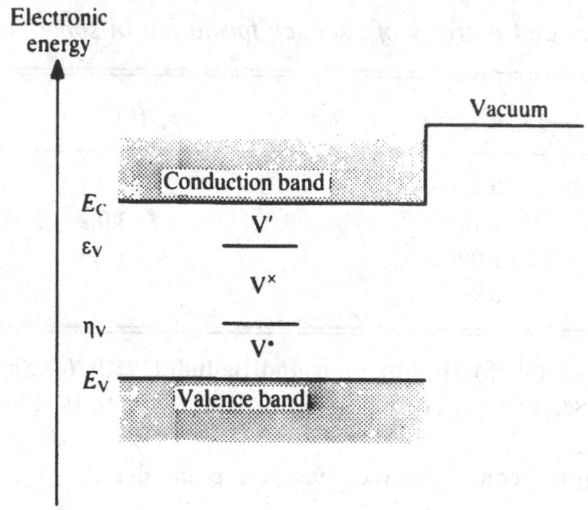


Fig. 3.1. The electronic energy level structure of the model semiconductor analysed in §3.2.2. The meaning of the local electronic levels associated with the vacancies can be seen by considering the charge state of vacancies present at $T = 0$, when the electron distribution will be a sharp Fermi–Dirac distribution (cut-off at the Fermi energy, $\varepsilon_F$). The vacancies will then be in the charge states indicated when the Fermi level lies in the corresponding part of the gap between conduction and valence states. The neutral vacancy $V^\times$ in this model is thus both a donor ($\eta_v$) and an acceptor ($\varepsilon_v$).

Our discussion is therefore really to be seen as a discussion of a model substance defined by Fig. 3.1. The essential differences from the system considered in the preceding section, 3.2.1, are twofold. Firstly, the vacancies may be in three possible charge states, with net electrical charges of $-e$, 0 and $+e$; these are denoted by $V'$, $V^\times$ and $V^\bullet$ respectively in Kröger–Vink notation (Kröger, 1974). We shall furthermore assume that for each of these charge states there is effectively only one electronic energy level available within the band gap. The corresponding electronic degeneracy of this level is $\omega_-$, $\omega_0$ and $\omega_+$ for $V'$, $V^\times$ and $V^\bullet$ respectively. The origin of this degeneracy, as usual, lies in either or both the spin and the orbital motion of the electrons (see e.g. Stoneham, 1975). The second essential difference from §3.2.1 is that we must explicitly consider the contribution to $G$ made by the electrons or holes in band states. Although this model differs in details from that needed for any one of diamond, Si and Ge our discussion of it will introduce the essential differences from metals and molecular solids as treated in the previous section.

We let there be $N$ atoms and $N_v$ vacancies altogether and also $n$ electrons in conduction states and $p$ holes in valence band states. Let the number of vacancies in the various charge states be as

| $V'$ | $V^\times$ | $V^\bullet$ |
|---|---|---|
| $N_1$ | $N_2$ | $N_3$ |

The total vacancy population will then be

$$N_v = N_1 + N_2 + N_3. \qquad (3.2.15)$$

As the reference state for expressing the Gibbs free energy we take a pure perfect crystal made up of $N_m \equiv N + N_v$ atoms with no sites vacant and with no free electrons or holes. Then the methods of statistical thermodynamics give us $G$ for the defective crystal as

$$G = G^0(N_m) + n\varepsilon_F - p\varepsilon_F + N_v g_v^\infty + (N_1\varepsilon_v - N_1 kT \ln \omega_-)$$
$$- (N_3\eta_v + N_3 kT \ln \omega_+) - N_2 kT \ln \omega_0 - kT \ln \left\{\frac{N_m!}{N!N_1!N_2!N_3!}\right\}. \qquad (3.2.16)$$

Here the terms immediately following $G^0$ on the right side are the standard Fermi–Dirac results for the Gibbs free energy of electrons and holes in band states; $\varepsilon_F$ is the Fermi energy or, in other words, the chemical potential of the electrons (Wilson, 1957, especially §6.3). The remaining terms on the right side of (3.2.16) all arise from the presence of the vacancies.

The quantity $g_v^\infty$ is the work required under conditions of constant $P, T$ to remove an atom from the lattice to a state of rest at infinity leaving the neutral vacancy so created in its relaxed ground state. The formation of a V′ vacancy requires the return also of an electron to the bound state provided by the vacancy; hence the formation of V′ requires energy $g_v^\infty + \varepsilon_v$ ($\varepsilon_v$ is negative relative to vacuum). Correspondingly the formation of V$^\bullet$ requires the removal of an additional electron and thus takes energy $g_v^\infty - \eta_v$ (Fig. 3.1). These additional 'electronic' terms, of course, also include the change in relaxation energy accompanying the change in electronic state.

To find the thermodynamic equilibrium concentrations of electrons, holes and vacancies in their several charge states we minimize $G$ with respect to these quantities, i.e. we set $\delta G = 0$. However, in this case the variation of $G$ is subject to the constraint of electroneutrality throughout the bulk of the crystal, i.e.

$$n + N_1 = p + N_3. \qquad (3.2.17)$$

This simply equates the total negative charge to the total positive charge in the crystal, since we can assume the crystal to be electrically neutral in the absence of electrons and holes. Thermal equilibrium conditions are thus obtained by setting $\delta G = 0$ subject to (3.2.17). By regarding $G$ as a function of $N_1$, $N_2$, $N_3$, $n$ and $p$ and using the Lagrange method of undetermined multipliers we then obtain as the conditions of internal equilibrium

$$\left(\frac{\partial G}{\partial N_1}\right) + \lambda = 0, \tag{3.2.18a}$$

$$\left(\frac{\partial G}{\partial N_2}\right) = 0, \tag{3.2.18b}$$

$$\left(\frac{\partial G}{\partial N_3}\right) - \lambda = 0, \tag{3.2.18c}$$

$$\left(\frac{\partial G}{\partial n}\right) + \lambda = 0, \tag{3.2.18d}$$

$$\left(\frac{\partial G}{\partial p}\right) - \lambda = 0. \tag{3.2.18e}$$

In each equation the partial derivative is to be taken with the other four variables constant. It must be remembered, however, in carrying out the indicated differentiation that $N_v$ in (3.2.16) is $N_1 + N_2 + N_3$ and that $N_m = N + N_v$. Each of the derivatives appearing in (3.2.18) may, like (3.2.5b), be referred to as the chemical potential, $\mu_i$, of the corresponding defect. From (3.2.16) these are then

$$\mu_1 = g_v + \varepsilon_v + kT\ln(c_1/\omega_-), \tag{3.2.19a}$$

$$\mu_2 = g_v + kT\ln(c_2/\omega_0), \tag{3.2.19b}$$

$$\mu_3 = g_v - \eta_v + kT\ln(c_3/\omega_+), \tag{3.2.19c}$$

$$\mu_4 = \varepsilon_F, \tag{3.2.19d}$$

$$\mu_5 = -\varepsilon_F, \tag{3.2.19e}$$

in which $c_i = N_i/N_m$ ($i = 1, 2, 3$) is the site fraction of the $i$th kind of vacancy while

$$g_v = \mu^0 + g_v^\infty \tag{3.2.20}$$

is the energy required to form a neutral vacancy $V^\times$ by setting the removed atom down on the surface or other defect source. The arguments of the natural logarithms in $\mu_1$, $\mu_2$ and $\mu_3$ are the thermodynamic *activities* of the corresponding defects. (N.B. In arriving at these expressions it is necessary to remember that $\varepsilon_F$ is an *intensive* quantity, i.e. $\partial\varepsilon_F/\partial n$, etc. are all zero.) Substitution of the $\mu_i$ from (3.2.19) for the derivatives in (3.2.18) then yields the equations of equilibrium. Equations (3.2.19d) and (3.2.19e) tell us that the Lagrange multiplier $\lambda = -\varepsilon_F$, while (3.2.19a) and (3.2.19c) tell us that

$$c_1 = \omega_- \exp\left(-\frac{1}{kT}(g_v + \varepsilon_v - \varepsilon_F)\right) \tag{3.2.21a}$$

and

$$c_3 = \omega_+ \exp\left(-\frac{1}{kT}(g_v - \eta_v + \varepsilon_F)\right), \tag{3.2.21b}$$

i.e. the concentrations of charged vacancies depend upon the position of the Fermi level. On the other hand, since $\lambda$ does not appear in (3.2.18b) the concentration of neutral vacancies $V^\times$ is independent of $\varepsilon_F$ and is an intrinsic property of the crystal, viz.

$$c_2 = \omega_0 \exp(-g_v/kT). \tag{3.2.21c}$$

This result is the analogue of (3.2.9) for metals.

Now, the above calculation might be deemed of limited interest in practice since it has taken the material to be pure, i.e. to be an intrinsic semiconductor, whereas one is more often concerned with semiconductors which are 'doped' to be strongly n-type or p-type. The calculation can easily be repeated by including donors and acceptors in their various charge states as additional structural elements. As long as there are no interactions between these additional species and the vacancies it will be found that (3.2.18) and (3.2.19) all remain the same. The effects of doping are all taken care of through the Fermi level parameter $\varepsilon_F$, which by (3.2.19d) is also the chemical potential of the electrons. Or course, the position of the Fermi level will depend upon the presence of donors or acceptors, but this dependence is well understood (see, e.g. Kittel (1986)). Certainly, in a semiconductor doped with shallow donors it will lie near the edge of the conduction band and in one doped with shallow acceptors it will lie near the upper edge of the valence band. In intrinsic material it will lie in the forbidden gap at a position dependent on the densities of conduction and valence states near these band edges. When the Fermi level is high the concentration of $V'$ is high (3.2.21a) and when it is low the concentration of $V^\bullet$ is high (3.2.21b). Since the concentration of neutral vacancies $V^\times$ is independent of the position of the Fermi level it follows that the total vacancy fraction $c_v = c_1 + c_2 + c_3$ will pass through a minimum as we go from n-type to intrinsic to p-type material. If there were no $V^\bullet$ state available to the vacancy in the band-gap then $c_v$ would simply fall monotonically to an asymptotic value as we go from n-type to p-type material.

As with metals (§3.2.1), the theory of interstitials in covalent semiconductors is very similar to that just given for vacancies. In particular, if the available charge states are like those given in Fig. 3.1 then the effect of doping upon the interstitial concentration will be similar. Such a model has, in fact, been advanced by Frank *et al.* (1984) to explain the diffusion properties of Si at high temperatures (see also Fahey, Griffin and Plummer, 1989). It now seems very likely that in this material the enthalpy of formation of interstitials is not much greater than that of vacancies, so that defects of both kinds are present simultaneously at high temperatures. But it may be noted that in absolute terms, these defect concentrations are much lower than in metals, being $<10^{-6}$ mole fraction at the melting point.

## 3.3 Compounds containing point defects

In this section we shall introduce the ideas of Schottky and Frenkel defects as they arise in compounds, e.g. ionic crystals, oxides, intermetallic compounds and compound semiconductors. As in the previous section, our aim is partly to derive certain basic results for the thermal equilibrium concentrations of these defects, but partly also to bring out clearly the nature of defect chemical potentials. In compounds, however, the constraints upon the statistical thermodynamic calculation demand some care. In particular, it is important to recognize that there are alternative ways to handle these constraints mathematically and not to confuse the elements of one approach with those of another. The significance of these remarks will be apparent in the particular calculations which follow. In these we have to consider three types of constraint, i.e. those of (i) structure, (ii) conservation of atoms and, for non-metals, (iii) overall electrical neutrality. All three are expressed through linear relations among the numbers of structural elements and, in the case of electroneutrality, free electrons and holes as well. Of the three the first exerts the strongest limitation on the formulation of the theory. We shall incorporate it from the beginning, essentially because in the general case we have no choice. The other constraints can be inserted by eliminating certain of the dependent variables, although the use of the method of Lagrange multipliers usually has the advantage of keeping the calculation in a more symmetrical form. For the most part that is the method we shall use.

### *3.3.1 Schottky and Frenkel defects in binary ionic crystals*

In this section we shall consider first the thermally created defects which may occur in strongly ionic crystals such as the alkali and silver halides, the alkaline earth fluorides and oxides. These materials are normally closely stoichiometric and have wide band-gaps so that thermally-produced electrons and holes are generally present in such minute concentrations that they can be ignored. Indeed these materials are the classic ionic or electrolytic conductors whose electrical conductivity arises from the presence and mobility of ionic defects, i.e. ion vacancies and interstitial ions. As such they are to be distinguished from the fast ion conductors (superionic conductors), such as $\alpha$-AgI or so-called $\beta$-alumina, in which a sizeable proportion of one of the ionic species is mobile. These fast ion conductors require different models, which are considered later (Chapter 13). As in the two preceding sections, §§3.2.1 and 3.2.2, we are here concerned with small concentrations of defects in an otherwise essentially perfect crystal. At the end of the section we shall briefly consider non-stoichiometric ionic compounds (especially oxides of transition metals) and the condition of equilibrium with a second phase. We take a broadly similar approach to that already used, i.e. the sequence of steps is (i) the specification of the ideal structure and the nature of the defects in it (ii) the expression of the Gibbs free energy as a function of the numbers of these defects (iii) the derivation of the (virtual) chemical potentials and their use for the expression of equilibrium conditions. To begin with we do not need to specify the electronic structure of the solid: for the time being therefore we refer to atoms rather than ions.

First, we suppose that the ideal stoichiometric compound has the chemical formula $A_aB_b$. For simplicity we shall further assume that all A sites are equivalent, and all B sites likewise, i.e. there are only two distinct sub-lattices overall. This will allow us to equate the number of distinct lattice cells to the number of (lattice) molecules. Other, more complex solids can be treated similarly after specifying the distinct sub-lattices and corresponding defects. Let us therefore assume that there are vacancies in both the A and B sub-lattices and also that certain numbers of A and B atoms are on interstitial sites. Furthermore we shall suppose that the interstitial atoms take up simple interlattice positions ($N'$ sites in all), rather than dumb-bell configurations about normal lattice sites, and that the same set of interlattice positions is available to both A and B atoms. The alternative dumb-bell model can be handled with no greater difficulty. These and the atoms on their proper sites ($A_A$, $B_B$) make up the structural elements. The corresponding numbers of these elements are $N_1 \ldots N_6$ as follows:

$$\begin{array}{cccccc} V_A & V_B & A_I & B_I & A_A & B_B \\ N_1 & N_2 & N_3 & N_4 & N_5 & N_6 \end{array}$$

We have the structural constraint that the total numbers of A-sites and B-sites must be in the correct proportion as given by the stoichiometric chemical formula, i.e.

$$N_m = (N_1 + N_5)/a, \tag{3.3.1a}$$

$$N_m = (N_2 + N_6)/b, \tag{3.3.1b}$$

in which $N_m$ is the total number of lattice unit cells. Thus

$$\frac{1}{a}(N_1 + N_5) = \frac{1}{b}(N_2 + N_6), \tag{3.3.2}$$

i.e. Proust's law of definite proportions.

If we can now neglect any interactions among the defects the Gibbs energy function can be written down in terms of $G^0(N_m)$, i.e. in terms of the Gibbs function for a perfect solid containing the same number of unit cells, $N_m$. Thus

$$G = G^0(N_m) + \sum_{r=1}^{4} N_r g_r^\infty - kT \ln \Omega \tag{3.3.3a}$$

where the index $r$ runs over the four types of defect and

$$\Omega = \frac{(aN_m)!}{N_1!N_5!} \cdot \frac{(bN_m)!}{N_2!N_6!} \cdot \frac{N'!}{N_3!N_4!(N' - N_3 - N_4)!} \tag{3.3.3b}$$

is the number of distinguishable arrangements of the six structural elements on the three sub-lattices (A, B and interstitial).

To obtain the conditions of thermodynamic equilibrium we want an expression for the change in $G$, $\delta G$, resulting from variations $\delta N_i$ in the numbers of the structural elements ($i = 1 \ldots 6$). Formally we can write this variation as

$$\delta G = \sum_{i=1}^{6} \left(\frac{\partial G}{\partial N_i}\right)_{N_{j\neq i}} \delta N_i \tag{3.3.4a}$$

$$\equiv \sum_{i=1}^{6} \mu_i \, \delta N_i. \tag{3.3.4b}$$

By the form of their definition these quantities may be called *virtual* chemical potentials – virtual because, for the following reasons, they are individually indeterminate to a certain extent. Firstly, in evaluating the partial derivatives of (3.3.3a) it must be remembered that $N_m$, the total number of lattice unit cells in the crystal, and $N' \equiv \zeta N_m$, the total number of interstitial sites, by eqns. (3.3.1) are linear functions of the $N_i$. However, since there are two equations for $N_m$ one is presented with a choice over which to use. There is nothing to constrain this choice, although in practice one would probably generally retain the maximum formal symmetry in the resulting set of expressions for the $\mu_i$. Thus one might use (3.3.1a) with the first factor in $\Omega$ for the A sub-lattice, (3.3.1b) with the second for the B sub-lattice and a simple average of the two for the factor for the interstitial lattice in $\Omega$ and for the leading term $G^0$. The set corresponding to this choice, in fact, is

$$\mu_1 = \frac{\mu^0}{2a} + g_1^\infty + kT \ln(c_1/a) + \cdots \tag{3.3.5a}$$

$$\mu_2 = \frac{\mu^0}{2b} + g_2^\infty + kT \ln(c_2/b) + \cdots \tag{3.3.5b}$$

$$\mu_3 = g_3^\infty + kT \ln(c_3/\zeta) + \cdots \tag{3.3.5c}$$

$$\mu_4 = g_4^{\infty} + kT\ln(c_4/\zeta) + \cdots \qquad (3.3.5d)$$

$$\mu_5 = \frac{\mu^0}{2a} + kT\ln(c_5/a) + \cdots \qquad (3.3.5e)$$

$$\mu_6 = \frac{\mu^0}{2b} + kT\ln(c_6/b) + \cdots \qquad (3.3.5f)$$

where $c_i \equiv N_i/N_m$ is the fractional concentration of structural element $i$. Quantities of this form, which arise naturally in the theory of complex systems, will be referred to generally as fractional concentrations. It should be noted that they may differ from the corresponding mole fractions, $x_i$, although for any particular system the relation between $c_i$ and $x_i$ can always be written down quite simply.

Returning now to eqns. (3.3.5) in each case $+\cdots$ indicates that there are additional terms of first and higher order in $c_3$ and $c_4$ which may often (though not always) be neglected. If they are negligible then the argument of each of the natural logarithms is the corresponding thermodynamic activity. All of these quantities (3.3.5a)–(3.3.5f), are *intensive* in the thermodynamic sense and thus one has the convenience that, as usual, one may immediately write

$$G = \sum_{i=1}^{6} \mu_i N_i, \qquad (3.3.6)$$

because $G$ is necessarily *extensive*, i.e. is mathematically a linear homogeneous form in the $N_i$.

Before we turn to the use of $G$ we should again refer to the fact that these virtual chemical potentials are not unique. In particular, we can see from (3.3.6) and the constraint (3.3.2) that we could add an arbitrary quantity $\xi/a$ to $\mu_1$ and $\mu_5$ and subtract $\xi/b$ from $\mu_2$ and $\mu_6$ and leave $G$ unaltered. These virtual chemical potentials may therefore only be used in conjunction with the constraining equations and the only physically significant quantities are those combinations of them which arise when these constraints have been incorporated. This non-uniqueness is the reason some authors (e.g. Schmalzried, 1981) avoid using them. On the other hand, to us they appear to be useful, at least formally, since they allow us to retain a more symmetrical overall mathematical structure which treats all the separate types of defects on an equal footing. This applies not only to the equations of equilibrium, but also to the flux equations (Chapter 4) for which there are homologous constraints.

Now let us use $G$ to determine the relations governing the defect concentrations under the condition of thermodynamic equilibrium. This condition demands that $G$ should be a minimum with respect to variations in the $N_i$ subject to the constraint (3.3.2) and to such other constraints as apply (e.g. in a closed system constant total numbers of atoms of the various kinds present).

Thus if we want the equilibrium numbers of vacancies and interstitials in a specimen of fixed composition we must add the constraining equations for constant total numbers of A and B atoms, viz.

$$N_A = N_3 + N_5, \tag{3.3.7a}$$

$$N_B = N_4 + N_6. \tag{3.3.7b}$$

If we introduce Lagrange multipliers $\lambda, \nu, \sigma$ for (3.3.2), (3.3.7a) and (3.3.7b) respectively, then the minimization of $G$ requires

$$\mu_1 + \frac{\lambda}{a} = 0, \tag{3.3.8a}$$

$$\mu_2 - \frac{\lambda}{b} = 0, \tag{3.3.8b}$$

$$\mu_3 + \nu = 0, \tag{3.3.8c}$$

$$\mu_4 + \sigma = 0, \tag{3.3.8d}$$

$$\mu_5 + \frac{\lambda}{a} + \nu = 0, \tag{3.3.8e}$$

$$\mu_6 - \frac{\lambda}{b} + \sigma = 0, \tag{3.3.8f}$$

whence, by elimination of $\lambda$, $\nu$ and $\sigma$,

$$a\mu_1 + b\mu_2 = 0, \tag{3.3.9}$$

$$\mu_1 + \mu_3 = \mu_5, \tag{3.3.10}$$

and

$$\mu_2 + \mu_4 = \mu_6. \tag{3.3.11}$$

The first of these gives the equation for *Schottky defects*, i.e. for the creation of $V_A$ and $V_B$ in the correct stoichiometric proportions according to the quasi-chemical reaction

$$aV_A + bV_B \rightleftharpoons 0, \tag{3.3.12}$$

in which the 0 signifies the sources and sinks for the vacancies. The second and third equations (3.3.10) and (3.3.11) give corresponding equations for *Frenkel defects* in the two sub-lattices, i.e. for the creation of an interstitial and a vacancy according to the quasi-chemical reactions

$$A_A \rightleftharpoons V_A + A_I \tag{3.3.13}$$

and

$$B_B \rightleftharpoons V_B + B_I. \tag{3.3.14}$$

By (3.3.5a) and (3.3.5b) the equation (3.3.9) for Schottky disorder becomes

$$(c_1/a)^a(c_2/b)^b = \exp(-g_S/kT) \tag{3.3.15a}$$

with

$$g_S = \mu^0 + ag_1^\infty + bg_2^\infty \tag{3.3.15b}$$

the free energy of formation of a complete Schottky defect by the removal of *a* A-atoms and *b* B-atoms from particular, but arbitrary, lattice sites and their subsequent replacement at correct positions at either an internal or an external surface. The quantities $c_1$ and $c_2$ are the fractional concentrations of $V_A$ and $V_B$. Eqn. (3.3.15a) is a typical equation of chemical equilibrium (*law of mass action*) for the reaction (3.3.12).

Likewise from (3.3.5a) and (3.3.5c) for A Frenkel defects and from (3.3.5b) and (3.3.5d) for B Frenkel defects we obtain

$$(c_1/a)(c_3/\zeta) = \exp(-g_{AF}/kT), \tag{3.3.16}$$

and

$$(c_2/b)(c_4/\zeta) = \exp(-g_{BF}/kT), \tag{3.3.17}$$

in which we have assumed that the defect fractions are all small so that $c_5 \simeq a$ and $c_6 \simeq b$. The Gibbs free energies of formation of Frenkel defects, $g_{AF}$ and $g_{BF}$, appearing in these relations are

$$g_{AF} = g_1^\infty + g_3^\infty \tag{3.3.18}$$

and

$$g_{BF} = g_2^\infty + g_4^\infty. \tag{3.3.19}$$

Eqns. (3.3.16) and (3.3.17) are the equilibrium equations for the reactions (3.3.13) and (3.3.14) respectively.

We now have sufficient equations to determine the equilibrium values of the six fractions $c_1 \ldots c_6$. The necessary six equations are the three equations of Schottky and Frenkel disorder, (3.3.15), (3.3.16) and (3.3.17), together with the three constraining equations of structure and composition, namely eqns. (3.3.1) re-expressed as

$$a = c_1 + c_5 \tag{3.3.20a}$$

$$b = c_2 + c_6 \tag{3.3.20b}$$

and the equation giving the chemical composition obtainable from (3.3.7)

$$\frac{c_A}{c_B} = \frac{c_3 + c_5}{c_4 + c_6}$$

$$= \frac{c_3 + a - c_1}{c_4 + b - c_2}, \quad (3.3.21)$$

by (3.3.20). The ratio $c_A/c_B$ alone is sufficient to express any departure from stoichiometric composition. Before moving on to more specific conclusions we make three comments on these results.

Firstly, we note that it is often said that the solutions (3.3.15)–(3.3.17) take this form on account of the condition of electroneutrality. However, in none of the equations of equilibrium so far derived have we used the condition of electroneutrality as such. While (3.3.21) can be viewed as an electroneutrality condition (by assigning ionic charges $q_A$ and $q_B$ to the atoms A and B) we would emphasize that it is really an expression of the structural and composition requirements. These constraints are quite rigid by comparison with the electroneutrality requirement which, although it cannot effectively be relaxed overall, may be relaxed locally in the vicinity of defect sources and sinks such as dislocations and surfaces (see, e.g., Whitworth, 1975 and Slifkin, 1989). In these regions the product relations (3.3.15)–(3.3.17), in fact, still hold. In any case, the requirements of structure and electroneutrality are quite different from one another when electronic defects are present, and in ternary and higher compounds.

Secondly, we would mention that similar remarks about their dependence on temperature apply to $g_S$, $g_{AF}$ and $g_{BF}$ and the corresponding enthalpies and entropies as were made about $g_v$ etc. for metals (§3.2.1). Some explicit calculations have been made within the framework of the quasi-harmonic model for several ionic crystals; see for example Catlow *et al.* (1981) and Harding (1985).

Thirdly, this set of equations has been obtained by using the virtual chemical potentials $\mu_1 \ldots \mu_6$ of (3.3.5a)–(3.3.5f). We emphasize that any alternative set which is compatible with the constraints on the system will lead to the same final equations (3.3.15)–(3.3.17). For any set of virtual chemical potentials, $\mu_i$, the combinations of them which appear in (3.3.9)–(3.3.11) are fully determined, as is easily verified. The other general approach, which avoids the use of virtual chemical potentials altogether, proceeds by first identifying the independent reactions, viz. (3.3.12)–(3.3.14), and then by minimizing $G$ with respect to the *advancement*, $\alpha$, of each reaction individually. Thus in the consideration of (3.3.12) we would set $\delta N_1 = \alpha a$ and $\delta N_2 = \alpha b$ and all other $\delta N_i = 0$ in the equation $\delta G = 0$. It is easily seen that this leads to the same final equation.

Before considering the solutions of these equations for the defect concentrations we derive the chemical potentials of the component atomic species (A, B). These, as well as the defect chemical potentials, are demanded by the formulation of the transport equations in the next chapter. Now, if we start with (3.3.6) and use the conditions of defect equilibrium (3.3.9)–(3.3.11) it is easy to show that $G$ can also be written as

$$G = \mu_3 N_A + \mu_4 N_B$$
$$\equiv \mu_A N_A + \mu_B N_B, \qquad (3.3.22)$$

in which $N_A$ and $N_B$ are respectively the total numbers of A and B atoms in the crystal (cf. eqns. 3.3.7). Hence $\mu_3$ can be identified with $\mu_A$, the chemical potential of species A; and $\mu_4$ with $\mu_B$ likewise. We now observe that, at no point in the derivation so far have we actually assumed the substance to be ionic in nature. The analysis applies equally well to a covalent insulator such as boron nitride. In this case $\mu_A$ would be the chemical potential of boron atoms and $\mu_B$ that of nitrogen atoms. However, the main application is to strongly ionic compounds and in these circumstances species A and B will be ions. The quantities $\mu_A$ and $\mu_B$ thus also refer to ionic species and would be appropriate for considering equilibrium with an external phase which also contained these ions (e.g. an electrolyte solution). Equilibrium with an external phase made up of atoms (e.g. gaseous alkali metal) would demand that we also considered the associated valence electrons which are removed when the ions are formed. An example of the change required is given below.

*Pure stoichiometric crystals*

Let us conclude this section now with some remarks on the nature of the solutions of these equations for $c_1 \ldots c_6$. First of all for crystals having the ideal stoichiometric composition $c_A/c_B = a/b$ (e.g., under normal conditions, most halides); by (3.3.21) we then have

$$\frac{(c_4 - c_2)}{b} = \frac{(c_3 - c_1)}{a}. \qquad (3.3.23)$$

Thus by (3.3.15)–(3.3.17) all four defect concentrations may be determined.

In most materials the energies $g_S$, $g_{AF}$ and $g_{BF}$ are sufficiently different that one type of disorder is far more extensive than the others. Thus in the (pure) alkali halides (where $a = b = 1$) the dominant disorder is of Schottky type

$$c_1 \simeq c_2 = \exp(-g_S/2kT), \qquad (3.3.24)$$

while $c_1$ and $c_2$ are $\gg c_3$ and $c_4$. Some values of the enthalpy and entropy of formation of Schottky defects, corresponding to $g_S$, are given in Table 3.2. Likewise,

Table 3.2. *Enthalpy and entropy of Schottky defect formation in some alkali halides*

| Compound | $h_S/(\text{eV})$ | $s_S/(k)$ | Ref. |
|---|---|---|---|
| NaCl | 2.44 | 9.8 | a |
| NaI | 2.00 | 7.6 | a |
| KCl | 2.54 | 9.0 | a |
| KBr | 2.53 | 10.3 | b |

(a) Bénière, Kostopoulos and Bénière (1980), (b) Chandra and Rolfe (1971). Other recent sources give values close to those quoted. Earlier measurements and analysis gave slightly lower values. For the calculation of these quantities by atomistic methods see Harding (1990).

Table 3.3. *Enthalpy and entropy of Frenkel defect formation in AgCl and AgBr*

| Compound | $h_F/(\text{eV})$ | $s_F/(k)$ | Refs. |
|---|---|---|---|
| AgCl | 1.45–1.55 | 5.4–12.2 | a, b, c, d |
| AgBr | 1.13–1.28 | 6.6–12.2 | b, c, e |

(a) Abbink and Martin (1966), (b) Weber and Friauf (1969), (c) Aboagye and Friauf (1975), (d) Corish and Mulcahy (1980), (e) Teltow (1949).

Theory indicates that in these substances $h_F$ and $s_F$ may vary significantly with temperature (Catlow *et al.*, 1981).

in AgCl and AgBr (also $a = b = 1$ but $\zeta = 2$) the dominant disorder is Frenkel disorder in the Ag sub-lattice and

$$c_1 \simeq c_3 = \sqrt{2}\exp(-g_F(\text{Ag})/2kT) \tag{3.3.25}$$

with $c_1, c_3 \gg c_2$ and $c_4$. (See Table 3.3.) Thirdly, in pure alkaline earth fluorides ($a = 1$, $b = 2$, $\zeta = 1$) it is Frenkel disorder in the fluorine sub-lattice which is dominant and

$$c_2 \simeq c_4 = \sqrt{2}\exp(-g_F(F)/2kT), \tag{3.3.26}$$

with $c_2, c_4 \gg c_1$ and $c_3$. The corresponding fractions of minority defects in these cases can be found by inserting these results into the appropriate remaining relations, e.g. for the alkali halides by inserting the result (3.3.24) for $c_1$ into (3.3.16) to get $c_3$, and that for $c_2$ into (3.3.17) to get $c_4$.

These results are for pure, stoichiometric crystals. The presence of impurity ions of a different charge or valence in the crystal exerts a strong influence on the defect concentrations, as does a departure from stoichiometric proportions. In both cases additional species are present. In the first it is the impurity ions. In the second it is defect electrons or holes, because only neutral atoms can be added or subtracted from the crystal, otherwise it would not remain electrically neutral overall. For example, AgBr in equilibrium with bromine vapour ($Br_2$) becomes $Ag_{1-\delta}Br$ by addition of $Br^-$ ions to the lattice and thus creation of additional cation vacancies ($V_{Ag}'$) together with electron holes ($h^{\bullet}$).

Now these new species (impurity ions, electron holes, etc.) can be added into the calculation and the sequence of operations repeated. It will, however, be found that, in thermodynamic equilibrium, the same product relations (3.3.15)–(3.3.17) are obtained when the defect fractions are all $\ll 1$ (i.e. so that small correction terms $O(c_i)$ are negligible). Naturally, there may also be additional relations corresponding to additional defect reactions, and these additional defects (impurity ions, defect electrons, etc.) may also enter into or modify the constraining relations (3.3.20) and (3.3.21), and may also introduce other constraining relations, most notably electroneutrality. A wide range of possibilities is thereby opened up (see e.g. Kröger, 1974 or Schmalzried, 1981). We consider just two examples which disclose the principles involved; (i) an alkali or silver halide containing substitutional divalent cations and (ii) a transition metal monoxide in equilibrium with an external oxygen phase (which for definiteness we can take to be gaseous).

*Crystals containing aliovalent solute ions*

Now it has long been known that alkali and silver halides (MX, say) will accept into solution small concentrations of halides of divalent metals ($NX_2$) without loss of halogen (X). The ions of the compound ($NX_2$) enter into the proper sub-lattices of the host (MX) but to every N ion entering the cation sub-lattice two X ions enter the anion sub-lattice. The structural constraints can therefore only be satisfied if simultaneously either a cation vacancy (or an anion interstitial) is created or else an anion vacancy (or a cation interstitial) is removed. Formally, the structural constraints (3.3.20) become (with $a = 1 = b$)

$$1 = c_1 + c_2 + c_N \tag{3.3.27a}$$

$$1 = c_2 + c_6, \tag{3.3.27b}$$

while the composition equation (by modification of eqns. 3.3.7) is

$$c_4 + c_6 = c_3 + c_5 + 2c_N, \tag{3.3.27c}$$

in which $c_N$ is the fractional concentration of $NX_2$ entering the lattice. From these

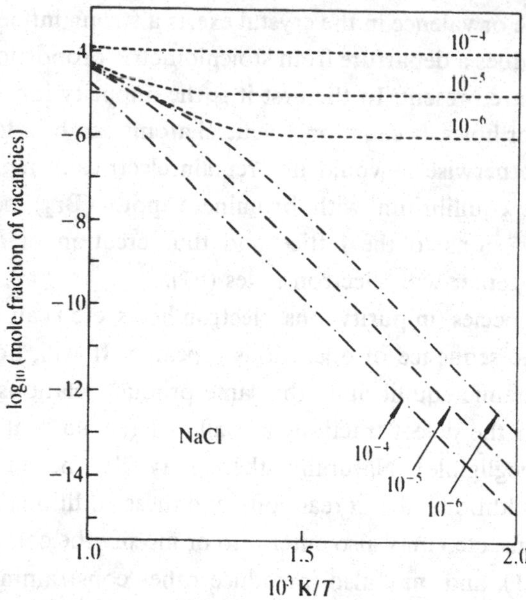


Fig. 3.2. The site fractions of cation and anion vacancies as a function of temperature for NaCl containing various fractions of substitutional divalent cations as indicated (cation vacancies ----, anion vacancies —·—·—).

equations we immediately obtain

$$c_1 - c_2 = c_3 - c_4 + c_N. \tag{3.3.28}$$

In an alkali halide only anion and cation vacancies are significant, so that the interstitial concentrations $c_3$ and $c_4$ can be dropped from (3.3.28) and the resulting equation taken with (3.3.15a) to yield $c_1$ and $c_2$. An example of the solution so obtained is shown in Fig. 3.2. From this we see that the addition of divalent cations enhances the concentration of cation vacancies and depresses the concentration of anion vacancies. In principle, the dissolution of divalent anions (e.g. O, S, $CO_3$, $SO_4$) will have the opposite effect, enhancing the anion vacancy concentration and depressing the cation vacancy concentration; but in practice the solubility of these proves to be considerably less than that of the divalent cations.

The corresponding result for a Ag-halide ($Ag^+$ Frenkel defects) is obtained with (3.3.16) by dropping $c_2$ and $c_4$ from (3.3.28). In these crystals the addition of divalent cations again enhances the cation vacancy concentration, but now it is the $Ag^+$ interstitial concentration that is decreased. Converse effects follow from the addition of divalent anions, but again their solubility is generally lower than that of divalent cations.

This ability to control the defect population by 'doping' with aliovalent ions is clearly of considerable practical value in the study of defect properties. Detailed reviews of its use have been given by Corish and Jacobs (1973a) and by Hayes (1974).

*Non-stoichiometric crystals*

Let us turn now to a different class of ionic compounds, the transition metal monoxides (e.g. NiO, CoO, etc.). These oxidize to a significant extent in the presence of oxygen gas, by the incorporation of oxygen in the anion sub-lattice (Schmalzried, 1981; Cox, 1992). This demands the creation of two electron holes (in the transition metal states, cf. §2.4) and the creation or removal of an appropriate point defect for every oxygen atom incorporated. To extend the calculation to these circumstances requires (i) the representation of the electron holes in the Gibbs function for the solid phase and (ii) the minimization of the sum of the Gibbs functions of the solid and the oxygen gas, subject to constraints for the conservation of the total numbers of atoms and for the structure of the solid ($a = b = 1$ in this case).* If we then repeat the calculation we will recover (3.3.9)–(3.3.11) as before, together with the additional relation

$$\tfrac{1}{2}\mu_{O_2,\,gas} = \mu_1 + \mu_6 - 2\varepsilon_F \qquad (3.3.29a)$$

$$= \mu_4 - 2\varepsilon_F, \qquad (3.3.29b)$$

being the expression of equilibrium in the reactions

$$\tfrac{1}{2}O_{2,gas} \rightleftharpoons O_O + V''_M + 2e^{\cdot}_h \qquad (3.3.30a)$$

$$\rightleftharpoons O_I + 2e^{\cdot}_h. \qquad (3.3.30b)$$

As with the doped crystals considered above, there are two ways of accommodating the excess oxygen, and, as there, the one which occurs depends on the relative importance of Schottky defects and anion Frenkel defects in the pure stoichiometric compound. In the transition metal monoxides (3.3.30a) is the important reaction: the oxygen sub-lattice remains largely perfect, whereas the concentration of cation vacancies increases as the oxygen chemical potential of the gas increases (i.e. high temperatures and pressures). In this case then it is the cation transport coefficients which are large. A contrasting example is provided by $UO_{2+x}$, for which the important reaction is (3.3.30b) corresponding to the dominance of anion Frenkel disorder in the pure stoichiometric compound. With this material it is the anion transport which is relatively large (Matzke, 1987, 1990). The difference

* A more comprehensive treatment would allow also for the possible trapping of the holes by the defects.

between these two examples, however, is not merely a formal one (cation for anion and vice versa), but has a strong effect on the rates at which equilibrium can be attained. Clearly this will be relatively rapid in $UO_2$ since oxygen is the more mobile species. In the transition metal monoxides it will be slow, since the dominant defects are the cation vacancies and the concentration of oxygen vacancies is depressed (via the Schottky equilibrium 3.3.15a). In fact, because of the relatively fast motion of the cation vacancies it is possible to build up such high local concentrations of $V_M''$ as to drive the reaction (3.3.30a) to the left, i.e. to cause local internal dissociation with production of oxygen gas (Schmalzried, 1986; Schmalzried and Pfeiffer, 1986).

### *3.3.2 Intermetallic compounds: Schottky and anti-site disorder*

In this section we shall consider the thermal creation of defects in intermetallic compounds. The defect and diffusion properties of such materials have been reviewed by Chang and Neumann (1982) and by Bakker *et al.* (1985). They include a variety of compounds based on the cubic B2(CsCl) structure (e.g. CoGa and NiAl) as well as structurally more complex materials such as $V_2Ga_5$. Their electronic properties are metallic in character. They normally exist in non-stoichiometric proportions and the dominant point defects are generally believed to be vacancies and also atoms on the wrong sub-lattice (e.g. V atoms on the Ga sub-lattice in $V_2Ga_5$), so-called anti-site disorder.

Such systems are physically very different from the ionic crystals considered above, although, as we shall see, the calculation of the thermal equilibrium concentration of these defects is formally not very different. In fact, the structure of the calculations follows similar lines. In turn, we shall see later (§3.10) how to apply the results obtained here to order–disorder and other alloys where the degree of anti-site disorder can reach extreme values.

We again suppose that the chemical formula of the stoichiometric compound is $A_aB_b$ and, as in the preceding section, again assume that there is just a single A sub-lattice and a single B sub-lattice. The structural elements are now vacancies in the A and B sub-lattices, A atoms on A and B sub-lattice sites and B atoms on A and B sub-lattice sites. The corresponding numbers of these elements $N_1 \ldots N_6$ are specified as follows

| $V_A$ | $V_B$ | $B_A$ | $A_B$ | $A_A$ | $B_B$ |
|---|---|---|---|---|---|
| $N_1$ | $N_2$ | $N_3$ | $N_4$ | $N_5$ | $N_6$ |

Let the solid be made up of $N_m$ lattice cells overall so that the structural constraints are

$$N_m = (N_1 + N_3 + N_5)/a, \tag{3.3.31a}$$

$$N_m = (N_2 + N_4 + N_6)/b, \tag{3.3.31b}$$

and thus

$$(N_1 + N_3 + N_5)/a = (N_2 + N_4 + N_6)/b. \tag{3.3.32}$$

These relations correspond to (3.3.1)–(3.3.2) for the model used in the preceding section.

Next we can write down the Gibbs energy function $G$ analogously to (3.3.3), although $\Omega$ is different.

$$G = G^0(N_m) + \sum_{r=1}^{4} N_r g_r^\infty - kT \ln \Omega, \tag{3.3.33a}$$

with

$$\Omega = \frac{(aN_m)!}{N_1!N_3!N_5!} \cdot \frac{(bN_m)!}{N_2!N_4!N_6!} \tag{3.3.33b}$$

for the number of distinct ways of assigning the $N_1$ $V_A$-, $N_3$ $B_A$- and $N_5$ $A_A$-structural elements to the $(N_m)$ A-sites and the $N_2$ $V_B$-, $N_4$ $A_B$- and $N_6$ $B_B$-structural elements to the $(bN_m)$ B-sites. As before, this equation for $G$ assumes the absence of interactions among the defects. $G^0(N_m)$ is the reference state Gibbs free energy of the perfect stoichiometric compound with the same total number of lattice cells.

We proceed as before to obtain a set of virtual chemical potentials $\mu_1 \ldots \mu_6$ (cf. (3.3.4a) and (3.3.4b)). This gives us the following set expressed in fractions, $c_i = N_i/N_m$,

$$\mu_1 = \frac{\mu^0}{2a} + g_1^\infty + kT \ln(c_1/a), \tag{3.3.34a}$$

$$\mu_2 = \frac{\mu^0}{2b} + g_2^\infty + kT \ln(c_2/b), \tag{3.3.34b}$$

$$\mu_3 = \frac{\mu^0}{2a} + g_3^\infty + kT \ln(c_3/a), \tag{3.3.34c}$$

$$\mu_4 = \frac{\mu^0}{2b} + g_4^\infty + kT \ln(c_4/b), \tag{3.3.34d}$$

$$\mu_5 = \frac{\mu^0}{2a} + kT \ln(c_5/a), \tag{3.3.34e}$$

and

$$\mu_6 = \frac{\mu^0}{2b} + kT\ln(c_6/b). \tag{3.3.34f}$$

Also

$$G = \sum_{i=1}^{6} \mu_i N_i. \tag{3.3.35}$$

To determine the equilibrium numbers of vacancies and anti-site defects in a specimen of fixed composition we must add the constraining equations for constant total numbers of A and B atoms, viz.

$$c_A = c_4 + c_5 \tag{3.3.36a}$$

$$c_B = c_3 + c_6, \tag{3.3.36b}$$

as well as the structural constraint (3.3.32). If we introduce Lagrange multipliers $\lambda$, $\nu$ and $\sigma$ for (3.3.32), (3.3.36a) and (3.3.36b) respectively then the minimization of $G$ requires

$$\mu_1 + \frac{\lambda}{a} = 0, \tag{3.3.37a}$$

$$\mu_2 + \frac{\lambda}{b} = 0, \tag{3.3.37b}$$

$$\mu_3 + \frac{\lambda}{a} + \sigma = 0, \tag{3.3.37c}$$

$$\mu_4 + \frac{\lambda}{b} + \nu = 0, \tag{3.3.37d}$$

$$\mu_5 + \frac{\lambda}{a} + \nu = 0, \tag{3.3.37e}$$

and

$$\mu_6 + \frac{\lambda}{b} + \sigma = 0. \tag{3.3.37f}$$

Elimination of $\lambda$ between (3.3.37a) and (3.3.37b) yields the Schottky product equation

$$a\mu_1 + b\mu_2 = 0, \tag{3.3.38}$$

or, equivalently,

$$(c_1/a)^a(c_2/b)^b = \exp(-g_S/kT), \tag{3.3.39a}$$

and

$$g_S = \mu^0 + ag_1^\infty + bg_2^\infty. \tag{3.3.39b}$$

The other equilibrium relations are obtained (i) by elimination of $\lambda$ and $\sigma$ from (3.3.37a), (3.3.37c) and (3.3.37f) and (ii) by elimination of $\lambda$ and $\nu$ from (3.3.37b), (3.3.37d) and (3.3.37e). This gives firstly

$$\mu_1 + \mu_6 = \mu_2 + \mu_3, \tag{3.3.40}$$

corresponding to the equilibrium in the reaction

$$V_A + B_B = V_B + B_A; \tag{3.3.41}$$

and secondly

$$\mu_2 + \mu_5 = \mu_1 + \mu_4, \tag{3.3.42}$$

corresponding to

$$V_B + A_A = V_A + A_B. \tag{3.3.43}$$

Explicitly, these equations (3.3.40) and (3.3.42) are

$$\frac{c_2 c_3}{c_1 c_6} = \exp(-g_{XB}/kT), \tag{3.3.44a}$$

where

$$g_{XB} = (g_2^\infty + g_3^\infty - g_1^\infty) \tag{3.3.44b}$$

is the free energy of B anti-site disorder. Also

$$\frac{c_1 c_4}{c_2 c_5} = \exp(-g_{XA}/kT) \tag{3.3.45a}$$

with

$$g_{XA} = (g_1^\infty + g_4^\infty - g_2^\infty) \tag{3.3.45b}$$

as the free energy of A anti-site disorder.

As before, we may solve for the equilibrium values of $c_1 \ldots c_6$ by using the three disorder equations (3.3.39), (3.3.44) and (3.3.45) together with the structure equations

$$a = c_1 + c_3 + c_5 \tag{3.3.46a}$$

$$b = c_2 + c_4 + c_6 \tag{3.3.46b}$$

and the composition equation

$$\frac{c_A}{c_B} = \frac{c_4 + c_5}{c_3 + c_6}. \tag{3.3.47}$$

The left side of (3.3.47) is determined just by the degree of non-stoichiometry. There will be different types of solutions of this set of equations, depending on the relative values of $g_{XA}$, $g_{XB}$ and $g_S$. Limiting forms corresponding to the dominance of one type of disorder are obtained easily, just as they were for ionic crystals.

This example of an inter-metallic compound has again illustrated the general principle that in compounds the concentrations of the individual defects are coupled together through the equivalents of mass-action equations for those defect reactions allowed by the structure. In particular, defects in one sub-lattice will be coupled in this way to those in the other sub-lattice. We must therefore expect a corresponding coupling between the atomic transport coefficients of the different components.

As in the preceding section we complete the calculation by finding the chemical potentials of the component species A and B. Again, we start from the equation for $G$ in terms of the $\mu_i$ (3.3.35) and then by using the equations of defect equilibrium (3.3.38), (3.3.40) and (3.3.42) obtain

$$\begin{aligned} G &= (\mu_5 - \mu_1)N_A + (\mu_6 - \mu_2)N_B \\ &\equiv \mu_A N_A + \mu_B N_B. \end{aligned} \tag{3.3.48}$$

Hence $\mu_A$ can be identified with $\mu_5 - \mu_1$ $(= \mu_4 - \mu_2)$ and $\mu_B$ with $\mu_6 - \mu_2$ $(= \mu_3 - \mu_1)$. These are the appropriate quantities to use to express the conditions of equilibrium with an external phase. (Since this system is assumed to be metallic we do not have to consider the electrons separately as with an ionic crystal: electronic terms are automatically included in the definition of the quantities $g_i^\infty$.) From the equation of equilibrium so obtained we can predict the dependence of composition on the external vapour pressure (or more precisely the activity) of the constituents.

In practice it appears that we may classify intermetallic compounds into those where anti-site defects occur on both sub-lattices in comparable concentrations (i.e. $g_{XA} \sim g_{XB}$) and those where anti-site defects occur to a significant concentration on only one sub-lattice (i.e. where $g_{XA}$ and $g_{XB}$ differ substantially). Well studied examples of the first group include AgMg, AuZn and AuCd (all with the CsCl structure). Examples of the second group include AlNi, CoGa and GaNi (also with the CsCl structure), where in each case only the first element gives rise to significant numbers of anti-site defects. As (3.3.39a), (3.3.44a) and (3.3.45a) show, these two groups respond differently to changes in stoichiometry. In the first group, an excess of either component will give rise primarily to an increase in the concentration of the corresponding anti-site defect. In the second group an excess of the component for which $g_X$ is larger (B, say) will lead directly to a corresponding increase in the population of vacancies on the opposite sub-lattice (A).

Lastly, we remark that throughout this section we have assumed that the degrees of disorder involved are small, so that the reaction constants in these mass-action equations are independent of the concentrations of vacancies and anti-site defects. If this is not the case a precise theory becomes enormously more difficult. However, an approximate theory is straightforwardly obtained by extending the analysis we have given by using the Bragg–Williams approximation. In this, the formation energies $g_r^\infty$ in (3.3.33a) are replaced by values appropriate to the average surroundings of the atom(s) involved in the formation of a defect of type $r$. We return to the matter in §3.10.

## 3.4 Interactions among defects

In §§3.2 and 3.3 we have considered four specific examples, two of which were models of monatomic substances while two were models of binary compounds. Obviously the number of model systems of interest is now very great. The object of our treatment of these four examples was thus to draw out the principles of a general approach valid for all similar systems. The essential limitation to our discussion in §§3.2 and 3.3 was the assumption that there were no interactions among the defects, so that in calculating the Gibbs free energy of the system we could assume that all configurations or arrangements of the structural elements on the lattice were of the same energy. For simplicity we also assumed that the materials were pure. This restriction can easily be removed as long as the first assumption is retained, as was indicated in §§3.2 and 3.3. In the next sections we shall turn to the task of performing similar calculations for systems in which interactions among the defects are present, but before doing so it is useful to summarize the approach and the inferences so far since this allows us to see clearly where modifications are required.

To calculate $G$ and other derived thermodynamic functions we follow this sequence.

(a) We specify the crystal structure and associated limitations of the model, e.g. the relation between the numbers of lattice sites belonging to different sub-lattices.

(b) We specify the structure elements (which define the point defects) and the electronic defects to be considered, and also whether these are in non-localized (Bloch) states or in localized (small polaron) states.

(c) We express $G$ for the defective solid as the sum of $G^0$ for a perfect stoichiometric solid with the same number of lattice cells ($N_m$) and appropriate defect terms.

(d) By the absence of interactions among the defects we can calculate the configurational entropy $S_{con}$ associated with the defects

$$S_{con} = k \ln \Omega \tag{3.4.1}$$

by a straightforward combinatorial evaluation of the number of distinct configurations, $\Omega$.

(e) From the Gibbs function so obtained we may obtain virtual chemical potentials for all the structural elements and defect species. These virtual chemical potentials can be put into the *standard* form

$$\mu_i = \mu_i^0 + kT \ln a_i$$
$$= \mu_i^0 + kT \ln(\gamma_i x_i), \tag{3.4.2}$$

in which $a_i$ is the activity of species $i$, $x_i$ is the mole fraction and $\gamma_i$ is the *activity coefficient*.

(f) Thermodynamic equilibrium corresponds to the minimization of $G$ subject to consistency with all the constraints which may hold (structure, constant numbers of atoms, electroneutrality). These constraints, being linear in the numbers $N_i$, are conveniently handled by the method of Lagrange multipliers.

(g) Elimination of the Lagrange multipliers from the equations so obtained gives quasi-chemical equations of mass-action corresponding to equilibrium in the possible defect reactions.

(h) Each of these mass-action equations when expressed in terms of activities remains valid in the presence of additional types of defect reaction.

However in §2.5 we referred to the extensive evidence, both experimental and theoretical, for the existence of significant interactions among defects and solute atoms. These have important effects on atomic transport properties and must therefore be included in the statistical mechanical calculations. The most direct and fundamental way of dealing with interacting systems would be to specify the form of the interactions (e.g. as a function of distance) and then to find the various thermodynamic functions and the pair and higher-order distribution functions needed in the calculation of transport coefficients (Chapter 6) by the general methods of statistical mechanics. Concentrated alloys and solid solutions (e.g. the order–disorder alloys introduced in §2.3.4) have been extensively studied in this way (e.g. Sato, 1970). Obviously it can also be followed for dilute systems, rather as in the Mayer theory of imperfect gases. However, the known physical existence of defect pairs and larger clusters allows us to short-circuit some of the formal theory, in effect by specifying the nature of the corresponding distribution functions.

Let us suppose then that we are concerned with a dilute system and that the interactions are of short range (as they are, for example, in metals). If we know, or suspect, that close neighbour pairs have an appreciable binding energy then the device is to treat such close neighbour pairs as a distinct type of defect or structural-element and follow the same path as before. In the simple case of a dilute alloy containing vacancies with attractions only between vacancies and solute atoms, the defect population would be made up of (a) vacancies which are not near to any solute atom (unpaired or *free* vacancies) (b) solute atoms which are not near to any vacancy (unpaired or *free* solute atoms) and (c) near-neighbour pairs each composed of a vacancy and a solute atom. These three types of defect may then be assumed to be mutually without interactions and the sequence (a)–(h) above followed. This approach has been extensively used and has been very productive. This idea can obviously be extended to include pairs of larger separations and also larger clusters (triplets, etc.) should that be necessary (e.g. Dorn and Mitchell, 1966 and, for application, Hehenkamp, 1986). However, the restriction that free solute atoms cannot be placed on sites near to free vacancies (for otherwise they would constitute a pair) can present difficulties to the straightforward evaluation of the configurational entropy, i.e. $\Omega$, by combinatorial arguments (step (d)). In the next section (§3.5) we shall therefore describe an alternative method which is based on the Mayer cluster expansion theory of imperfect gases. This is straightforward to apply to the approximation we need in dealing with dilute solid solutions (in effect to the inclusion of second virial terms).

So far we have concentrated on one type of defect interaction, namely a short-ranged interaction. This is a good representation in metals, where perturbations arising from changes in the ion-core charge (e.g. the missing ion-core of the vacancy) are screened out by the conduction electons. However, in ionic crystals and semiconductors effective charges on the defects are not screened in this way and there will be long-ranged Coulombic interactions to be considered, even though we still take explicit account of pairs or higher clusters. How these long-ranged interactions are to be dealt with is described in §3.9.

Lastly, there is the matter of interactions among the host atoms in concentrated alloys and solid solutions to be considered. To a degree we have already looked at this in §3.3.2 with our discussion of anti-site disorder in intermetallic compounds. In so-called order–disorder alloys (e.g. CuZn, $Fe_3Al$, etc.) however, the degree of anti-site disorder increases catastrophically at a critical temperature, $T_c$, above which distinct sub-lattices no longer exist. Such phenomena have been extensively researched for over 50 years and are usually interpreted on the basis of a model in which the energy of anti-site disorder is small although positive. The converse model in which this energy is negative would not, of course, have a stable ordered arrangement based on two distinct A and B sub-lattices. However, a disordered solid solution is again possible above a certain critical temperature. Below this, separation into two phases (mainly A and mainly B) takes place. Such solid solutions have also received a great deal of attention. We therefore turn to these systems in the context of point defects in §3.10.

### 3.5 Dilute solid solutions containing defect clusters

The object of this section is to present a general way of calculating the Gibbs free energy of a system in the presence of interactions among its defect components. The principal limitation is that the system should be dilute in defects, but all types of point defect are covered – vacancies, interstitials, solute atoms, anti-site defects and their simple aggregates. The host substance (solvent) may be a monatomic substance A or a compound, e.g. $A_aB_b$.

In setting up the statistical thermodynamic equations we distinguish between *simple* species (e.g. isolated vacancies, isolated interstitial atoms) and *complex* species (e.g. pairs formed from vacancies and substitutional solute atoms) which can be formed by bringing together appropriate simple species. A complex species is characterized by a distinct energy of complex formation (e.g. the binding energy of a vacancy–solute nearest-neighbour pair relative to an isolated solute atom and an isolated vacancy). We assume that all significant defect interaction energies are contained in the energies of formation of the various complex species from simple ones. Of course, only a system rather dilute in defects can be usually described in this manner. As in §§3.2 and 3.3 we consider a system specified by the independent variables $T$ and $P$ and the numbers $N_r$ of the various structural elements. Also as there, the essence of the calculation is the determination of the configurational entropy $S_{\text{con}}$. Since this is unaffected by the presence of free electrons and holes in non-localized states we leave these out of consideration: appropriate terms can be added to $G$ in the manner of §3.2.2 if desired. If the electronic defects are in localized states then these can be included as one of the types of simple defects in the calculation of $S_{\text{con}}$. The Gibbs energy function of the crystal is then

$$G(T, P) = G^0(N_{\text{m}}, T, P) + \sum_r N_r g_r^{\infty} - TS_{\text{con}}. \tag{3.5.1}$$

Here, $G^0(N_{\text{m}}, T, P)$ is the Gibbs function of a hypothetical perfect crystal of $N_{\text{m}}$ molecules of host substance having the same numbers of the various *lattice sites* as the crystal of interest. The second term gives the change in Gibbs function when the defects are introduced into this reference crystal at particular sites; $g_r^{\infty}$ is the change in Gibbs function accompanying the introduction of one defect of species $r$. The third term gives the contribution of the configurational entropy $S_{\text{con}}$ to the Gibbs function. As before, this entropy is to be calculated from the Boltzmann expression

$$S_{\text{con}} = k \ln \Omega \tag{3.5.2}$$

since all the configurations of defects on the lattice are, by definition, of equal energy.

In calculating the number of configurations $\Omega$ the essential difficulty is created by the 'excluded site' restrictions, i.e. by the fact that defects cannot be assigned to lattice sites independently of one another. For example, we may not place an isolated vacancy and an isolated impurity on sites so close that they should be counted as a pair. Further, we may not assign two defects to the same site or sites. These restrictions often make it impossible to obtain an exact expression for $\Omega$ in a closed form and most discussions are therefore approximate. The commonest approach is to use combinatorial arguments to construct expressions for $\Omega$. Although these methods are relatively easy to follow and have yielded useful approximate expressions for $\Omega$ for quite complex problems (e.g. de Fontaine, 1979: see also Miller, 1948; Guggenheim, 1952) they have the disadvantage that they do not generally provide successive approximations correct to a known order in concentration. As a result there may be no way of assessing the merits of one formulation of the method compared to another (Kidson, 1985).

In this section we shall therefore use a more systematic method to evaluate $\Omega$ which is based on the Mayer cluster-expansion technique originally devised for the theory of imperfect gases (Mayer and Mayer, 1940). The method allows one to write $\ln \Omega$ as the sum of an ideal lattice gas contribution (i.e. without excluded-site restrictions) plus the excluded-site corrections expressed as an infinite power series in defect concentrations. In the applications to dilute defect systems one needs only the leading corrections, i.e. second virial-coefficient terms. The method has the advantage that these terms can be written down rather easily without invoking the full formalism of the Mayer theory. The details of the method are given in Appendix 3.1. The result for the configurational entropy correct to second order in the defect fractions is

$$S_{\text{con}}/N_{\text{m}}k = -\sum_r c_r(\ln(c_r/z_r) - 1) + \frac{1}{2}\sum_r \sum_s c_r c_s B_{rs}, \tag{3.5.3}$$

in which $c_r$ is the fraction of defects of type $r$, i.e. $N_r/N_{\text{m}}$. The dimensionless coefficients $z_r$ and $B_{rs}$ are characteristic of the defects and the lattice. In the absence of any internal (electronic) degrees of freedom they are determined solely by geometrical characteristics. In the original Mayer notation, $z_r$ is formally defined by the equation

$$z_r = \frac{1}{N_{\text{m}}} \int \text{d}\{l_r\}, \tag{3.5.4}$$

where the integration over $\{l_r\}$ actually denotes a summation over all assignments of an individual defect $l$ of type $r$ to the empty lattice. Clearly, it is independent of which particular defect $l$ is taken. Thus, for a single lattice of available sites and a type of defect which occupies just one site at a time with no orientational degeneracy (e.g. a solute atom or a vacancy) we have $z_r = 1$. For a dumb-bell interstitial centred on the normal sites of a similar lattice $z_r$ is equal to the number of distinct orientations of the dumb-bell.

The quantity $B_{rs}$ is in the nature of a virial coefficient and is formally defined by

$$B_{rs} = \frac{1}{z_r z_s N_m} \int\int f_{rs}\, d\{l_r\}\, d\{m_s\}, \tag{3.5.5}$$

in which $f_{rs}$ embodies the restriction that certain sites may not be occupied by the particular defect $m$ of type $s$ when there is already a defect $l$ of type $r$ on the (otherwise empty) lattice. This quantity $f_{rs} = 0$ if there is no restriction and $f_{rs} = -1$ if the assignments are prohibited. For example, if $s$ is a solute–vacancy pair clearly there are two sites which may be occupied neither by a free vacancy nor by a free solute atom. Likewise, if $s$ is a free solute atom there are $z + 1$ sites which are not available to a free vacancy (i.e. the original site occupied by the solute plus its $z$ nearest neighbours, for, if any of these were vacant, we should have a bound solute–vacancy pair, not a free solute and a free vacancy). Other typical values of $z_r$ and $B_{rs}$ are given in Table 3.4.

By insertion of (3.5.3) into (3.5.1) we arrive at the Gibbs function $G$. We emphasize that this expression is unambiguous and rigorously correct to second order in the defect concentrations. However, when we use it to derive chemical potentials we may encounter the same indeterminacies that we found when handling compounds by the combinatorial method. We must therefore again distinguish monatomic hosts (i.e. a single lattice for the defects) from compound hosts with more than one distinct sub-lattice.

In the first case, by writing the configurational entropy in the form

$$S_{con} = N_m k\sigma(\ldots N_r/N_m \ldots), \tag{3.5.6}$$

we easily find that, to first order in the $c_s$,

$$\mu_r = \left(\frac{\partial G}{\partial N_r}\right)_{N_{s \neq r}}$$

$$= g_r + kT\left[\ln(c_r/z_r) - \sum_s B_{rs}c_s - y_r\left(\sum_s c_s\right)\right], \tag{3.5.7a}$$

in which

$$y_r = \frac{\partial N_m}{\partial N_r}, \tag{3.5.7b}$$

is found from the expression of $N_m$ as a summation over the structural elements. The corresponding expression for the chemical potential of the host atoms is

$$\mu_H = \left(\frac{\partial G^0}{\partial N_m}\right) - kT \sum_s c_s, \tag{3.5.8}$$

since evidently $y_H = 1$. If we want to express the defect chemical potential $\mu_r$ in the standard form (3.4.2) we can do so if we take the activity to be

$$a_r = \frac{c_r}{z_r}\left[1 - y_r \sum_s c_s - \sum_s B_{rs} c_s\right], \tag{3.5.9}$$

which is correct to first order in the $c_r$.

Thus in the case of the monatomic host there is no ambiguity. In the case of compounds we are faced again with the choice of using virtual chemical potentials or of finding minima of $G$ with respect to the degree of advancement of those defect reactions which are compatible with the structural constraint. The second of these routes can be used here just as easily as in §3.3.1. On the other hand, we could equally well take the alternative route and use virtual chemical potentials. Although somewhat greater care is required in general, it is immediately clear that we can use the above results for the monatomic lattice in essentially the same form for each sub-lattice of a compound when we are dealing with systems without defects of a sort which would link the sub-lattices together (e.g. anion–cation vacancy pairs in an alkali halide crystal). Other cases are probably best handled individually by following the sequence summarized in §3.4 and using (3.5.3) at the appropriate point. As was described at some length in §3.3.1 the essential point to remember is that in performing any necessary differentiations of $G$ the quantity $N_m$ enters via two distinct equations, each of which involves defects on one sub-lattice alone. How these two relations are deployed is arbitrary, although as always there is a gain in doing so in such a way as to obtain the maximum symmetry in the resulting virtual chemical potentials. By the form (3.5.6) of $S_{con}$, however, we always get the first two terms in square brackets on the right of (3.5.7a).

The calculation in this section obtained the chemical potentials in the absence of force fields and stress fields. In the presence of an electrical potential $\Phi$ the electrochemical potential of species $r$ (including host species)

$$\mu_r^{(e)} = \mu_r + q_r \Phi \tag{3.5.10}$$

takes the place of $\mu_r$, where $q_r$ is the charge of one unit of $r$. This modification is familiar in thermodynamics (Guggenheim, 1967). In the presence of a stress field the matter is not quite so simple. First of all it is clear that there will be additional stress-dependent terms in the Gibbs function (as discussed by Nye, 1985 for example). For dilute solid solutions containing either vacancy or interstitial defects it is straightforward to obtain these terms by specifying the dependence of the elastic strain of the solid upon the fractions of the various defect species, $r$. Then, by differentiating $G$ with respect to $N_r$ in the usual way, we find that the quantity taking the place of $\mu_r$ is then

$$\mu_r^{(s)} = \mu_r - v\lambda_{\alpha\beta}^{(r)}\sigma_{\alpha\beta} \tag{3.5.11}$$

in which $r$ may be either a solute or a defect, $v$ is the molecular volume, $\lambda_{\alpha\beta}^{(r)}$ is the elastic dipole strain tensor of $r$ (Nowick and Berry, 1972) and where we use the Einstein repeated index summation on the Cartesian components, $\alpha$, $\beta$, etc. There is no corresponding contribution to the chemical potentials of host species from the defects, which, in this case at least, is reasonable since the defect species alone should provide a complete description of the processes we are concerned with.

More generally, however, we must recognize a difficulty, or paradox, first pointed out one hundred years ago, by Gibbs; namely that, for a solid body under a uniform but non-hydrostatic stress, one cannot define absolute chemical potentials for all its chemical components. The difficulty is associated with the process of adding atoms to the body from a state of rest at infinity (equivalent to the differentiation of $G$). Different faces of the solid will necessarily be subject to different mechanical forces (unless the stress system is purely hydrostatic) so that the work required to add atoms to the solid will depend on which face they pass through, something which has been confirmed by an elegant experiment by Durham and Schmalzried (1987). This complication in the thermodynamics of stressed solids nevertheless need not trouble us here since we are only interested in internal changes and reactions and therefore only differences in chemical potentials are involved. Eqn. (3.5.11) therefore remains adequate for dilute systems.

Finally, we note that as the chemical potentials are *intensive* quantities the Gibbs energy function can be re-expressed in the usual way as

$$\begin{aligned} G &= \sum_{i=1}^{n} \mu_i N_i \\ &= \tilde{\boldsymbol{\mu}}\mathbf{N} \end{aligned} \tag{3.5.12}$$

where, in the first equality $i$ runs over all the structural elements, while, in the second, matrix notation has been introduced by forming the $\mu_i$ into a row matrix $\tilde{\boldsymbol{\mu}}$ and the $N_i$ into a column matrix $\mathbf{N}$. The $\mu_i$ can also be those given by (3.5.10) and (3.5.11) as appropriate.

## 3.6 Internal defect equilibria

In the previous section we have described a general method for finding (defect) chemical potentials in dilute solid solutions that contain small defect clusters formed as a result of attractive interactions among the primary defects. However, we have not yet considered thermodynamic equilibria in such systems. Basically there are two types. The first – which we shall refer to as *internal* equilibrium – relates to those reactions between defects which lead to the formation of pairs and larger clusters. Generally these reactions can be assumed to be fast since only short-range diffusion in small elements of volume is needed to establish equilibrium locally. In macroscopic diffusion processes one may therefore assume that such reactions are in equilibrium locally. The second type of equilibrium is the *overall* equilibrium in the numbers of primary defects, e.g. vacancies. This may require the migration of defects over distances comparable to the average separation of defect sinks such as grain boundaries and dislocations and thus takes longer to establish. Furthermore, in compounds, equilibrium with an external phase will require the migration of defects over the even larger distances to external surfaces (cf. §3.3.1). There can therefore be circumstances where such overall defect equilibrium is not achieved. We therefore deal with the two types of equilibrium – internal and overall – separately. In this section we deal with the conditions of internal equilibrium. In these circumstances the chemical potentials of the species involved in the internal equilibrium become linked together, the relations between them lead to the familiar mass-action equations relating the concentrations of the defect species taking part in the equilibrium. As before, the formal procedure for establishing such relations between the chemical potentials is to minimize the Gibbs function with respect to defect numbers subject to appropriate constraints for a closed system. Examples of such calculations abound in the literature on the statistical thermodynamics of defects. It is convenient to express the procedure by which the secondary defects are eliminated from the thermodynamic equations in a general compact form. The same ***reduced*** description enters naturally in the transport theory (Chapter 8), and it also maps directly on to a reduced description of the corresponding macroscopic flux equations, as we demonstrate in the next chapter (§4.4.2).

Let the labels $i = 1, \ldots, g$ be reserved for those structural elements which we shall identify as primary chemical and defect species, while the remaining labels $g + 1, \ldots, n$ refer to secondary defect species which can equilibrate with the primary, and which, by the criterion of local equilibrium, can be eliminated from a reduced description. The assignment of species to these two categories is dictated by the particular conditions of interest. Thus in the previous example of isolated defects and defect pairs, the defect pairs will be the natural choice of secondary species.

For convenience we shall now use a matrix notation. First we form all the chemical potentials $\mu_i$ ($i = 1 \ldots n$) into a row matrix and the numbers $N_i$ of all the $n$ species into a column matrix. We next partition these explicitly as

$$\tilde{\boldsymbol{\mu}} = [\tilde{\boldsymbol{\mu}}' : \tilde{\boldsymbol{\mu}}''] \tag{3.6.1}$$

$$\mathbf{N} = \{\mathbf{N}' : \mathbf{N}''\}, \tag{3.6.2}$$

where the prime and double prime refer to primary and secondary. Then we have for the Gibbs function

$$G = \tilde{\boldsymbol{\mu}}'\mathbf{N}' + \tilde{\boldsymbol{\mu}}''\mathbf{N}''. \tag{3.6.3}$$

Now, by assumption, some of the primary species can be combined to form secondary ones. Corresponding to each primary species $i$ we can define a number $M_i$, the total number of $i$ present either as primary $i$ or as a component of a secondary species. Formally, we can write

$$M_i = \sum_{j=1}^{n} A_{ij} N_j \qquad i = 1, \ldots, g$$

$$\mathbf{M} = \mathbf{A}'\mathbf{N}' + \mathbf{A}''\mathbf{N}'' \tag{3.6.4}$$

where the $g \times n$ rectangular matrix $\mathbf{A}$ has been partitioned into a $g \times g$ matrix $\mathbf{A}'$ and a $g \times (n - g)$ matrix $\mathbf{A}''$. The defects will equilibrate so that the defect numbers $N_j$ minimize the Gibbs function subject to the constraint that $M_1, \ldots, M_g$ are all constant. Therefore, with a row matrix of Lagrange undetermined multipliers

$$\tilde{\boldsymbol{\sigma}} = [\sigma_1, \ldots, \sigma_g] \tag{3.6.5}$$

we set

$$\delta(G + \tilde{\boldsymbol{\sigma}}\mathbf{M}) = 0 \tag{3.6.6}$$

and obtain

$$\tilde{\boldsymbol{\mu}}' + \tilde{\boldsymbol{\sigma}}\mathbf{A}' = 0 \tag{3.6.7a}$$

$$\tilde{\boldsymbol{\mu}}'' + \tilde{\boldsymbol{\sigma}}\mathbf{A}'' = 0. \tag{3.6.7b}$$

By eliminating the undetermined multipliers the relations between the chemical potentials are seen to be

$$\tilde{\boldsymbol{\mu}}'' = \tilde{\boldsymbol{\mu}}'(\mathbf{A}')^{-1}\mathbf{A}''. \tag{3.6.8}$$

Although not usually presented in this notation, eqns. (3.6.8) are, in fact, just the (quasi-)chemical equations of equilibrium for the $n - g$ reactions which we assume to be in balance. When expressions for the chemical potentials in terms of defect concentration and activity factors (as in 3.4.2) are inserted they will yield the usual mass-action relations (cf. Franklin, 1972; Kröger, 1974, vol. 2). By (3.6.8) we can now eliminate the chemical potentials of all the $n - g$ secondary species from the Gibbs function (3.5.12) and write it in the form

$$G = \tilde{\boldsymbol{\mu}}^{\mathrm{R}}\mathbf{M} \tag{3.6.9}$$

where the chemical potentials $\tilde{\boldsymbol{\mu}}^{\mathrm{R}}$ in the reduced description are defined by

$$\tilde{\boldsymbol{\mu}}^{\mathrm{R}} = \tilde{\boldsymbol{\mu}}'(\mathbf{A}')^{-1}. \tag{3.6.10}$$

This form (3.6.9) emphasizes that the $\mu_i^{\mathrm{R}}$ obtained in this way are the chemical potentials appropriate to the total numbers of each species in the reduced set whether in primary or secondary form.

The set of chemical potentials $\tilde{\boldsymbol{\mu}}^{\mathrm{R}}$ is sufficient to describe a system in which the secondary defects (e.g. pairs) are in equilibrium with the primary defects (e.g. vacancies, interstitials) but where these primary defects are not necessarily in overall thermodynamic equilibrium themselves. In particular, it is easy to show that (3.6.9) leads to a Gibbs–Duhem relation in the reduced description

$$\delta\tilde{\boldsymbol{\mu}}^{\mathrm{R}}\mathbf{M} = 0 \qquad (\text{constant } P, T). \tag{3.6.11}$$

The factor $(\mathbf{A}')^{-1}$ in (3.6.10) shows that these quantities may differ from those ($\boldsymbol{\mu}'$) provided by a statistical thermodynamical calculation for the complete set of structural elements, both primary and secondary. This is a rather important point and overlooking it may lead to error. For dilute substitutional alloys examples of the chemical potentials in the reduced scheme will be given in §3.8.

## 3.7 Overall defect equilibria

In the preceding section we allowed the concentrations of the primary defects to be arbitrary. That formulation will be directly useful later when we want to describe the fluxes of defects and atoms in systems containing non-equilibrium concentrations of primary defects (as, for example, under irradiation). However, there are many situations of experimental interest where the defects may be assumed to be present in concentrations appropriate to thermodynamic equilibrium (e.g. many macroscopic diffusion and ionic conductivity experiments are conducted under such conditions). In these cases the equilibrium concentrations of the primary defects enter directly into the expressions for the phenomenological coefficients $L$ to be presented in Chapters 6 and 8. In this section we therefore present the equations from which these concentrations may be obtained. Following the formulation (3.6.4) we shall denote the total number of species $i$ (structural elements), whether in primary or secondary form, by $M_i$.

Consider first the simple example of vacancy defects in a metal or alloy for which the host substance is a monatomic crystal. All the usual calculations, in which $G$ is obtained as a function of vacancy concentration and then minimized, in the present formulation are embraced by the equation

$$\mu_v^R = 0 \tag{3.7.1}$$

(corresponding to $\partial G/\partial M_v = 0$). A similar equation gives the equilibrium interstitial concentration

$$\mu_I^R = 0. \tag{3.7.2}$$

Appeal to (3.5.7) then yields expressions dominated by the usual exponential Boltzmann factors. Special cases of these relations for pure monatomic substances have already been given as (3.2.8) and (3.2.18b).

However, in the case of compound solid solutions the calculation of equilibrium defect concentrations is slightly more involved, for the reasons already described in detail in §3.3 in connection with pure compounds. In particular, there are the constraints of structure, conservation of numbers of each chemical species and electroneutrality to be included. These can all be handled by a simple extension of the formalism of §3.6. Thus each constraint is expressible as a linear equation among the $M_i$. We can gather them all together into a matrix equation of the form

$$\mathbf{BM} = \mathbf{\Gamma} \tag{3.7.3}$$

where $\mathbf{\Gamma}$ is a column matrix of constants. The equilibrium condition is that $G$ as given by (3.6.9) should be a minimum with respect to all variations $\delta M_i$ which are compatible with the constraints (3.7.3). Again resorting to the method of Lagrange multipliers, we see that this condition is equivalent to the unconditional minimum

$$\delta(G + \tilde{\boldsymbol{\rho}}\mathbf{BM}) = 0, \tag{3.7.4}$$

in which $\tilde{\boldsymbol{\rho}}$ is a row matrix of Lagrange multipliers. Thus

$$\tilde{\boldsymbol{\mu}}^R + \tilde{\boldsymbol{\rho}}\mathbf{B} = 0. \tag{3.7.5}$$

Elimination of the $\rho_i$ from this set of equations then leads to linear relations among the $\mu_i^R$ as appropriate conditions of overall equilibrium in these solid solutions. Each of these will then appear as the expression of defect equilibrium in a particular defect reaction. Particular examples of such relations are provided by the equations for Schottky defects (3.3.9 and 3.3.38), Frenkel defects (3.3.10 and 3.3.11), anti-site disorder (3.3.40 and 3.3.42) and equilibrium with an external phase (3.3.29), all of which were derived individually earlier.

## 3.8 Example of a dilute f.c.c. alloy

We now give examples to illustrate the generalities of the last two sections. Firstly we shall look at the example of a f.c.c. lattice of solvent (A) containing substitutional impurity atoms (B), vacancies and both A–A and A–B interstitial dumb-bells of the kind illustrated in Fig. 2.23. This complex situation is really only to be found in f.c.c. alloys under irradiation, in which overall thermodynamic equilibrium is not possible. Nevertheless, internal or local equilibria may exist and the conditions that they impose are needed for use in theories of atomic transport under irradiation (§11.7.2). We shall therefore take this example first in order to illustrate the formalism given in §3.6.

Secondly, we shall find the condition of overall defect equilibrium in dilute f.c.c. alloys. In such conditions the fraction of interstitials will be negligibly small (since $g_I$ is generally substantially greater than $g_V$). We shall therefore find the condition of overall equilibrium in the vacancy concentration.

### *3.8.1 Internal equilibria*

The complete description of the system thus involves the numbers of lattice A atoms that are not members of a dumb-bell interstitial ($N_A$), isolated substitutional B atoms ($N_B$), isolated vacant lattice sites ($N_V$), isolated A–A dumb-bell interstitials ($N_I$), mixed dumb-bell interstitials ($N_{IB}$), solute–vacancy pairs ($N_{BV}$), solute dumb-bell pairs in two distinct configurations ($N_{pa}$ and $N_{pb}$, cf. Fig. 2.23); and corresponding chemical potentials.

In applications we may generally assume that the pairing reactions and the internal conversion of solute–interstitial pairs to form mixed dumb-bells are in equilibrium, so that all we want to work with are the total numbers of A-atoms ($M_A$), B-atoms ($M_B$), vacancies ($M_V$) and interstitials ($M_I$). In other words, A, B, V and I are the labels $i = 1, \ldots, g$ in (3.6.1) and IB, BV, pa, pb are the labels of the $n$–$g$ secondary species. The equations corresponding to (3.6.4) in this particular example are

$$M_A = N_A + 2N_I + N_{IB} + 2(N_{pa} + N_{pb}) \tag{3.8.1a}$$

$$M_B = N_B + N_{IB} + N_{BV} + N_{pa} + N_{pb} \tag{3.8.1b}$$

$$M_V = N_V + N_{BV} \tag{3.8.1c}$$

$$M_I = N_I + N_{IB} + N_{pa} + N_{pb} \tag{3.8.1d}$$

whence

$$\mathbf{N}' = [N_A \quad N_B \quad N_V \quad N_I] \tag{3.8.2a}$$

$$\mathbf{A}' = \begin{bmatrix} 1 & 0 & 0 & 2 \\ 0 & 1 & 0 & 0 \\ 0 & 0 & 1 & 0 \\ 0 & 0 & 0 & 1 \end{bmatrix} \tag{3.8.2b}$$

$$\mathbf{N}'' = [N_{\mathrm{IB}} \quad N_{\mathrm{BV}} \quad N_{\mathrm{pa}} \quad N_{\mathrm{pb}}] \tag{3.8.2c}$$

$$\mathbf{A}'' = \begin{bmatrix} 1 & 0 & 2 & 2 \\ 1 & 1 & 1 & 1 \\ 0 & 1 & 0 & 0 \\ 1 & 0 & 1 & 1 \end{bmatrix} \tag{3.8.2d}$$

The expressions relating the chemical potentials of the secondary to those of the primary species are, by (3.6.8),

$$\mu_{\mathrm{IB}} = -\mu_{\mathrm{A}} + \mu_{\mathrm{B}} + \mu_{\mathrm{I}} \tag{3.8.3a}$$

$$\mu_{\mathrm{BV}} = \mu_{\mathrm{B}} + \mu_{\mathrm{V}} \tag{3.8.3b}$$

$$\mu_{\mathrm{pa}} = \mu_{\mathrm{B}} + \mu_{\mathrm{I}} \tag{3.8.3c}$$

$$\mu_{\mathrm{pb}} = \mu_{\mathrm{B}} + \mu_{\mathrm{I}}, \tag{3.8.3d}$$

and the row matrix of chemical potentials for use in the reduced scheme, by (3.6.10), is found to be

$$\tilde{\boldsymbol{\mu}}^{\mathrm{R}} = [\mu_{\mathrm{A}} \quad \mu_{\mathrm{B}} \quad \mu_{\mathrm{V}} \quad \mu_{\mathrm{I}} - 2\mu_{\mathrm{A}}]. \tag{3.8.4}$$

The chemical potentials in (3.8.3) and on the right side of (3.8.4) are those given by a calculation such as that described in §3.5, i.e. (3.5.7) and (3.5.8). The values of the constants $z_r$ and $y_r$ for all species and the virial coefficients $B_{rs}$ for the primary species are summarized in Table 3.4. The mass-action equations for the four internal equilibria found by substituting the appropriate chemical potentials in (3.8.3) are

$$c_{\mathrm{IB}}/c'_{\mathrm{B}}c'_{\mathrm{I}} = 6\exp[-(g^{\infty}_{\mathrm{IB}} - g^{\infty}_{\mathrm{B}} - g^{\infty}_{\mathrm{I}})/kT] \tag{3.8.5a}$$

$$c_{\mathrm{BV}}/c'_{\mathrm{B}}c'_{\mathrm{V}} = 12\exp[-(g^{\infty}_{\mathrm{BV}} - g^{\infty}_{\mathrm{B}} - g^{\infty}_{\mathrm{V}})/kT] \tag{3.8.5b}$$

$$c_{\mathrm{pa}}/c'_{\mathrm{B}}c'_{\mathrm{I}} = 24\exp[-(g^{\infty}_{\mathrm{pa}} - g^{\infty}_{\mathrm{B}} - g^{\infty}_{\mathrm{I}})/kT] \tag{3.8.5c}$$

$$c_{\mathrm{pb}}/c'_{\mathrm{B}}c'_{\mathrm{I}} = 12\exp[-(g^{\infty}_{\mathrm{pb}} - g^{\infty}_{\mathrm{B}} - g^{\infty}_{\mathrm{I}})/kT], \tag{3.8.5d}$$

to the neglect of the second virial terms in the chemical potentials. Here the fraction of defects of each species $r$ is $c_r = N_r/N_{\mathrm{m}}$ where $N_{\mathrm{m}}$ is the number of regular lattice sites in the crystal. The primes on $c'_{\mathrm{B}}$, $c'_{\mathrm{I}}$ and $c'_{\mathrm{V}}$ have been added to remind us

Table 3.4. *Constants for the chemical potentials, eqn.* (3.5.7*a*), *of the defects for the model of a dilute f.c.c. alloy containing both vacancies and* (100) *dumb-bell interstitials* ($z = 12$).

| $r$ | $z_r$ | $y_r$ |
|---|---|---|
| B | 1 | 1 |
| V | 1 | 1 |
| I | 3 | 1 |
| IB | 6 | 1 |
| BV | 12 ($=z$) | 2 |
| pa | 24 | 2 |
| pb | 12 | 2 |

| $r$ | $s$ | $-B_{rs}$ | $r$ | $s$ | $-B_{rs}$ |
|---|---|---|---|---|---|
| B | B | 1 | V | BV | 2 |
| B | V | 13 ($=z+1$) | V | pa | 2 |
| B | I | 13 ($=z+1$) | V | pb | 2 |
| B | IB | 1 | I | I | 1 |
| B | BV | 2 | I | IB | 1 |
| B | pa | 2 | I | BV | 2 |
| B | pb | 2 | I | pa | 2 |
| V | V | 1 | I | pb | 2 |
| V | IB | 1 | | | |

that these quantities represent the fractions of unpaired species. The quantity $(g_{IB}^{\infty} - g_{B}^{\infty} - g_{I}^{\infty}) \equiv \Delta g_{IB}$ in (3.8.5a) can be identified as the binding energy of a mixed dumb-bell relative to an unpaired solute atom and an isolated A–A dumb-bell, and there are similar interpretations for the quantities appearing in the other exponentials in (3.8.5). When there is an attraction between I and B then $\Delta g_{IB}$ is negative and the concentration of pairs is enhanced over that for a random distribution.

This example illustrates the point already made generally at the end of §3.6 that the use of the reduced scheme may require chemical potentials (e.g. the $\mu_I - 2\mu_A$ term) which differ from those provided by a statistical thermodynamical calculation for the wider set of defects. The reduced chemical potentials found from eqns. (3.5.7) and Table 3.4 are appropriate for transport calculations such as those described later in Chapter 8, in which the chemical potentials are required correct to terms linear in the concentrations. It may be noted that our expression for $B_{VI}$ (Table 3.4) assumes that a vacancy can be placed on any of the sites surrounding an interstitial. Alternatively we could have allowed for a number of sites $z'$ to be excluded because they would lead to spontaneous recombination (in which case $-B_{VI}$ would be $z' + 1$). The matter arises again when we consider the calculation of the $L$-coefficients for this model in later chapters.

### 3.8.2 *Overall equilibrium* (*vacancies only*)

We suppose that nearest-neighbour vacancy–solute pairs may form, as in the model of the preceding section (but in the absence of the interstitial defects included there). The expression for the chemical potentials of vacancies obtained from (3.5.7) and Table 3.4 is

$$\mu_V = g_V + kT \ln c_V + kT(c_P + zc_B). \tag{3.8.6}$$

The total mole fractions $x_V$ and $x_B$ of vacancies and solute may be expressed

$$x_V = c_V + c_P \tag{3.8.7a}$$

$$x_B = c_B + c_P. \tag{3.8.7b}$$

The concentration of pairs will normally be much less than the total concentration of solute and consequently the expression obtained by equating $\mu_V$ to zero can be written

$$x_V - c_P = (1 - zx_B)\exp(-g_V/kT). \tag{3.8.8}$$

The concentration of unpaired vacancies is therefore reduced by a factor $(1 - zx_B)$ relative to that of the pure crystal. When the mass-action expression (3.8.5b) for the formation of pairs is employed then the expression found from (3.8.8) for the total vacancy fraction correct to first order in $x_B$ is

$$x_V = \exp(-g_V/kT)[1 - zx_B + zx_B \exp(-\Delta g/kT)], \tag{3.8.9}$$

in which $\Delta g$ is the binding energy of a vacancy–solute pair. When there is binding ($\Delta g < 0$) the total concentration of vacancies is therefore greater than in the pure metal. In dilute alloys of the noble metals with solute elements lying to their right in the Periodic Table, $\Delta g$ is typically about $-0.2$ eV. (For a compilation of numerical values see e.g. Ullmaier, 1991). Equation (3.8.9) is often referred to as the Lomer expression (after Lomer, 1958). It provides a basis for the theory of the influence of solutes on self-diffusion in metals (§11.3). The calculation can be extended to include clusters involving more than one solute atom which form at higher solute concentrations.

## 3.9 Activity coefficients in ionic crystals

The method developed for dilute defect systems in §3.5 assumed that there were only short-range interactions among the defects and that the energies of interaction were absorbed into the energies of formation of the complex defects. This procedure is not wholly adequate for ionic solids, in which the defects may carry net electrical charges. At large enough separations the defect interactions are Coulombic and the energy varies as $q_r q_s/4\pi\varepsilon_0\varepsilon r_{sr}$ where $\varepsilon$ is the static relative permittivity (dielectric constant) of the crystal, while $q_r$ is the net charge on defects of type $r$. The system of defects is therefore analogous to an electrolyte solution. At the same time, in many systems there is ample evidence from paramagnetic resonance, dielectric and mechanical loss, ionic thermocurrents and other measurements for the formation of close pairs of oppositely charge defects at low temperature (i.e. generally well below the melting point), so that complex defects must still be included in the inventory. The simplest and by far the most commonly used procedure to take account of the long-range Coulombic interactions has been to add to the Gibbs function (3.5.1) an electrical term $G_{el}$, whose form is taken from the Debye–Hückel theory of electrolytes. This approach is modelled on the Bjerrum–Fuoss theory of electrolytes containing ion pairs and was first discussed in the applications to defects in ionic crystals by Teltow (1949) and Lidiard (1954). These applications have been described in greater detail by Fuller (1972), Corish and Jacobs (1973a), Laskar and Chandra (1989) and others, so we shall merely summarize the main features of the procedure.

The Debye–Hückel theory shows that the Coulombic interactions lead to an average distribution of charged defects about any particular charged defect which screens its Coulombic field. In the first-order approximation the mean electrical potential at distance $r$ from a defect of charge $q_s$ is proportional to $(q_s/4\pi\varepsilon_0\varepsilon r)\exp(-\kappa_D r)$ in which the screening parameter is given by

$$\kappa_D{}^2 = \sum_r n_r q_r{}^2/(\varepsilon_0 \varepsilon kT), \tag{3.9.1}$$

where $n_r$ is the concentration (number of defects per unit volume) of species $r$. The screening of the defect charges changes the chemical potential $\mu_r$ of defect species $r$ by the addition of an activity coefficient correction

$$kT \ln \gamma_r^{\mathrm{DH}} = -\frac{q_r{}^2 \kappa_D}{8\pi\varepsilon_0\varepsilon} \frac{1}{1+\kappa_D R}. \tag{3.9.2}$$

Here, $R$ is the 'distance of closest approach' at which a pair of oppositely charged defects are still counted as unpaired. As is evident from this formula, the chemical potential for a defect species bearing no net charge is unchanged.

The Bjerrum–Fuoss theory is known to be approximate and is limited to low defect concentrations, but a precise appraisal of its limitations and the formulation of more rigorous procedures which are applicable at relatively high concentrations both encounter formal difficulties that have not yet been satisfactorily overcome.

Allnatt and Loftus (1973) adapted the methods of Mayer ionic solution theory to obtain rigorous cluster expansions for the defect activity coefficients. They made a numerical study of a system of $Cd^{2+}$ impurity ions and cation vacancies in AgCl and concluded that that Teltow–Lidiard procedure was valid up to impurity fractions of about $2 \times 10^{-3}$ for temperatures within 200 K of the melting point and to lower concentrations at lower temperatures. At high defect concentrations (~1%) there will be clusters larger than pairs in this and many other ionic crystals, but the theory of Allnatt and Loftus (1973) is impracticable at these concentrations and no other rigorous method of calculating defect cluster concentrations is available. For the time being therefore there is little alternative but to use the approach of §3.5 with addition of a Debye–Hückel term $G_{el}$ to the Gibbs function to take account of the long-range interactions among defects bearing net charges.

The modifications to the chemical potentials arising from the Coulombic interactions are significant in a number of contexts. An important example is provided by the mass-action expressions for the concentrations of defects at overall thermodynamic equilibrium. For example, for Schottky defects in the NaCl structure one finds

$$c_1 c_2 = \exp(-g_S/kT)/\gamma_1^{DH}\gamma_2^{DH} \qquad (3.9.3)$$

in place of (3.3.15a). The qualitative effect of the activity coefficient corrections is to lead higher defect concentrations. A further example of the importance of the activity coefficient corrections is provided by the mass-action equations for the formation of nearest-neighbour pairs of oppositely charged defects. Both these examples are of importance in the analysis of the electrical conductivity of ionic crystals (Jacobs, 1983).

### 3.10 Concentrated solid solutions

We turn now to concentrated, binary, substitutional solid solutions. It is convenient when considering these to think of the two types of atom (A, B) as either (i) being indifferent to one another, or (ii) attracting one another or (iii) repelling one another. The first gives rise to ideal solid solutions in which the A and B atoms are distributed randomly relative to one another. In such systems the components are miscible in all proportions. Examples are provided by the alloys CuNi, AgAu and AuPt and by the pseudo-binary solid solutions which may be formed from MgO and the 3*d* transition metal monoxides MnO, FeO, CoO and NiO.

The second type of interaction gives rise to ordered solid solutions in which atoms of one kind are surrounded preferentially by those of the other. Intermetallic compounds (§3.3.2) provide examples in which the interaction is strong and the

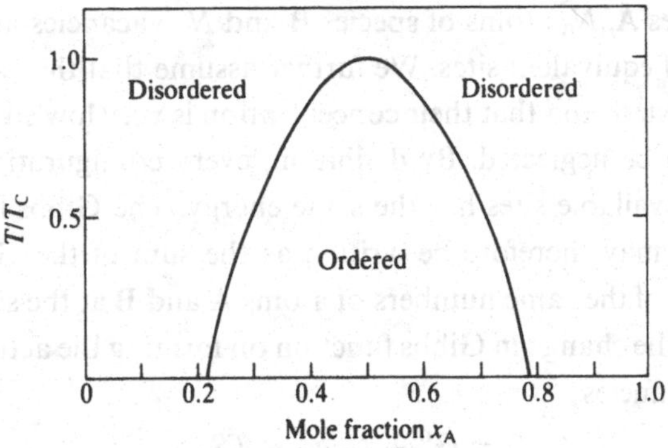


Fig. 3.3. Schematic phase diagram of a simple order–disorder alloy showing the boundary line for long-range order.

atoms are ordered on distinct sub-lattices at all temperatures up to the melting point. Other examples exist for which the ordering on to distinct sub-lattices only occurs beneath a certain critical temperature, $T_c$, e.g. $\beta$-CuZn, FeCo, $Fe_3Al$, etc. Many other alloys exist in which there is short-range order, i.e. a tendency for unlike atoms to be neighbours, but in which long-range ordering on separate sub-lattices is not observed.

The third type of interaction causes atoms of each type to be preferentially surrounded by atoms of the same type. This may lead to a separation into distinct phases at temperatures below a critical temperature, $T_c$.

These tendencies are, of course, strongly influenced by several factors. In all cases, high temperatures, by giving greater weight to the entropy term in $G$, favour the more disordered condition. Departures of composition from that stoichiometric ratio which favours maximum order likewise reduce the critical temperature of order–disorder alloys. These two effects are illustrated schematically in Fig. 3.3.

The thermodynamic behaviour of these alloys thus depends strongly upon the range and nature of the interatomic forces and is the subject of a large literature. By contrast, the analysis of atomic transport in concentrated solid solutions and alloys is still rather poorly developed (but see Chapter 13). In this section we shall therefore present only the rudiments of the theory of those types of system we have described. As in previous sections a main concern is with the concentration of defects, especially vacancies.

### *3.10.1 Ideal solid solutions*

The statistical thermodynamical theory of such solutions is a straightforward extension of that for a pure substance which was given in §3.2.1. We assume that

$N_A$ atoms of species A, $N_B$ atoms of species B and $N_V$ vacancies are distributed on a Bravais lattice of equivalent sites. We further assume that the vacancies occur in a single electronic state and that their concentration is very low so that interactions between them can be neglected. By definition, every configuration of atoms and vacancies on the available sites has the same energy. The Gibbs free energy, $G$, of the solid solution may therefore be written as the sum of the Gibbs free energy of a perfect crystal of the same numbers of atoms A and B at the same temperature and pressure plus the change in Gibbs function on forming the actual solid solution containing $N_V$ vacancies

$$G = G^0 + N_V g_V - TS_{con}. \tag{3.10.1}$$

Here $g_V$ is the Gibbs free energy of formation of one vacancy. The configurational entropy $S_{con}$ can be written

$$S_{con} = k \ln\left(\frac{(N_A + N_B + N_V)!}{N_A!N_B!N_V!}\right). \tag{3.10.2}$$

One then readily finds the following expressions for chemical potentials and thus the Gibbs energy

$$\mu_i = \mu_i^0 + kT \ln c_i, \qquad (i = \text{A, B}) \tag{3.10.3a}$$

$$\mu_V = g_V + kT \ln c_V \tag{3.10.3b}$$

and

$$G = N_A\mu_A + N_B\mu_B + N_V\mu_V. \tag{3.10.3c}$$

Here $c_i$ and $c_V$ are site fractions while $\mu_i^0$ denotes the chemical potential of pure crystalline $i$ at the same temperature and pressure.

Although this calculation provides a useful beginning to the theory of concentrated solid solutions, the assumption that the interactions among the atoms are independent of chemical type, of course, represents a limiting, ideal state of affairs. Even in nearly ideal systems there will always be at least small differences among the interactions of the different types of atom, and thus a tendency to short-range ordering (which will be stronger at lower temperatures). In nearly ideal systems such short-range order may be difficult to determine by direct diffraction methods, but in the case of metallic alloys it may still be apparent through its influence on sensitive properties such as the electrical resistivity. This provides a way to follow the rate of change of short-range order after abrupt changes of temperature; and thus, since this process depends upon the concentration and mobility of vacancies in the alloy, to obtain information about these vacancy properties (Schulze and Lücke, 1968, 1972). Elegant experiments of this kind have thus allowed the determination of vacancy properties in Ag–Au (Bartels *et al.*, 1987); Au–Pd (Kohl, Mais and Lücke, 1987) and other alloys. The technique has also been extended to allow the determination of interstitial properties from experiments on electron irradiated alloys (Bartels, 1987).

### *3.10.2 Order–disorder alloys and regular solutions*

The above calculation does not apply when the atoms A and B attract or repel one another. In such circumstances the calculation must allow for the different energies of the different configurations of atoms, but for this it is first necessary to specify the nature of the interactions. A simple model, which has long been used to represent the configuration-dependent energy term, assumes (non-saturable) interactions between pairs of atoms on nearest-neighbour sites, but no interactions at greater separations. We denote these interactions by $g_{AA}$, $g_{AB}$ and $g_{BB}$, according to the nature of the pair. (This model is referred to under various names – Ising, Bragg–Williams, lattice–gas, etc. – according to context; where we need a general term we shall refer to it as the *interacting lattice gas* model.) The energy of the crystal in a particular configuration therefore includes a term

$$N_{AA}g_{AA} + N_{AB}g_{AB} + N_{BB}g_{BB}, \tag{3.10.4}$$

in which $N_{AA}$, $N_{AB}$ and $N_{BB}$ are the numbers of nearest neighbour AA, AB and BB pairs respectively for the configuration concerned. This term is additional to the usual cohesive, vibrational and electronic terms going into $G^0$ for example. The accurate calculation of the thermodynamic properties of this model by the methods of statistical thermodynamics is difficult and has given rise to a huge literature. However, a simple approximation which is adequate for many purposes is that due originally to Bragg and Williams. In relation to $G$ that approximation retains the configurational entropy term in the form appropriate to ideal mixing (e.g. 3.10.2), but adds the additional energy of interaction (3.10.4) in which $N_{AA}$, etc. are replaced by their mean values, $\bar{N}_{AA}$, etc. for a random distribution. For example, in an alloy showing neither long-range order nor phase separation we should have $\bar{N}_{AB} = N_m z c_A c_B$, in which $N_m$ is the total number of sites and $z$ is the number of nearest neighbours. The theory of this approximation is given in many texts and, partly for this reason, we shall not reproduce it again here; partly also because we can easily obtain what we want from results already derived.

We demonstrate this by first considering a simple order–disorder alloy and appropriately modifying the treatment of §3.3.2 to allow for the dependence of the $g_i^\infty$ upon the degree of order. In this way we can derive equations for (i) the dependence of the degree of order upon temperature and composition (ii) the partition of the vacancies between the two sub-lattices and (iii) the vacancy chemical potential, as obtained previously by Bakker and van Winkel (1980).

### *Degree of order*

There are many distinct types of ordered alloys. Here we shall take the simplest, i.e. one consisting of two sub-lattices for each of which every site has $z$ equivalent nearest neighbours all lying on the other sub-lattice with none of its own. The best-studied example of this kind is provided by $\beta$-brass in the CsCl structure (for which $z = 8$ and each sub-lattice is simple cubic). Ideal ordering of such a system is possible at the stoichiometric composition AB and corresponds to all A atoms on one sub-lattice (conventionally denoted by $\alpha$) and all B atoms on the other ($\beta$). Lesser degrees of order are generated by putting A atoms on $\beta$-sites and B atoms on $\alpha$-sites. But this is just the same situation that was analysed in §3.3.2 when we considered intermetallic compounds. We can therefore use the same notation and call upon the results obtained there. The difference is that there the number of wrongly sited atoms was so small that the energies $g_i^\infty$ could be assumed to be given, whereas here we have to allow for their dependence on the degree of order. As already mentioned we shall, in fact, find that dependence by using the Bragg–Williams, or *mean-field*, approximation. First we define the degree of (long-range) order $s$ by setting

$$N_3 = \frac{N_B}{2} - \frac{N_B s}{2}, \tag{3.10.5a}$$

$$N_4 = \frac{N_A}{2} - \frac{N_B s}{2}, \tag{3.10.5b}$$

$$N_5 = \frac{N_A}{2} + \frac{N_B s}{2}, \tag{3.10.5c}$$

$$N_6 = \frac{N_B}{2} + \frac{N_B s}{2}, \tag{3.10.5d}$$

in which $N_3 \ldots N_6$ and $N_A$ and $N_B$ are as defined in §3.3.2. For convenience we take A to be the major component, i.e. $N_A > N_B$. We see that $s = 1$ corresponds to the maximum possible order: no B atoms are on A sites (3.10.5a) and the number of A atoms on A sites by (3.10.5c) is $(N_A + N_B)/2$, i.e. is equal to the number of A sites (in this case $a = b$). Likewise, $s = 0$ corresponds to maximum disorder: the A atoms and the B atoms are equally divided between the two sub-lattices. To obtain the equation for $s$ as a function of temperature and composition we combine (3.3.44a) and (3.3.45a) to give

$$\frac{N_3 N_4}{N_5 N_6} = \exp(-(g_{XA} + g_{XB})/kT), \tag{3.10.6}$$

in which $g_{XA} + g_{XB}$ is the energy required for the reaction

$$(\text{A on } \alpha) + (\text{B on } \beta) \rightleftharpoons (\text{A on } \beta) + (\text{B on } \alpha). \tag{3.10.7}$$

For this energy we now substitute the average change in the pair interactions on accomplishing the reaction (3.10.7). A simple calculation shows that this quantity is just

$$2zs(g_{AA} + g_{BB} - 2g_{AB})c_B \equiv 2zsEc_B, \tag{3.10.8}$$

in which $c_B$ is the site fraction of B ($=N_B/N_m$ with $N_m$ the total number of sites). The use of $E$ for $(g_{AA} + g_{BB} - 2g_{AB})$ is not particularly apposite, but is common. The equation (3.10.6) then gives as the equation for $s$

$$\frac{(1-s)(c_A - c_B s)}{(1+s)(c_A + c_B s)} = \exp\left(-\frac{2c_B zsE}{kT}\right). \tag{3.10.9}$$

For a stoichiometric alloy AB, i.e. $c_A = 1/2 = c_B$ this becomes

$$\left(\frac{1-s}{1+s}\right) = \exp\left(-\frac{c_B zsE}{kT}\right). \tag{3.10.10}$$

Although these equations cannot be solved for $s$ as a function of $T$ they can be inverted to give $T$ as a function of $s$. When $E > 0$ a solution $s > 0$ exists for $T < T_c$ with $kT_c/E = c_A c_B z$. Above $T_c$ the only solution is $s = 0$, i.e. no long-range order is possible. This critical temperature is highest for the stoichiometric composition. The solution of (3.10.10) is shown as the *mean-field* line in Fig. 2.18, which also shows the corresponding exact solution.

### *Vacancy concentrations*

These can also be found by adapting the analysis of §3.3.2. As there, the vacancy fractions on the two sub-lattices will differ in the presence of long-range order ($s > 0$). From (3.3.44a) and (3.3.45a) we obtain the ratio of the two populations

$$\frac{N_1}{N_2} = \left[\frac{(c_A + c_B s)(1-s)}{(c_A - c_B s)(1+s)}\right]^{1/2} \exp\left(-\frac{(g_{XA} - g_{XB})}{2kT}\right), \tag{3.10.11}$$

and then insert average quantities for the energies $g_{XA}$ and $g_{XB}$ of the reactions

$$V_\beta + A_\alpha \rightleftharpoons V_\alpha + A_\beta$$

and

$$V_\alpha + B_\beta \rightleftharpoons V_\beta + B_\alpha,$$

respectively. One easily finds that

$$g_{XA} - g_{XB} = 2zc_B(g_{AA} - g_{BB})s,$$

whence (3.10.11) becomes

$$\frac{N_1}{N_2} = \left[\frac{(c_A + c_B s)(1 - s)}{(c_A - c_B s)(1 + s)}\right]^{1/2} \exp\left(-\frac{zc_B(g_{AA} - g_{BB})s}{kT}\right), \qquad (3.10.12)$$

which gives the ratio of the vacancy populations in the two sub-lattices ($\alpha/\beta$).

Next we obtain the equation of Schottky equilibrium, which will yield the product $N_1N_2$. For this we need the average energies required to form $V_\alpha$ and $V_\beta$. Together these are

$$g_S = 2g_V' - z(g_{AA}c_A{}^2 + g_{BB}c_B{}^2 + 2g_{AB}c_Ac_B - c_B{}^2Es^2), \qquad (3.10.13)$$

in which $g_V'$ is the configuration-independent contribution to the vacancy formation energy (associated with the scattering of the free electrons). The product of the vacancy concentrations is then obtained from (3.3.39a). Solution of (3.10.12) and (3.3.39a) will yield the two concentrations separately. In the special case of a stoichiometric alloy with $g_{AA} = g_{BB}$ the two are equal by (3.10.12). The total vacancy fraction $c_V = (N_1 + N_2)/N_m$ is then

$$c_V = \exp\left(-\frac{1}{kT}\left(g_V'' + \frac{z}{8}Es^2\right)\right) \qquad (3.10.14a)$$

with

$$g_V'' = g_V' - \frac{z}{4}(g_{AA} + g_{AB}). \qquad (3.10.14b)$$

In other cases they are only equal above $T_c$, where $s = 0$. In all cases, however, the presence of the terms in $s$ and $s^2$ in (3.10.12) and (3.10.13) means that an Arrhenius plot of the total vacancy concentration will exhibit a kink at $T_c$. This is part of the reason for the similar feature observed in diffusion coefficients of order–disorder alloys (see Fig. 2.19).

### *Regular solutions*

When the parameter $E < 0$ ordering on to distinct sub-lattices is not possible, as (3.10.10) shows. In this case the system separates at low temperatures into two phases, one A-rich the other B-rich. At high temperatures there is a single homogeneous phase. The critical temperature dividing the two regions of the phase diagram depends on concentration in the same way as $T_c$ for the order–disorder alloy; in fact, apart from the replacement of $E$ by $|E|$ the two expressions are identical. This and other results can be obtained by the same mean-field arguments used above. In particular, the chemical potential of the vacancies in the single homogeneous phase follows from (3.10.13) with $s = 0$

$$\mu_V = g'_V - \frac{z}{2}(g_{AA}{c_A}^2 + g_{BB}{c_B}^2 + 2g_{AB}c_Ac_B) + kT\ln c_V. \qquad (3.10.15)$$

The equilibrium vacancy concentration, as before, is obtained by setting $\mu_V = 0$. It will be seen that this varies with composition.

In the same mean-field approximation the chemical potential of the A component in the standard form is

$$\mu_A = \mu_A^0 + kT\ln(c_A\gamma_A), \qquad (3.10.16a)$$

in which

$$kT\ln\gamma_A = -\frac{zE{c_B}^2}{2} \qquad (3.10.16b)$$

and $\mu_A^0$ is the chemical potential of pure A. There is a corresponding expression for $\mu_B$. The so-called thermodynamic factor for chemical diffusion ($\Phi$) which will be needed later (e.g. Chapter 5) then follows immediately from (3.10.16b);

$$\Phi = \left(1 + \frac{\partial\ln\gamma_A}{\partial\ln c_A}\right) = 1 + \frac{zEc_Ac_B}{kT}. \qquad (3.10.17)$$

Since $E$ is a negative quantity we see that (3.10.17) predicts $\Phi$ to be a simple parabolic function of $c_A$ with its minimum at $c_A = 1/2 = c_B$. Figure 5.4(b) shows an example of $\Phi$ for a real system (Au–Ni). Since $\partial\mu_A/\partial c_A$ must be $>0$ for thermodynamic stability $\Phi$ must also be $>0$. If $\Phi < 0$, then 'uphill' diffusion results and concentration fluctuations become steadily larger as the alloy separates into two phases (*spinodal decomposition*). The line $\Phi = 0$ thus defines the co-existence curve. From (3.10.17) we see that this yields the same relation between $c_A$ and $T_c$ with $|E|$ as we previously found between $c_A$ and $T_c$ for the order–disorder alloy with $E$.

To conclude this section we note that, despite the widespread convenience and usefulness of the mean-field approximation for qualitative predictions, its numerical accuracy is frequently poor. Various systematic approximation schemes for the thermodynamic properties of crystals in the absence of vacancies have therefore been proposed, generally yielding the mean-field approximation as the lowest-order theory. Among these the cluster variation scheme of Kikuchi (1951) has often been used (e.g. Sato, 1970; de Fontaine, 1979). Monte Carlo simulation has also provided much quantitative guidance on thermodynamic properties (Landau, 1979, 1984). More accurate approximations than that of Bragg and Williams have also been proposed for calculating the vacancy concentrations (see e.g. Schapink, 1965; Cheng, Wynblatt and Dorn, 1967). No way of checking such approximations by Monte Carlo simulation applicable to small vacancy contents has yet appeared.

## 3.11 Atomic mobilities

In the preceding sections we have used the methods of statistical thermodynamics to obtain expressions for the concentrations of defects and other defect-related thermodynamic quantities. For the theory of atomic transport we need not only these concentrations but also the rates at which the defects move through the crystal lattice. The basic view of this movement has already been presented in §2.3 as one in which random thermal fluctuations occasionally provide sufficient energy and momentum to cause the defect to jump from one position to another. The atoms surrounding the defect (cf. Fig. 2.1) spend most of their time vibrating about their positions in the lattice, occasionally to play their part in a defect jump of this type. This is the basic model which underlies most of the detailed analyses in later chapters. Since the atomic displacements are presumed to occur infrequently on the time scale of lattice vibrations ($\sim 10^{-13}$ s), we may further suppose that local thermal equilibrium will be re-established between defect jumps. We may therefore hope to use equilibrium statistical mechanics to give the frequency of those fluctuations in atomic positions and momenta necessary for the displacement of the defect to occur. This approach, which derives originally from the theory of chemical reaction rates, has been extensively used for many years. A comprehensive review of the subject taking in the applications to chemical reaction rates has recently been presented by Hänggi, Talkner and Borkovec (1990). Our interest here in migration rates in solids defines a more limited field from which many of the complexities of chemical reaction rate theory are absent. For this purpose the reviews of Franklin (1975) and of Jacucci (1984) are pertinent.

In this section we shall cover some of the basic ideas and then comment on a number of subsequent developments.

### *3.11.1 A simplified (single-particle) calculation*

The idea we have described of atoms which remain firmly held in their normal lattice positions until a sufficiently large thermal fluctuation displaces them to another position, implies an intermediate state (or states) where the potential energy of the lattice is a maximum along the path taken in the transition. This high-energy intermediate state is called the transition state. However, the energy of this transition state is not uniquely determined by the position of the moving atom in the average lattice since it evidently depends on the relative positions of the

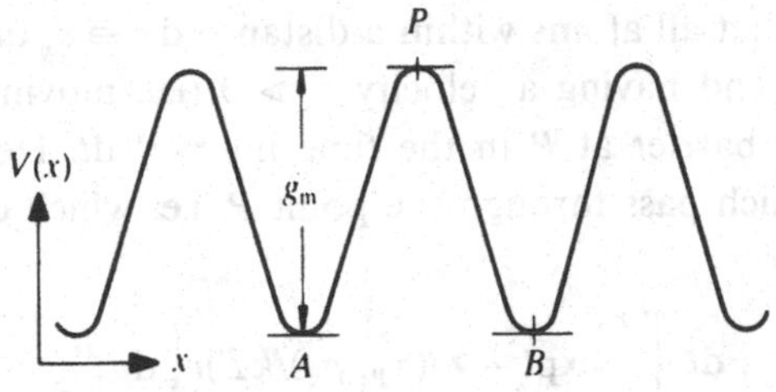


Fig. 3.4. Schematic diagram of the average potential energy of an interstitial atom as a function of position in one crystallographic direction, $x$.

neighbouring atoms, i.e. it depends on the vibrational state of the crystal and on the phases of the vibrations. Nevertheless the mean amplitude of these atomic vibrations will generally be small compared to the displacement required to take an atom from its initial position to the transitional position. Support for this assertion can be found in Lindemann's law relating to the melting of crystalline solids: this tells us that the r.m.s. displacement of atoms at the melting point is only about 1/10–1/5 of the nearest-neighbour distance. Therefore we can devise a simplified description of the displacement process by treating the motion of only one atom in the more-or-less well-defined potential field provided by its interaction with the other atoms of the lattice. We do this first to emphasize the principles involved; then we consider the corresponding many-body treatment.

For simplicity let us consider the example of simple interstitials (e.g. Figure 2.2(*a*)) which move directly from one interstitial site to another. The general form of the result for vacancies, dumb-bell interstitials, etc. will be the same, although minor details in the derivation will be different. We consider displacements of the interstitials in a particular crystal direction, $x$. The average potential, $V(x)$, in which they move will then be periodic as shown in Fig. 3.4. Now the classical motion of an atom in such a potential is described by the Hamiltonian function

$$\mathscr{H}(x, p_x) = (p_x{}^2/2m) + V(x), \tag{3.11.1}$$

in which $m$ is the mass of the atom and $p_x$ is its momentum in the direction $x$. Now from classical statistical mechanics (of systems of given temperature and volume), we know that the probability that at any one time the atom lies in the interval $x$ to $x + \mathrm{d}x$ of position and $p_x$ to $p_x + \mathrm{d}p_x$ of momentum is

$$\frac{\exp(-\mathscr{H}(x, p_x)/kT)\,\mathrm{d}x\,\mathrm{d}p_x}{\displaystyle\int_{L_x}\int_{-\infty}^{\infty} \exp(-\mathscr{H}(x, p_x)/kT)\,\mathrm{d}x\,\mathrm{d}p_x}. \tag{3.11.2}$$

in which the spatial integral in the denominator is taken over the length of the crystal in the $x$-direction, $L_x$.

We now observe that all atoms within a distance $\mathrm{d}x \equiv v_x\,\mathrm{d}t = (p_x/m)\,\mathrm{d}t$ to the left of $P$ in Fig. 3.4 and having a velocity $v_x > 0$ (i.e. moving to the right) will cross the top of the barrier at $P$ in the time interval $\mathrm{d}t$. Hence the number of interstitial atoms which pass through the point $P$, i.e. which cross the barrier, in time $\mathrm{d}t$ is

$$\frac{\mathrm{d}t \displaystyle\int_0^{\infty} \exp(-\mathscr{H}(x_P, p_x)/kT) v_x\,\mathrm{d}p_x}{\displaystyle\int_{L_x}\int_{-\infty}^{\infty} \exp(-\mathscr{H}(x, p_x)/kT)\,\mathrm{d}x\,\mathrm{d}p_x},$$

which by (3.11.1),

$$= \frac{\dfrac{\mathrm{d}t}{m}(mkT)\exp(-V(x_P)/kT)}{\displaystyle\int_{L_x} \exp(-V(x)/kT)\,\mathrm{d}x \int_{-\infty}^{\infty} \exp(-p_x{}^2/2mkT)\,\mathrm{d}p_x}$$

$$= \frac{\mathrm{d}t\left(\dfrac{kT}{2\pi m}\right)^{1/2} \exp(-V(x_P)/kT}{\displaystyle\int_{L_x} \exp(-V(x)/kT)\,\mathrm{d}x}. \qquad (3.11.3)$$

We now *assume* that all atoms crossing $P$ from $A$ towards $B$ are *deactivated* once they have done so, i.e. they have negligible chance of being reflected back to $A$ immediately. In other words any subsequent move of the atom back to $A$ is dynamically uncorrelated with the previous motion. With this assumption the quantity (3.11.3) can be identified with the product of the number of interstitial atoms on the plane $x$ ($Ns_x$) with the jump frequency ($w$) and $\mathrm{d}t$. Since the integrand in the denominator of (3.11.3) is periodic (Fig. 3.4) we can replace it with $(L_x/s_x)$ times the integral over one period $s_x$. Hence

$$w = \left(\frac{kT}{2\pi m}\right)^{1/2} \frac{\exp(-V(x_P)/kT)}{\displaystyle\int_{s_x} \exp(-V(x)/kT)\,\mathrm{d}x}. \qquad (3.11.4)$$

Now in most systems, fast ion conductors excepted, the height of the potential barrier, i.e. $V(x_P) - V(x_A)$, is much greater than $kT$. Most of the integral over $x$ in (3.11.4) therefore comes from the region around $A$ where $V(x)$ is lowest. Within the integral we therefore make the harmonic approximation about the original position $A$ and set

$$V(x) = V(x_A) + \tfrac{1}{2}Kx^2, \qquad (3.11.5)$$

whence by (3.11.4),

$$w = \frac{1}{2\pi}\left(\frac{K}{m}\right)^{1/2} \exp(-(V(x_P) - V(x_A))/kT)$$

$$\equiv \nu \exp(-e_{\mathrm{m}}/kT). \qquad (3.11.6)$$

From this simple calculation we can draw two immediate conclusions, apart from the obvious one that (3.11.6) shows that $w$ has an Arrhenius dependence upon temperature. The first concerns isotopic effects. Different isotopes of the same element will experience the same potential field $V(x)$ but will differ in mass $m$. Eqn. (3.11.6) predicts that $\nu$, and hence $w$, will be proportional to $m^{-1/2}$. Secondly, from the magnitudes of $e_m$ (as indicated by observed activation energies, i.e. $\sim 1$ eV) and of the displacement distance, we can infer from (3.11.6) that $\nu$ should often be $\sim 10^{13}\ s^{-1}$.

As we have emphasized, this calculation as presented omits the motions of the other atoms. However, these can be taken into consideration in an average way and $V(x)$ for the jumping atom is then obtained as a statistical mechanical average (potential of the mean force exerted on the jumping atom by all others). For a system at temperature $T$ under a given external pressure $P$ this has the effect of changing $e_m$ in (3.11.6) into a quantity with the character of a Gibbs energy, $g_m$, dependent on $T$ and $P$. Hence in place of (3.11.6) we have

$$w = \nu \exp(-g_m/kT), \tag{3.11.7}$$

as foreseen in §2.3. The entropy of activation, $s_m$, corresponds to a change in lattice vibrations associated with the displacement of the jumping atom from $A$ to $P$. This modification of the theory, however, uses the same criterion for a transition, namely passage in the direction $A$ to $B$ over the point $P$ of maximum mean potential. The predicted dependence of $w$ on isotopic mass will therefore be the same as before, i.e. as $m^{-1/2}$. In reality, however, we should suppose that there will be a correlation between the success of a given thrust by the jumping atom from $A$ towards $B$ and the motions of the other atoms in its vicinity. A many-body formulation of the problem to take account of this supposition will now be described.

### *3.11.2 Many-body calculation (Vineyard)*

In describing this generalization of the previous treatment we shall use the formulation of Vineyard (1957), although there are equivalent treatments by others. This is based upon the canonical ensemble of statistical mechanics for the distribution of atomic positions and velocities; this gives results for a closed system of fixed temperature and volume. In practice, of course, we are interested in systems at fixed temperature and pressure for which it would be appropriate to use a constant pressure ensemble (Hill, 1956; McQuarrie, 1976). However, the essentials of the calculation are more easily seen in the familiar method of the canonical ensemble, while the formal extension required by the constant pressure ensemble is easily made thereafter (see below).

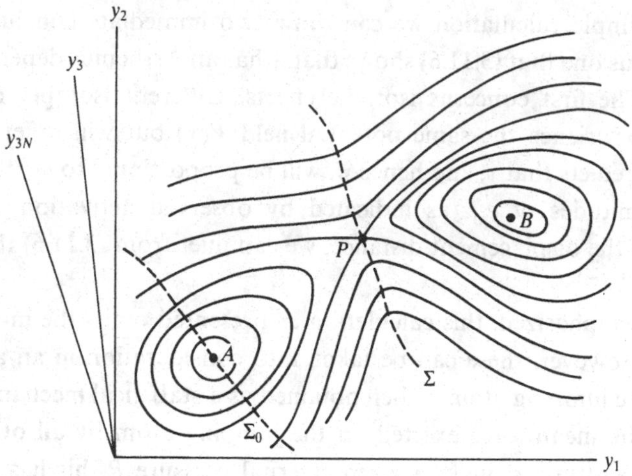


Fig. 3.5. Schematic diagram showing the contours of constant potential energy $\Phi(y_1 \ldots y_{3N})$ in the $3N$-dimensional co-ordinate space of the solid. The surface $\Sigma$ is the transition surface of local maxima while $P$ is a saddle point for which $\Phi$ takes its lowest value on the surface $\Sigma$.

We suppose that there are $N$ atoms in the solid and thus $3N$ spatial co-ordinates $x_i$. We shall actually use the mass-weighted co-ordinates $y_i = \sqrt{m_i} x_i$ (to allow contact with phonon theory where they give a symmetrical dynamical matrix). The potential energy function $\Phi(y_1 \ldots y_{3N})$ will then have minima corresponding to the various defect positions in the solid. Points $A$ and $B$ in Fig. 3.5 are such minima corresponding to two neighbouring configurations, i.e. configurations which differ by the movement of a defect from one position to a neighbouring one. We can now define reaction paths between these two minima, and there will be a saddle point $P$ on at least one of them. The difference in potential energy $\Phi(P) - \Phi(A)$ specifies the least energy necessary to pass from $A$ to $B$. Furthermore, there exists a unique surface $\Sigma$ of dimension $3N - 1$ which passes through $P$ and which is orthogonal to the contours of constant $\Phi$ everywhere else. The transition rate across this surface is the quantity we require to give us $w$. For this system of $3N$ co-ordinates, Vineyard made assumptions exactly analogous to those employed above for the model with one $x$-co-ordinate and so obtained

$$w = \left(\frac{kT}{2\pi}\right)^{1/2} \frac{\int_\Sigma \exp(-\Phi/kT)\, d\sigma}{\int_\Theta \exp(-\Phi/kT)\, d\tau)}, \tag{3.11.8}$$

where the integral in the numerator is over the $(3N - 1)$ dimensional surface $\Sigma$ while that in the denominator is over that region $\Theta$ of the $3N$ dimensional configuration space containing $A$ and bounded by $\Sigma$. The additional integrations contained in (3.11.8) compared to (3.11.4) take account of the different paths across $\Sigma$, while the apparent absence of the atomic mass is a consequence of the use of mass-weighted co-ordinates.

This expression, as we have mentioned, assumed that the probability distribution among states was that of the canonical ensemble for a system held at fixed volume and temperature. To obtain the corresponding result for the practical conditions of constant pressure and temperature we should instead use the statistical method of the constant pressure ensemble, in which the probability distribution contains an additional factor $\exp(-PV/kT)$ and thermodynamic average quantities are obtained by integrating over all volumes as well as summing over all states (Hill, 1956; McQuarrie, 1976). In place of (3.11.8) we therefore have the formal result

$$w = \frac{\left(\dfrac{kT}{2\pi}\right)^{1/2} \displaystyle\int_{V} \mathrm{d}V \int_{\Sigma} \exp(-(\Phi + PV)/kT)\,\mathrm{d}\sigma}{\displaystyle\int_{V} \mathrm{d}V \int_{\Theta} \exp(-(\Phi + PV)/kT)\,\mathrm{d}\tau}. \tag{3.11.9}$$

The simplest application of (3.11.9) is obtained by assuming that the dominant contributions to the integrals derive from the regions immediately around $P$ and $A$ respectively within which (separate) harmonic expansions of $\Phi$ in the atomic displacements can be used. The transition surface $\Sigma$ then becomes a hyperplane and (3.11.9) simplifies to

$$w = \bar{\nu} \exp(-h_{\mathrm{m}}/kT) \tag{3.11.10a}$$

with

$$h_{\mathrm{m}} = \Phi(P) - \Phi(A) + P(V_P - V_A)$$

$$\equiv \Phi(P) - \Phi(A) + Pv_{\mathrm{m}} \tag{3.11.10b}$$

and

$$\bar{\nu} = \left(\prod_{j}^{3N} \nu_j\right)_A \bigg/ \left(\prod_{j}^{3N-1} \nu'_j\right)_P. \tag{3.11.10c}$$

The form of $\bar{\nu}$ arising here (compared to $\nu$ in the 3.11.6) comes from integrating over the motions around $A$ and on the surface $\Sigma$. Thus the $\nu_j$ are the normal mode frequencies for vibrations about $A$ (the volume of the crystal being $V_A$) while $\nu'_j$ are the frequencies of the system when it is constrained to move within the surface $\Sigma$ (the volume of the crystal being $V_P$). Now the entropy of a harmonic solid is related to its vibration frequencies and (3.11.10c) therefore suggests defining an appropriate entropy of activation. We remove from the product in the numerator that frequency corresponding to motion normal to the surface $\Sigma_0$, i.e. in the direction of the reaction path at $A$. Call this frequency $\nu_r$ then

$$w = \nu_r \exp(-g_m/kT) \tag{3.11.11a}$$

with

$$g_m = h_m - Ts_m \tag{3.11.11b}$$

and

$$s_m = k\left\{\sum_{j\neq r}^{3N-1} \ln\left(\frac{h\nu_j}{kT}\right)_A - \sum_{j\neq r}^{3N-1} \ln\left(\frac{h\nu'_j}{kT}\right)_P\right\}, \tag{3.11.11c}$$

in which both summations are over $3N - 1$ modes.

We make the following comments on this treatment of the problem.

(1) By using the methods of equilibrium statistical mechanics it provides a representation of the basic idea that thermal fluctuations lead to a localized activation of atoms over an energy barrier and into a new configuration. It does not demonstrate, but assumes, that passage across the barrier necessarily leads to a new configuration. The rate of crossing the barrier is rigorously calculated but the assumption that all these crossings are successful means that (3.11.9) actually provides an upper limit to $w$, rather than an exact equation. Eqn. (3.11.9) thus provides a basis for variational calculations of $w$.

(2) At ordinary pressures the enthalpy of activation $h_m$ is indistinguishable from the internal energy of activation $u_m$. Eqn. (3.11.10b) then shows that this can be calculated from the potential energy function of a static lattice. Many atomistic calculations of this kind have been made. For the most part these have been carried out for systems held at constant volume. Thermodynamic relations which allow us to obtain the quantities for conditions of given $P$ and $T$ from the results of these constant-volume calculations together with examples are given in the review by Harding (1990).

(3) By the rule equating the product of the eigenvalues of a Hermitian matrix to the determinant of the matrix itself, it follows that $\bar{\nu}$ is proportional to the product of the magnitude of the (imaginary) frequency of the decomposition mode (mode through $P$ orthogonal to the surface $\Sigma$) and a function of the force constants. Thus with

$$\Phi = \Phi_0(A) + \frac{1}{2}\sum_{i,j} \frac{\beta_{ij} y_i y_j}{(m_i m_j)^{1/2}} \tag{3.11.12}$$

for the expansion about $A$ (with a similar expression about $P$) and the secular equation

$$\left|\frac{\beta_{ij}}{(m_i m_j)^{1/2}} - \lambda\delta_{ij}\right| = 0 \tag{3.11.13}$$

for the normal mode frequencies $\nu_j$ with $\lambda = (2\pi\nu)^2$ we see that

$$\prod_{j=1}^{3N} (2\pi\nu_j)^2 = |\beta_{ij}|_A \prod_{j=1}^{3N} m_j^{-1}, \tag{3.11.14a}$$

with a similar expression for the modes about $P$,

$$\prod_{j=1}^{3N} (2\pi\nu'_j)^2 = |\beta'_{ij}|_P \prod_{j=1}^{3N} m_j^{-1}, \tag{3.11.14b}$$

in which $|\beta_{ij}|$ and $|\beta'_{ij}|$ are the determinants of the force constant matrices for the motions about $A$ and $P$ respectively.

Thus, by (3.11.10c),

$$\bar{\nu} = \nu'_{\mathrm{d}}(|\beta_{ij}|_A/|\beta'_{ij}|_P)^{1/2}, \tag{3.11.15}$$

in which the frequency of the decomposition mode is written as $i\nu'_{\mathrm{d}}$. The dependence of $w$ upon the isotopic mass of the moving atom thus arises solely through $\nu'_{\mathrm{d}}$. One would generally expect the mass of the moving atom to be predominant in this frequency but, since other atoms certainly also move, the dependence will not be as fast as $m^{-1/2}$. Indeed, Le Claire (1966), considering two isotopes of masses $m_1$ and $m_2$, introduced the quantity

$$\Delta K = \frac{(\nu'_{\mathrm{d},1}/\nu'_{\mathrm{d},2}) - 1}{(m_2/m_1)^{1/2} - 1} \tag{3.11.16}$$

as a measure of this dependence and showed that it equalled the total kinetic energy in the decomposition mode divided into that part of the kinetic energy residing in the jumping atom: hence $\Delta K \leq 1$. A few detailed calculations of $\Delta K$ have been made. Some values inferred from experimental measurements of isotope effects are given later in Table 10.2. The values of $\Delta K$ obtained from (3.11.16) appear to be significantly closer to unity than the experimental values.

### *3.11.3 Other calculations of jump rates*

There have been many other discussions of this fundamental problem of jump rates, as may be seen from the reviews of Bennett (1975), Franklin (1975), Jacucci (1984) and Hänngi *et al.* (1990). Molecular dynamics computations are increasingly used to supplement and to check analytical theories. Some of the conclusions of this work are summarized below. Early indications of several of these were provided by calculations on a simple model due to McCombie and Sachdev (1975).

Their results (a) provided data on the fraction of unsuccessful passages through the transition state and (b) allowed comparisons of the three methods of calculation represented here by the results (3.11.7), (3.11.10) and a variational approximation to (3.11.9), (which they refer to as 'dynamical' and 'optimum surface' theories respectively). Their model was devised to give enhanced effects and it is not therefore surprising that they were, in fact, larger than those found subsequently for models based on detailed atomic force laws.

(1) The general rate theory expressions (3.11.9) and (3.11.10) do provide upper limits to $w$, but the indications are that the fraction of unsuccessful crossings (i.e. those counted in 3.11.9 and 3.11.10 but for which the atom is reflected back before it has a chance to become deactivated and settle into a new configuration $B$) is not more than about 10% at the melting temperature and is smaller still at lower temperatures. However, in cases where the activation energy is low (e.g. fast-ion conductors) the effects may be relatively more important (Funke, 1989).

(2) Possibly the use of the harmonic approximation to obtain (3.11.10) introduces greater errors. Although there are examples where there is a firm indication of harmonic behaviour at the saddle point (e.g. the oxygen vacancy in CoO, Harding and Tarento, 1986), there are others where some of the displacements are evidently anharmonic (e.g. vacancies in KCl, Harding, 1985). Of interest, however, are recent theories and simulations which indicate that successful calculations of the isotope effect $\Delta K$ require the inclusion of strongly anharmonic terms in the potential function for the transition state (Flynn, 1987).

(3) The molecular dynamics simulations also point to the possibility of multiple jumps, e.g. two atoms moving in rapid succession into the same vacancy. Possibly the most extreme example of this kind occurs in compounds with the fluorite structure at high temperatures ($T > 0.9T_m$) where sequences of several anion displacements are indicated (Gillan, 1989).

(4) All the above discussion has assumed that classical statistical mechanics is applicable, but, of course, diffusive atomic movements are observed at low temperatures where this may not be so ($T$ less than the Debye temperature for lattice vibrations). Notable examples are provided by the diffusion of hydrogen in metals and by the annealing of radiation-induced self-interstitials in a range of substances. The first of these has received considerable attention, partly because the comparative lightness of the hydrogen interstitial allows certain simplifications (e.g. the use of the adiabatic approximation, as in the treatment of Flynn and Stoneham, 1970). Quantum molecular dynamics has also been developed and applied successfully to this system (Gillan, 1988). These calculations explain the curvature observed in Arrhenius plots, but otherwise preserve the idea of jumps from one interstitial position to another.

### 3.11.4 Application to atomic transport theory

The theory of the rates at which atoms move from one lattice position to another may be rightly regarded as one of the fundamental aspects of atomic transport theory. As such it has received a great deal of attention, as the above discussion indicates. Nevertheless, for many practical purposes (including those of this book) only a few general conclusions are required. These we shall now underline.

First place must be given to support for the hypothesis that atomic migration is the result of a sequence of localized events or jumps from one site to another. Such support comes from both classical and quantum treatments, and in its most direct form from molecular dynamics simulations. The rate at which these elementary steps occur will thus be influenced by *local* conditions, e.g. the presence of solute atoms in the lattice may affect the movements of defects in their vicinity but not at remote positions. (Longer range effects may arise from Coulomb forces between charged species in insulators and semiconductors, but in practice their influence on jump rates is rather small. A representation of their effect on electrical mobility may be obtained from the theory of liquid electrolyte solutions, cf. Appendix 11.4.) Hence with jump rates we do not encounter the variety of considerations which arose in the earlier sections dealing with defect concentrations. These enter only later when we turn to the statistical description of atomic migration in terms of the elementary steps.

The second conclusion concerns the nature of these elementary steps. Molecular dynamics simulations mostly confirm that atoms jump directly from one site to a nearest-neighbour site and that multiple hops are rare (although their occurrence is indicated in some model substances). Subsequent theory allows for the possibility of multiple hops, although in particular illustrations and developments we shall assume only the simplest nearest-neighbour hops.

The third conclusion is the existence of the intermediate or transition state which is the same for the forward jump and the corresponding reverse jump. This will be useful when we consider migration in a force field. The conclusion clearly is an integral part of the treatments of §§3.11.1 and 3.11.2. In view of the evidence for the occurrence of a proportion of passages through the transition state which fail to lead to a new configuration, it might be doubted whether the conclusion can be used for exact calculations. However, use of the conclusion to give the ratio of the rates of forward and backward jumps is equivalent to the assumption of detailed balance, and is therefore exact. This is the purpose for which we shall use it.

Lastly, when a quasi-harmonic approximation can be used to describe atomic vibrations about both the initial configuration and the transition state, then the theory provides a basis on which to make atomistic calculations of activation energies and other parameters. The limitations on the use of the quasi-harmonic approximation are not severe enough to affect significantly the ability of atomistic calculations to distinguish likely from unlikely mechanisms of migration. On the other hand, the indications are that the quasi-harmonic approximation is not adequate for calculating isotope effects.

## 3.12 Summary

In this chapter we have used classical statistical mechanics to describe various aspects of the theory of crystals containing point defects (vacancies, interstitial atoms, impurity atoms and atoms on 'wrong' sites); in §§3.2–3.10 mainly the chemical potentials of the components and the concentrations of defects present under thermal equilibrium conditions and lastly in §3.11 the jump rates of atoms and defects. For the most part we have been concerned with systems in which the densities of such defects are low, as is the case in very many solids. In §§3.2 and 3.3 we discussed four examples in which the defects were without mutual interactions. For these we could use well-known and straightforward combinatorial methods to determine the configurational entropy and hence the all-important Gibbs free energy function. In turn this allowed us to introduce certain necessary but generally less familiar aspects of the theory, namely the notion of chemical potentials for defects and, in the case of compounds, virtual chemical potentials; and to do so in the absence of any extraneous mathematical complications. It is important not to misunderstand the nature of the virtual chemical potentials and the way in which they can be used. As we have already indicated in Chapter 1, and as will be clear in the next chapter, chemical potentials lie at the base of transport theory.

We then turned to the matter of defect interactions, dealing first with systems where these could be adequately handled by distinguishing defect pairs or other small clusters among the total defect population. For such systems the familiar combinatorial methods are less straightforward, but, when the defect concentrations are low, a systematic general method is available for evaluating the chemical potentials to the first order in these fractions (§3.5). Such results will be needed for transport theory later (see e.g. Chapter 8), but examples were considered here in §3.8. Further aspects of defect interactions were treated in §3.9 (long-range Coulomb interactions) and §3.10 (concentrated solid solutions). In all cases the object has been to convey an understanding of chemical potentials in defect solids sufficient for the appreciation of the phenomenological theory which is presented in the following chapter.

Lastly, in §3.11 we reviewed the theoretical description of the elementary, thermally-activated steps by which atoms move through crystals (the statistical description of the consequences of many such steps being the subject of later chapters). The basic description is provided by transition-state theory, but, although recent work is beginning to define the quantitative limitations to this theory, much remains to be done. Partly for this reason we have kept our discussion rather broad. Fortunately for our subsequent purposes only a few rather general conclusions (which we have given at the end of §3.11) are required.

# Appendix 3.1

## Configurational entropy of a dilute defect system by the Mayer method

In the usual formulation of the Mayer theory for imperfect gases one expresses the configurational partition function in terms of integrals over the positions of all the atoms (assumed to be of various types $1, 2 \ldots r \ldots$). Within the integrals the individual atoms are regarded as distinguishable, while the actual indistinguishability is represented by the usual $N_r!$ factors in the denominator of the partition function, $Z$. The theory also assumes that the energy function of the entire gas is a sum of pairwise interactions $u$. The starting point is thus the equation for the partition function

$$Z = \left(\prod_r N_r!\right)^{-1} \int \cdots \int \prod_{l_r, m_s} (1 + f(l_r, m_s)) \prod_r \prod_{n=1}^{N_r} \mathrm{d}\{l_r\}, \qquad \text{(A3.1.1)}$$

in which individual atoms of type $r$ are labelled $1_r, 2_r, \ldots l_r \ldots$ and the product of factors $(1 + f)$ is taken only over distinct pairs. The $\mathrm{d}\{l_r\}$ is an element of volume in the space available to atom $l_r$ while the integral is an integration over all that space. The factor

$$1 + f(l_r, m_s) = \exp(-u(l_r, m_s)/kT). \qquad \text{(A3.1.2)}$$

Mayer theory then obtains an expansion of $\ln Z$ in powers of the concentrations of the atoms by expanding the product $\Pi\,(1 + f)$ within the integral.

These equations can be transcribed for the circumstances of defects of various types $1, 2, \ldots r \ldots$ distributed on a lattice. The integrals over the multi-dimensional configuration space become summations over all assignments of the defects to the available lattice sites, i.e. over all configurations $C(\ldots l_r \ldots m_r \ldots)$. However, the formal steps in the argument actually follow somewhat more smoothly if we retain the integral notation, so we shall do so. We draw attention to the fact that it is necessary to include only the defects in (A3.1.1), since the host atoms merely define the medium or background in which the defects are arranged.

Now, from the way we have defined the systems of interest (§3.4), there are no interactions between any pair of defects at any separation. (The short-range interactions between solutes and vacancies, for example, have been taken care of by the definition of solute–vacancy pairs as a distinct defect which is without any further interaction with unpaired solutes or unpaired vacancies). This has two immediate consequences. First, $Z$ is simply $\Omega$ and, second, the role of the factors $1 + f(l_r, m_s)$ is to take care of the site restrictions, e.g. the requirement that no two defects can simultaneously occupy the same site and that no unpaired solute atom can be placed next to an unpaired vacancy (for, by definition, it would then no longer be unpaired). They are therefore just 1 (i.e. $f = 0$) for each allowed assignment of the pair $(l_r, m_s)$ and 0 (i.e. $f = -1$) for every assignment which violates any of the site restrictions. We can again follow the procedure for imperfect gases and obtain $\Omega$ by expanding the product of $(1 + f)$ factors as

$$\prod_{l_r, m_s} (1 + f(l_r, m_s)) = 1 + \sum_{l_r, m_s} f(l_r, m_s) + \cdots. \tag{A3.1.3}$$

The right side as written is correct to first order in the fractional concentrations of defects (not structural elements). Higher terms (represented by ...) would correspond to higher powers in the defect fractional concentrations.

$$\Omega = \Omega_0 + \Omega_1 + \cdots \tag{A3.1.4}$$

The leading term $\Omega_0$ corresponds to the omission of all site exclusion restrictions. It is easily seen that

$$\Omega_0 = \prod_r (N_m z_r)^{N_r} \Big/ \prod_r (N_r!), \tag{A3.1.5}$$

in which

$$N_m z_r \equiv \int 1 \, d\{l_r\}. \tag{A3.1.6}$$

This integral is evidently dependent just on the type of defect ($r$) and not on which particular defect $l_r$ of this type is considered. Furthermore, it must be extensive, i.e. proportional to the total number of sites available to it and thus proportional to $N_m$. It follows that $z_r$ is an intensive quantity characteristic just of the defect and the lattice.

In a like manner we find that

$$\Omega_1 = \frac{\Omega_0}{2} \sum_{r,s} \frac{N_r N_s}{(N_m z_r)(N_m z_s)} \int\int f(l_r, m_s)\, \mathrm{d}\{l_r\}\, \mathrm{d}\{m_s\}. \tag{A3.1.7}$$

The integral is independent of which particular defects of type $r$ and $s$ are involved and, since the site restriction function $f$ is zero unless the sites occupied by $l_r$ and $m_s$ are close together, it is again proportional to $N_m$. We therefore define an intensive quantity,

$$B_{rs} = \frac{1}{z_r z_s N_m} \int\int f(l_r, m_s)\, \mathrm{d}\{l_r\}\, \mathrm{d}\{m_s\}. \tag{A3.1.8}$$

Finally then, by Stirling's approximation,

$$\begin{aligned} \ln \Omega &= \ln \Omega_0 + (\Omega_1/\Omega_0) + \cdots \\ &= -\sum_r N_r(\ln(N_r/N_m z_r) - 1) + \frac{1}{2}\sum_r \sum_s \left(\frac{N_r}{N_m}\right) N_s B_{rs} + \cdots \end{aligned} \tag{A3.1.9}$$

and the configurational entropy $S_{\text{con}}$ is given by

$$\frac{S_{\text{con}}}{N_m k} = -\sum_r c_r (\ln(c_r/z_r) - 1) + \frac{1}{2}\sum_r \sum_s c_r c_s B_{rs} + \cdots \tag{A3.1.10}$$

in terms of the defect site fractions $c_r$.

# 4

# Non-equilibrium thermodynamics of atomic transport processes in solids

## 4.1 Introduction

In Chapter 1 we briefly introduced the extended macroscopic equations of atomic transport that are provided by non-equilibrium thermodynamics. In this chapter and the next we expand this description, in this chapter in general terms and then in the next by a series of applications. The objectives are (i) to show how to use the description provided by non-equilibrium thermodynamics and (ii) to focus attention on the corresponding transport coefficients (denoted by $L_{ij}$) and the expressions of various practical transport coefficients in terms of them. By working in terms of these basic $L$-coefficients we obtain a better understanding of the relations among the various practical coefficients and we provide a more sharply defined objective for the statistical atomic theories to be developed later.

The subject of non-equilibrium thermodynamics is now supported by a considerable body of statistical mechanical theory. Since our interest in this book lies primarily with the use of this formalism and with the calculation of the quantities appearing in it, it is not appropriate to go into this body of general theory widely. For this purpose we refer the reader to the books by de Groot and Mazur (1962) and by Haase (1969) for wide-ranging treatments, especially of the use of the macroscopic formulation, and to Kreuzer (1981) for a more recent account which emphasizes the general statistical mechanical foundations of this formalism. Here, a less extensive review will be given of those aspects pertinent to our interest in atomic transport in solids.

As we have already pointed out in §1.4 the immediately obvious features of the formalism are threefold: (i) it is a linear, local theory, relating the responses of a system (fluxes, $J_i$) linearly to the perturbing forces $X_j$ at the same place and time, (ii) it is more general than the elementary macroscopic laws such as Fick's law for diffusion and Ohm's law for conduction and (iii) it possesses a limited predictive power through the Onsager reciprocity theorem, which says that *for properly defined* forces and fluxes (see §4.2 below) the matrix of coefficients in the linear phenomenological expressions linking them together is a symmetric matrix (with a simple generalization in the presence of magnetic fields and mechanical rotation). These statements are expressed mathematically by equations like (1.4.1) *et seq.* This formalism has been extensively discussed for fluids (e.g. Fitts, 1962; de Groot and Mazur, 1962 and Haase, 1969) and, after a slow start, is now being used increasingly to represent atomic transport phenomena in solids (see e.g. the following references for milestones along the way: Bardeen and Herring, 1952; Le Claire, 1953; Howard and Lidiard, 1964; Adda and Philibert, 1966; Flynn, 1972; Anthony, 1975; Fredericks, 1975; Allnatt and Lidiard, 1987b). In this chapter we shall be concerned with this macroscopic theory in general terms and in the next with its application to particular systems and circumstances. It should, however, be explicitly noted that the diffusion and other transport processes we are concerned with are mostly sufficiently slow that the fluxes $J_i$ appearing in (1.4.2) at time $t$ depend only on the forces $X_j$ at the same time and not on their values at earlier times. This will be assumed throughout this and the next chapter. Particular examples where this assumption is not adequate (e.g. dielectric and mechanical relaxation phenomena, §§1.5 and 1.6) are described in later chapters.

We shall begin by summarizing the usual development (§§4.2 and 4.3). This often has fluids in mind; but, broadly speaking, the formalism is also the same for solids, the most obvious difference, apart from the lowered symmetry, being that, whereas for fluids it is natural to use the local centre of mass as the reference frame for diffusive flows, in crystalline solids it is more sensible to use the local crystal lattice as reference frame. The consequences of this and other differences are indicated in §4.4. Some specific relations between more usual phenomenological coefficients (e.g. diffusion coefficients, mobilities, etc.) and the quantities $L_{ij}$ introduced here are then used to guide the further development in Chapter 5.

### 4.2 Definition of fluxes and forces and Onsager's theorem

Since the characteristic feature of an irreversible process is the generation of entropy the calculation of the rate of entropy production is basic to the subject. Such calculations are based on known mechanical laws and the assumptions that the departures from thermodynamic equilibrium are sufficiently small for local values of the usual thermodynamic quantities (e.g. temperature, entropy, internal energy, etc.) to be definable and for the usual thermodynamic equations to hold between these local values.

The usual calculation of the rate of entropy production in *fluids* subject to thermodynamic forces (e.g. concentration gradients, potential gradients) thus assumes the various continuity equations for matter density, momentum density and energy density together with the thermodynamic equation expressing changes in entropy (per unit mass) in terms of corresponding changes in internal energy, volume and concentration. In this way a balance equation for entropy can be derived (see e.g. Kreuzer, 1981, chap. 2) and the rate of entropy production identified. One obtains thereby the rate of entropy production per unit volume, $\sigma$, in the form

$$T\sigma = \sum_{k}^{n} \mathbf{J}_k \cdot \mathbf{X}_k + \mathbf{J}_q \cdot \mathbf{X}_q + \text{viscosity terms.} \tag{4.2.1}$$

The various species of the system (atoms and molecules of various kinds) are here labelled by the index $k$ ($=1$ to $n$) and the flow vector $\mathbf{J}_k$ gives the number of atoms (or molecules, as appropriate) of species $k$ crossing a unit area, fixed relative to the local centre of mass, in unit time. The flow of heat vector $\mathbf{J}_q$ is also referred to the local centre of mass, i.e. $-\nabla \cdot \mathbf{J}_q$ gives the rate of addition of heat to a unit volume element moving with the local centre of mass. The thermodynamic vector forces $\mathbf{X}_k$ and $\mathbf{X}_q$ are measures of the imbalances generating these flows. If $\mathbf{F}_k$ is the external force per atom of $k$ and $\mu_k$ is the chemical potential of $k$, i.e. the partial derivative of the Gibbs free energy with respect to the number of atoms of $k$,

$$\mu_k = \left(\frac{\partial G}{\partial N_k}\right)_{T,P,N_{j \neq k}}, \tag{4.2.2}$$

then

$$\mathbf{X}_k = \mathbf{F}_k - T\nabla\left(\frac{\mu_k}{T}\right). \tag{4.2.3a}$$

(N.B. As in Chapter 3, we here follow the definition of chemical potential common in statistical mechanics, so that the derivative in (4.2.2) is taken with respect to the number of atoms (molecules) $N_k$, not with respect to the number of moles. This definition applies equally to fluids and to solids). The force $\mathbf{X}_q$ depends only on the temperature gradient

$$\mathbf{X}_q = -\frac{1}{T}\nabla T. \tag{4.2.3b}$$

We have not explicitly represented the viscosity terms since, although they may be significant for fluid phenomena, they are not relevant in the solids we are concerned with in this book. We may note that the occurrence of centre-of-mass flows $\mathbf{J}_k$ arises naturally in the demonstration of (4.2.1) from a separation of the work done by external forces into one part which increases the thermodynamic internal energy and another which increases the kinetic energy of the total mass; this second part is not of thermodynamic significance (unless or until it is degraded by viscous forces).

The importance of the derivation of equations (4.2.1)–(4.2.3) for transport theory is contained in the fact that linear laws of transport when re-expressed as relations between the $J$s and the $X$s are governed by a theorem known as the Onsager reciprocity theorem. These linear macroscopic laws are generalized so that they also include cross-phenomena, such as the influence of a gradient in concentration (activity) of one species upon the flow of another or the effect of a temperature gradient upon the flow of the various material species, both of which can be significant. They can then be written

$$J_k = \sum_i^n L_{ki} X_i + L_{kq} X_q \tag{4.2.4a}$$

$$J_q = \sum_i^n L_{qi} X_i + L_{qq} X_q, \tag{4.2.4b}$$

where for simplicity we drop the vector notation (the summations where necessary are to be over all Cartesian components as well as over all species). Again we omit viscous forces. The physical significance of these $L$-coefficients will emerge more fully as we proceed, but we can see immediately that each diagonal coefficient $L_{ii}$, since it expresses the contribution to $J_i$ per unit of force $X_i$, has the nature of the product of a concentration, $n_i$, and a mobility $M_i$. (It may be noted that in place of (4.2.4) some authors use equations in which a factor $T^{-1}$ is introduced explicitly on the r.h.s.: this merely changes the definition of the $L$-coefficients.)

The central theorem is that, in the absence of magnetic fields and of rotations, the matrix of phenomenological coefficients $\mathbf{L}$ is symmetric

$$L_{ik} = L_{ki}, \tag{4.2.5}$$

with like relations for $L_{iq}$. This is known as Onsager's theorem. Onsager's original demonstration of the relations, which is still often reproduced in the literature, was based on a discussion of the fluctuations $a_i$ in the thermodynamic parameters (e.g. local pressures, concentrations, etc.) of an adiabatically isolated system in equilibrium. For such a system the fluctuation in the entropy can be written in terms of the $a_i$s. The assumption was made that the regression of a fluctuation in this system followed the same laws as those which govern the return of a non-equilibrium system towards equilibrium. Use can be made of this assumption when the 'flows' $J_i$ which specify such a process are each the time derivative of an appropriate variable $a_i$, and if the 'forces' $X_i$ are suitable linear combinations of the $a_i$. This combination is that which makes the rate of entropy production during the regression of a fluctuation of the form (4.2.1). Then Onsager was able to show from the principle of time-reversal symmetry in mechanics that (4.2.5) followed. This demonstration is valid for those systems for which it can be shown that the $J$s and $X$s in question may be defined in this way in terms of the state variables $a_i$.

The original demonstration as it stands therefore does not apply to a continuous system in which the thermodynamic state variables are changing continuously with position. For then, the flows $J_i$ are vector quantities which are not representable as the time derivatives of scalar thermodynamic state variables, as the Onsager derivation requires. The demonstration can, however, be adapted by mentally 'discretizing' the system (Casimir, 1945; Flynn, 1972) or by related mathematical procedures (Sekerka and Mullins, 1980) so leading to the same result. The more usual demonstration these days, however, is via the fluctuation–dissipation hypothesis and the Kubo representation of transport coefficients in terms of dynamic correlation functions (Forster, 1975; de Groot and Mazur, 1962; Kreuzer, 1981). Unfortunately, such demonstrations are mathematically rather involved and lengthy, and it would not be appropriate to reproduce them here in their full generality. We can, however, be confident about using (4.2.5). For the solid systems which are the principal interest of this book we shall, in fact, verify (4.2.5) in the course of the derivation of statistical–mechanical expressions for the $L$-coefficients which is given in Chapter 6.

In connection with the use of this formalism in practice we insert the following remarks.

(i) Strictly speaking, each index $i$ in (4.2.4) and (4.2.5) is a compound of two indices, the first of which denotes a chemical species, while the second denotes a Cartesian component. For simplicity we shall throughout the text speak as though the index denoted just the chemical species in question, and take the extensions to include Cartesian compounds for granted. For simple fluids, which are necessarily isotropic, this simplification cannot lead to error, but in solids we have to recognize that there will generally be more than one independent Cartesian component to each $L_{ij}$ tensor, the actual number being determined by the crystal symmetry.

(ii) When eqns. (4.2.4) are substituted into (4.2.1) for the rate of entropy production we obtain a homogeneous quadratic function of the forces. By the second law of thermodynamics this function must be greater than or equal to zero (semi-positive definite), from which the following inequalities follow:

$$L_{ii} \geq 0 \qquad \text{(all } i\text{)} \tag{4.2.6a}$$

$$L_{qq} \geq 0 \tag{4.2.6b}$$

$$L_{ii}L_{kk} - L_{ik}L_{ki} \geq 0 \tag{4.2.7a}$$

$$L_{ii}L_{qq} - L_{iq}L_{qi} \geq 0 \tag{4.2.7b}$$

(all $i, k$). Such relations evidently provide a check on particular statistical theories of these coefficients, as we shall see later in particular cases.

(iii) We emphasize that these phenomenological coefficients $L_{ik}$ are kinetic quantities, which in general will be functions of the usual thermodynamic variables but which are independent of the forces $X_k$, etc. to which the system may be subjected.

(iv) It should be recognized that some formal subtleties arise in cases where there are linear relations among the fluxes (and among the forces); we return to this matter below in §4.3.

### *4.2.1 Alternative fluxes and forces*

There is an important degree of flexibility in the formalism which must be mentioned and that concerns alternative choices of fluxes and forces to represent the same phenomena. Let us assume that a set of forces and fluxes has been found such that the entropy production is given by (4.2.1) and that the Onsager relations are satisfied, i.e. that the $L$ matrix is symmetric,

$$\mathbf{L} = \tilde{\mathbf{L}}. \tag{4.2.8}$$

For some reason we may wish to choose new fluxes $J_i'$ which are linearly related to the original $J_i$. Let us arrange the fluxes into a column matrix so that this relation can be expressed in matrix form as

$$\mathbf{J}' = \mathbf{AJ}. \tag{4.2.9a}$$

Assume that $\mathbf{A}$ is non-singular, let us now introduce new forces $X_i'$ by the matrix equation

$$\mathbf{X}' = \tilde{\mathbf{A}}^{-1}\mathbf{X}. \tag{4.2.9b}$$

Then the rate of entropy production $\sigma$ is given by

$$T\sigma = \tilde{\mathbf{J}}\mathbf{X} = \tilde{\mathbf{J}}'\tilde{\mathbf{X}}', \tag{4.2.10}$$

and is invariant under the transformation (4.2.9). Furthermore, since the phenomenological equations (4.2.4) can be written as

$$\mathbf{J} = \mathbf{LX} \tag{4.2.11}$$

it follows that

$$\mathbf{J}' = \mathbf{AJ} = \mathbf{ALX} = \mathbf{AL}\tilde{\mathbf{A}}\mathbf{X}' \equiv \mathbf{L}'\mathbf{X}', \tag{4.2.12}$$

and if **L** is symmetric then so is **L**′. Thus, if the Onsager relations are valid for the original fluxes and forces they are also valid for the transformed quantities. Furthermore, it is clear that if the rate of entropy production, $\sigma$, is to remain invariant only transformations such as (4.2.9) are admissible in a linear theory. Lastly, we note that a transformation such as (4.2.12) will in general mean that the off-diagonal components of **L**′ will involve the diagonal components of **L**, and vice versa (in both senses).

A commonly used transformation of the type (4.2.9) which is significant when temperature gradients exist is

$$J'_k = J_k, \tag{4.2.13a}$$

$$J'_q = J_q - \sum_k^n h_k J_k, \tag{4.2.13b}$$

with

$$X'_k = X_k + h_k X_q, \tag{4.2.13c}$$

$$X'_q = X_q, \tag{4.2.13d}$$

in which $h_k$ is the partial enthalpy of component $k$. The flow $J'_q$ is called the *reduced* heat flow*: as (4.2.13b) shows, to obtain it we subtract from the total heat flow $J_q$ that part associated with the movement of material species. Explicitly, the new forces are

$$X'_k = F_k - (\nabla\mu_k)_T, \tag{4.2.14a}$$

$$X'_q = -\frac{1}{T}\nabla T, \tag{4.2.14b}$$

in which $(\nabla\mu_k)_T$ is that part of the gradient of $\mu_k$ due to gradients in pressure or concentration, but not temperature.

Various other general transformations of this kind were considered by de Groot (1952) and de Groot and Mazur (1962). A particular example which arises naturally in *solids* (see below) is a transformation from fluxes confined only to chemical species to a set which includes lattice defects as a species and eliminates the host or solvent species; see §4.4 below.

### *4.2.2 Heats of transport and effective charge parameters*

To conclude this brief introduction to non-equilibrium thermodynamics we introduce the definitions of heat of transport and of an analogous effective electric

* N.B. The specific meaning of the word 'reduced' in this connection is different from the sense in which it is used in §§3.6 and 4.4.2.

charge. The heat of transport is needed to represent the Soret effect and other phenomena arising when a temperature gradient exists. The formal definition of the heats of transport $Q_k^*$ is via the coefficients $L_{kq}$ appearing in (4.2.4), i.e.

$$L_{kq} = \sum_{i=1}^{n} L_{ki} Q_i^* . \tag{4.2.15}$$

Substitution of (4.2.15) into (4.2.4a) yields directly

$$J_k = \sum_{i=1}^{n} L_{ki}(X_i + Q_i^* X_q), \tag{4.2.16}$$

from which we see that $Q_i^* X_q$ specifies the thermal force contributing to the flux of species $i$. On the other hand, substitution into (4.2.4b) and use of the Onsager relation (4.2.5) gives

$$J_q = \sum_{i=1}^{n} Q_i^* J_i + \left( L_{qq} - \sum_{i=1}^{n} L_{iq} Q_i^* \right) X_q . \tag{4.2.17}$$

Thus in an isothermal system (where $X_q = 0$) $Q_i^*$ represents the heat flow consequent upon unit flux of component $i$. Whereas the experimental determination of $Q_i^*$ proceeds via (4.2.16), it is this last result which provides the more convenient approach to their theoretical calculation (e.g. by molecular dynamics) because the statistical mechanics of isothermal systems is far more tractable (Gillan, 1983).

Slightly different heats of transport, $Q_k^{*\prime}$, may also be defined in the manner of (4.2.15) when the phenomenological relations are expressed in terms of the $J'$ and $X'$ specified by equations (4.2.13). Equations of the same form as (4.2.16) and (4.2.17) again follow. The relation between the two kinds of heat of transport is

$$Q_k^{*\prime} = Q_k^* - h_k, \tag{4.2.18}$$

corresponding to the subtraction of $h_k J_k$ from $J_q$ in (4.2.13b) to give the reduced heat flow. The quantities $Q_k^{*\prime}$ are referred to as reduced heats of transport.

The heat flow in the above phenomenological equations is associated at a microscopic level with the thermal motion of the atoms and, in the case of metals and semiconductors, also with the conduction electrons. Thus for insulators the species $i$ will be just the chemically distinct species, while in electric conductors we may need also to include the conduction electron gas as one of them (denoted by $e$, say). Explicit allowance for the interplay of electron currents and atomic currents is particularly necessary in the consideration of the isothermal electromigration of atoms in metals (see e.g. Pratt and Sellors, 1973; Wever, 1973 and specifically for solids Huntington, 1975). The basic equations are still of the form

(4.2.4) but with $q$ now replaced by $e$. We may follow Doan (1971, 1972) and define quantities $z_i^*$ analogous to the $Q_i^*$ by

$$L_{ie} = \sum_{k=1}^{n} L_{ik} z_k^*. \tag{4.2.19}$$

Equations like (4.2.16) and (4.2.17) again follow:

$$J_k = \sum_{i=1}^{n} L_{ki}(X_i + z_i^* X_e) \tag{4.2.20}$$

$$J_e = \sum_{i=1}^{n} z_i^* J_i + \left(L_{ee} - \sum_{i=1}^{n} L_{ie} z_i^*\right) X_e. \tag{4.2.21}$$

From the first of these we see that a force $X_e$ acting on the electron gas implies a force $z_i^* X_e$ acting on the atoms of species $i$. The mechanism is the scattering of the electrons off the ion cores with consequent momentum transfer, sometimes called the *electron wind effect*. Furthermore, we see that if the only primary force is an external electric field, $E$, ($X_i = ez_iE$, $X_e = -eE$ where $ez_i$ is the ion core charge of atoms of type $i$) then $(z_i - z_i^*)e$ is the effective atomic charge for drift in this field. This difference $z_i - z_i^*$ is sometimes denoted by $Z_i^*$ (Doan, 1971, 1972; Huntington, 1975). From the second of these relations, we also see that $z_i^*$ would give the electron flux consequent upon unit flux of component $i$ when there is no direct force on the electrons ($X_e = 0$).

## 4.3 Linearly dependent fluxes and forces

It will be noted that, for the preceding definitions of heats of transport (4.2.15) and effective charges (4.2.19) to be unambiguous and to lead to unique quantities $Q_k^*$ and $z_k^*$, the matrix $\mathbf{L}$ of the phenomenological coefficients $L_{ik}$ must have an inverse. Given that the forces $X_k$ are linearly independent quantities this requires the $J_i$ also to be linearly independent; if they are not, the most straightforward procedure is to eliminate the dependent fluxes and then to employ the formalism (4.2.15) *et seq.* and (4.2.19) *et seq.* for the remaining independent quantities. A common example arises in the case of fluids where it is natural to refer the fluxes of all the component species to the local centre of mass. A linear relation between the fluxes then necessarily exists in the form

$$\sum_{j}^{n} m_j J_j = 0, \tag{4.3.1}$$

in which $m_j$ is the molecular weight of component $j$. If one is only interested in the diffusion of the solute species relative to the solvent, as will often be the case, it will then be convenient to eliminate the solvent species ($n$, say) from the set and to apply the formalism of §4.2 to the solute species. A similar situation arises in solids, as we consider in §4.4. In the past some authors have attempted in these circumstances to retain the full set of species but to introduce a corresponding linear relation among the heat of transport parameters. It must be said, however, that any such relation must be arbitrary; since only $n-1$ independent heats of transport can be defined from the phenomenological transport equations the choice of an $n$th heat of transport is wholly arbitrary.

The remarks so far have concerned the formal consequences of the existence of a linear relation among the fluxes when the forces are independent. However, in a system in mechanical equilibrium there is also a restriction on the forces as expressed by Prigogine's theorem (essentially a re-expression of the Gibbs–Duhem relation) viz.

$$\sum_{k=1}^{n} N_k X'_k = 0, \tag{4.3.2}$$

in which $N_k$ is the number of atoms of species $k$. The appropriateness of this arises because in diffusion problems generally one may assume that the adjustments to unbalanced mechanical forces are rapid compared to the adjustments to activity and temperature gradients. The situation where there are two restrictions, one like (4.3.1) and the other (4.3.2), has given rise to a certain amount of formal discussion in the literature (see e.g. de Groot and Mazur, 1962; Fitts 1962 especially section 4.7) the substance of which is often somewhat elusive, but which centres around the validity and significance of the Onsager relations (4.2.5) and on the ability to define the $Q^*$ and $z^*$ parameters under these conditions. Some of the points arising were reviewed by Howard and Lidiard (1964, §4.1.4), but for present purposes it will be sufficient to distinguish between relations of the type (4.3.1) which are a necessary and unavoidable consequence of the way the fluxes are defined and (4.3.2) which represents a condition the system may or may not be in. Since the whole basis on which the linear macroscopic relations are postulated is that $L$-coefficients are independent of the forces acting on the system, it follows that the existence of a relation such as (4.3.2) cannot lead to additional restrictions on the $L$-coefficients themselves. The statistical mechanical theory of these $L$-coefficients confirms this by showing them to be expressible as statistical averages of kinetic and dynamical quantities of the system in thermodynamic equilibrium, i.e. in the absence of any forces $X_k$.

## 4.4 Application to solids

As we have already indicated, the preceding formalism is for the most part introduced and justified in the context of fluids (de Groot and Mazur, 1962;

Kreuzer, 1981). However, it can be taken over without fundamental change to crystalline solids (as illustrated in the texts by Adda and Philibert, 1966; Flynn, 1972; Anthony, 1975; Fredericks, 1975 and Philibert, 1985). Five particular changes should, however, be noted, namely: (i) the local crystal lattice (rather than the local centre of mass) is the natural reference frame to employ in the definition of the various fluxes, $J_i$; (ii) this crystal lattice defines the geometrical symmetry of the system and this will always be lower than that for normal isotropic fluids; (iii) for systems showing small degrees of lattice disorder it is desirable to give explicit recognition of the role of vacancy and interstitial defects in causing atomic transport; (iv) for compound substances there will be structural constraints which have the effect of coupling certain defect populations together (cf. Chapter 3) and (v) a solid may sustain a general stress, $\sigma_{\alpha\beta}$.

The first point seems hardly to merit discussion since in crystalline solids the description of these migration processes at the atomic level necessarily involves the consideration of atomic movements relative to the local crystal lattice. Statistical theories of the sort we shall describe in later chapters certainly look at the processes in this way. It would therefore seem somewhat perverse to choose any other reference frame, although, of course, if we start with such a description in a lattice frame there is nothing to prevent us adding terms to allow for any inhomogeneous bulk movements of that lattice (cf. Howard and Lidiard, 1964, §4.4).

The second point has been touched on already in §1.4. In most extensively studied crystalline solids the symmetry is sufficient to allow the definition of principal axes relative to which the $L$-coefficient tensors become diagonal (in the Cartesian components). When we come to calculate these $L$-coefficients by statistical theory it will therefore be convenient to consider forces and gradients along a principal axis so that the problem becomes one-dimensional. Instead of a three-dimensional problem we therefore have to solve at most three one-dimensional problems to find the independent components. In the case of cubic solids, which we shall be much concerned with, there is only one independent component so that the $L$ tensor is isotropic and we can choose any axis for the one-dimensional analysis. In what follows therefore, we shall continue to write the fluxes, forces and transport coefficients as scalar quantities, but it should always be remembered that this means that we are considering one-dimensional flows which, except in cubic crystals, are along a principal axis.

The remaining three points require some formal elaboration and are therefore the subject of the following sections.

### 4.4.1 *Representation of vacancy and interstitial defects*

In this section we consider certain formal aspects of including the point defects as distinct species in the theory. Their representation in the statistical thermodynamic theory of chemical potentials and other thermodynamic functions was considered in Chapter 3. Here we look at the consequences for the forces, $X_k$, and the flux equations.

First suppose that we have a system of $n - 1$ atomic (or molecular) components ($i = 1 \ldots n - 1$) and simple vacancies (regarded as the $n$th thermodynamically independent component V) *on a single lattice*. One example would be a disordered alloy; others would be solid solutions of the noble gases or molecular solids. Then necessarily by the definition of $J_n$

$$\sum_{i=1}^{n} J_i = 0, \tag{4.4.1}$$

for fluxes defined relative to that lattice (in regions not containing sources and sinks of vacancies). This equation simply expresses the fact that every time an atom moves in one direction it does so by hopping to a vacant site, which thus moves in the opposite direction. This system thus provides another example where the chosen fluxes are linearly dependent, as already touched upon in §4.3.

Let us further suppose that the system is subject to a temperature gradient and, in the case of metallic and semiconducting alloys, also an electric field causing a conduction electron current $J_e$, as well as other forces contained in $X_i$ and $X_V$. Then by following the argument given originally by de Groot (1952, §46, also de Groot and Mazur, 1962, chap. VI), which assumes linearly independent forces, we obtain

$$\sum_{k=1}^{n} L_{ki} = 0, \qquad \sum_{k=1}^{n} L_{kq} = 0, \qquad \sum_{k=1}^{n} L_{ke} = 0 \tag{4.4.2a}$$

and

$$\sum_{i=1}^{n} L_{ki} = 0, \qquad \sum_{i=1}^{n} L_{qi} = 0, \qquad \sum_{i=1}^{n} L_{ei} = 0. \tag{4.4.2b}$$

We emphasize that these relations are the direct consequence of the restriction (4.4.1) on the fluxes of an $n$-component system. The basic flux equations (4.2.4 with additional representation of the electron terms) on elimination of $L_{ni}$ and $L_{kn}$

then become

$$J_k = \sum_{i=1}^{n-1} L_{ki}(X_i - X_V) + L_{kq}X_q + L_{ke}X_e \qquad (4.4.3a)$$

$$J_q = \sum_{i=1}^{n-1} L_{qi}(X_i - X_V) + L_{qq}X_q + L_{qe}X_e \qquad (4.4.3b)$$

$$J_e = \sum_{i=1}^{n-1} L_{ei}(X_i - X_V) + L_{eq}X_q + L_{ee}X_e. \qquad (4.4.3c)$$

Heats of transport, $Q_k^*$, and electron wind coupling constants, $z_k^*$, are then defined by

$$L_{iq} = \sum_{k=1}^{n-1} L_{ik}Q_k^*, \qquad (4.4.4a)$$

and

$$L_{ie} = \sum_{k=1}^{n-1} L_{ik}z_k^*. \qquad (4.4.4b)$$

Thus in place of (4.2.4) we have

$$J_k = \sum_{i=1}^{n-1} L_{ki}(X_i - X_V + Q_i^*X_q + z_i^*X_e), \qquad \text{(vacancies)}. \qquad (4.4.5)$$

Next suppose that instead of vacancy disorder we have interstitial disorder. We shall again suppose that the substitutional atoms occupy a single lattice; the interstitial defects may occupy a distinct lattice of interstitial sites or they may adopt dumb-bell configurations around the normal lattice sites (cf. Fig. 2.2). If we do not differentiate among the various kinds of interstitial possible in a multicomponent system (which, of course, implies that the various kinds are in local equilibrium with one another) and now let component $n$ refer to interstitials I generally then, on using the constraining equation

$$J_I = \sum_{k=1}^{n-1} J_k, \qquad (4.4.6)$$

(which expresses the assumption that atoms can only move by the action of interstitials) we shall similarly find in place of (4.4.5)

$$J_k = \sum_{i=1}^{n-1} L_{ki}(X_i + X_I + Q_i^*X_q + z_i^*X_e), \qquad \text{(interstitials)}. \qquad (4.4.7)$$

We shall have occasion to call on these two equations (4.4.5) and (4.4.7) at various times in the course of this book.

We note that if we had attempted to define heats of transport and electron wind coupling constants before eliminating the defect fluxes we should have encountered the difficulty that **L** was singular. As already indicated in §4.3 we have avoided the arbitrariness implicit in (4.2.15) by following de Groot's procedure and eliminating one of the dependent species first (*n* in this example).

However, it will be seen that the elimination of $J_V$ and $J_I$ is arbitrary and that we could have eliminated one of the other components instead. Indeed in dilute systems it often makes more sense to eliminate the major or 'host' element and thus to have explicit flux equations for the defects and all the minority or solute elements. We shall in fact, switch back and forth between these two descriptions as most convenient. Such a change can also be considered formally as a transformation of the type described by (4.2.9)–(4.2.12).

We indicate the nature of these changes by considering the example of a binary alloy (A, B) containing vacancies V. If we eliminate the vacancies then we have a *chemical* description in terms of $J_A$ and $J_B$. However, if the alloy is a dilute solution of B in A then a more natural *defect* description would be in terms of $J_B$ and $J_V$ with

$$J_V = -J_A - J_B. \tag{4.4.8}$$

In the chemical description we should have, in the absence of thermal gradients and electric fields,

$$J_A = L_{AA}(X_A - X_V) + L_{AB}(X_B - X_V), \tag{4.4.9a}$$

$$J_B = L_{BA}(X_A - X_V) + L_{BB}(X_B - X_V), \tag{4.4.9b}$$

by (4.4.5). The corresponding defect description is found from (4.2.9). Thus, by regarding the chemical fluxes as defining **J** in (4.2.9a) and the defect fluxes as defining **J**′, i.e.

$$\mathbf{J} = \begin{bmatrix} J_A \\ J_B \end{bmatrix}, \tag{4.4.10a}$$

$$\mathbf{J}' = \begin{bmatrix} J_V \\ J_B \end{bmatrix}, \tag{4.4.10b}$$

we see that the transformation matrix connecting them is

$$\mathbf{A} = \begin{bmatrix} -1 & -1 \\ 0 & 1 \end{bmatrix}. \tag{4.4.11}$$

Then by (4.2.9b) and (4.4.9) we have

$$\mathbf{X}' = \tilde{\mathbf{A}}^{-1}\mathbf{X} = \begin{bmatrix} -1 & 0 \\ -1 & 1 \end{bmatrix} \begin{bmatrix} X_A - X_V \\ X_B - X_V \end{bmatrix}$$

$$= \begin{bmatrix} X_V - X_A \\ X_B - X_A \end{bmatrix} \tag{4.4.12}$$

and by (4.2.12)

$$\mathbf{L}' = \begin{bmatrix} -1 & -1 \\ 0 & 1 \end{bmatrix} \begin{bmatrix} L_{AA} & L_{AB} \\ L_{BA} & L_{BB} \end{bmatrix} \begin{bmatrix} -1 & 0 \\ -1 & 1 \end{bmatrix}$$

$$= \begin{bmatrix} L_{AA} + L_{AB} + L_{BA} + L_{BB} & -(L_{AB} + L_{BB}) \\ -(L_{BA} + L_{BB}) & L_{BB} \end{bmatrix}$$

$$\equiv \begin{bmatrix} L_{VV} & L_{VB} \\ L_{BV} & L_{BB} \end{bmatrix}. \tag{4.4.13}$$

Thus the linear transport equations in the defect description corresponding to (4.4.9) are

$$J_V = L_{VV}(X_V - X_A) + L_{VB}(X_B - X_A) \tag{4.4.14a}$$

$$J_B = L_{BV}(X_V - X_A) + L_{BB}(X_B - X_A), \tag{4.4.14b}$$

where the coefficients $L_{VV}$, etc. are given in terms of the $L_{AA}$, etc. by (4.4.13).

With this defect description, parameters like electric charges and heats of transport must also be transformed. We show this for the heats of transport, but the calculation follows the same sequence for all. In the above example we have $Q_A^*$ and $Q_B^*$ in the chemical description. If we form these into a column matrix $\mathbf{Q}^*$ we evidently have

$$\mathbf{J} = \mathbf{L}(\mathbf{X} + \mathbf{Q}^* X_q) \tag{4.4.15}$$

for (4.4.5) and thus for the defect description

$$\mathbf{J}' = \mathbf{A}\mathbf{L}(\mathbf{X} + \mathbf{Q}^* X_q)$$

$$= \mathbf{A}\mathbf{L}\tilde{\mathbf{A}}\tilde{\mathbf{A}}^{-1}\mathbf{X} + \mathbf{A}\mathbf{L}\tilde{\mathbf{A}}\tilde{\mathbf{A}}^{-1}\mathbf{Q}^* X_q$$

$$\equiv \mathbf{L}'\mathbf{X}' + \mathbf{L}'\mathbf{Q}'^* X_q, \tag{4.4.16}$$

i.e. the heats of transport parameters in the defect description are

$$\mathbf{Q}'^* = \tilde{\mathbf{A}}^{-1}\mathbf{Q}^* \tag{4.4.17}$$

or explicitly

$$Q_V'^* = -Q_A^* \tag{4.4.18a}$$

$$Q_B'^* = Q_B^* - Q_A^*. \tag{4.4.18b}$$

Corresponding results will, of course, be obtained for the electron wind coefficients $z^*$. These results might have been anticipated since in the defect description we are dealing with effective quantities, i.e. with quantities relative to a perfect background provided by the host atoms A. The idea is most obvious in an ionic system in which the ions carry actual charges $q_A$ and $q_B$. The effective charges in the defect description are then evidently $-q_A$ for the vacancy and $q_B - q_A$ for a B ion which occupies an A-site, in agreement with one's intuition.

So far in this section we have considered the fluxes and the flux equations, but we have said nothing explicitly about the forces as they arise in solids. As would be expected, however, the chemical potentials that give the forces $X_i$, $X_V$ and $X_I$ in (4.4.5) and (4.4.7) are defined in the usual way by (4.2.2) but with the defects (vacancies or interstitials) as components of the system on the same footing as the material components. In particular applications one will generally obtain these quantities by means of statistical thermodynamics, as already described in Chapter 3. This approach is a natural one to take since the theory then remains in the same form as that commonly used for the thermodynamics of fluids, mixtures, alloys, etc.

However, certain alternatives offer themselves. For example, for a multi-component system containing vacancies ($n - 1$ material components plus vacancies) it is possible, since (4.4.5) involves a summation over the material components alone, to work instead in terms of different chemical potentials $\mu'_k$ obtained by the differentiation of $G$ at constant total number of lattice sites (as well as constant numbers of the material components $j \neq k$). In this case one would then find that the difference $X_i - X_V$ appearing in (4.4.5) is replaced by $X'_i$ as given by (4.2.3a) with $\mu'_i$ in place of our $\mu_i$. This representation is used for example by Cauvin (1981), Cauvin and Martin (1981) and Martin *et al.* (1983) in their discussion of the stability of alloys under irradiation. A related restriction to constant number of lattice sites is contained in the work of Larché and Cahn (1973, 1978a, 1978b and 1982). Such a scheme has its analogue in the theory of fluid mixtures when it is desired to express composition in terms of mole fractions (which necessarily add up to unity) rather than in terms of moles or numbers of atoms. It is important to recognize that these other approaches represent alternative choices that in the end lead to results equivalent to those presented here (see e.g. Cahn and Larché, 1983) but which should not be confused with that emphasized here.

In this section we have supposed that the atoms occupy a single lattice. Such an assumption is obviously appropriate for disordered alloys and solid solutions of monatomic substances. However, in compounds, including intermetallic compounds, the atoms of a particular species may be confined to a particular sub-lattice which is not available to atoms of other species. For example, in strongly ionic compounds there will be distinct anion and cation sub-lattices. The energy required to place a cation on an anion sub-lattice site (and vice versa) is so high that we can say that such transfers practically never occur in these materials (although they do occur in the less ionic III–V semiconducting compounds). These considerations have been incorporated in the statistical thermodynamical calculations of Chapter 3 and should obviously also be introduced into the description provided by the flux equations. Mostly, it will be physically obvious how this should be done and the formal changes required will be simple. For example, for a strongly ionic compound the equations for the flow of cations will refer to the cation sub-lattice and the vacancy term there will refer solely to the cation vacancies; likewise for the flow of anions. The vacancy populations in the two sub-lattices may, however, be coupled together through the equation expressing Schottky defect equilibrium (3.3.15a and 3.3.39a). To use the flow equations of non-equilibrium thermodynamics to describe processes taking place in compounds, the task is therefore primarily one of setting out the correct physical interpretation rather than of formal elaboration of the equations. We provide examples to illustrate this in the next chapter. We now return to another formal development which applies equally to all solids.

### *4.4.2 Allowance for internal equilibria*

When employing theoretical chemical potentials it is important to be clear which species are taken to be independent and also what internal (or local) equilibria are assumed. One example would be provided by a dilute alloy containing vacancies which may be attracted to the solute atoms. In the statistical thermodynamics we shall then want to distinguish isolated vacancies and isolated solute atoms from close solute–vacancy pairs. However, this pairing reaction will come to equilibrium locally very quickly, since relatively few vacancy jumps are required to make it so. From the point of view of transport over macroscopic distances it is clear that we can take the pairing reaction to be locally in equilibrium, which is to say that in the flux equations we do not need to distinguish isolated vacancies and solute atoms from pairs of such defects. Other examples are provided by systems which display Frenkel disorder (e.g. AgCl or $CaF_2$) where both interstitial and vacancy defects are present, but again in local equilibrium with one another. Metals under steady state irradiation may also provide an approximation to Frenkel disorder, although the 'local equilibrium' in this case is a dynamical steady state rather than a thermodynamic equilibrium.

The apparatus for dealing with these cases and 'reducing' the description from the full description necessary for the statistical thermodynamics to that adequate for the transport theory in the presence of local equilibrium has already been introduced in Chapter 3 (§3.6). We now use this to set out the formal expression of this reduction process, using the same notation.

We start from a full specification of the system as required for the statistical thermodynamics, i.e. with $n$ distinct species. We then proceed to eliminate all $n - g$ *secondary* defects which may be taken to be in local equilibrium from the macroscopic flux equations. That we can do so is a consequence of the similar mathematical structure of the equation of entropy production (4.2.1) and of the Gibbs equations for $G$ in terms of the chemical potentials (3.5.12).

By (4.2.1) the entropy production from matter transport alone can be expressed in terms of the same $n$ species initially employed to write the Gibbs function in §3.6

$$T\sigma = \sum_{i=1}^{n} J_i X_i = \tilde{\mathbf{X}}'\mathbf{J}' + \tilde{\mathbf{X}}''\mathbf{J}'', \tag{4.4.19}$$

where the prime and double prime have the significance defined there. New fluxes $J_i^{\mathrm{R}}$ (R = reduced) corresponding to the species whose numbers are given by $M_i$ can be defined in terms of the same matrix $\mathbf{A}$ as appears in (3.6.4),

$$J_i^{\mathrm{R}} = \sum_{j=1}^{n} A_{ij} J_j, \qquad i = 1, \ldots g$$

or

$$\mathbf{J}^{\mathrm{R}} = \mathbf{A}'\mathbf{J}' + \mathbf{A}''\mathbf{J}''. \tag{4.4.20}$$

The relation between $\mathbf{X}'$ and $\mathbf{X}''$ parallels that between the corresponding chemical potentials and when this is used to eliminate $\mathbf{X}''$ from the entropy production the result is

$$T\sigma = \tilde{\mathbf{X}}^{\mathrm{R}}\mathbf{J}^{\mathrm{R}} \tag{4.4.21}$$

where

$$\tilde{\mathbf{X}}^{\mathrm{R}} = \tilde{\mathbf{X}}'(\mathbf{A}')^{-1}. \tag{4.4.22}$$

The phenomenological flux equations, after the reactions involving the secondary defect species $g + 1, \ldots, n$ have equilibrated, are therefore

$$\mathbf{J}^{\mathrm{R}} = \mathbf{L}^{\mathrm{R}}\mathbf{X}^{\mathrm{R}}, \tag{4.4.23}$$

before the introduction of any restrictions which may exist among the $J_i^{\mathrm{R}}$. The phenomenological coefficients $\mathbf{L}^{\mathrm{R}}$ can be expressed in terms of the coefficients $\mathbf{L}$ of the unreduced, phenomenological equations. If we write the unreduced equation as

$$\mathbf{J}' = \mathbf{L}^{\prime,\prime}\mathbf{X}' + \mathbf{L}^{\prime,\prime\prime}\mathbf{X}'', \tag{4.4.24a}$$

$$\mathbf{J}'' = \mathbf{L}^{\prime\prime,\prime}\mathbf{X}' + \mathbf{L}^{\prime\prime,\prime\prime}\mathbf{X}'' \tag{4.4.24b}$$

we can then readily show that the relationship is

$$\mathbf{L}^{\mathrm{R}} = \mathbf{A}'\mathbf{L}'{}'\tilde{\mathbf{A}}' + \mathbf{A}'\mathbf{L}'{}''\tilde{\mathbf{A}}'' + \mathbf{A}''\mathbf{L}''{}'\tilde{\mathbf{A}}' + \mathbf{A}''\mathbf{L}''{}''\tilde{\mathbf{A}}''. \tag{4.4.25}$$

It follows that if the Onsager relations are valid for the original flux equations ($\mathbf{L}$ symmetric) then they are also valid for the flux equations in the reduced description ($\mathbf{L}^{\mathrm{R}}$ symmetric).

In summary, the reduced description is completely adequate and fits into the formalism of §4.4.1 as it stands without modification.

### 4.4.3 Stress fields

We now turn to the matter of an applied stress field. What does one expect the effect of a gradient in stress, $\sigma_{\alpha\beta}$, upon the thermodynamics to be? In answering this question we keep in mind three things, namely (i) the atomic lattice of solids (ii) the ensemble method of statistical thermodynamics and (iii) the hypothesis of non-equilibrium thermodynamics that all the usual thermodynamic functions can be defined locally for systems not too far from equilibrium. On this basis our expectations are the following. First, we expect eqns. (4.2.4) and such modified forms of them as arise in the presence of defects (eqns. 4.4.5 and 4.4.7) still to hold. The phenomenological coefficients $L_{ki}$ will be independent of the *gradient* of stress, although they will in general depend on the stress $\sigma_{\alpha\beta}$, as on the other thermodynamic variables. The presence of the stress field will manifest itself through an appropriate expression for the mechanical force $\mathbf{F}_k$ in (4.2.3a) and (4.2.14a). Since a spatially uniform stress cannot give rise to an atomic flux* (by Curie's principle, see de Groot and Mazur, 1962) we expect $\mathbf{F}_k$ to depend linearly on the spatial derivatives of the components of stress $\sigma_{\alpha\beta}$. The precise expression can be obtained by differentiation of (3.5.11). Remembering the use of the repeated index summation convention there, we then see that the Cartesian components of $\mathbf{F}_k$ are

$$F_{k\gamma} = v\frac{\partial}{\partial x_\gamma}(\lambda^{(k)}_{\alpha\beta}\sigma_{\alpha\beta}). \tag{4.4.26}$$

In cases where the sole interaction is with the hydrostatic component of stress (e.g. defects or solutes with full cubic symmetry, $\lambda_{\alpha\beta} = \lambda\delta_{\alpha\beta}$) (4.4.26) is equivalent to

$$\mathbf{F}_k = -\nabla(Pv_k), \tag{4.4.27}$$

* The phenomenon of Herring–Nabarro creep (see e.g. Nabarro, 1967) might be thought to contradict this statement. However, the flow of defects in a uniform stress in this case results not from a coupling of the flux directly to $\sigma_{\alpha\beta}$ but from differences between the equilibrium concentrations of defects at surfaces subject to different tractions. In other words, the stress changes the boundary conditions, not the flux relations. This is another example of the Gibbs paradox mentioned in §3.5.

in which $P$ is the pressure ($=\sigma_{\alpha\alpha}/3$) and $v_k$ ($=v\lambda$) is the partial excess volume of species $k$. A flux of defects or solute atoms in response to the force (4.4.27) is called the Gorsky effect (Gorsky, 1935).

## 4.5 Summary

In this chapter we have introduced the formalism provided by non-equilibrium thermodynamics for the description of atomic transport and diffusion processes. We did so by first summarizing the theoretical structure as usually presented for fluids and then introducing some of the changes required when we use it to describe these same processes in crystalline solids. An important feature of the flux equations in this structure is the explicit representation of 'cross-phenomena', i.e. of the generation of one sort of flow by a force normally associated directly with another. A particular example is the flux of an atomic species caused by the existence of a temperature gradient in the system, a force associated directly with heat flow.

The description of this formalism in this chapter has necessarily been rather abstract. In the next chapter we shall bring it into contact with the analysis of certain basic experiments and use it to introduce certain extensions of a 'physical' or intuitive kind.

# 5

# *Some applications of non-equilibrium thermodynamics to solids*

## 5.1 Introduction

The development in the previous chapter has necessarily been fairly abstract. In particular, although it is clear that the flux equations, when taken in conjunction with the corresponding continuity equations, in principle allow us to calculate the course of particular diffusion processes, we have not been able to give many indications of the physical characteristics of the $L$-coefficients appearing in these flux equations. This is, of course, to be expected at this stage since much of what follows in later chapters is about the theory of these quantities themselves. For the moment therefore they remain *phenomenological* coefficients, functions of thermodynamic variables ($T$, concentrations and stress) and governed by Onsager's theorem. It is helpful therefore at this stage to look at the relations between these $L$-coefficients and more familiar phenomenological coefficients adequate in special situations and which have been studied experimentally, e.g. chemical and isotopic diffusion coefficients, electrical ionic mobilities and ionic conductivity. Throughout we shall keep our assumptions about the nature of the systems being considered fairly broad. In particular, we shall not go out of our way to deal explicitly with dilute solid solutions, even though these are enormously important in fundamental studies. The more detailed results which are obtained for such systems are considered later after we have presented the atomic theory of the $L$-coefficients and evaluated them for particular models.

We begin by considering one of the simplest applications of the formalism, namely that to interstitial solid solutions in metals (§5.2). The simplicity of these systems allows us to describe several distinct phenomena with a minimum of mathematical manipulation. The next section then turns to isothermal diffusion in substitutional alloys, specifically isothermal chemical diffusion in a binary alloy

(§5.3) followed by migration in a temperature gradient and, by a simple formal substitution, in an electric field (§5.4). We then go on to consider isotopic diffusion ('self-diffusion') in the same system but in the absence of any gradients in chemical composition (§5.5). These considerations then lead us to approach the important question of the relation between chemical and isotopic diffusion coefficients (§§5.6–5.8). We then turn to the electrical mobilities of ions in strongly ionic solids (§5.9) and to the way electrical effects (the Nernst diffusion field) enter into chemical diffusion in this same class of solids. We illustrate these various discussions with experimental results, but space does not allow detailed reference to experimental methods nor to the associated phenomenological analysis.

Lastly, there are two points to be made which are common to all the applications of non-equilibrium thermodynamics made in this chapter, although they are in no way limited to them alone. The first is that, although the basic flux equations (4.4.5) and (4.4.7) apply to a lattice free from sources and sinks of defects (e.g. such as dislocations), they may still be used on a coarser scale for systems where the defects are maintained in (local) equilibrium concentrations by the operation of such sources and sinks, e.g. in the analysis of an experiment in which diffusion has taken place over a distance large compared to the separation between dislocations. The second point to be noted is that these experiments will be conducted under conditions in which the system is everywhere close to mechanical equilibrium. We say 'close to' rather than 'in' since, as we shall see, there are generally slow movements of one part of the system relative to another, as in the Kirkendall effect. This condition then allows us to employ the Prigogine relation (4.3.2) among the thermodynamic forces. In the absence of electrical or mechanical forces this relation is exactly equivalent to the Gibbs–Duhem relation.

## 5.2 Migration of interstitial solute atoms in metals

In this section we use the formalism of Chapter 4 to describe the migration of interstitial solute atoms in metals. The systems of practical concern are light elements, such as H, B, C, N and O, dissolved in Fe and other transition metals. These provide some of the simplest applications of the formalism, for two reasons. Firstly, although not all these solid solutions are necessarily extremely dilute, as a first approximation we may assume that the concentration is sufficiently low that the solute atoms move effectively independently of each other. Secondly, the tendency of these elements to take up interstitial, rather than substitutional, positions in the lattice is sufficiently strong that solute and solvent atoms do not exchange places. In other words the host metal merely provides a lattice structure through which the solute atoms move by thermally activated jumps from one interstitial site to another. It is thus reasonable to suppose that a gradient in the solute concentration will have a negligible influence on solvent diffusion so that we may neglect cross-terms in the general flux equations (i.e. the corresponding $L_{ij} = 0$).

With this simplification the equation for the flux of solute atoms (s) under the general conditions considered in §4.4 is

$$J_s = L_{ss}(X_s + Q_s^* X_q + z_s^* X_e), \tag{5.2.1}$$

in which, in the 'reduced' scheme, $X_s = -(\nabla\mu_s)_T + F_s$, $X_q = -\nabla T/T$ and $X_e = -eE$. The force $F_s$ on the solute atoms will be the sum of the force exerted by the electric field $E$ on the bare ion-core charge of the solute atom plus any mechanical forces acting (e.g. 4.4.26).

This simple equation provides a phenomenological description of several distinct phenomena. The first of these is diffusion in an isothermal metal in the absence of any electrical or mechanical forces. The flux $J_s$ is then simply

$$J_s = -L_{ss}\nabla\mu_s. \tag{5.2.2}$$

If we write the chemical potential in the standard form (3.4.2), i.e.

$$\mu_s = \mu_s^0(T, P) + kT\ln(\gamma_s x_s), \tag{5.2.3}$$

in which $x_s$ is the mole fraction of solute and $\gamma_s$ the activity coefficient, then (5.2.2) becomes

$$J_s = -\frac{kT}{n_s}L_{ss}\left(1 + \frac{\partial \ln \gamma_s}{\partial \ln x_s}\right)\nabla n_s, \tag{5.2.4}$$

in which $n_s$ is the number of solute atoms per unit volume ($\equiv nx_s$ with $n$ the total number of atoms per unit volume). This is just Fick's first law of diffusion (1.2.1), with the diffusion coefficient given by

$$D_s = \frac{kT}{n_s}L_{ss}\left(1 + \frac{\partial \ln \gamma_s}{\partial \ln x_s}\right). \tag{5.2.5}$$

Quantities like that in parentheses in (5.2.5) occur frequently in diffusion theory and are often referred to simply as *thermodynamic factors*. When the solution is very dilute (say $x_s < 1\%$) interactions among the solute atoms will be unimportant and the thermodynamic factor tends to unity and $D_s = kTL_{ss}/n_s$. An Arrhenius plot of such a limiting diffusion coefficient for $N$ atoms in $\alpha$-Fe is shown later as Fig. 8.1. The enthalpy of activation $h_m$ corresponding to such Arrhenius behaviour is mostly in the range 1–2 eV for the solutes C, N and O (Mehrer, 1990). Hydrogen diffuses much more rapidly and in many cases shows evidence of quantum effects leading to curved Arrhenius plots.

The second phenomenon represented by (5.2.1) is the migration or 'drift' of solute atoms in a stress field. If the concentration is uniform then the flux is simply

$$J_s = L_{ss}F, \tag{5.2.6}$$

in which the mechanical force $F$ is given by (4.4.26) or (4.4.27) as appropriate. The stress field involved could be generated externally (as in the Gorsky effect) or internally as a result of the presence of dislocations or loaded cracks. Since we should normally express the flux under these conditions as the product of solute concentration and a mobility ($u^M$) with the mechanical force it is clear that

$$L_{ss} = n_s u_s^M. \tag{5.2.7}$$

(We use the superscript M to distinguish this mobility from an electrical mobility from which the electric charge is always factored out.) In the dilute limit it follows that

$$D_s = kTu_s^M. \tag{5.2.8}$$

This relation is generally known as the Einstein relation. Clearly, if we are not near the dilute limit then by (5.2.5) the thermodynamic factor must be included on the right side of this equation. (In the context of ion motion in ionic substances we should replace $u_s^M$ by $u_s/|q_s|$ in which $q_s$ is the ion charge and $u_s$ the electrical mobility.)

The third phenomenon is the Ludwig–Soret effect, i.e. the establishment of a concentration gradient in a solution held in a temperature gradient (§1.2.4). A closed system held in a temperature gradient, i.e. $X_q \neq 0$, will come to a steady state specified by $J_s = 0$. By (5.2.1) with $X_e = 0$ we therefore obtain as the steady-state condition

$$-kT\frac{\nabla n_s}{n_s}\left(1 + \frac{\partial \ln \gamma_s}{\partial \ln x_s}\right) = Q_s^* \frac{\nabla T}{T}. \tag{5.2.9}$$

The concentration gradient is thus directly determined by the heat of transport parameter, $Q_s^*$. If the thermodynamic factor and $Q_s^*$ are both independent of $T$ we can immediately integrate (5.2.8) and infer that $\ln n_s$ will depend linearly on $1/T$. Experimental results illustrating this relation for C in α–Fe were shown in Fig. 1.5. These particular results yield a value of $Q_s^*$ of about $-1$ eV. Other systems of this class yield values in some cases positive in others negative, but mostly smaller in magnitude (Bocquet *et al.*, 1983). The atomic theory of these heats of transport is still very undeveloped, even though much important groundwork has been done (Gillan, 1983).

Lastly, we consider the electro-migration of these interstitial solute atoms. In a uniform isothermal system subject to an electric field, (5.2.1) predicts the flux of solute atoms to be

$$J_s = eL_{ss}(z_s - z_s^*)E \tag{5.2.10a}$$

$$\equiv eL_{ss}Z_s^*E, \tag{5.2.10b}$$

in which $e$ is the magnitude of the electronic charge while the ion-core charge of the solute atoms is written as $z_s e$. The electro-migration of the light elements H, C, N and O has been studied in a variety of metals. In a majority of these systems the solute atoms migrate towards the cathode, i.e. $Z^*$ is positive. Magnitudes range from a few times 0.1 up to about 6 (Bocquet *et al.*, 1983). In almost all cases, however, the direction of migration is determined by the host metal, independently of the interstitial solute. Elementary theories of $Z^*$ have been built on the idea that the contribution $z^*$ derives from the scattering of the conduction electrons by the migrating interstitial atoms (*electron wind*, e.g. Huntington, 1975). These predict that $z^*$ is proportional to the relaxation time of the conduction electrons and thus to the electrical conductivity (a prediction which has been confirmed experimentally for the substitutional solutes As and Sb in silver, see e.g. Hehenkamp, 1981). The same theories also predict a corresponding electron wind contribution to the heat of transport (because, in metals, heat is conducted overwhelmingly by the conduction electrons).

In this section we have described the application of the general flux equations to dilute interstitial solid solutions. These applications provided examples of relations between measureable and familiar quantities (e.g. $D_s$) and the more general transport coefficients introduced in Chapter 4 (e.g. $L_{ss}$). However, the simple nature of these interstitial systems meant that we had no occasion to introduce the off-diagonal transport coefficients. In the next section we consider substitutional solid solutions, for which the introduction of off-diagonal coefficients is unavoidable.

## 5.3 Isothermal chemical diffusion in a binary non-polar system

In this section we discuss the inter-diffusion of two distinct elements A and B in a single crystalline phase, as represented schematically in Fig. 5.1. We shall assume miscibility of the elements and a vacancy mechanism of transport. Since we wish to emphasize the basic principles we simplify the discussion by assuming that the mixing of the elements A and B has a negligible effect on the lattice parameter. Site fractions $c_i$ and number densities $n_i$ are thus effectively interchangeable, since the number of lattice sites per unit volume is independent of composition. The simplification is easily removed when we can assume Vegard's law (i.e. that the partial molar volumes of A and B are independent of composition). The changes enter mainly when we go from (gradients of) site fractions to (gradients of) number

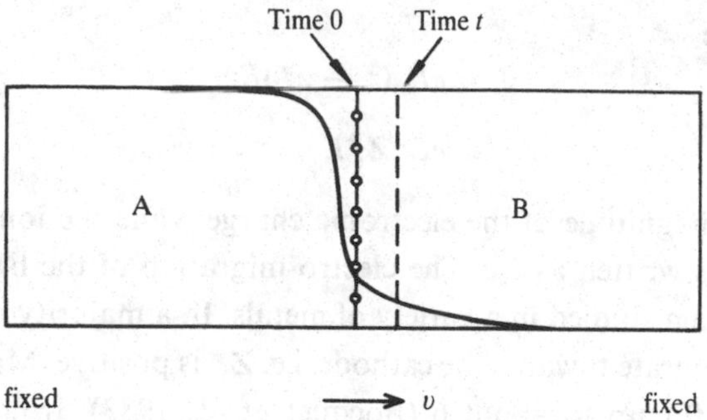


Fig. 5.1. Schematic diagram of a diffusion couple composed of two metals A and B. The circles represent wires or other inert markers inserted at the original interface, such as may be used to measure the velocity of the 'diffusion zone' relative to the fixed ends of the couple, i.e. to measure the Kirkendall velocity. The superimposed curved line is a graph of the fraction of A (vertical axis) v. position in the couple at some particular time, $t$. As drawn the atoms of B diffuse into A faster than those of A diffuse into B with the result that the interface moves to the right ($v > 0$).

densities. The immediate application of our discussion is to metal alloys, but the equations also hold for other binary and quasi-binary solid solutions, although in some such cases the correct interpretation of the formalism is only obtained once a more detailed, defect-based analysis has been made (see e.g. §5.10 below).

For an isothermal system in the absence of any electric fields $X_q = 0 = X_e$ in (4.4.5). Furthermore, if we assume that the vacancies are everywhere present in their (local) equilibrium concentration, i.e. that there is a sufficiently high density of sources and sinks of vacancies, we shall have $\mu_V = 0$ everywhere (cf. 3.7.1) and thus $X_V = 0$. Thus by (4.4.5) the flows of A and B atoms relative to the local crystal lattice are

$$J_A = L_{AA}X_A + L_{AB}X_B, \tag{5.3.1a}$$

and

$$J_B = L_{BA}X_A + L_{BB}X_B. \tag{5.3.1b}$$

We may further neglect any internal stresses in the diffusion couple, so that the Prigogine relation (4.3.2) reduces to the corresponding Gibbs–Duhem relation, i.e.

$$c_A X_A + c_B X_B = 0 \tag{5.3.2}$$

in which $c_A$ and $c_B$ are the site (mole) fractions $n_A/n$ and $n_B/n$ of A and B respectively. By (5.3.1) and (5.3.2) we can then write $J_A$ in the form

$$J_A = \left(L_{AA} - \frac{c_A}{c_B} L_{AB}\right) X_A. \tag{5.3.3}$$

If we now appeal to the form (3.4.2) for $\mu_A$, namely

$$\mu_A = \mu_A^0(T, P) + kT \ln(c_A \gamma_A)$$

where the activity coefficient is $\gamma_A$ we obtain

$$J_A = -\left(\frac{L_{AA}}{c_A} - \frac{L_{AB}}{c_B}\right) kT \left(1 + \frac{\partial \ln \gamma_A}{\partial \ln c_A}\right) \nabla c_A. \tag{5.3.4}$$

This has the form of Fick's first law (1.2.1)

$$J_A = -D_A \nabla n_A, \tag{5.3.5}$$

whence the diffusion coefficient of A is

$$D_A = \frac{kT}{n}\left(\frac{L_{AA}}{c_A} - \frac{L_{AB}}{c_B}\right)\left(1 + \frac{\partial \ln \gamma_A}{\partial \ln c_A}\right), \tag{5.3.6}$$

in which $n$ is the total number of lattice sites per unit volume. This diffusion coefficient is often called the *intrinsic diffusion coefficient* of A.

Of course, an exactly equivalent expression is obtained for B, namely

$$D_B = \frac{kT}{n}\left(\frac{L_{BB}}{c_B} - \frac{L_{BA}}{c_A}\right)\left(1 + \frac{\partial \ln \gamma_B}{\partial \ln c_B}\right). \tag{5.3.7}$$

(As here, we shall henceforth often retain both $L_{AB}$ and $L_{BA}$ to preserve the formal symmetry of equivalent expressions, even though, by the Onsager relations they are one and the same.) By the Gibbs–Duhem relation the last factors in (5.3.6) and (5.3.7) are seen to be equal to one another and hence the suffixes A and B can be dropped from them. (But this does not of course mean that $\gamma_A = \gamma_B$.) However, this is only true for binary systems and is not true generally for ternary and higher order alloys.

In the limit of a dilute alloy ($c_B \ll c_A$) we know that the activity coefficient will tend to a constant value as $c_B \to 0$. We also expect the diffusion coefficient of B, $D_B$, to be independent of $c_B$ in this limit, simply because as the atoms of B become more and more widely separated from one another any effect of their mutual interaction becomes smaller and smaller. Then, with the assumption that $L_{BA}/L_{BB}$ becomes independent of $c_B$ (which as we shall see later is justified by detailed calculations), we expect the limiting form of the relation (5.3.7) to be

$$D_B = \frac{kT}{n}\frac{L_{BB}}{c_B} = \frac{kT}{n_B} L_{BB}, \quad (c_B \to 0), \tag{5.3.8}$$

and that $L_{BB}$ becomes proportional to the concentration of B atoms, $n_B$. Although we have derived this result for a binary solid solution we can see that with the

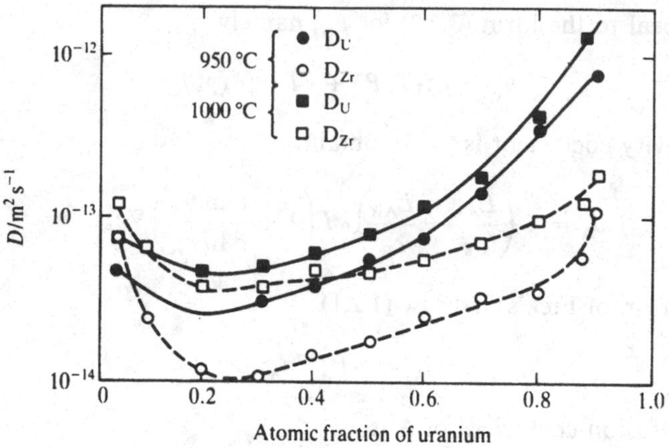


Fig. 5.2. Variation of the intrinsic diffusion coefficients $D_U$ and $D_{Zr}$ with the atomic fraction of U in U–Zr alloys at two different temperatures. (After Adda and Philibert, 1966.) For a compilation of such coefficients in other alloys see Mehrer (1990).

same assumptions it will be true as a limiting relation for a dilute component in any solid solution.

These two intrinsic diffusion coefficients, $D_A$ and $D_B$, are distinct and separately determinable experimentally. An example is shown in Fig. 5.2. The two quantities usually measured to obtain $D_A$ and $D_B$ are (i) the chemical interdiffusion coefficient, $\tilde{D}$, and (ii) the Kirkendall velocity, $v_K$ (§1.2.2), both of which relate to movements relative to the fixed ends of the couple where no diffusion is taking place. The analysis required to relate the basic equations (5.3.4) *et seq.* for atomic flows relative to the local crystal lattice to these measurable quantities may be complex. However, the simplest situation is sufficient to illustrate the nature of the relations between $D_A$ and $D_B$ on the one hand and $\tilde{D}$ and $v_K$ on the other. More complex cases are considered by Philibert (1985).

The fluxes $J_A$ and $J_B$ relative to the local crystal lattice are generally such that their sum is non-zero. This implies that the vacancy flux $J_V$ is also non-zero. Furthermore, since the concentration gradients and thus the fluxes are non-uniform we also have $\nabla \cdot \mathbf{J}_V \neq 0$. This requires vacancies to disappear or to be created in some regions, processes which, as we saw in Chapter 2, are made possible by the presence of dislocations. When this occurs a region where diffusion flows are large will move relative to those where they are small. The concentration distribution after a conventional diffusion experiment with a couple A–B, analysed with respect to the unaffected ends of the couple, will therefore yield a diffusion coefficient, the chemical interdiffusion coefficient $\tilde{D}$, linking the concentration gradient to the flows of atoms relative to the fixed parts of the lattice. In the simplest case these flows are

$$J'_A = J_A - c_A(J_A + J_B), \tag{5.3.9a}$$
$$J'_B = J_B - c_B(J_A + J_B), \tag{5.3.9b}$$

since the velocity of the local lattice relative to the fixed (non-diffusing) parts is $-v(J_A + J_B)$, if we assume (i) that the vacancies condense on to lattice planes perpendicular to the diffusion flow, i.e. there is no change in the cross-section or the shape of the specimen, and (ii) that the volume per lattice site, $v$, is constant. From (5.3.5) and (5.3.9) with $c_V \ll 1$ so that $c_A + c_B = 1$ we have

$$\begin{aligned} J'_A &= -(c_B D_A + c_A D_B)\nabla n_A \\ &\equiv -\tilde{D}\nabla n_A, \end{aligned} \tag{5.3.10}$$

and the same equation for B. Thus the chemical interdiffusion coefficient is the same for both species and equal to

$$\tilde{D} = c_A D_B + c_B D_A. \tag{5.3.11}$$

Likewise, the Kirkendall velocity, i.e. the motion of the diffusion zone relative to the fixed ends of the specimen as it results from the net non-zero flux of vacancies in the diffusion region, in the sense defined by positive atomic fluxes, is

$$v_K = (D_A - D_B)\nabla c_A. \tag{5.3.12}$$

These results show how we can determine the intrinsic diffusion coefficients $D_A$ and $D_B$ from measurements of $\tilde{D}$ and $v_K$. It is easily verified that the same results would be obtained if we supposed the defects to be interstitials rather than vacancies. On the other hand, more complex relations follow when we cannot make the simplifications specified in the course of deriving (5.3.10) and (5.3.11).

We note that, although such measurements have been made on a number of alloys, it is clear that the two quantities $D_A$ and $D_B$ are insufficient to obtain the three independent $L$-coefficients. As we shall see, this situation, in which the number of independently obtainable quantities is less than the number of independent $L$-coefficients entering into those quantities, is not uncommon and is one reason for the theoretical interest in the $L$-coefficients, especially the off-diagonal ones.

Lastly, we return to a point already made in Chapter 4, namely that, within the theory of non-equilibrium thermodynamics, different representations of the same system may be possible. The present example provides an opportunity to illustrate this and at the same time to draw a general conclusion. The details are given in Appendix 5.1, but the general conclusion is this: if equations expressing physical constraints, such as (5.3.2), are first used to reduce the number of independent variables in the expression for the rate of entropy production, then the consequent linear phenomenological equations are likewise fewer in number and can describe fewer phenomena. In this present example, use of (5.3.2) in this way leads to just one phenomenological equation, which can describe chemical diffusion but not the Kirkendall effect.

## 5.4 Drift experiments on binary non-polar systems

The calculation presented in §5.3 for diffusion in isothermal systems in the absence of any force fields is easily extended to more general conditions. Let us take first diffusion in a system in a temperature gradient. We again start from eqns. (4.4.5), but now we have $X_q \neq 0$. Another contrast with isothermal diffusion is that $X_V \neq 0$ even though the vacancies might also be locally in equilibrium. This is easily seen when it is recognized that the vacancy chemical potential $\mu_V$ is a function of vacancy fraction $c_V$, temperature and composition, i.e. $c_B$; and therefore that the meaning of $X_V = -(\nabla \mu_V)_T$ is

$$\begin{aligned} X_V &= -(\nabla \mu_V)_T \\ &= -\left(\frac{\partial \mu_V}{\partial c_V}\right)_{c_B, T} \nabla c_V - \left(\frac{\partial \mu_V}{\partial c_B}\right)_{c_V, T} \nabla c_B. \end{aligned} \tag{5.4.1}$$

When we evaluate (5.4.1) for the equilibrium vacancy fraction $\bar{c}_V$ defined by

$$\mu_V(\bar{c}_V, c_B, T) = 0, \tag{5.4.2}$$

(cf. 3.7.1) we easily see that

$$X_V = h_V X_q, \tag{5.4.3}$$

in which $h_V$ is the effective enthalpy of vacancy formation in the alloy as defined by $h_V = -\partial(\ln \bar{c}_V)/\partial(1/kT)$ (cf. 3.2.13).

The calculation now follows the same path as in §5.3 with the result that in place of (5.3.4) we have

$$J_A = -D_A \nabla n_A + [L_{AA}(Q_A^* - h_V) + L_{AB}(Q_B^* - h_V)]X_q \tag{5.4.4a}$$

and

$$J_B = -D_B \nabla n_B + [L_{BA}(Q_A^* - h_V) + L_{BB}(Q_B^* - h_V)]X_q, \tag{5.4.4b}$$

in which $D_A$ and $D_B$ are given by (5.3.6) and (5.3.7). The first terms on the right of these two relations are the diffusion terms, while the second are the drift terms. These equations give the fluxes relative to the local lattice. If we make the same simplifying assumptions that led to (5.3.9), then we find that the flux of B relative to the fixed ends of the specimen

$$J_B' = -\tilde{D}\nabla n_B - \nabla T\left(\frac{n_A n_B}{nkT^2}\right)\left\{\frac{D_B(Q_B^* - h_V) - D_A(Q_A^* - h_V)}{\left(1 + \dfrac{\partial \ln \gamma}{\partial \ln c}\right)}\right\}, \qquad (5.4.5)$$

with an equivalent expression for $J'_A$. Likewise, the velocity of the diffusion zone relative to the fixed ends is

$$v_K = (D_A - D_B)\nabla n_A/n$$
$$+ \left(\frac{\nabla T}{nT}\right)\{(L_{AA} + L_{BA})(Q_A^* - h_V) + (L_{BB} + L_{AB})(Q_B^* - h_V)\}. \qquad (5.4.6)$$

Various experiments have been carried out which allow the determination of the coefficients of $\nabla T$ in these expressions. For instance, the movement of the maximum of a bell-shaped concentration profile (i.e. $\nabla n_B = 0$) provides the second term in (5.4.5), although the strong dependence of the diffusion coefficients upon temperature means that the use of this fact in practice is not quite straightforward (Crolet and Lazarus, 1971). The steady-state Soret distribution is obtained by putting $J'_B = 0$. It is clear that in each case the measured quantity is fairly complicated.

A number of related drift experiments have been made on alloys subject to electric fields but under isothermal conditions. The corresponding calculation follows the same path, as one would expect from (4.4.5). The results are like those above except that $\nabla T/T$ is replaced by $eE$ while $(Q_A^* - h_V)$ and $(Q_B^* - h_V)$ are replaced by $Z_A^*$ and $Z_B^*$. Again therefore, the quantities obtainable experimentally for these substitutional alloys are complex. Their dissection into their component parts requires additional help from theory. In this connection considerable simplifications follow when the alloys are dilute and many basic experiments have therefore been made on such systems. We shall develop some of the necessary theory in later chapters. Here we shall continue without making that restriction.

## 5.5 Isothermal isotope diffusion (self-diffusion) in a binary non-polar system

The relative ease with which the fundamental mobilities of atoms can be obtained by following the migration of radioactive isotopes ('tracers') has allowed a vast number of such determinations to be made. For discussions of the experimental techniques used in such studies see Bénière (1983) and Rothman (1984). In the present section, therefore, we shall consider the diffusion of isotopes A* and B* in an (A, B) alloy or solid solution. These experiments are generally conducted as shown schematically in Fig. 5.3, i.e. the diffusion of small quantities of the isotope

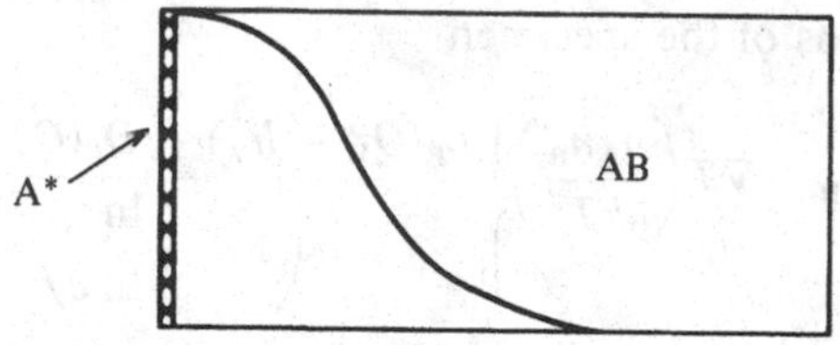


Fig. 5.3. Schematic diagram of an experiment to measure the isotopic ('tracer') diffusion coefficient $D_A^*$ in an alloy AB. The sensitivity generally afforded by radioactive isotopes allows the use of only a thin layer of isotope A*, which then diffuses into AB under conditions of negligible chemical concentration gradient. The curve illustrates the concentration profile after a particular time, $t$. For reviews of experimental techniques associated with the use of radiotracers see Bénière (1983) and Rothman (1984).

into the alloy is measured under conditions of essentially zero gradient in chemical composition. Although such experiments are capable of great refinement and precision the relation of the measured diffusion coefficients $D_A^*$ and $D_B^*$ to the intrinsic chemical diffusion coefficients is in no way trivial, as a moment's reflection will show. Thus it is obvious that if we use the formalism of eqns. (4.4.5) for example, a number of new $L$-coefficients, $L_{A^*A^*}$, $L_{A^*A}$, $L_{A^*B}$, etc. will enter. This number can be reduced by appeal to the Onsager relations and also to certain physical arguments ('Gedanken experiments') which we describe below, but not below two. Since these can be taken to be $D_A^*$ and $D_B^*$, it is clear that there is no simple formal argument based on what we have developed so far by which these quantities can be related to the quantities describing chemical diffusion, namely $L_{AA}$, $L_{AB}$ and $L_{BB}$. Either further intuitive physical arguments or analyses of detailed physical models are required. We describe some of the former in the next section, while later chapters deal with detailed models. Here we shall show how far we can get just with the basic phenomenology.

Let us begin with the formal analysis of isotope diffusion in an AB alloy. We set down the general equations for a system containing four species, A′, A″, B′ and B″, and again assume that the vacancies are everywhere in equilibrium, i.e. that $X_V = 0$. The isotopes A′ and A″ of element A are chemically indistinguishable; likewise for the isotopes B′ and B″ of B. The transport equations are

$$J_{A'} = L_{A'A'}X_{A'} + L_{A'A''}X_{A''} + L_{A'B'}X_{B'} + L_{A'B''}X_{B''} \tag{5.5.1a}$$

$$J_{A''} = L_{A''A'}X_{A'} + L_{A''A''}X_{A''} + L_{A''B'}X_{B'} + L_{A''B''}X_{B''} \tag{5.5.1b}$$

$$J_{B'} = L_{B'A'}X_{A'} + L_{B'A''}X_{A''} + L_{B'B'}X_{B'} + L_{B'B''}X_{B''} \tag{5.5.1c}$$

$$J_{B''} = L_{B''A'}X_{A''} + L_{B''A''}X_{A''} + L_{B''B'}X_{B'} + L_{B''B''}X_{B''} \tag{5.5.1d}$$

with Onsager's relation (4.2.5) among the off-diagonal $L$-coefficients. By the assumption that isotopes of the same element are chemically indistinguishable it follows that the chemical potentials, from which the forces, $X$, are to be obtained, are

$$\mu_{A'} = \mu_A^0(T, P) + kT\ln(c_{A'}\gamma_A) \tag{5.5.2a}$$

$$\mu_{A''} = \mu_A^0(T, P) + kT\ln(c_{A''}\gamma_A) \tag{5.5.2b}$$

$$\mu_{B'} = \mu_B^0(T, P) + kT\ln(c_{B'}\gamma_B) \tag{5.5.2c}$$

$$\mu_{B''} = \mu_B^0(T, P) + kT\ln(c_{B''}\gamma_B), \tag{5.5.2d}$$

in which the activity coefficients $\gamma_A$ and $\gamma_B$ depend on the total fractions of elements A and B, i.e. on $c_A \equiv c_{A'} + c_{A''}$ and $c_B \equiv c_{B'} + c_{B''}$ respectively but not on the proportions $c_{A'}/c_{A''}$ and $c_{B'}/c_{B''}$. (This statement, of course, neglects minute differences which would derive from the implied differences in nuclear mass of different isotopes, but these are far too small to be of concern at this point.)

These activity coefficients $\gamma_A$ and $\gamma_B$ of the two species are related by the Gibbs–Duhem equation, i.e.

$$\frac{\partial \ln \gamma_A}{\partial \ln c_A} = \frac{\partial \ln \gamma_B}{\partial \ln c_B} \equiv \frac{\partial \ln \gamma}{\partial \ln c}, \tag{5.5.3}$$

whence the forces are

$$X_{A'} = -kT\left(\frac{\nabla c_{A'}}{c_{A'}} + \frac{\partial \ln \gamma}{\partial \ln c}\frac{\nabla c_A}{c_A}\right) \tag{5.5.4a}$$

$$X_{A''} = -kT\left(\frac{\nabla c_{A''}}{c_{A''}} + \frac{\partial \ln \gamma}{\partial \ln c}\frac{\nabla c_A}{c_A}\right) \tag{5.5.4b}$$

$$X_{B'} = -kT\left(\frac{\nabla c_{B'}}{c_{B'}} + \frac{\partial \ln \gamma}{\partial \ln c}\frac{\nabla c_B}{c_B}\right) \tag{5.5.4c}$$

$$X_{B''} = -kT\left(\frac{\nabla c_{B''}}{c_{B''}} + \frac{\partial \ln \gamma}{\partial \ln c}\frac{\nabla c_B}{c_B}\right). \tag{5.5.4d}$$

An ideal measurement of the self-diffusion of A requires us to determine $J_{A'}$ (or $J_{A''}$) under conditions where no gradients exist in B′ and B″, i.e. when $X_{B'} = 0 = X_{B''}$. Under these conditions we shall also have $\nabla c_A = \nabla c_{A'} + \nabla c_{A''} = 0$. By (5.5.1a) and (5.5.4) we then obtain for the self-diffusion coefficient of A′.

$$D_{A'} = \frac{kT}{n}\left(\frac{L_{A'A'}}{c_{A'}} - \frac{L_{A'A''}}{c_{A''}}\right), \tag{5.5.5}$$

with a like expression for $D_{A''}$.

However, under the conditions postulated for this experiment there is no chemical mixing and we must therefore expect the total fluxes of both A atoms and B atoms to be zero, i.e.

$$J_{A'} + J_{A''} = 0, \tag{5.5.6}$$

as well as

$$J_{B'} = 0 = J_{B''}, \tag{5.5.7}$$

there being no gradient in the fractions of either of the B isotopes. Since these relations must be true whatever the gradient in the fraction of A isotopes we immediately see that

$$\frac{L_{A'A'}}{c_{A'}} - \frac{L_{A'A''}}{c_{A''}} = \frac{L_{A''A''}}{c_{A''}} - \frac{L_{A''A'}}{c_{A'}}, \tag{5.5.8}$$

$$\frac{L_{B'A'}}{c_{A'}} = \frac{L_{B'A''}}{c_{A''}}, \tag{5.5.9}$$

and

$$\frac{L_{B''A'}}{c_{A'}} = \frac{L_{B''A''}}{c_{A''}}. \tag{5.5.10}$$

Since the $L$-coefficients are explicitly independent of gradients these relations between quantities for the two isotopes A′ and A″ always hold. Similar relations between quantities for the two isotopes B′ and B″ likewise can be deduced.

$$\frac{L_{B'B'}}{c_{B'}} - \frac{L_{B'B''}}{c_{B''}} = \frac{L_{B''B''}}{c_{B''}} - \frac{L_{B''B'}}{c_{B'}}, \tag{5.5.11}$$

$$\frac{L_{A'B'}}{c_{B'}} = \frac{L_{A'B''}}{c_{B''}}, \tag{5.5.12}$$

$$\frac{L_{A''B'}}{c_{B'}} = \frac{L_{A''B''}}{c_{B''}}. \tag{5.5.13}$$

Furthermore by another thought experiment we can deduce a further six relations. These are like (5.5.8)–(5.5.13) but with the inequivalent indices on the $L$-coefficients transposed. We do so by (i) considering a situation in which the isotopic ratios $c_{A'}/c_{A''}$ and $c_{B'}/c_{B''}$ are everywhere the same but where there are gradients in the overall concentrations of A and B species, and then (ii) requiring that the flux ratios $J_{A'}/J_{A''}$ and $J_{B'}/J_{B''}$ equal $c_{A'}/c_{A''}$ and $c_{B'}/c_{B''}$ respectively for all possible values of the chemical forces $(X_A - X_V)$ and $(X_B - X_V)$. These additional relations taken with (5.5.8) and (5.5.11) then yield the Onsager equalities, $L_{A'A''} = L_{A''A'}$ and $L_{B'B''} = L_{B''B'}$. They do not, of course, yield the Onsager relations when the subscripts refer to chemically distinct species.

An important consequence of these relations, which follows from (5.5.8) and (5.5.5) is that $D_{A'}$ and $D_{A''}$ are equal. This is true whatever the ratio $c_{A'}$ to $c_{A''}$. The above argument is easily extended to a mixture of any number of A-isotopes, from which we conclude that there is only one distinct isotopic self-diffusion coefficient, which we shall denote by $D_A^*$. Furthermore, it is easily seen that this quantity is independent of the isotopic composition of the A component in the alloy. The same applies to $D_B^*$ and to the B component. In other words the usual conditions of a radiotracer diffusion experiment namely $c_{A'} \ll c_{A''}$ are not necessary for this equality to hold. However, the condition $\nabla c_B = 0$ is, and in practice it is much easier to get close to this requirement by using only a very small concentration of the isotope whose migration is followed.

Lastly, we consider how many of the original sixteen coefficients are left independent. Examination of the twelve new relations obtained from the thought experiments and of the Onsager relations shows that, in fact, we are left with five independent quantities. These could be chosen to be

$$L_{AA} \equiv \frac{c_A}{c_{A'}}(L_{A'A'} + L_{A''A'}) \tag{5.5.14a}$$

$$L_{AB} \equiv \frac{c_B}{c_{B'}}(L_{A'B'} + L_{A''B'}) = L_{BA} \tag{5.5.14b}$$

$$L_{BB} \equiv \frac{c_B}{c_{B'}}(L_{B'B'} + L_{B''B'}), \tag{5.5.14c}$$

together with two others – which, in view of the preceding conclusions, it is convenient to take to be $D_A^*$ and $D_B^*$ (although they could equally well be any independent pair of coefficients, e.g. $L_{A'A'}$ and $L_{B'B'}$). The quantities $L_{AA}$, etc. here are just those already introduced for chemical diffusion in the preceding section 5.3. Thus the above relations are easily arrived at by applying eqns. (5.5.1) to the situation in which the isotopic ratios $c_{A'}/c_{A''}$ and $c_{B'}/c_{B''}$ are everywhere the same but in which $\nabla c_A$ and $\nabla c_B$ are non-zero. Under these conditions the forces $X_{A'}$ and $X_{A''}$ are equal,

$$\begin{aligned} X_{A'} = X_{A''} &= -kT\frac{\nabla c_A}{c_A}\left(1 + \frac{\partial \ln \gamma}{\partial \ln c}\right), \\ &\equiv X_A, \end{aligned} \tag{5.5.15}$$

and similarly for $X_B$. At the same time, the fluxes $J_A = J_{A'} + J_{A''}$ and $J_B = J_{B'} + J_{B''}$ must be just those arising in the chemical diffusion experiment. In effect, we simply 'contract' the full description in terms of all the isotopes to one in terms of the chemical species alone by specifying constant isotopic ratios.

In summary then, a binary alloy AB is characterized by five independent transport coefficients which may be taken to be $L_{AA}$, $L_{AB}$ and $L_{BB}$, as for chemical diffusion, together with two further coefficients which may be the two isotopic self-diffusion coefficients $D_A^*$ and $D_B^*$ (or any independent pair of $L$-coefficients involving an isotope species). But we note that, when the alloy becomes very dilute ($c_B \to 0$), there is no longer any distinction between $D_B^*$ and the intrinsic coefficient $D_B$, i.e. we can substitute $D_B^*$ for $D_B$ on the left side of (5.3.8); there are then only four independent coefficients. Otherwise there is no rigorous general argument which allows us to reduce these five coefficients to any smaller number. There are, however, two approximate, intuitive arguments which have been devised and which are considered next. In view of the much greater precision and convenience of diffusion measurements with radioactive isotopes compared to chemical diffusion measurements these approximate relations are of considerable practical interest. The first argument is due to Darken (1948): the second derives from Manning (1968).

## 5.6 Darken's equation

As we have emphasized previously the coefficients $L_{AA}$, etc. appearing in (5.3.1) are functions of temperature, pressure and composition, but they do not depend on gradients of composition. One might therefore intuitively assume – as Darken did – that the average velocity of an isotope of A, say A′, per unit force $X_{A'}$ due to a gradient of A′ in a chemically homogeneous alloy would equal the average velocity of all A atoms per unit force $X_A$ due to chemical gradient, i.e. that

$$\frac{J_{A'}}{c_{A'}X_{A'}} = \frac{J_A}{c_A X_A}. \tag{5.6.1}$$

By (5.3.3) therefore

$$J_{A'} = c_{A'}\left(\frac{L_{AA}}{c_A} - \frac{L_{AB}}{c_B}\right)X_{A'}. \tag{5.6.2}$$

In the chemically uniform alloy only the entropy of mixing term is left in (5.5.2a) and we therefore have

$$J_{A'} = -\left(\frac{L_{AA}}{c_A} - \frac{L_{AB}}{c_B}\right)kT\nabla c_{A'}. \tag{5.6.3}$$

Hence the diffusion coefficient of A′ is

$$D_A^* = \frac{kT}{n}\left(\frac{L_{AA}}{c_A} - \frac{L_{AB}}{c_B}\right), \tag{5.6.4}$$

whence by (5.2.6)

$$D_A = D_A^*\left(1 + \frac{\partial \ln \gamma}{\partial \ln c}\right), \tag{5.6.5}$$

with a similar relation for $D_B$. The chemical interdiffusion coefficient, $\tilde{D}$, is thus

$$\tilde{D} = (c_A D_B^* + c_B D_A^*)\left(1 + \frac{\partial \ln \gamma}{\partial \ln c}\right). \tag{5.6.6}$$

This is the Darken equation. Figure 5.4 shows how well this equation is obeyed in the Au–Ni system.

This Darken relation between tracer diffusion coefficients and chemical diffusion coefficients is widely used in practice. Its simplicity provides an obvious convenience. We shall assess its accuracy later. The next step along these lines, due to Manning (1968), however, introduces an additional physical consideration related to atomic features of the diffusion process and to deal with that we first digress to consider transport in a pure substance A.

## 5.7 Self-diffusion in a pure substance

For a chemically pure material A composed of isotopes A′, A″, etc. we can go beyond the relations already obtained in §5.3 by considering a situation in which there is a gradient in vacancy fraction, $c_V$, and an associated vacancy flux $J_V$. We first introduce the vacancy diffusion coefficient $D_V$ through the relation

$$J_V = -D_V n \nabla c_V. \tag{5.7.1}$$

Since there is a corresponding net flux of A ($J_A = -J_V$) this situation is, in fact, a rather refined type of chemical diffusion experiment (which has nevertheless been realized in practice (Yurek and Schmalzried, 1975; Martin and Schmalzried, 1985, 1986)). Now we want (5.7.1) to be compatible with the flux equations for the isotopes. For simplicity we again assume that there are just two isotopes A′ and A″, although the argument is readily extended to any number. Since the vacancies are present in arbitrary concentration we need the flux equations in the form (cf. 4.4.5)

$$J_{A'} = L_{A'A'}(X_{A'} - X_V) + L_{A'A''}(X_{A''} - X_V), \tag{5.7.2a}$$

$$J_{A''} = L_{A''A'}(X_{A'} - X_V) + L_{A''A''}(X_{A''} - X_V). \tag{5.7.2b}$$

Since the isotopes A′, A″ and the vacancies constitute an ideal solution (we here neglect interactions among vacancies), each of the forces $X_i$ in (5.7.2) takes the simple form $-kT\nabla c_i/c_i$. We can establish contact between (5.7.1) and (5.7.2) by considering a system in which the isotopic ratio $c_{A'}/c_{A''}$ is everywhere the same. Then it is physically obvious that the flux ($J_{A'}$) will be

$$J_{A'} = -\frac{c_{A'} J_V}{(c_{A'} + c_{A''})} = \frac{n D_V c_{A'} \nabla c_V}{c_A} \tag{5.7.3}$$

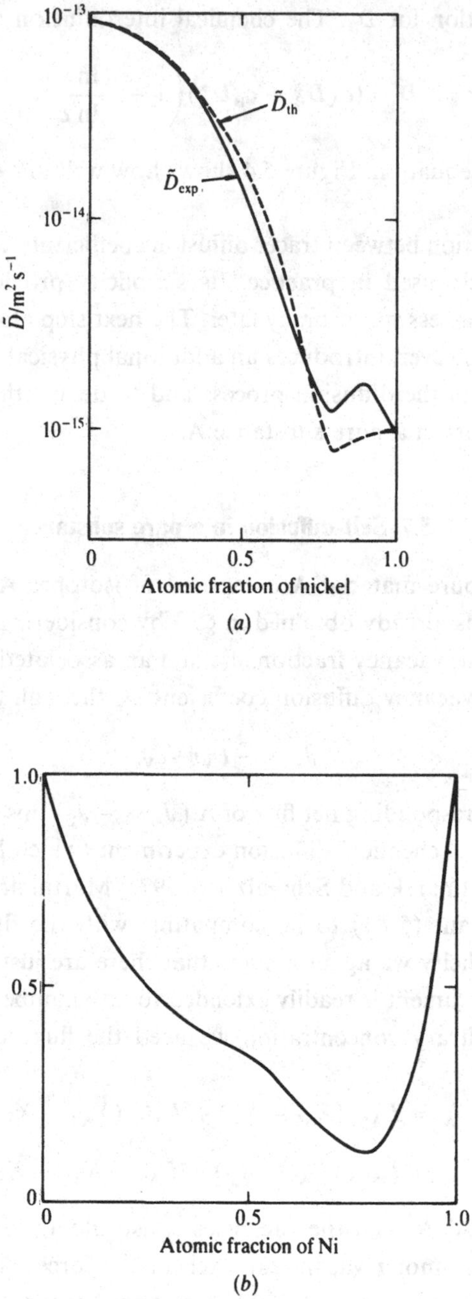


Fig. 5.4. (*a*) The chemical interdiffusion coefficient $\tilde{D}$ as a function of the atomic fraction of Ni in Au–Ni alloys at 900 °C. The full line represents the experimental results while the dashed line is as predicted by Darken's equation from the measured isotope diffusion coefficients $D^*_{\text{Ni}}$ and $D^*_{\text{Au}}$ and thermodynamic properties. (*b*) The contribution of the thermodynamic factor to the predicted $\tilde{D}$. (After Reynolds *et al.*, 1957.)

and likewise for $J_{A''}$. These relations are merely the expression of our assumption that the vacancies allow any particular A′ and A″ atoms to move with equal probability. At the same time, by (5.7.2)

$$J_{A'} = kT\left(\frac{L_{A'A'} + L_{A'A''}}{c_V c_A}\right)\nabla c_V \tag{5.7.4a}$$

and

$$J_{A''} = kT\left(\frac{L_{A''A''} + L_{A''A'}}{c_V c_A}\right)\nabla c_V. \tag{5.7.4b}$$

By (5.7.3) and (5.7.4) we have

$$L_{A'A'} + L_{A'A''} = \frac{nc_V c_{A'} D_V}{kT} \tag{5.7.5a}$$

and likewise

$$L_{A''A''} + L_{A''A'} = \frac{nc_V c_{A''} D_V}{kT}. \tag{5.7.5b}$$

The next step is to recognize that the self-diffusion coefficient $D^*_A$ will be proportional to the product $c_V D_V$ (§2.6). We let the constant of proportionality be $f$, i.e.

$$D^*_A = f c_V D_V. \tag{5.7.6}$$

Then from (5.5.5), (5.7.5) and (5.7.6) we obtain

$$L_{A'A'} = \frac{nc_{A'} D^*_A}{kT}\left[1 + \left(\frac{1-f}{f}\right)\frac{c_{A'}}{c_A}\right], \tag{5.7.7a}$$

$$L_{A'A''} = \frac{nc_{A'} D^*_A}{kT}\left(\frac{1-f}{f}\right)\frac{c_{A''}}{c_A} = L_{A''A'}, \tag{5.7.7b}$$

$$L_{A''A''} = \frac{nc_{A''} D^*_A}{kT}\left[1 + \left(\frac{1-f}{f}\right)\frac{c_{A''}}{c_A}\right], \tag{5.7.7c}$$

in which we may, if we choose, replace the (measurable) quantity $D^*_A$ by the expression (5.7.6). These relations apply whatever the relative proportions of A′ and A″. As already demonstrated, $D^*_A$, and thus $f$, are independent of the ratio $c_{A'}$ and $c_{A''}$ and so can be regarded as constants in the above relations. (Although eqns. (5.7.7) were derived for a pure substance we might expect them also to hold in alloys; in the final section of Chapter 6 we shall see why, in fact, this is so.)

It is sometimes useful to remember that

$$D_A^* = f\frac{kT}{n_A}L_{AA}, \tag{5.7.8}$$

which follows from (5.5.14a) and (5.7.7). Since these relations are exact under the conditions specified they can be regarded as limiting relations for more complex systems, and can thus provide useful checks on approximate theories or computer simulations of these more complex systems. Their extension to mixtures of more than two isotopes and to interstitial defects is easily obtained. In particular, the generalization of (5.7.5)

$$\sum_j L_{ij} = \frac{nc_i D_A^*}{kTf}, \tag{5.7.9}$$

in which the subscripts $i$ and $j$ stand for isotopes $A^i$ and $A^j$ in a general isotopic mixture, will be called upon shortly.

Before proceeding with the formal argument, however, let us pause to consider this coefficient $f$ which we have introduced. To understand this quantity we need a kinetic or other detailed statistical theory of both $D_A^*$ and $D_V$, or of the equivalent $L$-coefficients. These theories are the subject of later chapters. However, we note here that the older literature on solid state diffusion contains simplified kinetic treatments leading to the equality of $D_A^*$ to $c_V D_V$, i.e. to $f = 1$ and by (5.7.7b) to a zero cross-coefficient $L_{A'A''}$. It was first pointed out by Bardeen and Herring (1952) on the basis of a random walk treatment that this could not be true generally for migration via the action of vacancies. In fact, the relation $D_A^* = c_V D_V$ may be obtained by the method of random walks if we suppose that each atom of A′ makes a simple uncorrelated random walk from one lattice site to another. But since the atom A′ can move from its present site only if a neighbouring site is vacant, successive jumps of a particular atom A′ are not completely random, as we have already argued in §2.3.3. Successive displacements of A′ are therefore correlated with one another, even though the jumps of the vacancy are truly random. The effect of this correlation in the motion of A′ is to reduce $D_A^*$ from $c_V D_V$ to $fc_V D_V$ where $f < 1$. This aspect of atomic transport has undergone much subsequent development which we shall present in later chapters. Here we merely note that for vacancies, the quantity $f$ as introduced here is the same as the so-called Bardeen–Herring correlation factor $f_0$ (Chapter 10), but that for interstitialcy mechanisms or other mechanisms involving the movement of more than one atom in a single transition additional numerical factors must be included.

## 5.8 Manning's relations

Although Darken's equation remains a very useful approximation in many cases, it was recognized soon after the time of Darken's paper that, from a theoretical point of view, (5.6.5) was incomplete. As may be seen from (5.3.6), (5.5.5) and (5.5.14), the Darken equation only follows if the non-diagonal coefficients $L_{A'A''}$ and $L_{B'B''}$ are set equal to zero. However, it was recognized that these coefficients would in general be non-zero for a vacancy mechanism of diffusion (Bardeen and Herring, 1952). Subsequent studies of the kinetics of atomic migration via point defects led Manning to propose and develop model equations which provide corrections to the Darken equations (§10.8). These equations are nominally those for a 'random alloy', i.e. one in which the vacancies and the different atoms are presumed to be distributed randomly, although the rates at which the different atoms exchange with vacancies are allowed to be different. Also, in applications, the activity factors are generally included. Unfortunately, the arguments employed in the original derivations are difficult to follow and, despite various later discussions (e.g. Manning, 1971; Stolwijk, 1981), it is only recently that the essential approximations have begun to emerge clearly (see §§10.8 and 13.2) – although computer simulations of the model have shown Manning's equations to be remarkably accurate (Bocquet, 1974a; de Bruin *et al.*, 1975, 1977; Murch and Rothman, 1981; Murch, 1982b; Allnatt and Allnatt, 1984). However, the corresponding relations between the $L_{ij}$ for a multicomponent alloy and the various tracer diffusion coefficients can be derived, without appeal to the atomic model, from two physical postulates, which we now present.

We start with the equations for a multicomponent system (4.4.5) and (4.4.7) which we write in the shortened form

$$J_k = \sum_i L_{ki} X_i' \tag{5.8.1}$$

where the sum is over all chemical species, $i$, while $X_i'$ stands for that combination of forces between parentheses in (4.4.5) and (4.4.7). To be definite let us assume that vacancies are the dominant defects. The total vacancy flux is

$$J_V = -\sum_k J_k. \tag{5.8.2}$$

Next we consider circumstances in which the forces $X_i'$ are zero except for one pair, say $X_a'$ and $X_b'$. We can then write

$$J_b = \left[ L_{bb} - \frac{L_{ba} \sum_i L_{ib}}{\sum_i L_{ia}} \right] X_b' - \left( \frac{L_{ba}}{\sum_i L_{ia}} \right) J_V. \tag{5.8.3}$$

Now, effectively, Manning argues that the distinction between isotope diffusion and chemical diffusion is that in the former there is no vacancy flux, while in chemical diffusion there is. Thus if we set $J_V = 0$ in (5.8.3) we might suppose that the resulting equation will describe the diffusion of isotopes of $b$; or, in other words since then $X_b = -kT\nabla c_b/c_b$ and since the self-diffusion coefficient is independent of isotopic fraction

$$L_{bb} - \frac{L_{ba}\sum_i L_{ib}}{\sum_i L_{ia}} = \frac{N}{kT}c_b D_b^*, \qquad (b \neq a). \tag{5.8.4}$$

In other situations we may, following Manning, refer to the first term on the r.h.s. of (5.8.3) as the 'direct' term (because it results from the force $X_b'$ acting directly on the $b$ atoms) and to the second term as the 'vacancy wind' term. The direct term, being now of the form $(N/kT)c_b D_B^* X_b'$, is the exact analogue of Darken's assumption (5.6.1). The difference from Darken lies in the vacancy wind term in (5.8.3).

However, it is evident that for this view of the flux equations to be internally consistent (5.8.4) must hold whichever component we choose to experience the other non-zero force, here $X_a'$. Thus $L_{ba}/\sum_i L_{ia}$ must be independent of the index $a$ (as long as it is not $b$). Hence we can write

$$\frac{L_{ba}}{\sum_i L_{ia}} \equiv l_b, \qquad (\text{all } a \neq b). \tag{5.8.5}$$

We then see that the coupling to the vacancy flux is also independent of $a$, i.e. the vacancy wind contribution to $J_b$ depends on the magnitude of $J_V$ only and not on how it is generated. This self-consistency in the argument is encouraging.

However, eqns. (5.8.4) are insufficient to determine all the $L$-coefficients and we thus need a second assumption. Following Manning we assume that a certain mobility of the $i$-atoms in the alloy bears the same relation to $D_i^*$ as it does in a pure material (this mobility being obtained from the flux of $i$-atoms when an equal force $F$ is applied to all components as $J_i/Nc_iF$), i.e. by (5.8.1) and (5.7.8)

$$\sum_j L_{ij} = \frac{Nc_i D_i^*}{kTf}. \tag{5.8.6}$$

Eqns. (5.8.4) and (5.8.6), bearing in mind also (5.8.5), are sufficient to determine all the $L_{ij}$ coefficients (Lidiard, 1986a). The result is

$$L_{ik} = \frac{Nc_i D_i^*}{kT}\left[\delta_{ik} + \frac{(1-f)}{f}\frac{c_k D_k^*}{\sum_j c_j D_j^*}\right]. \tag{5.8.7}$$

This result is potentially of considerable value, for it tells us that from the $n$ isotopic (tracer) diffusion coefficients of an $n$-component alloy we can obtain *all* the $n^2$ atomic transport $L$-coefficients, both diagonal and non-diagonal. It is apparent that the argument leading to (5.8.7), although intuitive, is fairly general.

In particular, although (5.8.7) was derived originally by Manning (1970) with a specific model (the 'random alloy model') in mind the derivation given here is not obviously limited to that model. Furthermore, the argument is in no way limited to small vacancy fractions as long as we recognize that $f$, in a highly defective system, will be a function of the vacancy fraction $c_V$. Nor is it evidently limited to vacancy defects. If the defects were taken to be interstitials instead then we should have $J_I$ in place of $-J_V$ in (5.8.2) and (5.8.3), but otherwise the argument would be unaltered. That the result is compatible with certain general requirements is also encouraging. Firstly and most obviously, we observe that, although the Onsager relation has not been used in the derivation of (5.8.7), this relation follows directly from it,

$$L_{ik} = L_{ki}.$$

Secondly, the requirement that the $L$-matrix be positive definite ((4.2.6) and (4.2.7)) is easily seen to be satisfied by (5.8.7) when the condition $f < 1$ is inserted. Lastly, we recognize that since (5.8.7) applies equally well to a mixture of isotopes it is evidently compatible with the exact relations (5.5.8)–(5.5.10).

The results (5.8.7) can be employed to obtain chemical diffusion coefficients, Kirkendall shifts and other derived quantities in terms of tracer diffusion coefficients. In general, these are like those obtained by Darken but with vacancy wind corrections. For example, intrinsic diffusion coefficients and the chemical inter-diffusion coefficient for a binary alloy, by (5.3.6), (5.3.7), (5.3.11) and (5.8.7), are given by

$$D_A = D_A^*\Phi\left[1 + \left(\frac{1-f}{f}\right)\frac{c_A(D_A^* - D_B^*)}{(c_A D_A^* + c_B D_B^*)}\right], \tag{5.8.8}$$

with an equivalent expression for $D_B$, and

$$\tilde{D} = (c_B D_A^* + c_A D_B^*)\Phi\left[1 + \left(\frac{1-f}{f}\right)\frac{c_A c_B (D_A^* - D_B^*)^2}{(c_A D_A^* + c_B D_B^*)(c_A D_B^* + c_B D_A^*)}\right] \tag{5.8.9}$$

in which $\Phi$ is the thermodynamic factor

$$\Phi = \left(1 + \frac{\partial \ln \gamma}{\partial \ln c}\right). \tag{5.8.10}$$

The factors in square brackets are thus the Manning vacancy wind corrections to Darken's (5.6.5) and (5.6.6). In a similar way one finds that the Kirkendall velocity (5.3.12) is increased by the addition of a factor $f^{-1}$ to the Darken expression. It is evident from these results that the Manning corrections, while significant, will never be very large even when one of the self-diffusion coefficients is much greater

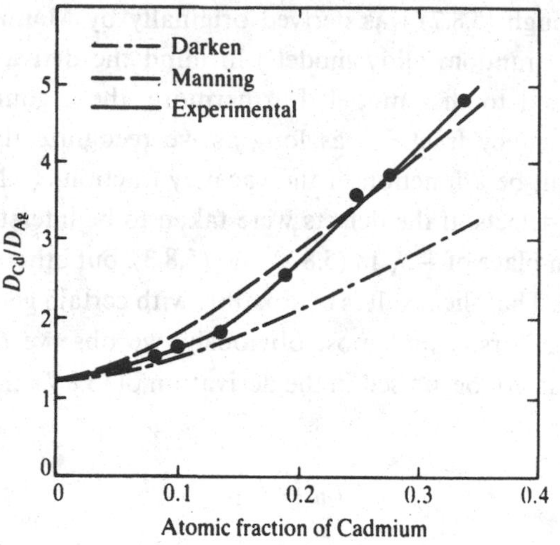


Fig. 5.5. A comparison of the ratio of intrinsic diffusion coefficients $D_{Cd}/D_{Ag}$ in Ag–Cd alloys as a function of the atomic fraction of cadmium. The experimental values of Iorio *et al.* (1973) are shown (———), together with curves obtained from Darken's equation (5.6.5) (—·—·—) and Manning's equations (5.3.6, 5.3.7 and 5.8.7) (— — —). (After Iorio *et al.*, 1973.)

than the others. The Kirkendall velocity for instance is increased by about 30% for a f.c.c. alloy.

There have been many experimental tests of the Manning relations via measurement of these quantities. The work of Lindstrom (1973a, b) on alkali halide couples, of Greene *et al.* (1971) and Dallwitz (1972) on Ag–Au couples, of Butrymowicz and Manning (1978) and Iorio *et al.* (1973) on Ag–Cd couples and of Damköhler and Heumann (1982) on Cu–Ni couples all provides evidence for the validity of the Manning corrections (Fig. 5.5). On the other hand, there are systems where the Manning corrections are evidently too small (e.g. V–Ti, Carlson, 1976 and Cu–Au, Heumann and Rottwinkel, 1978).

Some rather clear-cut tests of these relations are also possible via Monte Carlo computer simulations of model alloys, both random and interacting. For the random alloy models the relations prove to be remarkably accurate, while for the models in which there are interactions among the atoms they still provide good qualitative guides to the dependence of the *L*-coefficients upon composition and temperature. Details are given later in Chapter 13.

## 5.9 Electrical mobilities of ions

The considerations so far in this chapter have been for non-polar solids, specifically metals and alloys. We now turn to the strongly ionic, or ionic conducting, solids, well-known examples of which are the alkali and silver halides, some other halides with the fluorite structure, certain oxides, etc. We do so again with the object of relating some of the more familiar transport coefficients to the $L$-coefficients introduced by the non-equilibrium thermodynamics. First of all we consider the electrical mobilities of the ions.

In all these substances we are concerned with two or more distinct sub-lattices, e.g. the anion and the cation sub-lattices. We know that ions of one sub-lattice do not occupy sites on the other, because the energies involved would be prohibitively high. Vacancies on the two sub-lattices can therefore be regarded as distinct defects, although their concentrations may be coupled together through the conditions of Schottky equilibrium, and through the constraints of electro-neutrality and structure, as described in Chapter 3. Furthermore we shall assume that there is no significant electronic disorder and thus (i) that there can be no electron wind effects (i.e. all $z_i^* = 0$) and (ii) that the vacancies are present in only one charge state, as defined by the absence of the ion.

For definiteness consider a solid solution of chemical composition $M_aY_y$ displaying Schottky disorder and in which the metal sub-lattice is occupied by A ions of charge $q_A$ and B ions of charge $q_B$. The anions are of charge $q_Y$. Let us indicate vacancies on the cation sub-lattice by Vc and on the anion sub-lattice by Va. Then, recognizing that the cations can move only via the cation vacancies and the anions only via the anion vacancies, we can write the flux equations for our system as

$$J_A = L_{AA}(X_A - X_{Vc}) + L_{AB}(X_B - X_{Vc}), \tag{5.9.1a}$$

$$J_B = L_{BA}(X_A - X_{Vc}) + L_{BB}(X_B - X_{Vc}) \tag{5.9.1b}$$

and

$$J_Y = L_{YY}(X_Y - X_{Va}), \tag{5.9.1c}$$

(cf. §4.4.1). In keeping with the idea of two distinct sub-populations of vacancies we have supposed that any cross-coefficients $L_{AY}$, $L_{BY}$, etc. are zero. This will be a good approximation as long as the physical interactions between cation and anion vacancies are negligible in their effects. On the other hand, if the proportion of bound cation vacancy–anion vacancy pairs were comparable to or greater than those of the unpaired vacancies then this neglect of $L_{AY}$, $L_{BY}$, etc. would not be correct. Such cases can arise, but here we shall consider just the simpler situations.

We suppose that the system is subject to a uniform electric field, $E$, and that there are no concentration gradients. We then have

$$X_i = q_iE, \qquad (i = \text{A, B, Y}), \tag{5.9.2a}$$

and

$$X_{Vc} = 0 = X_{Va}. \tag{5.9.2b}$$

It follows that the electrical mobility of the A cations is

$$u_A = \frac{1}{n_A} |q_A L_{AA} + q_B L_{AB}|, \tag{5.9.3}$$

with a similar expression for the B ions. The presence of the cross-coefficient $L_{AB}$ in this expression comes from the assumption that both A and B ions are moved by the same cation vacancies. We observe that these electrical mobilities differ from the 'mobility' introduced in the discussion of Manning's relations (§5.8) by the presence of charges $q_A$ and $q_B$. These two types of mobility are essentially equivalent to one another when $q_A = q_B$ but not otherwise.

We can obviously also obtain diffusion coefficients for A and B ions on the cation sub-lattice. The relation of these to the electrical mobilities is of interest. The general treatment of diffusion coefficients in these ionic crystals is somewhat more complex than in metals and this is considered separately below. Some specific relations can be obtained easily, however. Firstly, for very dilute solid solutions where $c_B \ll c_{Vc}$, we can expect to use the limiting result (5.3.8) here as well as for metals. Hence by (5.9.3)

$$\frac{u_B}{D_B} = \frac{|q_B L_{BB} + q_A L_{BA}|}{kT L_{BB}}, \qquad (c_B \to 0). \tag{5.9.4}$$

This is a generalization of the well-known Nernst–Einstein relation. We see that the usual simplified form, namely,

$$\frac{u_B}{D_B} = \frac{|q_B|}{kT}, \tag{5.9.5}$$

only applies when $L_{BA}$ is zero, i.e. when there is no coupling between the flows of A and B ions.

In the special case where B is a particular isotope of A, $q_B$ is necessarily equal to $q_A$ and we can use (5.7.7) to obtain

$$\frac{u_A}{D_A^*} = \frac{|q_A|}{kTf}. \tag{5.9.6}$$

This relation is exact for these circumstances. However, it is also obtained, without restriction on the relative concentrations of elements A and B, when $q_A = q_B$ and when we can use the Manning relations (5.8.7). In these circumstances there will be an equivalent relation for B, but both relations will generally be approximate. However, (5.9.6) does not follow, even in these circumstances, when $q_B \neq q_A$.

The direct measurement of solute ion mobilities is not easy, but has been carried out in a few alkali halides (Chemla, 1956; Brébec, 1973). The results confirm the result (5.9.6) when $q_B = q_A$, but indicate considerable departures from it when $q_B \neq q_A$. Later in Chapter 11 we shall see that this contrast is understandable on the basis of more specific models.

The total ionic conductivity of a system, $\kappa$, is more easily measured. If the system is pure then by (5.9.6)

$$\frac{\kappa_A}{D_A^*} = \frac{nq_A^2}{kTf}, \tag{5.9.7}$$

with a similar expression for the anions Y, $\kappa_A$ and $\kappa_Y$ being the cation and anion contributions to the total ionic conductivity, $\kappa$. In practice it may be possible to separate the contributions $\kappa_A$ and $\kappa_Y$ to $\kappa$ by making classical transport number measurements, which yield the ratios $\kappa_A/\kappa$ and $\kappa_Y/\kappa$. Accurate tests of (5.9.7), however, are generally only possible when we can be sure that one of $\kappa_A$ and $\kappa_Y$ is negligibly small compared to the other (e.g. in the silver halides $\kappa_A \gg \kappa_Y$, while in the fluorite compounds like $CaF_2$ $\kappa_Y \gg \kappa_A$). In such cases determinations of the ratio $D^*/\kappa$ can give useful information about the underlying defect mechanism. (To this end the quantity $D^*/(kT\kappa/nq^2)$ is sometimes referred to as the Haven ratio $H_R$.) In analysing such measurements one often has to go beyond the simple form (5.9.7) on account of the complexities introduced when more than one type of defect is present as, e.g. with Frenkel defects where both vacancies and the corresponding interstitials are present. We return to this matter in Chapter 11.

## 5.10 Isothermal chemical diffusion in ionic solids

The discussion of diffusion in §5.3 assumed that no field was present. In strongly polar solids, such as the alkali halides where the ions carry electric charges, diffusion will, in general, give rise to an internal electric field, $E$, the so-called Nernst field. This field is just sufficient to ensure that the total electric current is zero in the absence of any external applied electric field. It is easily derived within the above formalism by setting $F_i = q_i E$ where $q_i$ is the electric charge of species $i$, and then determining $E$ from the condition

$$J_e \equiv \sum_i q_i J_i = 0. \tag{5.10.1}$$

We can proceed with this first step in a general way. Suppose the system contains $n$ independent mobile species (which could include electronic defects as well as mobile ions) and that we write the flux equations in matrix form as

$$\mathbf{J} = \mathbf{LX} \tag{5.10.2}$$

The forces $X_i$ appearing here will be written

$$X_i = X_i^{(0)} + q_i E \tag{5.10.3}$$

where $E$ is the Nernst electrical field, to be determined by the condition that no electrical current flows ('open circuit' condition), i.e. by (5.10.1). The part $X_i^{(0)}$ will generally be the diffusion force, but if we wish to include gradients of temperature it could also contain a thermal term $Q_i^* X_q$ (cf. §4.4.1). In that particular case the Nernst field would in fact yield the so-called homogeneous part of the thermopower (§1.3; also Howard and Lidiard, 1964).

The condition of zero electrical current (5.10.1) in matrix form and with (5.10.2) and (5.10.3) yields

$$\tilde{\mathbf{q}}\mathbf{J} = \tilde{\mathbf{q}}\mathbf{L}(\mathbf{X}^{(0)} + \mathbf{q}E) = 0 \tag{5.10.4}$$

and thus

$$\mathbf{J} = \mathbf{L}\left[\mathbf{X}^{(0)} - \mathbf{q}\frac{(\tilde{\mathbf{q}}\mathbf{L}\mathbf{X}^{(0)})}{(\tilde{\mathbf{q}}\mathbf{L}\mathbf{q})}\right] \tag{5.10.5a}$$

or, in component notation,

$$J_i = \sum_{j=1}^{n} \Lambda_{ij} X_j^{(0)} \tag{5.10.5b}$$

in which

$$\Lambda_{ij} = \sum_{r,s}^{n} q_r q_s \frac{(L_{ij}L_{rs} - L_{is}L_{rj})}{\sum_{r,s}^{n} q_r q_s L_{rs}}. \tag{5.10.5c}$$

We make two observations on this result. Firstly, we observe that the form (5.10.5b) is just that which would be suggested by the expression for the entropy production if we had first reduced that by eliminating $E$ from it via (5.10.4). Just as with the example treated in Appendix 5.1 the phenomenological coefficients arising in this representation, i.e. the $\Lambda_{ij}$, are distinct from the $L_{ij}$ appearing in the more general representation, although as (5.10.5c) shows they are expressible in terms of them. Secondly, we could say that, although the form of (5.10.5b) is like that of the starting equations (5.10.2), we see that the elimination of the Nernst field (giving $X_j^{(0)}$ in place of $X_j$) has 'mixed up' the initial set of $L$-coefficients to give the new $\Lambda$-coefficients. This is not a mere formal change since the statistical theories to be described in Chapters 8 to 11 will be derived in the absence of the constraint (5.10.1) and thus will yield directly expressions for the $L$-coefficients rather than the $\Lambda$-coefficients. The set of equations represented in (5.10.5) are now suitable for describing diffusion in ionic solids.

Before proceeding to any further development, however, we distinguish two types of system. Firstly, there are the semi-conducting ionic solids, exemplified by various oxides of the transition metals. Here mobile electronic defects will be present, electron holes in the above example: and this species, which must also be included in the above calculation, will almost always be far more mobile than the atomic species. In other words the corresponding coefficient, $L_{ee}$ say, will be much greater than all of the others, including $L_{ei} = L_{ie}$ $(i \neq e)$. In these circumstances (5.10.5a) simplifies to

$$\mathbf{J} = \mathbf{L}[\mathbf{X}^{(0)} - \mathbf{q}X_e^{(0)}/q_e]. \tag{5.10.6}$$

The second class of system, typified by the alkali and silver halides and other ionic conductors, is made up of those containing essentially no electronic defects. In this case there may not be any predominant $L$-coefficient, when it would be necessary to use (5.10.5). We see easily, however, that

$$\sum_{j=1}^{n} \Lambda_{ij} q_j = 0 \tag{5.10.7}$$

and thus if we wished to eliminate one of the species (say $n$) from explicit consideration (e.g. the ions of the host lattice) we can rewrite (5.10.5b) as

$$J_i = \sum_{j=1}^{n-1} \Lambda_{ij}(X_j^{(0)} - q_j X_n^{(0)}/q_n), \tag{5.10.8}$$

in analogy to (5.10.6).

A further difference between these two classes appears when we consider the conditions of local equilibrium. The strongly ionic compounds (e.g. ionic conductors) remain stoichiometric under a wide range of conditions and local equilibrium is attainable without change in the M:Y ratio, e.g. by internal adjustment of the Frenkel or Schottky defect population. This equilibrium can be achieved relatively quickly, depending on the defect type and the density of sources and sinks. By contrast, the defect population in the transition metal oxides and similar non-stoichiometric compounds is directly determined by equilibrium with a surrounding Y phase (cf. §3.3.1). Local equilibrium may therefore only be attained much more slowly, since either transport over macroscopic distances or internal decomposition is involved. The description of such processes may be complicated, although elegant experiments involving the transport of cations in specimens subject to a gradient in the chemical potential of Y (oxygen) have been devised to permit straightforward analysis (Schmalzried, Laqua and Lin, 1979; Schmalzried and Laqua, 1981). At this point therefore we analyse just the simpler situation presented by the more strongly ionic compounds.

Let us now show the explicit form of the results for the isothermal interdiffusion of cations A and B in an ionic conductor having the basic chemical structure $M_aY_y$ and displaying Schottky disorder. The basic flux equations for the model have already been set down in the preceding section (see eqns. (5.9.1)). The forces $X_i^{(0)}$ in (5.10.4) *et seq.* are to be identified with the terms $(X_A - X_{Vc})$, etc. in (5.9.1), while the Λ-coefficients, $\Lambda_{AA}$, etc., are obtained from (5.10.5c). Equation (5.10.5b) then yields the expressions for $J_A$ and $J_B$ in terms of $(X_A - X_{Vc})$, $(X_B - X_{Vc})$ and $(X_Y - X_{Va})$. These expressions then simplify further if (i) we can assume local defect equilibrium and thus the Gibbs–Duhem relation in the form

$$n_A(X_A - X_{Vc}) + n_B(X_B - X_{Vc}) + n_Y(X_Y - X_{Va}) = 0 \qquad (5.10.9)$$

and (ii) we insert the condition of overall electroneutrality for this model, namely

$$q_A n_A + q_B n_B + q_Y n_Y = 0. \qquad (5.10.10)$$

In this way one can straightforwardly show that the flux of A ions relative to the local crystal lattice is

$$J_A = \frac{1}{\kappa}[q_A(\nabla\mu_B - \nabla\mu_{Vc}) - q_B(\nabla\mu_A - \nabla\mu_{Vc})][q_B|L_c| - q_Y L_{YY}(n_B L_{AA} - n_A L_{AB})/n_Y] \qquad (5.10.11)$$

in which

$$|L_c| = L_{AA}L_{BB} - L_{AB}L_{BA}, \qquad (5.10.12)$$

and

$$\kappa = q_A^2 L_{AA} + q_A q_B(L_{AB} + L_{BA}) + q_B^2 L_{BB} + q_Y^2 L_{YY}, \qquad (5.10.13)$$

is actually the local electrical conductivity. The expression for $J_B$ is obtained from (5.10.11) by simply transposing A and B throughout.

Although (5.10.11) is more complicated than the corresponding expression for a binary non-polar system (5.3.4) we see that it too has the form of a product of a thermodynamic factor (in the $\nabla\mu$) and a kinetic factor (in the $L$-coefficients). We also see that the thermodynamic factor for $J_A$ is just the negative of that for $J_B$, as with the simpler example presented in §5.3 (see (5.3.4)). The more complex nature of each of these factors here is, of course, a consequence of the coupling together of the various fluxes by the Nernst internal electric field. To recast (5.10.11) in the form of Fick's first law and thus to infer a diffusion coefficient $D_A$ requires the re-expression of $\nabla\mu_A$, $\nabla\mu_B$ and $\nabla\mu_{Vc}$ in terms of $\nabla c_A$. It is here that different models may dictate somewhat different lines of development. However, in a general context we can see how to obtain the necessary re-expression. Thus we can conveniently regard the various chemical potentials as given functions of $c_A$, $c_B$ and $c_Y$. The fraction $c_Y$ is constrained by the electroneutrality condition (5.10.10), which thus also gives $\nabla c_Y$ in terms of $\nabla c_A$ and $\nabla c_B$. The relation between $\nabla c_A$ and $\nabla c_B$ is then obtained from the condition of Schottky equilibrium. In this way the thermodynamic factor in (5.10.11) is expressible as a multiple of $\nabla c_A$ alone, whence $D_A$ can be derived: likewise for $D_B$. A formal solution is easily obtained in this way. An equivalent solution is obtained if we work in terms of $c_{Vc}$, $c_B$ and $c_{Va}$ (the 'defect picture'), as is appropriate in dilute solutions where $c_B \ll c_A$. We remark that when $q_B \neq q_A$ the presence of the solute strongly affects the vacancy concentrations and thus $D_B$ may depend sensitively on $c_B$ (Fredericks, 1975). An example was presented in Fig. 1.2. Detailed analysis for such examples is given later in §11.8.

The above procedure is a natural one to follow when we have an adequate statistical theory of the component chemical potentials involved. However, from an electrochemical point of view it may seem desirable to avoid the use of separate chemical potentials for individual ionic species (on account of the coupling together enforced by the electroneutrality requirement). This can easily be done if we now specify $A_aY_y$ as the host material. Electroneutrality then requires

$$aq_A + yq_Y = 0, \tag{5.10.14}$$

while the chemical potential of $A_aY_y$ in the usual way will be

$$\begin{aligned} \mu(A_aY_y) &= a\mu_A + y\mu_Y \\ &= \frac{-yq_Y}{q_A}\mu_A + y\mu_Y, \end{aligned} \tag{5.10.15}$$

by (5.10.14). For the conditions already specified it is not difficult to show that the thermodynamic factor in square brackets in (5.10.11) is, in fact, just

$$-\frac{q_A n_Y}{y n_B}\nabla\mu(A_aY_y). \tag{5.10.16a}$$

Furthermore, if we now set down the chemical potential, $\mu(B_bY_y)$, of $B_bY_y$ when dissolved in the $A_aY_y$ structure (which requires us to recognize that $(q_B - q_A)/q_A$ cation vacancies are created for each molecule of $B_bY_y$ incorporated) we find that (5.10.16a) can equally well be written as

$$\frac{q_A n_Y}{y n_A}\nabla\mu(B_bY_y). \tag{5.10.16b}$$

This establishes the symmetry between the expressions for $J_A$ and $J_B$.

Although these results have been obtained by pursuing the microscopic formulation in terms of structural elements, we can now see that they fit into a mathematical structure exactly similar to that which we would have obtained by recognizing that our system is just a quasi-binary solid solution of $A_aY_y$ and $B_bY_y$ 'molecules' with the structure of $A_aY_y$ and then applying macroscopic thermodynamics. In particular, the equivalence of (5.10.16a) and (5.10.16b) is just the

Gibbs–Duhem relation in this representation. Despite the occurrence of the Nernst diffusion field there is therefore an equivalence between the thermodynamic description of this quasi-binary ionic system and the binary alloy treated in §5.3. A consequence of all this is that the thermodynamic factor in (5.10.11) in all cases necessarily relates to an electrically neutral combination of structural elements. Any theory which fails to comply with this conclusion is therefore wrong.

Let us now return to the defect picture to describe the Kirkendall effect. Thus when $q_B \neq q_A$, i.e. when the presence of B ions in the structure is 'compensated' by a change in the number of vacancies, part of the vacancy flux is the direct result of the flux of B ions. Formally, from the condition of zero charge current (5.10.1), we have

$$\frac{J_{Va}}{y} = \frac{1}{a}\left[J_{Vc} - \left(\frac{q_B - q_A}{q_A}\right)J_B\right], \tag{5.10.17a}$$

and equivalently

$$\frac{J_{Vc}}{a} = \frac{1}{y}\left[J_{Va} - \frac{y}{a}\left(\frac{q_A - q_B}{q_A}\right)J_B\right]. \tag{5.10.17b}$$

These equations may be interpreted as follows. When $q_B > q_A$ (5.10.17a) tells us that part of $J_{Vc}$ is tied to the motion of the B ions. Conversely, when $q_B < q_A$ (5.10.17b) tells us that part of $J_{Va}$ is tied to the motion of the B ions. Clearly in each case only the remaining part of the specified vacancy flux (corresponding to Vc and Va contributions in stoichiometric proportions) can lead to Schottky defect condensation and hence to a motion of the diffusion zone relative to the fixed ends of the system. Hence when $q_B > q_A$ the velocity of this Kirkendall shift in the direction of $J_B$ is

$$v_K = J_{Va}/yn, \qquad (q_B > q_A), \tag{5.10.18}$$

where the molecular volume has been taken as the inverse of $n$, the number of formula units of the structure $M_aY_y$ ('molecules') per unit volume. On the other hand, when $q_B < q_A$ the velocity of the Kirkendall shift in the direction of $J_B$ is

$$v_K = J_{Vc}/an, \qquad (q_B < q_A). \tag{5.10.19}$$

These expressions can be evaluated from the expressions for $J_A$ and $J_B$ (5.10.11). Obviously both cases involve the same thermodynamic factor, but the kinetic factor has a different form in the two cases.

Lastly, we note that as in Appendix 5.1 we could, quite legitimately, have eliminated dependent quantities from the expression for the rate of entropy generation and used the resultant form to determine the phenomenological flux equations. However, since there are several relations of structure, electroneutrality, vacancy equilibrium and Gibbs–Duhem defining dependent quantities it turns out that the final form of $T\sigma$ is sufficiently simple as to allow only one phenomenological equation, actually between $(c_B J_A - c_A J_B)$ and the same combination of forces as appears in (5.10.11). Hence it leads to just one phenomenological coefficient and only a partial representation of the system (as in the non-polar system considered in Appendix 5.1).

## 5.11 Summary

Our aim in this chapter has been to make the formal structure presented in Chapter 4 more readily understandable by establishing some of the connections with more familiar analyses. By examining some familiar processes, in particular chemical diffusion, isotope diffusion and ionic conductivity we have shown how the corresponding transport coefficients, such as diffusion coefficients and electrical mobilities of ions, are expressible in terms of the generalized transport coefficients, i.e. the $L$-coefficients introduced in Chapter 4, together with certain thermodynamic factors. In addition, we have shown how certain broad physical arguments and assumptions can lead to useful relations among some of these coefficients. Two principal types of argument have been illustrated. The first type may be called *thought experiments*, i.e. circumstances where we believe we know the precise outcome of the *experiment* on general physical grounds. We have used such arguments to obtain exact relations which apply to transport coefficients arising with isotope diffusion. The second type of argument involves more sweeping intuitive physical considerations which generally lead to approximate results. Examples are the Darken and the Manning relations between chemical diffusion coefficients and isotopic diffusion coefficients. These relations, by their simplicity, can be of considerable practical value, but their accuracy has to be determined by experiment and by the accurate theoretical analysis of specific models. In the next chapter we derive general statistical mechanical expressions for the $L$-coefficients in terms of detailed kinetic quantities which describe diffusive atomic movements in solids. These general results open the door to a whole range of subsequent calculations for specific systems.

# Appendix 5.1

## Isothermal chemical diffusion

In the conclusion to §5.3 we noted that an alternative phenomenological representation of chemical interdiffusion in a binary alloy could be given. This appendix provides the details.

We make the same assumptions as before, firstly that the vacancies are everywhere locally in equilibrium, $X_V = 0$. If we return to the basic equation for the rate of entropy production per unit volume, $\sigma$, we have

$$T\sigma = J_A X_A + J_B X_B. \tag{A5.1.1}$$

Now, if we insert the Gibbs–Duhem equation (5.3.2) into (A5.1.1) with the elimination of $X_B$, we have

$$T\sigma = \left( J_A - \frac{c_A}{c_B} J_B \right) X_A. \tag{A5.1.2}$$

This therefore leads to a phenomenological relation of the form

$$J_A - \frac{c_A}{c_B} J_B = \Lambda_{AA} X_A. \tag{A5.1.3a}$$

We use the symbol $\Lambda_{AA}$ for the phenomenological coefficient to indicate that it is necessarily distinct from $L_{AA}$ and should not be confused with it. Alternatively, if we had eliminated $X_A$ we should have obtained

$$J_B - \frac{c_B}{c_A} J_A = \Lambda_{BB} X_B. \tag{A5.1.3b}$$

However, (A5.1.3a) and (A5.1.3b) are not independent relations, i.e. $\Lambda_{BB}$ is not independent of $\Lambda_{AA}$. To see this multiply (5.1.3a) by $c_B/c_A$ throughout and add the result to (5.1.3b). Then with the Gibbs–Duhem relation between $X_A$ and $X_B$

we see that

$$\frac{c_B}{c_A}\Lambda_{AA} - \frac{c_A}{c_B}\Lambda_{BB} = 0, \qquad \text{(A5.1.4)}$$

i.e. there is only one independent phenomenological coefficient in this scheme, and we cannot solve (5.1.3a, b) for $J_A$ and $J_B$ separately.

In fact, the single coefficient corresponds to the chemical interdiffusion coefficient $\tilde{D}$. In the usual way the flux of A relative to the fixed ends of the diffusion couple is

$$J'_A = c_B J_A - c_A J_B = -J'_B. \qquad \text{(A5.2.5)}$$

Hence by (A5.1.3)

$$J'_A = c_B\left(J_A - \frac{c_A}{c_B}J_B\right)$$

$$= c_B \Lambda_{AA} X_A = -J'_B,$$

i.e.

$$\tilde{D} = \frac{kT}{n}\frac{c_B\Lambda_{AA}}{c_A}\left(1 + \frac{\partial \ln \gamma}{\partial \ln c}\right), \qquad \text{(A5.1.6a)}$$

$$= \frac{kT}{n}\frac{c_A\Lambda_{BB}}{c_B}\left(1 + \frac{\partial \ln \gamma}{\partial \ln c}\right). \qquad \text{(A5.1.6b)}$$

In conclusion then, we see that if we first reduce the expression for the entropy generation by using the Gibbs–Duhem relation and then use the reduced expression so obtained to give the phenomenological relations, there is in fact, only one such relation in the present example; this allows us to represent chemical interdiffusion. However, in this approach there is no way to represent the Kirkendall effect, since $J_A$ and $J_B$ cannot be independently represented. There is nothing wrong with the present reduced representation; it is simply that it deals with a more limited range of phenomena than the more general representation set out in §5.3. It is obvious that this must always be the case when the number of variables is reduced at the very beginning. Another example arises in §5.10.

# 6

# *Microscopic theories – the master equation*

## 6.1 Introduction

In this and following chapters we shall consider those theories which are based on microscopic models of atomic movements and which allow us to obtain expressions for the macroscopic phenomenological coefficients introduced in the preceding two chapters. The principle on which all these particular models are founded has already been introduced in §§2.3 and 3.11 and is that the motion of atoms (or ions) of the solid may be divided into (i) thermal vibrations about defined lattice sites and (ii) displacements or 'jumps' from one such site to another, the mean time of stay of an atom on any one site being many times both the lattice vibration period and the time of flight between sites. The justification of this principle is provided widely in solid state physics. In general terms it derives from the strong cohesion of solids and the associated strong binding of atoms to their assigned lattice sites. This leads to well-defined phonon spectra such as have now been observed, especially by neutron scattering methods, in a wide range of solids. Theories of solid structures also show that there are generally large energy barriers standing in the way of the displacement of atoms from one site to another, even in the presence of imperfections in the structure, with the consequence that such displacements must be thermally activated and will occur relatively infrequently compared to the period of vibrational motions. Thus even the highest diffusion rates (say $D \sim 10^{-9}\,\mathrm{m^2\,s^{-1}}$) correspond to a mean time of stay $\tau \sim 10^{-11}$ s, i.e. 100 times longer than the period of the lattice vibrations. It is more difficult to obtain information about the time of flight between one atomic configuration and another, but particular molecular dynamics simulations show this too to be much shorter than the time of stay. (For reviews of these molecular dynamics calculations see e.g. Bennett (1975) and Jacobs, Rycerz and Mościński (1991).)

The general principle is thus broadly founded, despite the possible existence of exceptions (e.g. some fast-ion conductors where the 'sites' involved may be rather indefinite, see Funke, 1989). In this chapter we shall therefore give the corresponding mathematical formulation, but without introducing other particular principles, as we shall want to do later (e.g. the existence of defects as the promoters of these transitions from one configuration to another). The immediate objective is to provide the general structure of these theories. Before setting out on that task we note that the picture of atomic motions and displacements which we have described is, in principle, derivable by applying the methods of statistical mechanics to a Hamiltonian representation of these atomic motions. Indeed, this is just the approach underlying computer simulations of the molecular dynamics type, and it is the basis of the transport theory of fluids generally (where one does *not* have the simplifying picture we have described). However, we shall not use the Hamiltonian formalism as our starting point here, because the validity of the physical picture of atomic movements in solids which we have just described means that a great deal that is contained within the full Hamiltonian description is superfluous from the present point of view (e.g. the vibrational atomic motions). We shall therefore follow Allnatt (1965) and base our discussion on the far less detailed description which is provided by the so-called 'master equation'. This we establish in the next section. In the following two sections we then trace two principal lines of development; (i) the evaluation of mean values of kinetic quantities and (ii) the derivation of expressions for the transport coefficients, $L_{ij}$, etc. None of this requires the lattice of sites which the atoms may occupy to be a regular lattice. It could be an irregular framework, as appropriate to a glassy substance, as long as the framework itself does not change with time. Furthermore, none of it requires the ideas of lattice defects (vacancies and interstitials) as the means by which atoms may move from one site to another. Such specific features are introduced in Chapters 7 *et seq.* However, although general in these respects, our discussion is limited to isothermal systems. The treatment of systems in which temperature gradients are present has not so far been carried out in the same generality.

The next two sections introduce the master equation and certain formal consequences of it. In §6.4 we use it to obtain expressions for the mean values of relevant kinetic quantities, while in §6.5 we develop the corresponding form of linear response theory. By these means we arrive at expressions for the transport coefficients $L_{ij}$ and obtain generalizations of the two elementary Einstein relations pertaining to Brownian motion (namely that between the diffusion coefficient, $D$, and the mean-square displacement of the particles and that between $D$ and their mobility in a force field). This last section also looks at certain implications, including those for the Darken and Manning approximations described in the previous chapter.

## 6.2 Formal development: the master equation for isothermal systems*

Since the mean time-of-stay of the atoms on the sites of the lattice is long compared to lattice vibration periods and to the flight-times of atoms between lattice sites, we may average out the vibrational motions and also may treat the jumps of atoms, when they occur, as accomplished effectively instantaneously. It follows that we can specify the states of the whole solid $\alpha, \beta, \gamma, \ldots$ by the way the atoms are distributed over the lattice sites, i.e. by just those configurations which we averaged over to get the configurational entropy in Chapter 3. We also introduce quantities, such as may be provided by transition-state theory (§3.11), which give the frequency, $w$, with which a system in contact with a 'heat-bath' (the lattice vibrations provide the contact) makes transitions between pairs of these accessible states. We denote the frequency of transition from an initial state $\alpha$ to a final state $\beta$ by $w_{\beta\alpha}$. (Note the order in which the subscripts appear.) The central hypothesis is that these transition frequencies at any one time depend only on the initial and final states of the whole system and are independent of its previous history. In other words, we assume that after a transition the thermal motions of the solid erase any memory of the preceding state, and that there is no possibility of an atom or group of atoms 'bouncing' dynamically through a series of states. (The system is thus Markovian: van Kampen, 1981.) These transition frequencies may, however, depend on time explicitly, for example when the system is subject to a time-dependent external field.

In the usual manner of statistical mechanics we replace the heat-bath to which the system is coupled by an ensemble of identical systems weakly coupled together and equate the observable macroscopic properties of the system to appropriate averages over the whole ensemble at any one time (see, e.g. McQuarrie, 1976). We thus introduce the probability $p_\alpha$ that any given system in the ensemble is in state $\alpha$. In general, this will be a function of time, $p_\alpha(t)$, as the system evolves.

With these definitions we can write down the equation governing the evolution of $p_\alpha(t)$, namely

$$\frac{\mathrm{d}p_\alpha}{\mathrm{d}t} = -\sum_{\beta \neq \alpha} w_{\beta\alpha} p_\alpha + \sum_{\beta \neq \alpha} w_{\alpha\beta} p_\beta. \tag{6.2.1}$$

The first term on the right side represents the rate at which systems are leaving state $\alpha$ for other states, while the second term gives the rate at which they are

* As already indicated we limit ourselves to isothermal systems. However, although the master equation is equally valid in both isothermal and anisothermal systems, there is no generally accepted description of the transition frequencies in anisothermal systems like that presented in §3.11 for isothermal systems. We therefore do not consider them further here.

arriving in $\alpha$ from other states, We write this in matrix form as

$$\frac{d\mathbf{p}}{dt} = -\mathbf{P}\mathbf{p}, \tag{6.2.2}$$

by forming the occupation probabilities into a column matrix (for convenience written horizontally between braces)

$$\mathbf{p} \equiv \{p_1\, p_2 \ldots\}, \tag{6.2.3}$$

while the matrix $\mathbf{P}$ is defined as follows:

$$P_{\beta\alpha} = -w_{\beta\alpha} \qquad (\alpha \neq \beta) \tag{6.2.4a}$$

$$P_{\alpha\alpha} = \sum_{\beta \neq \alpha} w_{\beta\alpha}. \tag{6.2.4b}$$

Equation (6.2.2) is known as the master equation. All the theories we are concerned with are based on (6.2.2) either explicitly or by implication.

Several characteristics of these equations should be noted. Firstly it is clear from the definition of the matrix $\mathbf{P}$ that

$$\sum_{\beta} P_{\beta\alpha} = 0, \qquad (\text{all } \alpha) \tag{6.2.5}$$

i.e. that $\mathbf{P}$ possesses a left eigenvector $\tilde{\boldsymbol{\tau}}$

$$\tilde{\boldsymbol{\tau}} = [1 \quad 1 \quad 1 \quad \ldots \quad 1], \tag{6.2.6}$$

each of whose elements is unity and for which the corresponding eigenvalue is zero. Equation (6.2.5) is equivalent to the statement that probability is conserved, i.e. that $d(\tilde{\boldsymbol{\tau}}\mathbf{p})/dt = 0$. It follows that the eigenvalues of $\mathbf{P}$ all have real parts greater than or equal to zero (known as Gershgorin's theorem; see, Oppenheim, Shuler and Weiss, 1977).

Furthermore, because any one row of $\mathbf{P}$ is equal to minus the sum of all the others, (6.2.5) also tells us that the determinant $|P|$ is zero, i.e. that the matrix $\mathbf{P}$ is singular. This warns us that matrix equations involving $\mathbf{P}$ cannot be straightforwardly inverted.

The matrix $\mathbf{P}$ under conditions of thermal equilibrium, which we denote by $\mathbf{P}^{(0)}$, is especially important. There are two points to be made. Firstly, the corresponding distribution among the states, $\mathbf{p}^{(0)}$, is given by statistical thermodynamics as

$$p_{\alpha}^{(0)} = \frac{\exp(-E_{\alpha}^{(0)}/kT)}{\sum_{\gamma} \exp(-E_{\gamma}^{(0)}/kT)}, \qquad (\text{all } \alpha), \tag{6.2.7}$$

where $E_{\gamma}^{(0)}$ is the energy of the system when it is in state $\gamma$. (Strictly, these energies are Gibbs energies, obtained by averaging over all vibrational and electronic states not directly relevant to atomic configurations and atomic displacements.) Secondly, under these same conditions we can invoke the principle of *detailed balance* which tells us that

$$w_{\beta\alpha}^{(0)}p_{\alpha}^{(0)} = w_{\alpha\beta}^{(0)}p_{\beta}^{(0)}, \qquad (\text{all } \alpha, \beta). \tag{6.2.8}$$

In words, this principle asserts that the rates at which forward and backward transitions occur in thermodynamic equilibrium are equal for every individual transition. From (6.2.7) and (6.2.8) we see that

$$\frac{w_{\beta\alpha}^{(0)}}{w_{\alpha\beta}^{(0)}} = \exp(E_{\alpha}^{(0)} - E_{\beta}^{(0)})/kT. \tag{6.2.9}$$

Computer simulations of systems in thermodynamic equilibrium by the Monte Carlo method (Chapter 13) will always impose this condition.

Now although $\mathbf{P}^{(0)}$ is not a symmetrical matrix (as 6.2.8 shows), it may be symmetrized as follows:

$$\mathbf{S} = \mathbf{N}^{-1/2}\mathbf{P}^{(0)}\mathbf{N}^{1/2}, \tag{6.2.10}$$

where $\mathbf{N}$ is a diagonal matrix whose elements are

$$N_{\rho\sigma} = p_{\rho}^{(0)}\,\delta_{\rho\sigma}. \tag{6.2.11}$$

That $\mathbf{S}$ is symmetrical, i.e. that $\mathbf{S} = \tilde{\mathbf{S}}$, follows from (6.2.8) and (6.2.11). The eigenvectors $\mathbf{a}^{(\nu)}$ of $\mathbf{S}$ are defined by

$$\mathbf{S}\mathbf{a}^{(\nu)} = \alpha^{(\nu)}\mathbf{a}^{(\nu)}. \tag{6.2.12}$$

It is easily seen that the $\alpha^{(\nu)}$ are also the eigenvalues of $\mathbf{P}^{(0)}$. But since $\mathbf{S}$ is symmetric we know that the $\alpha^{(\nu)}$ are all real. Furthermore, $\mathbf{S}$ may be shown to be positive semi-definite, i.e. its eigenvalues are positive or zero (Appendix 6.1). There is at least one eigenvalue of zero, i.e. that which corresponds to the equilibrium distribution $\mathbf{p}^{(0)}$ ($\mathbf{p}^{(0)}$, by (6.2.2), is an eigenvector of $\mathbf{P}^{(0)}$ of eigenvalue zero). The positive eigenvalues are the reciprocals of the relaxation times of the system ($\alpha^{(\nu)} \equiv 1/\tau^{(\nu)}$). Lastly, since $\mathbf{S}$ is real and symmetric we can assume that the $\mathbf{a}^{(\nu)}$ form an orthonormal set, i.e.

$$\tilde{\mathbf{a}}^{(\nu)*}\mathbf{a}^{(\mu)} = \delta_{\nu\mu}, \tag{6.2.13a}$$

and

$$\sum_{\nu} a_{\alpha}^{(\nu)*}a_{\beta}^{(\nu)} = \delta_{\alpha\beta}, \tag{6.2.13b}$$

and the set is complete for this class of problems. By these relations we can represent $\mathbf{P}$ in terms of the $\mathbf{a}^{(\nu)}$ by

$$\mathbf{P}^{(0)} = \mathbf{N}^{1/2} \sum_{\nu} \mathbf{a}^{(\nu)} \tilde{\mathbf{a}}^{(\nu)*} \alpha^{(\nu)} \mathbf{N}^{-1/2}. \tag{6.2.14}$$

This can be useful in formal analysis.

When the systems we are concerned with are crystalline, it follows that $\mathbf{P}^{(0)}$ will reflect their translational symmetry. The states $\alpha, \beta, \ldots$ of the system will be specified by the position vectors of particular atoms. For example, for the diffusion of isolated atoms on an interstitial lattice we need follow the movement of just one such atom and the state is therefore determined by the position of that atom alone. Furthermore, the eigenvectors take the form of plane waves, each of which is characterized by a wave-vector $\mathbf{k}$ and a 'branch' index $s$, i.e. $\nu \equiv (\mathbf{k}, s)$. It is appropriate also to use periodic boundary conditions to define the allowed wave-vectors. With these facts in mind it will not be surprising that it is often convenient to take Fourier transforms of the basic equations.

We conclude this section with two broader observations. Firstly, it will be evident that the detailed balance relations (6.2.8) are important; the possibility of symmetrizing $\mathbf{P}^{(0)}$ and the derivation of the properties of $\mathbf{S}$ depend on them. This importance is enhanced by the significance of $\mathbf{P}^{(0)}$, the transition matrix in thermal equilibrium, in the theory of the macroscopic transport coefficients – as will be shown in §6.5 and as might be expected from the Green–Kubo theory of transport coefficients more generally. These detailed balance relations appear to be necessary to the deduction of the Onsager reciprocal relations among the transport coefficients (4.2.5); they are the representation in this scheme of the dynamical principle of microscopic reversibility, which is usually employed in the derivation of these relations (e.g. Kreuzer, 1981).

The second observation is the formal similarity of (6.2.2) to the Liouville equation in the statistical mechanics of systems described by Hamiltonian dynamics, with $\mathbf{P}$ taking the place of $i\mathscr{L}$, where $\mathscr{L}$ is the Liouville operator, and with $\mathbf{p}$ replacing the probability density in the phase space of all the atoms in the system. It will not be surprising therefore to find that some of the steps in the formal development and the results are similar to those arising in the statistical mechanics of Hamiltonian systems.

## 6.3 Stochastic interpretation of the master equation

Let us return to the master equation (6.2.2). There is a useful stochastic interpretation of this equation which can be developed when the matrix of transition rates, $\mathbf{P}$, is constant in time. This limits the discussion to either thermal equilibrium or steady-state conditions, but both circumstances are important.

Formal integration of (6.2.2) gives

$$\mathbf{p}(t) = \exp(-\mathbf{P}t)\mathbf{p}(0) \equiv \mathbf{G}(t)\mathbf{p}(0), \tag{6.3.1}$$

where the matrix $\exp(-\mathbf{P}t)$ is defined in the usual way by the exponential series. This shows that the matrix element $[\exp -Pt]_{\rho\sigma}$ gives the probability that the system will be in state $\rho$ at time $t$ given that it was in state $\sigma$ at time zero. For reasons which are obvious from (6.3.1) the matrix $\mathbf{G}(t) = \exp(-\mathbf{P}t)$ may be called the *propagator* (as is the analogous quantity which arises in the use of the Liouville equation). The propagator for a system in thermodynamic equilibrium, $\mathbf{G}^{(0)}$, can be represented in terms of the eigenvectors and eigenvalues of $\mathbf{S}$ as

$$\mathbf{G}^{(0)}(t) = \mathbf{N}^{1/2} \sum_{\nu} \mathbf{a}^{(\nu)}\tilde{\mathbf{a}}^{(\nu)*} \exp(-\alpha^{(\nu)}t)\mathbf{N}^{-1/2}. \tag{6.3.2}$$

Now by expanding the exponential we also see that the probability that the system is still in a state $\sigma$ at time $t$ and that it has not passed through any other state in the meantime is simply

$$G^{(\mathrm{d})}_{\sigma\sigma} = \exp(-P_{\sigma\sigma}t). \tag{6.3.3}$$

It follows that the probability $\Pi^{(1)}_{\rho\sigma}$ that the system is in state $\rho$ after time $t$ having made just one transition from $\sigma$ to $\rho$ at some intermediate time $s$ is

$$\Pi^{(1)}_{\rho\sigma} = \int_0^t G^{(\mathrm{d})}_{\rho\rho}(t-s)(-P_{\rho\sigma})G^{(\mathrm{d})}_{\sigma\sigma}(s)\,\mathrm{d}s, \qquad (\rho \neq \sigma). \tag{6.3.4}$$

Likewise the probability $\Pi^{(2)}_{\rho\tau\sigma}$ that the system is in state $\rho$ after time $t$ having made two transitions $\sigma \to \tau$ and $\tau \to \rho$ at intermediate times $s_1$ and $s_2$ is

$$\Pi^{(2)}_{\rho\tau\sigma} = \int_0^t \mathrm{d}s_2 \int_0^{s_2} \mathrm{d}s_1 G^{(\mathrm{d})}_{\rho\rho}(t-s_2)(-P_{\rho\tau})G^{(\mathrm{d})}_{\tau\tau}(s_2-s_1)(-P_{\tau\sigma})G^{(\mathrm{d})}_{\sigma\sigma}(s_1), \qquad (\rho \neq \tau \neq \sigma) \tag{6.3.5}$$

and so on.

These quantities $\Pi^{(n)}_{\rho\ldots\sigma}$ are true probabilities. When added together and summed over all intermediate states ($\neq\rho$ or $\sigma$) they yield $G_{\rho\sigma}(t)$, i.e. the probability that the system is in state $\rho$ at time $t$ given that it was in state $\sigma$ initially. As a result they can be used to evaluate mean quantities of the system over periods of time $t$ (Allnatt and Lidiard, 1988). We turn to such applications next.

## 6.4 Mean values of additive variables

Classic relations like the Einstein relation between the diffusion coefficient and the mean-square displacement in time $t$ of a particle undergoing Brownian motion point to the interest in the calculation of mean values of kinetic variables. The usual treatment of this type of problem is based on a Hamiltonian description of the system whose time evolution is then expressed by the Liouville equation in either its classical or quantum form (McQuarrie, 1976). Here we do not have such a description, but only the master equation (6.2.2). Nevertheless it is still possible to obtain general expressions for quantities of interest.

Let the basic kinetic quantities of interest be denoted by $X$, $Y$, etc. A simple example would be the $x$-co-ordinate of a tracer atom, but, as we shall see, the most useful new results follow when $X, Y, \ldots$ are collective quantities. (They are, of course, quite distinct from the thermodynamic forces introduced in Chapters 4 and 5 and also denoted by $X$.)

We suppose that $X$ changes by an amount $X_{\beta\alpha}$ when the system makes a direct transition from state $\alpha$ to state $\beta$. Likewise the total change $\Delta X$ in the variable $X$ when the system undergoes a sequence of transitions $(\alpha \to \gamma, \gamma \to \delta, \ldots, \varepsilon \to \beta)$ from $\alpha$ ending in state $\beta$ is the sum $X_{\beta\varepsilon} + \cdots + X_{\gamma\alpha}$. Of course, for any variable whose value is determined uniquely by the state of the system we should have $X_{\beta\alpha} = X_\beta - X_\alpha$ and the change $\Delta X$ undergone in a sequence of transitions would depend only on the initial and final states. The analysis of mean values can be carried forward some way without making this simplification (see e.g. Allnatt and Lidiard, 1988), but as the important applications require it we shall not here deal explicitly with the more general case.

The averages of practical interest will be averages of simple functions of the chosen variables, e.g. $\langle \Delta X \rangle$, $\langle (\Delta X)^2 \rangle$, $\langle \exp(ik\,\Delta X) \rangle$, $\langle \Delta X\, \Delta Y \rangle$, etc., in which the angular brackets denote averages over all initial, intermediate and final states. For example

$$\langle (\Delta X)^n \rangle = \sum_{\beta,\alpha} \langle (\Delta X)^n \rangle_{\beta\alpha} p_\alpha(0) \tag{6.4.1}$$

in which $\langle (\Delta X)^n \rangle_{\beta\alpha}$ is the average of $(\Delta X)^n$ for systems which are in state $\beta$ at time $t$ having initially been in state $\alpha$. This conditional average $\langle (\Delta X)^n \rangle_{\beta\alpha}$ can be written in two ways, either as

$$\langle (\Delta X)^n \rangle_{\beta\alpha} = (X_{\beta\alpha})^n G_{\beta\alpha}(t) \tag{6.4.2}$$

(from 6.3.1) or as

$$\langle (\Delta X)^n \rangle_{\beta\alpha} = (X_{\beta\alpha})^n \Pi^{(1)}_{\beta\alpha} + \sum_{\gamma} (X_{\beta\gamma} + X_{\gamma\alpha})^n \Pi^{(2)}_{\beta\gamma\alpha}(t)$$
$$+ \sum_{\gamma,\delta} (X_{\beta\delta} + X_{\delta\gamma} + X_{\gamma\alpha})^n \Pi^{(3)}_{\beta\delta\gamma\alpha}(t) + \cdots \tag{6.4.3}$$

by (6.3.4) *et seq.*

With the transition rates, i.e. the $P_{\rho\sigma}$ factors, independent of time, subsequent development from either (6.4.2) or (6.4.3) is conveniently carried through with the aid of Laplace transformations. This development is not difficult, but is algebraically somewhat cumbersome and it is therefore given in Appendix 6.2. Here we restrict ourselves to a few particular results. These introduce new, non-diagonal matrices derived from the thermal equilibrium transition matrix $\mathbf{P}^{(0)}$ according to the definitions

$$(V_X)_{\beta\alpha} = -X_{\beta\alpha}P^{(0)}_{\beta\alpha}, \qquad \beta \neq \alpha \tag{6.4.4a}$$

$$(V_{XX})_{\beta\alpha} = -X_{\beta\alpha}{}^2P^{(0)}_{\beta\alpha}, \qquad \beta \neq \alpha \tag{6.4.4b}$$

$$(V_{XY})_{\beta\alpha} = -X_{\beta\alpha}Y_{\beta\alpha}P^{(0)}_{\beta\alpha}, \qquad \beta \neq \alpha \tag{6.4.4c}$$

and so on.

For averages in a system in thermodynamic equilibrium we find that

$$\langle \Delta X \rangle^{(0)}/t = \tilde{\tau}\mathbf{V}_X\mathbf{p}^{(0)} \tag{6.4.5}$$

$$\langle (\Delta X)^2 \rangle^{(0)}/t = \tilde{\tau}\mathbf{V}_{XX}\mathbf{p}^{(0)} + 2\tilde{\tau}\int_0^t \mathrm{d}s\left(1 - \frac{s}{t}\right)\mathbf{V}_X\mathbf{G}(s)\mathbf{V}_X\mathbf{p}^{(0)} \tag{6.4.6}$$

$$\langle \Delta X\, \Delta Y \rangle^{(0)}/t = \tilde{\tau}\mathbf{V}_{XY}\mathbf{p}^{(0)} + 2\tilde{\tau}\int_0^t \mathrm{d}s\left(1 - \frac{s}{t}\right)\mathbf{V}_X\mathbf{G}(s)\mathbf{V}_Y\mathbf{p}^{(0)}. \tag{6.4.7}$$

The superscript (0) has been added to the angular brackets to emphasize that these averages are for systems in thermal equilibrium. For these conditions we can employ the principle of detailed balance (6.2.8) and show that $\langle \Delta X \rangle^{(0)}$ by (6.4.5) is zero, an unsurprising result when we recall that $X$ is determined by the state of the system.

The results (6.4.5)–(6.4.7) are valid for all variables $X$, $Y$, etc. of the type specified when the averages are taken over a thermal equilibrium ensemble. The first terms on the right sides of (6.4.6) and (6.4.7) would be the sole contributions to the indicated averages on the left in the absence of any correlations between successive changes in the variables $X$ and $Y$. The second terms on the right with the time integral contain the correlations between successive changes. In these correlation terms the quantities $\mathbf{V}_X$ and $\mathbf{V}_Y$ play the role of stochastic analogues of the time rates of change of the variables $X$ and $Y$ as they would appear for a Hamiltonian dynamical system. For example, if $X$ and $Y$ represent the displacements of two particular atoms (X and Y, say) in a diffusive system then $\mathbf{V}_\mathrm{X}$ and $\mathbf{V}_\mathrm{Y}$ are the corresponding stochastic analogues of velocity. In appropriate circumstances we can use the results (6.4.6) and (6.4.7) to make contact with the linear response theory of systems subject to perturbations. We do that in the next section.

It is also possible to obtain results for a system in a steady state, i.e. when there is some external perturbation acting which changes $\mathbf{P}$ from its thermal equilibrium form $\mathbf{P}^{(0)}$ to $\mathbf{P}^{(0)} + \mathbf{P}^{(1)}$ but still leaves it constant in time. A steady electric field acting on an open system would be one such example. In these circumstances, however, additional considerations enter, in particular the linear reponse approximation, which we discuss in the next section. We therefore return to the matter of mean quantities later.

## 6.5 Linear response theory for isothermal systems

We conclude this chapter on the general stochastic theory with a presentation of the results of solving the master equation (6.2.2) in the linear response approximation. This linear approximation is evidently dictated by the aim of providing a microscopic interpretation of the linear macroscopic relations presented in Chapters 4 and 5. In view of the way the forces due to gradients in chemical potential and those having a mechanical or electrical origin combine together in these relations (see e.g. eqns. 4.2.3a and 4.2.14a), it would be desirable to develop this linear response theory for a similarly general type of perturbation. Unfortunately, however, in non-equilibrium statistical mechanics there is still no general, defined way of dealing with *thermal* or *boundary* transport problems (e.g. diffusion, heat transport, shear flow in liquids), i.e. those situations where the Hamiltonian function for the system as a whole is unchanged but where the boundary conditions dictate a departure from thermodynamic equilibrium. By contrast, the treatment of *mechanical* transport coefficients, i.e. those corresponding to transport in response to the imposition of a force field which does change the Hamiltonian function, is well defined. The same situation holds here.

One thus supposes that the system is subject to some mechanical perturbation specified by a conservative force field whose strength is measured by a parameter $\xi$, and one expands the transition frequencies $w_{\beta\alpha}$ and the occupation probabilities $p_\alpha$ in powers of $\xi$, thus

$$w_{\beta\alpha} = w_{\beta\alpha}^{(0)} + \xi w_{\beta\alpha}^{(1)} + \cdots, \tag{6.5.1}$$

$$p_\alpha = p_\alpha^{(0)} + \xi p_\alpha^{(1)} + \cdots, \tag{6.5.2}$$

but in subsequent equations retains only the terms of first order in $\xi$. Evidently the quantities $w_{\beta\alpha}^{(0)}$ and $p_\alpha^{(0)}$ are those pertaining to thermal equilibrium ($\xi = 0$). The quantities $w_{\beta\alpha}^{(1)}$ may be independently determined from a theory, or model, of the effect of the actual perturbation upon the transition frequency or, alternatively, obtained by assuming a local form of detailed balance. The perturbation in the probability, $\mathbf{p}^{(1)}$, then, by (6.2.2) with (6.5.1) and (6.5.2), satisfies the following equation

$$\frac{\mathrm{d}\mathbf{p}^{(1)}}{\mathrm{d}t} = -\mathbf{P}^{(0)}\mathbf{p}^{(1)} - \mathbf{P}^{(1)}(t)\mathbf{p}^{(0)}, \tag{6.5.3}$$

where the matrices $\mathbf{P}^{(0)}$ and $\mathbf{P}^{(1)}$ are defined by (6.2.4) in terms of the $w^{(0)}$ and the $w^{(1)}$ respectively, the strength parameter $\xi$ having now been absorbed into $\mathbf{p}^{(1)}$ and $\mathbf{P}^{(1)}$ for simplicity. In general, the $w^{(1)}$ and consequently $\mathbf{P}^{(1)}$ will depend on time (e.g. when the system is subject to an alternating electric field). We next consider the form of $\mathbf{P}^{(1)}$.

To obtain the matrix $\mathbf{P}^{(1)}$ we need to know the effect of the perturbation on the transition rates. For the greater number of systems of interest this can be obtained from classical transition state theory (§3.11). For isothermal systems this allows us to write

$$w_{\beta\alpha}^{(1)} = (w_{\beta\alpha}^{(0)}/kT)(\delta g_\alpha - \delta g_{\beta\alpha}^{+}), \tag{6.5.4}$$

in which $\delta g_\alpha$ is the change in the free energy of the system in state $\alpha$ (at constant $P$, $T$) as a result of the perturbation (correct to first order in the strength of that perturbation), while $\delta g_{\beta\alpha}^{+}$ $(=\delta g_{\alpha\beta}^{+})$ is the corresponding change when the system is in its saddle-point configuration for the transition from $\alpha$ to $\beta$. Eqns. (6.5.4) and (6.2.4) then allow us to write down $\mathbf{p}^{(1)}$.

The formal solution of (6.5.3) for time-dependent processes is then easily found to be

$$\mathbf{p}^{(1)}(t) = -\frac{1}{kT}\int_{-\infty}^{t} \exp[-\mathbf{P}^{(0)}(t-s)]\mathbf{P}^{(0)}\mathbf{N}\mathbf{g}(s)\,\mathrm{d}s, \tag{6.5.5}$$

in which $\mathbf{N}$ is the diagonal matrix defined by (6.2.11) while $\mathbf{g}$ is the column vector of the $\delta g_\alpha$, i.e.

$$\mathbf{g} = \{\delta g_1 \quad \delta g_2 \quad \ldots\}. \tag{6.5.6}$$

We have taken $\mathbf{p}^{(1)}(-\infty)$ to be zero. We observe that the terms in (6.5.4) relating to the transition state, i.e. $\delta g_{\beta\alpha}^{+}$, have cancelled out as a result of the use of (i) the detailed balance relations and (ii) the assumption that the same transition-state governs both the forward, $\alpha \to \beta$, and the backwards, $\beta \to \alpha$, transitions. This has the important practical consequence that all transport properties derivable from this analysis are independent of details of the paths by which the system makes transitions among its various states: they depend only on the states themselves, which transitions are allowed and the rates at which they occur.

An alternative form of (6.5.5) arises if we Fourier analyse the perturbation $\mathbf{g}(s)$ as

$$\mathbf{g}(s) = \int_{-\infty}^{\infty} \tilde{\mathbf{g}}(\omega)\,\mathrm{e}^{\mathrm{i}\omega s}\,\mathrm{d}\omega. \tag{6.5.7}$$

Insertion of this into (6.5.5) and use of (6.2.14) for $\mathbf{P}^{(0)}$ and of (6.3.2) for $\mathbf{G}(t)$

shows that

$$\mathbf{p}^{(1)}(t) = \int_{-\infty}^{\infty} \tilde{\mathbf{p}}^{(1)}(\omega)\, \mathrm{e}^{\mathrm{i}\omega t}\, \mathrm{d}\omega \tag{6.5.8}$$

with

$$\tilde{\mathbf{p}}^{(1)}(\omega) = -\frac{1}{kT}\mathbf{N}^{1/2} \sum_{v \neq 0} \frac{\alpha^{(v)}\mathbf{a}^{(v)}\tilde{\mathbf{a}}^{(v)*}}{(\alpha^{(v)} + \mathrm{i}\omega)} \mathbf{N}^{1/2}\tilde{\mathbf{g}}(\omega). \tag{6.5.9}$$

When we recall that $\alpha^{(v)} = 1/\tau^{(v)}$, it is clear that (6.5.9) has the appearance of a sum of Debye-like responses to the perturbing field (Fröhlich, 1958 and Scaife, 1989). In any particular case symmetry considerations will determine which modes $v$ to couple to the perturbation, i.e. which give non-zero $(\tilde{\mathbf{a}}^{(v)}\mathbf{N}^{1/2}\mathbf{g})$.

### *6.5.1 The transport coefficients*

So far this section has been concerned with the description of the microscopic evolution of the system. To make contact with the macroscopic description provided by non-equilibrium thermodynamics we need expressions for the fluxes in terms of microscopic variables, particularly the state vector **p**. In theories based on dynamical equations of motion such expressions will be provided as part of the dynamics (Forster, 1975; Kreuzer, 1981). Clearly this is not immediately possible with (6.2.2) as the starting point. Neither, since we are developing the theory at a fairly abstract level, is there any possibility of obtaining the atomic or defect fluxes by direct enumeration, as is done in elementary treatments (Shewmon, 1989; Philibert, 1985). Rather, as in the general development of non-equilibrium statistical mechanics, we must appeal to the equations of continuity or conservation. In fact, we shall follow this approach in detail in subsequent chapters when we develop more specific theories such as the random walk and kinetic theories. It is nevertheless possible to use it to develop general expressions for the macroscopic transport coefficients with only a minimum of specific physical assumptions. To do so we have to have regard to two points. First of all, we have to match the equations we have developed so far to a more specific representation of the atomic configurations hitherto specified only by the abstract state indices $\alpha$, $\beta$, $\gamma$, etc. Secondly, to proceed from a specification of atomic configurations in terms of assignments of atoms to discrete lattice points to a continuum representation (as e.g. in the continuity equation) it is necessary in principle to carry out averages over regions which are small in a macroscopic sense, but large enough to contain many lattice points. This is conveniently done by the device of Fourier transforming the equations. The argument is not difficult but is not especially compact and it is therefore given in Appendix 6.3.

The result is that the macroscopic transport coefficients for an isothermal system of volume $V$ containing atomic species $i, j$, etc. in numbers $N_i$, $N_j$, etc. are given by equilibrium ensemble averages of functions of the individual displacements, $\mathbf{r}_{\beta\alpha}$, of all the atoms in all possible transitions $\alpha \to \beta$ undergone by the system. More specifically,

$$\mathbf{L}_{ij} = \mathbf{L}_{ij}^{(0)} + \mathbf{L}_{ij}^{(1)}, \tag{6.5.10}$$

with

$$\mathbf{L}_{ij}^{(0)} = \frac{1}{2VkT} \sum_{\alpha,\beta} \sum_{m,n} \mathbf{r}_{\beta\alpha}(i_m)\mathbf{r}_{\beta\alpha}(j_n) w_{\beta\alpha}^{(0)} p_\alpha^{(0)}, \tag{6.5.11}$$

and

$$\mathbf{L}_{ij}^{(1)} = \int_0^\infty \mathbf{\Phi}_{ij}^{(1)}(t)\,\mathrm{d}t,$$

where

$$\mathbf{\Phi}_{ij}^{(1)}(t) = \frac{1}{VkT} \sum_{\alpha,\beta,\gamma,\delta} \sum_{m,n} (\mathbf{r}_{\delta\gamma}(i_m) w_{\delta\gamma}^{(0)}) G_{\gamma\beta}(t) (\mathbf{r}_{\beta\alpha}(j_n) w_{\beta\alpha}^{(0)}) p_\alpha^{(0)}. \tag{6.5.12}$$

These expressions are written as summations over all transitions of the system ($\alpha \to \beta$, etc.) and as summations over all atoms $m$, $n$ of species $i$ and $j$ respectively in the volume $V$. The vector displacement of atom $i_m$ in the transition $\alpha \to \beta$ is written as $\mathbf{r}_{\beta\alpha}(i_m)$ so that the $L$-coefficients are here presented in tensor notation, i.e. any particular Cartesian component, $L_{ijxy}$, of $\mathbf{L}_{ij}$ is obtained by taking the $x$-component of $\mathbf{r}(i_m)$ and the $y$-component of $\mathbf{r}(j_n)$ in (6.5.11) and (6.5.12). Both terms in $\mathbf{L}_{ij}$ are averages over a thermal equilibrium ensemble (cf. the factors $w_{\beta\alpha}^{(0)}$, $p_\alpha^{(0)}$, etc.). This reduction of transport coefficients to averages of kinetic and dynamic quantities over a thermal equilibrium ensemble is, of course, a general feature of linear response theory, albeit one which is usually demonstrated within the context of an underlying Hamiltonian mechanics (Kreuzer, 1981). In fact, the above result is seen to be the analogue of the transport coefficient for a dynamical system when we interpret summations such as $\sum_\beta \mathbf{r}_{\beta\alpha} w_{\beta\alpha}^{(0)}$ as the velocity of the corresponding atom, for (6.5.12) then takes on the form of a velocity–velocity time-correlation function. By using the principle of detailed balance, we can show that the Onsager relation (1.4.3) follows straightforwardly from (6.5.10)–(6.5.12), although this relation follows more immediately from (6.5.14) below. These particular results (first obtained in the present context by Allnatt, 1965) make no specific assumptions about defect type or structure and apply generally to both dilute and concentrated systems. They therefore provide the starting point for much of the development which follows.

### *6.5.2 Many-body form of the Einstein relations*

We now consider the generalization of the two elementary Einstein relations, namely (i) that between the mean-square displacement of a particle and its diffusion coefficient and (ii) that between its diffusion coefficient and its mobility. Let us return to the results of §6.4. First we see that if $\mathbf{G}(s)$ decreases sufficiently rapidly at long times $s$ (faster than $s^{-1}$) then at long times $t$ we can replace the integral in (6.4.7) by

$$\int_0^\infty (V_X)_{\delta\gamma} G_{\gamma\beta}(s)(V_Y)_{\beta\alpha} p_\alpha^{(0)}\, ds. \tag{6.5.13}$$

Evidently, long times $t$ here means times long compared to the internal relaxation times of the system. It follows that the averages $\langle(\Delta X)^2\rangle/t$ and $\langle \Delta X\, \Delta Y\rangle/t$ are then independent of $t$, i.e., the system is diffusive. If now we take the variables $X$ and $Y$ to be respectively the $x$ and $y$ components of $\sum_m \mathbf{r}(i_m)$ and $\sum_n \mathbf{r}(j_n)$, i.e. the sums of the positions of all atoms of types $i$ and $j$ respectively, then clearly by (6.5.10)–(6.5.12)

$$\langle \Delta X\, \Delta Y\rangle/2t = L_{ijxy} VkT \tag{6.5.14}$$

and similarly for $\mathbf{L}_{ii}$, $\mathbf{L}_{jj}$, etc. This result was obtained by Allnatt (1982), following a previous related discussion for liquids by Rice and Gray (1965, Chap 7).

This is the generalization of the elementary Einstein result that the diffusion coefficient $D$ for Brownian motion of a particle in a system in thermodynamic equilibrium is

$$D = \langle(\Delta x)^2\rangle/2t, \tag{6.5.15}$$

in which $x$ is the $x$-coordinate of the position of the particle. We do not need to emphasize that (6.5.14) is of far wider applicability than the elementary result, which can only be used for systems of independent particles. It has, however, a similar practical value, for it provides a convenient route to the evaluation of the $L$-coefficients by the technique of Monte Carlo computation (cf. Appendix 13.1). It also enables us to demonstrate the required positive definiteness of the matrix of $L$-coefficients (§4.2); see Appendix 6.4.

To obtain a generalization of the second Einstein relation we consider the mean quantity $\langle \Delta X\rangle$. For a system in thermodynamic equilibrium the equations of detailed balance show that (6.4.5) is zero for any additive variable such that $X_{\alpha\beta} = X_\alpha - X_\beta$. However, this will not generally be the case for steady-state conditions. Referring to Appendix 6.2, we see that to evaluate $\langle \Delta X\rangle$ for these conditions we need to develop an expression for $\bar{\mathbf{p}}(\lambda)$ for the system under the action of the perturbation (e.g. electric field). This can be done in the linear response approximation by appeal to (6.5.3) and by using (6.5.4) given by transition state theory. Details are given in Allnatt and Lidiard (1988) but the particular result of interest here is

$$\langle \Delta X\rangle = -\frac{1}{2kT}\langle \Delta X\, \Delta(\delta g)\rangle^{(0)} \tag{6.5.16}$$

which equates the mean change of $X$ over a time $t$, $\langle \Delta X \rangle$, in a system in a steady state to the mean correlation $\langle \Delta X \, \Delta(\delta g) \rangle^{(0)}$ in the same system in thermal equilibrium. (As before, the quantity $\delta g_\alpha$ is the change in Gibbs free energy of the system in state $\alpha$ due to the presence of the perturbing potential.)

This is the desired generalization of the elementary Nernst–Einstein relation between the mobility of a particle and its mean-square displacement or diffusion coefficient. For, if the energy of the system depends on the perturbing field only through the $x$-position of a particular particle as

$$\delta g_\alpha = -X_\alpha F, \tag{6.5.17}$$

then the right side of (6.5.16) gives $F\langle(\Delta X)^2\rangle^{(0)}/2kT$, while the left side is the mobility of the particle multiplied by $Ft$. With $\langle(\Delta X)^2\rangle^{(0)}/2t$ equal to the diffusion coefficient of the particle, we recover the elementary Nernst–Einstein relation (5.2.8). However, in solids generally the energy of a system subject to external forces (e.g. electrical, gravitational and centrifugal) will depend on the positions of all the atoms, with the result that, even if the $X$ variable refers to only a particular sub-group of atoms, the right side of (6.5.16) will contain correlations between the motion of the atoms in this sub-group and all the others (cf. 5.9.4). The elementary relation between mobility and a mean-square displacement is thus not universally valid. The form (6.5.16), however, applies to all systems within the stated linear response approximation.

Lastly, it should be remarked that there are systems where the function $\mathbf{G}(t)$ decays sufficiently slowly at long times $t$ that the integral (6.5.13) does not exist and there are, in fact, no linear macroscopic transport equations. One such example is provided by particles migrating in a partially blocked lattice at the percolation limit. In that case $\langle(\Delta X)^2\rangle$ is not proportional to $t$ but to some lower power of it (see e.g. Havlin and Ben-Avraham, 1987).

### 6.5.3 Some implications

We now draw out one or two implications of the above results, (6.5.10) *et seq.*, specifically for cubic systems for which there is only one distinct tensor component to each $\mathbf{L}_{ij}$; thus $L_{ijxx} = L_{ijyy} = L_{ijzz} \equiv L_{ij}$ while all off-diagonal $L_{ijxy}$, etc. $= 0$). Firstly, let us expand the expression (6.5.14) in terms of summations over the individual atoms. For the diagonal coefficient $L_{ii}$ we obtain

$$L_{ii} = \frac{1}{6VkT}\left\{\sum_m \frac{\langle(\Delta \mathbf{r}_i^{(m)})^2\rangle}{t} + \sum_{\substack{m,n \\ (m \neq n)}} \frac{\langle \Delta \mathbf{r}_i^{(m)} \cdot \Delta \mathbf{r}_i^{(n)} \rangle}{t}\right\}, \tag{6.5.18}$$

in which we have contracted the notation so that the (vector) displacement of atom $m$ of type $i$ in time $t$ is now denoted $\Delta\mathbf{r}_i^{(m)}$. By the nature of the thermodynamic ensemble average, each $\langle(\Delta\mathbf{r}_i^{(m)})^2\rangle$ is the same for every atom $m$ of species $i$, although it will, of course, be different for atoms of different chemical species. We can therefore drop the suffix $m$, so that the first term in $L_{ii}$ is just $(n_i/6kT)\langle(\Delta\mathbf{r}_i)^2\rangle/t$, with $n_i$ the number of atoms of type $i$ per unit volume. Likewise, in the second term $\langle\Delta\mathbf{r}_i^{(m)}\cdot\Delta\mathbf{r}_i^{(n)}\rangle$ will be the same for all pairs of two different atoms. The corresponding expression for $L_{ij}$ $(i \neq j)$ is

$$L_{ij} = \frac{1}{6VkT}\left\{\sum_{m,n}\frac{\langle\Delta\mathbf{r}_i^{(m)}\cdot\Delta\mathbf{r}_j^{(n)}\rangle}{t}\right\}, \tag{6.5.19}$$

in which $\langle\Delta\mathbf{r}_i^{(m)}\cdot\Delta\mathbf{r}_i^{(n)}\rangle$ is again the same for all pairs of different atoms.

With these observations in mind let us first consider the limiting case in which component $i$ is present only in a low concentration. It is then clear that the second term in $L_{ii}$ becomes of less and less significance compared to the first as the concentration is progressively reduced; for, as the average separation of a pair of atoms of type $i$ increases any correlation in their movements (caused by a single defect) must become very small. Hence in the dilute limit $(x_i \to 0)$

$$L_{ii} = \frac{n_i\langle(\Delta\mathbf{r}_i)^2\rangle}{6kTt}. \tag{6.5.20}$$

On the other hand, the same argument does not apply to the correlation $\langle\Delta\mathbf{r}_i\cdot\Delta\mathbf{r}_j\rangle$ between an atom of type $i$ and one of type $j$ (unless species $j$ is also dilute). There will always be some atoms of type $j$ sufficiently close initially to a given atom of type $i$ to give a significant correlation $\langle\Delta\mathbf{r}_i\cdot\Delta\mathbf{r}_j\rangle$. Thus $L_{ij}$ remains of order $n_i$ and the ratio $L_{ij}/L_{ii}$ is independent of the concentration of $i$ in the dilute limit. This is the fact we need to complete the argument leading to (5.3.8). Hence in the dilute limit, by (5.3.8) and (6.5.20), we have

$$D_i = \frac{\langle(\Delta\mathbf{r}_i)^2\rangle}{6t}. \tag{6.5.21}$$

A common example of a dilute species is provided by isotopic tracers, radioactive or otherwise. We can thus use (6.5.21) to give the self-diffusion coefficient $D_i^*$ of chemical species $i$, the self-correlation $\langle(\Delta\mathbf{r}_i)^2\rangle$ being that for the tracer atoms. But since the tracer atoms are chemically indistinguishable from the $i$-atoms in the solid this quantity is actually the same for all. Hence whether the system is dilute in species $i$ or not we can always rewrite (6.5.18) as

$$L_{ii} = \frac{n_i D_i^*}{kT} + \frac{n_i}{6kT}\sum_{n\neq m}\frac{\langle\Delta\mathbf{r}_i^{(m)}\cdot\Delta\mathbf{r}_i^{(n)}\rangle}{t}, \tag{6.5.22}$$

i.e. $L_{ii}$ is made up of a self-correlation term which can be replaced by the tracer diffusion coefficient plus a sum of 2-atom correlation terms. In other words the essential difference between the transport coefficient, $L_{ii}$, and the tracer diffusion coefficient, $D_i^*$, lies in the 2-atom correlation terms. The off-diagonal coefficient $L_{ij}$ is composed solely of 2-atom correlations (6.5.19).

In the light of this let us return for a moment to eqns. (5.7.7) for isotopic diffusion. From (6.5.22) and (5.7.7a) it is clear that the second part of $L_{A'A'}$, i.e. $n_{A'}D_A^*(1-f)c_{A'}/fc_AkT$, is determined by the correlation function of a pair of A′ atoms. Similarly for $L_{A''A''}$, while $L_{A'A''}$ is determined by the correlation function for a pair made up of one A′ and one A″ atom. But by the assumptions made in §5.7 (ideal thermodynamic mixing of the isotopes and equal jump frequencies) it follows that these pair correlation functions can differ only by the fraction of A′ or A″ atoms involved – as eqns. (5.7.7) confirm. But this conclusion follows equally in an alloy in which elements other than A are present. Hence the forms (5.7.7) and (5.7.9) apply as well in an alloy as in the pure substance A, although, of course, $f$ is now a property of the alloy (dependent on composition, interatomic interactions, etc.). Naturally, the argument applies only to the $L$-coefficients among the isotopes of individual chemical species.

The forms (6.5.19) and (6.5.22) also allow us to obtain additional understanding of the Darken and Manning approximations, discussed in §§5.6 and 5.8 respectively. The Darken relation (5.6.4) for a binary alloy is obviously equivalent to

$$L_{AA} - \frac{n_A D_A^*}{kT} = \frac{n_A}{n_B} L_{AB}. \tag{6.5.23}$$

We see that this relation is equivalent to the assertion that the 2-atom (AA) correlation part of $L_{AA}$ can be approximated by the 2-atom correlation function (AB) which is $L_{AB}$ scaled by the concentration of A atoms relative to that of B atoms. This will be exact when B is an isotope of A (cf. Eqns. 5.7.7), but otherwise the accuracy of the relation must depend on how differently the A and B atoms move.

The Manning vacancy wind argument led to a kindred relation. For a binary alloy (5.8.4) gives

$$L_{AA} - \frac{n_A D_A^*}{kT} = \frac{(L_{AA} + L_{BA})}{(L_{BB} + L_{AB})} L_{AB}, \tag{6.5.24}$$

with a similar relation for $L_{BB}$. In this case the 2-atom correlation part of $L_{AA}$ is represented by the 2-atom correlation function which is $L_{AB}$ scaled by a factor which is the ratio of the products of concentration and atom mobility (rather than just concentration as with Darken). This would lead us to suppose that (6.5.24) should be more accurate than (6.5.23), which, as we shall see later (§13.3), is often the case. However, we cannot expect either relation to be accurate when the alloy is composed principally of A atoms ($n_B \to 0$) because while the left sides necessarily depend on the properties of A atoms alone the right sides tend to a constant non-zero value which still depends on the properties of B atoms (even at $n_B = 0$!).

## 6.6 Summary

In this chapter we have set out the formal structure which provides the basis for the calculation of transport coefficients for particular model systems under isothermal conditions. We began with the master equation representation of the evolution of these (Markovian) systems and by means of a stochastic interpretation of this equation were able to obtain formal expressions for the mean values of kinetic variables appropriate to stationary conditions, i.e. thermal equilibrium and steady state conditions (§6.4). By solving the master equation in the linear response approximation we were able to obtain formal expressions for the linear transport coefficients (§6.5) and to verify the Onsager relations. The close similarity of the results of these two sections shows that in appropriate circumstances there is a generalized Einstein relation between mean values of kinetic quantities and corresponding transport coefficients. Although the formal structure of these results is known in other contexts it is noteworthy how few physical assumptions have been employed in the derivations given here. In particular, the system may be concentrated or dilute (in both atoms and defects). Furthermore, the lattice on which the atoms are arranged need not be regular, as for crystalline substances, but could be irregular as for glassy materials as long as the array of available atomic positions does not alter with time. It is the object of later chapters to apply these results to specific physical models.

# Appendix 6.1

## *Eigenvalues of the matrix* **S**

In this appendix we demonstrate that the eigenvalues of the matrix **S** are positive or zero, but never negative. We already know that they are real, because **S** is symmetric. Since all the elements of **S** are real and the $\alpha^{(\nu)}$ are real the eigenvectors may also be taken to be real without loss of generality. (Thus to any complex $\mathbf{a}^{(\nu)}$ there must be a conjugate $\mathbf{a}^{(\nu)*}$ with the same eigenvalue, hence in place of $\mathbf{a}^{(\nu)}$ and $\mathbf{a}^{(\nu)*}$ we could take their real and imaginary parts.)

From the eigenvalue eqn. (6.2.12) we form

$$\alpha^{(\nu)}\tilde{\mathbf{a}}^{(\nu)}\mathbf{a}^{(\nu)} = \tilde{\mathbf{a}}^{(\nu)}\mathbf{S}\mathbf{a}^{(\nu)}, \tag{A6.1.1}$$

and then show that the right side is positive or zero. Since the left side $\alpha^{(\nu)}\sum_\beta (a_\beta{}^{(\nu)})^2$ is then also positive or zero, it follows that $\alpha^{(\nu)} \geq 0$.

For simplicity we shall drop the superscript $(\nu)$ on the eigenvector and the superscript (0) on the quantities $\mathbf{p}^{(0)}$ and $\mathbf{P}^{(0)}$. Then

$$\begin{aligned}
\tilde{\mathbf{a}}\mathbf{S}\mathbf{a} &= \sum_{\alpha,\beta} a_\alpha S_{\alpha\beta} a_\beta \\
&= \sum_{\alpha\neq\beta} a_\alpha p_\alpha{}^{-1/2} P_{\alpha\beta} p_\beta{}^{1/2} a_\beta + \sum_\alpha a_\alpha p_\alpha{}^{-1/2} P_{\alpha\alpha} p_\alpha{}^{1/2} a_\alpha \\
&= -\sum_{\alpha\neq\beta} a_\alpha p_\alpha{}^{-1/2} w_{\alpha\beta} p_\beta{}^{1/2} a_\beta + \sum_{\alpha\neq\beta} a_\alpha w_{\beta\alpha} a_\alpha \\
&= -\sum_{\alpha\neq\beta} a_\alpha p_\alpha{}^{-1/2} w_{\alpha\beta} p_\beta{}^{1/2} a_\beta + \sum_{\alpha\neq\beta} a_\beta w_{\alpha\beta} a_\beta \\
&= -\sum_{\alpha\neq\beta} \{a_\alpha p_\alpha{}^{-1/2} w_{\alpha\beta} p_\beta{}^{1/2} a_\beta - a_\beta p_\beta{}^{-1/2} w_{\alpha\beta} p_\beta p_\beta{}^{-1/2} a_\beta\}
\end{aligned}$$

$$= -\sum_{\alpha\neq\beta} \{(a_\alpha p_\alpha{}^{-1/2} - a_\beta p_\beta{}^{-1/2})(w_{\alpha\beta} p_\beta) p_\beta{}^{-1/2} a_\beta\}$$

$$= -\sum_{\alpha,\beta} \{(a_\alpha p_\alpha{}^{-1/2} - a_\beta p_\beta{}^{-1/2})(w_{\alpha\beta} p_\beta)(p_\beta{}^{-1/2} a_\beta)\}$$

$$\equiv -\sum_{\alpha,\beta} (h_\alpha - h_\beta) g_{\alpha\beta} h_\beta, \tag{A6.1.2}$$

in which

$$g_{\alpha\beta} \equiv w_{\alpha\beta} p_\beta = g_{\beta\alpha} \tag{A6.1.3}$$

(by detailed balance, (6.2.8)) and

$$h_\alpha \equiv a_\alpha p_\alpha{}^{-1/2}. \tag{A6.1.4}$$

Hence the right side is

$$= -\frac{1}{2}\sum_{\alpha,\beta} \{(h_\alpha - h_\beta) g_{\alpha\beta} h_\beta + (h_\beta - h_\alpha) g_{\beta\alpha} h_\alpha\}$$

$$= -\frac{1}{2}\sum_{\alpha,\beta} \{(h_\alpha - h_\beta) g_{\alpha\beta} h_\beta + (h_\beta - h_\alpha) g_{\alpha\beta} h_\alpha\}$$

$$= \frac{1}{2}\sum_{\alpha,\beta} (h_\alpha - h_\beta)^2 g_{\alpha\beta}. \tag{A6.1.5}$$

Since $g_{\alpha\beta} \geq 0$ we see that the right side is necessarily $\geq 0$. The left side of (A6.1.2) is therefore also $\geq 0$, whence $\alpha$ is $\geq 0$, true for all eigenvalues $\alpha^{(\nu)}$. The equality applies when all the $h_\alpha$ are equal, i.e. when $a_\alpha = p_\alpha{}^{1/2}/(\sum_\alpha p_\alpha)^{1/2}$, where the numerator ensures that $a$ is normalized.

# Appendix 6.2

## Expressions for average kinetic quantities

In §6.4 we considered the representation of mean values of kinetic quantities for systems described by the master equation (6.2.2). In particular, we presented three relations, namely (6.4.5)–(6.4.7), for the mean values $\langle(\Delta X)^2\rangle^{(0)}$ and $\langle \Delta X\, \Delta Y\rangle^{(0)}$ for systems in thermal equilibrium. We here sketch their derivation. As in §6.4 we take the kinetic quantities to be wholly determined by the state the system is in. Then the change in $X$ in a transition from state $\alpha$ to state $\beta$ is $X_{\beta\alpha} = X_\beta - X_\alpha$.

We shall work with Laplace transforms of the conditional averages $\langle(\Delta X)^n\rangle_{\beta\alpha}$ determined by (6.4.2). We employ the notation

$$L\{f(t)\} \equiv \bar{f}(\lambda) = \int_0^\infty \mathrm{d}t \exp(-\lambda t) f(t) \qquad \text{(A6.2.1)}$$

for the Laplace transform of a function $f(t)$ of the time $t$. Then by (6.4.2)

$$L\{\langle(\Delta X)^n\rangle_{\beta\alpha}\} = (X_{\beta\alpha})^n \,\bar{G}_{\beta\alpha}(\lambda). \qquad \text{(A6.2.2)}$$

For $\mathbf{P}$ independent of time, i.e. thermal equilibrium or steady-state conditions, we can appeal to (6.3.1) to obtain

$$\bar{\mathbf{G}}(\lambda) = \int_0^\infty \mathrm{d}t \exp(-\lambda t) \exp(-\mathbf{P}t)$$

$$= (\mathbf{P} + \lambda \mathbf{1})^{-1}. \qquad \text{(A6.2.3)}$$

Let us first take the case $n = 1$. By starting from A6.2.2 and using the identity

$$\bar{\mathbf{G}} = \bar{\mathbf{G}}(\mathbf{P} + \lambda \mathbf{1})\bar{\mathbf{G}}, \qquad \text{(A6.2.4)}$$

which follows from (A6.2.3), and also by expanding $X_{\beta\alpha}$ as

$$X_{\beta\alpha} = X_{\beta\gamma} + X_{\gamma\delta} + X_{\delta\alpha}, \qquad \text{(A6.2.5)}$$

we find after a little algebra that

$$L\{\langle\Delta X\rangle_{\beta\alpha}\} = (\bar{G}(\lambda)V_X\bar{G}(\lambda))_{\beta\alpha}, \tag{A6.2.6}$$

where the matrix $\mathbf{V}_X$ is defined by (6.4.4a).

Next take the case $n = 2$. Again starting from (A6.2.2) but this time invoking the identity

$$\bar{\mathbf{G}} = \bar{\mathbf{G}}(\mathbf{P} + \lambda\mathbf{1})\bar{\mathbf{G}}(\mathbf{P} + \lambda\mathbf{1})\bar{\mathbf{G}} \tag{A6.2.7}$$

and expanding $X_{\beta\alpha}$ over four intermediate states, we find

$$L\{\langle(\Delta X)^2\rangle_{\beta\alpha}\} = (\bar{G}(\lambda)V_{XX}\bar{G}(\lambda))_{\beta\alpha} + 2(\bar{G}(\lambda)V_X\bar{G}(\lambda)V_X\bar{G}(\lambda))_{\beta\alpha}, \tag{A6.2.8}$$

in which the matrix $\mathbf{V}_{XX}$ is as defined by (6.4.4b) but with $\mathbf{P}^{(0)}$ replaced by $\mathbf{P}$.

In a similar way we find

$$L\{\langle\Delta X\,\Delta Y\rangle_{\beta\alpha}\} = (\bar{G}(\lambda)V_{XY}\bar{G}(\lambda))_{\beta\alpha} + (\bar{G}(\lambda)V_X\bar{G}(\lambda)V_Y\bar{G}(\lambda))_{\beta\alpha} + (\bar{G}(\lambda)V_Y\bar{G}(\lambda)V_X\bar{G}(\lambda))_{\beta\alpha}, \tag{A6.2.9}$$

with the matrix $\mathbf{V}_{XY}$ given by (6.4.4c) in which $\mathbf{P}^{(0)}$ is again replaced by $\mathbf{P}$.

These three equations (A6.2.6), (A6.2.8) and (A6.2.9) provide the basis from which to obtain equations (6.4.5)–(6.4.7). Before taking these final steps, however, we note that a more systematic development is possible from (6.4.3), This leads to a general formula for $L\{\langle(\Delta X)^n\rangle_{\beta\alpha}\}$. Details are given by Allnatt and Lidiard (1988).

To complete the calculation we need to insert these results for the transforms of the conditional mean values into the expression (6.4.1) for the corresponding unconditional mean values. The matrix products $\bar{\mathbf{G}}(\lambda)\mathbf{p}(0)$ and $\tilde{\boldsymbol{\tau}}\bar{\mathbf{G}}(\lambda)$ then appear at the right and left sides of the resulting expressions. But by (6.3.1) we have

$$\bar{\mathbf{G}}(\lambda)\mathbf{p}(0) = \bar{\mathbf{p}}(\lambda), \tag{A6.2.10}$$

while by (6.2.5) and (6.2.6)

$$\tilde{\boldsymbol{\tau}}\bar{\mathbf{G}}(\lambda) = \frac{1}{\lambda}\tilde{\boldsymbol{\tau}}. \tag{A6.2.11}$$

Thus for the Laplace transforms of the quantities of interest averaged over all initial and final states we then obtain

$$\begin{aligned} L\{\langle\Delta X\rangle\} &= \tilde{\boldsymbol{\tau}}\bar{\mathbf{G}}(\lambda)\mathbf{V}_X\bar{\mathbf{G}}(\lambda)\mathbf{p}(0) \\ &= \frac{1}{\lambda}\tilde{\boldsymbol{\tau}}\mathbf{V}_X\bar{\mathbf{p}}(\lambda), \end{aligned} \tag{A6.2.12}$$

and likewise

$$L\{\langle(\Delta X)^2\rangle\} = \frac{1}{\lambda}\tilde{\tau}\mathbf{V}_{XX}\bar{\mathbf{p}}(\lambda) + \frac{2}{\lambda}\tilde{\tau}\mathbf{V}_X\bar{\mathbf{G}}(\lambda)\mathbf{V}_X\bar{\mathbf{p}}(\lambda), \qquad \text{(A6.2.13)}$$

$$L\{\langle\Delta X\,\Delta Y\rangle\} = \frac{1}{\lambda}\tilde{\tau}\mathbf{V}_{XY}\bar{\mathbf{p}}(\lambda) + \frac{1}{\lambda}\tilde{\tau}\mathbf{V}_X\bar{\mathbf{G}}(\lambda)\mathbf{V}_Y\bar{\mathbf{p}}(\lambda) + \frac{1}{\lambda}\tilde{\tau}\mathbf{V}_Y\bar{\mathbf{G}}(\lambda)\mathbf{V}_X\bar{\mathbf{p}}(\lambda). \qquad \text{(A6.2.14)}$$

We have already emphasized that these results hold for both thermal equilibrium and steady-state conditions.

In the special case where the system is in thermal equilibrium $\mathbf{p}(t) = \mathbf{p}^{(0)}$, is independent of time, and $\bar{\mathbf{p}}(\lambda) = \mathbf{p}^{(0)}/\lambda$. By the convolution theorem on Laplace transforms we then obtain

$$\langle\Delta X\rangle^{(0)}/t = \tilde{\tau}\mathbf{V}_X\mathbf{p}^{(0)}, \qquad \text{(A6.2.15)}$$

$$\langle(\Delta X)^2\rangle^{(0)}/t = \tilde{\tau}\mathbf{V}_{XX}\mathbf{p}^{(0)} + 2\tilde{\tau}\int_0^t ds\left(1 - \frac{s}{t}\right)\mathbf{V}_X\mathbf{G}(s)\mathbf{V}_X\mathbf{p}^{(0)}, \qquad \text{(A6.2.16)}$$

and

$$\langle\Delta X\,\Delta Y\rangle^{(0)}/t = \tilde{\tau}\mathbf{V}_{XY}\mathbf{p}^{(0)} + 2\tilde{\tau}\int_0^t ds\left(1 - \frac{s}{t}\right)\mathbf{V}_X\mathbf{G}(s)\mathbf{V}_Y\mathbf{p}^{(0)}. \qquad \text{(A6.2.17)}$$

In (A6.2.17) the principle of detailed balance and the assumption $X_{\alpha\beta} = X_\alpha - X_\beta$ have been used to show the equality of the two contributions in $\mathbf{V}_X$ and $\mathbf{V}_Y$ which appear separately in (A6.2.14). The detailed balance relations also show that (A6.2.15) is zero for any additive variable such that $X_{\alpha\beta} = X_\alpha - X_\beta$, i.e.

$$\langle\Delta X\rangle^{(0)} = 0. \qquad \text{(A6.2.18)}$$

However, it should be noted that this will not generally be the case for steady-state conditions (cf. §6.5.2).

# Appendix 6.3

## Transport coefficients in the linear response approximation

In this appendix we shall derive the expressions for the transport coefficients in the linear response approximation as already set out in eqns. (6.5.10)–(6.5.12). As mentioned in §6.5, it is convenient in doing so to work with the Fourier transformed equations. We therefore begin by looking at the form of the transformed macroscopic flux equations. We shall suppose that the system is subject to an external field which leads to a thermodynamic force $\mathbf{F}_j(\mathbf{r}, t)$ on atoms of species $j$. We also go beyond the phenomenology introduced in Chapter 4 by supposing that the flux of species $i$ in these circumstances can be written as

$$\mathbf{J}_i(\mathbf{r}, t) = \sum_j \int \mathrm{d}\mathbf{r}' \int_{-\infty}^{t} \mathrm{d}t' \boldsymbol{\Phi}_{ij}(\mathbf{r}, \mathbf{r}' : t, t') \cdot \mathbf{F}_j(\mathbf{r}', t'). \tag{A6.3.1}$$

Here $i$ and $j$ stand for any of the distinct atomic species present in the system. As here expressed the transport coefficients allow for non-local effects (i.e. for the flux at $\mathbf{r}$ to depend on the force acting at some other position $\mathbf{r}'$) and for time-dependent relaxation effects, as are relevant in dielectric and anelastic relaxation phenomena. Although for the most part we shall deal with static or slowly-varying force fields we shall here retain the indicated dependence on $t$ and $t'$. We shall return to the matter in later chapters.

We now proceed to the Fourier transformed equations. The spatial variable $\mathbf{r}$ is to be continuous so the transform of a function $f(\mathbf{r})$, which for simplicity we shall write as $\tilde{f}(\mathbf{k})$, is

$$\tilde{f}(\mathbf{k}) = \int f(\mathbf{r})\, \mathrm{e}^{\mathrm{i}\mathbf{k}\cdot\mathbf{r}}\, \mathrm{d}\mathbf{r}, \tag{A6.3.2}$$

where the integration is taken over a periodicity volume $V$. The inverse transform is a sum over all the **k**-points allowed by the periodic boundary conditions, viz.

$$f(\mathbf{r}) = \frac{1}{V}\sum_k \tilde{f}(\mathbf{k})\,\mathrm{e}^{-\mathrm{i}\mathbf{k}\cdot\mathbf{r}}. \tag{A6.3.3a}$$

If so desired, one can replace the sum $\sum_k$ by an integral by introducing the density of states factor $V/(2\pi)^3$

$$f(\mathbf{r}) = \frac{1}{(2\pi)^3}\int \tilde{f}(\mathbf{k})\,\mathrm{e}^{-\mathrm{i}\mathbf{k}\cdot\mathbf{r}}\,\mathrm{d}\mathbf{k}, \tag{A6.3.3b}$$

the integration being now over the Brillouin zone.

Now, as general statistical theory tells us that $L_{ij}$ is obtained by averaging certain correlated quantities over an equilibrium ensemble, we can assume that it depends only on the difference $(\mathbf{r} - \mathbf{r}')$ (and $t - t'$). The Fourier transformed equation thus becomes

$$\tilde{\mathbf{J}}_i(\mathbf{k}, t) = \int_0^\infty \sum_j \tilde{\mathbf{\Phi}}_{ij}(\mathbf{k}, -\mathbf{k}{:}t')\cdot\tilde{\mathbf{F}}_j(\mathbf{k}, t - t')\,\mathrm{d}t'. \tag{A6.3.4}$$

That only the $\tilde{\mathbf{\Phi}}_{ij}$ $(\mathbf{k}, -\mathbf{k}{:}t')$ enter is a consequence of the requirement of spatial invariance, i.e. that $\mathbf{\Phi}_{ij}(\mathbf{r}, \mathbf{r}')$ should depend only on $(\mathbf{r} - \mathbf{r}')$ and not on $\mathbf{r}$ and $\mathbf{r}'$ separately. As $\mathbf{k} \to 0$ the response function $\tilde{\mathbf{\Phi}}_{ij}(\mathbf{k}, -\mathbf{k}{:}t')$ must become independent of **k**, corresponding to spatially uniform systems. The equation governing behaviour in long-wavelength, slowly varying fields is thus

$$\tilde{\mathbf{J}}_i(\mathbf{k}, t) = \sum_j \mathbf{L}_{ij}\cdot\tilde{\mathbf{F}}_j(\mathbf{k}, t) \tag{A6.3.5a}$$

in which the quantity

$$\mathbf{L}_{ij} = \lim_{\mathbf{k}\to 0}\int_0^\infty \tilde{\mathbf{\Phi}}_{ij}(\mathbf{k}, -\mathbf{k}{:}t)\,\mathrm{d}t \tag{A6.3.5b}$$

is the macroscopic response tensor for uniform fields which we seek. We shall proceed to calculate $\tilde{\mathbf{\Phi}}_{ij}(\mathbf{k}, \mathbf{k}')$ and thus, via (A6.3.5b), to obtain $\mathbf{L}_{ij}$. We do so by first setting down an equation for the density, $n_i$, of atoms of type $i$ under conditions where the system is perturbed by the set of forces $\mathbf{F}_j$. Appeal to the equation of conservation of atoms $i$ then allows us to infer $J_i$ thus

$$\frac{\partial n_i}{\partial t} + \nabla\cdot\mathbf{J}_i = 0, \tag{A6.3.6a}$$

the Fourier transformed form of which is

$$\frac{\partial \tilde{n}_i(\mathbf{k}, t)}{\partial t} + i\mathbf{k}\cdot\tilde{\mathbf{J}}_i(\mathbf{k}, t) = 0. \tag{A6.3.6b}$$

We shall assume that the system is made up of $N_i$ atoms of type $i$ where $i$ runs over all atomic species present. Then if the volume of the system is $V$ the local density of atoms of type $i$ at position $r$ when the system is in state $\alpha$ may be written

$$n_\alpha^{(i)}(\mathbf{r}) = \sum_{m=1}^{N_i} v(\mathbf{r} - \mathbf{r}(i_m, \alpha)) \tag{A6.3.7}$$

where the function $v$ represents the distribution of atomic positions around the assigned lattice point as caused by thermal vibrations. This function $v(\mathbf{r})$ is thus sharply peaked around $\mathbf{r} = 0$ and is normalized to unity

$$\int v(\mathbf{r})\, \mathrm{d}\mathbf{r} = 1. \tag{A6.3.8}$$

The position $\mathbf{r}(i_m, \alpha)$ is the lattice site to which the particular $i$-atom $i_m$ is assigned to the state $\alpha$. The ensemble average density of $i$ atoms at $\mathbf{r}$ is thus

$$n_i(\mathbf{r}, t) = \sum_\alpha n_\alpha^{(i)}(\mathbf{r}) p_\alpha(t). \tag{A6.3.9}$$

The density $n_i(\mathbf{r}, t)$ depends on time through the probabilities $p_\alpha$. Hence, by using the equation for $\mathbf{p}$ in the form (6.5.3) we easily find that

$$\frac{\partial n_i(\mathbf{r}, t)}{\partial t} = \sum_{\alpha,\beta} (n_\alpha^{(i)}(\mathbf{r}) - n_\beta^{(i)}(\mathbf{r}))(w_{\alpha\beta}^{(1)} p_\beta^{(0)} + w_{\alpha\beta}^{(0)} p_\beta^{(1)}). \tag{A6.3.10}$$

In setting down this equation we have omitted the time-dependence of $v$ since the period of the lattice vibrations is so short compared to the mean time between atomic jump that only the mean density is required.

From (6.5.4) for $w_{\alpha\beta}^{(1)}$ and (6.5.5) for $p_\beta^{(1)}$ together with the principle of detailed balance (6.2.8) we then find

$$\frac{\partial n_i(\mathbf{r}, t)}{\partial t} = \frac{1}{2kT} \sum_{\alpha,\beta} (n_\alpha^{(i)}(\mathbf{r}) - n_\beta^{(i)}(\mathbf{r})) w_{\alpha\beta}^{(0)} p_\beta^{(0)} (\delta g_\beta(t) - \delta g_\alpha(t))$$

$$+ \frac{1}{kT} \sum_{\alpha,\beta,\gamma,\delta} (n_\alpha^{(i)}(\mathbf{r}) - n_\beta^{(i)}(\mathbf{r})) w_{\alpha\beta}^{(0)}$$

$$\times \int_{-\infty}^{t} G_{\beta\gamma}^{(0)}(t - s) w_{\gamma\delta}^{(0)} p_\delta^{(0)} (\delta g_\delta(s) - \delta g_\gamma(s))\, \mathrm{d}s \tag{A6.3.11}$$

in which

$$\mathbf{G}^{(0)}(t) = \exp(-\mathbf{P}^{(0)}t) \tag{A6.3.12}$$

is the propagator for the system in thermal equilibrium (cf. 6.3.1). Before proceeding to the Fourier transform of (A6.3.11) we note that the **r**-dependence of these expressions comes solely through the factors $(n_\alpha^{(i)} - n_\beta^{(i)})$ on the r.h.s. while any time-dependence enters through the factors $(\delta g_\beta - \delta g_\alpha)$. These last factors are properties of the system as a whole, not of some local part of it. However, since the perturbing field may be assumed to interact locally with the atomic density these factors can be written as an integral over the volume of the system, thus

$$\delta g_\alpha(t) - \delta g_\beta(t) = \sum_j \int_V (n_\alpha^{(j)}(\mathbf{r}) - n_\beta^{(j)}(\mathbf{r}))\chi^{(j)}(\mathbf{r}, t)\,\mathrm{d}\mathbf{r}, \tag{A6.3.13}$$

where $\chi^{(j)}$ is the potential of the force $\mathbf{F}_j$ introduced in (A6.3.1).

These factors are evidently the important ones in generating the response of the system. However, we do not expect to be dealing with fields whose strength varies significantly over interatomic distances, while as already noted the factor $(n_\alpha^{(i)}(\mathbf{r}) - n_\beta^{(j)}(\mathbf{r}))$ is a sum of terms each of which is sharply localized about individual lattice sites. We shall therefore expand $\chi^{(j)}$ in a Taylor series. From (A6.3.7) and (A6.3.13)

$$\begin{aligned}\delta g_\alpha(t) - \delta g_\beta(t) &= \sum_j \sum_{n=1}^{N_j} \int [v(\mathbf{r} - \mathbf{r}(j_n, \alpha)) - v(\mathbf{r} - \mathbf{r}(j_n, \beta))]\chi^{(j)}(\mathbf{r}, t)\,\mathrm{d}\mathbf{r} \\ &= \sum_j \sum_{n=1}^{N_j} \int v(\mathbf{r}')[\chi^{(j)}(\mathbf{r}' + \mathbf{r}(j_n, \alpha), t) - \chi^{(j)}(\mathbf{r}' + \mathbf{r}(j_n, \beta), t)]\,\mathrm{d}\mathbf{r}'.\end{aligned} \tag{A6.3.14}$$

We now expand $\chi^{(j)}$ in the integrand in a Taylor series about **r**′. The first term in the series obviously gives nothing, in agreement with our expectation that the flux cannot depend on the absolute value of the potential but only on the fields derived from it. The leading term for the expression in square brackets in (A6.3.14) is thus

$$-(\mathbf{r}(j_n, \alpha) - \mathbf{r}(j_n, \beta))\cdot\mathbf{F}_j(\mathbf{r}'). \tag{A6.3.15}$$

We shall discard the higher terms as we do not expect quadrupolar terms and those associated with higher multipoles to contribute significantly to the fluxes. In place of (A6.3.14) we therefore have

$$\delta g_\alpha(t) - \delta g_\beta(t) = -\sum_j \sum_{n=1}^{N_j} (\mathbf{r}(j_n, \alpha) - \mathbf{r}(j_n, \beta))\cdot\int_V v(\mathbf{r}')\mathbf{F}_j(\mathbf{r}', t)\,\mathrm{d}\mathbf{r}'. \tag{A6.3.16}$$

Next, by using (A6.3.3a) we insert the Fourier representations of $v(\mathbf{r}')$ and $\mathbf{F}_j(\mathbf{r}')$

into (A6.3.16):

$$v(\mathbf{r}') = \frac{1}{V}\sum_{\mathbf{k}'} \tilde{v}(\mathbf{k}')\,\mathrm{e}^{-\mathrm{i}\mathbf{k}'\cdot\mathbf{r}'},$$

$$\mathbf{F}_j(\mathbf{r}', t) = \frac{1}{V}\sum_{\mathbf{k}''} \tilde{\mathbf{F}}_j(\mathbf{k}'', t)\,\mathrm{e}^{-\mathrm{i}\mathbf{k}''\cdot\mathbf{r}'}.$$

On carrying out the spatial integration over $\mathbf{r}'$ in (A6.3.16) we see that the only contributions are from the terms with $\mathbf{k}'' = -\mathbf{k}'$. The result is

$$\delta g_\alpha(t) - \delta g_\beta(t) = -\frac{1}{V}\sum_j \sum_{n=1}^{N_j} \sum_{\mathbf{k}'} \tilde{v}(\mathbf{k}')\mathbf{r}_{\alpha\beta}(j_n)\cdot\tilde{\mathbf{F}}_j(-\mathbf{k}', t), \qquad \text{(A6.3.17)}$$

in which we have used $\mathbf{r}_{\alpha\beta}(j_n)$ to denote the displacement of atom $j_n$ in the transition $\beta \to \alpha$.

Having thus obtained a Fourier representation of the $\delta g_\alpha - \delta g_\beta$ factors in (A6.3.11) we now return to the task of obtaining the Fourier transform of $\partial n_i/\partial t$. As already remarked the spatial dependence of $\partial n_i/\partial t$ comes from the factor $(n_\alpha^{(i)}(\mathbf{r}) - n_\beta^{(i)}(\mathbf{r}))$ in (A6.3.11). The Fourier transform of this factor by (A6.3.7) is

$$\sum_{m=1}^{N_i} [\mathrm{e}^{\mathrm{i}\mathbf{k}\cdot\mathbf{r}(i_m,\alpha)} - \mathrm{e}^{\mathrm{i}\mathbf{k}\cdot\mathbf{r}(i_m,\beta)}]\tilde{v}(\mathbf{k}),$$

which can be written in the form

$$\sum_{m=1}^{N_i} (\mathrm{i}\mathbf{k}\cdot\mathbf{r}_{\alpha\beta}(i_m))\eta(\mathrm{i}\mathbf{k}\cdot\mathbf{r}(i_m,\alpha), \mathrm{i}\mathbf{k}\cdot\mathbf{r}(i_m,\beta))\tilde{v}(\mathbf{k}), \qquad \text{(A6.3.18)}$$

in which $\eta(0, 0) = 1$.

We now use (A6.3.18) and (A6.3.17) to obtain $\partial n_i(\mathbf{k})/\partial t$, and thus from (A6.3.6b) also $\tilde{\mathbf{J}}_i(\mathbf{k})$. We then compare this expression with (A6.3.4) and thus infer $\tilde{\mathbf{\Phi}}_{ij}(\mathbf{k}, -\mathbf{k}\!:\!t)$ from which finally the macroscopic $\mathbf{L}_{ij}$ is obtained from (A6.3.5b). Since in the limit $\mathbf{k} \to 0$ both $\eta$ and $\tilde{v}(\mathbf{k}) \to 1$ we thus obtain the following expressions:

$$\lim_{\mathbf{k}\to 0} \mathbf{\Phi}_{ij}(-\mathbf{k}, \mathbf{k}\!:\!t) = \mathbf{\Phi}_{ij}^{(0)}(t) + \mathbf{\Phi}_{ij}^{(1)}(t) \qquad \text{(A6.3.19)}$$

with

$$\mathbf{\Phi}_{ij}^{(0)}(t) = \frac{1}{2VkT}\sum_{\alpha,\beta}\sum_{m,n} \mathbf{r}_{\beta\alpha}(i_m)\mathbf{r}_{\beta\alpha}(j_n)w_{\beta\alpha}^{(0)}p_\alpha^{(0)}\,\delta(t), \qquad \text{(A6.3.20)}$$

and

$$\mathbf{\Phi}_{ij}^{(1)}(t) = \frac{1}{VkT}\sum_{\alpha,\beta,\gamma,\delta}\sum_{m,n} (\mathbf{r}_{\delta\gamma}(i_m)w_{\delta\gamma}^{(0)})G_{\gamma\beta}^{(0)}(t)(\mathbf{r}_{\beta\alpha}(j_n)w_{\beta\alpha}^{(0)})p_\alpha^{(0)}. \qquad \text{(A6.3.21)}$$

To obtain the transport coefficients for a static force field one just integrates over all time $t$, because $\mathbf{F}_j$ is constant and so can be removed from the integral in (A6.3.4). This results in the expressions given as eqns. (6.5.10)–(6.5.12).

There is nothing in this deviation to limit the calculation to crystalline solids. As elsewhere in this chapter the analysis is also valid for irregular lattice structures, as long as these do not change with time, i.e. as long as the meaning of the states $\alpha, \beta, \ldots$ does not change.

# Appendix 6.4

## Positive definiteness of $\mathbf{L}$

In §4.2 we drew attention to the requirement that the matrix of $L$-coefficients should be positive definite, and it is obviously desirable to verify that that condition is satisfied by specific forms of the $L$-coefficients. This can be conveniently done here by using the relation (6.5.14). For simplicity, we assume that all $L_{ij}$ have been diagonalized in the Cartesian components $x, y, z$ and deal just with the $xx$-component. We then have

$$L_{ij} = \frac{1}{2kTV} \frac{\langle \Delta X_i \, \Delta X_j \rangle^{(0)}}{t}, \tag{A6.4.1}$$

in which $\Delta X_i$ is the sum over all atoms of type $i$ of the displacement of each in the $x$-direction. By definition

$$\langle \Delta X_i \, \Delta X_j \rangle^{(0)} = \sum_{\alpha, \beta} X_{\beta\alpha}(i) X_{\beta\alpha}(j) G^{(0)}_{\beta\alpha}(t) p^{(0)}_{\alpha}, \tag{A6.4.2}$$

with

$$X_{\beta\alpha}(i) = X_{\beta}(i) - X_{\alpha}(i), \tag{A6.4.3}$$

where $X_{\alpha}(i)$ is the sum of the $x$-co-ordinates of atoms of species $i$ when the system is in state $\alpha$. $G^{(0)}_{\beta\alpha}(t)$, as before, is the conditional probability of the transition $\alpha \to \beta$ occurring in time $t$, while $p^{(0)}_{\alpha}$ is the probability that the system is in state $\alpha$, in both cases for a system in thermodynamic equilibrium.

First, we observe that by the interpretation of $p^{(0)}_{\alpha}$ and $G^{(0)}_{\beta\alpha}(t)$ as probabilities $G^{(0)}_{\beta\alpha}(t) \, p^{(0)}_{\alpha} \geq 0$. By (A6.4.1) and (A6.4.2) it then follows immediately that

$$L_{ii} \geq 0, \tag{A6.4.4}$$

as demanded by (4.2.6a).

It remains to show that (4.2.7a) is satisfied. For this it will be sufficient to show that

$$\alpha \equiv F_{ii}F_{jj} - F_{ij}F_{ji}, \tag{A6.4.5}$$

where

$$F_{ij} = \langle \Delta X_i \, \Delta X_j \rangle, \qquad \text{all } i, j, \tag{A6.4.6}$$

satisfies

$$\alpha \geq 0. \tag{A6.4.7}$$

By (A6.4.2) we have

$$\alpha = \sum_{\alpha,\beta,\gamma,\delta} \{(X_{\beta\alpha}(i))^2(X_{\delta\gamma}(j)^2) - (X_{\beta\alpha}(i))(X_{\beta\alpha}(j))(X_{\delta\gamma}(i))(X_{\delta\gamma}(j))\}G^{(0)}_{\beta\alpha}(t)p^{(0)}_{\alpha}G^{(0)}_{\delta\gamma}(t)p^{(0)}_{\gamma}.$$

$$= \sum_{\alpha,\beta,\gamma,\delta} \{(X_{\beta\alpha}(i)X_{\delta\gamma}(j))(X_{\beta\alpha}(i)X_{\delta\gamma}(j) - X_{\beta\alpha}(j)X_{\delta\gamma}(i))\}G^{(0)}_{\beta\alpha}(t)p^{(0)}_{\alpha}G^{(0)}_{\delta\gamma}(t)p^{(0)}_{\gamma}. \tag{A6.4.8}$$

Likewise by interchanging $i$ and $j$ we obtain

$$\alpha = -\sum_{\alpha,\beta,\gamma,\delta} \{(X_{\beta\alpha}(j)X_{\delta\gamma}(i))(X_{\beta\alpha}(i)X_{\delta\gamma}(j) - X_{\beta\alpha}(j)X_{\delta\gamma}(i))\}G^{(0)}_{\beta\alpha}(t)p^{(0)}_{\alpha}G^{(0)}_{\delta\gamma}(t)p^{(0)}_{\gamma}. \tag{A6.4.9}$$

By adding 1/2 of (A6.4.8) to 1/2 of (A6.4.9) we obtain then

$$\alpha = \frac{1}{2}\sum_{\alpha,\beta,\gamma,\delta} (X_{\beta\alpha}(i)X_{\delta\gamma}(j) - X_{\beta\alpha}(j)X_{\delta\gamma}(i))^2 G^{(0)}_{\beta\alpha}(t)p^{(0)}_{\alpha}G^{(0)}_{\delta\gamma}(t)p^{(0)}_{\gamma}. \tag{A6.4.10}$$

Since the probabilities $G^{(0)}_{\beta\alpha}$, $p^{(0)}_{\alpha}$, etc. are $\geq 0$, it follows from (A6.4.10) that $\alpha \geq 0$, as required.

In summary, we have verified the positive definiteness of the matrix of $L$-coefficients for isothermal conditions.

# 7

# *Kinetic theory of relaxation processes*

## 7.1 Introduction

The theory presented in the last chapter has already provided important general insight into atomic transport coefficients, as well as specific results which define routes to the evaluation of these coefficients for particular systems, both dilute and concentrated. We could therefore go straight to these evaluations. However, there are good reasons for delaying the matter. One is that there are alternative approaches to dilute systems which convey particular insights into the processes of atomic transport and which define additional quantities of experimental interest, most notably the so-called *correlation factor* (Chapters 8 and 10). Two of the most important of these alternative approaches are the kinetic and random-walk theories: the former are the subject of this and the next chapter while the latter are dealt with in Chapters 9–10. Although these theories are very different in appearance we shall show that in the end they can lead to the same expressions for measurable quantities in terms of specific atomic features of the system under study. Ultimately therefore they may use the same techniques for the evaluation of these expressions (as we shall show in Chapter 11 especially).

Although we shall henceforth be much more concerned with the physical details of the systems of interest, both approaches relate rather closely to what we have presented in Chapter 6. We shall take kinetic theory first because it offers a rather direct representation of many of the basic equations of Chapter 6 together with a number of easily obtained approximate results for particular systems. At the same time some of its notions are useful in the development of the random-walk theories.

It may also be remarked that in the theory of transport phenomena in *fluids*, kinetic theories (e.g. the Boltzmann transport equation) have been fundamental to the evaluation of the properties of particular systems. By setting down particular representations of the master equations (6.2.2) and (6.5.3) in terms of the spatial distributions of atoms and defects we shall establish the basis of analogous theories for atomic transport in solids. Although these have been developed to only a small extent relative to the great mass of kinetic theory for fluids, they have nevertheless proved similarly useful and fruitful. They have been used to discuss a range of physical phenomena (e.g. relaxation processes, electrical conductivity, diffusion, the Soret effect, the dynamical structure factor, Mössbauer line width, etc.) in different defect systems and have been shown capable of providing all the required phenomenological $L$-coefficients in dilute systems. Such theories have also provided certain basic ideas about the solute–vacancy pair distribution functions in diffusion which have been used (notably by Manning, 1968) to extend the capability of the random-walk approach to deal with systems in which there is a net defect flux, and, most successfully, to deal with diffusion in concentrated alloys via the random alloy model.

We shall introduce kinetic theories of diffusion in the next chapter. To provide the necessary basis we shall here first consider the simpler situation presented by the mechanical (anelastic) and dielectric relaxation of cubic solids containing distributions of defects of lower than cubic symmetry. Widely studied examples of this kind are provided by (i) solute–vacancy pairs in crystals with the NaCl structure, as illustrated in Fig. 7.1, (ii) pairs formed from trivalent solute ions (e.g. rare earth ions) and interstitial fluorine ions in $CaF_2$ and similar compounds, as shown in Fig. 7.2, and (iii) interstitial solute atoms (e.g. C, N and O) in b.c.c. metals such as $\alpha$-Fe, as represented in Fig. 7.3. In the first two examples the lowered symmetry arises from a physical attraction and consequent pairing of defects with solute atoms; in the third it results from the tetragonal symmetry associated with the interstitial site itself and the consequent anisotropic displacement of the neighbours to the interstitial site. In each case it is clear that several distinct orientations of the defect exist; for the defects shown in the figures there are respectively 12, 6 and 3 geometrically distinct orientations. In the absence of any external perturbation these orientations nevertheless all have the same energy, and thus in thermodynamic equilibrium the defects will be distributed equally among them.

The imposition of a perturbation such as a stress field, however, will remove this energy equivalence (at least partially) and the distribution of defects among the various orientations will tend to become non-uniform. The rate at which it does so is determined by the various jump rates of the defects. If the applied stress field is periodic or otherwise dependent upon time, the adjustment of the distribution of defects will lag behind the variations in the field, thus giving rise to an absorption of energy from the field. When the stresses are generated by setting the specimen into vibration this energy absorption can be measured by observing the damping of these vibrations.

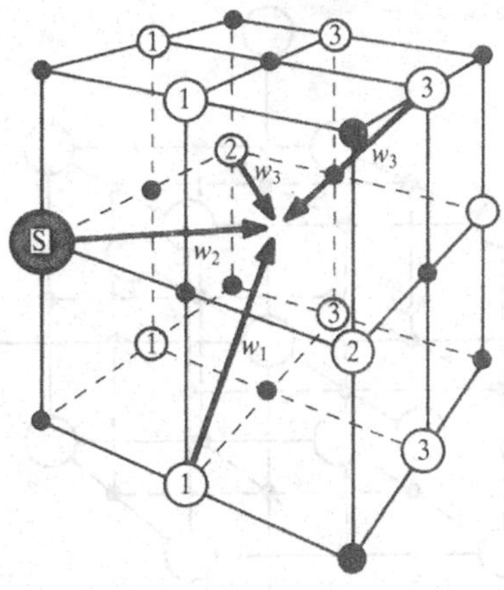


Fig. 7.1. Schematic diagram of a nearest-neighbour solute–vacancy pair in a NaCl lattice. Open circles represent one sub-lattice (f.c.c.) to which the solute atom (S) and the vacancy are confined. The filled circles represent the other sub-lattice (also f.c.c.). First neighbours (110) of the solute (S) are marked 1, second neighbours (200) are marked 2 and third neighbours (211) are marked 3. The solute atom may jump into the vacancy at a rate $w_2$, first neighbour atoms may do so at a rate $w_1$ and second neighbour atoms at a rate $w_3$. An extension of this model will be required later for the theory of diffusion: in this, vacancies may jump back to first neighbour positions at a rate $w_4$ while jumps of vacancies between positions remote from any solute atom occur at a rate $w_0$.

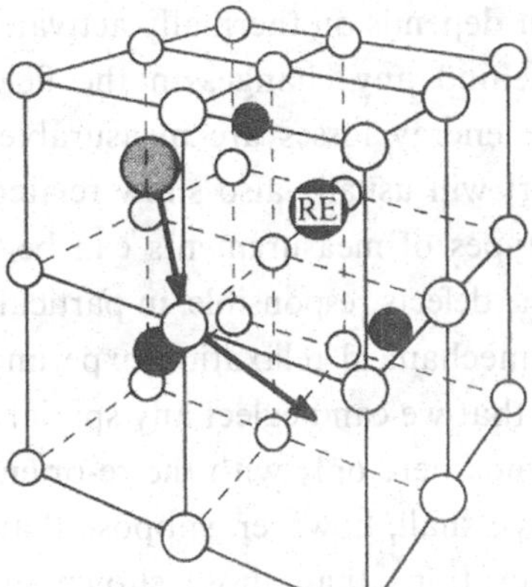


Fig. 7.2. Schematic diagram of a $CaF_2$ lattice containing a nearest-neighbour defect pair formed from a rare-earth solute cation (RE) and a $F^-$ interstitial (shaded). The open circles represent the anions (on a simple cubic sub-lattice) and the filled circles the cations (on a f.c.c. sub-lattice). The interstitial anion is believed to move by the interstitialcy mechanism indicated. The rare-earth cation is not moved by these anion interstitial movements.

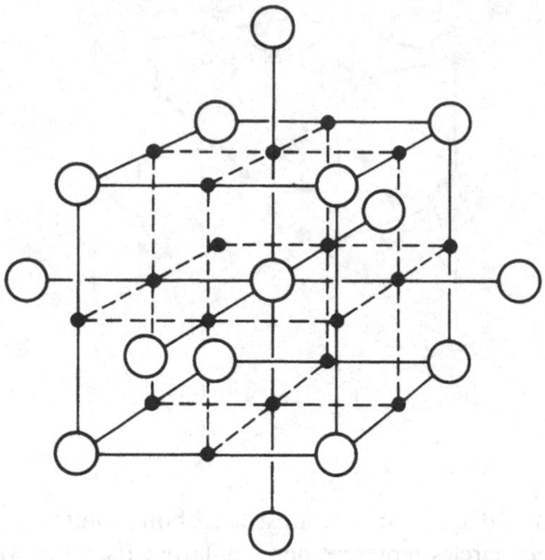

Fig. 7.3. Schematic diagram of the positions available to an interstitial solute atom (filled circles) in a b.c.c. metal lattice such as α-Fe (open circles). Note the tetragonal symmetry defined by the solute atom and the arrangement of its neighbours.

There is an analogous situation with dielectrics containing defects possessing electrical dipole moments. The distribution of dipoles among the available orientations will be perturbed by an external electric field, but the adjustment of the distribution, because it depends on thermally activated reorientation jumps of the defects, again lags behind any changes in the field. As a result dielectric relaxation and associated energy losses are measurable. Materials which show dielectric losses of this sort will usually also show related mechanical losses. The combination of the two types of measurements can be very useful in providing basic information about the defects responsible, in particular about their symmetry.

In both dielectric and mechanical relaxation experiments the frequencies and other conditions are such that we can neglect any spatial variations in the applied fields. We are thus concerned here only with the re-orientation of the defects and not with their diffusion. We shall, however, suppose that the defects may also be in more extended configurations than those shown in Figs. 7.1 and 7.2. For example, it may be necessary to allow the solute–vacancy pairs of Fig. 7.1 to be at the next nearest-neighbour separation (or even at more distant separations) in addition to the nearest-neighbour separation shown. We shall turn to the role of such defects in diffusion in the next chapter. Later in Chapter 13 we shall present a more formal way of generating kinetic theories which is particularly useful for concentrated alloys and solid solutions.

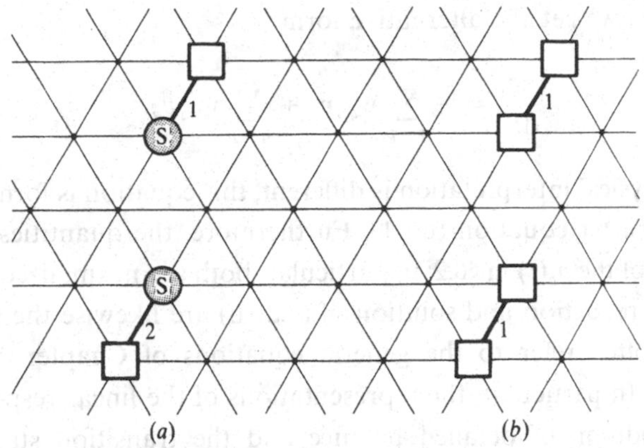


Fig. 7.4. Schematic diagram to illustrate the meaning of *distinct* configurations of (*a*) an impurity–vacancy pair and (*b*) a divacancy. Since the elements of the impurity–vacancy pair are distinguishable, configurations (1) and (2) are physically distinct. On the other hand, for the vacancy pair, where the two vacancies are indistinguishable from one another, there is only one corresponding distinct configuration.

## 7.2 Relaxation of complex defects: general theory

We label the configurations of the complex defect (e.g. a solute–defect pair) as $p$, $q$, etc. (By 'configuration' we here mean any distinct orientation or relative arrangement of the components of the complex defect: see Fig. 7.4.). We suppose that a complex defect changes from configuration $p$ to $q$ as a result of thermal activation at a rate $w_{qp}$. These transitions are taken to be Markovian, i.e. the $w_{qp}$ depend on the initial and final configurations but are independent of all previous transitions. We denote the number density of defects which are in configuration $p$ at time $t$ by $n_p(t)$. For a closed set of configurations $p$ the rate equations for the densities, $n_p(t)$ are

$$\frac{\mathrm{d}n_p}{\mathrm{d}t} = -\sum_{q \neq p} w_{qp} n_p + \sum_{q \neq p} w_{pq} n_q. \tag{7.2.1a}$$

The first term on the right side gives the rate of loss of defects by jumps out of configuration $p$ into all the others, while the second gives the corresponding rate of gain of defects by jumps into configuration $p$. However, we can make direct use of the formalism of Chapter 6 by introducing the fraction of all defects which are in configuration $p$ at time $t$, $p_p(t)$, such that $n_p(t) = n_\mathrm{d} p_p(t)$ where $n_\mathrm{d}$ is the number density of defects irrespective of configuration. By dividing (7.2.1a) throughout by $n_\mathrm{d}$ we get the alternative form

$$\frac{\mathrm{d}p_p}{\mathrm{d}t} = -\sum_{q\neq p} w_{qp}p_p + \sum_{q\neq p} w_{pq}p_q. \tag{7.2.1b}$$

Although the physical interpretation is different, this equation is formally identical to the general master equation (6.2.1). Furthermore, the quantities $p_p(t)$ are the exact analogues of the $p_\alpha(t)$ in §6.2; in particular, both are normalized to unity. The formal stages of reduction and solution of (7.2.1b) are likewise the same, and we shall therefore often refer to the general equations of Chapter 6 rather than re-express them. In particular, the representations of the linear response approximation, the condition of detailed balance and the transition state theory are formally identical. Hence in matrix form (7.2.1b) becomes

$$\frac{\mathrm{d}\mathbf{p}^{(1)}}{\mathrm{d}t} = -\mathbf{P}^{(0)}\mathbf{p}^{(1)} - \mathbf{P}^{(1)}(t)\mathbf{p}^{(0)} \tag{7.2.2}$$

with $\mathbf{P}^{(0)}$ and $\mathbf{P}^{(1)}$ given by the analogues of (6.2.4) in $w_{qp}^{(0)}$ and $w_{qp}^{(1)}$ respectively (the indices $p$, $q$, etc. replacing $\alpha$, $\beta$, etc.). As before, the superscripts (0) and (1) imply respectively an unperturbed quantity under thermal equilibrium conditions and the first-order perturbation of that quantity. The analogue of (6.5.4) then leads to

$$\mathbf{P}^{(1)}\mathbf{p}^{(0)} = \frac{1}{kT}\mathbf{P}^{(0)}\mathbf{N}\mathbf{g} \tag{7.2.3a}$$

$$= \frac{1}{kT}\mathbf{W}\mathbf{p}^{(0)}, \tag{7.2.3b}$$

in which $\mathbf{N}$ and $\mathbf{g}$ are defined by (6.2.11) and (6.5.6) respectively, while $\mathbf{W}$ is a wholly diagonal matrix whose elements are

$$W_{pq} = W_{pp}\,\delta_{pq} \tag{7.2.4}$$

and

$$W_{pp} = \sum_q (\delta g_p - \delta g_q) w_{qp}^{(0)}. \tag{7.2.5}$$

As in the general theory, all details of the transition state (*saddle-point configuration*) have cancelled out and the inhomogeneous term in (7.2.2) contains only the perturbations in the ground-configuration energies, $\delta g_p$. Here this has the practical consequence that these relaxation phenomena are independent of any details of the *path* taken by the defect in moving from one configuration to another, although, of course, they depend upon which configurations are actually accessible.

If we can now specify the perturbations of these energies we complete the microscopic basis of the theory of dielectric and mechanical relaxation due to these defects. For dielectric relaxation in an electric field $\mathbf{F}(t)$ we have (in vector notation)

$$\delta g_p = -\boldsymbol{\mu}_p \cdot \mathbf{F}(t) \tag{7.2.6}$$

and for anelastic relaxation in a uniform stress $\sigma_{\alpha\beta}$ (Cartesian components $\alpha$, $\beta$, etc.)

$$\delta g_p = -v\lambda_{\alpha\beta}^{(p)}\sigma_{\alpha\beta}(t), \tag{7.2.7}$$

in which $\boldsymbol{\mu}_p$ is the electric dipole moment of the defect in configuration $p$ and $\lambda_{\alpha\beta}{}^{(p)}$ is its elastic dipole tensor (see §4.4.3 and Nowick and Berry, 1972). Here, $v$ is the molecular volume and we use the convention that a repeated index is to be summed over.

In the analysis of relaxation phenomena on the basis of eqns. (7.2.2)–(7.2.7) the eigenvectors and eigenvalues of $\mathbf{P}^{(0)}$, or rather of its symmetrized form (cf. 6.2.10)

$$\mathbf{S} = \mathbf{N}^{-1/2}\mathbf{P}^{(0)}\mathbf{N}^{1/2}, \tag{7.2.8}$$

in which

$$N_{pq} = p_p^{(0)}\,\delta_{pq}, \tag{7.2.9}$$

play an important part.

For example, if, as in the experiments of Dreyfus (1961), the applied field is held constant up to $t = 0$ and then suddenly removed, so that

$$\mathbf{g} = \mathbf{g}_0\theta(-t), \tag{7.2.10}$$

where $\theta$ is the Heaviside step function,* the decay of $\mathbf{p}^{(1)}$ will be described by

$$\mathbf{p}^{(1)}(t) = \sum_{\nu}{}' c_\nu \mathbf{a}^{(\nu)} \exp(-t/\tau^{(\nu)}), \tag{7.2.11}$$

in which the eigenvectors $\mathbf{a}^{(\nu)}$ and the corresponding eigenvalues $\alpha^{(\nu)} \equiv 1/\tau^{(\nu)}$ are given by (6.2.12) with (7.2.8). The prime on the summation indicates that the mode with zero eigenvalue is excluded (because, in the absence of the field, $\mathbf{p}^{(1)}$ is zero in thermodynamic equilibrium). The coefficients $c_\nu$ are found from $\mathbf{p}^{(1)}(0)$ by using the orthogonality property (6.2.13),

$$c_\nu = \tilde{\mathbf{a}}^{(\nu)}\mathbf{p}^{(1)}(0). \tag{7.2.12}$$

If the system was in thermodynamic equilibrium in the field up to the time (zero) at which the field was removed, then $\mathbf{p}^{(1)}(0)$ is easily found from the equilibrium Boltzmann distribution. To first order in the field we obtain

$$p_p^{(1)}(0) = -\frac{1}{kT}p_p^{(0)}\frac{\sum_q(\delta g_p - \delta g_q)\exp(-g_q^{(0)}/kT)}{\sum_q \exp(-g_q^{(0)}/kT)}, \tag{7.2.13}$$

* $\theta(x) = 1$ for $x > 0$ and $\theta(x) = 0$ for $x < 0$.

in which $g_q^{(0)}$ is the energy of the defect in configuration $q$ in the absence of the field. This could also be written

$$p_p^{(1)}(0) = -\frac{1}{kT} p_p^{(0)} \sum_q (\delta g_p - \delta g_q) p_q^{(0)}. \tag{7.2.14}$$

It is easy to verify that (7.2.13) satisfies (7.2.2) with $\mathrm{d}p_p^{(1)}/\mathrm{d}t = 0$, as required in thermodynamic equilibrium.

In many cases where spectroscopic techniques (optical, infra-red, elecron spin resonance) allow one to 'see' defects in particular orientations, it has been possible to follow changes in the distribution $\mathbf{p}^{(1)}(t)$ directly, and thus to obtain information about some of the relaxation times $\tau^{(\nu)}$. Usually a non-equilibrium distribution is prepared by some means and the relaxation from this initial state to a thermal equilibrium distribution is followed. Which relaxation times can be obtained will depend on the means available for preparing the initial state (e.g. uniaxial stress, optically-induced re-orientation, sudden change of temperature following an initial equilibration, etc.).

The commoner situation, however, is that one must study these defect distributions less directly through macroscopic relaxation measurements. To complete the theory for these circumstances we need (i) to solve the kinetic equations for a general time-dependent field $\mathbf{F}(t)$ and (ii) to specify observable macroscopic quantities in terms of $\mathbf{p}^{(1)}$. For the first of these we can use the solution (6.5.5). Likewise, in terms of the Fourier transformed quantities we can use eqns. (6.5.7–6.5.9) as they stand.

As regards the second point, the contribution to dielectric polarization will be

$$\Delta P = n_{\mathrm{d}} \tilde{\boldsymbol{\mu}} \mathbf{p}^{(1)} \tag{7.2.15}$$

in which $n_{\mathrm{d}}$ is the number of defects per unit volume and $\tilde{\boldsymbol{\mu}}$ is the row matrix of dipole moments

$$\tilde{\boldsymbol{\mu}} = [\mu_1 \quad \mu_2 \quad \ldots]. \tag{7.2.16}$$

Correspondingly, the contribution to the elastic strain tensor will be the summation over all configurations $p$ of the product of $\lambda_{\alpha\beta}^{(p)}$ for the defects in configuration $p$ with the mole fraction of such defects, i.e.

$$\Delta e_{\alpha\beta} = n_{\mathrm{d}} v \tilde{\boldsymbol{\lambda}}_{\alpha\beta} \mathbf{p}^{(1)}, \tag{7.2.17}$$

where $v$ is the molecular volume and $\tilde{\boldsymbol{\lambda}}_{\alpha\beta}$ is the row matrix of elastic dipole tensors,

$$\tilde{\boldsymbol{\lambda}}_{\alpha\beta} = [\lambda_{\alpha\beta}^{(1)} \quad \lambda_{\alpha\beta}^{(2)} \quad \ldots], \tag{7.2.18}$$

(see §4.4.3; also Nowick and Berry, 1972).

In connection with dielectric relaxation we should also note that long-range dipolar interactions among the defects will mean that the average internal field **F** acting on the defects will differ in principle from the electric **E**-field of electromagnetic theory. This internal field will not be the simple Lorentz internal field (because the defects are extended and not point dipoles; Fröhlich, 1958) but may differ from **E** by a modest factor. The experimental work of Burton and Dryden (1970) on alkali halide crystals indicated that this factor could not be very different from 1. Ways of estimating it theoretically were discussed by Boswarva and Franklin (1965), but do not seem to have been much considered since. If we call this factor $\eta$ then we can use (7.2.6), (6.5.9) and (7.2.15) to give the contribution of the relaxing defects to the complex electric permittivity, in tensor notation and SI units, as

$$\Delta\varepsilon_{\alpha\beta}(\omega) = \sum_{\nu} \frac{\Delta\varepsilon_{\alpha\beta}^{(\nu)}}{1 + i\omega\tau^{(\nu)}}, \tag{7.2.19}$$

in which

$$\Delta\varepsilon_{\alpha\beta}^{(\nu)} = \frac{n_d \eta}{kT} (\tilde{\boldsymbol{\mu}}_\alpha \mathbf{N}^{1/2}\mathbf{a}^{(\nu)})(\tilde{\mathbf{a}}^{(\nu)}\mathbf{N}^{1/2}\boldsymbol{\mu}_\beta). \tag{7.2.20}$$

Comparison with (1.5.9) shows that each eigenvalue of **S** gives a Debye relaxation mode.

There is no corresponding internal field for anelastic relaxation and the contribution of the defects to the complex elastic compliance tensor by (7.2.6), (6.5.9) and (7.2.17) is

$$\Delta s_{\alpha\beta\lambda\mu}(\omega) = \sum_{\nu} \frac{\Delta s_{\alpha\beta\lambda\mu}^{(\nu)}}{1 + i\omega\tau^{(\nu)}}, \tag{7.2.21}$$

where,

$$\Delta s_{\alpha\beta\lambda\mu}^{(\nu)} = \frac{v}{kT} (\tilde{\boldsymbol{\lambda}}_{\alpha\beta} \mathbf{N}^{1/2}\mathbf{a}^{(\nu)})(\tilde{\mathbf{a}}^{(\nu)}\mathbf{N}^{1/2}\boldsymbol{\lambda}_{\lambda\mu}). \tag{7.2.22}$$

It is clear that the evaluation of these expressions requires the determination of the relaxation times $\tau^{(\nu)}$ and the corresponding modes of relaxation, i.e. the determination of the eigenvalues ($\alpha^{(\nu)} \equiv 1/\tau^{(\nu)}$) and the eigenvectors $\mathbf{a}^{(\nu)}$ of the matrix **S** defined by (7.2.8) in accordance with (6.2.12). Although it does not contribute to $\Delta\varepsilon_{\alpha\beta}(\omega)$ nor to $\Delta s_{\alpha\beta\lambda\mu}(\omega)$, there is here, as in the general case described in Chapter 6, at least one eigenvalue $\tau^{-1}$ which is zero; this has the corresponding eigenvector $\mathbf{N}^{-1/2}\mathbf{p}^{(0)}$ (with $\mathbf{p}^{(0)}$ the thermal equilibrium distribution) since necessarily then

$$\mathbf{P}\mathbf{p}^{(0)} = 0,$$

and thus

$$\mathbf{S}(\mathbf{N}^{-1/2}\mathbf{p}^{(0)}) = 0. \tag{7.2.23}$$

Furthermore, all the other eigenvalues are real and $\geq 0$, as was proved in Appendix 6.1. In practice they will be positive unless the system is overspecified by the inclusion of zero jump-frequencies which make it impossible for the system to relax in certain modes.

In contrast to the general case, however, here the eigenvectors for any particular system can be classified by their spatial symmetry, and this symmetry determines the nature of the fields to which the modes can couple, i.e. it tells us which modes $\nu$ contribute to the summations in (7.2.19) and (7.2.21). Such symmetry information is important to the experimental analysis of particular systems and we therefore summarize the principal results for defects in cubic solids in the next section.

## 7.3 Symmetry of relaxation modes

The results obtained in the previous section form the basis of an extensive body of analysis of dielectric and anelastic relaxation processes in a variety of solids. As already remarked, symmetry analysis plays an important part in both theoretical and experimental investigation of these systems. The basic techniques are common to much solid state and molecular theory and are described in many texts (see e.g. Cotton, 1971). For their application to the processes of concern here there are the extensive reviews by Nowick (1967) and Nowick and Heller (1963, 1965) and the book by Nowick and Berry (1972). In this section we shall therefore first summarize some of the more commonly used results and then illustrate them with three particular examples.

### *7.3.1 Summary of results for defects in a cubic lattice*

We refer here only to cubic crystals, although symmetry analyses have been made for most other classes as well. We may suppose that, although the defects in their various configurations are individually of lower symmetry (e.g. tetragonal, trigonal or orthorhombic), the entire set of these configurations has the cubic symmetry of the crystal, and for definiteness we take this to be $O_h$. Each of the relaxation modes (eigen-solutions of 6.2.12) of these defects then belongs to one of the irreducible representations of the symmetry group $O_h$. The particular irreducible representation specifies the spatial symmetry of the relaxation mode. The cubic group $O_h$ has 10 irreducible representations conventionally designated $A_{1g}$, $A_{2g}$, $E_g$, $T_{1g}$, $T_{2g}$, $A_{1u}$, $A_{2u}$, $E_u$, $T_{1u}$ and $T_{2u}$. (For full definitions of this notation we refer to Cotton, 1971, especially Chap. 4, but some idea of the symmetry implied by these descriptions may be obtained from the particular illustrations in Fig. 7.5.) Which of these actually may arise in any particular case is determined by the symmetry of the configurations available to the defect. However, only four types of

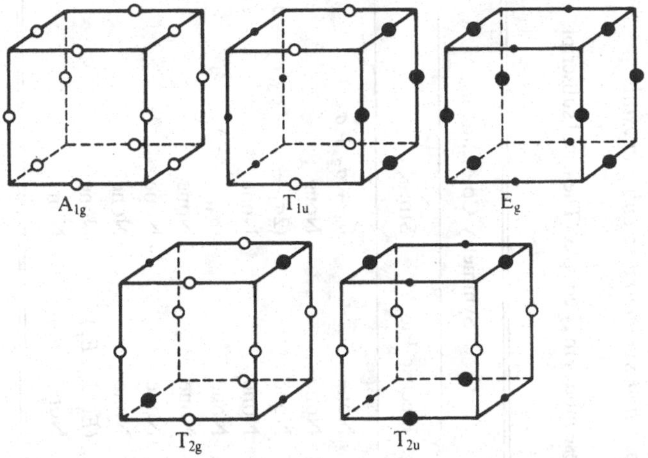


Fig. 7.5. A Haven–van Santen diagram showing the five types of relaxation mode of the nearest-neighbour solute–vacancy pair in the NaCl lattice. The circles (both filled and open) represent the 12 nearest-neighbour positions to the solute which is located at the centre of the cube. The open circles, large filled circles and the small filled circles represent distributions of vacancies on sites which are respectively equal to, greater than and less than that for thermodynamic equilibrium. (After Haven and van Santen, 1958.)

mode, namely, $A_{1g}$, $E_g$, $T_{2g}$ and $T_{1u}$, are really relevant here, since it is only modes of these symmetries that can couple to uniform elastic and electric fields. More particularly, only $T_{1u}$ modes can couple to a uniform electric field, while only $A_{1g}$, $E_g$ and $T_{2g}$ can couple to uniform stress fields.

Underlying these statements is the fact that any uniform electric or stress field can be written as a linear combination of symmetry-adapted combinations of the field components, each of which also belongs to an appropriate irreducible representation. Such *symmetry co-ordinates* of the fields are given in Table 7.1 Only relaxation modes having the same designation can couple to these combinations of field components. Thus each of the three Cartesian components of a uniform electric field along the ⟨100⟩ crystal directions can couple only to the similarly oriented $T_{1u}$ relaxation mode, while each of the three pure shear components $\sigma_{xy}$, $\sigma_{yz}$, $\sigma_{zx}$ (i.e. $\sigma_4$, $\sigma_5$, $\sigma_6$ in Voigt notation) of a uniform elastic field can couple only to the corresponding $T_{2g}$ mode and to no other. Lastly, uniaxial stresses, which are far more convenient experimentally than these symmetrized stresses, can be resolved into combinations of the symmetrized stresses, e.g. a uniaxial stress along ⟨100⟩ is a sum of symmetrized stresses $\sigma(A_{1g})$ and $\sigma(E_g)$ while one along ⟨111⟩ is a sum of $\sigma(A_{1g})$ and $\sigma(T_{2g})$.

We shall now illustrate these general remarks by summarizing several particular results. We do so first for the three simplest cases of defects which are either

Table 7.1. *Possible relaxation normal modes and their symmetry type for defects of specified symmetry in cubic crystals*

Also given are the symmetry co-ordinates of the electric field and stress components to which these modes may couple. (The usual subscript notation of Voigt is used for the stress components).

| Mode | Possible for defects of symmetry: | | | Symmetry Co-ordinates | |
|---|---|---|---|---|---|
| Symmetry type and degeneracy | Tetragonal | ⟨110⟩ Orthorhombic | Trigonal | Electric Field | Stress, $\sigma$ |
| $A_{1g}$ ($\times 1$) | Yes | Yes | Yes | None | $\sigma_1 + \sigma_2 + \sigma_3$ |
| $A_{2g}$ ($\times 1$) | No | No | No | None | None |
| $E_g$ ($\times 2$) | Yes | Yes | No | None | $(2\sigma_1 - \sigma_2 - \sigma_3, \sigma_2 - \sigma_3)$ |
| $T_{1g}$ ($\times 3$) | No | No | No | None | None |
| $T_{2g}$ ($\times 3$) | No | Yes | Yes | None | $(\sigma_4, \sigma_5, \sigma_6)$ |
| $A_{1u}$ ($\times 1$) | No | No | No | None | None |
| $A_{2u}$ ($\times 1$) | No | No | Yes | None | None |
| $E_u$ ($\times 2$) | No | No | No | None | None |
| $T_{1u}$ ($\times 3$) | Yes | Yes | Yes | $(E_x, E_y, E_z)$ | None |
| $T_{2u}$ ($\times 3$) | No | Yes | No | None | None |

tetragonal, trigonal or ⟨110⟩ orthorhombic in all orientations and configurations.

*Tetragonal defects*

There may be solutions of the eigenvalue equation (6.2.12) corresponding to the $A_{1g}$ (non-degenerate), $E_g$ (doubly degenerate) and $T_{1g}$ (triply degenerate) representations, but not all these will necessarily occur in any particular case. The degeneracy specifies the number of distinct modes having the same eigenvalue, $\tau^{-1}$. Of these, only $T_{1u}$ (which is odd) can couple to a uniform electric field, i.e. only $T_{1u}$ relaxation modes can be studied in dielectric loss or I.T.C. experiments (cf. §1.5.2). The doubly degenerate $E_g$ modes can couple to uniaxial stresses applied along ⟨100⟩ or ⟨110⟩ directions but *not* along ⟨111⟩. In principle, the totally symmetric $A_{1g}$ mode can couple to a hydrostatic stress, though for this to give rise to an observable relaxation it would be necessary for the defect to be able to extend as well as to re-orient (Nowick, 1970a; Lidiard, 1986b).

*Trigonal defects*

There may be solutions of (6.2.12) corresponding to the $A_{1g}$ (non-degenerate), $T_{2g}$ (triply degenerate), $A_{2u}$ (non-degenerate) and $T_{1u}$ (triply degenerate) representations. As before only $T_{1u}$ modes can be seen in dielectric relaxation and I.T.C experiments. The $T_{2g}$ modes can be studied in mechanical relaxation experiments when uniaxial stresses are applied in ⟨110⟩ or ⟨111⟩, but not ⟨100⟩, directions. The remarks made above about the symmetric $A_{1g}$ mode apply also here.

*Orthorhombic ⟨110⟩ defects*

Here there may be solutions belonging to the $A_{1g}$, $E_g$, $T_{2g}$, $T_{1u}$ and $T_{2u}$ representations. The coupling of these to uniaxial stress fields and to uniform electric fields is the same as in the two previous examples. In particular, defects of this symmetry should give rise to mechanical relaxation effects for uniaxial fields along all three principal cubic directions. The preceding remarks about the symmetric $A_{1g}$ mode apply also here.

These differences in response to uniaxial stress between the three types of defects are useful in determining the underlying symmetry of the defects giving rise to the relaxation. Dielectric relaxation, by contrast, tells us only that the defect has an electric dipole moment but cannot tell us anything about the axial symmetry of the defect, essentially because the dielectric constant tensor reduces to a scalar in cubic crystals (cf. §1.5). When the nature of the defect is reasonably well known, however, a combinaton of dielectric and mechanical relaxation measurements, possibly supplemented by other kinetic information (e.g. lifetime broadening of electron paramagnetic resonance lines, diffusion data, etc.), may allow the accurate determination of several jump frequencies and corresponding activation energies. We then need the particular solutions of (6.2.12) in terms of the various jump frequencies which characterize the defect. These must be worked out for each case as appropriate. Some of these practically important cases show that we need to include defect jumps which change the symmetry of the complex; for example, in the case of solute–vacancy pairs in alkali halides with the rock salt structure there may be only small differences in energy between nearest- and next-nearest neighbour separations, i.e. between $\langle 110 \rangle$ orthorhombic and tetragonal configurations. In these cases the range of symmetries of the allowed solutions of (6.2.12) is correspondingly enlarged; in this example we obtain solutions belonging to all of $A_{1g}$, $E_g$, $T_{2g}$, $T_{1u}$ and $T_{2u}$. In each of the cases of $A_{1g}$, $E_g$ and $T_{1u}$ there are two distinct eigensolutions with this same symmetry but having different eigenvalues, $\tau^{-1}$. Below we illustrate these remarks by considering this particular example of solute–vacancy pairs in the NaCl lattice in greater detail.

Before doing so, however, we append a few further remarks about the implications of these results for the observable tensor quantities $\Delta\varepsilon_{\alpha\beta}$ and $\Delta s_{\alpha\beta\lambda\mu}$ of (7.2.19) and (7.2.21). Firstly, since these are calculated in the linear response approximation they will have the same overall symmetry properties as the usual dielectric constants and the elastic compliances of the crystals in question. For cubic crystals this means that $\Delta\varepsilon_{\alpha\beta}$ will be a simple scalar, $\Delta\varepsilon$, while $\Delta s_{\alpha\beta\lambda\mu}$ will have just three independent elements, which in Voigt notation will be denoted $\Delta s_{11}$, $\Delta s_{12}$ and $\Delta s_{44}$. Secondly, in the simplest situation where only one distinct mode of the appropriate kind contributes, the expressions (7.2.19) and (7.2.21) take simple explicit forms. Thus, when there is only one electrically active mode ($T_{1u}$), (7.2.19) becomes

$$\Delta\varepsilon = \frac{n_d \eta \mu^2}{3kT} \frac{1}{(1 + i\omega\tau(T_{1u}))}, \tag{7.3.1}$$

in which $\mu$ is the *magnitude* of the dipole moment borne by the defects. In the same circumstances the equations of the technique of ionic thermocurrents are also obtained (cf. eqns. 1.5.15 and 6.3.2).

Corresponding results when there is just one elastically active mode of either $E_g$ or $T_{2g}$ symmetry can be set down as follows. In these equations, the elastic dipole tensor $\lambda_{\alpha\beta}$ of the defects, as it appears in (7.2.22), has been re-expressed in terms of the principal components defined relative to the principal axes of the defect itself. In each case the $\lambda_{\alpha\beta}$ tensor becomes diagonal (by definition), with components $\lambda_1$, $\lambda_2$ and $\lambda_2$ for tetragonal and trigonal defects and components $\lambda_1$, $\lambda_2$ and $\lambda_3$ for orthorhombic defects. One obtains the following results.

(i) Tetragonal defects

$$\Delta(s_{11} - s_{12}) = \frac{n_d v^2(\lambda_1 - \lambda_2)^2}{3kT[1 + i\omega\tau(E_g)]} \tag{7.3.2a}$$

$$\Delta(s_{44}) = 0. \tag{7.3.2b}$$

(ii) Trigonal defects

$$\Delta(s_{11} - s_{12}) = 0, \tag{7.3.3a}$$

$$\Delta s_{44} = \frac{4n_d v^2(\lambda_1 - \lambda_2)^2}{9kT[1 + i\omega\tau(T_{2g})]}. \tag{7.3.3b}$$

(iii) ⟨110⟩ Orthorhombic defects

$$\Delta(s_{11} - s_{12}) = \frac{n_d v^2(\frac{1}{2}(\lambda_1 + \lambda_2) - \lambda_3)^2}{3kT[1 + i\omega\tau(E_g)]}, \tag{7.3.4a}$$

$$\Delta s_{44} = \frac{n_d v^2(\lambda_1 - \lambda_2)^2}{3kT[1 + i\omega\tau(T_{2g})]}. \tag{7.3.4b}$$

Each of these changes in moduli (7.3.1)–(7.3.4) correspond to a Debye type of loss (dielectric or mechanical as appropriate). The phenomenological equations of §§1.5 and 1.6 are therefore followed; in particular the tangent of the loss angle will be given by (1.5.13) and its mechanical counterpart.

In systems where there is more than one mode of the necessary symmetry then the total defect contribution to the electric permittivity and the elastic compliances will be a sum of terms like (7.3.1)–(7.3.4) and the relative strengths of these terms must be determined by solving the eigenvalue equation (6.2.10). Once the symmetry analysis is employed then this equation is reduced to block diagonal form, each block corresponding to one irreducible representation.

We shall now discuss three examples for which we can write down explicit expressions for the relaxation times in terms of appropriate jump frequencies.

### *7.3.2 Interstitial solute atoms in a body-centred cubic lattice*

The behaviour of light elements such as C, N and O in b.c.c. metals such as α-Fe and other transition metals is of importance in metallurgy: aspects of this have already been mentioned in Chapters 2 and 5. These elements dissolve interstitially and are generally believed to take up the octahedral interstitial sites shown in Fig. 7.3. Since two of the solvent neighbours are closer to the interstitial site than the other four there is a $\langle 100 \rangle$ tetragonal distortion associated with each interstitial atom. This can couple to applied stress fields in accordance with the above results. The corresponding anelastic relaxation is called Snoek relaxation after its discoverer. Corresponding to the three distinct $\langle 100 \rangle$ crystal directions, there are therefore three distinct groups of interstitial atoms to be distinguished in the total population when the system is subjected to a stress field. If we suppose that jumps occur between neighbouring interstitial sites only and that the jump frequency for a transition from one site to any one of its four neighbours is $w$, then eqns. (7.3.2) apply with

$$\tau^{-1}(\mathrm{E_g}) = 6w, \tag{7.3.5}$$

where for simplicity we have dropped the superscript (0) indicating thermal equilibrium.

This result is easily confirmed by writing out (7.2.1) and finding the eigenvalues of the $3 \times 3$ matrix $\mathbf{P}^{(0)}$. The $\mathrm{E_g}$-mode (7.3.5) is doubly degenerate. The remaining mode (of $\mathrm{A_{1g}}$ symmetry) has eigenvalue zero, corresponding to the fact that there is no way this interstitial population can respond to a uniform hydrostatic stress. The model can be tested by comparing the magnitude of $|\lambda_1 - \lambda_2|$ obtained from the strength of the Snoek peak (7.3.2) with the same quantity as inferred from diffuse X-ray and neutron scattering measurements of the strain field around the interstitial atom (Krivoglaz, 1969). However, such comparisons generally involve lattice statics calculations and thus depend upon details of the interatomic forces assumed to be acting (e.g. Rebonato, 1989). A more direct confirmation of the model is obtained by taking jump frequencies inferred from measured Snoek relaxation times in conjunction with corresponding measurements of the interstitial diffusion coefficients. In this way the validity of the model is confirmed (see §8.2 and in particular Fig. 8.1 of the next chapter).

### *7.3.3 Solute–defect pairs in the $CaF_2$ lattice*

Although compounds with the fluorite structure undergo a transition to a fast-ion-conducting state at high temperatures ($\geq 85\%$ of $T_\mathrm{m}$), below the transition temperature they display normal Frenkel disorder in the anion sub-lattice. They may be doped with cations of both higher and lower charge (e.g. trivalent rare-earth ions, $RE^{3+}$, and monovalent alkali ions). Structural and charge compensation is effected by corresponding numbers of interstitial anions and anion vacancies respectively (Hayes, 1974). In both cases the opposite effective charges of the solute cation and the compensating anion defect lead to the formation of solute defect pairs, e.g. of $RE^{3+}$–interstitial $F^-$ pairs in the one case and of $Na^+$–$F^-$ vacancy pairs in the other. If the $RE^{3+}$-interstitial pairs are at the closest separation allowed by the lattice (Fig. 7.2) then the defect has a ⟨100⟩ axis and thus tetragonal symmetry. This would yield only one, triply degenerate $T_{1u}$ mode (electrically active) and one, doubly degenerate $E_g$ mode (elastically active). A number of experiments on fluorite compounds doped with various trivalent cations, however, indicate that in practice more modes than this occur, the assignment of which may not yet be settled.

By contrast, the situation appears to be simpler when the solute ions are of lower charge than the host, and there is little doubt that nearest-neighbour solute–anion vacancy pairs are formed. These have a ⟨111⟩ axis and thus trigonal symmetry. There are eight distinct orientations of the pair. Reorientation occurs by direct ⟨100⟩ jumps of the vacancy (three possibilities for any one orientation, each with frequency $w$). By using the results of the above symmetry analysis we then easily find the eigenvalues of $\mathbf{P}^{(0)}$, i.e. the relaxation times:

$$\tau^{-1}(A_{1g}) = 0 \tag{7.3.6a}$$

$$\tau^{-1}(T_{2g}) = 4w \tag{7.3.6b}$$

$$\tau^{-1}(A_{2u}) = 6w \tag{7.3.6c}$$

$$\tau^{-1}(T_{1u}) = 2w. \tag{7.3.6d}$$

There is no $A_{1g}$ mode because we have assumed that the anion vacancy is confined to the first-neighbour positions around the solute cation. As Table 7.1 shows, the $T_{2g}$ mode will be responsive to uniaxial stress in ⟨111⟩ and ⟨110⟩ directions but not in ⟨100⟩ directions. The $T_{1u}$ mode will respond to electric fields. The $A_{2u}$ mode will not couple to any uniform stress or electric fields.

These predictions are borne out by various experimental results. For example, Johnson *et al.* (1969) confirmed the absence of relaxation under a ⟨100⟩ uniaxial stress applied to single crystals of $CaF_2$ doped with $Na^+$ ions. Also Wachtman (1963), who studied Ca-doped $ThO_2$ and Lay and Whitmore (1971), who studied Ca-doped $CeO_2$, found that the ratio of relaxation times $\tau(T_{1u})/\tau(T_{2g})$ was close to 1/2, as required by eqns. (7.3.6).

### *7.3.4 Solute–vacancy pairs in the NaCl lattice*

There have been many studies of alkali halides containing small concentrations of divalent cations such as $Ca^{2+}$, $Cd^{2+}$, $Mn^{2+}$, etc. The presence of such ions results in the introduction of an equivalent number of cation vacancies, essentially for the structural reason treated formally in §3.3.1. It was recognized many years ago that the opposite effective charges on the cation vacancy ($-e$) and the solute ion ($+e$) would give an appreciable binding energy to nearest-neighbour and other close pairs of such defects and that pairs so formed would possess an electric dipole moment (cf. §2.5.3). The thermally activated re-orientation of such pairs was then studied experimentally via measurements of dielectric and mechanical relaxation.

If the vacancy is confined to be at one of the 12 nearest-neighbour (110) positions to the solute ion there are evidently 12 distinct orientations of the pair. If the pair is allowed to extend by having the vacancy at the next nearest-neighbour positions (200) there are a further 6 orientations to be considered. And so on. Clearly, as more extended pairs are included, the straightforward algebraic solution of the eigenvalue equation (6.2.12) rapidly becomes very tedious, if not unmanageable. As elsewhere the task is greatly simplified by making use of the results of symmetry analysis. Such an analysis allowing for this possibility of extended pairs was made by Franklin, Shorb and Wachtman (1964), and Table 7.2, taken from that paper, lists the corresponding basis vectors (out to the third neighbour shell, 211 sites). Figure 7.5 shows these graphically for the nearest-neighbour (110) configurations. Given this information, the most straightforward way to solve the eigenvalue equation is to assume that the eigenvector has the indicated symmetry when the order of the equation to be solved will be immediately reduced.

Thus if only nearest-neighbour pairs exist we shall find in each case that the equation becomes of first order so that each relaxation time is immediately obtained in terms of the jump frequencies defined in Fig. 7.1, viz.

$$\tau^{-1}(A_{1g}) = 0 \tag{7.3.7a}$$

$$\tau^{-1}(E_g) = 6w_1 \qquad (\times 2) \tag{7.3.7b}$$

$$\tau^{-1}(T_{2g}) = 4w_1 \qquad (\times 3) \tag{7.3.7c}$$

$$\tau^{-1}(T_{1u}) = 2(w_1 + w_2) \qquad (\times 3) \tag{7.3.7d}$$

Eqn. (7.3.7d) gives the relaxation time to be inserted in (7.3.1) for dielectric relaxation, while (7.3.7b) and (7.3.7c) give those for insertion in (7.3.4a) and (7.3.4b) respectively. For this model of nearest-neighbour pairs there can be no active $A_{1g}$ mode simply because $A_{1g}$ would be a wholly symmetric, or 'breathing', mode and there is no way in this model for the pairs to extend or contract.

On the other hand, if we also include the six (200) next-nearest-neighbour positions for the vacancy we shall obtain second-order secular equations for three of the symmetries, viz. $A_{1g}$, $E_g$, and $T_{1u}$. The corresponding eigenvalues are as follows.

$A_{1g}$ *modes:*

$$\tau_{\pm}^{-1} = 0, \qquad 2w_3 + 4w_4. \tag{7.3.8a}$$

## Table 7.2. *Basis vectors for solute–vacancy pairs in NaCl lattice (Franklin et al., 1964)*

| Site | xyz | $A_{1g}$ | | | $A_{2u}$ | $E_g$ | | | | | | $E_u$ | | $T_{1g}$ | | | $T_{1u}$ | | | | | | | | | | | | $T_{2g}$ | | | | | | | | | $T_{2u}$ | | | | | |
|---|---|---|---|---|---|---|---|---|---|---|---|---|---|---|---|---|---|---|---|---|---|---|---|---|---|---|---|---|---|---|---|---|---|---|---|---|---|---|---|---|---|---|---|
| 1 | $\bar{1}\bar{1}0$ | 1 | ... | ... | ... | −2 | ... | ... | 0 | ... | ... | ... | ... | ... | ... | ... | −1 | ... | ... | ... | −1 | ... | ... | ... | 0 | ... | ... | ... | 1 | ... | ... | 0 | ... | ... | 0 | ... | ... | 0 | ... | −1 | ... | 1 | ... |
| 2 | $0\bar{1}1$ | 1 | ... | ... | ... | 1 | ... | ... | −1 | ... | ... | ... | ... | ... | ... | ... | 0 | ... | ... | ... | −1 | ... | ... | ... | 1 | ... | ... | ... | 0 | ... | ... | −1 | ... | ... | 0 | ... | ... | −1 | ... | 0 | ... | −1 | ... |
| 3 | $1\bar{1}0$ | 1 | ... | ... | ... | −2 | ... | ... | 0 | ... | ... | ... | ... | ... | ... | ... | 1 | ... | ... | ... | −1 | ... | ... | ... | 0 | ... | ... | ... | −1 | ... | ... | 0 | ... | ... | 0 | ... | ... | 0 | ... | 1 | ... | 1 | ... |
| 4 | $0\bar{1}\bar{1}$ | 1 | ... | ... | ... | 1 | ... | ... | −1 | ... | ... | ... | ... | ... | ... | ... | 0 | ... | ... | ... | −1 | ... | ... | ... | −1 | ... | ... | ... | 0 | ... | ... | 1 | ... | ... | 0 | ... | ... | 1 | ... | 0 | ... | −1 | ... |
| 5 | $\bar{1}01$ | 1 | ... | ... | ... | 1 | ... | ... | 1 | ... | ... | ... | ... | ... | ... | ... | −1 | ... | ... | ... | 0 | ... | ... | ... | 1 | ... | ... | ... | 0 | ... | ... | 0 | ... | ... | −1 | ... | ... | 1 | ... | 1 | ... | 0 | ... |
| 6 | $101$ | 1 | ... | ... | ... | 1 | ... | ... | 1 | ... | ... | ... | ... | ... | ... | ... | 1 | ... | ... | ... | 0 | ... | ... | ... | 1 | ... | ... | ... | 0 | ... | ... | 0 | ... | ... | 1 | ... | ... | 1 | ... | −1 | ... | 0 | ... |
| 7 | $10\bar{1}$ | 1 | ... | ... | ... | 1 | ... | ... | 1 | ... | ... | ... | ... | ... | ... | ... | 1 | ... | ... | ... | 0 | ... | ... | ... | −1 | ... | ... | ... | 0 | ... | ... | 0 | ... | ... | −1 | ... | ... | −1 | ... | −1 | ... | 0 | ... |
| 8 | $\bar{1}0\bar{1}$ | 1 | ... | ... | ... | 1 | ... | ... | 1 | ... | ... | ... | ... | ... | ... | ... | −1 | ... | ... | ... | 0 | ... | ... | ... | −1 | ... | ... | ... | 0 | ... | ... | 0 | ... | ... | 1 | ... | ... | −1 | ... | 1 | ... | 0 | ... |
| 9 | $\bar{1}10$ | 1 | ... | ... | ... | −2 | ... | ... | 0 | ... | ... | ... | ... | ... | ... | ... | −1 | ... | ... | ... | 1 | ... | ... | ... | 0 | ... | ... | ... | −1 | ... | ... | 0 | ... | ... | 0 | ... | ... | 0 | ... | −1 | ... | −1 | ... |
| 10 | $011$ | 1 | ... | ... | ... | 1 | ... | ... | −1 | ... | ... | ... | ... | ... | ... | ... | 0 | ... | ... | ... | 1 | ... | ... | ... | 1 | ... | ... | ... | 0 | ... | ... | 1 | ... | ... | 0 | ... | ... | −1 | ... | 0 | ... | 1 | ... |
| 11 | $110$ | 1 | ... | ... | ... | −2 | ... | ... | 0 | ... | ... | ... | ... | ... | ... | ... | 1 | ... | ... | ... | 1 | ... | ... | ... | 0 | ... | ... | ... | 1 | ... | ... | 0 | ... | ... | 0 | ... | ... | 0 | ... | 1 | ... | −1 | ... |
| 12 | $01\bar{1}$ | 1 | ... | ... | ... | 1 | ... | ... | −1 | ... | ... | ... | ... | ... | ... | ... | 0 | ... | ... | ... | 1 | ... | ... | ... | −1 | ... | ... | ... | 0 | ... | ... | −1 | ... | ... | 0 | ... | ... | 1 | ... | 0 | ... | 1 | ... |
| 13 | $\bar{2}00$ | ... | 1 | ... | ... | ... | −1 | ... | ... | 1 | ... | ... | ... | ... | ... | ... | ... | −1 | ... | ... | ... | 0 | ... | ... | ... | 0 | ... | ... | ... | ... | ... | ... | ... | ... | ... | ... | ... | ... | ... | ... | ... | ... | ... |
| 14 | $002$ | ... | 1 | ... | ... | ... | 2 | ... | ... | 0 | ... | ... | ... | ... | ... | ... | ... | 0 | ... | ... | ... | 0 | ... | ... | ... | 1 | ... | ... | ... | ... | ... | ... | ... | ... | ... | ... | ... | ... | ... | ... | ... | ... | ... |
| 15 | $200$ | ... | 1 | ... | ... | ... | −1 | ... | ... | 1 | ... | ... | ... | ... | ... | ... | ... | 1 | ... | ... | ... | 0 | ... | ... | ... | 0 | ... | ... | ... | ... | ... | ... | ... | ... | ... | ... | ... | ... | ... | ... | ... | ... | ... |
| 16 | $00\bar{2}$ | ... | 1 | ... | ... | ... | 2 | ... | ... | 0 | ... | ... | ... | ... | ... | ... | ... | 0 | ... | ... | ... | 0 | ... | ... | ... | −1 | ... | ... | ... | ... | ... | ... | ... | ... | ... | ... | ... | ... | ... | ... | ... | ... | ... |
| 17 | $0\bar{2}0$ | ... | 1 | ... | ... | ... | −1 | ... | ... | −1 | ... | ... | ... | ... | ... | ... | ... | 0 | ... | ... | ... | −1 | ... | ... | ... | 0 | ... | ... | ... | ... | ... | ... | ... | ... | ... | ... | ... | ... | ... | ... | ... | ... | ... |
| 18 | $020$ | ... | 1 | ... | ... | ... | −1 | ... | ... | −1 | ... | ... | ... | ... | ... | ... | ... | 0 | ... | ... | ... | 1 | ... | ... | ... | 0 | ... | ... | ... | ... | ... | ... | ... | ... | ... | ... | ... | ... | ... | ... | ... | ... | ... |
| 19 | $\bar{2}\bar{1}1$ | ... | ... | 1 | 1 | ... | ... | −1 | ... | ... | 1 | −1 | 1 | 1 | −1 | 0 | ... | ... | −1 | 0 | ... | ... | 0 | −1 | ... | ... | 0 | 1 | ... | 0 | 1 | ... | −1 | 0 | ... | 0 | −1 | ... | 1 | ... | 0 | ... | 1 |
| 20 | $\bar{1}\bar{1}2$ | ... | ... | 1 | 1 | ... | ... | 2 | ... | ... | 0 | 2 | 0 | 0 | 1 | 1 | ... | ... | 0 | −1 | ... | ... | 0 | −1 | ... | ... | 1 | 0 | ... | 1 | 0 | ... | 0 | −1 | ... | 0 | −1 | ... | 0 | ... | 1 | ... | −1 |
| 21 | $\bar{1}\bar{2}1$ | ... | ... | 1 | 1 | ... | ... | −1 | ... | ... | −1 | −1 | −1 | −1 | 0 | −1 | ... | ... | 0 | −1 | ... | ... | −1 | 0 | ... | ... | 0 | 1 | ... | 0 | 1 | ... | 0 | −1 | ... | −1 | 0 | ... | −1 | ... | −1 | ... | 0 |
| 22 | $\bar{2}11$ | ... | ... | 1 | −1 | ... | ... | −1 | ... | ... | 1 | 1 | −1 | −1 | −1 | 0 | ... | ... | −1 | 0 | ... | ... | 0 | 1 | ... | ... | 0 | 1 | ... | 0 | −1 | ... | 1 | 0 | ... | 0 | −1 | ... | 1 | ... | 0 | ... | −1 |
| 23 | $\bar{1}21$ | ... | ... | 1 | −1 | ... | ... | −1 | ... | ... | −1 | 1 | 1 | 1 | 0 | 1 | ... | ... | 0 | −1 | ... | ... | 1 | 0 | ... | ... | 0 | 1 | ... | 0 | −1 | ... | 0 | 1 | ... | −1 | 0 | ... | −1 | ... | −1 | ... | 0 |
| 24 | $\bar{1}12$ | ... | ... | 1 | −1 | ... | ... | 2 | ... | ... | 0 | −2 | 0 | 0 | 1 | −1 | ... | ... | 0 | −1 | ... | ... | 0 | 1 | ... | ... | 1 | 0 | ... | −1 | 0 | ... | 0 | 1 | ... | 0 | −1 | ... | 0 | ... | 1 | ... | 1 |
| 25 | $121$ | ... | ... | 1 | 1 | ... | ... | −1 | ... | ... | −1 | −1 | −1 | −1 | 0 | 1 | ... | ... | 0 | 1 | ... | ... | 1 | 0 | ... | ... | 0 | 1 | ... | 0 | 1 | ... | 0 | 1 | ... | 1 | 0 | ... | −1 | ... | 1 | ... | 0 |
| 26 | $211$ | ... | ... | 1 | 1 | ... | ... | −1 | ... | ... | 1 | −1 | 1 | 1 | 1 | 0 | ... | ... | 1 | 0 | ... | ... | 0 | 1 | ... | ... | 0 | 1 | ... | 0 | 1 | ... | 1 | 0 | ... | 0 | 1 | ... | 1 | ... | 0 | ... | −1 |
| 27 | $112$ | ... | ... | 1 | 1 | ... | ... | 2 | ... | ... | 0 | 2 | 0 | 0 | −1 | −1 | ... | ... | 0 | 1 | ... | ... | 0 | 1 | ... | ... | 1 | 0 | ... | 1 | 0 | ... | 0 | 1 | ... | 0 | 1 | ... | 0 | ... | −1 | ... | 1 |
| 28 | $1\bar{2}1$ | ... | ... | 1 | −1 | ... | ... | −1 | ... | ... | −1 | 1 | 1 | 1 | 0 | −1 | ... | ... | 0 | 1 | ... | ... | −1 | 0 | ... | ... | 0 | 1 | ... | 0 | −1 | ... | 0 | −1 | ... | 1 | 0 | ... | −1 | ... | 1 | ... | 0 |
| 29 | $2\bar{1}1$ | ... | ... | 1 | −1 | ... | ... | −1 | ... | ... | 1 | 1 | −1 | −1 | 1 | 0 | ... | ... | 1 | 0 | ... | ... | 0 | −1 | ... | ... | 0 | 1 | ... | 0 | −1 | ... | −1 | 0 | ... | 0 | 1 | ... | 1 | ... | 0 | ... | 1 |
| 30 | $1\bar{1}2$ | ... | ... | 1 | −1 | ... | ... | 2 | ... | ... | 0 | −2 | 0 | 0 | −1 | 1 | ... | ... | 0 | 1 | ... | ... | 0 | −1 | ... | ... | 1 | 0 | ... | −1 | 0 | ... | 0 | −1 | ... | 0 | 1 | ... | 0 | ... | −1 | ... | −1 |
| 31 | $\bar{1}\bar{2}\bar{1}$ | ... | ... | 1 | −1 | ... | ... | −1 | ... | ... | −1 | 1 | 1 | −1 | 0 | 1 | ... | ... | 0 | −1 | ... | ... | −1 | 0 | ... | ... | 0 | −1 | ... | 0 | 1 | ... | 0 | 1 | ... | 1 | 0 | ... | 1 | ... | −1 | ... | 0 |
| 32 | $\bar{1}\bar{1}\bar{2}$ | ... | ... | 1 | −1 | ... | ... | 2 | ... | ... | 0 | −2 | 0 | 0 | −1 | −1 | ... | ... | 0 | −1 | ... | ... | 0 | −1 | ... | ... | −1 | 0 | ... | 1 | 0 | ... | 0 | 1 | ... | 0 | 1 | ... | 0 | ... | 1 | ... | −1 |
| 33 | $\bar{2}\bar{1}\bar{1}$ | ... | ... | 1 | −1 | ... | ... | −1 | ... | ... | 1 | 1 | −1 | 1 | 1 | 0 | ... | ... | −1 | 0 | ... | ... | 0 | −1 | ... | ... | 0 | −1 | ... | 0 | 1 | ... | 1 | 0 | ... | 0 | 1 | ... | −1 | ... | 0 | ... | 1 |
| 34 | $\bar{2}1\bar{1}$ | ... | ... | 1 | 1 | ... | ... | −1 | ... | ... | 1 | −1 | 1 | −1 | 1 | 0 | ... | ... | −1 | 0 | ... | ... | 0 | 1 | ... | ... | 0 | −1 | ... | 0 | −1 | ... | −1 | 0 | ... | 0 | 1 | ... | −1 | ... | 0 | ... | −1 |
| 35 | $\bar{1}1\bar{2}$ | ... | ... | 1 | 1 | ... | ... | 2 | ... | ... | 0 | 2 | 0 | 0 | −1 | 1 | ... | ... | 0 | −1 | ... | ... | 0 | 1 | ... | ... | −1 | 0 | ... | −1 | 0 | ... | 0 | −1 | ... | 0 | 1 | ... | 0 | ... | 1 | ... | 1 |
| 36 | $\bar{1}2\bar{1}$ | ... | ... | 1 | 1 | ... | ... | −1 | ... | ... | −1 | −1 | −1 | 1 | 0 | −1 | ... | ... | 0 | −1 | ... | ... | 1 | 0 | ... | ... | 0 | −1 | ... | 0 | −1 | ... | 0 | −1 | ... | 1 | 0 | ... | 1 | ... | −1 | ... | 0 |
| 37 | $21\bar{1}$ | ... | ... | 1 | −1 | ... | ... | −1 | ... | ... | 1 | 1 | −1 | 1 | −1 | 0 | ... | ... | 1 | 0 | ... | ... | 0 | 1 | ... | ... | 0 | −1 | ... | 0 | 1 | ... | −1 | 0 | ... | 0 | −1 | ... | −1 | ... | 0 | ... | −1 |
| 38 | $11\bar{2}$ | ... | ... | 1 | −1 | ... | ... | 2 | ... | ... | 0 | −2 | 0 | 0 | 1 | 1 | ... | ... | 0 | 1 | ... | ... | 0 | 1 | ... | ... | −1 | 0 | ... | 1 | 0 | ... | 0 | −1 | ... | 0 | −1 | ... | 0 | ... | −1 | ... | 1 |
| 39 | $12\bar{1}$ | ... | ... | 1 | −1 | ... | ... | −1 | ... | ... | −1 | 1 | 1 | −1 | 0 | −1 | ... | ... | 0 | 1 | ... | ... | 1 | 0 | ... | ... | 0 | −1 | ... | 0 | 1 | ... | 0 | −1 | ... | −1 | 0 | ... | 1 | ... | 1 | ... | 0 |
| 40 | $2\bar{1}\bar{1}$ | ... | ... | 1 | 1 | ... | ... | −1 | ... | ... | 1 | −1 | 1 | −1 | −1 | 0 | ... | ... | 1 | 0 | ... | ... | 0 | −1 | ... | ... | 0 | −1 | ... | 0 | −1 | ... | 1 | 0 | ... | 0 | −1 | ... | −1 | ... | 0 | ... | 1 |
| 41 | $1\bar{1}\bar{2}$ | ... | ... | 1 | 1 | ... | ... | 2 | ... | ... | 0 | 2 | 0 | 0 | 1 | −1 | ... | ... | 0 | 1 | ... | ... | 0 | −1 | ... | ... | −1 | 0 | ... | −1 | 0 | ... | 0 | 1 | ... | 0 | −1 | ... | 0 | ... | −1 | ... | −1 |
| 42 | $1\bar{2}\bar{1}$ | ... | ... | 1 | 1 | ... | ... | −1 | ... | ... | −1 | −1 | −1 | 1 | 0 | 1 | ... | ... | 0 | 1 | ... | ... | −1 | 0 | ... | ... | 0 | −1 | ... | 0 | −1 | ... | 0 | 1 | ... | −1 | 0 | ... | 1 | ... | 1 | ... | 0 |
| Vector No. | | 1 | 2 | 3 | 4 | 5 | 6 | 7 | 8 | 9 | 10 | 11 | 12 | 13 | 14 | 15 | 16 | 17 | 18 | 19 | 20 | 21 | 22 | 23 | 24 | 25 | 26 | 27 | 28 | 29 | 30 | 31 | 32 | 33 | 34 | 35 | 36 | 37 | 38 | 39 | 40 | 41 | 42 |

As before there is a zero eigenvalue, $\tau^{-1} = 0$, which corresponds to the thermal equilibrium distribution among the available configurations.

$E_g$ *modes*:

$$\tau_{\pm}^{-1} = (3w_1 + w_3 + 2w_4) \pm \{(3w_1 + w_3 - 2w_4)^2 + 2w_3w_4\}^{1/2}. \quad (7.3.8b)$$

$T_{2g}$ *modes*:

$$\tau^{-1} = 4w_1 + 2w_3. \quad (7.3.8c)$$

$T_{1u}$ *modes*:

$$\tau_{\pm}^{-1} = (w_1 + w_2 + w_3 + 2w_4) \pm \{(w_1 + w_2 + w_3 - 2w_4)^2 + 4w_3w_4\}^{1/2}. \quad (7.3.8d)$$

We notice that only the electrically active modes depend upon $w_2$, the vacancy–impurity exchange rate. This fact can be useful in the inference of jump frequencies from experimental information when data on both dielectric and mechanical relaxation is available.

The relative strengths of the contributions from modes of like symmetry to the total $\Delta\varepsilon$ or $\Delta s$ can also be written down when we know the electric dipole moments and the elastic-dipole tensors for the paired defects in their available configurations. In the case of the two $T_{1u}$ electrically active modes we may suppose that the electric dipole moments of the second and first neighbour pairs are in the ratio of the corresponding solute–vacancy separations, i.e. $\sqrt{2}:1$, and obtain

$$\frac{\Delta\varepsilon_+}{\Delta\varepsilon_-} = \frac{2(w_1 + w_2 + w_3) - \tau_-^{-1}(1 + w_3/w_4)}{\tau_+^{-1}(1 + w_3/w_4) - 2(w_1 + w_2 + w_3)}, \quad (7.3.9)$$

in which $\tau_{\pm}^{-1}$ are given by (7.3.8d) above. In practice, the mode with the longer relaxation time, i.e. $\tau_-$, often is dominant, i.e. $\Delta\varepsilon_- \gg \Delta\varepsilon_+$ and the curve of $\Delta\varepsilon$ $(=\Delta\varepsilon_+ + \Delta\varepsilon_-)$ v. $\omega$ is very close to a single Debye curve. In some cases of alkali halides doped with divalent cations, however, information about the weaker mode has also been obtained (e.g. Dreyfus, 1961; Burton and Dryden, 1970).

One of the better studied examples of this class is NaCl doped with $MnCl_2$, which contains $Mn^{2+}$-vacancy pairs. For this system a combination of dielectric loss, mechanical loss and e.p.r. measurements has yielded data on all the modes of relaxation represented by eqns. (7.3.8). Analysis based on these equations has yielded results for the four jump frequencies which can be represented by Arrhenius expressions

$$w_i = A_i \exp(-Q_i/kT), \quad (7.3.10)$$

Table 7.3. *Arrhenius parameters for $Mn^{2+}$-vacancy pairs in NaCl*

| $w_i$ | Vacancy jump | $A_i/(s)$ | $Q_i/(eV)$ |
|---|---|---|---|
| $w_1$ | n.n. → n.n. | $8.410 \times 10^{12}$ | 0.659 |
| $w_2$ | $Mn^{2+}$ ↔ vacancy | $1.927 \times 10^{12}$ | 0.601 |
| $w_3$ | n.n. → n.n.n. | $1.059 \times 10^{14}$ | 0.671 |
| $w_4$ | n.n.n. → n.n. | $7.357 \times 10^{13}$ | 0.632 |

(It should be noted that our definitions of $w_3$ and $w_4$ are transposed with respect to those used by Symmons, 1971.)

with the parameters $A_i$ and $Q_i$ given in Table 7.3 (Symmons, 1971). However, it should be noted that this may not be a very reliable way to determine the solute–vacancy exchange frequency, $w_2$. But as we shall see in the next chapter the rate of diffusion of the solute depends directly on this quantity. Inclusion of diffusion data into the analysis should therefore lead to more secure values for $w_2$.

It is possible to extend the above analysis to include solute–vacancy pairs at the third nearest-neighour separation (i.e. the (211) separation in the NaCl lattice) and this has been done by Franklin *et al.* (1964) and by Franklin and Young (1982). Nevertheless, because it is generally difficult to see the separate modes experimentally, in practice, the calculation of the total $\Delta\varepsilon$ and $\Delta s$ may actually be better done in these cases, not by mode analysis, but by a more direct method due to Okamura and Allnatt (1985) which is based on results such as (A6.3.19)–(A6.3.21). We shall describe this method in §11.2.

## 7.4 Ionic conductivity and the total electrical response

The above theory of the relaxation of complex defects provides only part of the description of real ionic conductors. Thus in ionic substances, where the complex defects will contain vacancies or interstitial ions as components, it is well known that dielectric relaxation phenomena are generally accompanied by an ionic electrical conductivity that is due to the drift of those vacancies and/or interstitial ions which are not bound to solute ions or other defects (see any of the reviews by Lidiard, 1957; Fuller, 1972; Flynn, 1972; Corish and Jacobs, 1973a; Jacobs, 1983). That there are in effect two distinct populations, paired and unpaired defects, is also suggested by the general success of the idea in the statistical thermodynamics of defect populations (see Chapter 3) and by spectroscopic

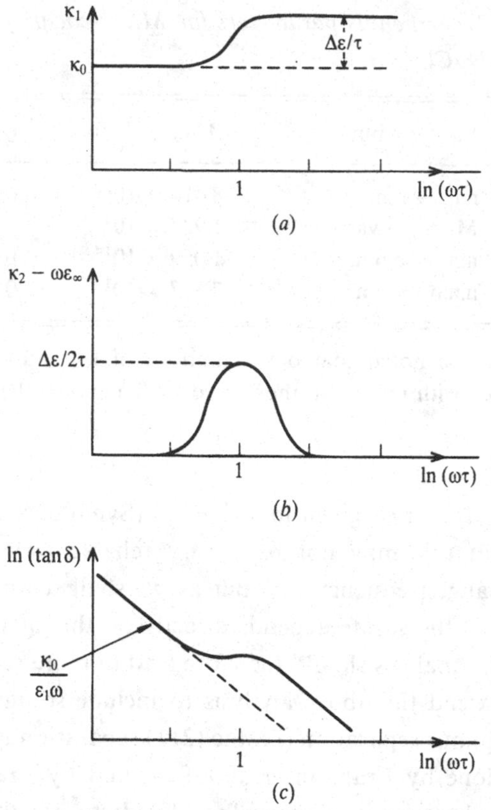


Fig. 7.6. Schematic diagram showing the variation of the (*a*) real and (*b*) imaginary parts of the ionic conductivity of an alkali halide crystal containing divalent solute cations as a function of ln $\omega$. Likewise (*c*) represents a log-log plot of the tangent of the loss angle, tan $\delta$, as a function of $\omega$. The frequency-independent ionic conductivity due to the unassociated defects is $\kappa_0$. This gives a contribution to tan $\delta$ of $\kappa_0/\varepsilon_1\omega$, in which $\varepsilon_1$ is the real part of the total electric permittivity. The remainder is due to the paired defects.

techniques of observation which allow us to distinguish isolated solute atoms from those in combination with defects (e.g. Hayes, 1974; Henderson and Hughes, 1976). The theory of electrical response in such systems therefore generally simply adds together the separate responses of the paired, or complex, defects and of the unpaired, or isolated, defects. In so far as the ionic conductivity of the unpaired defects should be independent of frequency while the paired defects yield a Debye-like contribution, the overall dependence of the complex conductivity on frequency is thus expected to be as shown in Fig. 7.6.

This approach has been widely, and successfully, used in the interpretation of experimental results. The obvious question of the influence of the exchange of defects between these two populations has received rather little attention. It turns out, however, that it is important to the theory of atomic transport in two ways. Firstly, when we calculate the ionic conductivity more rigorously we discover that there are additional terms in the d.c. conductivity which are proportional to the number of defect pairs, even though these pairs may carry no net electrical charge. These terms result from the way the vacancy movements are changed in the vicinity of the solute ions (e.g. because $w_1$, $w_2$, $w_3$ and $w_4$ differ from the free vacancy jump rate $w_0$). Secondly, there is a subtler effect which also emerges from these more rigorous calculations. Thus we might suppose that we could handle both populations together by indefinitely extending the definition of pairs in the formalism of §7.2 so that it embraced defects and solutes at greater and greater separations. If we do that, however, we find that all four phenomenological coefficients $L_{SS}$, $L_{SD}$, $L_{DS}$ and $L_{DD}$ (S = solute, D = defect) are always equal (cf. Chapter 8). This shows that such an approach fails even though it may not assume any physical interaction between the defect and the solute at large separations. It is the failure to liberate the defect from association with a particular solute atom which is the essential inadequacy. With this in mind we turn in the next chapter to the formulation of kinetic theory in a form that describes atomic transport and associated relaxation phenomena together.

## 7.5 Summary

In this chapter we have shown how simple kinetic theory describes dielectric and mechanical relaxation in terms of the thermally activated movements of solute–defect pairs and other defects of low symmetry. Of course, this does not constitute a comprehensive discussion of relaxation processes as a whole. There are various others having different physical origins from those considered here, e.g. anelastic relaxation associated with dislocation movements (Nabarro 1967). Our specific purpose here has been to introduce the kinetic theory appropriate to a number of well-characterized systems. Although the physical justification of this development is quite clear and specific, it can also be viewed as a direct representation of the general microscopic equations of Chapter 6. We have seen how, with the aid of symmetry analysis, experimental measurements can provide direct information about the symmetry and thermally activated movements of these complex defects. Such information is clearly directly relevant to the theory of diffusion and migration of solute atoms. Beyond that, however, it would seem natural – since the theory treats solute atoms and defects on a more or less equal footing – to extend it to provide a general kinetic theory of atomic transport. In doing so, however, we must pay explicit attention to the exchange of defects between the sub-populations of unpaired and paired defects. This is dealt with in the next chapter.

# 8

# *Kinetic theory of isothermal diffusion processes*

## 8.1 Introduction

In the previous chapter we showed how the thermally activated movements of defects of low symmetry (e.g. solute–defect pairs) may give rise to dielectric and anelastic relaxation processes. To do so we employed a particular example of the master equation (already introduced in Chapter 6) to represent the movement of these defects among the possible configurations and orientations open to them. These systems were, by assumption, spatially uniform, but it is natural to seek to extend such a theoretical approach to spatially non-uniform systems in which necessarily there will be diffusion processes taking place. Such extensions are the subject of this chapter. We call them kinetic theories because in them we are concerned with the changes in time of the distributions of the solute atoms, defects, solute–defect pairs, etc. in space and among available configurations. We shall use such approaches to find expressions for the diffusion coefficients and the more general transport coefficients of non-equilibrium thermodynamics for dilute alloys and solid solutions. Both interstitial and substitutional solid solutions are considered. In doing so we shall give quantitative expression to the correlation effects anticipated in atomic migration coefficients when defect mechanisms are active (cf. §§2.5.3 and 5.5). The treatments are mostly limited to isothermal systems, although they can be extended to systems in a thermal gradient.

Kinetic theories come in a variety of forms, depending on the characteristics of the systems of interest and on the physical quantities to be represented (e.g. transport coefficients, quasi-elastic scattering functions, Mössbauer cross-sections, etc.). As elsewhere, the aim is to achieve a useful level of generality and this is achieved for systems dilute in both defects and solute atoms. For concentrated systems the task is more difficult and so far only physically simplified models have been treated successfully.

The development of the present chapter is therefore as follows. We begin (§8.2) with the simplest problem, the motion of particles on an empty lattice. This represents certain interstitial solid solutions (mostly light atoms in b.c.c. and f.c.c. metals) and also the motions of intrinsic defects (vacancies or interstitials) when present in low concentration. This example then guides our extension of the treatment of Chapter 7 to give a theory of the diffusion of substitutional solutes (§8.3). Further extension and generalization for the purpose of inferring all the $L$-coefficients is the subject of §8.4. The expressions which we derive (§8.5) will be found to agree with those to be obtained from the many-body forms of §6.5 when we particularize these to dilute systems. In the next section (§8.6) we consider more highly defective systems, or rather *lattice gas models* of such systems. These are used as models of fast-ion-conductors and metallic hydrides. In both systems a large number of mobile ions (e.g. $Na^+$ ions in $\beta$-alumina) or atoms (e.g. H in metals) move among an even greater number of (interstitial) sites; the excess of available sites ensures that the atoms in question are mobile, but since many atoms are present they get in one another's way. This presents a more difficult statistical problem. §8.6 briefly considers the way kinetic theories have tackled it. A more systematic and powerful approach to this particular problem is the subject of Chapter 13.

## 8.2 Particles moving on an empty lattice

We shall begin the description of kinetic theories by considering a simple case, namely that of particles which migrate in a series of jumps on an empty lattice. Realizations of this model are to be found in metals containing low concentrations of interstitial solutes, such as C or N in $\alpha$-Fe (cf. Fig. 7.3). In this example the C atoms are the particles, while the lattice is the lattice of available interstitial sites; the Fe atoms need not be considered explicitly in the C migration, although obviously they define the lattice and supply the impetus for the thermally activated movements of the C atoms (§5.2). Another realization of the model is provided by the motion of intrinsic defects, either vacancies or interstitials, present in low concentrations in an otherwise perfect lattice.

Since the concentration of interstitial atoms or defects is low, each one may be assumed to move without hindrance from the others. As in Chapter 6 we therefore proceed by considering a statistical ensemble of identical systems each containing the same number of particles. The states of the system ($\alpha$, $\beta$, etc. of Chapter 6) are thus each specified by an assignment of the particles to the sites $l$, $m$, etc. of the lattice. The master equation is then an equation for the occupancy of these sites. With $w_{lm}$ as the jump frequency for a particle on site $m$ going to site $l$ this equation is evidently of the form

$$\frac{\partial p_l}{\partial t} = -\sum_{m \neq l} w_{ml} p_l + \sum_{m \neq l} w_{lm} p_m, \tag{8.2.1}$$

in which $p_l$ is defined to be the (ensemble) average number of particles on site $l$. (This definition of $p_l$ means that it is not precisely analogous to the $p_\alpha$ of Chapter 6 because the $p_l$ are not normalized to unity, the summation over all sites $\sum_l p_l$ being the total number of particles rather than unity. It is, however, more convenient for the purpose of this chapter.)

Although (8.2.1) is formally identical to (6.2.1) the physical simplicity of the present model and the specific physical interpretation in terms of the movements of isolated particles on a lattice allow more direct treatments. We shall contrast one or two of these, partly to illuminate some of the general results already presented in Chaper 6 and partly to prepare the ground for the treatment of more complex examples later.

### 8.2.1 Fick's law

When we are interested in diffusion over macroscopic distances (8.2.1) invites us to go to a continuum form. To this end let us introduce the vector variable $\mathbf{r}$ for the position of the particle on the lattice and write $vp(\mathbf{r}, t)$ for $p_l$ when $\mathbf{r}$ corresponds to site $l$ and $v$ is the volume per lattice site. Further let the position of the $z$ neighbours to the site at $\mathbf{r}$ be at positions $\mathbf{r} + \mathbf{s}_j, j = 1, 2 \ldots z$. Then, in the absence of external force fields and for particle jumps between nearest-neighbours only, all non-zero $w_{lm} = w^{(0)}$ and (8.2.1) becomes

$$\frac{\partial p(\mathbf{r}, t)}{\partial t} = w^{(0)} \sum_{j=1}^{z} [p(\mathbf{r} + \mathbf{s}_j, t) - p(\mathbf{r}, t)]. \tag{8.2.2}$$

For movements over macroscopic distances ($\gg \mathbf{s}_j$) we can treat $\mathbf{r}$ as a continuous variable and expand (8.2.2) by Taylor's theorem to give

$$\frac{\partial p(\mathbf{r}, t)}{\partial t} = w^{(0)} \sum_{j=1}^{z} \left[ \sum_{\alpha} s_{j,\alpha} \frac{\partial p}{\partial x_\alpha} + \frac{1}{2} \sum_{\alpha,\beta} s_{j,\alpha} s_{j,\beta} \frac{\partial^2 p}{\partial x_\alpha \, \partial x_\beta} + \cdots \right]. \tag{8.2.3}$$

We here impose a general symmetry condition; namely, that the lattice on which the particles are free to move is either (a) a Bravais lattice (i.e. one lattice point per primitive unit cell, which implies inversion symmetry about each site) or (b) such that the several sites in the primitive unit cell available to the particles are equivalent with respect to the point-group operations of the crystal lattice and have at least tetrahedral point symmetry ($T_d$). The lattice of octahedral interstitial sites in the f.c.c. lattice (e.g. H interstitials in Pd) satisfies the first condition (because it is also f.c.c.), while the lattice of octahedral interstitial sites in the b.c.c. lattice (e.g. C in α-Fe), which is not a Bravais lattice, satisfies the second. Then the first term on the right side of (8.2.3) is zero because

$$\sum_{j=1}^{z} s_{j,\alpha} = 0 \qquad (\text{all } \alpha).$$

If we now truncate (8.2.3) at the second term on the right side, we see that we have just Fick's second law

$$\frac{\partial p(\mathbf{r}, t)}{\partial t} = \sum_{\alpha,\beta} D_{\alpha\beta} \frac{\partial^2 p(\mathbf{r}, t)}{\partial x_\alpha \, \partial x_\beta}, \tag{8.2.4a}$$

where

$$D_{\alpha\beta} = \tfrac{1}{2} w^{(0)} \sum_{j=1}^{z} s_{j,\alpha} s_{j,\beta}. \tag{8.2.4b}$$

Since the local concentration of particles is $np(\mathbf{r}, t)$ (where $n$ is the number of lattice sites per unit volume), $D_{\alpha\beta}$ is evidently the diffusion coefficient tensor. Referred to its principal axes this becomes diagonal, i.e.

$$D_{\alpha\alpha} = \tfrac{1}{2} w^{(0)} \sum_{j=1}^{z} s_{j,\alpha}{}^2. \tag{8.2.5}$$

In a cubic lattice all three diagonal components are equal and we obtain the well-known elementary result

$$D = \tfrac{1}{6} w^{(0)} z s^2, \tag{8.2.6}$$

where $s$ is the magnitude of the jump distance between neighbouring sites.

As already indicated, one particular example to which (8.2.6) applies is that of interstitial solutes occupying the octahedral interstitial positions in b.c.c. transition metals. This was considered in connection with Snoek relaxation in §7.3.2. In this case $z = 4$ and by (8.2.6) and (7.3.5) $D$ can be calculated from measured Snoek relaxation times. From a comparison of measured values of $D$ with those obtained from Snoek relaxation times in this way the underlying model has been confirmed; see Fig. 8.1 for an example of this.

Another way of regarding (8.2.4a) which we shall use in more complex examples below is obtained via the equation of continuity, viz.

$$\frac{\partial}{\partial t}(np(\mathbf{r}, t)) + \nabla \cdot \mathbf{J} = 0, \tag{8.2.7}$$

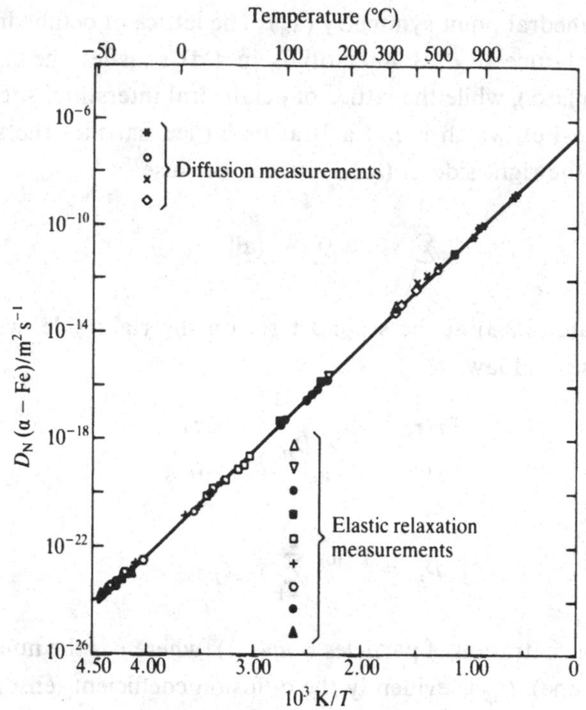


Fig. 8.1. An Arrhenius plot of the diffusion coefficient $D$ of interstitial $N$ atoms in b.c.c. iron ($\alpha$ and $\delta$ phases). The plotted values derive from direct diffusion measurements and from Snoek relaxation times via eqns. (7.3.5) and (8.2.6). The two sets are seen to be wholly consistent. (After Beshers, 1973.)

which expresses the conservation of particles, $\mathbf{J}$ being the flux of particles at position $\mathbf{r}$ and time $t$. By writing (8.2.4a) as

$$\frac{\partial p(\mathbf{r}, t)}{\partial t} = \sum_{\alpha} \frac{\partial}{\partial x_{\alpha}} \left( \sum_{\beta} D_{\alpha\beta} \frac{\partial p(\mathbf{r}, t)}{\partial x_{\beta}} \right)$$

and comparing this with (8.2.7) we see that

$$J_{\alpha} = -n \sum_{\beta} D_{\alpha\beta} \frac{\partial p(\mathbf{r}, t)}{\partial x_{\beta}} \tag{8.2.8}$$

(apart from the mathematically possible addition of a term of the form $\nabla \times \mathbf{S}$, which nevertheless must be zero since $\mathbf{J}$ as a flux of particles must be a polar vector whereas $\nabla \times \mathbf{S}$ is necessarily an axial vector). Since (8.2.8) is in the form of Fick's first law the significance of $D_{\alpha\beta}$ as the diffusivity tensor is confirmed.

### 8.2.2 *Mobilities*

It is not difficult to repeat the calculation which led to (8.2.8) in the presence of a uniform force field which biases the jumps of the particle. The starting equation (8.2.1) is the same, but the jump frequencies $w_{lm}$ now depend on the relation of the displacement $\mathbf{r}_l - \mathbf{r}_m$ to the force field $\mathbf{F}$. If we assume that the force field is conservative then this dependence can be obtained from transition state theory, i.e. from

$$w_{lm} = w^{(0)} \exp\left[\frac{1}{kT}(\delta g_m - \delta g_{lm}^{+})\right], \tag{8.2.9}$$

in which $\delta g_m$ is the change in Gibbs energy of a particle at site $m$ due to the imposition of the force field, while $\delta g_{lm}^{+}$ is the corresponding change when the particle is in its saddle-point configuration for its jump from site $m$ to $l$. For linear effects we need (8.2.9) just to first order in these changes, i.e. we can set

$$w_{lm} = w^{(0)}\left[1 + \frac{1}{kT}(\delta g_m - \delta g_{lm}^{+})\right]. \tag{8.2.10}$$

In fact, we only need the difference $w_{lm} - w_{ml}$ and by (8.2.10) the saddle-point term $\delta g_{lm}^{+}$ $(= \delta g_{lm}^{+}$ by hypothesis) then drops out. The same result can also be obtained without appealing to transition state theory but by assuming that the principle of detailed balance holds in the presence of the force field – which again demands that it is conservative. Such an argument gives

$$\frac{w_{lm}}{w_{ml}} = \exp\left[-\frac{1}{kT}(\delta g_l - \delta g_m)\right], \tag{8.2.11}$$

which, in the linear approximation, yields the same results.

If now we return to (8.2.1) and make the same Taylor expansion of $p$ as before, but allow for the changes in transition rates caused by the applied field, we obtain in place of (8.2.4)

$$\begin{aligned}\frac{\partial p}{\partial t} &= \frac{w^{(0)}}{2}\sum_{\alpha,\beta}\sum_{j=1}^{z} s_{j,\alpha}s_{j,\beta}\frac{\partial^2 p}{\partial x_\alpha\,\partial x_\beta} - \frac{w^{(0)}}{2kT}\sum_{\alpha,\beta}\sum_{j=1}^{z} s_{j,\alpha}s_{j,\beta}\frac{\partial p}{\partial x_\alpha}F_\beta \\ &= \sum_{\alpha,\beta} D_{\alpha\beta}\frac{\partial^2 p}{\partial x_\alpha\,\partial x_\beta} - \frac{1}{kT}\sum_{\alpha,\beta} D_{\alpha\beta}\frac{\partial p}{\partial x_\alpha}F_\beta,\end{aligned} \tag{8.2.12}$$

with $D_{\alpha\beta}$ as before given by (8.2.4b). Correspondingly, the expression for the particle current obtained from (8.2.12) and (8.2.7) is

$$J_\alpha = -n\sum_\beta D_{\alpha\beta}\frac{\partial p}{\partial x_\beta} + \frac{n}{kT}\sum_\beta D_{\alpha\beta}F_\beta p. \tag{8.2.13}$$

The second (drift) term on the right side clearly defines a mobility tensor $u^M_{\alpha\beta}$ such that

$$\frac{u^M_{\alpha\beta}}{D_{\alpha\beta}} = \frac{1}{kT}. \tag{8.2.14}$$

This is evidently just the Nernst–Einstein relation (5.2.8). (Electrical mobility, however, is defined as the coefficient of $E$ rather than of $F$, which is charge multiplied by $E$.)

### 8.2.3 *Quasi-elastic incoherent neutron scattering*

Quasi-elastic incoherent neutron scattering allows the possibility of studying atomic movements over small distances, the relevant quantity being the incoherent scattering function $S_s(\mathbf{q}, \omega)$ introduced in §1.8 as the space–time Fourier transform of the self-correlation function, $G_s(\mathbf{r}, t)$. We shall now use $G_s(\mathbf{r}_l, t)$ to denote the probability that the atom which was originally at the origin is to be found the lattice point $\mathbf{r}_l$ at time $t$: it is thus $v$ times the quantity denoted by the same symbol within the continuum representation used in §1.8. It is clear then that $G_s(\mathbf{r}_l, t)$ also satisfies (8.2.1), with the initial condition

$$G_s(\mathbf{r}_l, 0) = \delta_{l,0}, \tag{8.2.15}$$

corresponding to the certainty that the atom was at the origin at $t = 0$. If we take the spatial Fourier transform of the equation (8.2.1) we obtain an equation for the transform of $G_s(\mathbf{r}_l, t)$, i.e. for the intermediate scattering function $I_s(\mathbf{q}, t)$ defined by (1.9.3). By (8.2.1) this yields

$$\frac{\partial I_s(\mathbf{q}, t)}{\partial t} = -w^{(0)} I_s(\mathbf{q}, t) \sum_{j=1}^{z} (1 - e^{i\mathbf{q}\cdot\mathbf{s}_j}) \tag{8.2.16}$$

with the initial condition corresponding to (8.2.15), i.e.

$$I_s(\mathbf{q}, 0) = 1. \tag{8.2.17}$$

If we now write

$$\Gamma(\mathbf{q}) = w^{(0)} \sum_{j=1}^{z} (1 - e^{i\mathbf{q}\cdot\mathbf{s}_j}), \tag{8.2.18}$$

the solution for $I_s(\mathbf{q}, t)$ is evidently

$$\begin{aligned} I_s(\mathbf{q}, t) &= I_s(\mathbf{q}, 0) \exp(-\Gamma(\mathbf{q})t) \\ &= \exp(-\Gamma(\mathbf{q})t). \end{aligned} \tag{8.2.19}$$

The quantity $\Gamma(\mathbf{q})$ is real by the assumed Bravais nature of the lattice, i.e. to every $\mathbf{s}_j$ there is another neighbour at $-\mathbf{s}_j$.

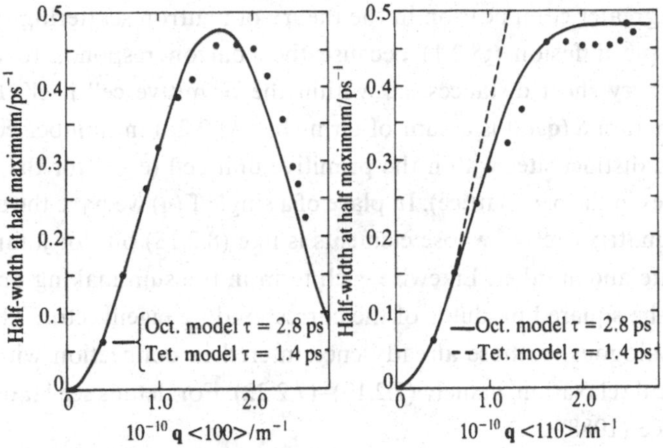


Fig. 8.2. Half-width at half-maximum of the quasi-elastic lines for the scattering of neutrons by H in Pd, as measured by Rowe *et al.* (1972). The lines are calculated from (8.2.18) for the scattering vectors **q** as indicated. The full lines are for octahedral interstitial sites while the dashed line is for tetrahedral interstitial sites. For **q** in the ⟨100⟩ directions both asumptions lead to the same Γ(**q**).

Finally, the *incoherent scattering function* for these translational motions $S_s(\mathbf{q}, \omega)$, being the temporal Fourier transform of $I_s(\mathbf{q}, t)$, is

$$S_s(\mathbf{q}, \omega) = \frac{1}{\pi} \mathrm{Re} \int_0^\infty I_s(\mathbf{q}, t)\, e^{i\omega t}\, dt$$

$$= \frac{1}{\pi} \frac{\Gamma(\mathbf{q})}{(\Gamma(\mathbf{q}))^2 + \omega^2}, \qquad (8.2.20)$$

as foreshadowed in eqn. (1.8.7). Allowance for thermal vibrations of the atoms about their lattice sites will fold in a Debye–Waller factor $\exp(-2W(\mathbf{q}))$ so that the observable quantity is (8.2.20) multiplied by this factor. For the scattering of neutrons in a particular direction **q**, (8.2.20) shows that the variation with change of energy $\hbar\omega$ is Lorentzian in form with a half-width at half-maximum given by Γ(**q**). By such measurements the migration of H atoms in Pd has been shown to be by jumps between nearest-neighbour octahedral sites of the (f.c.c.) interstitial lattice; see Fig. 8.2.

Hydrogen also dissolves interstitially in the b.c.c. transition metals and its migration has also been studied by quasi-elastic neutron scattering. The theory in this case is more complicated, however, because neither the lattice of octahedral interstitial sites nor that of tetrahedral sites is a Bravais lattice and the occupancy of each inequivalent site in the primitive unit cell must be separately represented.

This leads to greater complication in the theory of neutron scattering than in that for macroscopic diffusion (§8.2.1) because the neutron responds to the atomic motion over very short distances, i.e. within the primitive cell itself. In fact, the scattering function $S_s(\mathbf{q}, \omega)$ is a sum of terms like (8.2.20) in number equal to the number, $m$, of distinct sites within the primitive unit cell (e.g. 3 for the octahedral interstitial sites in the b.c.c. lattice). In place of a single $\Gamma(\mathbf{q})$ we have the eigenvalues of an $m \times m$ matrix each of whose elements is like (8.2.18) but for jumps between one *type* of site and another. Likewise each term in the sum making up $S_s(\mathbf{q}, \omega)$ is weighted by the squared modulus of the corresponding eigenvector. These results are rather analogous to those already encountered in connection with dielectric and mechanical relaxation, namely (7.2.19)–(7.2.22). For details see Haus and Kehr (1987) and Bée (1988).

Although Mössbauer atoms are rarely interstitial, essentially the same calculation (1.9.2) shows us that for such cases the cross-section for $\gamma$-ray emission follows a Lorentzian line-shape about $\omega = \omega_0$, but with an energy width at half-height of $\gamma + 2\hbar\Gamma(\mathbf{q})$ in place of the natural width $\gamma$.

## 8.3 Solute diffusion via vacancies

In the previous section we considered the theory of diffusion of interstitial solute atoms moving through an otherwise empty interstitial lattice. Successive jumps of a solute atom take place randomly as the result of thermal activation. The same broad model also describes the diffusion of vacancies in a pure substance, because we can treat the vacancies as the independently moving particles. However, it does not in general describe the diffusion of substitutional solute atoms. We have already pointed out in §§2.5.3 and 5.5 that when solute atoms move by jumping into vacant sites successive displacements are correlated with one another, both in space and in time. To describe this situation requires us to represent the movement of both the solute atoms and the vacancies. In dilute solid solutions it is sufficient to follow the movements of one solute atom and one vacancy in interaction with one another, and then to average their behaviour statistically.

### *8.3.1 Basic kinetic equations*

We shall begin by assuming that there is an attraction between a solute atom and a vacancy, as is indeed often the case (§2.5.3). It is then natural to extend the mathematical structure presented in the previous chapter (§7.2) for relaxation processes associated with such solute–vacancy pairs. This is easily done by defining the distinct configurations $p, q, \ldots$ available to the solute–vacancy pair in the same way as before. Now, however, we must replace the fraction $p_p(t)$ by a function of

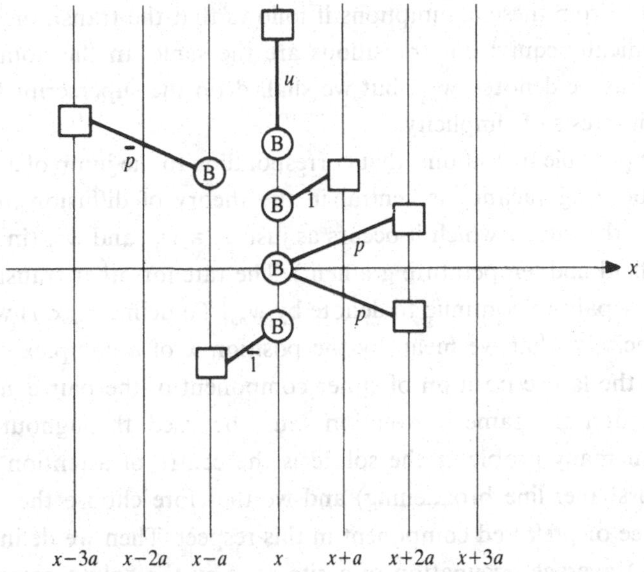


Fig. 8.3 Schematic diagram showing the planes of the lattice normal to one of the principal axes of the crystal (*x*-direction). Pairs of solute atoms, B, and vacancies are shown explicitly, but the solvent atoms A are not represented. The rotational symmetry of the crystal about the *x*-axis will ensure that there are several equivalent defect configurations of each type (1, $\bar{1}$, *p*, *q*, etc.). Only *types* of configuration in this sense are distinguished in the equations. By the definitions introduced later in §8.3.2 the configuration labelled here as *p* is a member of the (+)-set while $\bar{p}$ belongs to the (−)-set. The configurations labelled *u* belong to the (0)-set. In all cases the defects are assumed to be uniformly distributed across the planes normal to the *x*-direction.

position as well as time, i.e. $p_p(x, t)$ for diffusion in one dimension *x*. We concentrate here on the diffusion of the solute atoms (which, when we want to be specific, we shall label B, as distinct from the solvent atoms which are of species A). Looking at Fig. 8.3 and assuming that the *x*-direction is a principal axis of the crystal, we can see that there are two types of configurations, 1 and $\bar{1}$ which are particularly important when only nearest-neighbour jumps occur; for these are the only configurations whereby the solute atom has the chance to move directly along the *x*-axis without having to wait for the vacancy to make jumps which bring it into a nearest-neighbour position of the solute.

We now set down the necessary extension of the treatment of §7.2. In this section we are interested in pure isothermal diffusion, i.e. we may suppose that no external forces are acting on the system and that, apart from the small gradient in the solute concentration, there is no other source of inhomogeneity present. In dilute solid solutions we can ignore any dependence of the transition rates upon concentration. From these assumptions it follows that the transition rates for all crystallographically equivalent transitions are the same. In the notation of §7.2 they would thus be denoted $w_{pq}^{(0)}$, but we shall drop the superscript for the time being in the interests of simplicity.

Among the possible transitions, that corresponding to the jump of a solute atom into a neighbouring vacancy is central in the theory of diffusion and we shall therefore label the rate at which it occurs as just $w_s \equiv w_{1\bar{1}}$ and $w_{\bar{1}1}$ (in the absence of any force field and temperature gradient). The rate for other transitions of the solute–vacancy pair we continue to denote by $w_{pq}$. To define $p_p(x, t)$ we need first to specify precisely what we mean by the position $x$ of a complex defect like a pair. Clearly, the lattice position of either component of the pair can be used to do this, although the same convention must be used throughout any given calculation. In many problems the solute is the centre of attention (e.g. solute diffusion, Mössbauer line broadening) and we therefore choose the solute atom as the reference or preferred component in this respect. Then we define $p_p(x, t)$ as the (ensemble) average occupation of a site at $x$ by the solute component of a pair in configuration $p$. As such, it is a quantity of the same sort as $p_l$ in the preceding section.

Since we have singled out the forwards (1) and backwards ($\bar{1}$) nearest-neighbour configurations, there will be three equations in place of the single equation (7.2.1), one for each of 1 and $\bar{1}$ and one for all, more extended configurations $p$, $q$, etc. Let us take that for $p_1(x, t)$ first. From our definitions it is clear that

$$\frac{\partial p_1(x, t)}{\partial t} = -\sum_{p \neq 1} w_{p1} p_1(x, t) + \sum_{p \neq 1, \bar{1}} w_{1p} p_p(x, t) + w_s p_{\bar{1}}(x + a, t). \quad (8.3.1)$$

The first term on the right side represents the rate of loss of pairs from configuration 1 and the last two the corresponding rate of gain. All possible vacancy movements are thus included. The essential difference from (7.2.1), apart from the specific changes in notation, comes from the last term, $w_s p_{\bar{1}}(x + a, t)$, which is for position $x + a$ of the solute while all the other terms refer to position $x$. Likewise the corresponding equation for $\partial p_{\bar{1}}(x, t)/\partial t$ will contain a term $w_s p_1(x - a, t)$. It is through these terms that the effects of concentration gradients enter into the microscopic description. For the description of diffusion on the macroscopic scale (as e.g. by Fick's law) it is sufficient to expand these terms in Taylor series about $x$ and to retain only terms up to the second derivatives. From (8.3.1) we then get

$$\frac{\partial p_1}{\partial t} = -\sum_{p \neq 1} w_{p1} p_1 + \sum_{p \neq 1} w_{1p} p_p + a w_s \frac{\partial p_{\bar{1}}}{\partial x} + \frac{a^2 w_s}{2} \frac{\partial^2 p_{\bar{1}}}{\partial x^2}, \quad (8.3.2a)$$

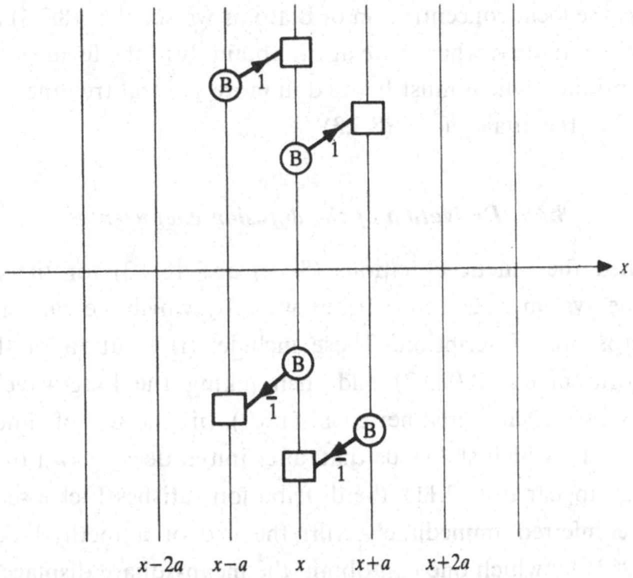


Fig. 8.4. Schematic diagram showing the transitions which contribute to the flux of solute atoms $J_B$ across the lattice plane at $x$. Each transition shown contributes one of the four terms on the right side of (8.3.3).

$$\frac{\partial p_p}{\partial t} = -\sum_{q \neq p} w_{qp} p_p + \sum_{q \neq p} w_{pq} p_q, \qquad (p \neq 1, \bar{1}), \tag{8.3.2b}$$

and

$$\frac{\partial p_{\bar{1}}}{\partial t} = -\sum_{p \neq \bar{1}} w_{p\bar{1}} p_{\bar{1}} + \sum_{p \neq 1} w_{\bar{1}p} p_p - a w_s \frac{\partial p_1}{\partial x} + \frac{a^2 w_s}{2} \frac{\partial^2 p_1}{\partial x^2}. \tag{8.3.2c}$$

The argument of all the quantities $p$ is here the same, namely $(x, t)$, and therefore is no longer shown explicitly. In (8.3.2a) the leading term in the Taylor expansion of $p_{\bar{1}}(x + a, t)$ in (8.3.1) has been combined with the second term in (8.3.1) by removing the restriction $p \neq \bar{1}$ from the summation. Correspondingly with (8.3.2c).

To complete the basic equations we need an expression for the flux of B-atoms in terms of the $p_p$. From Fig. 8.4 we see directly that this is

$$\begin{aligned} J_B &= \tfrac{1}{2} n a w_s (p_1(x, t) + p_1(x - a, t) - p_{\bar{1}}(x, t) - p_{\bar{1}}(x + a, t)) \\ &= n a w_s \left( (p_1 - p_{\bar{1}}) - \frac{a}{2} \frac{\partial p_1}{\partial x} - \frac{a}{2} \frac{\partial p_{\bar{1}}}{\partial x} \right), \end{aligned} \tag{8.3.3}$$

to first-order terms in the Taylor expansion. An alternative is to add up all the eqns. (8.3.2) to give

$$\frac{\partial}{\partial t}\left( \sum_q p_q \right) = a w_s \left( \frac{\partial p_1}{\partial x} - \frac{\partial p_{\bar{1}}}{\partial x} \right) + \frac{a^2 w_s}{2} \frac{\partial^2}{\partial x^2} (p_1 + p_{\bar{1}}). \tag{8.3.4}$$

Since $n \sum_q p_q$ is the local concentration of B atoms we see that (8.3.4) is effectively the equation of continuity, whence we again obtain $J_{\mathrm{B}}$ in the form (8.3.3). It is the equation of continuity which must be used in more general treatments where it is impossible to find the analogue of (8.3.3).

### *8.3.3 Derivation of the diffusion coefficient*

Having obtained the kinetic equations (8.3.2) and (8.3.3) for the microscopic evolution of the system, there are various ways by which we can make contact with the macroscopic description. These include: (i) solution of the Fourier-transformed form of eqns. (8.3.2) and then taking the long-wavelength low-frequency limit in the same manner as in (1.8.9), (ii) the use of time-dependent perturbation theory, which shows us that, after initial decays with the same time constants $\tau^{(\nu)}$ as appear in (7.2.11), the distribution satisfies Fick's second law so that $D$ can be inferred immediately, (iii) the use of a method described by McCombie (1962) by which one can obtain the mean-square displacement of the solute atoms and so infer $D$ from the Einstein relation (6.5.15), and (iv) the consideration of eqns. (8.3.2) and (8.3.3) under uniform steady state conditions which allow one to obtain the flux $J_{\mathrm{B}}$ in the form of Fick's first law and so to identify $D$ immediately. Of these, the last is the simplest and we shall therefore use it here.

We therefore consider time-independent diffusion under a uniform concentration *gradient*. Then the terms in $\partial/\partial t$ and in $\partial^2/\partial x^2$ in eqns. (8.3.2) and (8.3.4) are all zero. As in §7.2 we write

$$p_p(x, t) \equiv p_p^{(0)} + p_p^{(1)}(x, t), \tag{8.3.5}$$

in which $p_p^{(0)}$ is the thermal equilibrium value, while $p_p^{(1)}(x, t)$ is the disturbance brought about by the gradient in overall concentration. Since eqns. (8.3.2) are linear they are also satisfied by the $p_p^{(1)}$ alone (the $p_p^{(0)}$ satisfy (8.3.2) in the absence of any gradients). In matrix notation these equations become

$$-\mathbf{P}\mathbf{p}^{(1)} + \mathbf{V}\frac{\partial \mathbf{p}^{(1)}}{\partial x} = 0. \tag{8.3.6}$$

Here the $p_p^{(1)}$ have been formed into a column matrix $\mathbf{p}^{(1)}$, while the matrix $\mathbf{P}$ is given by the analogue of (6.2.4) in the jump frequencies $w_{pq}$. For a given system it is identical with the matrix $\mathbf{P}^{(0)}$ introduced in §7.2. The matrix $\mathbf{V}$ has elements

$$\left.\begin{aligned} V_{1\bar{1}} &= aw_{\mathrm{s}}, \\ V_{\bar{1}1} &= -aw_{\mathrm{s}}, \\ V_{pq} &= 0, \quad \text{otherwise.} \end{aligned}\right\} \tag{8.3.7}$$

We seek a solution to the set of linear first-order equations (8.3.6) of the form

$$\mathbf{p}^{(1)} = \boldsymbol{\phi} + \boldsymbol{\gamma} x \tag{8.3.8}$$

where $\boldsymbol{\phi}$ and $\boldsymbol{\gamma}$ are column matrices independent of $x$. We substitute (8.3.8) into (8.3.6) and see that if the result is to be true for all $x$ we must have

$$\mathbf{P}\boldsymbol{\gamma} = 0, \tag{8.3.9}$$

i.e. $\gamma$ is an eigenvector of $\mathbf{P}$ of eigenvalue zero. Since there can be only one uniform stationary solution of (8.3.2), namely that for thermal equilibrium, this means that $\boldsymbol{\gamma}$ must be a simple multiple of the thermal equilibrium eigenvector $\mathbf{p}^{(0)}$. The multiplying factor, $\kappa$ say, is found by using (8.3.8) to obtain $\partial p_p^{(1)}/\partial x = \gamma_p$; whence summing over $p$ gives

$$\kappa = \left( \frac{1}{p^{(0)}} \frac{\partial p}{\partial x} \right), \tag{8.3.10}$$

with $p$ as the total fraction of B atoms

$$p = \sum_p (p_p^{(1)} + p_p^{(0)}) \tag{8.3.11}$$

and

$$p^{(0)} = \sum_p p_p^{(0)}. \tag{8.3.12}$$

We now insert this result into (8.3.6) to obtain an equation for $\phi$, namely

$$-\mathbf{P}\boldsymbol{\phi} + \mathbf{V}\mathbf{p}^{(0)} \left( \frac{1}{p^{(0)}} \frac{\partial p}{\partial x} \right) = 0. \tag{8.3.13}$$

Unfortunately, we cannot solve this for $\boldsymbol{\phi}$ simply by multiplying through by $\mathbf{P}^{-1}$ because $\mathbf{P}$ is singular by reason of the defining eqn. (6.2.4). This difficulty can, however, be circumvented by making use of symmetry. Firstly, for the lattices we are concerned with, the set of pair configurations can be divided into three sub-sets as shown in Fig. 8.3. By the assumed symmetry of the lattice we see that to each configuration, $p$, of the $(+)$-set there will be a corresponding configuration, $\bar{p}$, in the $(-)$-set obtainable from $p$ by one of the following symmetry operations: (i) reflection in a plane normal to the $x$-axis, (ii) inversion through a centre on the $x$-axis or (iii) such inversion in combination with rotation about the $x$-axis, i.e. an improper rotation. Clearly then the thermal equilibrium values of $p_p^{(0)}$ satisfy

$$p_p^{(0)} = p_{\bar{p}}^{(0)} \tag{8.3.14a}$$

and in particular

$$p_1^{(0)} = p_{\bar{1}}^{(0)}. \tag{8.3.14b}$$

Secondly, for perturbations of the system having the form of the gradient of a scalar field we shall have

$$p_p^{(1)}(x, t) = -p_{\bar{p}}^{(1)}(-x, t), \tag{8.3.15}$$

as derived in Appendix 8.1. When, as for steady state conditions, $\mathbf{p}^{(1)}$ has the form (8.3.8) it follows that

$$\phi_p = -\phi_{\bar{p}} \tag{8.3.16a}$$

and in particular

$$\phi_1 = -\phi_{\bar{1}}. \tag{8.3.16b}$$

In other words, the column matrix $\boldsymbol{\phi}$ is antisymmetric with respect to the pair configurations. In particular, for configurations, $p$, in the (0)-set, $\phi_p = 0$.

We shall now make use of these results, firstly in an analysis which shows the relation of diffusion to dielectric relaxation and then in one which leads to more explicit results for $D_{\mathrm{B}}$. For both we need the equation for the flux $J_{\mathrm{B}}$. By (8.3.3), (8.3.8), (8.3.10) and (8.3.16b) we easily obtain

$$J_{\mathrm{B}} = naw_{\mathrm{s}}(2\phi_1 - a\kappa p_1^{(0)}). \tag{8.3.17}$$

*Solution in terms of relaxation modes*

For the first form of solution we introduce a small positive quantity $\lambda$ and then formally re-express (8.3.13) as

$$\boldsymbol{\phi} = \lim_{\lambda \to 0} \bar{\mathbf{G}}(\lambda)\mathbf{V}\mathbf{p}^{(0)}\left(\frac{1}{p^{(0)}}\frac{\partial p}{\partial x}\right), \tag{8.3.18}$$

in which $\bar{\mathbf{G}}(\lambda) \equiv (\mathbf{P} + \lambda\mathbf{1})^{-1}$ can be expressed in terms of the eigenvalues and eigenvectors of the symmetrized matrix

$$\mathbf{S} = \mathbf{N}^{-1/2}\mathbf{P}\mathbf{N}^{1/2} \tag{8.3.19}$$

via Laplace transformation of (6.3.2). Recalling now the definition (8.3.7) of the matrix $\mathbf{V}$, we see that (8.3.18) yields

$$\phi_1 = \lim_{\lambda \to 0} aw_{\mathrm{s}}(\bar{G}_{11} - \bar{G}_{1\bar{1}})\left(\frac{1}{p^{(0)}}\frac{\partial p}{\partial x}\right). \tag{8.3.20}$$

We can now use (6.3.2) to obtain $\phi_1$ $(= -\phi_{\bar{1}})$ as

$$\phi_1 = \lim_{\lambda \to 0} \tfrac{1}{2} a w_s \left( \frac{1}{p^{(0)}} \frac{\partial p}{\partial x} \right) \sum_{\nu} \frac{|a_1^{(\nu)} - a_{\bar{1}}^{(\nu)}|^2}{(\alpha^{(\nu)} + \lambda)}. \tag{8.3.21}$$

Now the eigenvectors $\mathbf{a}^{(\nu)}$ of $\mathbf{S}$ will either be symmetric or antisymmetric with respect to the symmetry operations which we introduced above to define the relation between configurations $p$ and $\bar{p}$ in general and that between 1 and $\bar{1}$ in particular. For the symmetric modes $a_1 = a_{\bar{1}}$ and there is therefore no contribution to the sum in (8.3.21). In particular, the (thermal equilibrium) mode with zero eigenvalue is symmetric, so that we can take the limit $\lambda \to 0$ in (8.3.21) with impunity. For the antisymmetric modes we will have $a_1^{(\nu)} = -a_{\bar{1}}^{(\nu)}$ and thus

$$\phi_1 = 2 a w_s \left( \frac{1}{p^{(0)}} \frac{\partial p}{\partial x} \right) \sum_{\nu}^{\text{A.S.}} \tau^{(\nu)} |a_1^{(\nu)}|^2, \tag{8.3.22}$$

in which the summation is over antisymmetric modes only. Lastly therefore, from (8.3.10) and (8.3.17) we obtain $J_B$ and hence $D_B$.

$$D_B = a^2 w_s \left( \frac{p_1^{(0)}}{p^{(0)}} \right) \left[ 1 - 4 w_s \sum_{\nu}^{\text{A.S.}} \tau^{(\nu)} |a_1^{(\nu)}|^2 \right]. \tag{8.3.23}$$

We observe that $D_B$ is the product of two parts. The first part $a^2 w_s (p_1^{(0)}/p^{(0)})$ is the expression we would have obtained for $D_B$ if we had assumed that the vacancies were available randomly, irrespective of the movements of the solute atoms. This would have implied in turn that successive moves of a solute atom were themselves random, i.e. uncorrelated with one another. The second part, i.e. that factor in (8.3.23) in [ ] results from the fact that successive moves of a solute atom are correlated with one another. As argued earlier in Chapter 2 this is a general feature of diffusion by vacancy mechanism. This factor is referred to as the *correlation factor*,

$$f_B = \left[ 1 - 4 w_s \sum_{\nu}^{\text{A.S.}} \tau^{(\nu)} |a_1^{(\nu)}|^2 \right], \tag{8.3.24}$$

for B-atom diffusion. We see immediately that the same anti-symmetric modes as enter into the dielectric relaxation of solute–vacancy pairs also enter into the solute diffusion coefficient, but that none of the other relaxation modes do. Since the relaxation times $\tau^{(\nu)} > 0$ (Appendix 6.1) while $|a_1^{(\nu)}|^2 > 0$ it follows that $f_B < 1$. (This conclusion also follows under the more general conditions considered in §§8.4 and 8.5.) We also see that the faster the re-orientation rate, i.e. the shorter $\tau^{(\nu)}$ relative to the mean-time between solute–vacancy exchanges, $w_s^{-1}$, the closer $f_B$ comes to 1. In practice (8.3.24) has been rather little used for calculations, although one example is provided by Franklin (1965). We shall therefore turn to a more explicit solution.

*Solution in terms of reduced jump-frequency matrix*

To obtain this we return to (8.3.13) and again make use of the results (8.3.14)–(8.3.16) coming from the symmetry of the system. In doing so we also recognize that the product $\mathbf{Vp}^{(0)}$ possesses the same antisymmetry as $\boldsymbol{\phi}$ on account of (8.3.7) and (8.3.14a). We can thus write out (8.3.13) using just the indices of the (+)-set:

$$\sum_{q}^{(+)} Q_{pq}\phi_q = 0, \qquad (p \neq 1), \tag{8.3.25a}$$

and

$$\sum_{q}^{(+)} Q_{1q}\phi_q = aw_s\kappa p_1^{(0)} \tag{8.3.25b}$$

in which

$$Q_{pq} = P_{pq} - P_{p\bar{q}}. \tag{8.3.26}$$

Although $\mathbf{P}$ is singular (cf. 6.2.5) there is no general necessity for the summation

$$\sum_{p}^{(+)} Q_{pq}$$

to be zero, i.e. there is no reason not to invert $\mathbf{Q}$ as required. From (8.3.25a) we thus get

$$\sum_{q\neq 1}^{(+)} Q_{pq}\phi_q = -Q_{p1}\phi_1, \qquad (i \neq 1), \tag{8.3.27}$$

whence

$$\phi_q = -\sum_{p\neq 1}^{(+)} (\hat{Q}^{-1})_{qp}Q_{p1}\phi_1, \qquad (q \neq 1), \tag{8.3.28}$$

in which $\hat{\mathbf{Q}}$ is the matrix of the complementary minor of the element $Q_{11}$, i.e. the matrix obtained by removing the 1-row and the 1-column from $\mathbf{Q}$. We now treat (8.3.25b) similarly by separating the term in $\phi_1$ from those in $\phi_q$ $(q \neq 1)$. On using (8.3.28) to eliminate $\phi_q$ we then obtain as the equation for $\phi_1$

$$\phi_1\left[Q_{11} - \sum_{p,q\neq 1}^{(+)} Q_{1q}(\hat{Q}^{-1})_{qp}Q_{p1}\right] = aw_s\kappa p_1^{(0)}. \tag{8.3.29}$$

Lastly, we insert this solution for $\phi_1$ together with (8.3.10) into (8.3.17) and thus infer the diffusion coefficient $D_{\mathrm{B}}$ as

$$D_{\mathrm{B}} = a^2 w_s\left(\frac{p_1^{(0)}}{p^{(0)}}\right) f_{\mathrm{B}} \tag{8.3.30}$$

as before, but now with the correlation factor $f_B$ in the form

$$f_B = 1 - \frac{2w_s}{\Omega} \tag{8.3.31}$$

with

$$\Omega = Q_{11} - \sum_{p,q \neq 1}^{(+)} Q_{1q}(\hat{Q}^{-1})_{qp}Q_{p1}. \tag{8.3.32}$$

Although the inverse of $\hat{\mathbf{Q}}$ must be found, (8.3.31) and (8.3.32) give an explicit prescription by which $f_B$ may be calculated. The order of $\hat{\mathbf{Q}}$ is one less than the number of types of configuration in the (+)-set. Clearly then if the number of configurations is not too large we can obtain $D_B$ explictly for any specified model of the configurations. We shall illustrate its use with a simple example, namely solute–vacancy pairs in a f.c.c. lattice, as already considered in §7.3.3.

*Example of solute–vacancy pairs in a f.c.c. lattice*

Firstly consider just nearest-neighbour pairs so that the only available jumps are the $w_1$ and $w_2$ jumps shown in Fig. 7.1. Consideration of the geometry of the lattice then quickly shows us that the matrix $\mathbf{P}$ is

$$\mathbf{P} = \begin{bmatrix} (2w_1 + w_2) & -2w_1 & -w_2 \\ -2w_1 & 4w_1 & -2w_1 \\ -w_2 & -2w_1 & (2w_1 + w_2) \end{bmatrix}, \tag{8.3.33}$$

the elements being arranged in order of the configurations 1, 0 and $\bar{1}$ as shown in Fig. 8.5. Evidently, $\mathbf{P}$ is singular in accordance with (6.2.5). We now reduce $\mathbf{P}$ to $\mathbf{Q}$ by (8.3.26). There is only one configuration in the (+)-set, namely 1, so that $\mathbf{Q}$ contains just the one element $2(w_1 + w_2)$. Correspondingly there is no $\hat{\mathbf{Q}}$. Thus by (8.3.31) the correlation factor $f_B$ is simply $w_1/(w_1 + w_2)$. The fraction $p_1^{(0)}/p^{(0)}$ is 1/3 in this case, whence by (8.3.30)

$$D_B = a^2 w_1 w_2 / 3(w_1 + w_2). \tag{8.3.34}$$

This simple result for the diffusion coefficient may be taken with the corresponding relaxation times for dielectric and mechanical relaxation (eqns. 7.3.7) as the basis for the determination of $w_1$ and $w_2$ separately. In this connection it will be seen that the dielectric relaxation time (7.3.7d) is determined by the faster of $w_1$ and $w_2$ while the diffusion coefficient is determined by the slower. Therefore the characteristic Arrhenius activation energies for these two processes need not be the same.

The model for which (8.3.34) was deduced is, of course, very simple and, as we saw in §7.3, further solute–vacancy configurations may have to be considered.

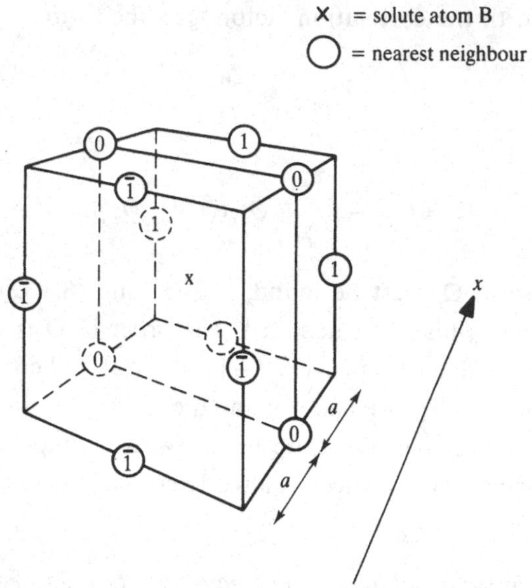


Fig. 8.5. A section of the f.c.c. lattice showing the three distinct types of nearest-neighbour configurations of solute-vacancy pairs with respect to the $x$-direction, taken to be along one of the $\langle 100 \rangle$ axes.

For example, second nearest-neighbour (200) configurations can easily be included to yield results corresponding to eqns. (7.3.8). We shall not, however, pursue this or similar examples at this point since there is a more significant consideration to be included. We have already touched on this in §7.4: it is the recognition that the dissociation of pairs into separate solute atoms and vacancies should also be allowed for.

### 8.3.3 *Inclusion of association and dissociation reactions in the expression for $D_B$*

The physical motivation for extending the calculation of the preceding section is simply that it is no longer sensible to think of solute–vacancy pairs as bound once the separation between the solute and the vacancy becomes large. Thus, atomistic calculations generally show that interaction energies beyond third neighbours are rarely more than $kT$ at the usual temperatures of experimentation. In addition to the population of bound pairs in their various configurations we now wish to include free, or isolated, solute atoms and vacancies. We have already used this distinction between pairs and free component defects in §§3.4–3.8 when dealing with equilibrium thermodynamic properties. It is natural therefore to take it over into the treatment of non-equilibrium properties.

We shall reserve the results of the full extension of the theory until §8.4. Here we shall indicate the rather simple extensions of the results (8.3.30)–(8.3.32) which follow when we may assume that the free vacancy population is locally in equilibrium. First we specify the configurations of the pair which are to be treated as bound. Then for each such configuration $p$ we allow dissociation to occur at a rate $w_{dp}$. (Of course, for many of the closer configurations $w_{dp}$ may be zero). Symmetry considerations of the same kind as in the last section tell us that the elements of $\boldsymbol{\phi}$ corresponding to the free solutes and free vacancies are zero because they are contained in the (0)-set. However, the gradients in their concentrations will in general not be zero, although when we can assume the free vacancy population to be locally in equilibrium then the concentration is constant and its gradient is zero. Under these conditions the kinetic equations are just like (8.3.13) except that the matrix $\mathbf{P}$ is now replaced by

$$\mathbf{P}' = \mathbf{P} + \mathbf{D}, \tag{8.3.35a}$$

where $\mathbf{D}$ is a diagonal matrix of dissociation rates

$$D_{pq} = \delta_{pq} \sum_{d} w_{dp}. \tag{8.3.35b}$$

We then find that the analysis goes through as before and leads again to the results (8.3.30)–(8.3.32), the only changes being (i) that $\mathbf{Q}$ is obtained from $\mathbf{P}'$ rather than $\mathbf{P}$ (8.3.26) and (ii) that the quantity $p^{(0)}$ appearing in the denominator of (8.3.30) now includes the unpaired solute atoms as well as those in pairs. Thus even when exchanges between the populations of paired and free species are allowed the structure of the results is the same as before, providing only that the vacancy population can be taken as locally in equilibrium. Of course, if the vacancies were not in local equilibrium it would not be possible to define $D_B$, since $J_B$ would depend not only on the gradient of solute concentration but also on the gradient of vacancy concentration.

*Example of solute–vacancy pairs in a f.c.c. lattice*

We have already touched on this example, but inclusion of dissociation jumps by the vacancies even from nearest-neighbour positions makes the results much more useful. As Fig. 7.1 shows there are seven dissociation jumps by which a vacancy can move from a first neighbour position to more distant positions. If we suppose these all occur at the same rate $w_3$ then $\mathbf{P}'$ is the sum of $\mathbf{P}$ and $7w_3$ times the unit matrix. The consequence is that

$$D_B = a^2 w_2 \left( \frac{p_1^{(0)}}{p^{(0)}} \right) f_B \tag{8.3.36}$$

with

$$f_B = \frac{w_1 + 7w_3/2}{(w_1 + w_2 + 7w_3/2)}. \tag{8.3.37}$$

Since $p^{(0)}$ is now the total solute fraction, the quantity $(p_1^{(0)}/p^{(0)})$ is 1/3rd the fraction of impurity present in the form of pairs (obtainable from the equations of local pairing equilibrium, e.g. (3.8.5b)). This result is applicable to a much wider set of circumstances than is (8.3.33). In particular, if we evaluate (8.3.37) when there is no significant physical distinction between the solute and the solvent atoms and all the jump frequencies are the same (solvent tracer diffusion) we obtain $f_B = 9/11$, which is quite a good approximation to the exact result, $f_B = 0.781$, for this limiting case (Chapter 10). Thus here we have a model based on the idea of a physical attraction between solute and vacancy which gives results accurate to a few per cent, even when there is no attraction at all.

The method described in this section has been used previously (under the title *pair-association* method) to calculate diffusion coefficients in a variety of model systems. Each of these calculations has, however, been carried through in terms specific to the system studied. All the results can be obtained from the general results (8.3.30)–(8.3.32) once the specific form of the matrix $\mathbf{Q}$ has been set down. Use of these general results makes it easier to avoid algebraic mistakes associated with the manipulation of cumbersome expressions. Nevertheless the limitations of this section will be apparent. As it stands, it deals only with the diffusion of substitutional solutes via vacancies, and does not make contact with the more extended phenomenology offered by non-equilibrium thermodynamics (Chapters 4 and 5). In the next section we review a more complete treatment which is not limited in these ways. It will also be shown that this treatment allows us to address certain basic questions about this approach. For example, must there be a physical attraction between solute and vacancy in order to define the paired configurations $p$, $q$, etc.; and, if there need not be such an interaction for the approach to be valid, is there not an unsatisfactory arbitrariness about the distinction between paired and unpaired species? As we shall see, a physical interaction is *not* necessary for the method to be valid and the results which we obtain for the $L$-coefficients can be transformed so as to remove any distinction between paired and unpaired species when there is no interaction between them.

## 8.4 General kinetic theory of dilute isothermal systems

In this section we address the task of generalizing the treatment of §8.3 in order to obtain expressions for all the appropriate $L$-coefficients. This generalization was developed in a series of papers (Franklin and Lidiard, 1983, 1984; Lidiard, 1984, 1985, 1986b, 1987). Unfortunately, space does not allow us to provide a full derivation of the results here. On the other hand, the principal steps in their derivation are all understandable from what has gone before. The following sketch of the derivation should therefore be adequate for the understanding and use of the results. We begin with some minor extensions of notation and then review the basic kinetic equations, the flux equations, the solution under steady-state conditions, comparison with the flux equations of non-equilibrium thermodynamics and the inference of the $L$-coefficients and other results. (In any comparison between this chapter and the original papers it should be recognized that where here we write $w_{qp}$ for the rate of transition from $p \rightarrow q$ in the original paper this would be written as $w_{pq}$. This has the consequence that all equations in matrix form must be transposed in going from one to the other system. There are some other minor differences as well.) The principal complications which we have to contend with in this extension come from the association and dissociation reactions, i.e. from the formation and break-up of pairs. One or two rather detailed arguments are required to handle these correctly in general circumstances. These are given in the original papers but are omitted from the account here for simplicity: the results are, however, complete. The main point is that in this extension we are analysing an assemblage of solute atoms, point defects (vacancies or interstitials) and solute–defect pairs and treating each of these components on an equal footing. We have here the kinetic analogue of §§3.5–3.8. As there, it will be convenient to refer to the components of the assemblage collectively as defects; when the term defect is used specifically to mean vacancy or interstitial this will be clear from the context.

### *8.4.1 General analysis*

We begin with some extensions in notation. As before we ignore the solvent atoms (A) and consider the distributions of paired and free solute atoms (B) and defects (D) (which may now be either vacancies or interstitials). We use the same labelling for the pairs, namely $p$, $q$, $u$, etc.; unpaired species we label just B or D as appropriate. At this stage, the dividing line between separations at which the B atom and the defect are to be regarded as paired (i.e. as constituting a single, complex entity) and those at which they are taken to be distinct entities (i.e. free) may be fixed arbitrarily, although one would generally take it to enclose the range of any significant physical interaction. Later we shall see how this arbitrariness is removed (§8.5.2). As in §8.3, we simplify matters by supposing that all gradients of potential and concentration are along a particular crystal principal axis, the

$x$-axis (but note that this restriction can be straightforwardly removed to obtain a fully three-dimensional formulation, Lidiard, 1984). We also need a convention to specify the $x$-position of a paired defect (e.g. the $x$-coordinate of the solute or of the defect), but this is arbitrary and we must expect to find results which are invariant with respect to the particular choice. Once this convention is settled we define the corresponding displacements, $a_{qp}$, of defects in particular transitions, e.g.

$$D_p(x) \underset{w_{pq}}{\overset{w_{qp}}{\rightleftharpoons}} D_q(x + a_{qp}), \tag{8.4.1a}$$

for a transition within the set of paired configurations. There will be corresponding quantities for reactions in which a pair dissociates into its elements (e.g. p → d) or by which it is formed from them (e.g. d → p). Thus

$$D_p(x) \underset{w_{pd}}{\overset{w_{dp}}{\rightleftharpoons}} D_{\mathrm{B}}(x + a^{\mathrm{B}}_{dp}) + D_{\mathrm{D}}(x + a^{\mathrm{D}}_{dp}), \tag{8.4.1b}$$

where $d$ is used to specify any configuration which may be reached in a dissociation jump of the pair $D_p$ and from which a pair $D_p$ can be formed by an association jump. These various displacements depend, of course, on the convention chosen to define position; for example, for a solute–vacancy pair whose position is defined by that of the solute atom (as in §8.3) the only non-zero $a_{qp}$ are those for solute–vacancy exchange transitions, whereas if defined by that of the vacancy, all transitions have non-zero $a_{qp}$. Lastly, we must allow for the movement of the free components (B and D) from one site to another

$$D_{\mathrm{R}}(x) \underset{w^{\mathrm{R}}_{0r}}{\overset{w^{\mathrm{R}}_{r0}}{\rightleftharpoons}} D_{\mathrm{R}}(x + a^{\mathrm{R}}_{r0}). \tag{8.4.1c}$$

Here $r$ denotes a site that can be reached in one jump by component R (=B or D) from an initial site 0 lying in plane $x$, it being understood that none of these sites bring the moving species into association with another. Of course, with substitutional solutes the isolated solute atom cannot move (the required intervention of a defect necessarily requires it to be part of pair) so that $w^{\mathrm{B}}$ in (8.4.1c) is zero. However, the theory also deals with interstitial solutes when we need (8.4.1c) for the solute B as well as for the defect D.

As in §8.3 we again specify the states of the system in terms of the occupancy of the lattice sites by the paired and unpaired defects, and introduce the average site occupancies $p_p(x)$ for the paired defects in the various configurations, $p$, together with corresponding quantities, $p_{\mathrm{S}}(x)$, for unpaired species S (=B, D). The basic kinetic equations then take the form

$$\frac{\partial p_p}{\partial t} = -\left(\sum_{q\neq p} w_{qp}(x) + \sum_{d} w_{dp}(x)\right)p_p(x) + \sum_{q\neq p} w_{pq}(x + a_{qp})p_q(x + a_{qp})$$

+ bimolecular terms in $p_{\mathrm{B}}$ and $p_{\mathrm{D}}$ representing the formation of solute–defect pairs $D_p(x)$, (8.4.2)

with corresponding equations for $\partial p_{\mathrm{B}}/\partial t$ and $\partial p_{\mathrm{D}}/\partial t$.

An important difference between these equations and (8.3.2) lies in the presence of (i) dissociation terms (those in $w_{dp}$) and (ii) association terms which are second order ('bimolecular') in $p_{\mathrm{B}}$ and $p_{\mathrm{D}}$. Another difference is that in (8.4.2) we have also allowed for the defect jump frequencies to depend on position as indicated, so that we can represent systems subject to external fields (e.g. electric and centrifugal fields) or internal stress fields. Equation (8.4.2) can therefore be used for the circumstances of Chapter 7 as well for those of §8.3, or for the two in combination.

We next reduce (8.4.2) to the form of a Fokker–Planck equation by making the several assumptions already described in Chapter 6, namely (i) the linear response approximation (ii) the principle of detailed balance and (iii) transition-state theory, as well as making a multipole expansion (about $x$) of the change in defect energy $\delta g_p$ brought about by the perturbation, together with a Taylor expansion of the $p_p$, to second order about $x$. The result is then a set of parabolic partial differential equations for the perturbed part of $p_p$, i.e. $p_p^{(1)}$, which for isothermal conditions and in matrix notation takes the particular form

$$\frac{\partial \mathbf{p}^{(1)}}{\partial t} = -\mathbf{P}^{(1)}\mathbf{p}^{(0)} - \mathbf{P}^{(0)}\mathbf{p}^{(1)} + \mathbf{V}\frac{\partial \mathbf{p}^{(1)}}{\partial x} + \mathbf{T}\frac{\partial^2 \mathbf{p}^{(1)}}{\partial x^2}. \tag{8.4.3}$$

In this equation $\mathbf{p}^{(1)}$ is the column matrix

$$\mathbf{p}^{(1)} = \{p_1^{(1)} \quad p_2^{(1)} \quad p_3^{(1)} \quad \cdots \quad p_{\mathrm{B}}^{(1)} \quad p_{\mathrm{D}}^{(1)}\} \tag{8.4.4}$$

while $\mathbf{p}^{(0)}$ as before denotes the column matrix of thermal equilibrium quantities $p_p^{(0)}$. Among these we have the detailed balance relations corresponding to reactions (8.4.1a, b), viz.

$$w_{qp}^{(0)}\, p_p^{(0)} = w_{pq}^{(0)}\, p_q^{(0)} \tag{8.4.5a}$$

and

$$w_{dp}^{(0)}\, p_p^{(0)} = w_{pd}^{(0)}\, p_{\mathrm{B}}^{(0)} p_{\mathrm{D}}^{(0)}. \tag{8.4.5b}$$

The matrix elements of $\mathbf{P}^{(0)}$, $\mathbf{V}$ and $\mathbf{T}$ are

$$P_{pq}^{(0)} = -w_{pq}^{(0)}, \qquad p \neq q \tag{8.4.6a}$$

$$P_{qq}^{(0)} = \sum_{q \neq p} w_{pq}^{(0)} + \sum_{d} w_{dq}^{(0)} \tag{8.4.6b}$$

plus further elements in ($p$, B), ($p$, D), (B, D), etc.

$$V_{pq} = -a_{pq} w_{pq}^{(0)}, \qquad p \neq q \tag{8.4.7a}$$

$$V_{qq} = 0 \tag{8.4.7b}$$

plus further elements in ($p$, B), ($p$, D), (B, D), etc.

$$T_{pq} = \tfrac{1}{2} a_{pq}{}^{2} w_{pq}^{(0)}, \qquad p \neq q \tag{8.4.8a}$$

$$T_{qq} = 0 \tag{8.4.8b}$$

plus further elements in ($p$, B), ($p$, D), (B, D), etc. It will be seen that $\mathbf{P}^{(0)}$ is the same as the matrix $\mathbf{P}'$ encountered already towards the end of §8.3. The elements of $\mathbf{P}^{(1)}$ in the inhomogeneous term all derive from the first-order perturbation of the jump frequencies by the impressed fields (cf. 6.5.4) and are each made up of three terms, one in the field potential (or the stress), one in the $x$-derivative of the potential (or stress) and one in the second $x$-derivative of the potential (or stress). It turns out that the corresponding matrices, when multiplied by $\mathbf{p}^{(0)}$ as in the first term on the right side of 8.4.3, are closely related to $\mathbf{P}^{(0)}$, $\mathbf{V}$ and $\mathbf{T}$ respectively, all transition state terms having cancelled out. In effect, the relation – which is the same for all three parts of $\mathbf{P}^{(1)}$ – represents a generalization of the Nernst–Einstein relation and is necessary if we are to obtain flux equations like (4.2.4), (4.4.5) and (4.4.7) in which the phenomenological $L$-coefficients are independent of the nature of the forces acting. The details can be found in Franklin and Lidiard (1983, 1984) and in Lidiard (1984, 1985). (We again note that (8.4.3) is transposed with respect to the corresponding equations in these papers and that the matrix elements in (8.4.3) are thus the transposes of the quantities denoted by the same symbols there.)

Equation (8.4.3) describes the microscopic evolution of the system. In the absence of external or internal forces, i.e. in pure diffusion ($\mathbf{P}^{(1)} = 0$), this evolution is determined by the matrices $\mathbf{P}^{(0)}$, $\mathbf{V}$ and $\mathbf{T}$. On the other hand, if the system is spatially homogeneous but is prepared in some non-equilibrium state then its relaxation to equilibrium is governed by $\mathbf{P}^{(0)}$ alone, the separate modes of relaxation being defined by the eigenvectors of $\mathbf{P}^{(0)}$ (cf. §7.2). Lastly, the influence of spatially uniform force fields is governed by $\mathbf{P}^{(0)}$ and $\mathbf{P}^{(1)}$ alone, but, as we have already noted, $\mathbf{P}^{(1)}$ is composed of terms related to $\mathbf{P}^{(0)}$, $\mathbf{V}$ and $\mathbf{T}$. It remains to relate the microscopic evolution described by (8.4.3) to the macroscopically observable relations between the fluxes and the thermodynamic forces.

Since in the kinetic treatment we have eliminated the host atoms and are explicitly considering the movement of solutes and defects (B, D, etc.), it is convenient to do the same in the macroscopic relations by eliminating the host species from (4.2.4) with the aid of eqns. (4.4.2) and/or their equivalents for interstitials, as carried out in the later part of §4.4.1. We have already shown in Chapter 3 how the chemical potentials, $\mu$, and thus the forces, $X$, are calculated. It remains therefore to obtain equations for the fluxes in terms of the $\mathbf{p}^{(1)}$. In all the explicit uses of the pair-association method which have been made the flux expressions have been obtained by direct enumeration, i.e. by a straightforward extension of the usual elementary approach as used already to get (8.3.3). However, this is not possible in a general formulation. The most direct general method is to appeal to the continuity equation that relates the local density, $n_S$, of the species, S, considered to the corresponding flux, viz.

$$\frac{\partial n_S}{\partial t} + \frac{\partial J_S}{\partial x} = 0. \tag{8.4.9}$$

This approach proves to be especially convenient when taken in conjunction with the device, already used in §8.3, of seeking solutions of (8.4.3) for uniform steady-state conditions. Then in effect the left-hand side of (8.4.3) by appropriate summation gives $\partial n_B/\partial t$ and $\partial n_D/\partial t$ directly. By integrating the resulting expressions with respect to $x$, we thus obtain the corresponding total fluxes $J_B$ and $J_D$ in terms of the steady-state solution, i.e. of $\mathbf{p}^{(1)}$ as given by

$$0 = -\mathbf{P}^{(1)}\mathbf{p}^{(0)} - \mathbf{P}^{(0)}\mathbf{p}^{(1)} + \mathbf{V}\frac{\partial \mathbf{p}^{(1)}}{\partial x}. \tag{8.4.10}$$

The result contains two parts, one deriving from the homogeneous terms in (8.4.3), the other from the inhomogeneous terms:

$$J_S = J_S^{\text{hom}} + J_S^{\text{inhom}}. \tag{8.4.11}$$

The homogeneous part, which relates to pure diffusion in the absence of any external force fields, is

$$-\frac{1}{n}J_S^{\text{hom}} = \tilde{\boldsymbol{\tau}}_S\mathbf{V}\mathbf{p}^{(1)} + \tilde{\boldsymbol{\tau}}_S\mathbf{T}\frac{\partial \mathbf{p}^{(1)}}{\partial x}, \tag{8.4.12}$$

in which $\tilde{\boldsymbol{\tau}}_S$, which performs the operation of summation over all elements corresponding to defect configurations containing an S-species, is the row matrix

$$\tilde{\boldsymbol{\tau}}_B = [1 \quad 1 \quad 1 \quad 1 \ldots \quad 1 \quad 0], \qquad S = B, \tag{8.4.13a}$$

the 0 entry corresponding to the unpaired D species, or

$$\tilde{\tau}_D = [1 \quad 1 \quad 1 \quad 1 \quad \ldots \quad 0 \quad 1], \qquad S = D, \tag{8.4.13b}$$

the 0 entry this time corresponding to the unpaired B atoms. (If we were considering a system containing more than two species any two of which could form pairs, other zeros would enter into $\tilde{\tau}_S$; but the end result will be simply additive as (4.2.4), etc. demand). Equation (8.4.12) shows that $J_S^{hom}$ is composed of two terms, a 'direct' term $\tilde{\tau}_S \mathbf{T}\, \partial \mathbf{p}^{(1)}/\partial x$ and an 'indirect' term $\tilde{\tau}_S \mathbf{V} \mathbf{p}^{(1)}$. The direct term alone would correspond to the uncorrelated diffusion of the species (cf. §8.3.3). The indirect term represents the effects of correlation in the motion of B and D species. The component $J_S^{inhom}$, which is present when there are force fields to be considered, introduces additional direct terms (for details see Franklin and Lidiard, 1984 and Lidiard, 1984).

By these means and by application of the expressions for the chemical potentials derived in §§3.5–3.8 one can obtain general expressions for the $L$-coefficients and for the macroscopic response functions for time-varying electric and elastic fields We shall now briefly describe how these are obtained and some of the results.

## 8.5 Some results of kinetic theory

We begin by specifying the same symmetry conditions as in §8.3. There are three aspects to these: (i) the direction $x$ of diffusion or of force fields is a principal crystal axis (ii) all configurations that are equivalent with respect to a crystal rotation about the $x$-axis are equally occupied and represented by a single term in all equations and (iii) to each distinct configuration, $p$, of the B–D pair in the (+)-set there is an equivalent configuration $\bar{p}$ in (−)-set obtainable from $p$ by one of the symmetry operations specified in §8.3, i.e. reflection in a plane normal to $x$, inversion through a centre on the $x$-axis or such inversion in combination with rotation about the $x$-axis. As before, configurations for which $p = \bar{p}$ must lie in, or symmetrically about, a plane normal to $x$ and are referred to as belonging to the (0)-set. The unpaired species will be in the (0)-set.

### *8.5.1 The transport coefficients, L*

The procedure for extracting the $L$-coefficients from the kinetic theory contains three steps. Firstly, we solve the steady-state equation (8.4.10). Secondly, we insert this solution into the flux equations (8.4.11), (8.4.12), etc. This gives us the fluxes in terms of the gradients of concentration of the *free* solute atoms and defects. Thirdly, these gradients are replaced by corresponding thermodynamic forces (negative gradients of chemical potential, electrochemical potential, etc.) by using the results of the statistical thermodynamics analysis given in Chapter 3, especially §§3.5 *et seq.*

By this procedure one finds equations for the fluxes $J_B$ and $J_D$ in terms of the thermodynamic forces, which are of precisely the form required by non-equilibrium thermodynamics. That is to say, the fluxes of solute atoms and defects are given by equations of the form of (4.4.14), in which the forces are defined by (4.2.3a) or, equivalently, when, as here, $\nabla T = 0$, by (4.2.14a). This enables us to identify the $L$-coefficients given by the kinetic theory and, as expected, they are found to be functions only of the equilibrium thermodynamic variables (concentrations, temperature, pressure or other uniform stresses). It should be noted that in deriving, and in using, these $L$-coefficients one needs (some of) the second virial-coefficients as well as the leading $\ln c_r$ terms in the chemical potentials.

The mechanical part $F_S$ of the force is defined by the nature of the applied field, but in all cases it involves a coupling constant that is characteristic of the *unpaired* species (e.g. unpaired solute atom or unpaired defect). In the case of an electric field $E$, $F_S$ equals the net charge of S in the lattice, $q_S$, times $E$; for a centrifugal field it is the net mass of S in the lattice times the centrifugal acceleration; and for a stress field, $\sigma$, it is the gradient of a tensor product $v\,\partial(\lambda^{(S)}_{\alpha\beta}\sigma_{\alpha\beta})/\partial x$ involving the elastic dipole strain tensor $\lambda^{(S)}_{\alpha\beta}$ of the unpaired defect. That the forces should involve only the characteristics of the unpaired species is perhaps not immediately obvious. However, a return to the discussion of chemical potentials in §§3.6 and 4.4.2 shows that it should be so. For, our assumption of detailed balance implies local thermodynamic equilibrium between paired and unpaired defects in just the way assumed there. It is only necessary to enlarge the discussion given there by considering systems in the appropriate potential or stress field and to change each of the chemical potentials $\mu$ appropriately; this does not affect the structure of the arguments leading to the eqns. (4.4.21)–(4.4.23). The interactions between different species (as shown, for example, by the tendency to form close pairs) all appear in the form of thermodynamic averages in the $L$-coefficients. Again, (4.4.23) shows that this should be so.

Each phenomenological $L$-coefficient is made up of two parts, the first of which involves a (double) sum over all configurations, while the second one can be written as a (quadruple) sum over configurations of the (+)-set alone. With subscripts $R$ and $S$, each of which may be either B or D the essence of the results is contained in the following curtailed expressions.

$$L_{RS} = L'_{RS} + L''_{RS}, \quad \text{(all } R, S\text{)}, \tag{8.5.1}$$

with

$$\left(\frac{kT}{n}\right) L'_{RS} = \frac{1}{2}\sum_{p,q} a_{qp}{}^2 w_{qp}{}^{(0)} p_p{}^{(0)} + \begin{array}{l}\text{dissociation terms and free–free}\\ \text{terms dependent on } (R, S),\end{array} \tag{8.5.2}$$

and

$$\left(\frac{kT}{n}\right)L''_{RS} = 2\sum_{q,u}{}^{+}\left(\sum_{v} a_{vu}w^{(0)}_{vu}\right)(Q^{-1})_{uq}\left(\sum_{p} a_{qp}w^{(0)}_{qp}\,p^{(0)}_{p}\right)$$

$$+ \text{ dissociation terms dependent on } (R, S). \qquad (8.5.3)$$

As elsewhere, $n$ is the number of lattice sites per unit volume and the superscript + on the second summation indicates that only configurations in the (+)-set are to be included. The matrix $\mathbf{Q}$ is obtained by partitioning $\mathbf{P}^{(0)}$ according to the above classification of configurations (+, 0 and −),

$$\mathbf{Q} = \mathbf{P}^{(0)}_{++} - \mathbf{P}^{(0)}_{+-}. \qquad (8.5.4)$$

The matrix introduced previously in §8.3 is an example of (8.5.4); cf. (8.3.26). The form of the result (8.5.3) uses the symmetry of $\mathbf{P}^{(0)}$ which follows from the classification of configurations. The full results which can be used to obtain particular expressions for specific models are given in Appendix 8.2. The first part, $L'_{RS}$, in all cases can be obtained by simple enumeration, but the second part $L''_{RS}$, like the corresponding correlated term in $D$ in §8.3, requires appropriate elements of $\mathbf{Q}^{-1}$. We shall now comment on three aspects of these results.

*Onsager reciprocal relations*

It is a straightforward matter to verify the correctness of the Onsager reciprocal relation (4.2.5) from the complete expressions for $L_{RS}$ (A8.2.1–A8.2.3). This relation is a consequence of the assumption of detailed balance, which in turn is the expression of microscopic reversibility (Kreuzer, 1981). However, we should recall at this point our assumption that the symmetry of the system is such as to allow the definition of principal axes and that we have only considered transport along such directions. In the general case $L_{RS}$ is a second-rank tensor $L_{RS\alpha\beta}$ and detailed balance, or equivalently microscopic reversibility, leads us to the following expression of the Onsager reciprocal theorem

$$L_{RS\alpha\beta} = L_{SR\beta\alpha}, \qquad (8.5.5)$$

(Nye, 1985; Lidiard, 1985). Symmetry in the Cartesian indices $\alpha, \beta$ alone is a consequence of the symmetry of the crystal structure and does not hold in triclinic, monoclinic and certain trigonal, tetragonal and hexagonal classes (Nye, 1985). In such crystals then we cannot derive the simple form (4.2.5), but only (8.5.5).

*Significance of the populations of free solute atoms and defects*

By the discussion of the solute diffusion coefficient given in §§8.3.2 and 8.3.3 we have seen that the inclusion of dissociation transitions leads to a more accurate expression for $D_B$. There is, however, a broader significance to the inclusion of the populations of free solute atoms and defects into the theory which is visible from the expressions (8.5.1)–(8.5.3). Thus suppose we drop the unpaired species from consideration, then in (8.5.2) and (8.5.3) we are left with just those terms explicitly given. But these terms do not depend on the choice of $R$ and $S$. In other words all the $L$-coefficients given by these equations are equal, i.e. for diffusion by vacancies $L_{BB} = L_{VV} = L_{BB} = L_{VB} \equiv L$ or, equivalently by (4.4.10) *et seq.*, $L_{AA} = 4L$, $L_{AB} = L_{BA} = -2L$ and $L_{BB} = L$. The reason for this result is quite plain, for it is equivalent to having $J_V = J_B$; because the B–V pairs never dissociate (or re-form from free solute atoms and vacancies) the macroscopic fluxes of solute atoms and vacancies must be equal, however extended the B–V pairs are allowed to become on a microscopic scale. Hence a correct theory demands the inclusion of exchanges between the populations of the paired and the unpaired species. The representation of the association reactions by a bimolecular reaction term introduces a degree of statistical averaging and curtails the spatial correlation between the motions of a solute and a defect implied when they are treated as a pair, i.e. as a single distinct entity. There is a broad analogy here to the celebrated Stosszahlansatz in Boltzmann gas kinetic theory, although there it is the distribution of momenta which is assumed to be averaged in the general population following a collision beween a pair of atoms. Returning to defect theory, we will see later (§§9.5, 12.5 and 12.6) that the same idea is sometimes referred to as the 'encounter model', especially in connection with the theory of nuclear magnetic relaxation, (although the term 'model' is not really appropriate). Even though the theory might appear still to contain an arbitrary element by its distinction between paired and unpaired species, we shall see in the following section §8.5.2 that this can be removed by a purely mathematical transformation. All that remain are the physical interactions which led us to the idea of pairs in the first place.

*Relation to calculation of solute diffusion coefficient in §8.3*

The equivalence of these results to those obtained in §8.3 for solute diffusion via vacancies can be established by obtaining $L_{BB}$ from (A8.2.1)–(A8.2.3) and choosing a reference system for the solute–vacancy pairs which is based on the position of the solute atom. This then specifies the displacements $a_{pq}$ as defined at the beginning of §8.4.1. We then easily see that the $a_{pq}$ entering into (A8.2.2) and (A8.2.3) are zero except for those corresponding to solute–vacancy exchange. If for definiteness we assume a cubic crystal with $x$ along $\langle 100 \rangle$ then; (i) the only non-zero $a_{qp}$ are $\pm a$; (ii) all $w_{qp}^{(0)}$, etc. in the non-zero terms in (A8.2.2) and (A8.2.3) are $\equiv w_s$; (iii) in (A8.2.2) $p$ is either the forwards or backwards nearest-neighbour configuration; (iv) in (A8.2.3) the restriction on the summation means that all the displacements are $+a$ and that $p$ can be the forwards nearest-neighbour configuration, 1, only. Lastly, all other terms are zero for the solute coefficient $L_{BB}$. The result is then

$$\frac{kT}{n} L_{BB} = a^2 w_s p_1^{(0)} (1 - 2w_s (Q^{-1})_{11}). \tag{8.5.6}$$

This is evidently equivalent to (8.3.31) and (8.3.32) when the relation (5.2.8) between $D_B$ and $L_{BB}$ is inserted.

*Example of solute–vacancy pairs in a f.c.c. lattice*

We already considered this example towards the end of §8.3 when we presented $D_B$ in the approximation where only nearest-neighbour pairs were considered. By using the results (A8.2.1)–(A8.2.3) it is easy to establish all three $L$-coefficients in the same approximation. The detailed arguments are set out in the second part of Appendix 8.2. Expressed in terms of the site (mole) fractions of (a) solute–vacancy pairs, $c_p$ ($p_1^{(0)} = c_p/3$ in this lattice) and (b) all vacancies (paired and unpaired), $c_V$, the expressions are

$$L_{BB} = \frac{na^2 c_p}{3kT} \frac{w_2(w_1 + 7w_3/2)}{(w_1 + w_2 + 7w_3/2)}, \tag{8.5.7a}$$

$$L_{BV} = L_{VB} = \frac{na^2 c_p}{3kT} \frac{w_2(w_1 - 13w_3/2)}{(w_1 + w_2 + 7w_3/2)} \tag{8.5.7b}$$

and

$$L_{VV} = \frac{na^2 c_p}{3kT} \frac{[w_2(w_1 + 47w_3/2) + 40w_3(w_1 + w_3)]}{(w_1 + w_2 + 7w_3/2)} - \frac{na^2 c_p}{3kT}\left(\frac{7w_3 w_0}{w_4}\right) + \frac{4na^2 w_0}{kT}(c_V - c_p). \tag{8.5.7c}$$

The vacancy jump frequencies are as defined in Fig. 7.1, while $a$ is the separation between (100) planes, i.e. nearest-neighbour jump distances are $\sqrt{2}a$. These are the coefficients for use with the flux equations (4.4.14). If instead we wanted to use (4.4.9) then the coefficients $L_{AA}$, $L_{AB}$ and $L_{BB}$ can be obtained easily from (8.5.7) by using (4.4.13). Although these expressions (8.5.7) are only approximate and the exact results for this model are known (Chapter 11), in practice uncertainties about the jump frequencies, and indeed about the physical correctness of the model in detail, mean that eqns. (8.5.7) are probably adequate in most circumstances. In the limit when the solute is just an isotope of the host (i.e. when all five jump frequencies are the same) (8.5.7b) and (8.5.7c) are exact (cf. (5.5.5)), while, as previously noted, (8.5.7a) is in error by only a few percent as it yields 9/11 instead of 0.78 for the tracer correlation factor.

Approximate results like (8.5.7) can also be obtained relatively easily from (8.5.1)–(8.5.3) for other similar models, including those with interstitial defects, and for other lattices. For example results analogous to (8.5.7) have been obtained (Allnatt *et al.*, 1983) for dumb-bell interstitials in the f.c.c. lattice (Fig. 2.23). These are presented and considered along with their application in Chapter 11.

### *8.5.2 Relation to general expressions for L-coefficients given in §6.5*

Let us return to the general expressions (8.5.1)–(8.5.3) for the $L$-coefficients. We observe that the division of $L_{RS}$ into $L'_{RS}$ and $L''_{RS}$ in (8.5.1) is, in fact, a division into (i) a summation over independent transitions $L'_{RS}$ and (ii) a summation over pairs of transitions coupled together through the $\mathbf{Q}^{-1}$ factor ($L''_{RS}$). We may refer to these as the uncorrelated and the correlated parts of $L_{RS}$ respectively. Furthermore, there is a suggestive similarity to the general formulae for $L$-coefficients obtained in Chapter 6 ((6.5.10)–(6.5.12)). (N.B. In making this comparison it will be helpful to recognize that the symmetry considerations set out in §§8.3 and 8.5, which by (8.5.4) take us from the singular matrix $\mathbf{P}$ to the non-singular matrix $\mathbf{Q}$, effectively allow us to evaluate the time integral of $\mathbf{G}(t)$ in (6.5.12) and to replace it by $\mathbf{Q}^{-1}$.) However, the correspondence between the two sets of results is not so straightforward as it first appears. Firstly, the division of the kinetic theory expressions into correlated and uncorrelated terms is not unique on account of the arbitrariness of the $a_{pq}$ in this formulation. Secondly, the separation of defects and their configurations into paired and free means that the full expressions corresponding to (8.5.2) and (8.5.3) are somewhat inelegant and increasingly troublesome to evaluate as more and more paired configurations are included. In other words (8.5.2) and (8.5.3) are not well suited as they stand to the calculation of precise results, although, as we have seen, they do provide a simple and convenient way to obtain approximate results.

As a result the method has only been used for approximate calculations in which at most a few near-neighbour configurations have been included, all more distant configurations being treated as dissociated. Although such approximate calculations are undoubtedly adequate for most practical purposes, one should recognize that this is not an intrinsic limitation of the method. We have already noted that the convention employed for the definition of the $a_{pq}$ is arbitrary and therefore that the expressions for the $L$-coefficients must be invariant with respect to changes from one convention to another. In fact, the invariance is even more extensive. Surprisingly, we can not only change the $a_{qp}$ according to

$$a_{qp} \rightarrow a_{qp} - \zeta_p + \zeta_q,$$

where the $\zeta_p$ are any set of parameters, and leave the $L$-coefficients invariant, but, taking $L_{RS}$ as a whole, we can also separately transform the two $a_{qp}$ factors in (8.5.2) and the $a_{qp}$ and the $a_{vu}$ factors in (8.5.3) as

$$a_{qp} \rightarrow a_{qp} - \xi_p + \xi_q \equiv a^R_{qp} \tag{8.5.8}$$

and

$$a_{qp} \rightarrow a_{qp} - \eta_p + \eta_q \equiv a^S_{qp} \tag{8.5.9}$$

and likewise for the displacements in the dissociation transitions and still leave $L_{RS}$ invariant. This invariance comes about by the transfer of terms between the two parts of (8.5.1): it is in this sense that there is no absolute division of $L_{RS}$ into correlated and uncorrelated terms. This result is conveniently written as

$$\frac{kT}{n} L_{RS} = \frac{1}{2} \sum_{p,q} a^R_{qp} a^S_{qp} w^{(0)}_{qp} p^{(0)}_p + \sum_{d,p} a^R_{dp} a^S_{dp} w^{(0)}_{dp} p^{(0)}_p$$

$$+ \tfrac{1}{2} \delta_{RS} p^{(0)}_R (1 - z_{Rf} p^{(0)}_R) \sum_r (a^R_{r0})^2 w^{R(0)}_{r0}$$

$$-2 \sum_{p,q}^{(+)} v^R_q (Q^{-1})_{qp} v^S_p p^{(0)}_p, \tag{8.5.10a}$$

in which

$$v^S_p = \sum_u a^S_{up} w^{(0)}_{up} + \sum_p a^S_{dp} w^{(0)}_{dp}. \tag{8.5.10b}$$

Apart from the changes to the displacements defined by (8.5.8) and (8.5.9) the notation is as explained in Appendix 8.2. The first term on the right side is the uncorrelated contribution of transitions from one paired configuration to another, the second term gives the sum of the two corresponding contributions from dissociation and association transitions (equal by detailed balance, hence no factor 1/2 as in the first term), while the third term is the uncorrelated contribution from the free–free transitions (corrected by the term in $z_{Rf}$ for the fact that some movements of an unassociated pair may result in the formation of an associated pair, the contributions of which have already been accounted for in the second term). This free–free term is unchanged by the transformations (8.5.8) and (8.5.9). Correlated movements are represented in the fourth term. (We note that the summation is over only those pair configurations in the (+)-set, although the summations contained in the velocity analogue $v_p^S$ (8.5.10b) are over all configurations which can be reached in one transition from a configuration $p$ lying in the (+)-set.)

The convenience of (8.5.8)–(8.5.10) is now apparent, for we can choose the $\xi_p$ so that the $a_{qp}^R$ refer directly to the displacements of the $R$ species and the $\eta_p$ likewise for the $S$ species. The apparent difficulty over the convergence of the sums is thereby removed (essentially by cancellation between the correlated and uncorrelated parts). Furthermore, we can also see from (8.5.10) that, in any model in which the effect of the solute atom upon the jump frequency of the vacancy is localized to just a few surrounding sites, the vectors $\mathbf{v}^R$ and $\mathbf{v}^S$ pre- and post-multiplying $\mathbf{\Omega}^{-1}$ will be similarly limited. This follows because, for the more distant unperturbed sites, (a) all the jump frequencies and occupation probabilities are equal and (b) $\sum_v a_{vu} = 0 = \sum_p a_{qp}$. Hence, the only non-zero elements are those close in. Likewise, only these same elements of $\mathbf{\Omega}^{-1}$ are needed. Since we can take the division between paired and unpaired defects outside this region, it follows that (8.5.10) will simply appear as a sum of contributions from the separate populations of paired and unpaired defects, details of the exchange between the two populations having been absorbed into the summations. The use of (8.5.10) is illustrated for the f.c.c. five-frequency model in Appendix 8.2.

Lastly, we return to the point of the relation to the general many-body expressions for the $L$-coefficients given in §6.5. By re-expressing these in terms of the same displacements and jump frequencies as defined for (8.5.10) we can easily establish the equivalence of (8.5.10) to (6.5.10)–(6.5.12).

### *8.5.3 Dielectric and anelastic relaxation*

Since the kinetic description of transport as presented in this section grew out of the theory of relaxation in uniform fields (Chapter 7) it is a trivial matter to recover this theory when the association–dissociation reactions are omitted (i.e. all $w_{dp} = 0$). However, as the consideration of these reactions proved to be necessary in transport theory it is desirable to examine the consequences of including them in the relaxation theory. This matter has been examined within the context of kinetic theory by Nowick (1970a, b), Franklin and Lidiard (1984) and Lidiard (1985, 1986b).

Now in dielectric relaxation the complex defects responsible will mostly be composed of separate charged defects (e.g. a solute ion with an ion vacancy or with an interstitial ion); in this situation one may expect an influence of the association–dissociation reaction upon both the d.c. ionic conductivity and the a.c. dielectric relaxation (§7.2). One finds, in fact, that the total electrical response function, or equivalently the total complex conductivity, is still composed of two terms, one corresponding to the d.c. ionic conductivity, $\sigma_0$, and the other to a set of Debye relaxation modes (Franklin and Lidiard, 1984). More formally, the response function $\phi(\tau)$, which gives the total electric current $J$ when the time-varying electric field $E$ is known,

$$J(t) = \int_0^\infty E(t-\tau)\phi(\tau)\,\mathrm{d}\tau, \tag{8.5.11}$$

is given by

$$\phi(\tau) = \sigma_0\,\delta(\tau) + 2(n/kT)\,\delta(\tilde{\boldsymbol{\mu}}'\mathbf{N}\boldsymbol{\Phi}(\tau)\boldsymbol{\mu}'), \tag{8.5.12}$$

in which

$$\boldsymbol{\Phi}(\tau) = \mathbf{Q}\,\delta(\tau) - \mathbf{Q}^2\exp(-\mathbf{Q}\tau) \tag{8.5.13}$$

and the row matrix $\boldsymbol{\mu}'$ and the diagonal matrix $\mathbf{N}$ have elements for the (+)-set of configurations only. The matrix $\mathbf{Q}$, given by (8.3.26), is similarly restricted. (N.B. the internal field factor, $\delta$, introduced in §7.2 appears before the relaxing term only; it cannot enter into $\sigma_0$.) Hence the division generally assumed in the analysis of experimental measurements of electrical response is maintained. The ionic conductivity is just as expected from (5.8.12) and the expressions for the $L$-coefficients already obtained. The remainder is in the form expected from (7.2.19) and (7.2.20) but with a change in the dipole-moment row matrix from the actual $\boldsymbol{\mu}$, as in (7.2.20), to an effective quantity $\boldsymbol{\mu}'$. Details of the differences between $\boldsymbol{\mu}$ and $\boldsymbol{\mu}'$ are discussed by Franklin and Lidiard (1984) and in the context of generalized linear response theory (Chapter 6) by Okamura and Allnatt (1985). If we evaluate the corresponding dielectric loss in the Lidiard–Le Claire approximation we find that it could be substantially smaller than would be predicted from the actual dipole moment $\mu$; but this approximation, while useful for the transport coefficients $L$, may apparently be misleading in these circumstances (Okamura and Allnatt, 1985).

Since we found that the static $L$-coefficients, and thus the d.c. conductivity, were invariant with respect to the changes in the displacements ((8.5.8) and (8.5.9)) it is natural to examine the remainder of the electrical response function (8.5.12) for a similar invariance. We do indeed find such a general invariance. As with the correlated term in $L_{RS}$, (i.e. 8.5.3) we find that we can transform one of the $\boldsymbol{\mu}'$ factors in (8.5.12) according to (8.5.8) and the other according to (8.5.9), and still leave the response function unaltered in value. By appropriate choice of the $\xi_p$ and $\eta_p$ this again allows us to establish the equivalence with the equations of linear response theory (Okamura and Allnatt, 1985).

The above remarks have concentrated upon the response to electrical forces. There is rather less to be said in relation to anelastic relaxation (Lidiard, 1986b). In particular, there is no effect on the elastic dipole strain tensor analogous to the change from $\boldsymbol{\mu}$ to $\boldsymbol{\mu}'$; hence (7.2.21) and (7.2.22) remain valid as they stand, although the inverse relaxation times, $1/\tau^{(\nu)}$ will be changed by the inclusion of the association–dissociation terms in $\mathbf{P}^{(0)}$. The reason for the difference between the two cases is that a complex defect having an electric dipole moment will dissociate into elementary defects each bearing a net electrical charge, which therefore also contribute to the electrical current; while there is no corresponding defect current in a *uniform* stress field.

Although the presence of association and dissociation terms in $\mathbf{P}^{(0)}$ will change all the eigenvalues, the only eigenvectors to be affected are those for modes which are invariant under the operations of the crystal point group, i.e. which have the full point group symmetry (Nowick, 1970a). The modes that couple strongly to uniaxial stresses are normally of lower symmetry, so that the corresponding $\mathbf{a}^{(\nu)}$ will be unaffected.

### *8.5.4 Self-correlation function*

In §8.2 we saw that the same microscopic kinetic equation from which we obtained the macroscopic transport coefficients of interstitial solute atoms is also satisfied by the self-correlation function. It was therefore possible to use it to obtain the incoherent scattering function for quasi-elastic neutron scattering and the line shape for Mössbauer $\gamma$-ray emission. In principle much the same can be said here. The basic microscopic equations (8.4.2) *et seq.* describe a dilute system in which the motions of the solute atoms (B) are essentially independent of one another. They can therefore be used to describe the motion of a single B atom if we choose the initial conditions appropriately. In particular, if we let the position $x$ of a solute–defect pair be fixed by that of the solute then the matrix product $\tilde{\boldsymbol{\tau}}_{\mathbf{B}}\mathbf{p}^{(1)}$ (cf. (8.4.3) and (8.4.13a)) will give us the required probability that a B atom is at position $x$ at time $t$ if we solve the kinetic equations in the absence of external forces with the initial condition that the atom was at $x = 0$ at $t = 0$. The initial vacancy distribution about the B atom will be that for thermal equilibrium. Fourier transformation of the kinetic equations will then allow us to obtain the intermediate scattering function $I_s(\mathbf{q}, t)$, the incoherent scattering function $S_s(\mathbf{q}, \omega)$ and the Mössbauer line shape. Use of (8.4.3) gives the macroscopic small $\mathbf{q}$ limit and recovers the result (1.8.6) and its equivalents. The formalism has not been used for general $\mathbf{q}$ (which requires 8.4.2) but alternative analyses using the same physical ideas have been given by Gunther (1976, 1986), by Gunther and Gralla (1983) and by Dattagupta and Schroeder (1987).

## 8.6 Diffusion in highly defective systems

In the discussion of the diffusion of interstitial H atoms in metals which we gave in §8.2, we mentioned that in some circumstances (e.g. H in Pd) the concentration of solute atoms can become so high that they may hinder one another in their motion through the interstitial sites. Analogous situations arise with a number of fast ion conductors, e.g. the so-called $\beta$-aluminas containing high concentrations of alkali or similar ions (Chandra, 1981; Laskar and Chandra, 1989). This has stimulated the theory of transport in 'lattice-gas' models, i.e. systems in which a certain number of atoms are assumed to be able to jump among the (larger number of) sites of a specified lattice, subject only to the restriction that each site can be empty or occupied by just one atom. Given that a neighbouring site is vacant the jump-frequency of an atom is assumed to depend only on its nature and to be independent of the way other neighbouring sites are occupied. Such models have been studied by a variety of techniques, some of which have been borrowed from other branches of many-body theory. We shall study them in a systematic way in Chapter 13, but by way of introduction it is appropriate here to review briefly the approach via kinetic theory. As in §§8.3–8.5 attention has been focussed on dilute solutions of B in A, but with the difference that the vacancy concentration is no longer assumed to be small. The object of the calculation is the diffusion coefficient $D_B$. We shall give attention to those theories which first obtain the self-correlation function for a B atom, or rather its transform $S_s(\mathbf{q}, \omega)$, and then derive $D_B$ by use of (1.8.9).

Let the frequency with which the B atom jumps into a vacant neighbouring site by $w_B$ and let the corresponding jump-frequency of the host atoms be $w_A$. Let us also introduce discrete site occupation variables for each site $l$, $\rho_l^B$ for the solute atom and $\rho_l^A$ for the host atoms, each of which may take the value 0, if the site is not occupied by an atom of that type, or 1, if it is. The configurations $\alpha, \beta, \ldots$ of Chapter 6 are thus specified by the sets $\{\rho_l^A, \rho_l^B\}$ for all sites $l$. If we suppose that the solute atom occupied the origin site o at time zero, i.e. that $\rho_l^B(0) = \delta_{lo}$ then the desired self-correlation function is the ensemble average $\langle\rho_l^B\rangle$. By straightforward physical arguments or by appeal to the stochastic master equation (6.2.1) we can write down an equation for the time rate of change of this quantity, thus

$$\frac{\mathrm{d}}{\mathrm{d}t}\langle\rho_l^B\rangle = \sum_m w_B[\langle\rho_m^B(1-\rho_l^A-\rho_l^B)\rangle - \langle\rho_l^B(1-\rho_m^A-\rho_m^B)\rangle]$$

$$= \sum_m w_B[\langle\rho_m^B(1-\rho_l^A)\rangle - \langle\rho_l^B(1-\rho_m^A)\rangle]. \tag{8.6.1a}$$

This equation differs in form from (8.2.2) by the factors $(1-\rho_l^A-\rho_l^B)$ and $(1-\rho_m^A-\rho_m^B)$ in the first line of this equation which express the restriction that an atom can only jump to a new site ($l$ or $m$ respectively) if that site is vacant. Clearly if either $\rho^A$ or $\rho^B$ is unity the site is occupied and no jump to that site is possible. In §8.2 it was assumed that the concentration of atoms was so low that such hindrance to movement was negligible. There is a corresponding equation also for $\langle\rho_l^A\rangle$, viz.

$$\frac{d}{dt}\langle\rho_l^A\rangle = \sum_m w_A[\langle\rho_m^A(1-\rho_l^B)\rangle - \langle\rho_l^A(1-\rho_m^B)\rangle]. \tag{8.6.1b}$$

We see that the right side introduces a correlation function of higher order. Likewise, further appeal to the master equation or to appropriate physical arguments gives an equation for $d\langle\rho_l^A\rho_m^B\rangle/dt$, etc. which introduces third-order correlation functions $\langle\rho_n^B\rho_l^A\rho_m^A\rangle$, $\langle\rho_n^A\rho_l^A\rho_m^B\rangle$, etc. And so on. This is a familiar situation in many-body theory and requires some 'decoupling' procedure to truncate the otherwise infinite hierarchy of equations.

Before considering ways of doing this we make a small digression by looking at the case where $w_A = w_B \equiv w$, e.g. where B is an isotope of A and where only nearest neighbour jumps $\mathbf{l} \to \mathbf{l} + \mathbf{s}_j$ $(j = 1 \ldots z)$ occur. By addition of (8.6.1a) and (8.6.1b) we see that

$$\frac{d}{dt}(\langle\rho_l^A + \rho_l^B\rangle) = w\sum_{j=1}^{z}[\langle\rho_{l+j}{}^A + \rho_{l+j}{}^B\rangle - \langle\rho_l^A + \rho_l^B\rangle]. \tag{8.6.2a}$$

Equivalently, if we introduce the vacancy 'occupation' variable $\rho_l^V = 1 - \rho_l^A - \rho_l^B$ (all $l$) we have

$$\frac{d}{dt}\langle\rho_l^V\rangle = w\sum_{j=1}^{z}[\langle\rho_{l+j}{}^V\rangle - \langle\rho_l{}^V\rangle]. \tag{8.6.2b}$$

But these equations are formally identical to (8.2.2). In particular, (8.6.2b) shows that the vacancies diffuse as though they were independent particles. Equivalently (8.6.2a) shows that as long as we are not interested in the *mixing* of the two isotopes, the atoms also diffuse as though they were independent particles, irrespective of their concentration. We also note that this conclusion is independent of the nature of the lattice as long as all sites are energetically equivalent and we have just a single jump frequency, $w$. In particular, it applies to such a lattice in which a proportion of sites are 'blocked' through being occupied by immobile atoms of another kind. These results provide useful tests of the correctness and accuracy of numerical methods of simulating diffusion (e.g. Monte Carlo methods). In fact, they appear to have been first inferred from the results of such simulations (Bowker and King, 1978; Murch, 1980) and later demonstrated analytically (Kutner, 1981). We emphasize that they apply only to the overall transport of the atoms and not to the mixing of isotopes, i.e. they apply to chemical diffusion and not to tracer diffusion. That problem demands separate solutions for $\langle\rho_l^B\rangle$ and

$\langle \rho_l^A \rangle$ and to obtain these we need a suitable (approximate) decoupling procedure, just as in the general case, $w_A \neq w_B$.

Let us therefore now return to the treatment of the model on the basis of eqns. (8.6.1). The first step towards obtaining a satisfactory approximation, is to introduce a variable $\Delta\rho_l^A \equiv \rho_l^A - \langle \rho_l^A \rangle$ which describes the fluctuations in occupancy of site $l$ by host atoms about its mean value $\langle \rho_l^A \rangle$, and to re-express the various correlation functions in terms of $\Delta\rho_l^A$ rather than $\rho_l^A$. The lowest order of approximation is the mean-field approximation in which we simply neglect fluctuations in $\rho_l^A$ about $\langle \rho_l^A \rangle$ and drop all $\langle \rho_m^B \Delta\rho_l^A \rangle$. In the mean-field approximation the vacancies are effectively smeared out uniformly and the motion of the solute will be a random-walk without correlation, as described in §8.2.

An advance on the mean-field approximation is clearly necessary if correlation effects are to be described. Such a step forward was taken by Tahir-Kheli and Elliott (1983) who retained all second-order functions but dropped all third-order functions $\langle \rho_l^B \Delta\rho_m^A \Delta\rho_n^A \rangle$ in which $l$, $m$, $n$ are three different sites. The resulting equations were then solved by Fourier transformation to give $S_s(\mathbf{q}, \omega)$ and, by (1.8.9), the solute diffusion coefficient, $D_B$, was obtained.

This approximation is quite successful. Thus in the limit of low vacancy concentration the correlation factor, $f_B$, is given exactly (for the cubic lattices considered). At arbitrary vacancy fraction $c_V$ it is predicted to be

$$f_B = \left(1 + \frac{(1-f_0)}{f_0}\frac{w_B(1-c_V)}{(w_A + w_B c_V)}\right)^{-1}, \tag{8.6.3}$$

in which $f_0$ is the correlation factor for isotope diffusion in the limit $c_V \to 0$ (Table 10.1). A comparison of $f_B$ as predicted by (8.6.3) with the Monte Carlo estimates of Kehr *et al.* (1981) is shown in Fig. 8.6 for the case $w_A = w_B$. The agreement is remarkably good. However, the predictions are not accurate as we go towards the limit $w_A \to 0$, i.e. where the host atoms become immobile. In this case we know that there is a lower limit to the vacancy fraction $c_V$ below which the solute atom cannot diffuse (percolation limit). No such limit is predicted by (8.6.3).

A way round this difficulty was conjectured by Tahir-Kheli (1983) as a result of an examination of the structure of the equations giving the transform of $\langle \rho_l^B \Delta\rho_m^A \rangle$. It was suggested that the quantity $(w_A + w_B c_V)$ appearing in (8.6.3) should be replaced by $(w_A + w_B f_B c_V)$; i.e.

$$f_B = \left(1 + \frac{(1-f_0)}{f_0}\frac{w_B(1-c_V)}{(w_A + w_B f_B c_V)}\right)^{-1}. \tag{8.6.4}$$

If we set $w_A = 0$ (static background) we see that (8.6.4) predicts a percolation

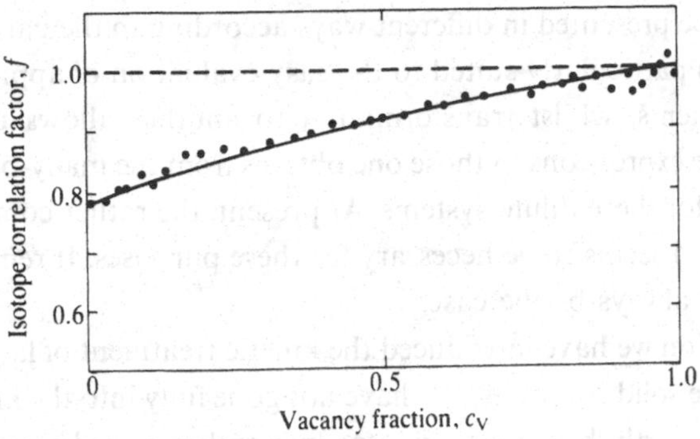


Fig. 8.6. The correlation factor $f$ for self-diffusion in a pure substance having the f.c.c. structure as a function of the vacancy fraction $c_v$. The full curve is as predicted by (8.6.3) for $w_A = w_B$ while the circles represent the Monte Carlo estimates of Kehr *et al.* (1981).

limit

$$c_{V,\,\mathrm{perc}} = (1 - f_0), \qquad (8.6.5)$$

which is close to known percolation limits obtained by the Monte Carlo method (Stauffer, 1979) and in other ways. A way of rationalizing (8.6.5) in terms of an approximation to the third-order correlations $\langle \rho_l^B \, \Delta\rho_m^A \, \Delta\rho_n^A \rangle$ was obtained by Holdsworth and Elliott (1986). In Chapter 13 we shall see how these ideas can be embodied in a more powerful theory which enables us to calculate all the $L$-coefficients for this and other more complex models.

## 8.7 Summary

In this chapter we have described the kinetic theory of atomic transport in mainly dilute interstitial and substitutional alloys and solid solutions. The limitation to dilute systems means that the theory is essentially a theory of the average motion of a single atom (as e.g. in interstitial solid solutions like that of C in α-Fe) or of a single solute atom in interaction with a vacancy or interstitial defect. In this second, more complex case it is necessary to represent the statistical interplay of the solute and defect movements, even when the physical interaction between the two is of very short range or even non-existent (as when the solute is just an isotope of the host element). This has been done here by describing that variant of kinetic theory known as the pair-association method. Although in the literature this approach has been used in rather specific forms we have been able to make the analysis generally and to derive general expressions for the transport coefficients $L$ which can then be evaluated for any specific model. Furthermore, these expressions can be presented in different ways according to the particular purpose. Thus one form is particularly suited to the easy evaluation of approximate results for the $L$-coefficients, whilst transformation to another allows us to show the equivalence of the expressions to those one obtains from the many-body expressions (6.5.10)–(6.5.12) for these dilute systems. At present, the rather complex apparatus we have assembled seems to be necessary for these purposes. It remains to be seen whether this will always be the case.

In the last section we have introduced the kinetic treatment of lattice-gas models of highly defective solid solutions. We have not gone fully into the kinetic approach to these systems, partly because they introduce rather complex mathematics, but mainly because they can be handled more completely by the methods described in Chapter 13. We have, however, demonstrated one exact result, namely that the chemical diffusion coefficient of a system containing just one mobile component is independent of the vacancy concentration. Such a result can provide a useful test of methods devised for more general systems.

In the next two chapters we turn to another branch of the theory of dilute systems, namely random-walk theory. We shall again emphasize the general structure and the way it relates to the many-body treatment of Chapter 6. In Chapter 10 we shall also see how the random-walk theory for dilute systems has been extended to give a theory of diffusion in model concentrated alloys.

# Appendix 8.1

## Symmetry of the pair distribution

Let us denote by $\bar{p}$ the configuration of the impurity–vacancy pair which is obtained by reflecting configuration $p$ in the plane $x$ through the impurity atom (Fig. 8.3). We consider that the system is subject to a vectorial perturbation along the $x$-axis (e.g. electric field, $E$; concentration gradient, $\nabla c$; thermal gradient $\nabla T$).

Then suppose that, in place of the $x$-co-ordinate system, we change to a new system $x' = -x$. Then as the physical properties of the system must be independent of this choice we must have

$$p_p^{(1)}(x; E, \nabla c, \nabla T) = p_{\bar{p}}^{(1)}(-x; -E, -\nabla c, -\nabla T), \qquad \text{(A8.1.1)}$$

for, in the reversed system $\bar{p}$ takes the place of those configurations labelled $p$ with $x$, $+E$ etc.

But now we use the fact that we are working in the linear response approximation so that the dependence of $p_p^{(1)}$ on $E$, $\nabla c$, $\nabla T$ is linear and homogeneous. We therefore have

$$p_{\bar{p}}^{(1)}(-x; -E, -\nabla c, -\nabla T) = -p_{\bar{p}}^{(1)}(-x; E, \nabla c, \nabla T). \qquad \text{(A8.1.2)}$$

If we combine (A8.1.1) and (A8.1.2) we therefore obtain

$$p_p^{(1)}(x; E, \nabla c, \nabla T) = -p_{\bar{p}}^{(1)}(-x; E, \nabla c, \nabla T)$$

or, for short,

$$p_p^{(1)}(x) = -p_{\bar{p}}^{(1)}(-x), \qquad \text{(A8.1.3)}$$

it being understood that the perturbations are the same on both sides of (A8.1.3).

*Example 1. Electric field only*

For a uniform electric field there is no dependence of $p_p^{(1)}$ on $x$ and (A8.1.3) becomes

$$p_p^{(1)} = -p_{\bar{p}}^{(1)}. \tag{A8.1.4}$$

This is an example of the $T_{1u}$ solutions obtained in §7.3. In this case then

$$\sum_p p_p^{(1)} = 0,$$

i.e. the total number of pairs is unaffected by the presence of the electric field.

*Example 2. Diffusion (including thermal diffusion or $\nabla T \neq 0$)*

In this case (A8.1.3) holds as it stands. But the solution (8.3.8) with

$$\gamma_p = \partial p_p^{(1)}/\partial x \tag{A8.1.5}$$

then gives

$$p_p^{(1)}(x) = \phi_p + \gamma_p x, \tag{A8.1.6}$$

$$p_{\bar{p}}^{(1)}(-x) = -\phi_{\bar{p}} - \gamma_{\bar{p}} x. \tag{A8.1.7}$$

But because

$$p_p^{(0)} = p_{\bar{p}}^{(0)} \tag{A8.1.8}$$

and thus that

$$\gamma_p = \gamma_{\bar{p}},$$

it follows from (A8.1.3), (A8.1.6) and (A8.1.7) that

$$\phi_p = -\phi_{\bar{p}}. \tag{A8.1.9}$$

This is the result used in §8.3. We notice that

$$\sum_p p_p^{(1)} = \left(\sum_p \gamma_p\right) x \neq 0,$$

i.e. that the total number of pairs *is* affected by the perturbation in this case – as is physically obvious it must be.

# Appendix 8.2

## Expressions for the L-coefficients obtained by the pair-association method

In this appendix we give the complete expressions for the coefficients $L_{RS}$ as derived from Franklin and Lidiard (1983) obtained by the pair-association method. The original paper should be referred to for details of the derivation, but when doing so it should be remembered that the notation used here differs in detail from that used originally, as explained in the text. As shown by (8.5.1) there are two parts to $L_{RS}$. For the first part $L'_{RS}$ we have

$$\left(\frac{kT}{n}\right)L_{RS}' = \frac{1}{2}\sum_{p,q} a_{qp}{}^2 w_{qp}^{(0)} p_p^{(0)} + \sum_{d,p} a_{dp}^R a_{dp}^S w_{dp}^{(0)} p_p^{(0)}$$

$$+ \frac{\delta_{RS}}{2} p_{\bar{R}}^{(0)}(1 - z_{Rf} p_{\bar{R}}^{(0)}) \sum_r (a_{r0}^R)^2 w_{r0}^{R(0)}. \qquad \text{(A8.2.1)}$$

Here the first summation is over all pair-configurations $p$ and $q$, while the second summation is over all pair-configurations $p$ and all dissociated configurations $d$ which may be reached from them. The last term, which is only present when $R = S$, is the contribution of the unpaired $R$ species making free–free jumps from site o to $r$: $\bar{R}$ signifies 'not $R$' (i.e. V if $R =$ B and vice versa) while $z_{Rf}$ is the number of sites available to $\bar{R}$ that are within range for association (pairing) with $R$ on a site $r$ but not within range from the (neighbouring) origin site o.

For the second part of $L_{RS}$ we have

$$\left(\frac{kT}{n}\right)L''_{RS} = -2\sum_{p,q}^{(+)} v_q^R (Q^{-1})_{qp} v_p^S p_p^{(0)}, \qquad \text{(A8.2.2)}$$

in which

$$v_p^S = \sum_p a_{up} w_{up}^{(0)} + \sum_d a_{dp}^S w_{dp}^{(0)}. \qquad \text{(A8.2.3)}$$

The summation in (A8.2.2) is over all pair configurations $p, q$ lying in the $(+)$-set. The first summation in (A8.2.3) is over all pair configurations $u$ which can be reached from $p$ without restriction, while the second is over all dissociated configurations $d$ which can be reached from $p$. There are no contributions to $L''_{RS}$ from free–free transitions. If will also be noted that, although dissociation transitions enter into both $L'_{RS}$ and $L''_{RS}$, there might appear to be no contributions from the converse association transitions: in fact, there are such contributions, but the principle of detailed balance, as expressed in eqns. (8.4.5), has been used to re-express all of them in terms of the dissociation transitions so that only these have to be enumerated. For this reason there is no factor $1/2$ in the second term.

### Example. The five-frequency model ($f.c.c.$) in the 1-shell approximation

For any model having a localized defect-solute interaction the evaluation of $L_{RS}$ via (A8.2.1)–(A8.2.3) is straightforward in principle, the main task, as in all similar calculations, being the evaluation of the inverse matrix $\mathbf{Q}^{-1}$. Beyond that, one only needs orderly book-keeping to write down $\mathbf{Q}$ and to enumerate the quantities $L'_{RS}$, $v^R$ and $v^S$. We shall illustrate the procedure for the five-frequency model specified by Fig. 2.22 (and again for the NaCl lattice by Fig. 7.1). We do so in the 1-shell approximation in which a solute atom and a vacancy are taken to be unassociated, or free. This is the same approximation as was considered in §8.3. We choose the $x$-direction to be the $\langle 100 \rangle$-direction shown in Fig. 8.5. We shall spell out the argument in words, although in practice it can be convenient to draw up tables of distinct configurations, transitions, displacements, etc. and to add up contributions systematically.

Let us first take $R = B$ and $S = B$ and turn to $L_{BB}'$ (A8.2.1). The first term on the right side is simply $p_1^{(0)}a^2w_2$, because the only transition which gives a non-zero $a_{qp}$ by the definition (8.4.1a) is the exchange of the solute atom and the vacancy when the pair is initially in either a 1-orientation (gives $+a$) or a $\bar{1}$-orientation (gives $-a$), and, of course, $p_{\bar{1}}^{(0)} = p_1^{(0)}$. The second term on the right side is zero because all dissociation jumps leave the solute atom where it was, i.e. $a_{dp}^B = 0$ by the definition (8.4.1b). The third term is zero because unpaired solute atoms cannot move at all. Hence $(kT/n)L'_{BB} = p_1^{(0)}a^2w_2 = c_pa^2w_2/3$ where $c_p$ is the site fraction of pairs (4 of the total of 12 orientations being in the $(+)$-set).

Next take $R = B$ and $S = V$ and turn to $L_{BV}'$ (A8.2.1). The first term on the right side is the same as before, $p_1^{(0)}a^2w_2$. The second term is again zero because $a_{dp}^B$ is zero for all dissociation transitions (even though $a_{dp}^V$ is not). The third term is not present because $R \neq S$. Hence $(kT/n)L'_{BV} = c_pa^2w_2/3 = (kT/n)L'_{VB}$.

The case $R = S = V$ is a little more involved. The first term on the right side of (A8.2.1) is $p_1^{(0)}a^2w_2$ as before. To obtain the second term we observe that from each $(+)$-configuration (e.g. 110) there are $4w_3$ jumps where the vacancy ends on the plane $x = 2a$ (relative to the B atom), $2w_3$ where it ends on the plane $x = a$ and $w_3$ where it ends at $x = 0$, i.e. on the same $x$-plane as the solute (Fig. 8.5). Likewise for each $\bar{1}$-configuration – except that the displacements are now of the opposite sign (although this makes no difference as they enter (A8.2.1) squared in this case). We further observe that from each 0-configuration (e.g. 011) there are $2w_3$ dissociation jumps where the vacancy ends on the plane $x = a$, $3w_3$ where it ends on plane $x = 0$ and $2w_3$ on plane $x = -a$. The total contribution of the second term is thus $36p_1^{(0)}a^2w_3 + 4p_0^{(0)}a^2w_3 = 40p_1^{(0)}a^2w_3$ since $p_0^{(0)} = p_1^{(0)}$. Since each free vacancy is assumed to jump with a frequency $w_0$ per neighbouring site, the third term in (A8.2.1) is

$$p_V^{(0)}(1 - 7p_B^{(0)})(4a^2w_0),$$

there being 7 sites which are first neighbours to a neighbour of a particular site o which are not also neighbours of o. Hence

$$\left(\frac{kT}{n}\right)L'_{VV} = \frac{c_p a^2}{3}(w_2 + 40w_3) + p_V^{(0)}(1 - 7p_B^{(0)})(4a^2w_0).$$

It remains to calculate the corresponding parts $L''_{RS}$. Since only first-shell configurations are counted as associated, the matrix $\mathbf{Q}$ is found from (8.5.4) to possess the single element $Q_{11} = (2w_1 + 2w_2 + 7w_3)$. Correspondingly there is only one term in the summation (A8.2.2), namely that for $p = q = 1$. The quantity $v_{B1} = aw_2$, the only contribution coming from the exchange of position of the solute and the vacancy. Dissociation of the pair leaves the solute at $x = 0$, i.e. $a_{d1}^B = 0$. On the other hand, the quantity $v_{V1} = a(w_2 + 10w_3)$ because $4w_3$ dissociative jumps leave the vacancy at $x = 2a$, $2w_3$ leave it at $x = a$ and $w_3$ leave it at $x = 0$. Hence by (A8.2.2)

$$\left(\frac{kT}{n}\right)L''_{BB} = \frac{-p_1^{(0)}a^2w_2^{\ 2}}{\left(w_1 + w_2 + \dfrac{7w_3}{2}\right)}$$

$$\left(\frac{kT}{n}\right)L''_{BV} = \frac{-p_1^{(0)}a^2w_2(w_2 + 10w_3)}{\left(w_1 + w_2 + \dfrac{7w_3}{2}\right)}$$

$$= \left(\frac{kT}{n}\right)L''_{VB}$$

$$\left(\frac{kT}{n}\right)L''_{VV} = \frac{-p_1^{(0)}a^2(w_2 + 10w_3)^2}{\left(w_1 + w_2 + \dfrac{7w_3}{2}\right)}$$

Then by gathering $L'_{RS}$ and $L''_{RS}$ together we arrive at the expressions given in the text as eqns. (8.5.7).

It is, of course, necessary that we obtain the same result if we start from the transformed expression (8.5.10a). This we now verify. As before we refer to Fig. 8.5 and begin with $L_{BB}$. The first term is the same as the first term of $L'_{BB}$ (eqns. A8.2.1) since the B atom can only move by making a $w_2$ exchange of position with the neighbouring vacancy. The second and third terms are both zero for the same reason. The velocity analogue $v_1^B$ is likewise as before (A8.2.3). Hence $L_{BB}$ is also as before (8.5.7a).

But now when we turn to $L_{BV}$, the change in the definition of the displacements does have some effect on the component terms. In particular, the first term of (8.5.10a) is $-a^2 w_2 p_1^{(0)}$, because the vacancy displacement is equal and opposite to the B-atom displacement in the $w_2$ jumps from the 1- and $\bar{1}$-configurations. There are no other contributions to the first term, since all other transitions have $a^B = 0$. Likewise, the second term in (8.5.10a) is zero and the third is absent (because $R \neq S$). For the fourth term we need $v_1^V$ as well as $v_1^B$ which we already have ($aw_2$). From (8.5.10b) we find

$$v_1^V = -a(w_2 + 2w_1 - 3w_3)$$

because, for jumps from configuration 1 (i.e. 110), $a^V$ is $-a$ for the $w_2$ jump and for two of the four $w_1$ jumps (the other two giving zero, cf. Fig. 8.5), while it is $+a$ for four of the $w_3$ dissociative jumps, 0 for two others and $-a$ for the seventh. In the 1-shell approximation $(Q^{-1})_{11}$ is $(2w_1 + 2w_2 + 7w_3)^{-1}$ as before. By putting these parts together we find that (8.5.10a) leads to (8.5.7b), as required.

As before, $L_{VV}$ is slightly more complicated. Reference to Fig. 8.5 and recognition that $p_1^{(0)} = p_0^{(0)} = p_{\bar{1}}^{(0)}$ shows that the first two terms in (8.5.10a) in this case yield

$$p_1^{(0)} a^2 (w_2 + 4w_1 + 14w_3).$$

We already have $v_1^V$, so the fourth term is

$$-2a^2 p_1^{(0)} \frac{(w_2 + 2w_1 - 3w_3)^2}{(2w_1 + 2w_2 + 7w_3)}.$$

The third term is unaltered by the transformations (8.5.8) and (8.5.9). Adding these parts together we then arrive at (8.5.7c) as before.

This example has been particularly straightforward partly because no matrix inversion was required, $\mathbf{Q}$ having just one element. Nevertheless, more complex models with paired configurations extending out to several shells (necessary in more open lattices) are manageable in this approach.

# 9

# *The theory of random walks*

## 9.1 Introduction

In the last two chapters we have studied kinetic theories of relaxation and diffusion as specific representations of the general master equation (6.2.1). Such analyses allowed us to obtain insight into a number of aspects of these processes, especially in dilute systems, e.g. the $L$-coefficients, the way correlations in atomic movements enter into diffusion coefficients, the relation of diffusion to relaxation rates and so on. In this chapter and the next we turn to another well-established body of theory, namely the theory of *random walks*. This too can be presented in the context of the general analysis of Chapter 6 and additional insights obtained.

The basic model or system which is analysed in the mathematical theory of discrete random walks is that of a particle (or 'walker') which moves in a series of random jumps or 'steps' from one lattice site to another. It can be used to represent physical systems such as interstitial atoms (e.g. C in $\alpha$-Fe) or point defects moving through crystal lattices under the influence of thermal activation, as long as the concentrations of these species are low enough that their movements do not interfere with one another. The mathematical theory of such random walks has received considerable attention and is well recorded in many books and articles (recent examples include Barber and Ninham, 1970 and Haus and Kehr, 1987). For this reason it would be superfluous (and impractical) to go over all the same ground again here. Nevertheless there are various results which are useful in the theory of atomic transport either directly (e.g. in the evaluation of transport coefficients accordings to eqns. (6.5.10)–(6.5.12)) or for the ideas they give rise to. They are the subject of this chapter, while their uses are dealt with in the following three. We shall derive these necessary results from the structure already developed in Chapter 6. This is possible because the random walk model is a particular example of the general model analysed there and it is only necessary to interpret the states $\alpha, \beta, \ldots$ of the general system as the sites $\mathbf{l}, \mathbf{m}, \ldots,$ of the lattice which may be occupied by the particle executing the random walk in the same way as we have done already in §8.2.

The structure of this chapter is therefore as follows. The next section presents formal preliminaries which follow from Chapter 6 and which provide the basic relations needed here. Section 9.3 then derives results for unrestricted random walks of a single particle while §9.4 deals with walks subject to restrictions (e.g. those which must avoid a certain site). Some of the results for restricted walks are needed for the description of atomic diffusion when this is effected by defects (which is the subject of the next chapter). In particular they provide necessary background to the so-called 'encounter model' introduced in §9.5. Section 9.6 concludes the chapter.

## 9.2 Formal preliminaries

At various points in Chapter 6 we have employed the propagator $\mathbf{G}(t)$, the elements $G_{\beta\alpha}(t)$ of which give the conditional probability that a system which is initially in state $\alpha$ will be in state $\beta$ after time $t$ (6.3.1). At the same time we have seen that the diagonal matrix $\mathbf{G}^{(d)} = \exp(-\mathbf{P}^{(d)}t)$ (6.3.3) also plays a part and that the Laplace transforms of both $\mathbf{G}(t)$ and $\mathbf{G}^{(d)}(t)$, i.e. $\bar{\mathbf{G}}(\lambda)$ and $\bar{\mathbf{G}}^{(d)}(\lambda)$ are also useful quantities. By the definition of $\mathbf{P}^{(d)}$ we see that the matrix $\mathbf{G}^{(d)}(t)$ corresponds to a set of uncoupled decays and that each diagonal matrix element gives the decay of the probability that the system remains in the corresponding state. In this section we shall take one or two further formal steps which prove to be useful for the random walk theory particularly.

We begin by defining a matrix $\mathbf{U}(\lambda)$ by writing

$$\bar{\mathbf{G}}(\lambda) \equiv \bar{\mathbf{G}}^{(d)}(\lambda)\mathbf{U}(\lambda). \tag{9.2.1}$$

In fact, and as we shall see, from $\mathbf{U}(\lambda)$ we can obtain the generating functions for restricted as well as unrestricted random walks (§§9.3 and 9.4 below), and also measures of correlation effects in the random walk description of defect-induced diffusion (Chapter 10). It is therefore an important quantity.

The so-called Dyson expansion of $\bar{\mathbf{G}}(\lambda)$ gives

$$\begin{aligned}
\bar{\mathbf{G}}(\lambda) &= (\mathbf{P} + \lambda\mathbf{1})^{-1} \\
&= (\mathbf{P}^{(d)} + \mathbf{P}^{(nd)} + \lambda\mathbf{1})^{-1} \\
&= \bar{\mathbf{G}}^{(d)}(\lambda) - \bar{\mathbf{G}}^{(d)}(\lambda)\mathbf{P}^{(nd)}\bar{\mathbf{G}}^{(d)}(\lambda) + \bar{\mathbf{G}}^{(d)}(\lambda)\mathbf{P}^{(nd)}\bar{\mathbf{G}}^{(d)}(\lambda)\mathbf{P}^{(nd)}\bar{\mathbf{G}}^{(d)}(\lambda) - \cdots \\
&= \bar{\mathbf{G}}^{(d)}(\lambda)\{\mathbf{1} - \mathbf{P}^{(nd)}\bar{\mathbf{G}}^{(d)}(\lambda) + \mathbf{P}^{(nd)}\bar{\mathbf{G}}^{(d)}(\lambda)\mathbf{P}^{(nd)}\bar{\mathbf{G}}^{(d)}(\lambda) - \cdots\},
\end{aligned} \tag{9.2.2}$$

in which $\mathbf{P}$ has been separated into its diagonal and its non-diagonal parts, and where $\bar{\mathbf{G}}^{(d)}(\lambda) = (\mathbf{P}^{(d)} + \lambda\mathbf{1})^{-1}$ with $\mathbf{1}$ as the unit matrix. It is implicit in the use of these expressions for $\bar{\mathbf{G}}(\lambda)$ and $\bar{\mathbf{G}}^{(d)}(\lambda)$ that the matrix of transition rates $\mathbf{P}$ is independent of time, which means that the systems considered must either be in thermodynamic equilibrium or in a non-equilibrium steady state. Apart from that restriction we could, of course, expand $\bar{\mathbf{G}}(\lambda)$ in this way for any convenient separation of $\mathbf{P}$ into two parts. In this chapter we shall use it only for the separation into diagonal and non-diagonal parts indicated. This expansion is equivalent to that in §6.3, eqns. (6.3.3) *et seq.*, and the significance of the successive terms is the same.

Next, we observe that the product $\mathbf{P}^{(nd)}\bar{\mathbf{G}}^{(d)}(\lambda)$ appearing in (9.2.2) can be rewritten as

$$\mathbf{P}^{(nd)}\bar{\mathbf{G}}^{(d)}(\lambda) = \mathbf{P}^{(nd)}(\mathbf{P}^{(d)})^{-1}\mathbf{P}^{(d)}\bar{\mathbf{G}}^{(d)}(\lambda)$$
$$\equiv -\mathbf{A}\mathbf{Y}, \tag{9.2.3}$$

in which $\mathbf{A}$ is the non-diagonal matrix $-\mathbf{P}^{(nd)}(\mathbf{P}^{(d)})^{-1}$, i.e. from the definition (6.2.4)

$$A_{\rho\sigma} = w_{\rho\sigma}/P_{\sigma\sigma}, \qquad \sigma \neq \rho,$$
$$= 0, \qquad \rho = \sigma, \tag{9.2.4a}$$

or, in the display form,

$$\mathbf{A} = \begin{bmatrix} 0 & w_{12}/P_{22} & w_{13}/P_{33} & \cdots \\ w_{21}/P_{11} & 0 & w_{23}/P_{33} & \cdots \\ w_{31}/P_{11} & w_{32}/P_{22} & 0 & \\ \vdots & \vdots & \vdots & \end{bmatrix}; \tag{9.2.4b}$$

while $\mathbf{Y}$ is the diagonal matrix $\mathbf{P}^{(d)}\bar{\mathbf{G}}^{(d)}(\lambda)$, i.e.

$$Y_{\rho\sigma} = Y_{\rho\rho}\delta_{\rho\sigma}, \tag{9.2.5a}$$

and

$$Y_{\rho\rho} = P_{\rho\rho}/(P_{\rho\rho} + \lambda). \tag{9.2.5b}$$

Thus (9.2.2) becomes

$$\bar{\mathbf{G}}(\lambda) = \bar{\mathbf{G}}^{(d)}(\lambda) \sum_{n=0}^{\infty} (\mathbf{A}\mathbf{Y})^n \tag{9.2.6}$$

and (9.2.1) yields

$$\mathbf{U}(\lambda) = \sum_{n=0}^{\infty} (\mathbf{AY})^n$$

$$= (\mathbf{1} - \mathbf{AY})^{-1}, \qquad (9.2.7)$$

for $\lambda > 0$. This result can also be written in a form analogous to an integral equation, viz.

$$\mathbf{U}(\lambda) = \mathbf{1} + (\mathbf{AY})\mathbf{U}(\lambda). \qquad (9.2.8)$$

It should be recognized that $\bar{\mathbf{G}}(\lambda) = (\mathbf{P} + \lambda\mathbf{1})^{-1}$ has a singularity at $\lambda = 0$ when $\mathbf{P}$ is defined by (6.2.4). Hence $\mathbf{U}(\lambda)$ also does. However, from the mathematical point of view (9.2.1) and (9.2.7) are effectively just a factorization based on the separation of $\mathbf{P}$ into two parts $\mathbf{P}^{(d)}$ and $\mathbf{P}^{(nd)}$; the paarticular form (6.2.4) of the diagonal elements has not been employed. In later sections in this and the next chapter we shall, in fact, discover quantities of the same type as $\bar{\mathbf{G}}(\lambda)$ and $\mathbf{U}(\lambda)$ which do not have this singularity at $\lambda = 0$, but which are still related by (9.2.1) and (9.2.7). In any case as we shall see below, it turns out that the singularity arising from (6.2.4) is not of direct significance in the present chapter and that it may be removed from our considerations, although in some later applications there may be mathematical subleties associated with the limit $\lambda \to 0$. Lastly, it should be remembered that (as in Chapter 6) this formalism can be used with different definitions of the *state* of the system. In the present chapter we develop the theory of the movements of a single particle: hence the states are defined by its position alone. By contrast, later applications to the theory of nuclear spin relaxation require states to be specified by the simultaneous positions of two spins.

## 9.3 Unrestricted random walk on a cubic Bravais lattice

In this context the states $\alpha, \beta, \ldots$ entering into the general formulation of Chapter 6 are simply the positions $\mathbf{l}, \mathbf{m}, \ldots$ of a single specific particle making a random walk on a perfect lattice. A practical realization would be an interstitial atom moving under thermal activation among the available interstitial sites in a crystal lattice, e.g. H atoms in metals, as already treated in §§7.3.2 and 8.2. Another would be a vacancy defect moving through a perfect crystal lattice. We assume that the particle jumps directly only between sites that are nearest neighbours of one another and that it does so with a single jump rate $w$. We limit the treatment to cubic lattices of the Bravais type and to such systems in the absence of any perturbing force field (for, otherwise, $w$ would be differently biased in the 'forward' and 'backward' directions). The implication then is that the random walk is conducted in a system in thermodynamic equilibrium.

For this model the elements of **P** are then very simple, as follows:

$$P_{\mathbf{lm}} = wz, \qquad \mathbf{m} = \mathbf{l}$$
$$= -w\theta(\mathbf{l} - \mathbf{m}), \qquad \mathbf{m} \neq \mathbf{l} \tag{9.3.1a}$$

in which

$$\theta(\mathbf{l} - \mathbf{m}) = 1 \text{ if } \mathbf{m} \text{ and } \mathbf{l} \text{ are nearest neighbours}$$
$$= 0 \text{ otherwise} \tag{9.3.1b}$$

and $z$ is the number of nearest neighbours for the lattice in question. This defines a matrix $\boldsymbol{\theta}$. For simplicity we have dropped the superscript $^{(0)}$ which would otherwise be attached to **P** under conditions of thermodynamic equilibrium.

From this simple **P** it is possible to deduce a number of the properties of the propagator $\mathbf{G}(t)$ and of the associated function $\mathbf{U}(\lambda)$ introduced by (9.2.1). We consider these two types of function separately.

### *9.3.1 The propagator for unrestricted random walks*

As already indicated, it is convenient to introduce the Laplace transform $\bar{\mathbf{G}}(\lambda)$ of $\mathbf{G}(t)$ and the associated quantity $\mathbf{U}(\lambda)$. From (9.2.2) we see that

$$\bar{\mathbf{G}}(\lambda) = [\lambda\mathbf{1} + zw(\mathbf{1} - \boldsymbol{\gamma})]^{-1} \tag{9.3.2a}$$

in which

$$\boldsymbol{\gamma} = \boldsymbol{\theta}/z. \tag{9.3.2b}$$

Since the diagonal matrix $\mathbf{G}^{(\mathrm{d})}(\lambda)$ derived from $\mathbf{P}^{(\mathrm{d})}$ is just $(\lambda + zw)^{-1}$ times the unit matrix, the matrix $\mathbf{U}(\lambda)$ is given by

$$\bar{\mathbf{G}}(\lambda) = (\lambda + zw)^{-1}\mathbf{U}(\lambda) \tag{9.3.3}$$

together with (9.3.2a).

Several useful and well-known results follow from (9.3.2). First, in order to illustrate the time dependence of $\mathbf{G}(t)$ in a less formal manner than in (6.3.1), we expand (9.3.2) as follows

$$\bar{\mathbf{G}}(\lambda) = (\lambda + zw)^{-1}\mathbf{U}(\lambda)$$
$$= (\lambda + zw)^{-1} \sum_{n=0}^{\infty} (zw/(\lambda + zw))^n \boldsymbol{\gamma}^n. \tag{9.3.4}$$

In this case the matrix **Y** defined by (9.2.5) is simply $zw/(\lambda + zw)$ times the unit matrix. It is clear both physically from the translational invariance of the Bravais lattice and formally from (9.3.4) and (9.3.2b) and (9.3.1b) that the elements of the propagator $G_{\mathbf{lm}}(t)$ will depend only on the separation $\mathbf{l}$–$\mathbf{m}$ between the sites. We therefore write it more briefly in non-matrix notation as $G_{\mathbf{l}}(t)$, i.e. we take the initial site to be the origin. We use a similar notation for $\mathbf{U}$ and elsewhere as appropriate. This feature is an important simplification.

By inverting the Laplace-transformed equation (9.3.4) we see that

$$G_{\mathbf{l}}(t) = \sum_{n=0}^{\infty} \left( \frac{(zwt)^n}{n!} \exp(-zwt) \right) p_n(\mathbf{l}), \tag{9.3.5}$$

in which $p_n(\mathbf{l})$ is the $(\mathbf{l}, 0)$ element of the matrix $\boldsymbol{\gamma}^n = (\boldsymbol{\theta}/z)^n$. The matrix product $(\gamma^n)_{\mathbf{l}0}$ generates a sum over all the paths made up of $n$ nearest-neighbour steps between the origin 0 and site $\mathbf{l}$ and in consequence we find that

$$p_n(\mathbf{l}) = N_n(\mathbf{l})/z^n. \tag{9.3.6}$$

Here $N_n(\mathbf{l})$ is simply the number of possible $n$-step random walks by which the particle can reach site $\mathbf{l}$ from the origin. Since each step could be made in $z$ distinct directions it is clear that $p_n(\mathbf{l})$ is the probability than an $n$-step random walk starting at the origin finishes at site $\mathbf{l}$. However, we know that the coefficient of $p_n(\mathbf{l})$ in the sum (9.3.5) is, in fact, the probability that exactly $n$ steps are made in time $t$ for a Poisson process having a mean number, $zw$, of transitions per unit time (see e.g. van Kampen, 1981). The sum (9.3.5) over all $n$ is thus the total probability that at time $t$ the particle is at $\mathbf{l}$ given that it was initially at the origin, as it should be by the general definition of $\mathbf{G}(t)$ given in §6.3.

Another view of the space–time dependence of the propagator for a particle executing a random walk is obtained from an examination of its spatial Fourier transform, viz.

$$\tilde{G}_{\mathbf{q}}(t) = \sum_{\mathbf{l}} \exp(i\mathbf{q}\cdot\mathbf{r}_{\mathbf{l}})G_{\mathbf{l}}(t). \tag{9.3.7}$$

We shall assume that periodic boundary conditions are applied to $G_{\mathbf{l}}(t)$, so that the allowed values of $\mathbf{q}$ are discrete. Also since $G_{\mathbf{l}}(t)$ is only defined at discrete points in real space, these $\mathbf{q}$-values are confined to the first Brillouin zone. The Laplace transform of this quantity $\tilde{G}_{\mathbf{q}}(t)$ can be found directly by Fourier transforming (9.3.4) and using the fact that $\gamma_{\mathbf{lm}}$ is determined by the difference $\mathbf{l}$–$\mathbf{m}$ alone. A short calculation then shows that

$$\begin{aligned} \bar{\tilde{G}}_{\mathbf{q}}(\lambda) &= (\lambda + zw)^{-1}\bar{U}_{\mathbf{q}}(\lambda) \\ &= (\lambda + zw(1 - \tilde{\gamma}_{\mathbf{q}}))^{-1}, \end{aligned} \tag{9.3.8}$$

in which $\tilde{\gamma}_{\mathbf{q}}$ is the Fourier transform of $\gamma_{\mathbf{l}}$, i.e.

$$\tilde{\gamma}_{\mathbf{q}} = \sum_{\mathbf{l}} \exp(i\mathbf{q}\cdot\mathbf{r}_{\mathbf{l}})\gamma_{\mathbf{l}}.$$

Since $\gamma_{\mathbf{l}}$ is non-zero only when $\mathbf{l}$ is a nearest neighbour of the origin, such transforms can be found straightforwardly. For the basic cubic lattices (simple cubic, body-centred cubic and face-centred cubic) the expressions are respectively

$$\tilde{\gamma}_{\mathbf{q}} = (c_x + c_y + c_z)/3, \qquad \text{(s.c.)}, \qquad (9.3.9a)$$

$$\tilde{\gamma}_{\mathbf{q}} = c_x c_y c_z, \qquad \text{(b.c.c.)}, \qquad (9.3.9b)$$

$$\tilde{\gamma}_{\mathbf{q}} = (c_x c_y + c_y c_z + c_z c_x)/3, \qquad \text{(f.c.c.)}, \qquad (9.3.9c)$$

where

$$c_i \equiv \cos(q_i a) \qquad (i = x, y, z) \qquad (9.3.10)$$

and $a$ is the lattice constant such that the nearest-neighbour distance, $s$, is $a$ (s.c.), $\sqrt{3}a$ (b.c.c.) and $\sqrt{2}a$ (f.c.c.).

We now recall the definition of the incoherent scattering factor $S_s(\mathbf{q}, \omega)$ given by (1.8.4). We then see immediately that, for the system considered here, namely particles hopping unhindered among the sites of a cubic lattice, this quantity – apart from the Debye–Waller factor – is just $(1/\pi)\,\mathrm{Re}\,\bar{\tilde{G}}_q(i\omega)$, i.e. by (9.3.8)

$$S_s(\mathbf{q}, \omega) = \frac{1}{\pi}\frac{\Gamma(\mathbf{q})}{\omega^2 + (\Gamma(\mathbf{q}))^2} \qquad (9.3.11a)$$

with

$$\Gamma(\mathbf{q}) = zw(1 - \tilde{\gamma}_{\mathbf{q}}). \qquad (9.3.11b)$$

This form (9.3.11) is sometimes referred to as the Chudley–Elliott form (Chudley and Elliott, 1961) and is often used in the interpretation of measurements of neutron scattering by dilute solid solutions of hydrogen in metals. In concentrated solid solutions the individual interstitial hydrogen atoms get in one another's way and the problem becomes more difficult. The theory can be extended to give an approximate description of concentrated solutions by simply replacing the jump frequency $w$ by its product with the fraction of vacant sites, $c_v$, i.e. by $wc_v$. This is called the *mean-field approximation.* More accurate theories are the subject of Chapter 13.

In the limit when $\mathbf{q} \to 0$ the factor $(1 - \tilde{\gamma}_{\mathbf{q}})$ is proportional to $q^2$ in all cubic lattices (cf. 9.3.9) and (9.3.8) thus assumes the form

$$\bar{\tilde{G}}_{\mathbf{q}}(\lambda) = (\lambda + Dq^2)^{-1} \qquad (9.3.12a)$$

with

$$D = zws^2/6, \qquad \text{(cubic lattices)}, \qquad (9.3.12b)$$

in which $s$ is the nearest-neighbour distance. This expression (9.3.12a) may easily be seen to be the Laplace–Fourier transform of the fundamental solution or

'Green's function' $G(\mathbf{r}, t)$, of Fick's second equation, viz.

$$\frac{\partial G(\mathbf{r}, t)}{\partial t} = D\nabla^2 G(\mathbf{r}, t), \tag{9.3.13a}$$

with

$$G(\mathbf{r}, 0) = \delta(\mathbf{r}). \tag{9.3.13b}$$

Hence for small $\mathbf{q}$, i.e. large lattice distances $\mathbf{l}$, the macroscopic diffusion equation describes the conditional probability that the particle is at position $\mathbf{l}$ at time $t$ and $D$ may therefore be identified as the diffusion coefficient of such particles. By (9.3.11) this result may be regarded as an example of (1.8.9).

This expression (9.3.12b) giving the diffusion coefficient for particles executing unbiassed, uncorrelated random walks agrees with the equation (8.2.6) obtained already by kinetic theory. It is, of course, well known and is the basis of countless discussions of specific systems. Under the same circumstances we also have the Einstein relation between mobility and diffusion coefficient as arrived at previously in §§5.2, 6.5 and 8.2. (Correlation effects brought about by defect mechanisms of atomic transport modify both relations and are dealt with in the next chapter.)

### *9.3.2 $U(\lambda)$ as a generating function for unrestricted random walks*

For many applications a knowledge of $U_{\mathbf{l}}(\lambda)$ turns out to be central. A more conventional notation for this quantity would be $U(\mathbf{l}, y)$ in which

$$y = zw/(\lambda + zw) \tag{9.3.14}$$

so that the formal expression (9.3.4) becomes

$$U(\mathbf{l}, y) = \sum_{n=0}^{\infty} y^n p_n(\mathbf{l}). \tag{9.3.15}$$

This form shows that the coefficient of $y^n$ in the expansion of $U(\mathbf{l}, y)$ is just the probability that the particle making the random walk will be at site $\mathbf{l}$ after taking $n$ steps from the origin. In other words $U(\mathbf{l}, y)$ is the *generating function* for this probability (Barber and Ninham, 1970).

These generating functions $U(\mathbf{l}, y)$ (sometimes also called *lattice Green functions* and denoted by the symbol $P$) are most conveniently calculated from the integral representation which is obtained by inversion of the Fourier transform (9.3.8). To do this we bear in mind the remarks following (9.3.7) about the use of periodic boundary conditions (over a volume $V$ containing $N_m$ unit cells), and first use the inverse Fourier transform of (9.3.8) corresponding to that employed in (9.3.7) to obtain $U(\mathbf{l}, y)$ from (9.3.4), thus

$$U(\mathbf{l}, y) = \frac{1}{N_m} \sum_{\mathbf{q}}^{\text{B.Z.}} \frac{\exp(-\mathrm{i}\mathbf{q}\cdot\mathbf{r}_l)}{(1 - y\gamma_{\mathbf{q}})}. \tag{9.3.16a}$$

Then, by introducing the density-of-states factor $V/(2\pi)^3$, we pass from this sum over discrete points $\mathbf{q}$ to an integral, the result being

$$U(\mathbf{l}, y) = \frac{v}{(2\pi)^3} \iiint \frac{\exp(-\mathrm{i}\mathbf{q}\cdot\mathbf{r}_l)\,\mathrm{d}\mathbf{q}}{(1 - y\gamma_{\mathbf{q}})} \tag{9.3.16b}$$

Here $v$ is the lattice cell volume $(V/N_m)$ and the integral is taken over the first Brillouin zone.

These functions are of interest to us both for real values of $y$, as in the applications of the theory of random walks to diffusion in the remainder of this chapter and in Chapter 10, and also for imaginary values ($\lambda = \mathrm{i}\omega$), as in the theory of nuclear magnetic relaxation times in Chapter 12. In fact, they also occur in various other areas of solid state physics, with the result that they have been extensively studied for cubic (and other) lattices and many of their analytical and numerical values are known. Although no comprehensive review exists, particular aspects have been reviewed by Joyce (1972) and Schroeder (1980). A few of them can be calculated analytically; the earliest examples were the functions $U(0, 1)$ for the cubic lattices, the so-called Watson integrals (Barber and Ninham, 1970). Other recent analytical results which have been useful in defect calculations are given in the papers by Montet (1973) and Sholl (1981b). There are also tables of values of $U(\mathbf{l}, y)$ for the principal lattices. As an example, Table 9.1 gives $U(l, 1)$ for the first ten neighbour distances, $l$, in both b.c.c. and f.c.c. lattices.

We conclude with four particular remarks.

(1) In the limit of large $l$, corresponding to the macroscopic diffusion equation limit, the dominant contributions to the integral (9.3.16b) come from small $\mathbf{q}$ (because at large $\mathbf{q}$ the factor $\exp(-\mathrm{i}\mathbf{q}\cdot\mathbf{r}_l)$ undergoes rapid oscillations). In this limit $\gamma_{\mathbf{q}} \to 1 - q^2s^2/6$. By substituting this into the integral and extending the range of integration over all $\mathbf{q}$-space, we obtain

$$U(l, y) = \frac{3v}{2\pi s^2} \frac{1}{y r_l} \exp(-\kappa r_l), \tag{9.3.17a}$$

where

$$\kappa^2 = 6(1 - y)/ys^2. \tag{9.3.17b}$$

Table 9.1. *The lattice Green function $U(l, 1)$ for the first ten neighbour distances for b.c.c. and f.c.c. lattices*

Accurate numerical values (Koiwa and Ishioka, 1983b) are compared with the asymptotic form (9.3.17). It may be noted that these authors also give $U(l, 1)$ for the diamond lattice.

| Body-centred cubic | | | Face-centred cubic | | |
|---|---|---|---|---|---|
| Coordinate | Accurate | Asymptotic | Coordinate | Accurate | Asymptotic |
| (000) | 1.393 204 | — | (000) | 1.344 661 | — |
| (111) | 0.393 204 | 0.367 553 | (110) | 0.344 661 | 0.337 619 |
| (200) | 0.290 901 | 0.318 310 | (200) | 0.229 936 | 0.238 732 |
| (220) | 0.229 599 | 0.225 079 | (211) | 0.195 467 | 0.194 924 |
| (222) | 0.190 927 | 0.183 776 | (220) | 0.170 889 | 0.168 809 |
| (311) | 0.188 599 | 0.191 948 | (222) | 0.138 363 | 0.137 832 |
| (331) | 0.147 995 | 0.146 051 | (310) | 0.149 680 | 0.150 988 |
| (333) | 0.124 429 | 0.122 518 | (321) | 0.127 953 | 0.127 608 |
| (400) | 0.154 800 | 0.159 155 | (330) | 0.112 971 | 0.112 540 |
| (420) | 0.141 304 | 0.142 353 | (332) | 0.102 080 | 0.101 796 |

By taking the limit $y \to 1$ we find

$$U(l, 1) = \frac{3v}{2\pi s^2} \frac{1}{r_l}, \tag{9.3.17c}$$

as an approximation valid at large $l$.

As Table 9.1 shows this is a useful approximation for third or more distant neighbours in the f.c.c. lattice, although for the more open b.c.c. lattice it is only satisfactory at somewhat larger separations, $l$.

(2) We previously remarked on the existence of a singularity in $\mathbf{U}(\lambda)$ at $\lambda = 0$ in cases where $\bar{\mathbf{G}}(\lambda)$ derives from (6.2.4) – as it does here. This implies a singularity in $U(l, y)$ at $y = 1$ (cf. (9.3.14)). Yet no singularity is evident among the results which we have just given for $U(l, 1)$. This paradox is resolved by looking more closely at the step from (9.3.16a) to (9.3.16b) and thus recognizing that we have effectively taken the thermodynamic limit ($N_m$, $V \to \infty$, but with $V/N_m = v$ the lattice cell volume) in the course of it. To be more rigorous we should have first separated out the $\mathbf{q} = 0$ term, namely $1/N_m(1 - y)$ from (9.3.16a). This term clearly is responsible for the singularity in question at $y = 1$, because all remaining terms have $\gamma_q < 1$ ($\mathbf{q}$ within the Brillouin zone). It is now only necessary to take the thermodynamic limit $N_m \to \infty$ first, when this term goes to zero and the remainder gives (9.3.16b). The way in which the singularity disappears from this has an

analogue in the lattice-statics theory of the displacement field caused by point defects. In that case the $\mathbf{q} = 0$ term corresponds to a uniform translation of the lattice – which clearly plays no part in the displacement field around a defect (see e.g. Tewary, 1973).

(3) As we shall see later it is sometimes useful to regard the generating functions as solutions of a set of equations of the following form

$$U(\mathbf{l}, y) = \delta(\mathbf{l}) + y \sum_{m} \gamma(\mathbf{l} - \mathbf{m}) U(\mathbf{m}, y). \tag{9.3.18}$$

This relation, which follows from (9.2.8), is useful because, starting from the left side, it enables us to 'work outwards' from the origin ($\mathbf{l} = 0$) to the first neighbours ($\mathbf{m} = \mathbf{s}$) all of which have the same $U(\mathbf{s}, y)$ by symmetry, then to second neighbours and so forth. Another useful form of the same relation is

$$U(\mathbf{l}, y) = \delta(\mathbf{l}) + y \sum_{m} U(\mathbf{l} - \mathbf{m}, y) \gamma(\mathbf{m}). \tag{9.3.19}$$

(4) The formalism of this section has been extended to lattices more complicated than the cubic lattices; for example, Ishioka and Koiwa (1978) have provided the necessary extension for two of the most important non-Bravais lattices, namely the diamond and the hexagonal close-packed lattices.

## 9.4 Restricted random walks on a perfect lattice

Before discussing applications of the results so far obtained it is necessary to supplement them by the properties of random walks in which there are restrictions on the path of the walker. We are particularly interested in situations where a site or group of sites is excluded from the walk because we seek conditional probabilities rather than the unconditional $p_n(\mathbf{l})$ probabilities which arose with (9.3.5). A practical example, which is important in applications to solute diffusion in dilute mixtures, is provided by the need to evaluate the movements of a vacancy in only that region of the lattice around the solute atom which is unperturbed by it.

We first show in general terms how the generating function for these restricted walks can be related to those for unrestricted walks (9.4.1) and then make three particular applications of the general results.

### *9.4.1 Generating functions for conditional walks*

The formalism introduced in §9.2 encompasses both unrestricted and restricted random walks because the set of transition rates $w_{\beta\alpha}$, i.e. the matrix $\mathsf{P}$, was unspecified. We now establish the relation between $\mathsf{U}(\lambda)$ for an unrestricted walk and the corresponding quantity for a walk restricted to avoid a certain set of sites. Let these excluded sites be grouped together collectively as 'region I' and all the permitted sites as 'region II'. In principle, any set of sites can be chosen to constitute region I, in particular they need not be adjoining sites. We then partition the matrix **AY** in (9.2.7) as follows

$$\mathbf{AY} = \begin{bmatrix} (\mathbf{AY})_{11} & (\mathbf{AY})_{12} \\ (\mathbf{AY})_{21} & (\mathbf{AY})_{22} \end{bmatrix}. \tag{9.4.1}$$

For the restricted walk the summation in (9.2.7) is to be evaluated without including any intermediate sites from region I, i.e.

$$U_{\beta\alpha}{}^{(\mathrm{r})} = \delta_{\beta\alpha} + (AY)_{\beta\alpha} + \sum_{\gamma}{}' (AY)_{\beta\gamma}(AY)_{\gamma\alpha} + \sum_{\gamma}{}'\sum_{\delta}{}' (AY)_{\beta\gamma}(AY)_{\gamma\delta}(AY)_{\delta\alpha} + \cdots \tag{9.4.2}$$

where $\sum'$ means that no site in region I appears in the sum. To relate $\mathbf{U}^{(\mathrm{r})}$ to $\mathbf{U}$ we introduce a projection operator type of matrix $\mathbf{V}$ such that

$$\begin{aligned} V_{\sigma\rho} &= 0 \text{ if either or both of } \sigma \text{ and } \rho \in \text{region I} \\ &= \delta_{\sigma\rho} \text{ otherwise.} \end{aligned} \tag{9.4.3a}$$

Displayed as a partitioned matrix, this has the form

$$\mathbf{V} = \begin{bmatrix} \mathbf{0} & \mathbf{0} \\ \mathbf{0} & \mathbf{1} \end{bmatrix}. \tag{9.4.3b}$$

Then we may re-express (9.4.2) as

$$\begin{aligned} \mathbf{U}^{(\mathrm{r})} &= \mathbf{1} + (\mathbf{AY}) + (\mathbf{AY})\mathbf{V}(\mathbf{AY}) + (\mathbf{AY})\mathbf{V}(\mathbf{AY})\mathbf{V}(\mathbf{AY}) + \cdots \\ &= \mathbf{1} + (\mathbf{AY})(\mathbf{1} - \mathbf{VAY})^{-1}. \end{aligned} \tag{9.4.4}$$

But $\mathbf{U}$ for the unrestricted random walk over both regions I and II is given by

$$\mathbf{U} = (\mathbf{1} - \mathbf{AY})^{-1}. \tag{9.4.5}$$

A little re-arrangement then allows us to obtain from (9.4.4) and (9.4.5)

$$(\mathbf{U} - \mathbf{1})(\mathbf{1} - \mathbf{V})(\mathbf{U}^{(\mathrm{r})} - \mathbf{1}) = \mathbf{U} - \mathbf{U}^{(\mathrm{r})}, \tag{9.4.6}$$

which is the desired relation between $\mathbf{U}$ and $\mathbf{U}^{(\mathrm{r})}$. This matrix relation is most useful when written in terms of the components of the matrices partitioned as in (9.4.1), as follows

$$U_{11}(U_{1j}^{(\mathrm{r})} - \delta_{1j}) = U_{1j} - \delta_{1j}, \qquad j = 1, 2 \tag{9.4.7a}$$

i.e.

$$U_{1j}^{(r)} = (U_{11})^{-1}(U_{1j} + U_{11}\delta_{1j} - \delta_{1j}), \qquad j = 1, 2 \tag{9.4.7b}$$

and

$$U_{2j}^{(r)} = -U_{2j} - U_{21}(U_{1j}^{(r)} - \delta_{1j}), \qquad j = 1, 2 \tag{9.4.8}$$

which by (9.4.7a) gives

$$U_{2j}^{(r)} = U_{2j} - U_{21}(U_{11})^{-1}(U_{1j} - \delta_{1j}), \qquad j = 1, 2. \tag{9.4.9}$$

In the same way that $\mathbf{U}(\lambda)$ in §9.3.2 gave the generating function for unrestricted random walks so (9.4.7) will yield the generating function for walks beginning in region $j$ (=I or II) and ending in region I without having passed through region I at any intermediate stage. Likewise (9.4.9) will yield the generating function for walks beginning in $j$ (=I or II) and ending in region II without having passed through region I at any intermediate stage.

We next demonstrate how these formal expressions can be used. In particular, we shall consider two particular cases and then in §9.5 further application of one of them. These examples, although distinct by their definitions, are closely related to one another, as we shall show.

### *9.4.2 First passage probabilities*

In this section we consider the probability $p_n^{(f)}(\mathbf{l})$ that the particle starting from the origin reaches site $\mathbf{l}$ for the first time in step $n$. There will be a corresponding generating function $U^{(f)}(\mathbf{l}, y)$ (cf. (9.3.15)) which we shall relate to the generating function for unrestricted walks $U(\mathbf{l}, y)$ by using (9.4.7). We can do this by designating $\mathbf{l}$ (alone) as region I and the rest as region II so that, when $\mathbf{l}$ is not the origin, the origin lies in region II. For this case then we can use (9.4.7a) by recognizing that $U_{11}$ has only one element, equal to $U(0, y)$, and that $U_{12}$ is a row matrix whose elements are labelled by the site in region II from which the random walk begins; and likewise for $U_{12}^{(r)}$. We are interested in the element corresponding to the origin and this is therefore just what we have designated as $U(\mathbf{l}, y)$; and similarly for $U^{(f)}(\mathbf{l}, y)$. However, in this case the matrix $\mathbf{U}^{(r)}$ differs slightly from the $\mathbf{U}^{(f)}$ we want because $\mathbf{U}^{(r)}$ includes the possibility that the particle does not actually move away from the origin at all (cf. the term $\delta_{\beta\alpha}$ in (9.4.2)), whereas the definition of the first-passage probability demands that the particle does leave the site. Thus $\mathbf{U}^{(f)} = \mathbf{U}^{(r)} - \mathbf{1}$. Hence, taking just the origin element of both sides of (9.4.7a) yields

$$U(0, y)U^{(f)}(\mathbf{l}, y) = U(\mathbf{l}, y) - \delta_{\mathbf{l}0}. \tag{9.4.10}$$

We could have obtained this equation more directly by noting that if the particle is at $\mathbf{l}$ after the $n$th step then it must have arrived there for the first time in a number of steps $m \leq n$ and then ended up at $\mathbf{l}$ after the remaining $n - m$ steps. This observation is expressed by the equation

$$p_n(\mathbf{l}) = \sum_{m=1}^{n} p_{n-m}(0)\, p_m^{(f)}(\mathbf{l}) + \delta_{n0}\, \delta_{\mathbf{l}0}. \tag{9.4.11}$$

(N.B. $p_0(0) = 1$ by definition.) We now recall the definition (9.3.15) and then multiply this equation by $y^n$ and sum over $n$ while also inverting the order to summation in the double sum on the right side. This gives us just (9.4.10). However, although this direct approach can be used for other restricted walks (see, e.g. Montroll, 1964 or Barber and Ninham, 1970) the arguments differ in each case, whereas the formal development given in §9.4.1 covers them all. We shall therefore rely on that henceforth.

Before turning to another type of restricted walk, however, we note one or two useful results which the first arrival probabilities provide access to. First, the probability $\pi_{\mathbf{l}}$ that a particle visits $\mathbf{l}$ at least once in the course of a random walk which begins at the origin can be simply related to the generating function $U(\mathbf{l}, 1)$ (section 9.3.2 above and Table 9.1). Evidently

$$\begin{aligned} \pi_{\mathbf{l}} &= \sum_{n=1}^{\infty} p_n^{(f)}(\mathbf{l}) = U^{(f)}(\mathbf{l}, 1) \\ &= (U(\mathbf{l}, 1) - \delta_{\mathbf{l}0})/U(0, 1), \end{aligned} \tag{9.4.12}$$

where we have used (9.4.10) in the second line. Of particular interest are (i) the probability, $\pi_0$, that the particle will at some point return to the origin from which it started and (ii) the probability, $\pi_e$, that the particle will never return to the origin ('escape probability'). These can be obtained from (9.4.12) as

$$\pi_0 = (U(0, 1) - 1)/U(0, 1) \tag{9.4.13a}$$

and

$$\pi_e = 1 - \pi_0 = [U(0, 1)]^{-1}. \tag{9.4.13b}$$

Some values of $\pi_0$ calculated from $U(0, 1)$ are collected in Table 9.2. In a two-dimensional square lattice the return probability $\pi_0$ is unity, i.e. the particle will certainly revisit the origin at some time, but in three-dimensional lattices this is not the case and $\pi_0 < 1$. The results for the cubic lattices in Table 9.2 (Montroll, 1964) show that $\pi_0$ diminishes as the coordination number of the lattice increases, essentially because there are then more ways to escape. The calculations have been extended to non-Bravais lattices by Ishioka and Koiwa (1978) and by Koiwa and Ishioka (1979) who obtained the last three entries in Table 9.2. Although these

Table 9.2. *The probability $\pi_0$ that a random walker starting from the origin will subsequently return there, the probability $\pi'$ that a tracer which has exchanged sites with a vacancy will later exchange with it again, and the mean number $\nu$ of exchanges in a complete encounter of a tracer with a vacancy are shown for various lattices.*

| Lattices | $\pi' = \pi_0$ | $\nu = (1 - \pi_0)^{-1}$ |
|---|---|---|
| Square planar | 1 | $\infty$ |
| Simple cubic | 0.3405 | 1.5164 |
| Body-centred cubic | 0.2822 | 1.3932 |
| Face-centred cubic | 0.2563 | 1.3447 |
| Diamond | 0.4423 | 1.7929 |
| B.C.C., octahedral interstices | 0.4287 | 1.7504 |
| B.C.C., tetrahedral interstices | 0.4765 | 1.9102 |

three lattices all have the same number of nearest neighbours (4), they have slightly different values of $\pi_0$.

Lastly, we can obtain another result which shows the physical significance of $U(\mathbf{l}, 1)$. Thus the mean number of times $\nu_\mathbf{l}$ that a particle starting from the origin subsequently visits the site $\mathbf{l}$ is

$$\nu_\mathbf{l} = \pi_\mathbf{l}[\pi_e + 2\pi_0\pi_e + 3\pi_0^2\pi_e + \cdots] \tag{9.4.14}$$

in which successive terms are the mean number of times the particle visits site $\mathbf{l}$ just once, twice, three times, and so on (see Fig. 9.1). Hence

$$\begin{aligned} \nu_\mathbf{l} &= \pi_\mathbf{l}\pi_e/(1 - \pi_0)^2 \\ &= \pi_\mathbf{l}/\pi_e \\ &= U(\mathbf{l}, 1) - \delta_{\mathbf{l}0}, \end{aligned} \tag{9.4.15}$$

by (9.4.12) and (9.4.13b).

### *9.4.3 Probability of reaching site* l *without passing through site* m

A second example of a restricted walk arises when we seek the probability $p_n(\mathbf{l}, \tilde{\mathbf{m}})$ that the particle reaches site $\mathbf{l}$ at the $n$th step *without* having passed through the site $\mathbf{m}$. (We use the tilde to denote an excluded site; some authors use a bar instead, but we wish to avoid confusion with the bar notation for the symmetry-related

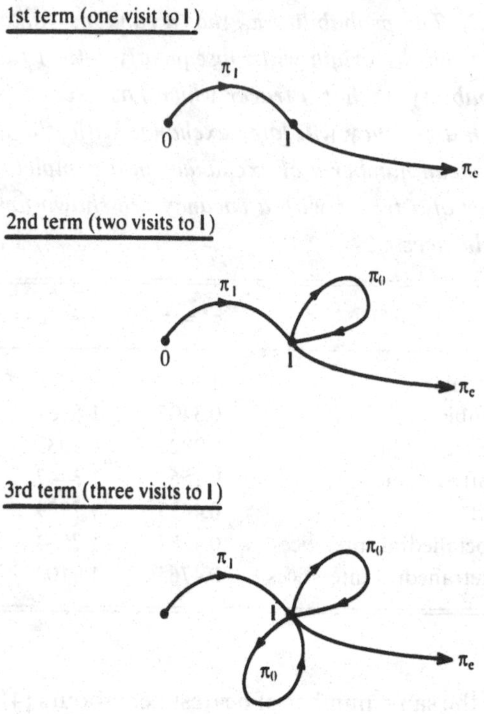


Fig. 9.1.

configurations which was introduced in Chapter 8.) The corresponding generating function for this probability

$$U(\mathbf{l}, \tilde{\mathbf{m}}, y) = \sum_{n=0}^{\infty} y^n p_n(\mathbf{l}, \tilde{\mathbf{m}}), \tag{9.4.16}$$

may be obtained directly from (9.4.9) for $U_{22}^{(r)}$ by identifying region I with the site **m** alone (with region II as all other sites) and by taking the $(\mathbf{l}, 0)$ matrix element of both sides. The left side is then just the $U(\mathbf{l}, \tilde{\mathbf{m}}, y)$ we require and (9.4.9) becomes

$$U(\mathbf{l}, \tilde{\mathbf{m}}, y) = U(\mathbf{l}, y) - \frac{U(\mathbf{l} - \mathbf{m}, y)(U(\mathbf{m}, y) - \delta_{\mathbf{m}0})}{U(0, y)}. \tag{9.4.17}$$

This result will be used in the following section on solute diffusion caused by defects, where we shall need certain characteristics of those random walks executed by the defect which avoid the site occupied by the solute atom.

## 9.5 Encounter between a vacancy and a tracer atom

In the last chapter we made use of the idea that, after an exchange of places between a solute atom and a vacancy, the vacancy may cause further movements of the atom before it eventually leaves the vicinity to rejoin the (randomized) population of free vacancies. This idea is sometimes referred to as the encounter model. In this section we shall use the preceding results to characterize the idea more quantitatively for the case where the solute is a tracer atom (isotope) of the host. This section therefore differs from the preceding ones in that it considers the correlated walk of one particle (the tracer) as caused by the intervention of the vacancy which moves randomly.

We shall therefore calculate (i) the probability $\pi'$ that a given exchange between the tracer atom and the vacancy is followed by a second exchange, and (ii) the mean number $\nu$ of exchanges between the tracer and vacancy in a single encounter. We then characterize the behaviour of the tracer in one encounter by considering the probability $\varpi(\mathbf{l})$ that the tracer undergoes a displacement from the origin to the site $\mathbf{l}$ in a single encounter. We may suppose that after the initial exchange the tracer is at site $\mathbf{l}_0$ and the vacancy is at the origin. The probability that the *vacancy* will subsequently arrive at one of the $z$ sites neighbouring the tracer, say $\mathbf{l}_j$, without having exchanged with the tracer in the course of its walk to $\mathbf{l}_j$ is clearly just $\sum_{n=0}^{\infty} p_n(\mathbf{l}_j, \tilde{\mathbf{l}}_0)$. Furthermore, the chance that the next jump of the vacancy at $\mathbf{l}_j$ results in an exchange with the tracer is just $1/z$. Hence the total probability of a second tracer–vacancy exchange is

$$\pi' = \frac{1}{z} \sum_{j} \sum_{n=0}^{\infty} p_n(\mathbf{l}_j, \tilde{\mathbf{l}}_0). \tag{9.5.1}$$

By (9.4.16) and then (9.4.17) this becomes

$$\begin{aligned}\pi' &= \frac{1}{z} \sum_{j=1}^{z} U(\mathbf{l}_j, \tilde{\mathbf{l}}_0, 1) \\ &= \frac{1}{z} \sum_{j=1}^{z} \left( U(\mathbf{l}_j, 1) - \frac{(U(\mathbf{s}_j, 1))^2}{U(0, 1)} \right)\end{aligned} \tag{9.5.2}$$

where the $\mathbf{s}_j$ are the nearest-neighbour position vectors $\mathbf{l}_j - \mathbf{l}_0$. Reverting now to (9.3.18) and (9.3.19), we see from the first that all the $U(\mathbf{s}_j, 1)$ are equal to $U(0, 1) - 1$ and from the second that,

$$\begin{aligned}\frac{1}{z} \sum_{j} U(\mathbf{l}_j, 1) &= U(\mathbf{s}, 1) \\ &= U(0, 1) - 1.\end{aligned} \tag{9.5.3}$$

Finally therefore

$$\pi' = (U(0, 1) - 1)/U(0, 1)$$
$$= \pi_0, \qquad (9.5.4)$$

by (9.4.13a). Hence the probability that a given exchange is followed by a second is exactly equal to $\pi_0$, the probability that the vacancy will revisit a given site at least once. Numerical values have already been given in Table 9.2. They show that for the three-dimensional cubic lattices $\pi_0$ is less than 1/2 and that it diminishes as the co-ordination number of the lattice increases.

Now we can calculate the mean number of exchanges per encounter, $\nu$. This is simply related to the probability $\pi_0$ of a subsequent exchange which we have just calculated. Thus the probability $\pi^{(n)}$ of exactly $n$ exchanges occurring in one encounter is

$$\pi^{(n)} = \pi_0^{n-1}(1 - \pi_0), \qquad (9.5.5)$$

where $(1 - \pi_0)$ is the probability that there is no further tracer–vacancy exchange after any such exchange. The mean number of exchanges is therefore

$$\nu = \sum_{n=1}^{\infty} n\pi^{(n)} = (1 - \pi_0)^{-1}$$
$$= \nu_0 + 1, \qquad (9.5.6)$$

where in the second line we have used (9.4.13a) and (9.4.15). Hence the mean number of exchanges between the tracer and a vacancy in a single encounter is one plus the mean number of times that a vacancy starting from the origin will subsequently return to the origin. The results for $\nu$ for cubic lattices given in Table 9.2 show that it too decreases as the coordination number increases. The numbers are relatively small, e.g. 1.5 for a simple cubic lattice.

Finally, we can characterize the effect on the tracer atom of an encounter with a vacancy through the probability $\varpi(\mathbf{l})$ that the tracer undergoes a displacement from the origin to lattice site $\mathbf{l}$. The calculation of this quantity is facilitated by the results of §9.4. Sholl (1992) has shown that it may be reduced to an integral form which is a generalization of (9.3.16b). Numerical evaluation of this expression leads to the results for the b.c.c. and f.c.c. lattices given in Table 9.3. (These values agree with those obtained previously by a less direct method, Sholl 1981b). They show that $\varpi(\mathbf{l})$ decreases rather rapidly with increasing distance $r_l$, being only 0.001 or less for a fourth neighbour site. We also observe that the probability that the tracer ends up at the origin where it started is close to $z^{-1}$, rather as might be expected. These calculations can be extended to deal with cases where instead of a host tracer atom we have a chemically distinct solute atom (§12.6.3).

Table 9.3. *The probability $\varpi(\mathbf{l})$ that a host tracer undergoes a displacement from the origin to lattice site* $\mathbf{l}$ *in a single encounter with a vacancy for the b.c.c. and f.c.c. lattices, as calculated by Sholl* (1981*b*; 1992).

| Body-centred cubic | | Face-centred cubic | |
|---|---|---|---|
| Coordinate, $\mathbf{l}$ | $\varpi(\mathbf{l})$ | Coordinate, $\mathbf{l}$ | $\varpi(\mathbf{l})$ |
| 000 | 0.120 89 | 000 | 0.081 78 |
| 111 | 0.096 44 | 110 | 0.070 83 |
| 200 | 0.009 00 | 200 | 0.003 65 |
| 220 | 0.003 02 | 211 | 0.001 49 |
| 222 | 0.001 04 | 220 | 0.000 62 |
| 311 | 0.000 29 | 222 | 0.000 05 |
| 331 | 0.000 06 | 310 | 0.000 08 |
| 333 | 0.000 01 | 321 | 0.000 02 |
| 400 | 0.000 02 | 330 | 0.000 01 |
| 420 | 0.000 01 | | |
| 422 | 0.000 01 | | |

In closing this section on the idea of encounters between tracer atoms and vacancies we emphasize that the results we have obtained apply only to three-dimensional lattices, where the vacancy escape probability $1 - \pi_0$ is $>0$. The idea is not valid in two-dimensional lattices since here $\pi_0 = 1$ (cf. Table 9.2).

## 9.6 Summary

In this chapter we have shown how the formal structure presented in Chapter 6 allows us to obtain in a rather direct way many of the results of the theory of uncorrelated random walks, including those pertaining to the commonly employed generating functions for the probabilities of various kinds of random walk. Those for unrestricted random walks of this kind may be used directly to describe the diffusion of point defects and interstitial atoms, provided only that these are present in the crystal lattice in sufficiently low concentration that their interactions or interference with one another can be neglected. The results of this application are simple and widely known. In particular, there are no off-diagonal transport coefficients, $L$, diffusion coefficients have the simple Einstein form, and the elementary Nernst–Einstein relation between mobility and diffusion coefficient also holds.

By contrast, when we consider the movement of solute atoms and tracers by vacancies or interstitialcies, as we do in the next chapter, we are necessarily concerned with correlations between successive moves of the atoms and these lead to non-zero off-diagonal transport coefficients, among other things. The evaluation of these effects then requires the probabilities for certain restricted random walks of the types considered in §9.4. The application of the results given here is pursued in the next chapter. Further use of the results of this chapter will be made in Chapter 12.

# 10

# *Random-walk theories of atomic diffusion*

## 10.1 Introduction

In this chapter we show how to use some aspects of the random-walk theory to describe the macroscopic diffusion of solute atoms (and isotopic atoms) caused by point defects. In doing so we have to handle solute movements and defect movements together, rather than the random walk of just one type of 'walker' or 'particle', as in the preceding chapter. This presents a more complex problem; nevertheless the approach we shall describe has provided a major part of the subject for many years. It was started by Bardeen and Herring (1952) who drew attention to the fact that the movement of solute atoms by the action of point defects was such that the directions of successive jumps of the solute atom were necessarily correlated with one another for the reasons already presented in Chapters 2 and 5. In Chapter 8 we showed how these correlated movements are handled in kinetic theories. However, it is the calculation of these correlation effects by random-walk theory which has engaged the attention of theoreticians to a considerable extent. Although the consequence of these effects for isotopic self-diffusion may be to introduce only a simple, numerical factor into the expression for the diffusion coefficient, the consequences for solute diffusion can be qualitatively significant, particularly when the motion of the defect is substantially affected in the vicinity of the solute. Under these same conditions qualitatively significant correlations can also appear in the self-diffusion of the solvent. The calculation of these correlation effects is thus an important part of this chapter.

We begin (§10.2) by deriving the Einstein–Smoluchowski equation, for this often forms the starting point for the calculation of diffusion coefficients. This equation provides an alternative derivation of the Einstein expressions for the diffusion coefficient and the mobility already obtained in Chapter 6. It is interesting in that this derivation does not make, nor require, the assumption of sudden transitions which is the basis of the master equation used there, but applies equally well if we have a dynamical (Hamiltonian) formulation in mind. Furthermore, this approach via the Einstein–Smoluchowski equation has featured sufficiently frequently in the development of the subject that its inclusion is appropriate, even though the results of Chapter 6 provide all that we shall need. Subsequent development, on the basis of the Einstein relation between diffusion coefficient and mean-square displacement of the diffusing atoms, is limited to dilute solid solutions containing low concentrations of defects. The condition that the system be dilute in solute atoms ensures that we can neglect their mutual interactions. Of course, the results can also be specialized to isotopic self-diffusion, although, as we saw in §5.3, the results will be valid irrespective of the proportions of isotope actually present. Further development of the Einstein equation leading to the definition of the correlation factor is then given in §10.3. Sections 10.4 to 10.7 then pursue the evaluation of the correlation factor for various types of dilute solid solutions and for self-diffusion in pure solids. Section 10.4 presents certain preliminaries for both vacancy and interstitial defects. Section 10.5 then considers the evaluation of the correlation factor for self-diffusion in pure solids while §§10.6 and 10.7 deal with correlation factors for dilute alloys. We emphasize the inter-relations among various detailed methods of evaluation. In this part of the chapter we are able to make use of some of the methods and results presented in the previous chapter. Section 10.8 then describes an approximate (though rather accurate) extension of the scheme which allows the calculation of correlation factors for (isotopic) self-diffusion in a certain model concentrated alloy (the 'random alloy model'). After this, §10.9 discusses two ways in which experimental measurements may come close to giving direct information about correlation factors alone (as distinct from just the whole diffusion coefficient).

It will be evident from the above that this chapter is principally about the diffusion coefficient $D_B$ (B = solute) in the dilute limit, i.e. about $L_{BB}$ (see (5.3.8)). In the language of §6.5 it is thus about self-correlation functions and not about correlation functions for pairs of different atoms. Although the chapter contains a number of particular results (e.g. correlation factors for self-diffusion), it will also be seen that the main emphasis here is on the general structure and methodology of the theory rather than on detailed predictions for specific physical models. In this respect it is similar to Chapter 8 on kintic theories. Significant results include the mutual equivalence of different variants of the random-walk approach and the overall equivalence of the general formulae for $D_B$ to those already obtained by the kinetic theory of Chapter 8. (In contrast to that theory the random-walk approach has not proved convenient for $L_{AA}$ (A = solvent) nor for $L_{AB}$, essentially because these require two-particle correlation functions as well as self-correlation functions.) Having shown the equivalence of the various results for $D_B$, we are then in a position to obtain explicit results for particular models. This we do in the following chapter as part of a coherent presentation of results for all the transport coefficients.

Lastly, we remark that some readers may find it helpful, at a first reading, to concentrate on the essential structure of the theory and on the nature of the principal results rather than to attempt to follow all the detailed arguments.

## 10.2 Einstein–Smoluchowski equation

We suppose that we are primarily interested in the motion, and thus the distribution, of only one type of atom or defect (B, say). We therefore take averages over the motions of all the others (which we shall only specify when needed). We shall suppose that we are interested in diffusion and drift along a principal crystal direction ($x$) and that any perturbing fields and concentration gradients are along that axis too. This allows us to use a one-dimensional notation (but again one should remember that there are crystal classes for which principal crystal directions cannot be defined; Nye, 1985).

We now introduce the probability $p(X, t; x)$ that a B atom which was on plane $x$ at time zero is on plane $x + X$ at a time $t$ later. (Other quantities of the same kind have already been introduced, e.g. Van Hove's self-correlation function in neutron scattering (§1.8) and, more generally, the functions $\mathbf{G}(t) = \exp(-\mathbf{P}t)$ in §6.3.) We shall now use this to obtain the flux equation. Let $n(x)$ be the concentration, i.e. the number per unit volume, of B atoms at $x$ at time zero; see Fig. 10.1. Then the number of B atoms which at time zero were at positions $x < x_0$ and which at time $t$ are to be found at positions $x > x_0$, i.e. which have crossed each unit area of the plane $x = x_0$ in a forward direction, is

$$\int_{-\infty}^{x_0} n(x)\,\mathrm{d}x \int_{x_0 - x}^{\infty} p(X, t; x)\,\mathrm{d}X. \tag{10.2.1a}$$

Likewise, the number of B atoms which at time zero were at $x > x_0$ and which at time $t$ are at $x < x_0$ is

$$\int_{x_0}^{\infty} n(x)\,\mathrm{d}x \int_{-\infty}^{x_0 - x} p(X, t; x)\,\mathrm{d}X. \tag{10.2.1b}$$

We now invoke the condition of the conservation of B atoms and set the difference between (10.2.1a) and (10.2.1b) equal to $J_x t$, so obtaining the flux $J_x$ of B atoms in the $x$-direction. Now, if the spatial variation of $n(x)$ is slow compared to that of $p(X, t; x)$ – as will generally be the case in macroscopic diffusion

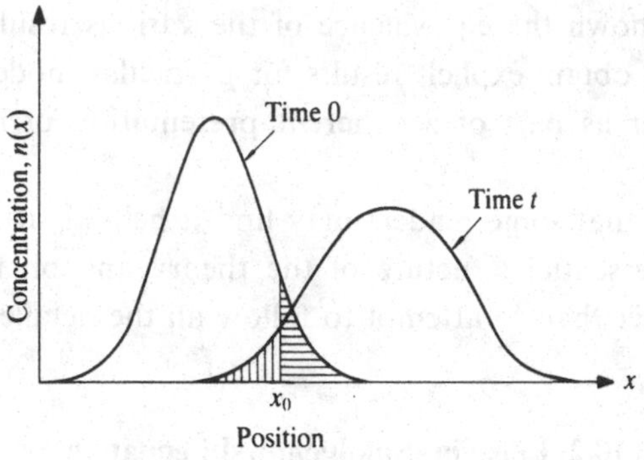


Fig. 10.1. Schematic diagram of the concentration of B atoms, $n(x)$, as a function of position $x$. Two distributions are shown, one for $t = 0$, and one for a later time $t$. The plane $x = x_0$ across which the flux is calculated is also shown.

problems – we can expand $n(x)$ in a Taylor series about $x_0$. We truncate this series at the first-order term and re-arrange the integrals in (10.2.1) by using integration by parts. The result is that

$$J_x = n\frac{\langle X\rangle}{t} - \frac{\partial n}{\partial x}\left(\frac{\langle X^2\rangle}{2t}\right) - n\frac{\partial}{\partial x}\left(\frac{\langle X^2\rangle}{2t}\right), \tag{10.2.2}$$

in which

$$\langle X^m\rangle = \int X^m p(X, t; x)\,\mathrm{d}X, \tag{10.2.3}$$

it being assumed that the probability $p(X, t; x)$ is normalized to unity – since the atom considered must be somewhere. This result is referred to as the Einstein–Smoluchowski equation (although it is a particular example of the Fokker–Planck type of equation, see van Kampen, 1981). Equations (10.2.2) and (10.2.3) are valid in the presence of concentration gradients, temperature gradients and force fields. However, it is only useful as a macroscopic flux equation when we can obtain the average quantities $\langle X\rangle/t$ and $\langle X^2\rangle/2t$ and their dependence on the perturbations acting on the system (which in general arises from the dependence of $p(X, t; x)$ upon these perturbations).

For dilute solid solutions of B in some host (A) one can appeal to physical intuition and obtain a useful result, namely the Einstein relation, immediately. This has provided the basis of much subsequent analysis, some of which we review in the remainder of this chapter, and also of direct numerical simulations as described in Chapter 13. From this point on therefore and unless specifically indicated we shall have in mind solutions dilute in B.

In systems which are not subject to force fields and which are homogeneous apart from a dilute concentration of B atoms, we must have $\langle X \rangle = 0$, while we also expect $\langle X^2 \rangle$ to be independent of position. It follows that the diffusion coefficient of the B atoms (in the $x$-direction) in this dilute limit is

$$D_{\text{B}} = \frac{\langle X^2 \rangle}{2t}, \tag{10.2.4a}$$

which is the well-known Einstein formula for Brownian motion. By (5.3.8) we thus also have $L_{\text{BB}}$. Strictly, these quantities should be written $D_{\text{B}xx}$ and $L_{\text{B}xx}$. There are corresponding expressions in the other principal directions $y$ and $z$. Of course, for cubic crystals there is only one distinct $D_{\text{B}}$ and one $L_{\text{BB}}$ and for these we can write equivalently to (10.2.4a)

$$D_{\text{B}} = \frac{\langle R^2 \rangle}{6t}. \tag{10.2.4b}$$

One or two general observations on this result (10.2.4) are in order. The first is that it is a special example of the general relation (6.5.14). The requirement that $\langle X^2 \rangle / 2t$ be evaluated for a thermal equilibrium ensemble, which is emphasized by this relation, is met here by the specification that (a) the system is free from force fields and (b) the atomic transition rates are independent of the concentration of B atoms. An important corollary to this is that $\langle X^2 \rangle / 2t$ as defined by (10.2.3) may not be identified with $D_{\text{B}}$ if these conditions (a) and (b) are not satisfied and it would be incorrect to substitute (10.2.4) into (10.2.2) in the general case.

If we now consider the same system but subject to a uniform force field, $F$, then the last term in (10.2.2) will still be zero (by the assumption of uniformity) but the first term will not. In fact, the first term defines the mechanical mobility, $u_{\text{B}}^{\text{M}}$, of the species B considered

$$u_{\text{B}}^{\text{M}} = \frac{\langle X \rangle}{Ft} \tag{10.2.5}$$

The mobility, $u_{\text{B}}^{\text{M}}$, so defined should be distinguished from the electrical mobility, $u_{\text{B}}$, which is defined with the electric field rather than the force per particle in the denominator. One would use $u_{\text{B}}^{\text{M}}$, for example, when calculating the drift of solutes in an internal stress field around a dislocation or a crack. Of course, for a particle bearing an electric charge $q_{\text{B}}$, $u_{\text{B}}$ is simply $u_{\text{B}}^{\text{M}} / |q_{\text{B}}|$.

There are three principal ways by which $D_{\text{B}}$ and $u_{\text{B}}^{\text{M}}$ can be obtained from (10.2.4) and (10.2.5), namely (i) computer simulation by molecular dynamics, (ii) computer simulation by Monte Carlo methods and (iii) analytic methods. The first can be used since our derivation of the Einstein–Smoluchowski equation was independent of any assumption about the dynamical description of the system at

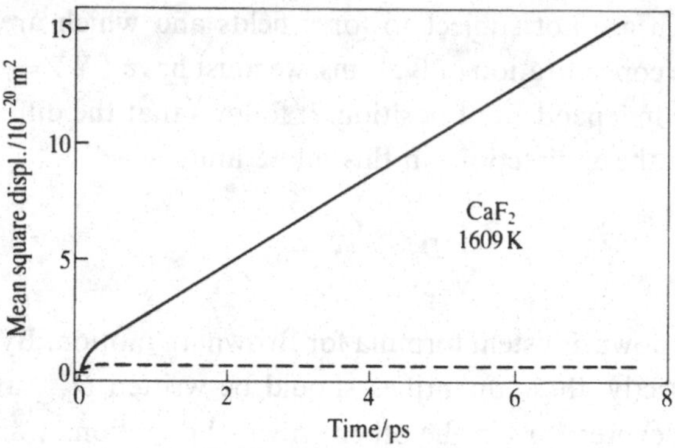


Fig. 10.2. The mean square displacement of anions (full line) and cations (dashed line) as a function of time, as obtained from a molecular dynamics simulation of model $CaF_2$ at 1609 K. By (10.2.4b) one-sixth of the slope of the line gives the self-diffusion coefficient of the corresponding ionic species. The self-diffusion coefficient of the $Ca^{2+}$ ions is thus effectively zero at this temperature, while that of the $F^-$ ions is very large ($3 \times 10^{-9}$ $m^2/s$). (After Gillan, 1985).

the molecular level. Simulations can therefore be based on a classical Hamiltonian description of the system, as in the example of Fig. 10.2. However, this is not a practicable method for solids unless diffusion is exceptionally fast; usually the jump frequencies are too small for a simulation of a reasonable length to produce enough jumps for useful averages to be obtained. In practice then, most computer simulations are based on the master equation description, i.e. Monte Carlo methods. We shall return to these again in Chapter 13. Here we shall continue with analytic methods. Up to a point these also define the most economical way to carry out the Monte Carlo simulations.

## 10.3 Further development of the Einstein equation: the correlation factor

It is possible to go substantially further in the discussion of the Einstein equation (10.2.4). There are three stages in the discussion. Firstly, there is the introduction of the *correlation factor*; that is the task of this section. Secondly, as part of the analysis of this factor, there is the determination of the relation between the correlation between *any* two movements of the diffusing atom and that between two *consecutive* movements (§10.4). Thirdly, there is the expression of this quantity in terms of details of the system, i.e. in terms of the geometry and frequency of jumps as defined by the defect involved, the lattice structure, etc. (§§10.5 and 10.6).

Let us label the successive jumps of a particular B atom (*not* the sites visited by it) by $r = 1, 2, \ldots, n$. Then we can set

$$X = \sum_{r=1}^{n} x_r \tag{10.3.1}$$

in which $x_r$ is the $x$-displacement of the atom in jump $r$. Hence the ensemble average of $X^2$ over the motions of a large number of equivalent B atoms as given by (10.2.3) is

$$\langle X^2 \rangle = \sum_{r=1}^{n} \langle x_r^2 \rangle + 2 \sum_{s=1}^{n-1} \sum_{r=1}^{n-s} \langle x_{r+s} x_r \rangle. \tag{10.3.2}$$

It will be noted that in equating the average in (10.2.4) for fixed $t$ to that in (10.3.2) for a fixed number of jumps $n$ we have omitted any dispersion in the number of jumps which an atom will undergo in a time $t$. This is justified because $n$ is very large and thus the statistical distribution of $n$ about it means value ($\Gamma t$) is effectively a Dirac $\delta$-function. More fundamentally, it should be said that we are here concerned with one aspect of the equivalence of discrete random walks (i.e. those represented in terms of discrete steps on the lattice) and continuous-time random walks (i.e. those represented by a master equation in the continuous variable $t$). For further discussion of this equivalence see Haus and Kehr (1987, especially section 2.3).

By (10.3.2) we can therefore re-write (10.2.4) as

$$D_B = D_B^{(0)} f \tag{10.3.3}$$

with

$$D_B^{(0)} = \frac{\sum_{r=1}^{n} \langle x_r^2 \rangle}{2t}, \tag{10.3.4}$$

and

$$f = \frac{\langle X^2 \rangle}{\sum_{r=1}^{n} \langle x_r^2 \rangle} \tag{10.3.5a}$$

or

$$f = 1 + \frac{2 \sum_{s=1}^{n-1} \sum_{r=1}^{n-s} \langle x_{r+s} x_r \rangle}{\sum_{r=1}^{n} \langle x_r^2 \rangle}. \tag{10.3.5b}$$

In cubic crystals, in which $D$ is necessarily a scalar (Nye, 1985), it is often more convenient to write (10.3.4) and (10.3.5) in their equivalent three-dimensional forms, i.e. as

$$D_B^{(0)} = \frac{\sum_{r=1}^{n} \langle r_r^2 \rangle}{6t} \tag{10.3.6}$$

and

$$f = 1 + \frac{2 \sum_{s=1}^{n-1} \sum_{r=1}^{n-s} \langle \mathbf{r}_{r+s} \cdot \mathbf{r}_r \rangle}{\sum_{r=1}^{n} \langle r_r^2 \rangle}, \tag{10.3.7}$$

in which $\mathbf{r}_r$ is the vector displacement of the solute in jump $r$.

Before proceeding to further mathematical development we comment on the separation of $D_B$ into factors $D_B^{(0)}$ and $f$. Although this separation may seem rather formal at first sight, it appears less so when we recognize certain physical features. There are several points to be made.

(1) As will be more apparent shortly, both factors $D_B^{(0)}$ and $f$ are independent of $n$ as $n \to \infty$.

(2) The separation is the same as that already effected in §8.3, as will be shown in §10.6.1.

(3) If, after any one jump of an atom B, all possible directions for its next jump were equally probable, then the factor $f$ would be unity because all the $\langle x_{r+s} x_r \rangle$ would be zero. It will be clear from earlier discussions (Chapter 2, and §5.7) that this cannot apply generally to atomic migration via vacancy, interstitialcy and dumb-bell migration mechanisms, although it would be correct for a simple interstitial mechanism. The extent to which $f$ departs from unity (in fact is less than unity) is a measure of the correlations between successive displacements of an atom, i.e. of the $\langle x_{r+s} x_r \rangle$. This dimensionless factor, $f$, is therefore usually known as the *correlation factor*. Its value lies between 0 and 1 (see §8.3.3). Correspondingly, $D_B^{(0)}$ would be the diffusion coefficient if no account were taken of the occurrence of these correlations.

(4) It is easy to write down expressions for $D_B^{(0)}$. The simplest case to deal with is one in which there is just one crystallographically or energetically distinct type of jump. Jumps between nearest-neighbour positions on a cubic lattice provide one such example. Then it follows immediately that $D_B^{(0)}$ is simply $\Gamma s^2/6$ where $\Gamma = n/t$ is the average number of jumps made by a B atom in unit time. Evidently $\Gamma$ will be different for different mechanisms. For example for interstitial solutes, B, it is simply the product of a jump frequency, $w_B$, and the number, $z$, of distinct sites which can be reached in one jump (cf. (8.2.6)). In other cases it will involve the product of a jump frequency with a suitable probability that there is a defect

nearby to effect the jump. This probability in turn depends on the defecty concentration and on any interactions between the solute atoms and the defects.
(5) In practice then, the dependence of $D_B$ on temperature and pressure and on specific features of the host lattice and the solute (e.g. the interatomic forces) comes mainly through the factor $D_B^{(0)}$, i.e. through $\Gamma$. Compared to this rather rich field of variation associated with $D_B^{(0)}$, the correlation factor $f$ may at first sight appear rather pallid and uninteresting. However, this factor is specific to the mechanism by which the B atoms move, and this, coupled with the possibilities of isolating it in one or two types of experiment, accounts for much of the theoretical and experimental attention which it has received. Furthermore, in appropriate circumstances it too may depend on defect concentration, temperature, pressure, etc., and when very small can be limiting.
(6) Although the analysis of $D_B^{(0)}$ is straightforward, the calculation of $f$ is more difficult. The form (10.3.7) is suited to Monte Carlo computations (see e.g. the review by Murch, 1984a), but further analytical development of (10.3.7) is also possible, and, at least at low defect concentrations, this is more accurate. Monte Carlo calculations are of particular value for the more difficult circumstances presented by systems with high defect concentrations.
(7) Lastly, it is important to distinguish the correlation factor $f$ from other similarly denoted quantities which have been introduced into the scientific literature. When we want to emphasize that we are concerned with $f$ as defined by eqns. (10.3.5) and (10.3.7) we shall refer to the Bardeen–Herring correlation factor, but in general this is to be understood.

In the next section we show that in all cases of low defect concentrations we can relate the $\langle x_{r+s}x_r\rangle$ for all $s$ to just the $\langle x_{r+1}x_r\rangle$, i.e. that to obtain $f$ we only need to evaluate the extent of correlations between one jump and the next. Then in §§10.5 and 10.6 we consider the determination of the $\langle x_{r+1}x_r\rangle$.

## 10.4 The correlation factor in terms of the correlations between consecutive jumps

In this section we consider the further reduction of (10.3.7) for the correlation factor. An exact and largely analytical development is possible when the defect concentration is sufficiently low that the movements of the solute atoms result from their individual involvement with one defect at a time. Correlations between successive movements of a solute atom thus result from its direct involvement over a limited period of time with a single defect. Movements of this atom caused by successive encounters with different defects are uncorrelated, simply because different defects arrive in the vicinity of the solute atom randomly. The mathematical problem of calculating the correlation factor becomes much more difficult when the defect concentration is high enough that the sequence of solute movements due to a single defect is curtailed at an early stage by the intervention of others. We shall deal with such cases in Chapter 13. Here we consider systems containing low concentrations of defects, firstly cubic crystals containing vacancies. This section (10.4.1) contains the essential ideas. Section (10.4.2) and Appendix 10.1 give the generalization to crystals and vacancy defects of lower symmetry; this is presented largely for theoretical completeness and may be omitted at a first reading. Section 10.4.3 deals with interstitial defects. There are differences here from vacancy defects.

### *10.4.1 Cubic crystals containing low concentrations of vacancies*

As an introduction to the more general development we shall consider the simple but important example of the diffusion of a dilute solute B in cubic crystals containing vacancies. We can then make the basic simplification that all B-atom jumps are energetically equivalent. (Actually this restriction to just one type of jump is the real limitation on what follows and, of course, this admits a wider range of applications than the usual cubic Bravais lattices; for example, it applies equally well to the diamond lattice and to the two-dimensional honeycomb and triangular lattices.) It is convenient then to use (10.3.7) for the correlation factor. The first point to notice is that $\langle \mathbf{r}_{r+s} \cdot \mathbf{r}_r \rangle$ is independent of $r$; after each jump of the solute atom, i.e. after solute–vacancy exchange, the vacancy and solute stand in an equivalent relation defined by $\mathbf{r}_r$, but the scalar product is invariant with respect to rotations and translations from one lattice displacement vector, $\mathbf{r}_r$, to another. We can therefore re-express (10.3.7) by carrying out the summation over $r$

$$f = 1 + 2 \sum_{s=1}^{n-1} (1 - s/n)\langle \cos \theta_s \rangle, \tag{10.4.1}$$

where $\langle \cos \theta_s \rangle$ is the average cosine of the angle between one solute jump and the $s$th jump later and we take the limit $n \to \infty$.

We next consider the quantities $\langle \cos \theta_s \rangle$. Since we assume that the system is dilute in both solute atoms and vacancies we need to focus only on individual pairs of solute atoms and vacancies, i.e. we consider individual encounters of one solute atom with one vacancy. During the course of these encounters successive movements of the solute atom will be correlated with one another and it is this effect which we want to represent. After each solute movement (resulting from an exchange of position with the vacancy) there will be a whole range of possible paths open to the vacancy before it next exchanges with the solute (Fig. 10.3). However, the rotational invariance of the lattice and the consequential equivalence of the situation immediately after the $s$th solute jump to that immediately after all others means that the correlation between the directions of the $s$th jump and

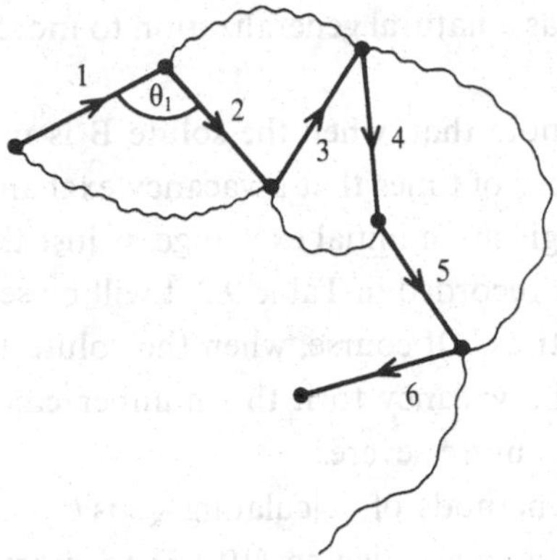


Fig. 10.3. A schematic diagram showing a sequence of steps taken by a solute atom. The wavy lines joining alternate vertices along this path represent the paths taken by the vacancy between the successive displacements of the solute caused by this vacancy.

the first only arises from (a) the correlation between the $s$th jump and the $(s-1)$th immediately preceding and (b) that between the $(s-1)$th and the first. Formally then

$$\langle \cos \theta_s \rangle = \langle \cos \theta_1 \rangle \langle \cos \theta_{s-1} \rangle \tag{10.4.2a}$$

and hence

$$\langle \cos \theta_s \rangle = \langle \cos \theta_1 \rangle^s. \tag{10.4.2b}$$

If we now insert (10.4.2b) into (10.4.1) and pass to the limit of large $n$ we obtain

$$f = \frac{1 + \langle \cos \theta_1 \rangle}{1 - \langle \cos \theta_1 \rangle} = \frac{1+T}{1-T}, \tag{10.4.3a}$$

in which

$$T = \langle x_2 x_1 \rangle / a^2 \tag{10.4.3b}$$

and $a$ is the projection of the nearest-neighbour displacement distance along the (100) cube axis. The calculation of $f$ is then completed by evaluating $\langle \cos \theta_1 \rangle$, or equivalently $T$, by following all subsequent movements of the vacancy (which has just exchanged position with the solute atom) and adding up the products of the chance at each vacancy jump that the solute atom is again displaced multiplied by the corresponding value of $\cos \theta$. For cubic lattices, where the projection of each displacement on to a cube axis is either $+a$, $-a$ or 0, this quantity $T$ as so defined is simply the difference $(p^+ - p^-)$, where $p^+$ is the probability that the next jump of the *solute* is in the same sense as the first (i.e. $x_2 = x_1$) while $p^-$ is the corresponding probability that it is in the opposite sense $(x_2 = -x_1)$, i.e.

$$T = p^+ - p^-. \tag{10.4.4}$$

This form of expression has a natural generalization to more complex cases (§10.4.2 and Appendix 10.1).

At this point we may note that, when the solute B is merely an isotope of the host A, the average number of times that a vacancy exchanges with a solute atom in a given encounter, i.e. given an initial exchange, is just the quantity $\nu$ evaluated in §9.4.2. From the values recorded in Table 9.2 it will be seen that $\nu$ is quite small for three-dimensional lattices. Of course, when the solute B is chemically distinct and is such as to bind the vacancy to it this number can be very much greater and the correlation effects more severe.

We shall describe the methods of calculating $\langle\cos\theta_1\rangle$ in §§10.5 and 10.6, but first we extend the analysis which led to (10.4.3) to more complex examples of vacancy-induced diffusion (e.g. in lattices of lower than cubic symmetry or by more complex vacancy defects) and to interstitialcy migration.

### *10.4.2 The general case of several types of jump: vacancy defects*

The discussion in §10.4.1 above was limited to cases where there was only one distinguishable type of solute–atom jump. Before turning to the methods by which we calculate $\langle\cos\theta_1\rangle$ or $T$ for such cases, we must generalize (10.4.4) to those other cases where there are several, crystallographically and/or energetically inequivalent types of solute jump. Let us define jumps, $r$, to be of the same type if they have the same value of $x_r^2$ and of $\sum_s \langle x_{r+s}x_r\rangle$. Since jumps for which $x_r = 0$ contribute nothing to $D_{Bxx}$ we can omit them in enumerating the different types of jump.

An example is shown in Fig. 10.4 which represents the possible jumps of a solute

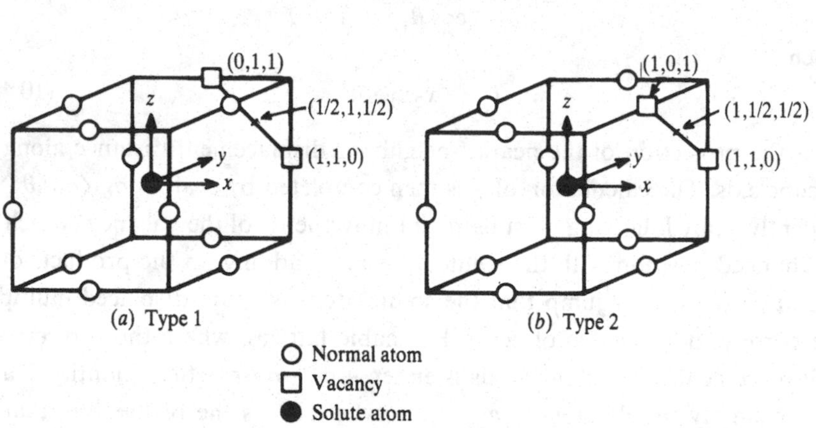


Fig. 10.4. The two types of jump (with non-zero $x_r$) for a solute atom which is migrating by means of a bound vacancy pair in a face-centred cubic crystal. In both cases the solute atom shown has made a $\langle 110\rangle$ jump to the origin site, in (a) of type 1 and in (b) of type 2.

atom into a bound vacancy pair in a f.c.c. lattice. In this example there are just two types of jump; type 1 caused by pairs both of whose vacancies have different $x$-co-ordinates from the solute atom in its initial position, and type 2 caused by pairs which have one vacancy with the same $x$-co-ordinate as the solute atom in its initial position. In other cases also one will normally be able to classify the possible jumps by inspection of the geometrical and mechanistic features. The following analysis of such systems is due originally to Mullen (1961a, b) and to Howard (1966). We again suppose the vacancy concentration to be low.

Now suppose that there are $\sigma$ different types of jump and that $\Gamma_\alpha$ ($\alpha = 1, \ldots, \sigma$) is the average number of jumps of type $\alpha$ made by a solute atom in unit time. By definition, all jumps $r$ of type $\alpha$ have the same value of $|x_r|$ so that (10.3.4) can be written

$$D_B^{(0)} = \frac{1}{2}\sum_\alpha \Gamma_\alpha x_\alpha^2. \tag{10.4.5}$$

Turning next to the summations in the expression for the correlation factor (10.3.5), we may similarly write

$$\frac{\sum_{s=1}^{n-1}\sum_{r=1}^{n-s}\langle x_{r+s}x_r\rangle}{\sum_{r=1}^{n}\langle x_r^2\rangle} = \frac{\sum_{s=1}^{n-1}\sum_\alpha (n-s)C_\alpha\langle x_{\alpha s}x_\alpha\rangle}{\sum_\alpha nC_\alpha x_\alpha^2}, \tag{10.4.6}$$

assuming $n$ to be large. Here $C_\alpha$ ($=\Gamma_\alpha/\sum_\alpha \Gamma_\alpha$) is the fraction of all jumps for which $x_r$ is non-zero of kind $\alpha$ in any long sequence of jumps, while $x_{\alpha s}$ is the $x$-component of the $s$th displacement following an initial jump of type $\alpha$. (In writing this expression we have also taken $(n - s)$ to be large, but since $\langle x_{\alpha s}x_\alpha\rangle$ is expected to decrease rapidly with increasing $s$ this will be the case for all terms of any significance). With (10.4.6) and by taking the limit $n \to \infty$ we obtain for $f$

$$f = 1 + 2\left(\sum_\alpha C_\alpha \sum_{s=1}^{\infty}\langle x_{\alpha s}x_\alpha\rangle\right)\Bigg/\left(\sum_\alpha C_\alpha x_\alpha^2\right). \tag{10.4.7}$$

In complex cases it is often more convenient to express $D_B$ in terms of partial correlation factors, thus in place of (10.3.3) we write

$$D_B = \frac{1}{2}\sum_\alpha \Gamma_\alpha x_\alpha^2 f_\alpha \tag{10.4.8}$$

where

$$f_\alpha = 1 + 2\left(\sum_{s=1}^{\infty}\langle x_{\alpha s}x_\alpha\rangle/x_\alpha^2\right). \tag{10.4.9}$$

In cases where $|x_\alpha|$ is the same for all $\alpha$ (e.g. movement via vacancy defects bound to solute atoms in a cubic lattice) we have the further relation

$$f = \sum_{\alpha} C_\alpha f_\alpha. \tag{10.4.10}$$

After these preliminaries the next step is the analysis of $\langle x_{\alpha s} x_\alpha \rangle$, which is in fact related to $\langle x_{\alpha 1} x_\alpha \rangle$ by a generalization of (10.4.3). This analysis introduces the probability that the $s$th jump following an initial jump of type $\alpha$ is a jump of type $\beta$ having an $x$-displacement in the same sense as the initial jump. We call this conditional probability $p_{\beta\alpha}{}^{s+}$. Likewise, there will be another quantity $p_{\beta\alpha}{}^{s-}$ giving the probability that this $s$th jump of type $\beta$ following an initial jump of type $\alpha$ is in the opposite sense to the initial displacement. (Note the implied assumption in these definitions that only the *relative* sense of the two indicated jumps needs to be specified, which in turn demands that the inverse of an $\alpha$-jump is also of type $\alpha$, and likewise for a $\beta$-jump. This is not the case for interstitialcy mechanisms, which are considered separately below.)

The result of the analysis, details of which are given in Appendix 10.1, is then that the correlation factor (10.4.10) can be expressed as a simple matrix product, viz.

$$f = 1 + 2\tilde{\mathbf{d}}\mathbf{T}(\mathbf{1} - \mathbf{T})^{-1}\mathbf{b}. \tag{10.4.11}$$

Here the matrix elements are labelled by the jump types: so that $\mathbf{b}$ is the column matrix with elements $C_\alpha |x_\alpha| / (\sum_\alpha C_\alpha x_\alpha{}^2)$, $\tilde{\mathbf{d}}$ is the row matrix of displacement magnitudes $|x_\alpha|$ while $\mathbf{T}$ is the square matrix whose elements are

$$T_{\beta\alpha} = p_{\beta\alpha}{}^{1+} - p_{\beta\alpha}{}^{1-}. \tag{10.4.12}$$

This result (10.4.11) is the required generalization of (10.4.4). Evidently it reduces to (10.4.4) when there is only one type of jump available to the solute atom.

Equation (10.4.11) provides the starting point for calculations of more complex systems, e.g. self-diffusion by bound vacancy pairs, the effect of solutes on the self-diffusion of the host ('solute enhancement factors'), self-diffusion and solute diffusion in anisotropic crystals and so on. In some of these cases all the non-zero $x$-displacements are of equal length. An example would be provided by vacancy pairs or other vacancy complexes in a cubic lattice. Certain convenient simplifications are then possible. In particular, the expression for $f$ in terms of partial correlation factors (10.4.10) is convenient. It can then be written in the compact form

$$f = \tilde{\mathbf{C}}\mathbf{f}, \tag{10.4.13}$$

in which $\tilde{\mathbf{C}}$ is the row matrix with elements $C_\alpha$, while $\mathbf{f}$, the column matrix of partial correlation factors, is to be calculated from

$$\mathbf{f} = \boldsymbol{\tau} + 2\mathbf{T}(\mathbf{1} - \mathbf{T})^{-1}\boldsymbol{\tau}, \tag{10.4.14}$$

where $\boldsymbol{\tau}$ is the unit column matrix defined by (6.2.6).

This completes our formulation of the correlation factor for vacancy mechanisms in terms of correlations between successive jumps alone.

### *10.4.3 Migration by the interstitialcy mechanism*

Although vacancy defects are the commonest type of defect in systems in or near thermal equilibrium, there are nevertheless some compounds where the predominant defects in such conditions are Frenkel defects, i.e. vacancies and interstitials. Examples are the silver halides, AgCl and AgBr, which contain Ag Frenkel defects, and compounds with the fluorite structure (e.g. the alkaline earth fluorides, $SrCl_2$, etc.) which contain anion Frenkel defects. In these examples there are strong indications that the interstitial ion moves by the so-called interstitialcy mechanism, illustrated for various lattices in Fig. 10.5. In this an interstitial ion is presumed to move by jumping on to a neighbouring lattice site at the same time as the ion on that site moves away to another interstitial position. Any particular atom therefore moves in an alternating sequence, lattice (l) → interstitial (i) → lattice (l), and so on.

When such a mechanism operates the jumps i → l and l → i are always of different types, so that there is always an even number of jump types. Now, because the interstitial will jump with equal probability to any one of the surrounding lattice sites it follows that

$$\langle x_{r+s} x_r \rangle = 0 \text{ for } s \geq 1 \text{ if } r \text{ is an l} \rightarrow \text{i jump.} \qquad (10.4.15)$$

However, if $r$ is an i → l jump the next move of the solute atom $(r + 1)$ is correlated with it through the presence of the interstitial ion created by the move of the solute. Thus for i → l jumps $\langle x_{r+1} x_r \rangle \neq 0$. However, such jumps must be followed by l → i jumps with the consequence $\langle x_{r+s} x_r \rangle = 0$ for $s \geq 2$ if $r$ is an i → l jump. If now we label the types of i → l jumps with odd numbers and the types of l → i jumps with even numbers then we obtain directly from (10.3.5) with (10.4.15)

$$f = 1 + 2 \sum_{\substack{\alpha \\ \text{odd}}} \sum_{\substack{\beta \\ \text{even}}} |x_\beta| \, T_{\beta\alpha} \, |x_\alpha| \, C_\alpha \Bigg/ \left( \sum_\alpha C_\alpha x_\alpha^{\,2} \right). \qquad (10.4.16)$$

In the terms of Appendix 10.1 we see that, for $s > 1$, $p_{\beta\alpha}^{s+}$ equals $p_{\beta\alpha}^{s-}$ so that the only non-zero $T_{\beta\alpha}^{(s)}$ appearing here is that for $s = 1$.

The simplest cases are those where there is only a single type of interstitialcy jump in a cubic crystal, e.g. one of the three kinds of interstitialcy jump illustrated in Fig. 10.5 for the rocksalt structure. Here, since $C_1 = C_2 = 1/2$ and $x_1 = x_2$, $f$ reduces simply to

$$f = 1 + T_{21}. \qquad (10.4.17)$$

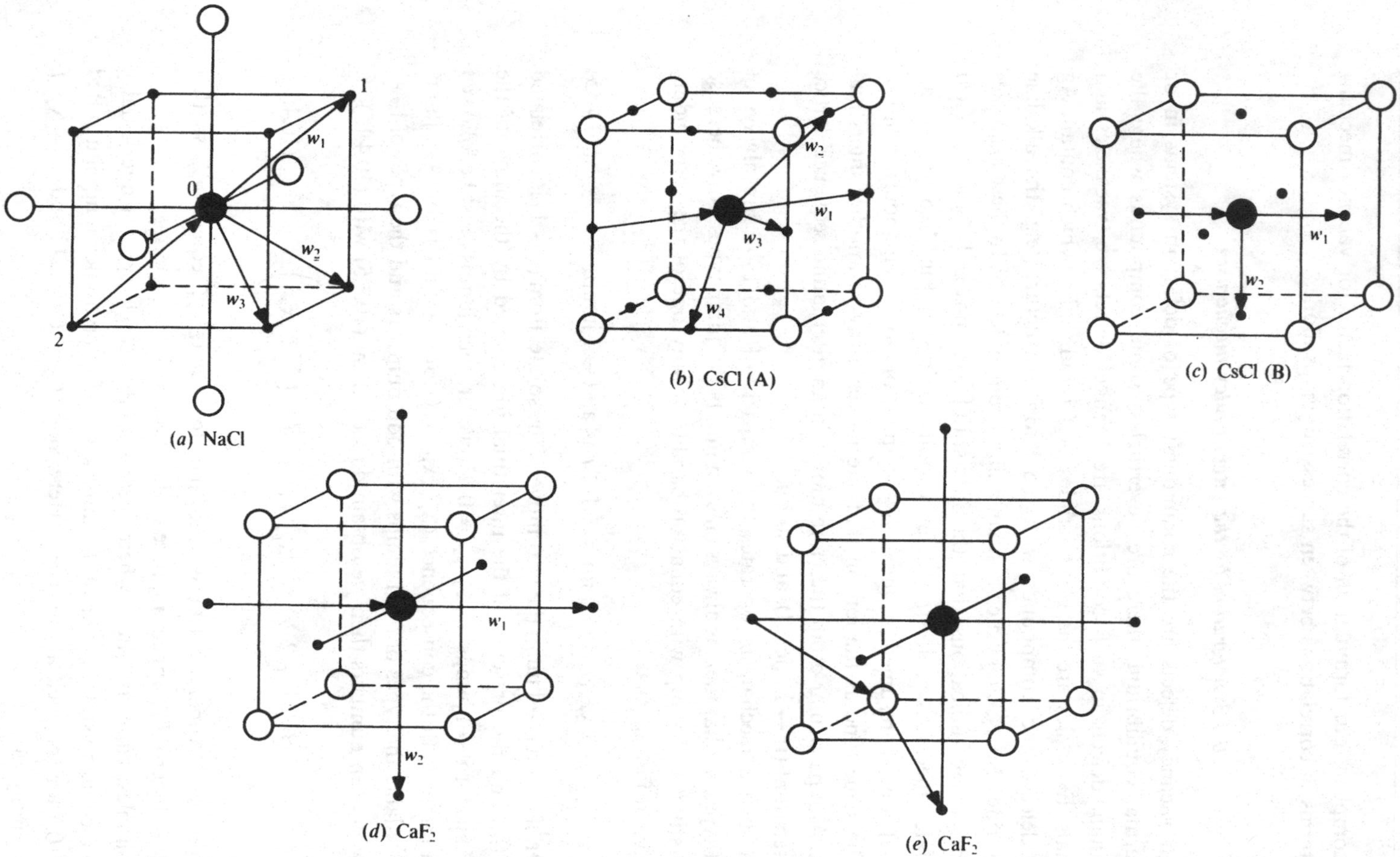


Fig. 10.5. Possible types of interstitialcy movement in three different cubic lattices (*a*) NaCl, (*b*) and (*c*) CsCl and (*d*) and (*e*) $CaF_2$. The two different cases in the CsCl lattice correspond to two possible stable interstitial positions, while in the $CaF_2$ lattice the two cases correspond to possible interstitialcy movements for cations (*d*) and anions (*e*). In all cases the large filled circles represent normal cations, the large open circles represent normal anions while the small filled circles indicate possible interstitial sites.

In this section we have seen that the problem of calculating the correlation factor for diffusion can be reduced to the evaluation of correlations between two successive jumps of the atom, e.g. $\langle \cos \theta_1 \rangle$ or $\langle x_2 x_1 \rangle$ in (10.4.4) or more generally $\mathbf{T}$ in (10.4.11) and (10.4.16). In the next two sections we turn to the evaluation of these quantities, firstly for isotopic self-diffusion (§10.5) and secondly for solute diffusion (§10.6). For systems containing vacancy defects we shall need to obtain the probabilities that the second movement of the solute atom is caused by the same vacancy as its first. We must thus expect to sum over all movements of the vacancy that follow the initial solute–vacancy exchange up until the time that there is a second exchange – the relation of still later solute movements being covered by the relations (10.4.3) and (A10.1.6). From this we can expect to make contact with some of the conditional probabilities and other quantities introduced in Chapter 9. In fact, there are various ways of formulating the problem. Earlier methods have been reviewed by Manning (1968) and Le Claire (1970). We shall describe these and later methods in ways which show their relation to one another and to general results already presented.

## 10.5 The correlation factor for isotopic self-diffusion

Although methods designed to provide correlation factors for solute diffusion can be specialized to give those for self-diffusion (by giving the various jump frequencies their unperturbed values), the generating functions for random walks already obtained in Chapter 9 are especially suited to giving these directly. We therefore treat self-diffusion first.

### *10.5.1 Vacancy defects*

We begin with the evaluation of the correlation factor from (10.4.3a) for the vacancy mechanism in a cubic lattice. We have to calculate $\langle \cos \theta \rangle$, the average of the cosine of the angle between successive exchanges of the tracer with a vacancy. We suppose that after an initial exchange of sites the tracer is at $\mathbf{l}_0$ and the vacancy is at the origin and we denote the nearest neighbours of $\mathbf{l}_0$ by $\mathbf{l}_j$, $j = 1$ to $z$. As in §9.4.3 we denote by $p_n(\mathbf{l}_j, \tilde{\mathbf{l}}_0)$ the probability that the vacancy reaches $\mathbf{l}_j$ at the $n$th step from the origin without having passed through $\mathbf{l}_0$. The probability that the vacancy at $\mathbf{l}_j$ will exchange with the tracer at its next jump is $z^{-1}$ and we may therefore write

$$\langle \cos \theta \rangle = z^{-1} \sum_{j=1}^{z} \cos \theta_j \sum_{n=0}^{\infty} p_n(\mathbf{l}_j, \tilde{\mathbf{l}}_0)$$

$$= z^{-1} \sum_{j=1}^{z} \cos \theta_j U(\mathbf{l}_j, \tilde{\mathbf{l}}_0, 1). \tag{10.5.1}$$

Here $\theta_j$ is the angle between the jump vectors of the tracer in its initial jump from the origin to $\mathbf{l}_0$ and in its next jump to $\mathbf{l}_j$. In the second line we have identified the sum of the $p_n$ as the generating function for a restricted random walk as introduced in (9.4.16). Now this function may be expressed in terms of the generating function for unrestricted walks by (9.4.17). By the symmetry of the lattice implied by the restriction to just one type of jump

$$\sum_j \cos\theta_j U(\mathbf{l}_j - \mathbf{l}_0, 1) = 0, \tag{10.5.2}$$

so that the second term on the right side of (9.4.17) gives no contribution to $\langle\cos\theta\rangle$ and we obtain simply

$$\langle\cos\theta\rangle = z^{-1}\sum_{j=1}^{z}\cos\theta_j U(\mathbf{l}_j, 1), \tag{10.5.3}$$

showing that the restriction on the generating functions in (10.5.1) can in fact be ignored in the cubic lattices. Hence $\langle\cos\theta\rangle$ and the correlation factor for isotopic self-diffusion, $f_0$, can be obtained immediately once the generating functions $U(\mathbf{l}_j, 1)$ have been calculated by one of the methods described in Chapter 9. Take, for example, the case of the f.c.c. lattice (where $z = 12$). If we choose $\mathbf{l}_0$ to be (110) then examination of the lattice structure shows that the 12 neighbours, $\mathbf{l}_j$, fall into five groups, namely (i) (220) with $\cos\theta = 1$, (ii) (211) and three equivalent sites with $\cos\theta = \frac{1}{2}$, (iii) (200) and (020) with $\cos\theta = 0$, (iv) (101) and three equivalent sites with $\cos\theta = -\frac{1}{2}$ and (v) the origin site (000) with $\cos\theta = -1$. Hence

$$\langle\cos\theta\rangle = \tfrac{1}{12}(U(220, 1) + 2U(211, 1) - 2U(101, 1) - U(000, 1)),$$

which, by Table 9.1, yields $\langle\cos\theta\rangle = -0.1227$ and thus by (10.4.3a), $f_0 = 0.7815$. This equation, (10.5.3) and equivalent expressions for the matrix elements of $\mathbf{T}$ appearing in the more general expressions (10.4.11) for the correlation factor have been quite widely used, notably by Montet (1973), Benoist, Bocquet and Lafore (1977) and by Sholl (1981b). Results for the vacancy mechanism are collected in Table 10.1.

It will be noticed from Table 10.1 that for the honeycomb and diamond lattices the tracer correlation factors for the vacancy mechanism are rational fractions. These particular results can be obtained by using the difference equations (9.3.18) without the need to calculate particular generating functions. For example, for the honeycomb lattice shown in Fig. 10.6 one finds from (10.4.12) and (9.4.17) that

$$T = [U(\mathbf{l}_2, 1) - U(\mathbf{l}_1, 1)]/3 \tag{10.5.4}$$

where the origin of the vacancy random walks is site $\mathbf{l}_1$. Consideration of Fig. 10.6 then shows that $U(\mathbf{l}_5, 1) = U(\mathbf{l}_4, 1) = U(\mathbf{l}_0, 1)$ and that $U(\mathbf{l}_3, 1) = U(\mathbf{l}_2, 1)$,

Table 10.1. *Some calculated tracer correlation factors for self-diffusion in pure lattices at infinitely low defect concentrations*

| Lattice | Mechanism | $f_0$ | Reference |
|---|---|---|---|
| Honeycomb (2D) | Vacancy | 1/3 | 1 |
| Square (2D) | Vacancy | $1/(\pi - 1) = 0.4669$ | 1 |
| Triangular (2D) | Vacancy | $(\pi + 6\sqrt{3})/(11\pi - 6\sqrt{3})$ | 1 |
| Diamond | Vacancy | 1/2 | 1 |
| s.c. | Vacancy | 0.6531 | 1 |
| b.c.c. | Vacancy | 0.7272 | 1 |
| f.c.c. | Vacancy | 0.7815 | 1 |
| f.c.c. | Vacancy pair | 0.46 | 2 |
| f.c.c. | ⟨100⟩ dumb-bell interstitial | 0.4395 | 3 |
| NaCl | Cation or anion interstitialcy (as defined in Fig. 10.5) | | |
| | collinear $w_1$ | 2/3 | 4 |
| | non-collinear $w_2$ | 32/33 | 4 |
| | non-collinear $w_3$ | 0.9643 | 4 |
| CsCl | Cation or anion interstitialcy (A, as defined in Fig. 10.5) | | |
| | collinear $w_1$ | 2/3 | 4 |
| | non-collinear $w_2$ | 0.9323 | 4 |
| | non-collinear $w_3$ | 1 | 4 |
| | non-collinear $w_4$ | 0.9120 | 4 |
| CsCl | Cation or anion interstitialcy (B, as defined in Fig. 10.5) | | |
| | collinear $w_1$ | 0 | 4 |
| | non-collinear $w_2$ | 1 | 4 |
| $CaF_2$ | Cation interstitialcy (as defined in Fig. 10.5) | | |
| | collinear $w_1$ | 4/5 | 4 |
| | non-collinear $w_2$ | 1 | 4 |
| $CaF_2$ | Anion interstitialcy (non-collinear, as defined in Fig. 10.5) | 68/69 | 4 |

References to Table 10.1: (1) Montet (1973); (2) Murch (1984c); (3) Benoist *et al.* (1977); (4) Compaan and Haven (1958).

whence (9.3.18) yields

$$U(\mathbf{l}_1, 1) = 1 + U(\mathbf{l}_0, 1) \tag{10.5.5a}$$

and

$$U(\mathbf{l}_0, 1) = [2U(\mathbf{l}_2, 1) + U(\mathbf{l}_1, 1)]/3. \tag{10.5.5b}$$

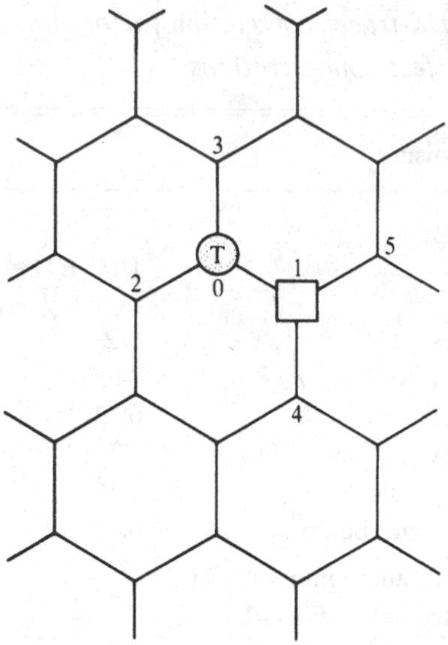


Fig. 10.6. Schematic diagram for the calculation of the tracer correlation factor for self-diffusion on the (two-dimensional) honeycomb lattice. The diagram shows the state after an initial exchange which has left the tracer at site 0 and the vacancy at site 1. The vacancy subsequently makes a random walk and, from one of sites 1 to 3, then makes a second exchange with the tracer. The coordinates of sites 0–3 are denoted $\mathbf{l}_j, j = 0, \ldots, 3$ in the text.

From these three equations one then readily finds $T = -1/2$ and hence $f_0 = 1/3$. Even when such simple results are unavailable, the difference equations can still be useful in the calculation of approximate results. One may repeatedly replace the $U$-function for the shortest random walk in the expression for $T$ in terms of $U$-functions for longer walks. The process can be continued until the value of $f_0$ calculated by neglecting the remaining $U$-functions is stable to the required number of figures.

We conclude this section by referring to a subtlety associated with the calculation of $\langle \cos \theta \rangle$ as it appears in (10.4.3). This concerns the fact that in formulating the expression (10.5.1) for $\langle \cos \theta \rangle$ we were concerned with probabilities which allow for the escape of the vacancy from the tracer. The generating functions $U(\mathbf{l}_j, \bar{\mathbf{l}}_0, 1)$, $j = 1, \ldots, z$, appearing there, although sometimes referred to as probabilities, when summed over all nearest neighbours therefore do not add up to unity. We have already seen in §9.5 that the probability, $\pi'$, that the vacancy will make a second exchange with the tracer is, in fact, relatively small; for example, it is just over 1/4 for an f.c.c. lattice. The implications of such small values of $\pi'$ for the calculation of the correlation factor were pointed out and analysed by Kidson (1978) and Koiwa (1978). The result of their analyses was to modify (10.4.3a) for the correlation factor by the substitution of $\pi'(\cos\theta_1)$ for $\langle\cos\theta_1\rangle$ where the average $(\cos\theta_1)$ is to be calculated from probabilities which are normalized to unity. In the present framework it is clear from (10.5.1) that we can write the same form of expression for the correlation function if we define the average $(\cos\theta_1)$ by

$$(\cos\theta_1) = \sum_{j=1}^{z} q_j \cos\theta_j \tag{10.5.6a}$$

where

$$q_j = U(\mathbf{l}_j, \tilde{\mathbf{l}}_0, 1)/\pi' z. \tag{10.5.6b}$$

It is readily seen from (9.5.2) that the functions $q_j$ are normalized to unity so that the end result of the analyses of Kidson (1978) and Koiwa (1978) is the same as that based on (10.5.1), as one would expect.

### *10.5.2 Interstitial defects*

We turn now to the correlation factors for self-diffusion via interstitial defects. As is already clear, for direct interstitial movements from one interstitial site to another there are no correlations between successive moves and $f_0 = 1$. More appropriate to the consideration of self-diffusion, however, are the dumb-bell interstitial and the interstitialcy mechanisms of migration (Chapters 2 and 8), particularly the latter. Some of the results are given in Table 10.1. We consider the calculation of those for the interstitialcy mechanism first.

For models, such as those defined in Fig. 10.5 the calculation of the correlation factor differs in one way from those so far described. We may illustrate the point for the calculation of the correlation factor from (10.4.17) for the particular case of the collinear cation interstitialcy mechanism in the rocksalt structure. Fig. 10.5(*a*) shows the configuration after an initial tracer jump from interstitial site 2 to substitutional site 0 has left an interstitial at site 1. The subsequent interstitialcy random walk prior to the next tracer jump will take the interstitial over a diamond lattice of sites, there being only four directions in which an interstitial can move by collinear jumps. By inspection of the lattice we can then see that, if we start with an interstitial ion at site 1, interstitial ions can only appear subsequently at sites 1 and 2 among those neighbouring the tracer. The next tracer jump can therefore be only from site 0 to site 2 or from site 0 to site 1 and we therefore obtain from (10.4.17)

$$f_0 = 1 + (U_{21}^{(\mathrm{r})} - U_{11}^{(\mathrm{r})})/4, \tag{10.5.7}$$

in which the generating functions $U^{(\mathrm{r})}$ are labelled by the site of the interstitial on the diamond lattice (and are evaluated at $y = 1$, in the notation of §9.3). The random walks of the interstitial described by these functions have the restriction that the interstitial may not jump from site 1 to site 2, nor the reverse, since these jumps would displace the tracer. This is an illustration of the point that restrictions on the generating functions for interstitialcy mechanisms are generally such that jumps between certain nearest-neighbour pairs of sites are forbidden rather than, as with vacancies, that a certain site is not visited.

It is straightforward to relate restricted to unrestricted generating functions in such cases. In the matrix notation of (9.2.7) we have

$$\mathbf{U}^{(r)} = (1 - \mathbf{A}^{(r)})^{-1} \tag{10.5.8a}$$

and

$$\mathbf{U} = (1 - \mathbf{A})^{-1} \tag{10.5.8b}$$

whence by defining $\Delta\mathbf{A}$ as $\mathbf{A}^{(r)} - \mathbf{A}$ we obtain the Dyson equation

$$\mathbf{U}^{(r)} = \mathbf{U} + \mathbf{U}\,\Delta\mathbf{A}\,\mathbf{U}^{(r)}. \tag{10.5.9}$$

Now the matrix $\Delta\mathbf{A}$ has only a small number of non-zero elements corresponding to the forbidden jumps and it is therefore conveniently partitioned in the form

$$\Delta\mathbf{A} = \begin{bmatrix} \Delta\mathbf{a} & \mathbf{0} \\ \mathbf{0} & \mathbf{0} \end{bmatrix} \tag{10.5.10}$$

where $\Delta\mathbf{a}$ is a small matrix with elements $\Delta a_{\mathbf{ml}}$ equal to $-z^{-1}$ if one jump from site $\mathbf{l}$ to site $\mathbf{m}$ is forbidden and equal to zero otherwise. If the matrices $\mathbf{U}^{(r)}$ and $\mathbf{U}$ are partitioned in the same block structure and $\mathbf{u}^{(r)}$ and $\mathbf{u}$ denote the block matrices spanning the same sites as $\Delta\mathbf{a}$, then we obtain

$$\mathbf{u}^{(r)} = (\mathbf{1} - \mathbf{u}\,\Delta\mathbf{a})^{-1}\mathbf{u}. \tag{10.5.11}$$

The unrestricted generating functions $\mathbf{u}$ may be evaluated by the analytic and numerical methods described in §9.3 (see e.g. Benoist *et al.* (1977) for examples).

As with vacancies, some exact results can be obtained for interstitialcy mechanisms without numerical work by exploiting the difference equations (9.3.18). Let us return to the example of the collinear interstitialcy mechanism in the rocksalt structure already referred to. We need the combination of generating functions appearing in (10.5.7) for $f_0$. Firstly, by starting from (10.5.11) and remembering that $u_{21} = u_{12}$ and $u_{11} = u_{22}$ (Fig. 10.5(a)) we find after a short calculation that

$$u_{21}^{(r)} - u_{11}^{(r)} = (u_{21} - u_{11})/(1 + (u_{21} - u_{11})/4). \tag{10.5.12}$$

We next use (9.2.8) for the unrestricted generating function. With $\lambda = 0$, corresponding to $y = 1$ (cf. 9.3.14), and the recognition that $u_{j1} = u_{21}$ for all four

$j \neq 1$ (cf. Fig. 10.5(a)) we obtain $u_{21} - u_{11} = -1$. Hence by (10.5.12) and (10.5.7) $f_0 = 2/3$, as shown in Table 10.1.

A decimal approximation to this particular result was first obtained by McCombie in the context of the self-diffusion of Ag in AgCl (McCombie and Lidiard 1956). Shortly afterwards the exact result and other results for the interstitialcy mechanism in a variety of other lattices were obtained by Compaan and Haven (1958). Their method, in effect, replaced the difference equations for the generating functions by analogous equations of current flow in a resistance network. This allowed them to obtain both exact results, like that just described, and also numerical results in other cases. Their numerical results were obtained by direct measurements of electrical currents in a resistance network in the laboratory. The method has been subsequently reviewed by Adda and Philibert (1966) and Le Claire (1970). Franklin (1965) has pointed out a further analogy in terms of the charge and potential distributions in a capacitance network. These analogues are of less interest today now that the numerical work required by the methods of this section can be done so conveniently by digital computers.

Lastly, we comment on the dumb-bell mechanism. Although the result for the f.c.c. dumb-bell given in Table 10.1 was first obtained by Benoist *et al.* (1977) by a special method, it can also be obtained in the way described here. Several types of jump must be recognized, but all but one of the matrix elements $T_{\alpha\beta}$ can be determined by inspection and (10.4.11) then reduces to the form given by Benoist *et al.* (1977), namely

$$f_0 = \frac{1+T}{2+T}, \tag{10.5.13}$$

in which $T$ is a matrix element which must be determined numerically. Their value $T = -0.2160$ then yields $f_0 = 0.4395$, as in Table 10.1.

Having considered the calculation of correlation factors for self-diffusion, we turn next to those for dilute alloys and solid solutions. In these cases we must take account of the changes to the defect movements which arise in the vicinity of the solute atoms. These are represented by models of the sort we have encountered already in Chapters 7 and 8.

## 10.6 The correlation factor for dilute alloys: one type of vacancy jump

There are two principal methods to be considered, the *matrix method* (a name suggested by Manning, 1968) and the *diffusion of probability method*. We deal with these separately.

### 10.6.1 *The matrix method*

Consider the situation already schematized in Fig. 8.3 where we suppose that all solute–vacancy nearest-neighbour exchanges occur with the same jump rate (which we shall later designate as $w_s$). We shall suppose that the $x$-axis passing through the site occupied by the solute atom is an axis of rotational symmetry; and we shall group rotationally equivalent sites together (e.g. those nearest-neighbour sites marked 1 etc. in Fig. 8.5). Let us label these sets of equivalent sites $i, j, k, \ldots$, etc., and denote the jump rate for vacancy jumps from a particular site of type $i$ to one of type $j$ by $w_{ji}$. For simplicity we omit the superscript which we would otherwise use to indicate that the jump rates are for thermodynamic equilibrium. (In general, these quantities $w_{ji}$ will include appropriate numerical factors when there is more than one way of accomplishing the indicated transition.) Now suppose that the vacancy has made $n$ jumps since the time of the initial solute–vacancy exchange *without exchanging again with the solute*, and denote the probability that it is on the sites $i$ (i.e. equally on all equivalent sites of type $i$) at this stage by $p_i^{(n)}$. It is then apparent from Fig. 8.3 that the contribution of the $(n+1)$th vacancy jump to $\langle x_2 x_1 \rangle$ is simply

$$-a^2(p_1^{(n)} - p_{\bar{1}}^{(n)})w_s \bigg/ \left(\sum_i w_{i1}\right),$$

in which $w_{i1}$ is the vacancy jump frequency from a site of type 1 to a site of type $i$ (including the solute site $s$). Here we have used the equivalence of the sites $i$ accessible from 1 with the sites $\bar{i}$ accessible from $\bar{1}$ and set $w_{s1} = w_{s\bar{1}} \equiv w_s$ and $\sum_i w_{i1} = \sum_i w_{\bar{i}\bar{1}}$. It follows therefore that

$$\langle x_2 x_1 \rangle = -a^2 \frac{w_s}{(\sum_k w_{k1})} \sum_{n=0}^{\infty} (p_1^{(n)} - p_{\bar{1}}^{(n)}), \tag{10.6.1}$$

where the summation over $k$ in the denominator includes the site occupied by the solute s. The relation of the $p_i^{(n)}$ to the $p_i^{(n-1)}$ can be expressed by a matrix equation.

$$\mathbf{p}^{(n)} = \mathbf{A}\mathbf{p}^{(n-1)}, \tag{10.6.2}$$

where the element $A_{ji}$ of the matrix $\mathbf{A}$ is the probability that when a vacancy initially at a site $i$ jumps it does so to a site $j$ ($\neq$s), i.e.

$$A_{ji} = w_{ji} \bigg/ \left(\sum_k w_{ki}\right), \qquad (i, j \neq \text{s}). \tag{10.6.3}$$

We notice that this matrix $\mathbf{A}$ belongs to that class of matrices defined by equation (9.2.4). However, we must note that, although $\mathbf{A}$ is defined for all sites $i, j$ other than the solute site s, the jumps represented in the denominator in (10.6.3) do include jumps to the solute site, i.e. the sum over $k$ includes s. This means that the matrix $\mathbf{1} - \mathbf{A}$ is non-singular and thus $(\mathbf{1} - \mathbf{A})^{-1}$ exists. Lastly, if we introduce the row vector $\boldsymbol{\xi}$ whose 1-component is $+1$, $\bar{1}$-component is $-1$ and all others are zero, i.e.

$$\boldsymbol{\xi} = [1 \quad -1 \quad 0 \; 0 \quad \cdots \quad 0] \tag{10.6.4}$$

then (10.6.1) with (10.6.2) yield

$$\langle x_2 x_1 \rangle = -a^2 w_s \boldsymbol{\xi}(\mathbf{1} - \mathbf{A})^{-1}\mathbf{p}^{(0)} \Big/ \left( \sum_i w_{i1} \right). \tag{10.6.5}$$

The vector $\mathbf{p}^{(0)}$ describing the vacancy distribution immediately following the initial solute–vacancy exchange but before any further vacancy jumps have taken place has only zero elements except for the 1-component, which is unity,

$$\mathbf{p}^{(0)} = \{1 \quad 0 \quad 0 \quad \cdots \quad 0\}. \tag{10.6.6}$$

If we denote the matrix $(\mathbf{1} - \mathbf{A})^{-1}$ by $\mathbf{U}^{(1)}$, then by (10.6.4)–(10.6.6) it is clear that

$$T = \langle x_2 x_1 \rangle / a^2 = -w_s (U_{11}^{(1)} - U_{\bar{1}1}^{(1)}) \Big/ \left( \sum_i w_{i1} \right) \tag{10.6.7}$$

The matrix $\mathbf{U}^{(1)}$ introduced here is clearly an example of (9.2.7) for $\lambda = 0$, when $\mathbf{Y}$ becomes just the unit matrix.

The argument and method leading to (10.6.7) is essentially that suggested originally by Bardeen and Herring (1952), as slightly modified by later workers (Compaan and Haven, 1956; Mullen, 1961b). Many correlation factors have been calculated in this way (see e.g. Le Claire, 1970). However, a second formulation has also been widely used. We may call this the *diffusion of probability method* because it follows the changes in vacancy occupation probability continuously in time rather than jump by jump as in the matrix method. (N.B. Manning (1968) uses the same name in a different connection.)

### *10.6.2 The diffusion of probability method*

If we measure time, $t$, from the instant of the first solute–vacancy exchange and let $p_i(t)$ be the probability that the vacancy is at site $i$ at time $t$ then

$$\langle x_2 x_1 \rangle = -a^2 w_s \int_0^\infty \boldsymbol{\xi} \mathbf{p}(t) \, dt, \tag{10.6.8}$$

which is the continuous time equivalent of (10.6.1). As before, $\mathbf{p}(t)$ must be calculated by terminating all vacancy paths which lead to an exchange with the solute, i.e. by treating the solute site s as a vacancy sink. The evolution of $p_i(t)$ under these conditions is given by

$$\frac{\mathrm{d}p_i}{\mathrm{d}t} = -\sum_{j \neq 1} w_{ji} p_i + \sum_{j \neq i}' w_{ij} p_j. \tag{10.6.9}$$

The prime on the second sum – which represents vacancy jumps *to* site $i$ – indicates that when $i = 1$ the term in $w_{1\bar{1}}$ is omitted and when $i = \bar{1}$ the term in $w_{\bar{1}1}$ is omitted, thus expressing the restriction that only vacancy movements between two consecutive vacancy–solute exchanges are to be considered or, in other words, that the solute site represents a perfect sink for the vacancy. Equation (10.6.9) can be written in matrix form

$$\frac{\mathrm{d}\mathbf{p}}{\mathrm{d}t} = -\mathbf{R}\mathbf{p} \tag{10.6.10}$$

in which

$$R_{ii} = \sum_{j \neq i} w_{ji}, \tag{10.6.11a}$$

$$R_{ij} = -w_{ij}, \qquad (i \neq j), \tag{10.6.11b}$$

except for $(i, j) = (1, \bar{1})$ when

$$R_{1\bar{1}} = 0 = R_{\bar{1}1}. \tag{10.6.11c}$$

If we take the solution of (10.6.10) to be

$$\mathbf{p}(t) = \mathrm{e}^{-\mathbf{R}t}\mathbf{p}^{(0)}, \tag{10.6.12}$$

with $\mathbf{p}^{(0)}$ given by (10.6.6), and insert it into (10.6.8), we obtain

$$T = \langle x_2 x_1 \rangle / a^2 = -w_s(G^{(1)}_{11} - G^{(1)}_{\bar{1}1}), \tag{10.6.13}$$

in which

$$\bar{\mathbf{G}}^{(1)} = \mathbf{R}^{-1}. \tag{10.6.14}$$

The notation $\bar{\mathbf{G}}^{(1)}$ is chosen to emphasize the formal similarity with some of the analysis of Chapter 6. Thus (10.6.10) and (10.6.12) are similar to (6.2.2) and (6.3.1) respectively, so that the quantity (10.6.14) is the analogue of the resolvent $\bar{\mathbf{G}}(\lambda)$ defined by (A6.2.3) for $\lambda = 0$ (except that $\mathbf{R}$, unlike $\mathbf{P}$, is non-singular by virtue of (10.6.11c)).

The equivalence of (10.6.13) to the expression (10.6.7) obtained by the matrix method follows by noting that, from the definitions of $\mathbf{A}$ and $\mathbf{R}$,

$$\mathbf{R} = (\mathbf{1} - \mathbf{A})\mathbf{R}_{\mathrm{d}}, \tag{10.6.15}$$

in which $\mathbf{R}_d$ is the diagonal part of $\mathbf{R}$; therefore

$$\bar{\mathbf{G}}^{(1)} = \mathbf{R}_d^{-1}(\mathbf{1} - \mathbf{A})^{-1} = \mathbf{R}_d^{-1}\mathbf{U}^{(1)}, \tag{10.6.16}$$

and the equivalence of (10.6.7) and (10.6.13) follows from (10.6.11a). This relation (10.6.16) is an example of (9.2.1).

### *10.6.3 Further reduction of both methods by use of symmetry*

In practice the accuracy of calculations based on either the matrix method (10.6.7) or the diffusion of probability method (10.6.13) will be limited by the size of the matrix $(\mathbf{1} - \mathbf{A})$ or $\mathbf{R}$ which is finally inverted. Reducing the size of the matrix means neglecting certain random walks of the vacancy by which it would return to the solute atom and exchange with it again. The further away the vacancy wanders the more nearly will the return to the solute site be random in direction and the smaller will be the contribution of such paths to $\langle x_2 x_1 \rangle$. It is therefore important to use symmetry to reduce the dimensions of the matrices so that as many distant sites as possible are included when the matrix is truncated.

In cases where the plane through the solute atom normal to the $x$-axis is a plane of symmetry (either of reflection or of reflection plus rotation about $x$) further reduction of (10.6.13) is straightforward. For then we can match sites $i$ in the forward half-space to those $\bar{i}$ in the backward half-space so that

$$w_{ij} = w_{\bar{i}\bar{j}}, \tag{10.6.17a}$$

and

$$w_{\bar{i}j} = w_{i\bar{j}}. \tag{10.6.17b}$$

We then re-write (10.6.9) as an equation for $(p_i - p_{\bar{i}}) \equiv \pi_i$ ($i$ in the forward half-space only) and find that the equation for the reduced vector $\mathbf{p}$ is of the form (10.6.10) but with $\mathbf{R}$ replaced by $\mathbf{R}'$ which is such that

$$R'_{ij} = R_{ij} - R_{i\bar{j}}. \tag{10.6.18}$$

(The pairs of configurations $j$ and $\bar{j}$ appearing in this definition of $R'_{ij}$ are sometimes referred to as *negatively equivalent configurations* (e.g. Manning, 1968; 1972).)

Since $i$ is necessarily identical with $\bar{i}$ for sites lying on the reflection plane, $\pi_i$ is zero for such sites, which are thereby removed from the calculation entirely. Hence the order of $\mathbf{R}'$ is never greater than half that of $\mathbf{R}$ and for many lattices is less than half, which can be a valuable saving. If we define

$$\bar{\mathbf{G}}^{(2)} = (\mathbf{R}')^{-1} \tag{10.6.19}$$

we then get in place of (10.6.13)

$$T = \langle x_2 x_1 \rangle / a^2 = -w_s \bar{G}_{11}^{(2)}. \tag{10.6.20}$$

In a similar way (10.6.7) for the matrix method can be replaced by

$$T = \langle x_2 x_1 \rangle / a^2 = -w_s U_{11}^{(2)} \left( \sum_i w_{i1} \right)^{-1} \tag{10.6.21}$$

in which

$$\mathbf{U}^{(2)} = (\mathbf{R}' \mathbf{R}_d^{-1})^{-1}. \tag{10.6.22}$$

Of course, the two expressions (10.6.20) and (10.6.21) remain exactly equivalent, as were (10.6.7) and (10.6.13).

### *10.6.4 Relation to kinetic theory*

In the preceding section we showed the equivalence of the two principal ways of evaluating $T$ in (10.4.3a) for $f$. We are now in a position to demonstrate the further equivalence of the correlation factor so obtained to that arising in the kinetic theory presented in Chapter 8 (8.3.32). This will then provide one of the principal conclusions of this chapter.

First of all we see that the matrix $\mathbf{Q}$ of §8.3 is related to the matrix $\mathbf{R}'$ defined by (10.6.18) in the following way:

$$Q_{ij} = R'_{ij}, \tag{10.6.23a}$$

except when $i = j = 1$ when

$$Q_{11} = R'_{11} + w_s. \tag{10.6.23b}$$

It is then easy to see that with

$$\mathbf{G}^{(3)} = \mathbf{Q}^{-1} \tag{10.6.24}$$

we obtain

$$G_{11}^{(2)} = G_{11}^{(3)} / (1 - w_s G_{11}^{(3)}) \tag{10.6.25}$$

and therefore by (10.4.3) and (10.6.20) that the correlation factor

$$f = 1 - 2 w_s G_{11}^{(3)}. \tag{10.6.26}$$

The quantity $G_{11}^{(3)}$ obtained from (10.6.24) thus describes the correlation effects in the solute motion directly. Furthermore, since we only require the (1, 1) element of $\mathbf{Q}^{-1}$ it is instructive to write $G_{11}^{(3)}$ in a slightly expanded form as

$$G_{11}^{(3)} = \left[ Q_{11} - \sum_{p,r} Q_{1p} (\hat{\mathbf{Q}}^{-1})_{pr} Q_{r1} \right]^{-1}, \tag{10.6.27}$$

in which the sites $r$ and $p$ are those which can be reached in a single jump from the first-neighbour shell site 1 (but will not themselves contain any first shell sites). The matrix $\hat{\mathbf{Q}}$ is obtained from $\mathbf{Q}$ by removing the 1-row and 1-column; $(\hat{\mathbf{Q}}^{-1})_{pr}$ thus links together only second-shell sites $r$ and $p$ via vacancy trajectories which do *not* enter the first shell. Equation (10.6.27) is thus capable of a physical interpretation. From the definitions (10.6.11a), (10.6.18) and (10.6.23b) we see that $Q_{11}$ is the total frequency of vacancy jumps from a 1-site to other sites, except that the frequency of vacancy–solute exchange is doubled in this sum. The second term in (10.6.27) represents a reduction in the overall frequency of vacancy jumps from a first-neighbour-site to second shell sites as the result of vacancies returning to the first-neighbour-sites at the end of trajectories which have proceeded through the second and more distant shells.

One feature of the exact expression for the solute correlation factor which turns out to be of importance later is the form of the dependence of $f$ upon the solute jump frequency $w_s$. In the denominator of (10.6.27) the first term $Q_{11}$ contains $2w_s$ plus terms independent of $w_s$, while the second term is evidently completely independent of $w_s$. It therefore follows from (10.6.26) and (10.6.27) that the correlation factor may be written in the form

$$f = \frac{H}{2w_s + H}, \tag{10.6.28}$$

in which the quantity $H$ is independent of the solute jump frequency. This particular structure holds for transport by single vacancies, but it often does not hold for other migration mechanisms.

### *10.6.5 A simple approximation*

It is sometimes a useful first approximation to $G_{11}^{(3)}$ to neglect the second term in the square brackets in (10.6.27), i.e. to neglect those vacancy paths which bring the vacancy back to the first shell of neighbours once it has left it. Then

$$f = 1 - \frac{2w_s}{Q_{11}}, \tag{10.6.29}$$

which, for any particular model, can be written down by inspection. For example, for the five-frequency model for solute diffusion in a f.c.c. lattice (in which the solute–vacancy exchange frequency $w_s$ is denoted $w_2$ while the other vacancy jump frequencies are as defined in Figs 2.22 and 7.1) we see that

$$Q_{11} = 2w_2 + 2w_1 + 7w_3, \tag{10.6.30}$$

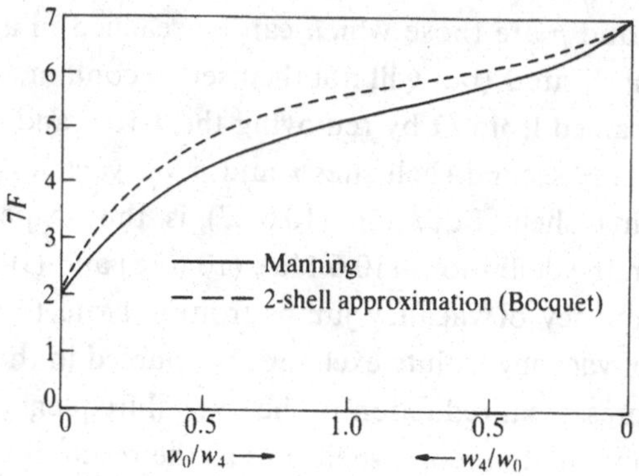


Fig. 10.7. The quantity $7F$ for the f.c.c. five-frequency model defined by Fig. 2.22, shown as a function of the ratio of the jump frequencies $w_0$ and $w_4$.

whence

$$f = \frac{(2w_1 + 7w_3)}{(2w_2 + 2w_1 + 7w_3)}. \tag{10.6.31}$$

The approximation embodied in (10.6.29) is sometimes referred to as the Lidiard–Le Claire approximation. The term omitted from (10.6.27) to get (10.6.29) defines the parameter $F$ introduced by Manning to preserve the form of the expressions resulting from (10.6.29). For example, in place of (10.6.31) we would have

$$f = \frac{(2w_1 + 7Fw_3)}{(2w_2 + 2w_1 + 7Fw_3)}, \tag{10.6.32}$$

in which $F$ is a function of the ratio of jump frequencies $(w_4/w_0)$ as defined in Fig. 2.22. For this model it is a purely numerical quantity which has been calculated by Manning (1964) and others subsequently by following the motion of the vacancy among the distant sites of the lattice (see Chapter 11). The result for $F$ for the five-frequency model is shown in Fig. 10.7. (Analogous results for similar models of diffusion in other lattices have also been obtained; see Chapter 11.)

The expression (10.6.32) for the correlation factor for the five-frequency model naturally has the structure anticipated in (10.6.28) for the dependence of $f$ on the solute jump-frequency. If we were to neglect those vacancy paths that bring the vacancy back to the first shell of neighbours once it has left it ($F = 1$) then $H$ would be simply the total escape frequency of the vacancy from a particular position neighbouring the solute atom. When we allow for the possibility that the vacancy may return to the sites neighbouring the solute atom ($F \neq 1$) then the value of $H$ is reduced and it can then be interpreted as an *effective* frequency of escape of the vacancy from a particular position neighbouring the solute atom. The evaluation of $F$ and $H$ is considered further in §11.2.

## 10.7 The correlation factor for dilute alloys: more than one type of vacancy jump

We have already referred to several important examples where there is more than one type of distinct vacancy jump. In such cases one has to calculate all the elements $T_{\beta\alpha}(\alpha, \beta = 1, \ldots, \sigma)$ defined in (10.4.12) and the formalism of the preceding section requires elaboration. One important example is that of host tracer diffusion in a dilute f.c.c. alloy, where we need to consider those movements of the host tracer atom A* which are caused by a vacancy in the vicinity of a solute atom, B. (A later figure (11.2) defines the jump types when the five-frequency model is employed.) We supply the details of this calculation in Appendix 10.2, and take up the topic again in §11.2.

## 10.8 Tracer diffusion in concentrated alloys – the random alloy model

As we have emphasized already, the analysis of the preceding sections has been for dilute solid solutions. It thus yields the diffusion coefficients of two distinct systems: (i) of solutes themselves, (ii) of isotopes of the host (either in its pure form or as the solvent in a dilute alloy). The essential limitation of this analysis is the assumption that it is adequate to examine encounters between one defect and one solute atom (or isotopic atom of the chemical species considered), while the consequences of this encounter are evaluated by following the movements of the defect – which may be complex (e.g. a solute–vacancy pair) – through a perfect lattice.

By using the insights obtained from these calculations Manning (1968, 1970, 1971) proposed a theory of tracer diffusion in a simplified model of a *concentrated* alloy. In effect, he retained the emphasis upon the encounter between one isotopic atom and one vacancy but the movements of the vacancy in the model alloy were averaged out in a self-consistent way. The model was defined by assuming (i) that the arrangement of the atoms of the different chemical species on the lattice was random and, correspondingly, that the vacancies had no preference for association with atoms of any one component over the others, and (ii) that the rates of atom–vacancy exchange $w_i$ were characteristic of the atomic species, $i$, and were independent of the detailed configuration of atoms around the vacancy. This we shall refer to as the *random alloy model.*

Manning's analysis of this model is approximate, but turns out to be rather accurate. The simple mathematical form of its results makes this accuracy all the more remarkable. It is therefore an important theory. Nevertheless the arguments deployed are complex and rather particular, so that it has not proved at all easy to extend the theory to other systems. We shall show how it fits into a more rigorous and systematic analysis in Chapter 13. Here it is sufficient to note that the arguments reduce to the following three assumptions (Manning, 1971; Stolwijk, 1981; Allnatt and Allnatt, 1984) which, however, remain intuitive in character.

The first assumption is that the diffusion coefficient, $D_i^*$, for a tracer atom of species $i$ in an alloy of uniform composition preserves the form (10.3.3) derived for solutes in a dilute alloy and, in particular, that the correlation factor is likewise of the form

$$f_i = H_i/(2w_i + H_i) \tag{10.8.1}$$

(cf. (10.6.28) in which $w_s$ takes the place of $w_i$). Like $H$ in the expression (10.6.28) for dilute alloys, the quantity $H_i$ is interpreted as an effective frequency of escape of the vacancy from a particular position neighbouring the tracer atom in question. This equation is the expression of the idea that the diffusion of the tracer of $i$ results from a series of encounters of each tracer atom with one vacancy at a time. The remaining two assumptions provide a self-consistent way of obtaining $H_i$.

These remaining assumptions require the notion of the average uniform alloy in which vacancies move with the average jump frequency

$$w = \sum_i c_i w_i. \tag{10.8.2}$$

In such a system the vacancy diffusion coefficient would be

$$D_V^{(0)} = \tfrac{1}{6} z w s^2, \tag{10.8.3}$$

where $z$ is the co-ordination number of the lattice and $s$ is the nearest-neighbour jump distance. In the actual alloy, due to the differing rates of exchange of the vacancy with the various types of atom, the vacancy will actually diffuse more slowly than (10.8.3) predicts – by a factor $f_V$, referred to as the vacancy correlation factor. The second assumption is then that $H_i$ for species $i$ is simply $f_V$ times the value of $H$ in the uniform average alloy obtained from (10.8.1) with $w_i$ replaced by $w$, i.e. that

$$H_i = f_V H_0 = \frac{2wf_0 f_V}{(1 - f_0)}, \tag{10.8.4}$$

in which $H_0$ is the value of $H$ in the uniform average alloy and $f_0$ is the isotope correlation factor for the pure lattice (Table 10.1).

The third assumption has already been presented in one form, namely (5.8.6), which states that the mobility $u_i^M$ of $i$ atoms bears the same relation to $D_i^*$ as it does in the pure material. When this assumption is employed in the formal expression for $D_V$ in terms of the $L_{ij}$, namely $D_V = kT\sum_{i,j} L_{ij}/nc_V(1 - c_V)$ (cf. eqns. (5.7.5)) then the corresponding expression for the vacancy correlation factor $f_V = D_V/D_V^{(0)}$ is found to be

$$f_V = \frac{\sum_j c_j w_j f_j}{w f_0 \sum_j c_j}. \tag{10.8.5}$$

It may also be noted that by this equation and (10.8.1) and (10.8.4) we also obtain

$$\sum_i c_i f_i = f_0 \sum_i c_i, \tag{10.8.6}$$

so that $f_V$ can alternatively be written as a sort of mean of the atom correlation factors

$$w f_V = \frac{\sum c_j w_j f_j}{\sum c_j f_j}. \tag{10.8.7}$$

The form of this expression is notable since it approximates a quantity which depends on all the $L_{ij}$ of the chemical components, namely $f_V$, by a function of only the tracer correlation factors. That a useful approximation of this kind should exist is not intuitively obvious.

For an $n$-component alloy there are therefore $(n + 1)$ equations for the $(n + 1)$ correlation factors $f_j$ $(j = 1, \ldots, n)$ and $f_V$. The equation for $f_V$ becomes an $n$th order polynomial. In the special case of a binary alloy AB one may easily obtain the explicit solution

$$H_i = h + [h^2 + 4w_A w_B f_0/(1 - f_0)]^{1/2}, \qquad (i = \mathrm{A, B}), \tag{10.8.8a}$$

with

$$h = w/(1 - f_0) - (w_A + w_B). \tag{10.8.8b}$$

The predictions of (10.8.1) and (10.8.8) for $f_A$ and $f_B$ have been tested by computer simulation of the model and found to be remarkably accurate (Bocquet, 1974a; de Bruin *et al.*, 1975, 1977; Murch and Rothman, 1981): see Fig. 10.8.

An important practical consequence of Manning's theory is that, for those systems for which the random alloy model is applicable, the tracer correlation factors can be calculated if all the tracer diffusion coefficients are known. The expression which allows this, namely,

$$f_A = 1 - \frac{D_{A*}(1 - f_0)\left(\sum_i c_i\right)}{\sum_i c_i D_i^*}, \tag{10.8.9}$$

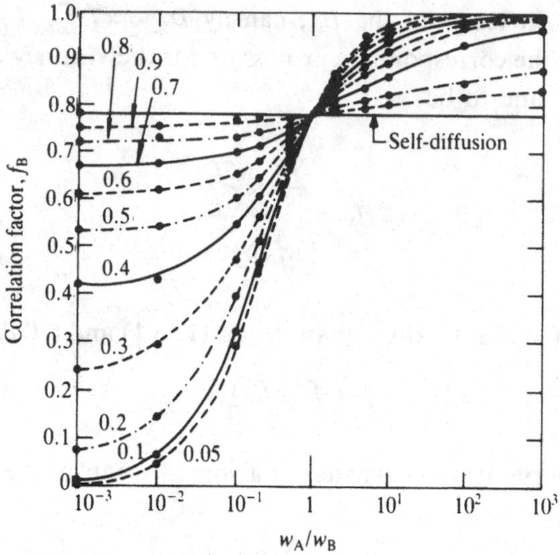


Fig. 10.8. The correlation factor $f_B$ for a tracer atom of chemical species B in a random f.c.c. alloy $A_xB_{1-x}$ as a function of jump frequency ratio $w_A/w_B$ for various compositions $x$ as indicated. The lines are the predictions of eqns. (10.8.1) and (10.8.8) while the circles are the results of Monte Carlo simulations by de Bruin *et al.* (1975).

is readily derived from (10.3.3), (10.8.1) and (10.8.6). The jump frequency ratio for any two components can then be determined from the corresponding correlation factors since from (10.8.1) one has

$$\frac{w_A}{w_B} = \frac{(1-f_A)f_B}{(1-f_B)f_A}. \tag{10.8.10}$$

The individual jump frequencies can only be determined if the vacancy concentration is also known.

Several experimental tests of the theory have been made. Good confirmations of the theory were found in CoO–NiO and CoO–MgO solid solutions (Dieckmann and Schmalzried, 1975 and Schnehage *et al.*, 1982) and in olivine solid solutions $(Fe, Mg)_2SiO_4$ (Hermeling and Schmalzried, 1984). In each of these systems the two metallic species form a pseudo-binary system on the cation sub-lattice and their tracer diffusion coefficients vary strongly with the composition of the solid solutions, as illustrated for example by Fig. 10.9 for the CoO–MgO solid solutions. Nevertheless, the jump frequency ratio $w_{Co}/w_{Mg}$ determined from these results by means of Manning's equations is practically independent of the composition, as it should be if the model is strictly obeyed. The correlation factors determined from the analysis are, of course, sensitive to composition: see Fig. 10.10.

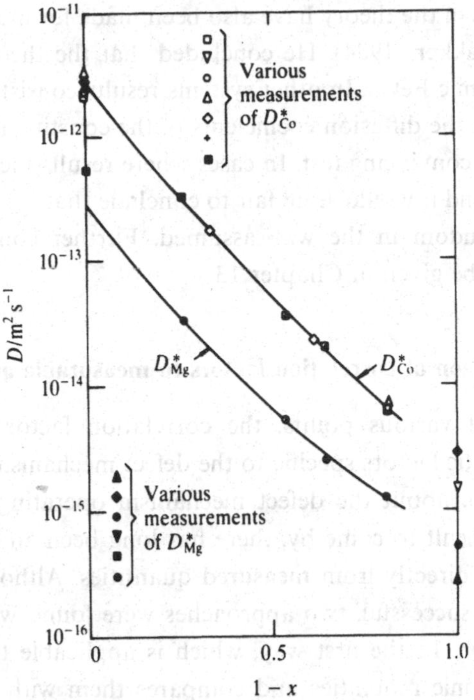


Fig. 10.9. The composition dependence of Co and Mg tracer diffusion coefficients in $(Co_xMg_{1-x})O$ at 1573 K in air. (After Schnehage *et al.*, 1982.)

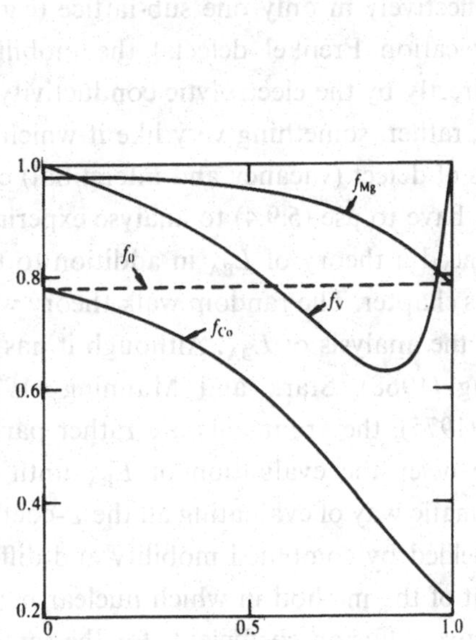


Fig. 10.10. The correlation factors for cations and vacancies in $(Co_xMg_{1-x})O$ as a function of $x$ calculated with $w_{Co}/w_{Mg} = 12.8$ for the same system as Figure 10.9. (After Schnehage *et al.*, 1982.)

Experimental tests of the theory have also been made for metals and these have been reviewed by Bakker (1984). He concluded that the theory is applicable to α-CuZn and equiatomic FeCo. In other systems results consistent with the theory have been found, but the diffusion coefficients of the constituents are not different enough to provide a convincing test. In cases where results incompatible with the theory have been found it would seem fair to conclude that these alloys are locally ordered and not random in the way assumed. Further consideration of such ordering effects will be given in Chapter 13.

## 10.9 Relation of correlation factors to measurable quantities

As we have seen at various points, the correlation factor, $f$, depends upon geometrical and kinetic factors specific to the defect mechanism considered. Since definitive information about the defect mechanism operating in any particular solid is generally difficult to come by, there has long been an interest in possible ways of obtaining $f$ directly from measured quantities. Although this quest has not been completely successful, two approaches were found which have received considerable attention. In the first way, which is applicable to ionic solids, one measures electrical ionic mobilities and compares them with the corresponding isotope diffusion coefficients. We can see how this works from (5.9.7), which applies to one species (A) of a pure compound (AY, say). In those electrolytic conductors which have defects effectively in only one sub-lattice (e.g. the Ag halides, which show predominantly cation Frenkel defects) the mobility of the ions in this sub-lattice is given directly by the electrolytic conductivity. As a result (5.9.7) can be used directly – or, rather, something very like it which allows for the possible presence of two types of defect (vacancy and interstitial) can be so used. In more general cases we may have to use (5.9.4) to analyse experimental information and for that purpose we need a theory of $L_{BA}$ in addition to the theory of $D_B$ (i.e. of $L_{BB}$) developed in this chapter. The random walk theory which we have presented is not convenient for the analysis of $L_{BA}$. Although it has been extended for that purpose by Manning (1968), Stark and Manning (1974), Stark (1974) and Manning and Stark (1975), the arguments are rather particular and not easy to follow. We therefore defer the evaluation of $L_{BA}$ until Chapter 11 where we describe a more systematic way of evaluating all the $L$-coefficients. We shall return to the information yielded by combined mobility and diffusion measurements in Chapter 11. A variant of the method in which nuclear magnetic relaxation times are used to give effective diffusion coefficients for the same sort of comparison is discussed in Chapter 12.

The second experimental approach to the correlation factor is via the measurement of isotope effects in diffusion. Although, for many purposes, it is adequate to ignore differences in diffusion brought about by the different masses of different isotopes of the same element, such differences are measurable by suitably refined techniques (see e.g. Rothman, 1984). When we consider the diffusion of two different isotopes of the same solute species we expect the greatest effect to come through the influence of the isotopic mass on the jump frequency $w_s$ which enters into both $D_B^{(0)}$ and $f$ in (10.3.3). $D_B^{(0)}$ is directly proportional to $w_s$ while $f$ depends on $w_s$ only through $Q_{11}$ (cf. (10.6.27) and (10.6.28)). Simple algebra then shows that, so long as the other jump frequencies, which do not involve the solute, arc not affected by the change in solute mass, the relative difference in diffusion coefficients is

$$\frac{D' - D''}{D'} \equiv \frac{\Delta D}{D}$$

$$= f\,\frac{(w_s' - w_s'')}{w_s'} = f\,\frac{\Delta w_s}{w_s}. \tag{10.9.1}$$

Elementary transition-state theory would lead us to expect that $w_s$ was proportional to $m_s^{-1/2}$. If this were the case, then from (10.9.1) we could deduce $f$ from a measurement of $\Delta D/D$ for two isotopes of known mass. Many such measurements have in fact been made (see e.g. Peterson, 1975). Unfortunately, $w_s$ depends on $m_s$ more slowly than $m_s^{-1/2}$ because the transition path over the saddle-point configuration involves the coupled movement of the jumping atom and a number of neighbours. Likewise, those atomic jumps in the vicinity of the solute may also depend to some extent on the mass of the solute atom. These complications make (10.9.1) less useful than it otherwise might be. Nevertheless in practice it seems often adequate to use (10.9.1) with

$$\frac{\Delta w_s}{w_s} = \left(1 - \sqrt{\frac{m'}{m''}}\right)\Delta K, \tag{10.9.2}$$

in which the factor $\Delta K$ is largely independent of the chemical nature of the solute. An analysis of $\Delta K$ on the basis of Vineyard's formulation of transition-state theory (§3.11) was given by Le Claire (1966) and shows it to be the fraction of kinetic energy in the reaction mode which is associated with the motion of the jumping atom. As such, $0 < \Delta K < 1$. Of course, in circumstances where one is confident of the defect mechanism responsible for the atomic transport, so that $f$ is known, a measurement of $\Delta D/D$ allows us to infer $\Delta K$. Some values of $\Delta K$ obtained in that way are given in Table 10.2. Few atomistic calculations of $\Delta K$ have been made, so that little guidance is available from that direction. In practice, it is often assumed that, for a given host, $\Delta K$ is the same for solute diffusion as for self-diffusion: this seems to lead to conclusions consistent with other inferences.

Table 10.2. *Illustrative values of $\Delta K$ for vacancy migration as determined from measurements of isotope effects in self-diffusion*

| Structure | Substance | $\Delta K$ | Reference |
|---|---|---|---|
| f.c.c. | NaCl (cation) | 0.9 | 1 |
| f.c.c. | AgBr (cation) | 0.66 | 2 |
| f.c.c. | Cu | 0.87 | 3 |
| f.c.c. | Ag | 0.92–0.96 | 3, 4 |
| f.c.c. | Au | 0.9–1.0 | 4 |
| f.c.c. | Pd | 1.02 | 3, 4 |
| f.c.c. | $\gamma$-Fe | 0.68 | 3, 4 |
| b.c.c. | Na | 0.6 | 3, 4 |
| b.c.c. | $\alpha$-Fe | 0.6 | 5 |
| b.c.c. | $\delta$-Fe | 0.46 | 5 |
| Hexagonal | Zn | 0.95 | 3, 4 |
| Diamond | Ge | 0.5–0.6 | 3, 4 |

References to Table 10.2: (1) Rothman *et al.* (1972); (2) Peterson *et al.* (1973); (3) Peterson (1975); (4) Peterson (1978); (5) Walter and Peterson (1969).

## 10.10 Summary

In this chapter we have described the structure of random-walk theories of solute and (isotopic) self-diffusion. In a rigorous form this structure is limited to dilute solid solutions containing a low concentration of defects (vacancies or interstitials). The principal subject of the discussion has been cubic systems containing simple defects, for which there is only one class of atomic movements. One important result (§10.6.4) is the equivalence of the formulation presented here to the kinetic theory already given in Chapter 8. The appropriate generalization of random-walk theory to more complex cases (e.g. non-cubic lattices, vacancy pair defects, etc.) where there is more than one class of atomic movements is given in Appendices 10.1 and 10.2. Numerical values for the correlation factors for self-diffusion in pure lattices were given in Table 10.1 for those circumstances where they are pure numbers. Results for more complicated circumstances where these factors are functions of several jump frequencies (e.g. self-diffusion in anisotropic lattices, solute diffusion, etc.) are reserved to the next chapter. We do so because, as will be shown in §11.2, for some lattices it is possible, in one and the same evaluation, to obtain not only $D_B$, and hence $f_B$, but also other transport coefficients $L_{AA}$, $L_{AB}$, etc. at the same time. In view of the rather difficult nature of the calculations, this is clearly a valuable result pointing to a useful saving of effort.

Although most of the chapter has thus been concerned with dilute solid solutions, an approximate extension of these methods which describes self-diffusion in a particular model of a concentrated alloy has also been presented in §10.9. This is due to Manning (1968) and assumes that the defects are vacancies present in a low concentration. This theory, although approximate, is surprisingly accurate and we have thus presented its essentials here. Nevertheless, in its original form as an intuitive derivative of the theory of dilute alloys, it has not proved convenient for further development, the work of Bocquet (1986, 1987) on its extension to interstitial defects notwithstanding. We shall therefore return to it as part of a more rigorous and systematic development to be given in Chapter 13.

# Appendix 10.1

## The correlation factor in the presence of several types of jump

In this appendix we present the argument leading to the result (10.4.11). We assume that the definitions of jump types as given in §10.4.2 are unambiguous and that the probabilities $p^{s\pm}_{\beta\alpha}$ are unique. Then we may write

$$\langle x_{\alpha s} x_\alpha \rangle = \sum_\beta |x_\beta|\ T^{(s)}_{\beta\alpha}\ |x_\alpha|, \qquad \text{(A10.1.1)}$$

with

$$T^{(s)}_{\beta\alpha} = p^{s+}_{\beta\alpha} - p^{s-}_{\beta\alpha}. \qquad \text{(A10.1.2)}$$

A compact expression for the correlation factor is then immediately obtained by substituting (A10.1.1) into (10.4.7). In matrix notation this is

$$f = 1 + 2\sum_s \tilde{\mathbf{d}}\mathbf{T}^{(s)}\mathbf{b}. \qquad \text{(A10.1.3)}$$

Here the matrix elements are labelled by the jump types; so that $\mathbf{b}$ is the column matrix with elements $C_\alpha|x_\alpha|/(\sum_\alpha C_\alpha x_\alpha{}^2)$, $\tilde{\mathbf{d}}$ is the row matrix of displacement magnitudes $|x_\alpha|$ while $\mathbf{T}^{(s)}$ is the matrix of the coefficients $T_{\beta\alpha}{}^{(s)}$.

We now show that $\mathbf{T}^{(s)}$ can be expressed simply in terms of $\mathbf{T}^{(1)}$, i.e. of the matrix for successive jumps. First we have the recursion formula for the conditional probabilities

$$p^{(s+1)+}_{\beta\alpha} = \sum_\gamma (p^{1+}_{\beta\gamma}\, p^{s+}_{\gamma\alpha} + p^{1-}_{\beta\gamma}\, p^{s-}_{\gamma\alpha}). \qquad \text{(A10.1.4a)}$$

This expresses the fact that for the $(s+1)$th jump (of type $\beta$) following the initial jump (of type $\alpha$) to be in the same sense as the initial jump, either the $s$th jump and the next jump following it must be in the same sense along the $x$-axis as the initial jump, or the $s$th jump must be opposed to the initial jump and the next jump following it must be opposed to the $s$th jump itself. There is a corresponding relation when the $(s + 1)$th jump is opposed to the initial jump, namely

$$p_{\beta\alpha}^{(s+1)-} = \sum_{\gamma} (p_{\beta\gamma}^{1-} p_{\gamma\alpha}^{s+} + p_{\beta\gamma}^{1+} p_{\gamma\alpha}^{s-}). \qquad \text{(A10.1.4b)}$$

These relations also express the Markovian property and are examples of the Chapman–Kolmogoroff equation (van Kampen, 1981). It is then easy to see from (A10.1.2) and (A10.1.4) that

$$T_{\beta\alpha}^{(s+1)} = \sum_{\gamma} T_{\beta\gamma}^{(1)} T_{\gamma\alpha}^{(s)} \qquad \text{(A10.1.5)}$$

and hence that

$$\mathbf{T}^{(s)} = (\mathbf{T}^{(1)})^{s}. \qquad \text{(A10.1.6)}$$

This form of this relation allows us now to simplify the notation and to write just $\mathbf{T}$ in place of $\mathbf{T}^{(1)}$ and $\mathbf{T}^{s}$ in place of $\mathbf{T}^{(s)}$.

By substitution of this simple result into (A10.1.3), we then obtain for the correlation factor

$$f = 1 + 2\tilde{\mathbf{a}}\mathbf{T}(\mathbf{1} - \mathbf{T})^{-1}\mathbf{b}. \qquad \text{(A10.1.7)}$$

This is the required generalization of (10.4.3).

# Appendix 10.2

## Evaluation of the correlation factor in the presence of several types of jump

In this appendix we supply details of the way in which the elements $T_{\beta\alpha}$ may be calculated in terms of the characteristic jump frequencies of the system. We do so for the more general class of model defined in §10.4.2, as exemplified by the case of solute enhanced self-diffusion. The method is due to Howard and Manning (1967). We use the same notation as before.

We suppose that an $\alpha$-type of jump of the tracer atom A* has occurred. We then have to calculate the probability that the next jump of this atom is of type $\beta$ in a particular direction relative to the initial jump, i.e. $p_{\beta\alpha}{}^{1\pm}$. Just as with the simpler systems considered in §10.6.1 we can identify sets of configurations which have the same probability of occurrence after the initial jump and which are therefore equivalent. In our example, configurations must be specified by the position of both the vacancy and the solute atom, B, relative to the tracer A* at the origin. Configurations obtained from one another by a crystal lattice symmetry rotation about the $x$-axis or by a reflection in a mirror symmetry plane which contains the $x$-axis are equivalent.

Now, as with the calculation of $T$ for systems with only one type of jump (eqns. (10.4.3) and (10.4.4)), we may introduce the probability $p_i^{(n)}$ that the configuration will be $i$ after the *vacancy* has made $n$ jumps since the initial tracer–vacancy exchange but without exchanging with the tracer again. The relation of $p_i^{(n)}$ to $p_i^{(n-1)}$ is again expressed by the matrix equations (10.6.2) and (10.6.3), except that now $i, j$, etc. relate to the configurations identified above. Next we define a vector $\mathbf{p}$ by the equation

$$\mathbf{p} = \sum_{n=0}^{\infty} \mathbf{p}^{(n)}. \tag{A10.2.1}$$

The components of $\mathbf{p}$ give the total probabilities that each of the configurations is occupied at some time between those consecutive tracer jumps giving rise to displacements along the $x$-axis. Clearly, from (10.6.2) and (A10.2.1)

$$\mathbf{p} = \mathbf{U}^{(1)}\mathbf{p}^{(0)} \tag{A10.2.2}$$

with

$$\mathbf{U}^{(1)} = (\mathbf{1} - \mathbf{A})^{-1}. \tag{A10.2.3}$$

Now let $\mathbf{p}_\alpha^{(0)}$ be the initial distribution appropriate to an initial tracer jump of type $\alpha$. This allows us to write

$$\mathbf{p}_\alpha = \mathbf{U}^{(1)}\mathbf{p}_\alpha^{(0)}. \tag{A10.2.4}$$

It remains only to combine these probabilities of particular configurations with the relative rates of tracer jumps to obtain the $T_{\beta\alpha}$ quantities. To this end we select from the $\mathbf{A}_{ji}$ (defined by 10.6.3) the relative rates of transition for vacancy jumps from configuration $i$ which cause a $\beta$-type jump of the tracer in either the positive or the negative $x$-direction and form these into the row matrices $\tilde{\mathbf{q}}_\beta^\pm$ in order of the index $i$. Then

$$\begin{aligned} T_{\beta\alpha} &= \tilde{\mathbf{q}}_\beta^+ \mathbf{p}_\alpha - \tilde{\mathbf{q}}_\beta^- \mathbf{p}_\alpha \\ &= (\tilde{\mathbf{q}}_\beta^+ - \tilde{\mathbf{q}}_\beta^-)\mathbf{U}^{(1)}\mathbf{p}_\alpha^{(0)}. \end{aligned} \tag{A10.2.5}$$

This is the appropriate generalization of the expression (10.6.7) for systems having only a single type of jump.

In practical evaluations it is again important to reduce the size of the matrix to be inverted as far as possible. In lattices where the plane through the tracer normal to the $x$-axis is a plane of symmetry, so that there are pairs of 'negatively equivalent' configurations, a further reduction can be made along the lines of (10.6.20) and (10.6.21). The enumerated configurations may then be restricted to the positive half-space and $T_{\beta\alpha}$ becomes

$$T_{\beta\alpha} = (\tilde{\mathbf{q}}_\beta^+ - \tilde{\mathbf{q}}_\beta^-)\mathbf{U}^{(2)}\mathbf{p}_\alpha^{(0)} \tag{A10.2.6}$$

where

$$\mathbf{U}^{(2)} = (\mathbf{R}'')^{-1} \tag{A10.2.7}$$

and

$$\mathbf{R}''_{ij} = -(\mathbf{A}_{ij} - \mathbf{A}_{i\bar{j}}). \tag{A10.2.8}$$

The $T_{\beta\alpha}$ and hence also the partial correlation factors $f_\alpha$ are functions of the various ratios of jump rates. These functions can usually only be obtained as numerical tabulations. Details of a particular case are given in Chapter 11.

# 11

# Transport coefficients of dilute solid solutions – results and applications

## 11.1 Introduction

In Chapters 6–10 we have dealt with the general structure of the linear response, kinetic theory and random-walk approaches to the calculation of the phenomenological coefficients and to the dielectric and anelastic response functions. We gave some straightforward examples of approximate results that can be obtained from the general expressions. The mathematical inter-relations between the three approaches demonstrated in those chapters allow the use of common techniques for the evaluation of the expressions for the transport coefficients of particular models.

In the present chapter, which, like Chapters 7–10, is concerned with dilute alloys and solid solutions, we first consider these techniques and then go on to present some results and applications of those results. It divides therefore into three more or less distinct parts. Techniques are the subject of §§11.2 and 11.3, the resulting transport coefficients are the subject of §§11.4–11.6 while various applications are reviewed in §§11.7–11.10.

In Chapters 7 and 8 general expressions were derived for the phenomenological coefficients and response functions from kinetic and linear response theories while consistent expressions for diffusion coefficients were obtained from random walk theory in Chapter 10. The techniques for the evaluation of these expressions are reviewed here in §§11.2 and 11.3. In the first of these sections the techniques are limited to calculations for which, in the terminology introduced in Chapter 10, only one *type* of jump occurs in the formal analysis. Important examples of such calculations are furnished by evaluations of the three independent phenomenological coefficients $L_{AA}$, $L_{AB}$, $L_{BB}$ for dilute binary alloys of cubic structure with transport by single vacancies, simple interstitials or dumb-bell interstitials. Our description, in §11.3, of techniques for calculations in which more than one type of jump must be distinguished is less general. At the present time the extension of the random walk formulation of §§10.4.2, 10.4.3 and 10.7 to the calculation of all the $L$-coefficients is less well established. We shall therefore not give a general discussion. Rather, we concentrate on working out the particular example of the enhancement of solvent self-diffusion by solutes (§10.7 and Appendix 10.2).

To a considerable extent we can approach these tasks for all types of solid together, since, at least in the case of vacancies, similar models are often used for a given structure irrespective of the type of bonding. Differences between different classes of solid enter most noticeably through the dependence of the concentrations of defects upon composition (as Chapter 3 demonstrated) and, in ionic crystals and semiconductors, through the occurrence of internal electric fields (Nernst fields, §5.10). However, to keep the arguments as simple as possible we develop the techniques in the particular context of metals and alloys, with brief indications of the changes which are necessary for other substances.

When we turn to the results of using these techniques and to the corresponding applications we are faced with an enormous amount of detail – different substances, different measurements, different defect regimes, etc. We can therefore only present results and applications very selectively. To be comprehensive would require further chapters and would inevitably repeat much that is already covered in the literature. We therefore shall not go into full details of these better known applications and analyses, confining ourselves instead to a review of essentials. On the other hand we shall give greater attention to those applications which rely more heavily on the theoretical results developed in the earlier part of this chapter and in preceding chapters.

## 11.2 Techniques for dilute alloys

In considering the evaluation of transport coefficients in dilute binary alloys (AB) there are several distinct considerations. Firstly, there are the different transport coefficients – the three independent ('chemical') $L$-coefficients ($L_{AA}$, $L_{AB}$ and $L_{BB}$) and the two tracer diffusion coefficients $D_A^*$ and $D_B^*$. However, since the system is dilute in B, $D_B^*$ is equal to the intrinsic diffusion coefficient $D_B$, which to zero order in solute fraction ($x_B$) is given directly by $L_{BB}$ (cf. (5.3.8)). (N.B. These equalities may not hold in systems where Nernst diffusion fields arise, although the necessary modifications are generally obtainable without difficulty; see e.g. §§11.8 and 11.9.) On the other hand the tracer diffusion coefficient of the host species, $D_A^*$, must be obtained to first order in $x_B$ since to zero order it is simply that of the pure solvent. This forces us to consider the motion of tracer atoms A* not only by unperturbed defects but also by those paired with B atoms or otherwise perturbed by them.

Secondly, there is the matter of lattice symmetry and defect symmetry (which may be lower than the point symmetry of the lattice). Formally, these come together in the classification by *types of jump* given in §10.4, where we calculated correlation factors for tracer diffusion, i.e. $D_A^*$ and $D_B^*$. The simplest of the calculations given there were for cubic lattices containing vacancy (or interstitial) defects which retain the point symmetry of the lattice (or interstitial lattice) and which thus give rise to only a single type of jump. If the lattice symmetry is lower than cubic there will be more than one type of jump. The problem is then more complex and requires the introduction of the matrix $\mathbf{T}$ in place of the single quantity $T$. The same device allows us to treat the diffusion of tracer atoms which are moved by defects of low symmetry (e.g. solute–vacancy pairs, dumb-bell interstitials, etc.), although in practice the mathematical complexity of the calculation may be somewhat greater.

For both types of system (i.e. low lattice symmetry and low defect symmetry) it now appears to be possible to express the three $L$-coefficients in terms of three matrices $\mathbf{T}_{AA}$, $\mathbf{T}_{AB}$ and $\mathbf{T}_{BB}$ each of which is a generalization of the matrix $\mathbf{T}$ defined in §10.4.2. Each of the three expressions has a structure analogous to (10.4.11). However, although such a structure has been used to calculate the $L$-coefficients for transport via dumb-bell interstitials (Chaturvedi and Allnatt, 1992), further experience may be needed before we can be fully confident of having a truly general formalism.

For these reasons we shall here pursue the more limited aims of exposing the principles involved in the calculation of $L_{AA}$, $L_{AB}$, $L_{BB}$ and $D_A^*$ in cubic lattices. The $L$-coefficients are considered in the remainder of this section (§11.2.1) while $D_A^*$ is treated in the next (§11.3).

### *11.2.1 Techniques for $L_{AA}$, $L_{AB}$ and $L_{BB}$: one type of defect jump*

In this section we turn to the evaluation of eqns. (8.5.10) for the $L$-coefficients. For a dilute alloy of a solute B in host A we are concerned with three independent phenomenological coefficients $L_{AA}$, $L_{AB} = L_{BA}$ and $L_{BB}$. We shall consider their evaluation with the five-frequency model of vacancy–solute interactions in a dilute f.c.c. alloy in mind (see Fig. 2.22). In that model vacancy jump frequencies are assumed to be affected by the presence of the solute only in its immediate vicinity, as suggested by various atomistic calculations for metals (as summarized, for example, by March, 1978). The assumption that the solute causes only a localized disturbance of the vacancy (or interstitial) movements is essential for the use of eqns. (8.5.10) in the form given. It is present in almost all models which have been analysed in detail and appears to be adequate, certainly in metals and in other cases where there are no long-range Coulomb interactions. Even when such

Table 11.1. *Pair types and velocity functions for the five-frequency model*

The diffusion is in a ⟨100⟩ direction and the planes perpendicular to this direction have a spacing $a$.

| Pair type, $p$ | Vacancy sites (B at origin) | $v_p^{(A)}/a$ | $v_p^{(B)}/a$ |
|---|---|---|---|
| 1 | $(1,1,0), (1,\bar{1},0), (1,0,1)\ (1,0,\bar{1})$ | $(2w_1 - 3w_3)$ | $w_2$ |
| 2 | $(2,0,0)$ | $4(w_4 - w_0)$ | 0 |
| 3 | $(2,1,1), (2,1,\bar{1}), (2,\bar{1},1), (2,\bar{1},\bar{1})$ | $2(w_4 - w_0)$ | 0 |
| 4 | $(1,2,1), (1,2,\bar{1}), (1,\bar{2},1), (1,\bar{2},\bar{1})$ $(1,1,2), (1,1,\bar{2}), (1,\bar{1},2), (1,\bar{1},\bar{2})$ | $(w_4 - w_0)$ | 0 |
| 5 | $(2,2,0), (2,\bar{2},0), (2,0,2), (2,0,\bar{2})$ | $(w_4 - w_0)$ | 0 |

long-range interactions have to be considered, as will commonly be the case in ionic crystals, we may separate the long- and the short-range interactions. We do so by using the above models for the short-range effects, while representing the effects of the long-range Coulomb forces in the way indicated by the theory of electrolyte solutions – rather as we did in §3.9 for equilibrium properties. For the time being we concentrate on the more specific effects of the localized interactions.

For these models of localized interaction the uncorrelated terms in (8.5.10) may be evaluated by straightforward enumeration. It is the correlated (fourth) term which demands attention. The principal mathematical task involved is the determination of those elements of the inverse of the matrix $\mathbf{\Omega}$ which are needed in (8.5.10). These are selected by the velocity functions (8.5.10b) which also enter into the fourth term of (8.5.10a).

We begin by reference to the five-frequency model for which the labelling of the different species of solute–vacancy pair is given in Table 11.1. As eqns. (8.5.10) show we need to consider (i) the probability that there is a vacancy at the various positions around the solute atom, (ii) the velocity functions for these positions and (iii) matrix elements of $\mathbf{\Omega}^{-1}$ for pairs of these positions. We take these in order. In this model the solute and vacancy are taken to be physically bound only when they are nearest neighbours. The site occupancies $p_l$ are therefore readily expressed in terms of the mole fractions of physically bound first-neighbour pairs, $c_P$, and of unbound defects, $c'_V$ and $c'_B$:

$$\begin{aligned} p_1 &= c_P/3, \\ p_l &= \zeta_l c'_B c'_V, \qquad l > 1. \end{aligned} \tag{11.2.1}$$

Here $\zeta_l$ is the number of sites of type $l$. The defect concentrations are related through the law of mass action (eqn. (3.8.5)).

We turn next to the velocity functions. Now, although vacancy jumps to the bound first-neighbour positions from more distant positions are assumed to occur with frequency $w_4$, all other vacancy jumps from these and other more distant positions are assumed to occur with frequency $w_0$ (Fig. 2.22). It follows that the velocity function for the host atoms, $v_m^{(A)}$, is zero for every site $m$ from which all vacancy jumps occur with frequency $w_0$ (because the vector sum of the nearest neighbour displacements is zero in this, and any other Bravais, lattice). This velocity function $v_m^{(A)}$ is thus non-zero only for first-neighbour sites $m$ and for those sites which can be reached in one jump from a first-neighbour site (the non-zero values are in Table 11.1). Since the solute atom B can move only when there is a vacancy next to it the velocity function for B atoms $v_m^{(B)}$ is zero unless $m = 1$.

The elements of $(Q^{-1})_{ml}$ that are needed are therefore $m, l = 1$ for $L_{BB}$, $m = 1$–$5$ with $l = 1$ for $L_{AB}$, and $m, l = 1$–$5$ for $L_{AA}$. The number of non-zero terms in (8.5.10) will evidently always be small for any model with short-range defect interactions. However, when the particular combinations of sums of products of jump frequency times elements of $\mathbf{Q}^{-1}$ appearing in $L_{AB}$ and $L_{AA}$ are examined for the five-frequency model, it is found that all the required elements of $\mathbf{Q}^{-1}$, except $(Q^{-1})_{11}$ can be eliminated. This good fortune arises from the identities

$$\sum_{q=1}^{5} Q_{1q}(Q^{-1})_{qr} = \delta_{1r} = \sum_{q=1}^{5} (Q^{-1})_{1q}Q_{qr}. \tag{11.2.2}$$

Take, for example, $L_{AB}$. By (8.5.10) we need the quantity

$$\sum_{q,r} v_q^{(A)}(Q^{-1})_{qr}v_r^{(B)},$$

which reduces to

$$\sum_{q=1}^{5} v_q^{(A)}(Q^{-1})_{q1}v_1^{(B)}.$$

From the velocity functions listed in Table 11.1 this becomes

$$a^2(2w_1 - 3w_3)w_2(Q^{-1})_{11} + a^2(w_4 - w_0)w_2(4(Q^{-1})_{21} + 2(Q^{-1})_{31} + (Q^{-1})_{41} + (Q^{-1})_{51}). \tag{11.2.3}$$

If we now substitute the values of $Q_{1q}$ (see e.g. (11.2.7) below) into the left hand equation of the pair (11.2.2) we obtain

$$(2w_2 + 2w_1 + 7w_3)(Q^{-1})_{11} + w_4(4(Q^{-1})_{21} + 2(Q^{-1})_{31} + (Q^{-1})_{41} + (Q^{-1})_{51}) = 1. \tag{11.2.4}$$

We see therefore that the terms in (11.2.3) other than that in $(Q^{-1})_{11}$ can be eliminated *en bloc*, i.e. $L_{AB}$ is expressible in terms of $(Q^{-1})_{11}$ alone. The same identities (11.2.2) also allow us to reduce $L_{AA}$ to $(Q^{-1})_{11}$ alone, although in this case the algebraic manipulations are a little longer. Thus, as already stated above, only $(Q^{-1})_{11}$ is needed for all three coefficients $L_{AA}$, $L_{AB}$ and $L_{BB}$.

For the models so far studied this kind of elimination step has proved possible for the five-frequency model, the corresponding model for the b.c.c. lattice, and very similar models with simple interstitials. In other cases, for example models with second- as well as first-neighbour binding or models with dumb-bell interstitials, no similar elimination can be made and one is left with a small number of separate $\mathbf{Q}^{-1}$ elements to calculate.

We must now consider the actual evaluation of the required matrix elements of $\mathbf{Q}^{-1}$. We shall describe only the calculation of $(Q^{-1})_{11}$ since any other elements which are required for certain models can readily be treated by minor variants of the methods described. We also recall that many of the calculations of the impurity correlation factor (equivalently $L_{BB}$) evaluate the expressions for $\langle \cos\theta_1 \rangle$ or $\langle x_1 x_2 \rangle$ by the matrix method or by the diffusion of probability method as described in Chapter 10. As we saw there, in such cases one requires the (11)-element of the inverse of some other matrix which is always closely related to $\mathbf{Q}$ (see e.g. (10.6.23)) and therefore no essentially different problems arise. The method and approximations used to evaluate $(Q^{-1})_{11}$ can be understood in terms of the block structure of $\mathbf{Q}$. This is defined in terms of the coordination shells of the vacancy sites of the pairs about the solute atom. The $n$th coordination shell consists of those sites from which the vacancy can reach the solute site in a minimum of $n$ nearest-neighbour jumps. The block structure of the matrix $\mathbf{Q}$ can then be written:

$$\begin{bmatrix} Q_{11} & Q_{12} & 0 & 0 & \cdots \\ Q_{21} & Q_{22} & Q_{23} & 0 & \cdots \\ 0 & Q_{32} & Q_{33} & Q_{34} & \cdots \\ 0 & 0 & Q_{43} & Q_{44} & \cdots \\ \vdots & \vdots & \vdots & \vdots & \end{bmatrix} \tag{11.2.5}$$

Here, for the elements $Q_{rq}$ of block $Q_{mn}$, $r$ refers to a site in the $m$th coordination shell and $q$ to a site in the $n$th coordination shell. In row $m$ the only blocks whose elements are not all zero are $Q_{m\,m-1}$, $Q_{m\,m}$ and $Q_{m\,m+1}$.

A plausible first approximation for defect pairs which are strongly bound only when they are nearest-neighbours is to retain only block $Q_{11}$ in $Q$. In the particular case of the five-frequency model this is a single element $Q_{11} = (2w_1 + 2w_2 + 7w_3)$. This 'one-shell' approximation is the Lidiard–Le Claire approximation of §10.6.5 ((10.6.29) *et seq.*). We recall from the earlier discussion that is is convenient for this model to write

$$(Q^{-1})_{11} = (2w_1 + 2w_2 + 7w_3F)^{-1}, \tag{11.2.6}$$

which defines a quantity $F$, which is 1 for the Lidiard–Le Claire approximation. This approximation assumes in effect that a vacancy which has exchanged with the solute and which subsequently leaves the first coordination shell does not return (or does so from a random direction). The quantity $(1 - F)$ is evidently the fractional reduction in the overall frequency of jumps from a first-shell site to a second-shell site caused by returns of the vacancy to first-shell sites. A similar formulation is readily made for other models.

The second general approximation is to retain the four blocks $Q_{11}$, $Q_{12}$, $Q_{21}$ and $Q_{22}$ in $\mathbf{Q}$. For the five-frequency model this is the approximation of Bocquet (1974b) which reduces $\mathbf{Q}$ to a $5 \times 5$ matrix by entirely neglecting that part of $\mathbf{Q}$ associated with vacancy jumps in unperturbed crystal but retaining all the perturbed jumps. This matrix is

$$\begin{bmatrix} (2w_2 + 2w_1 + 7w_3) & -4w_4 & -2w_4 & -w_4 & -w_4 \\ -w_3 & (4w_4 + 8w_0) & -w_0 & 0 & 0 \\ -2w_3 & -4w_0 & (2w_4 + 10w_0) & -w_0 & -2w_0 \\ -2w_3 & 0 & -2w_0 & (2w_4 + 9w_0) & -2w_0 \\ -w_3 & 0 & -2w_0 & -w_0 & (w_4 + 11w_0) \end{bmatrix} \tag{11.2.7}$$

In effect, this two-shell approximation assumes that a vacancy which jumps from the second to the third shell never returns (or does so from a random direction). By using the inversion formula to obtain $(Q^{-1})_{11}$ and then obtaining $F$ from the definition (11.2.6) we find that the result for the two-shell approximation, or indeed any approximation retaining more shells, can be reduced exactly to

$$7(1 - F) = \frac{10\xi^4 + B_1\xi^3 + B_2\xi^2 + B_3\xi}{2\xi^4 + B_4\xi^3 + B_5\xi^2 + B_6\xi + B_7}, \tag{11.2.8}$$

in which $\xi \equiv w_4/w_0$ and the coefficients $B_i$ are pure numbers defined in terms of determinants of $\mathbf{Q}$ with certain rows and columns omitted. These numbers are readily calculated for the two-shell approximation, but for approximations retaining more shells an electronic computer is useful. It turns out that the convergence of the $B_i$ as the number of shells is increased is rather slow.

Methods which extrapolate the results of successive approximations to those for an infinite matrix are described in Appendix 11.1. The extrapolation method used by Manning (1964) to obtain the first accurate results for the five-frequency

Table 11.2. *Coefficients in the expression for F for the five-frequency model according to various authors*

In the approximation of Bocquet (1974b) and Bakker (1970) a vacancy which migrates outside the second or fifth shell, respectively, is assumed not to return to the impurity. The other results refer to the accurate calculations described in §11.2.

|         | Bocquet (1974b) | Bakker (1970) | Manning (1964) | Koiwa and Ishioka (1983b) |
|---------|----------------|---------------|----------------|---------------------------|
| $B_1$   | 190            | 181           | 180.5          | 180.3                     |
| $B_2$   | 1031           | 936           | 927            | 924.3                     |
| $B_3$   | 1594.5         | 1362          | 1341           | 1338.1                    |
| $B_4$   | 45             | 41            | 40.2           | 40.1                      |
| $B_5$   | 328            | 261           | 254            | 253.3                     |
| $B_6$   | 930.5          | 627           | 597            | 596.0                     |
| $B_7$   | 855.5          | 471           | 436            | 435.3                     |

and other models employs a reformulation of the problem and is so efficient that paper and pencil calculations suffice. Koiwa and Ishioka (1983a,b) have shown how such extrapolations can be avoided and exact results obtained by application of perturbation theory in a manner similar to that employed in other defect problems. This method, which is summarized in Appendix 11.2, reduces the problem to the numerical calculation of a set of lattice Green functions for unrestricted vacancy random walks on a perfect lattice, just as in Montet's calculation of the tracer correlation factor described in §10.5.1. To illustrate the results obtained by these various methods the coefficients $B_i$ for two- and five-shell approximations are compared with accurate results of Manning (1964) and Koiwa and Ishioka (1983b) in Table 11.2. The dependence of $F$ on $(w_4/w_0)$ has already been shown graphically in Fig. 10.7 where the dashed line gives the two-shell approximation, while the full line gives the accurate results. Evidently the two-shell approximation is far more accurate than the one-shell Lidiard–Le Claire approximation ($F = 1$). Nevertheless since the accurate computation of $F$ by the methods we have described can be time consuming it is worth noting that simple analytical expressions like those resulting from the one- and two-shell approximations can be adequate for many purposes.

For models in which more than one element of $\mathbf{Q}^{-1}$ is required one has several quantities like $F$ to calculate, each one being a function of a single jump frequency ratio and defined formally in terms of cofactors of $\mathbf{Q}$. Similar methods of calculation can be used.

The preceding discussion has been concerned with the transport coefficients for slow processes. However, we noted in Appendix 6.3 that time-dependent response coefficients, as required for dielectric and mechanical relaxation processes, can be obtained by a slight formal extension of the same basic theory. In essence, instead of $\mathbf{Q}^{-1}$, what is now required is the matrix $(\mathbf{Q} - i\omega\mathbf{1})^{-1}$; and for cases where only the $(1, 1)$ element was needed we now require just $[(\mathbf{Q} - i\omega\mathbf{1})^{-1}]_{11}$. The previous methods can be straightforwardly adapted for this purpose. For the f.c.c. five-frequency model we obtain

$$[(\mathbf{Q} - i\omega\mathbf{1})^{-1}]_{11} = (2w_1 + 2w_2 + 7w_3F(\omega) - i\omega)^{-1}, \qquad (11.2.9)$$

in place of (11.2.6). The quantity $F(\omega)$ is a generalization of the vacancy escape factor $F$, being a complex function of both $\omega/w_0$ and $w_4/w_0$. An analytic expression for $F$ can be obtained in the two-shell approximation or it can be computed numerically in higher approximation (Okamura and Allnatt, 1985); in the one-shell approximation $F$ remains equal to unity.

In this section we have concentrated on $L_{AA}$, $L_{AB}$ and $L_{BB}$. There are, however, two other transport coefficients to be considered, namely the two tracer diffusion coefficients $D_A^*$ and $D_B^*$. Since the system is assumed dilute in B, the coefficient $D_B^*$ is equal to the intrinsic diffusion coefficient $D_B$, which, to zero order in solute fraction, is given directly by $L_{BB}$ (5.3.8). So no further calculation is required, unless we want $D_B^*$ to first order in $x_B$. On the other hand, the tracer diffusion coefficient $D_A^*$ in the alloy must be obtained to first order in $x_B$ since to zero order it is simply that of the host and so tells us nothing about the alloy. To obtain $D_A^*$, or equivalently $L_{A^*A^*}$, thus requires us to consider the motion of tracer atoms A* as caused by vacancies which are in the vicinity of solute atoms B as well as in the unperturbed lattice. For those in the vicinity of B atoms there is more than one type of jump in the sense defined in §10.4.2. We therefore now turn to this extension.

## 11.3 Techniques for dilute alloys: more than one type of defect jump

In §10.4 we drew attention to various examples where, in calculating diffusion coefficients, we needed to distinguish more than one type of jump. These examples included anisotropic lattices and defects of low symmetry in cubic lattices. The discussion in §§10.4.2 and 10.7 has already shown that there are four essential steps in the calculation: (i) the identification of the jump types of the diffusing atom, (ii) the evaluation of $\Gamma_\alpha$, the mean number of jumps of type $\alpha$ made by the diffusing atom per unit time, and hence the calculation of $C_\alpha = \Gamma_\alpha/(\sum_\alpha \Gamma_\alpha)$, (iii) the evaluation of the quantities $T_{\beta\alpha}$ as given for example by equation (A10.2.6) and (iv) the calculation of $f$ from these results through (10.4.11) (or equivalently through (10.4.13) and (10.4.14) where appropriate). Usually steps (i), (ii) and (iv) are straightforward while (iii) is the main part of the calculation. Anisotropic lattices containing vacancies (e.g. tetragonal or hexagonal lattices) provide examples where step (iii) is more complicated than for, say, cubic lattices (in which **T** has only one element). Low symmetry defects (e.g. vacancy pairs and solute–vacancy pairs) usually result in even greater complexity.

For the reasons already given in the previous section we shall not attempt to deal with this problem in general terms. Rather we shall concentrate on the specific task of evaluating the influence of solute atoms upon the self-diffusion of the solvent A in dilute alloys AB. Although we carry the calculation through consistently for a metal, the central (kinetic) part of the calculation can also be used to describe corresponding influences in other types of solid. We shall indicate the nature of the required changes at appropriate points.

Now because we are only interested in the lowest, linear approximation we can calculate the effect by using the same solute–vacancy pair models as were described in the preceding section. The vacancy jump frequencies are changed in the vicinity of the solute atoms so that as long as these remain dispersed as single atoms (and do not cluster) we can be confident that the effect upon the diffusion of the host atoms will be proportional to the fraction $c_B$, i.e.

$$D_A^*(c_B) = D_A^*(0)(1 + b_{A*}c_B). \tag{11.3.1}$$

We shall calculate $b_{A*}$ in terms of the properties of the solute–vacancy models for the particular case of the five-frequency model of the f.c.c. lattice.

Before setting out on that task, however, we note that in practice (11.3.1) may be limited to quite dilute alloys, say $<1\%$. Experimental studies, even those nominally on dilute alloys, may well be made at higher concentrations than this and such results are thus often fitted to a polynomial containing terms in $c_B{}^2$ and $c_B{}^3$ as well. Many systems actually fit an exponential dependence

$$D_A^*(c_B) = D_A^*(0)\exp(b_{A*}c_B) \tag{11.3.2}$$

(see e.g. Hagenschulte and Heumann, 1989b). This tendency of $D_A^*$ to increase more and more rapidly with $c_B$ is understandable in terms of the occurrence of pairs and larger clusters of solute atoms and associated vacancies which are attracted to them more strongly than to isolated solute atoms. However, we shall here pursue only the simple model of solute–vacancy pairs which leads to (11.3.1); but when we need experimental values of $b_A^*$ we shall take these as the coefficient of the first-order term in the exponential or polynomial representation of the experimental values of $D_A^*(c_B)$.

The enumeration of the types of tracer atom jumps is generally straightforward and is illustrated for the f.c.c. model system in Fig. 11.1. The jump types in this model are determined by the position of the solute B relative to the solvent tracer A* and to the vacancy. There are twelve distinct types ($\alpha = 1, \ldots, 12$) of tracer

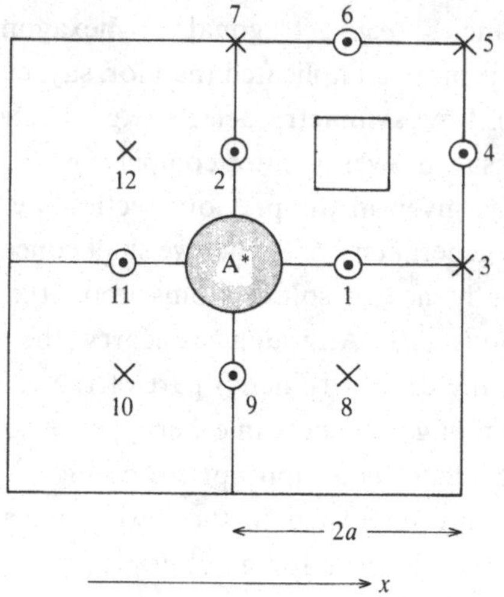


Fig. 11.1. Schematic diagram to define the types of jump (α) for a tracer A* migrating by the vacancy mechanism in a dilute f.c.c. solid solution of B in A (viewed along a 001 axis). The positions of the tracer A* and the vacancy □ are shown explicitly, while the other sites indicated are possible positions for the B atom. Those sites marked × are all in the plane of the paper whereas the sites marked ○ represent equivalent pairs of sites half a cell-side $a$ above and below the plane of the paper. The tracer jump frequency is $w_1$ if B is at one of the four sites 1 and 2, is $w_3$ when B is at one of the seven sites 3–7, and is $w_4$ if B is at one of the seven sites 8–12. If the B atom is on any other site the tracer movement is unaffected ($w_0$). The site labels are also the labels of the distinct types of jump α as used in §11.3.

jumps with frequencies $w_1$, $w_3$ or $w_4$, and there are further types for the various jumps with frequency $w_0$. In order to keep the computations within bounds the number of types of $w_0$ jumps has to be restricted by some approximation. For example, in the pioneering calculation by Howard and Manning (1967) it was assumed that all $w_0$ jumps could be treated as a single type (type 0) and that the partial correlation factor for these jumps is the usual tracer correlation factor ($f_0 = 0.781$; cf. Table 10.1). In higher approximations (e.g. Ishioka and Koiwa, 1984) some types of $w_0$ jumps are treated as distinct and the rest assigned to type 0 as before.

The mean jump rates $\Gamma_\alpha$ can be written down by inspection. For the five-frequency model the expressions, correct to first order in the site fractions of free vacancies and solute atoms and of solute–vacancy pairs, are

$$\Gamma_1 = \Gamma_2 = \tfrac{2}{3}c_P(2w_1), \tag{11.3.3a}$$

$$\Gamma_3 = \Gamma_5 = \Gamma_7 = \tfrac{1}{2}\Gamma_4 = \tfrac{1}{2}\Gamma_6 = \tfrac{2}{3}c_P w_3, \tag{11.3.3b}$$

$$\Gamma_8 = \Gamma_{10} = \Gamma_{12} = \tfrac{1}{2}\Gamma_9 = \tfrac{1}{2}\Gamma_{11} = \tfrac{2}{3}(12c'_V c'_B)w_4. \tag{11.3.3c}$$

The suffixes on these mean jump rates relate to the sites 1–12 defined in Fig. 11.1. The factor 2/3 is the fraction of all tracer jumps with non-zero $x$-components. For type-0 jumps, in the approximation of Howard and Manning, we have

$$\Gamma_0 = \tfrac{2}{3}12c''_V(1 - 7c'_B)w_0. \tag{11.3.4}$$

The quantity $c''_V$ is the probability that an unassociated vacancy is at a particular site neighbouring a tracer, given the presence of the tracer.

The various fractions $c'_V$, $c'_B$, $c_P$ and $c''_V$ are obtained from the statistical thermodynamic theory of Chapter 3. It is here that differences between different substances may enter. In metals the binding of solute atoms and vacancies is relatively weak, so that appreciable effects will only be apparent for solute fractions of the order of one per cent; hence $c_B \gg c_V$ and $c_P$. Some other substances may be similar in this respect (e.g. molecular solids). However, ionic crystals, with their requirements of structure and charge neutrality, may provide a different situation. This is most obviously so with aliovalent solute ions which can strongly affect the overall vacancy fraction (§3.3) and radically change the dependence of $c_P$ upon $c_B$. Such differences affect the form in which the predictions are most usefully presented.

We proceed as for f.c.c. metals for which the statistical theory of dilute alloys gives

$$c'_V = (1 - 12c_B)c_V(0), \tag{11.3.5a}$$

$$c''_V = (1 - 11c_B)c_V(0), \tag{11.3.5b}$$

and

$$c'_V c'_B = c_P(w_3/12w_4), \tag{11.3.5c}$$

where $c_V(0)$ is the site fraction of vacancies in a pure crystal.

The present calculation allows one to calculate the self-diffusion coefficient $D^*_A$ to terms linear in solute concentration and we therefore write

$$D^*_A(c_B) = D^*_A(0)[1 + b_{A*}c_B], \tag{11.3.6}$$

which defines the *solvent enhancement factor* $b_{A*}$. From (10.4.8) and (11.3.3)–(11.3.6) one readily obtains

$$D^*_A(0) = 4a^2c_V(0)f_0w_0 \tag{11.3.7}$$

and, in the Howard and Manning approximation for $w_0$ jumps,

$$b_{A*} = -18 + (4w_4/f_0w_0)(\chi_1w_1/w_3 + 7\chi_2/2), \tag{11.3.8}$$

where

$$\chi_1 = \tfrac{1}{2}(f_1 + f_2), \tag{11.3.9a}$$

and

$$\chi_2 = \tfrac{1}{14}(f_3 + 2f_4 + f_5 + 2f_6 + f_7 + f_8 + 2f_9 + f_{10} + 2f_{11} + f_{12}). \quad (11.3.9b)$$

The partial correlation factors $f_\alpha$ may be obtained from (10.4.14) once the quantities $T_{\beta\alpha}$ defined in (10.4.12) have been found. The matrix $\mathbf{T}$ is a square $n \times n$ matrix where $n$ is the number of jump types. It has elements $T_{\beta 0} = 0$ for $\beta \neq 0$ and $T_{00}$ equal to the value of $T$ for the single jump type appropriate to self-diffusion in a pure solvent. The evaluation of the remaining elements of $\mathbf{T}$ is the central part of the calculation.

Before proceeding with those considerations, we note that when the solute exerts only a weak effect on the vacancy movements we may reasonably approximate the $f_\alpha$, and thus $\chi_1$ and $\chi_2$, simply by $f_0$. We then obtain

$$b_{A*} = -18 + \frac{4w_4}{w_3 w_0}(w_1 + \tfrac{7}{2}w_3). \quad (11.3.10)$$

Taken in combination with (8.3.36) and (8.3.37) this expression shows that we may infer approximate values of the solute correlation factor $f_B$ from measurements of $D_A^*(0)$, $b_A^*$ and $D_B$ (Lidiard, 1960). See §11.5.3 later.

Returning now to the evaluation of $\mathbf{T}$, we recall that it was shown in Appendix 10.2 that an expression, (A10.2.6), can be derived for the elements $T_{\beta\alpha}$ which is a generalization of (10.6.20) for $T \equiv \langle x_1 x_2 \rangle / a^2$ for the case of one jump type. The essential feature is that its evaluation requires the enumeration of the configurations occurring between successive tracer-atom jumps in the $x$-direction. The configurations are defined by the relative positions of the three defects (A*, B, V) and may be grouped into sets of configurations such that all members of a set are *equivalent* because they have the same probability of occurrence after the initial tracer jump. In order to keep the calculation tractable the number of sets of configurations must be restricted by neglecting those configurations in which the vacancy has crossed a certain boundary, in effect assuming that if it crosses the boundary either it will not return or it will do so from a random direction. For example, Howard and Manning (1967) retained only the 21 sets of configurations in which at least one defect is nearest neighbour to the other two, plus 98 sets of configurations obtained from these by one vacancy jump, i.e. a total of 119 sets. In general, if there are $m$ sets, the calculation of $T_{\beta\alpha}$ requires the inversion of an $m \times m$ matrix (see e.g. (A10.2.6), (A10.2.7)). Finally, the calculation of partial correlation factors from (10.4.14) requires a further $n \times n$ matrix inversion. Because the matrices are large the calculations must be done by computer. For the five-frequency model $\chi_1$ and $\chi_2$ (11.3.9) were obtained as numerical tables as functions of $w_2/w_1$, $w_3/w_1$ and $w_4/w_0$. Very similar calculations have been made by Jones and Le Claire (1972) for the corresponding model for the b.c.c. lattice.

Ishioka and Koiwa (1984) have also made calculations for these same f.c.c. and b.c.c. models. In these they retain more jump types and configurations and use a slightly modified method, which is outlined in Appendix 11.3. In the extreme when the solute jump frequency is small compared with the solvent–atom jump frequencies there are significant differences between their numerical values of the solvent enhancement factor $b_A^*$ and those from the earlier calculations. However, in the circumstances defined by the random alloy model (i.e. $w_1 = w_3 = w_4 = w_0 \equiv w_A$ while $w_2 \equiv w_B$) the earlier results predict that $D_A^*$ increases with increasing $w_B/w_A$, as is physically reasonable, whereas the values of Ishioka and Koiwa show the opposite trend (Allnatt and Lidiard, 1987a). There may therefore be some as yet unresolved difficulties in making satisfactory calculations for certain combinations of jump frequencies.

In concluding this section we note that, although eqns. (11.3.6)–(11.3.8) cannot be employed as they stand to describe corresponding effects in ionic compounds, the correlation factors $\chi_1$ and $\chi_2$ are immediately transferable. For aliovalent solutes B giving rise to a fraction of free vacancies $c_V'$ and a fraction of solute–vacancy pairs $c_P$ in the A sub-lattice (f.c.c.), the appropriate expression for $D_A^*$ is

$$D_A^* = 4w_0 a^2 f_0 c_V' + \tfrac{4}{3} w_3 a^2 c_P \left( \frac{\chi_1 w_1}{w_3} + \frac{7\chi_2}{2} \right), \tag{11.3.11}$$

to first order in the site fractions $c_V$, $c_B$ and $c_P$. The quantities $c_V'$ and $c_P$ are obtained in terms of $c_B$ from the equations of Schottky equilibrium, pairing equilibrium (11.3.5c) and electroneutrality. Equation (11.3.11) taken in conjunction with these other equations shows us that, in contrast to metals, $D_A^*$ may depend non-linearly upon $c_B$ at quite low solute fractions.

## 11.4 Atomic transport in metals and alloys

From the previous two sections it will be clear that we are now in a position to work out transport coefficients for specific models of alloys and solid solutions. Before doing so, however, it will be useful to survey some of the information obtainable from experiment and from atomistic calculations about the nature of the defects involved. We can use this to point to appropriate models for alloys and solid solutions. We have, of course, already done this to some extent in Chapter 2. But in this and succeeding sections we shall give more attention to the analysis of diffusion and transport data in the light of theoretical results already obtained in previous chapters. (For extensive compilations of data see Mehrer, 1990 and Ullmaier, 1991.)

### *11.4.1 Diffusion mechanisms – evidence from pure metals*

In the present context metals are commonly divided into *normal* metals and the others. Normal metals are defined empirically as those whose tracer self-diffusion coefficients $D^*$ have the following characteristics:

(i) $D^*$ is closely represented by the Arrhenius expression (1.2.4) with a constant activation energy $Q$ and pre-exponential factor, $D_0$.

(ii) $D_0$ lies in the range $5 \times 10^{-6}$ to $10^{-3}\ \mathrm{m^2\,s^{-1}}$, while $Q/k$ scales with the melting temperature, being approximately $18T_m$/K.

Examples of metals in this class are the f.c.c. metals Al, Ag, Au, Cu, Ni and Pt and the b.c.c. metals Cr, Li, Na, Nb, Ta and V.

Metals whose properties do not conform to (i) and (ii) are called *anomalous.* These are generally particular phases of allotropic metals for which several distinct solid phases exist. Examples include the b.c.c. metals $\delta$-Ce, $\beta$-Gd, $\beta$-Hf, $\gamma$-La, $\beta$-Pr, $\varepsilon$-Pu, $\beta$-Ti, $\gamma$-U, $\gamma$-Yb and $\beta$-Zr. Arrhenius plots of the self-diffusion coefficients of these metals show a clearly visible curvature while the pre-exponential factor $D_0$ and the activation energy $Q$ in any limited range are considerably lower than those of normal metals, although they both increase as one moves upwards in temperature. The most striking examples are provided by Ti and Zr which, as $\beta$-Ti and $\beta$-Zr, exist in the b.c.c. phase over a wide range temperature.

The normal metals have been investigated by many refined measurements of several different kinds in addition to diffusion (e.g. differential dilatometry, positron annihilation, residual electrical resistivity after quenching, direct observation in the field ion microscope, etc.). The results are fully consistent with single vacancies as the dominant defects in self-diffusion (e.g. Siegel, 1982). The inferences from the studies of the anomalous metals are less definite, mainly because their study is made more difficult by the existence of other phases, so that, for example, quenching techniques may be less effective, the preparation of suitable specimens may be more difficult and so on. Nevertheless it may still be correct to suppose that vacancies are the defects responsible for self-diffusion. The anomalous behaviour is then to be interpreted as resulting from a strong temperature dependence of the vacancy energies and other properties associated with large thermal effects in the lattice vibrations ('soft' modes, see, e.g. Herzig, 1990). Such an idea is consistent with the existence of allotropy in these metals.

From the conclusion that vacancies are the dominant defects in both the normal and possibly also the anomalous metals, it is natural to suppose that the diffusion of substitutional solutes in these metals is also due to vacancies, although the possibility of interactions must be allowed for. Atomistic calculations of various kinds provide guides to the magnitude of these interactions and indicate that they are generally small (of the order of 0.1 eV; see, e.g. March, 1978; Alonso and March, 1989) and of short range. In other words we should not expect solute diffusion coefficients to differ greatly from those for self-diffusion. In fact, there is a wide range of dilute alloys for which the solute and the self-diffusion coefficients in the pure solvent do not differ by more than a factor of 10 at the melting point, although the difference may be greater at lower temperatures. We shall use the term normal solute diffusion in these circumstances. By contrast to these normal systems there is a smaller group of so-called anomalously fast diffusers which, in certain (normal) solvents (e.g. Pb), have diffusion coefficients which may be several orders of magnitude greater than the corresponding self-diffusion coefficients and for which the Arrhenius energy $Q$ is only one half or one third of that for the solvent. We shall deal with these separately after the normal systems.

Normal solute diffusion can therefore be reasonably represented by models of the same sort as the five-frequency model for f.c.c. lattices. Indeed several such models have been proposed for other lattices. These are all based on the idea that (single) vacancies are the dominant defects and they all assume that (i) the solute–vacancy interaction is of short-range and (ii) the vacancy jump frequency is perturbed only in the immediate vicinity of the solute. We shall analyse the transport coefficients for these models in §11.5 below.

Before leaving this section however we note that, as measurements of diffusion coefficients and related quantities have become more precise and have extended over wider ranges of temperature, they have indicated that even in normal metals the defect picture may be more complicated than we have indicated. These indications arise from observations of small deviations from linearity in Arrhenius plots (of $\ln D$ v. $T^{-1}$), of a dependence of isotope effects on temperature, of the magnitude and temperature dependence of the activation volume and so on. At present two alternative interpretations of these results are advanced; one that the vacancy parameters (e.g. enthalpy of formation and migration) vary with temperature as a result of increasing lattice vibrations and the associated expansion, the other that in addition to single vacancies there is a second defect with a somewhat higher Arrhenius energy, which thus becomes relatively more important at higher temperatures. Most commonly, the second defect is taken to be a vacancy pair, although in some b.c.c. metals dumb-bell interstitials may be possible. These matters cannot, however, be regarded as settled and we shall therefore set them aside from what follows.

## 11.5 Transport coefficients of normal dilute alloys and solid solutions

In §§11.2 and 11.3 we were concerned with ways to work out transport coefficients for specific models. Although we focussed on the five-frequency model of f.c.c.

alloys, the same approaches may be applied generally to all systems in which atomic migration involves the motion of vacancies or interstitials or more complex defects formed from them (e.g. solute–vacancy pairs). Indeed, transport coefficients have now been worked out for a number of different types of model. The results are the subject of this and later sections of this chapter. Before proceeding, however, we note that all the dilute alloy models to be described in this chapter share two characteristic features with the five-frequency model. The first of these is the short range of the solute–defect interactions and the smallness of the region in which the jump frequency of the defect is perturbed by the solute. The second is that certain crystallographically inequivalent jumps are assigned the same frequency, partly in order to reduce the number of parameters in the model. Both these features are supported to a greater or lesser extent by detailed atomistic calculations (e.g. March, 1978).

We turn now to the presentation of explicit expressions for the transport coefficients. We shall also summarize some of the information obtained by using them to analyse the corresponding quantities as measured experimentally. We shall do this for each of several types of alloy and solid solution. We begin with those we have referred to as *normal*, i.e. those where straightforward vacancy models are believed to be correct.

### *11.5.1 Models*

We shall limit the discussion to alloys sufficiently dilute that we may neglect the presence of pairs or larger clusters of solute atoms. No rigorous extension of transport theory to include such clusters has yet been provided, although a semi-quantitative theory based on the models of Dorn and Mitchell (1966) and Bérces and Kovács (1983) has been offered by Faupel and Hehenkamp (1986) and shown to be quite successful for noble metal alloys containing electropositive solutes.

Our attention therefore focusses on solute–vacancy pairs. For f.c.c. lattices the five-frequency model of solute–vacancy pairs (Fig. 2.22) is standard. The analogues for b.c.c., s.c. and diamond lattices are four-frequency models. In the f.c.c. lattice the nearest neighbours of a given site may also be nearest neighbours of one another, but in none of the b.c.c., s.c. and diamond lattices is this possible. Hence in these there are no analogues of the $w_1$ jump in the f.c.c. lattice. The other four frequencies can be labelled in the same way as for the five-frequency model: namely $w_2$ (the $w_s$ of §8.3) for solute exchanges, $w_3$ for all solvent exchanges that carry the vacancy from a first neighbour to a more distant position relative to the solute, $w_4$ for the reverse of $w_3$ jumps and $w_0$ for all other solvent exchanges. The last are assumed to be the same as in the pure host lattice. We shall refer to these four- and five-frequency models as the first-neighbour binding models.

In the b.c.c. and diamond lattices the second-neighbour distance is relatively close to the first-neighbour distance and binding at both positions should probably be allowed for. A useful model of this kind for the b.c.c. lattice which involves six distinct jump frequencies was defined by Le Claire (1970): see also Jones and Le Claire (1972). Vacancy jumps from first to second and from first to more distant neighbour sites are assigned different frequencies, $w_3$ and $w_3'$ respectively, and their inverses have frequencies $w_4$ and $w_0$, respectively, where $w_0$ is the unperturbed frequency. Jumps from second to more distant neighbours are assigned the frequency $w_5$ and all other solvent exchanges are assumed to occur at the unperturbed frequency $w_0$. The sixth frequency $w_2$ is the solute–vacancy exchange rate. An analogue of this model can be defined for the diamond lattice, although here there are no $w_3$ jumps since the sites concerned are not nearest neighbours. We shall refer to models of this kind which allow for vacancy–solute binding at both first- and second-neighbour positions as second-neighbour models.

Lastly, we note that no models of these types seem to have been proposed for h.c.p. metals, despite a considerable number of atomistic calculations of the properties of point defects in such lattices (see, e.g. Bacon 1988; 1991).

### *11.5.2 The transport coefficients $L_{AA}$, $L_{AB}$ and $L_{BB}$*

It is again convenient to begin with the example of the five-frequency model. It is clear, both from physical considerations and from the analysis developed in Chapter 8, that we shall need both the site fraction $c_P$ of solute atoms which have a vacancy among their $z$ nearest-neighbour sites and also the fractions of unbound vacancies $c_V' = c_V - c_P$ and of unbound solute atoms $c_B' = c_B - c_P$. These are, of course, related through the mass action equation (3.8.5b), namely

$$\frac{c_P}{c_V' c_B'} = z\exp(-\Delta g/kT) = \frac{zw_4}{w_3}. \tag{11.5.1}$$

Here the binding energy $\Delta g$ has been related to $w_4/w_3$ by the use of detailed balance.

By using the techniques described in §11.2 we can now obtain the transport coefficients. In terms of the division of each $L$-coefficient into an uncorrelated part $L^{(0)}$ and a correlated part $L^{(1)}$ (6.5.10) we then obtain for the five-frequency model

$$L_{AA}^{(0)} = (ns^2/6kT)(12c_V'(1-7c_B')w_0 + c_P A_{AA}^{(0)}), \tag{11.5.2a}$$

$$L_{AA}^{(1)} = (ns^2/6kT)c_P A_{AA}^{(1)}, \tag{11.5.2b}$$

$$L_{AB}^{(0)} = L_{BA}^{(0)} = 0, \tag{11.5.3a}$$

$$L_{AB}^{(1)} = L_{BA}^{(1)} = (ns^2/6kT)c_P A_{AB}^{(1)}, \tag{11.5.3b}$$

$$L_{BB}^{(0)} = (ns^2/6kT)c_P w_2, \tag{11.5.4a}$$

$$L_{BB}^{(1)} = -2(ns^2/6kT)c_P {w_2}^2/\Omega, \tag{11.5.4b}$$

in which $s$ is the nearest-neighbour spacing $(=\sqrt{2}a)$ and

$$A_{AA}^{(0)} = 4w_1 + 14w_3, \tag{11.5.5}$$

$$A_{AA}^{(1)} = \left[-2(3w_3 - 2w_1)^2 + 28w_3(1-F)(3w_3 - 2w_1)\left(\frac{w_0 - w_4}{w_4}\right) - 14w_3(1-F)(2w_1 + 2w_2 + 7w_3)\left(\frac{w_0 - w_4}{w_4}\right)^2\right]\bigg/\Omega \tag{11.5.6}$$

$$A_{AB}^{(1)} = w_2\left[2(3w_3 - 2w_1) + 14w_3(1-F)\left(\frac{w_0 - w_4}{w_4}\right)\right]\bigg/\Omega \tag{11.5.7}$$

and

$$\Omega = 2w_1 + 2w_2 + 7w_3 F. \tag{11.5.8}$$

Here $F$ is the vacancy escape factor discussed in §11.2 (eqns. 11.2.6 and 11.2.8). We recall that it is a function of $w_4/w_0$. Approximations as well as the accurate description of it were given in §11.2. The one-shell (Lidiard–Le Claire) approximation, $F = 1$, gives the $L_{ij}$ coefficients of Howard and Lidiard (1963) (cf. §8.5) and the more accurate two-shell approximation gives the $L_{ij}$ coefficients of Bocquet (1974b). The complete set of equations (11.5.2) *et seq.* were first obtained in this form by Allnatt (1981). We make several comments on them as follows.

(1) Once the various jump frequencies and the vacancy concentration are known, these expressions allow us to calculate $L_{AA}$, $L_{AB}$ and $L_{BB}$ as functions of temperature, pressure and solute concentration. Conversely, an empirical knowledge of the $L$-coefficients may allow us to infer the jump frequencies entering into (11.5.2)–(11.5.8).

(2) The expressions (11.5.2) are correct to first order in the fractions $c_V$, $c_B$ and $c_P$. Because they are correct to first order in $c_P$ they are therefore also correct to first order in the product $c_V c_B$. But they are not correct to terms in ${c_V}^2$ or ${c_B}^2$. This should be kept in mind when proceeding to the limit of very weak or zero binding.

(3) When we wish to use these $L$-coefficients to describe a diffusion process (e.g. in eqns. 4.2.4) it is important to represent the forces to the correct order in the

Table 11.3. *Models of a dilute alloy of solute B in host A for which $L_{AA}$, $L_{AB}$ and $L_{BB}$ have all been calculated*

| Migration mechanism | Lattice | Bound defect pairs | References |
|---|---|---|---|
| Vacancy | f.c.c. | first neighbour B–V | Howard and Lidiard (1963, 1964), Bocquet (1974b), Allnatt (1981) |
| Vacancy | f.c.c. | first and second neighbour B–V | Okamura and Allnatt (1983b) |
| Vacancy | b.c.c. | first and second neighbour B–V | Serruys and Brébec (1982b), Okamura and Allnatt (1983b) |
| Vacancy | diamond | first and second neighbour B–V | Okamura and Allnatt (1983b) |
| A and B interstitials | f.c.c. | first neighbour A–A, A–B and B–B interstitial pairs | Lidiard and McKee (1980), Okamura and Allnatt (1983a) |
| Interstitial B and lattice vacancy | f.c.c. | first neighbour interstitial B–lattice vacancy | Okamura and Allnatt (1984) |
| Interstitial B and lattice vacancy | b.c.c. | first neighbour interstitial B–lattice vacancy | Serruys and Brébec (1982a) |
| AA and AB dumb-bells[a] | f.c.c. | first neighbour (AA dumb-bell) substitutional B | Allnatt, Barbu, Franklin and Lidiard (1983), Okamura and Allnatt (1986a) |
| AA and AB dumb-bells[b] | f.c.c. | none | Chaturvedi and Allnatt (1992) |

[a] Only the AB dumb-bell can rotate on its lattice site.
[b] Both AA and AB dumb-bells can rotate on their lattice sites.

site fractions too. In practice this requires the retention of some of the terms linear in $c_s$ in (3.5.7) and (3.5.8).

(4) If we take the limit in which A and B are merely isotopes of the same chemical element (i.e. all $w = w_0$) we should regain eqns. (5.7.7). In particular, we expect that $L_{AA} + L_{AB} + L_{BA} + L_{BB}$ will tend to $Nc_V D_V/kT$ to first order in $c_V$. It is easily verified from (11.5.2) that this is so (with $D_V = 2w_0 s^2$).

(5) Results analogous to eqns. (11.5.2) have been obtained for both first- and second-neighbour models of other lattices. Table 11.3 provides references to these other cases. (N.B. The level of approximation in these references varies.)

(6) The same general methods can be used to obtain corresponding coefficients for second-neighbour models, i.e. those which postulate a binding between solute atoms and vacancies at second- as well as first-neighbour separations. There will now be two relations like (11.5.1) for the fractions of first- and second-neighbour pairs. Expressions for $L_{AA}$, $L_{AB}$ and $L_{BB}$ for the second-neighbour models have been determined for the b.c.c. lattice (Serruys and Brébec, 1982b; Okamura and Allnatt 1983b) and for the diamond and f.c.c. lattices (Okamura and Allnatt, 1983b). The structure of these results with respect to the defect fractions $c'_V$ and $c_P$ is like that for the five-frequency model. From a practical point of view these more elaborate models have the drawback of containing a relatively large number of unknown jump frequencies. To be useful they therefore normally require strong backing by particular atomistic calculations.

(7) Lastly, we compare these accurate solutions with approximate ones. The approximate solutions are, of course, much easier to obtain and thus from a practical point of view it is important to know whether the approximations made are accurate enough. If we compare eqns. (11.5.2) *et seq.* for the five-frequency model with the corresponding results in the Lidiard–Le Claire approximation (eqns. 8.5.7) equivalent to $F = 1$, then we might expect that the approximation will not make much difference unless $w_0$ and $w_4$ are very different from one another; for, when $w_0 = w_4$ the quantity $F$ appears only in the denominator $\Omega$ of the correlated terms. We can test this expectation by looking at the ratio $L_{AB}/L_{BB}$, which is of direct interest in the analysis of various drift experiments. This quantity is independent of $w_2$ and, when $F = 1$, it depends only on the ratio $w_1/w_3$. With more accurate $L$-coefficients it depends also on $w_4/w_0$ as shown in Fig. 11.2. In this representation the Lidiard–Le Claire one-shell approximation gives a horizontal straight line, while the Bocquet, two-shell approximation gives results very similar to the exact ones. A rather similar pattern emerges for the diamond and b.c.c. lattices (Allnatt and Okamura, 1986). Clearly, the one-shell approximation does not give this ratio at all precisely outside the range where $w_4$ is comparable to $w_0$ (say, $\frac{1}{2} < w_4/w_0 < 2$) but fortunately, as will be seen later, a number of the systems investigated fall into this range. On the other hand the two-shell approximation is adequate for most purposes. Finally, we note that the differences between first- and second-neighbour models are likely to be small for many purposes and therefore it is usually not possible to distinguish between them experimentally.

### *11.5.3 Self-diffusion coefficients, $D^*_A$ and $D^*_B$*

As we saw in Chapter 5, the complete specification of a binary alloy AB requires the self-diffusion coefficients $D^*_A$ and $D^*_B$ as well as $L_{AA}$, $L_{AB}$ and $L_{BB}$. That our

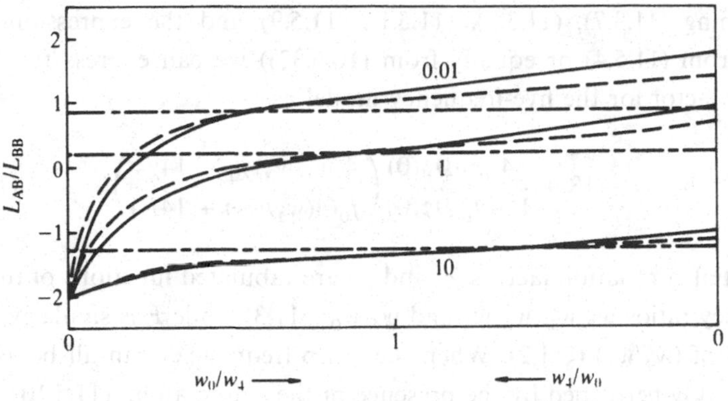


Fig. 11.2. The ratio $L_{AB}/L_{BB}$ for the f.c.c. five-frequency model of solute–vacancy interactions as a function of the ratio $(w_4/w_0)$ for values of $(w_1/w_3)$ as indicated. The chain curves give the one-shell (Lidiard–Le Claire) approximation, the broken curves the two-shell (Bocquet) approximation and the full curves the accurate results. Rather similar curves are obtained for the f.c.c. second neighbour model. (After Allnatt and Okamura, 1986.)

models omit pairs or larger clusters of solute atoms here implies that we consider $D_A^*$ to first order in $c_B$, but $D_B^*$ only to zero order. Taking $D_B^*$ first, we see furthermore that in this dilute limit $D_B^*$ is identical to the intrinsic diffusion coefficient $D_B$ given by (5.3.8) and the expression for $L_{BB}$ (e.g. (11.5.4)), i.e.

$$D_B^* = D_B = \tfrac{1}{6}c_V^{(0)}zs^2 w_2 f_B \exp(-\Delta g/kT) \qquad (11.5.9)$$

independent of the concentration of B. Here $c_V^{(0)}$ is the fraction of vacancies in the pure solvent, $f_B$ is the correlation factor for B atom diffusion and $\Delta g$ is the free energy of binding of vacancies to solute atoms in nearest-neighbour positions. This result has been obtained by considering the encounters between individual B (or B*) atoms and free vacancies. To get the next higher term in $c_B$ we would need to consider encounters between vacancies and pairs of B atoms (or B–B* pairs). This has not yet been done exactly (although see Faupel and Hehenkamp (1986) and Hehenkamp (1986) for an approximate, but quite successful, treatment).

By similar reasoning it is clear that models of the type described in §11.4.1 will allow us to obtain $D_A^*$ to first order in $c_B$ by consideration of the movement of A* atoms by those vacancies attached to solute B atoms as well as by free vacancies. This evaluation has been demonstrated in §11.3 and leads to results for the solvent tracer enhancement factor $b_{A*}$ defined by (11.3.6). In general, $b_{A*}$ is a function of the partial correlation factors of the solvent jumps as, for example, in (11.3.8) for the five-frequency model. As we saw in §11.3 these have been numerically tabulated as functions of jump frequencies for a few of the models cited.

By combining (11.3.7), (11.3.8), (11.5.1), (11.5.9) and the expression for $f_B$ (obtainable from (11.5.4) or equally from (10.6.32)) we can express the solvent enhancement factor for the five-frequency model as

$$b_{A*} = -18 + \frac{4f_0}{1 - f_B} \frac{D_B^*(0)}{D_A^*(0)} \left( \frac{4(w_1/w_3)\chi_1 + 14\chi_2}{f_0(4(w_1/w_3) + 14F)} \right). \qquad (11.5.10)$$

Here the partial correlation factors $\chi_1$ and $\chi_2$ are tabulated functions of the three jump frequency ratios $w_2/w_1$, $w_3/w_1$ and $w_4/w_0$ (§11.3) while $F$ is similarly known as a function of $(w_4/w_0)$ (§11.2). When the jump frequencies can all be assumed to be only weakly perturbed by the presence of the solute atom, (11.5.10) takes a convenient form consistent with the Lidiard–Le Claire approximation (10.6.31): thus with $\chi_1 = \chi_2 = f_0$ and $F = 1$ we obtain

$$b_{A*} = -18 + \frac{4f_0}{1 - f_B} \frac{D_B^*(0)}{D_A^*(0)}. \qquad (11.5.11)$$

When B is an isotope of A, i.e. $D_B^*(0) = D_A^*(0)$ we expect $b_{A*} = 0$ and $f_B = f_0$; (11.5.11) then gives $f_0 = 9/11$ as does (10.6.31). In other cases this approximate form (11.5.11) is convenient as it will provide a fair estimate of the solute correlation factor $f_B$ from $D_A^*(0)$, $D_B^*(0)$ and $b_{A*}$. The reason it does so is that in practical cases where the five-frequency model holds, the term in braces in (11.5.10) is still of order of magnitude unity, even if it is not exactly unity.

Equation (11.5.10) shows that fast diffusing solutes ($D_B^*(0) > D_A^*(0)$) should tend to enhance diffusion rates. Since values of $D_B^*(0)/D_A^*(0)$ range up to 15–20 while typical values of $f_B$ are about 1/2, it follows that positive values of $b_{A*}$ should be in the range 0–100 if the vacancy mechanism holds. Although slow diffusers may be expected to slow down solvent diffusion, the coefficient $b_{A*}$ cannot be less than $-18$ (for this would require $f_B > 1$). These expectations are in accord with experimental results for suitable systems as shown in Tables 11.4 and 11.5. It should be remarked that when $D_B^*(0)$ and $D_A^*(0)$ are comparable, then the particular values of the correlation factors in (11.5.10) can be important; so that, for example a fast diffuser may give a negative solvent enhancement, as does Fe in Cu.

Equation (11.5.10) is also valuable in providing a minimum possible value for $b_{A*}$ consistent with a given experimental value of $D_B^*(0)/D_A^*(0)$. The minimum will occur when the vacancy exchanges readily with a solute to which it is strongly bound so that solvent jumps are rare. In the appropriate limit ($w_3/w_1 \to 0$, $w_2/w_1 \to \infty$, and hence also $f_B \to 0$) the result is

$$b_{A*}(\text{min}) = -18 + 1.945 D_B^*(0)/D_A^*(0). \qquad (11.5.12)$$

Table 11.4. *Jump frequency ratios according to the first-neighbour (five-frequency) model of solute–vacancy interactions in f.c.c. alloys as determined from solute and solvent tracer diffusion coefficients, linear solvent-enhancement factors and solute correlation factors found by isotope effect measurements*

| Alloy A | Alloy B | $T/K$ | $\frac{D^*_B(0)}{D^*_A(0)}$ | $b_{A*}$ | $f_B$ | $\frac{w_2}{w_1}$ | $\frac{w_3}{w_1}$ | $\frac{w_4}{w_0}$ |
|---|---|---|---|---|---|---|---|---|
| Cu | Mn | 1199 | 4.2 | ~5 | 0.36 | 2.0–8.0 | 0.15–1.0 | 0.6–1.3 |
| Cu | Fe | 1293 | 1.06 | −5 | 0.79 | 0.35 | 0.095 | 0.3 |
| Cu | Co | 1133 | 0.61 | 0 | 0.88–0.91 | 0.1–0.3 | 0.02–0.4 | 0.1–0.7 |
| Cu | Zn | 1168 | 3.56 | 7.23 | 0.47 | 2.5 | 0.5 | 1.2 |
| | | 1220 | 3.39 | 8.78 | 0.47 | 3.6 | 0.91 | 1.5 |
| Cu | Cd | 1076 | 10.2 | 35 | 0.22 | 10 | 1 | 3 |
| Cu | In | 1089 | 12.0 | 43 | 0.07 | 20.0–500.0 | 0.4–9.5 | 2.4–5.5 |
| Cu | Sn | 1014 | 17 | 40 | 0.15 | 6.0–20.0 | 0.1–0.4 | 0.9–3.2 |
| | | 1089 | 14.1 | 48 | 0.15 | 6.0–40.0 | 0.2–0.8 | 1.9–5.7 |
| Cu | Au | 1133 | 1.15 | 8.1 | 0.87–0.90 | 0.12–0.26 | 0.05–0.20 | 0.40–0.80 |
| Ag | Zn | 1010 | 4.1 | 12.6 | 0.52 | 1.53 | 0.27 | 1.15 |
| | | 1153 | 3.9 | 12.7 | 0.57 | 1.54 | 0.39 | 1.30 |
| Ag | Cd | 1060 | 3.8 | ~4 | 0.41 | 1.9–5.8 | 0.2–1.0 | 0.7–1.2 |
| | | 1133 | 3.28 | 9.2 | 0.71 | 0.49 | 0.07 | 0.52 |
| | | 1197 | 2.96 | 13.7 | 0.62 | 1.7 | 0.8 | 1.7 |
| Ag | Sn | 1043 | 5.8 | 20.2 | 0.46 | 2.00–2.72 | 0.29–0.47 | 1.45–1.81 |
| Ag | In | 1065 | 5.7 | 17.6 | 0.35 | 3.45–5.63 | 0.45–0.86 | 1.63–2.0 |
| Au | Zn | 1058 | 6.2 | 24 | 0.15 | 18.0–900.0 | 1.5–82.0 | 2.4–3.5 |
| | | 1117 | 5.7 | 23 | 0.15 | 18.0–950.0 | 5.0–95.0 | 2.4–3.4 |
| Au | In | 1075 | 8.6 | 71 | 0.26 | 8.0–300.0 | 1.4–80.0 | 6.0–6.5 |
| | | 1175 | 7.5 | 49 | 0.26 | 7.0–400.0 | 1.0–90.0 | 3.0–5.3 |
| Au | Sn | 1059 | 16.4 | 130 | 0.16 | | No solution | |

Table adapted (with corrections) from Bocquet *et al.* (1983) except for entries for Ag Sn and Ag In which are from Köhler, Neuhaus and Herzig (1985).

This relation allows an important test of the model, since it predicts that the observed $b_{A*}$ must be greater than the value given by (11.5.12): see Fig. 11.3. Corresponding expressions for $b_{A*}(\text{min})$ for b.c.c. first and second neighbour models can also be obtained; see e.g. Le Claire (1978).

Table 11.5. *Jump frequency ratios according to the first neighbour (five-frequency) model of solute-vacancy interactions in f.c.c. alloys as determined from solute and solvent tracer diffusion coefficients, linear solvent-enhancement factors and* $L_{AB}(0)/L_{BB}(0)$ *either determined from the solvent intrinsic diffusion coefficient and* (11.5.13) *or from the enhancement factor of the solvent drift velocity in electromigration* $b_{A*}'$, (11.5.18)

| Alloy A | Alloy B | $T/K$ | $\frac{D_{B*}(0)}{D_{A*}(0)}$ | $b_{A*}$ | $\frac{L_{AB}(0)}{L_{BB}(0)}$ or $b_{A*}'$ | a / b | $\frac{w_2}{w_1}$ | $\frac{w_3}{w_1}$ | $\frac{w_4}{w_0}$ |
|---|---|---|---|---|---|---|---|---|---|
| Cu | Ni | 1273 | 0.36 | −5 | 0.07 | a | 0.2 | 1 | 1 |
| | | 1273 | 0.36 | −5.3 | 0.12 | a | 0.27 | 0.42 | 0.53 |
| Cu | Zn | 1168 | 3.3 | 8 | −0.22 | a | 3 | 0.5 | 1 |
| Cu | Cd | 1076 | 10.2 | 35 | −0.7 | a | 0.1 | 1 | 3 |
| Cu | In | 1005 | 13.3 | 42 | −0.71 | a | 18 | 0.5 | 3 |
| | | 1089 | 11.4 | 43 | −0.57 | a | 11 | 1 | 4 |
| Cu | Sn | 1014 | 15.5 | 40 | −1.06 | a | 13 | 0.2 | 2 |
| | | 1089 | 13.6 | 48 | −0.84 | a | 7 | 0.33 | 3 |
| Cu | Sb | 1005 | 24.1 | 79 | −1.2 | a | 15 | 0.40 | 5 |
| Ag | Sb | 1048 | 6.24 | 27 | −0.76 | a | 1.85 | 0.38 | 2.02 |
| | | 952 | 8.69 | 43 | −0.67 | a | 3.36 | 0.70 | 3.50 |
| | | 926 | 9.60 | 47 | −0.66 | a | 4.35 | 0.84 | 3.96 |
| | | 890 | 10.85 | 51 | −0.64 | a | 6.44 | 0.99 | 4.40 |
| Ag | Zn | 1153 | 3.9 | 12.6 | 6 | b | 1.25 | 0.26 | 1.12 |
| Ag | Cd | 1153 | 3.18 | 6.5 | −12 | b | 0.5 | 0.07 | 0.46 |

Table adapted from Bocquet *et al.* (1983) except for the entries for Ag Sb which are from Hagenschulte and Heumann (1989a).

As Fig. 11.3 shows, for those solutes which exhibit normal diffusion, the experimental values of $D_A^*(0)$, $D_B^*(0)$ and $b_{A*}$ are compatible with the theoretical constraints just outlined and they therefore provide support for the assumption that migration is by the monovacancy mechanism. For the anomalously fast diffusing solutes, as we shall see later, this turns out not to be the case.

### *11.5.4 The Kirkendall effect in dilute alloys*

The intrinsic diffusion coefficient of the solvent, $D_A$, as already indicated (§5.3), may be determined from measurements of interdiffusion and Kirkendall effect in

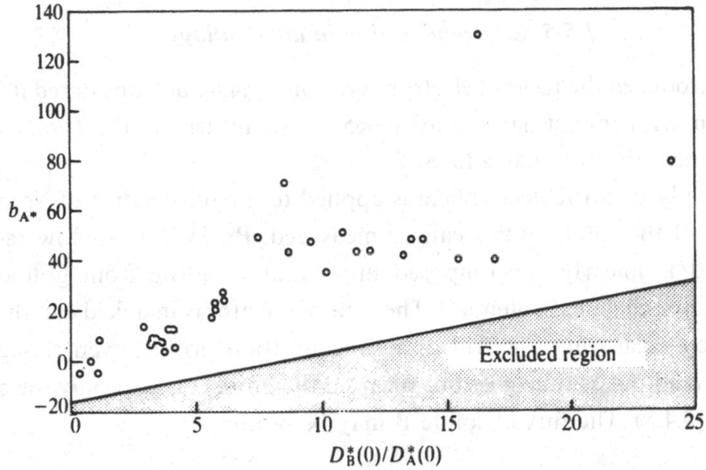


Fig. 11.3. A plot of the solvent-enhancement coefficient, $b_{A*}$, for self-diffusion v. $D_B^*(0)/D_A^*(0)$ in dilute f.c.c. alloys AB, using the data in Table 11.4. The line is given by (11.5.12): for the five-frequency model to be applicable the data points should lie above this line.

couples A/B. By extrapolating to the limit of extreme dilution ($c_B \to 0$) to obtain $D_A(0)$, one can infer the ratio $L_{AB}/L_{BB}$ in the same limit, so long as we also know the self-diffusion coefficients $D_A^*(0)$ and $D_B^*(0)$. In turn this may allow us to test the models used to calculate these quantities. The method uses the following relation due to Heumann (1979)

$$\lim_{c_B \to 0} \left(\frac{L_{AB}}{L_{BB}}\right) = \frac{D_A^*(0)}{D_B^*(0)}\left[\frac{1}{f_0} - \frac{D_A(0)}{D_A^*(0)}\frac{\bar{V}_B}{\bar{V}_A}\right], \tag{11.5.13}$$

in which $\bar{V}_A$ and $\bar{V}_B$ are the partial molar volumes of A and B respectively. To derive this we extend (5.3.6) to the case where $\bar{V}_A$ is different from $\bar{V}_B$ (but assuming Vegard's law) and then rewrite it as an equation for $L_{AB}$. We also use the corresponding expression for $D_B$ and the fact that, in the limit $c_B \to 0$, $D_B$ and $D_B^*$ become identical. Lastly, we also have (5.7.8) in this limit (pure A). The usefulness of this relation (11.5.13) comes from the fact that the quantities on the right side are either measurable on pure A or are obtainable by extrapolation to $c_B \to 0$ from Kirkendall measurements, while the left side is a property of dilute alloys since both $L_{AB}$ and $L_{BB}$ are proportional to $c_B$ (§11.5.2). Practical aspects are discussed by Heumann and Rottwinkel (1978). The use of this information to test and analyse models is considered shortly, after we have reviewed another method which can be similarly employed to yield $L_{AB}/L_{BB}$.

### *11.5.5 Electromigration in dilute alloys*

We briefly introduced the topic of electromigration in §4.4 and considered it further in connection with interstitial solutes in §5.2. We return to the topic now in connection with substitutional solutes.

When a steady uniform electric field is applied to an alloy a drift of isotopes of the host and of the solute atoms can be measured. By (4.4.5) the flow resulting from the field is linearly superimposed upon that resulting from gradients in concentration (or chemical potential). The drift of the atoms in a field can therefore be followed by measuring the drift of a concentration profile, even though this profile will spread out as time goes on. We need therefore consider only the electric field term in (4.4.5). The flux of solute B may be written as

$$J_B = L_{BB} Z_B^{**} eE = D_B^* n c_B Z_B^{**} eE/kT \tag{11.5.14}$$

where

$$Z_B^{**} = Z_B^* + Z_A^*(L_{BA}/L_{BB}). \tag{11.5.15}$$

Here $Z_i^*$ denotes $(z_i - z_i^*)$, in which $z_i$ is the ion core charge of atoms of kind $i$ and $z_i^*$ the coefficients introduced by (4.4.4b). If B is merely an isotope of the host A then by eqns. (5.7.7) we have $Z_A^{**} = Z_A^*/f_0$ where $f_0$ is the correlation factor for tracer self-diffusion in the host. Both $Z_A^{**}$ (and thus $Z_A^*$) and $Z_B^{**}$ may be determined from the drift of diffusion profiles in an electric field. However, to determine $Z_B^*$ we need to know $L_{AB}/L_{AA}$. This can, of course, be calculated from the appropriate jump-frequency ratios if these are known (see e.g. Tables 11.4, 11.6 below). An interesting alternative is to determine $L_{AB}/L_{AA}$ from further electromigration experiments.

A suitable experiment for this purpose is a measurement of the effect of solute atoms upon the drift velocity $v_{A*}$ of a tracer A* of the host species. This is characterized by a coefficient $b'_{A*}$. Thus

$$v_{A*}(c_B) = v_{A*}(0)(1 + b'_{A*} c_B). \tag{11.5.16}$$

Since $v_A$ is proportional to $J_A/c_A$ (all isotopic components of A having the same mobility) we can use the expression for A corresponding to (11.5.14) and, by re-expressing $L_{AA}$ as

$$L_{AA}(c_B) = L_{AA}(0)(1 + b_{AA} c_B), \tag{11.5.17}$$

we find

$$b'_{A*} = 1 + b_{AA} + \frac{f_0 D_{B*}(0)}{D_{A*}(0)} \left[ \frac{Z_B^{**}}{Z_A^*} - \left( \frac{L_{AB}}{L_{BB}} \right)_0 \right] \left( \frac{L_{AB}}{L_{BB}} \right)_0. \tag{11.5.18}$$

All the quantities $b'_{A*}$, $D_{B*}(0)/D_{A*}(0)$, $Z_B^{**}$ and $Z_A^*$ are measurable, while both $(L_{AB}/L_{BB})$ and $b_{AA}$ are known as functions of the jump-frequency ratios. In combination with the solvent enhancement factor $b_{A*}$, this data may therefore be used to infer values of the jump-frequency ratios, and also $(L_{AB}/L_{BB})$ itself, without the need to know the solute isotope effect.

Although we here emphasize the use of electromigration experiments for testing and parameterizing models, we should also note the interest in the $Z^*$ quantities themselves (Huntington, 1975; Ho and Kwok, 1989). The theory of these quantities emphasizes the importance of the scattering of the conduction electrons by the moving atoms and so relates them to other properties such as the electrical conductivity. Such predictions have been confirmed experimentally for the substitutional solutes As and Sb in Ag (Hehenkamp, 1981). The same theories also predict a corresponding contribution to the heats of transport in metals (e.g. Gerl, 1967).

### *11.5.6 Analysis of some dilute alloys*

A first step towards testing models of the type analysed in the preceding sections is to infer the various jump frequencies (or ratios of jump frequencies) appearing in the expressions for experimental quantities. Quantities used for this purpose include solvent and solute tracer diffusion coefficients, isotope effects (to yield correlation factors cf. §10.9), $L_{AB}/L_{BB}$ from Kirkendall shifts, electromigration experiments, etc. The alloys we consider have normal metals as solvents.

We begin by looking at f.c.c. alloys as represented by the five-frequency model. Our results show that the three ratios $w_2/w_1$, $w_3/w_1$ and $w_4/w_0$ can be obtained from the three quantities $D_B^*(0)/D_A^*(0)$, $b_{A*}$ and $f_B$. Results of such analyses are given in Table 11.4. Alternatively, one can analyse $D_B^*(0)/D_A^*(0)$, $b_A^*$ and $L_{AB}/L_{BB}$, the last of these being obtained from either Kirkendall experiments or electromigration experiments. Table 11.5 shows the results of these alternative analyses. For four alloys – CuCd, CuIn, CuSn and CuZn – there is sufficient data for both methods of analysis to be used. By comparing the entries in Tables 11.4 and 11.5 it will be seen that the inferences are generally consistent with one another (the obvious exception being the ratio $w_2/w_1$ for CuCd). Bearing in mind that the inferred ratios are likely to be sensitive to errors in the experimental quantities, this consistency can be taken as evidence for the correctness of the five-frequency model.

For b.c.c. alloys the approach is slightly different, for in this case both the first- and second-neighbour models show that two frequency ratios may be determined from the two quantities $D_B^*(0)/D_A^*(0)$ and $b_{A*}$. The results of such analyses are given in Table 11.6. This table also contains solute correlation factors

Table 11.6. *Jump frequency ratios according to the first-neighbour (model II) and second-neighbour (model I) models of solute-vacancy interactions in b.c.c. alloys as determined from solute and solvent tracer diffusion coefficients and linear solvent-enhancement factors*

The solute correlation factors are calculated from the jump frequency ratios

| Alloy | | | | | Model I | | | Model II | | |
|---|---|---|---|---|---|---|---|---|---|---|
| A | B | $T/K$ | $\frac{D_{B*}(0)}{D_{A*}(0)}$ | $b_{A*}$ | $w_2/w_3'$ | $w_3/w_3'$ | $f_B$ | $w_2/w_3$ | $w_4/w_0$ | $f_B$ |
| Fe | Si | 1400 | 2.1 | 24.4 | 1.86 | 6.8 | 0.82 | 0.61 | 3.2 | 0.78 |
| Fe | Co | 1168 | 0.74 | −1.1 | 0.69 | 0.80 | 0.78 | 0.74 | 0.92 | 0.79 |
| Fe | Sn | 1168 | 8.3 | 98 | 7.6 | 27 | 0.79 | 0.7 | 12 | 0.71 |
| | | 1093 | 6.9 | 63 | 6.8 | 17 | 0.74 | 1.1 | 7.5 | 0.64 |
| | | 1009 | 8.6 | 63 | 9.4 | 17 | 0.67 | 1.5 | 7.4 | 0.55 |
| Ti | V | 1673 | 1.26 | −4.3 | 1.73 | 0.34 | 0.53 | 2.64 | 0.67 | 0.52 |
| | | 1373 | 0.94 | −3.7 | 1.01 | 0.43 | 0.67 | 1.42 | 0.72 | 0.67 |
| V | Ti | 1673 | 1.66 | 16 | 1.49 | 4.6 | 0.81 | 0.64 | 2.4 | 0.78 |
| V | Fe | 1573 | 4.06 | 18.7 | 5.33 | 4.89 | 0.554 | 2.51 | 2.48 | 0.474 |
| | | 1757 | 3.68 | 10 | 6.65 | 2.76 | 0.402 | 4.38 | 1.71 | 0.357 |
| | | 2001 | 3.04 | 12.6–15.6 | 3.40–3.56 | 3.37–4.10 | 0.566–0.621 | 2.15–1.64 | 1.96–2.23 | 0.523–0.571 |
| V | Ta | 1960 | 0.74 | 1.68 | 0.66 | 1.21 | 0.814 | 0.605 | 1.09 | 0.813 |
| Zr | V | 1167 | 0.89 | −4.14 | 0.97 | 0.34 | 0.67 | 1.43 | 0.68 | 0.66 |
| | | 1476 | 1.14 | 13.60 | 0.97 | 3.92 | 0.85 | 0.46 | 2.17 | 0.83 |

Table adapted from Le Claire (1978, 1983) except for the entries for Fe Sn, which are from Kumugai, Iijima and Hirano (1983), for Zr V which is from Pruthi and Agarawala (1982) and for V Ta which is from Ablitzer, Haeussler and Sathyraj (1983).

calculated from these ratios for first- and second-neighbour models. The differences are too small for the models to be distinguished by isotope effect measurements, although the magnitudes are reasonable.

Lastly, we turn to some dilute alloys which show striking departures from the normal behaviour just considered. Table 11.7 lists many of these (as classified into three groups by Le Claire, 1978). Possibly the most thoroughly studied are the (f.c.c.) Pb-based alloys, especially those containing Ag, Au, Cu, Ni, Pd or Zn as solute. For these the solute diffusion coefficients are as much as $10^3$ to $10^5$ times that for Pb self-diffusion; see Fig. 11.4. (Indeed, it was the very rapid diffusion of Au in Pb that enabled Roberts-Austen in 1896 to make the first direct observations

Table 11.7. *Dilute alloy systems in which anomalously fast solute diffusion occurs*

| Solutes | Solvents |
|---|---|
| (1) Low valency groups I and II<br>Cu, Ag, Au, Be, Zn, Cd, Hg, Pd | High valency groups III and IV<br>Pb, Sn, Tl, In |
| (2) Later transition metals<br>Ni, Co, Fe, Cr, Mn, Pd | Early members of d-transition groups<br>$\beta$-Ti, $\beta$-Zr, $\beta$-Hf, Nb, La, Ce, Pr, Nd, $\gamma$-U, Pu |
| (3) Noble metals | Li, Na, K |

Table adapted from Le Claire (1978).

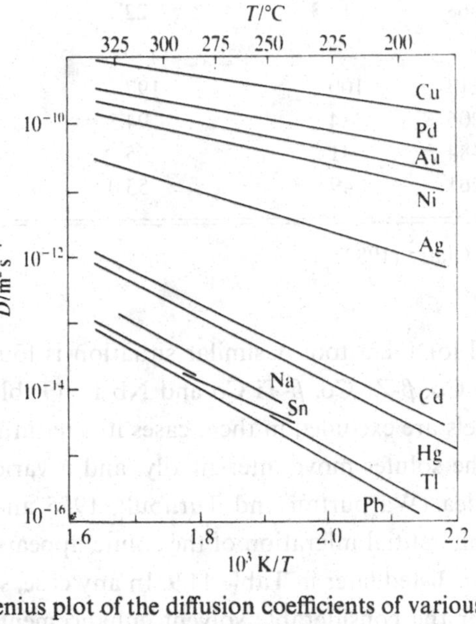


Fig. 11.4. Arrhenius plot of the diffusion coefficients of various solute elements in Pb. (After Warburton and Turnbull, 1975.)

of diffusion in the solid state.) Now, for the vacancy mechanisms already analysed it is clear that such rapid solute diffusion demands a strong solute–vacancy binding and both a rapid solute–vacancy exchange rate ($w_2$) and a rapid vacancy jump rate around the solute ($w_1$). These requirements can be expressed quantitatively through the minimum value of the enhancement factor $b_{A*}$ given by (11.5.12). As Table 11.8 shows, the known enhancement factors for Pb Ag and Pb Au are so much less than these minimum values that the usual vacancy mechanism is excluded. Furthermore, since the enhancement factors for Pb Cd and Pb Hg are so close to the corresponding allowed minimum values, the vacancy mechanism

Table 11.8. *Measured solvent-enhancement factors for anomalously fast diffusing solutes and the minimum values consistent with $D_B^*(0)/D_A^*(0)$ for diffusion by the vacancy mechanism*

For the f.c.c. alloys the five-frequency model and for b.c.c. alloys model I or model II are assumed

| System | | | | | | |
|---|---|---|---|---|---|---|
| A | B | $T$/K | $\dfrac{D_{B*}(0)}{D_{A*}(0)}$ | $b_{A*}$ | $b_{A*}$ (min) | |
| Pb | Ag | 573 | 870 | 137 | 1675 | |
| Pb | Au | 488 | $8.9 \times 10^4$ | $4.3 \times 10^3$ | $1.7 \times 10^5$ | |
| Pb | Cd | 573 | 19.9 | 19.1 | 20.8 | |
| Pb | Hg | 568 | 14.3 | 22.1 | 9.8 | |
| | | | | | I | II |
| U | Co | 1213 | 100 | 197 | 277 | 369 |
| Zr | Co | 1206 | 384 | 94.3 | 1095 | 1407 |
| Ti | Co | 1484 | 41 | 5 | 109 | 153 |
| Nb | Fe | 2261 | 49 | 55.0 | 144 | 191 |

Table adapted from Le Claire (1983).

must appear doubtful for them too. A similar situation is found in several b.c.c. alloys, namely in $\gamma$-U Co, $\beta$-Zr Co, $\beta$-Ti Co and Nb Fe (Table 11.8).

Since vacancy models are excluded in these cases it is natural to turn to models which suppose that the solutes move interstitially, and a variety of experimental results support the idea (Warburton and Turnbull, 1975 and Le Claire, 1978). Simple uncorrelated interstitial migration of the solute appears to be excluded by the small isotope effects listed later in Table 11.9. In any case, such an assumption would not account for the considerable solvent enhancement factors associated with these solutes (Table 11.8). These observations have thus led to proposals for more complex interstitial mechanisms which can explain them. We turn to their analysis in the next section.

## 11.6 Transport coefficients for dilute alloys and solid solutions: interstitial mechanisms

Atomistic calculations show that atomic relaxation around vacancies is generally small and such that it often leaves the point symmetry of the site unaltered. In this sense therefore all vacancy mechanisms of diffusion are much the same. By contrast, atomistic calculations for interstitials disclose a variety of structures and possible ways of moving. Further possibilities may also be suggested by more intuitive considerations. The transport coefficients must be evaluated for each of these in turn. However, although the details differ, the general principles of these evaluations remain as set out in §§11.2 and 11.3 as long as the perturbing effects of the solute on the interstitial movements are still local.

In §11.5.6 we saw that there was good evidence that the solutes in certain alloys moved by an interstitial mechanism. One of the earlier models proposed for f.c.c. alloys of this kind is due to Miller (1969). The idea is that, although the solute atoms may be present substitutionally, they are mostly present as interstitial solute vacancy pairs: both components of the pair can jump to new sites, but dissociation of the pair occurs infrequently.

The pair can 'collapse' to give a substitutional solute atom. Simplified versions of this model were proposed and analysed later by Warburton (1973) and McKee (1977). An elaboration of the model to include binding between the vacancy and the interstitial solute at second-neighbour positions was suggested by Huntley (1974). By using the basic formal expressions given in §§6.5 and 8.5 and the methods of evaluation described in §11.2, Okamura and Allnatt (1984) obtained the transport coefficients $L_{AA}$, $L_{AB}$ and $L_{BB}$ in sufficient generality to embrace all these variants. As before, $L_{BB}$ gives $D_B = D_B^*$ and thus also the correlation factor $f_B$. The self-diffusion coefficient for the solvent, $D_A^*$, and thus also the enhancement coefficient $b_{A*}$ were obtained by Le Claire (1978, correcting an earlier result of Miller) and by Warburton (1973) for certain special assumptions. When we use these results to relate $b_{A*}$ to the ratio $D_B^*/D_A^*$ we find that they predict that $b_{A*}$ should be roughly equal to $D_B^*/D_A^*$. Table 11.8 shows that this is true for Pb Cd and Pb Hg but not for Pb Ag and Pb Au where $b_{A*}$ is much smaller than predicted. These small values of $b_{A*}$ could be accommodated if the diffusion of the solute is dominated by the migration of free interstitials and the solvent enhancement is ascribed to the pairs. However, the free interstitials cannot be isolated interstitials undergoing uncorrelated diffusion since that would be inconsistent with the small solute isotope effect in these systems.

Warburton and Turnbull (1975) therefore suggested that in these systems the interstitials may exist as 'mixed' (solute–solvent) dumb-bells each centred on a substitutional site in the manner shown in Fig. 2.23(*c*). Migration of a dumb-bell (which they called a 'diplon') is assumed to occur by a mixture of rotation and displacement as shown in Fig. 2.23. In the displacement jump the solute dissociates from its host partner and forms a new mixed dumb-bell with a neighbouring host atom. This motion is sometimes called 'caging' since it can only transport the solute among an octahedron of sites (Fig. 2.23). In the second process the dumb-bell rotates on its lattice site and so allows the transport of the solute to a new octahedron. Both processes are evidently required for long range transport.

No theoretical calculations are available to provide guidance on the relative jump frequencies for these alloys and Warburton and Turnbull (1975) did not, in fact, make a detailed development of the kinetic theory of their model, nevertheless several features make the model plausible.

(i) If transport occurs in the way suggested by Warburton and Turnbull then the solute is, in effect, transported via interstitial positions with a diffusion coefficient dependent on the concentration of mixed dumb-bells and hence on their binding energy. Since this may be rather variable for different solutes, a wide spread of diffusion coefficients is possible. The host atom of the mixed dumb-bell may, however, also exchange with a vacancy so that the small solvent enhancement factors for Pb due to the solutes Au and Ag are in principle explicable.

(ii) Since three atoms are directly involved in the migration of a dumb-bell the isotope effect is likely to be small, as observed for these solutes in Pb.

(iii) The model is similar to models adopted for other systems where the dumb-bells are produced by irradiation. We shall therefore now consider the results for f.c.c. lattices in detail.

### *11.6.1 Dumb-bell interstitial mechanism*

When an alloy is irradiated by high energy particles which displace atoms from their lattice sites the interstitial defects created play an important role in the subsequent transport of matter. In this section we shall first define an appropriate model of interstitial migration in the f.c.c. lattice as originally proposed by Barbu (1980) and then outline the results for the phenomenological coefficients which have been determined for it. The justification of this model is provided both by atomistic calculations of Dederichs *et al.* (1978) and by the consideration of various kinds of experimental observations as reviewed for example by Wiedersich and Lam (1983).

The interstitial defects introduced by irradiation in the f.c.c. lattice assume a dumb-bell form centred on a substitutional lattice site and with $\langle 100 \rangle$ orientation. Both A–A and A–B dumb-bells are supposed to occur, but B–B dumb-bells are neglected because the concentration of solute B is taken to be low. The migration mechanism of a dumb-bell combines a translation to a nearest-neighbour site with a $\pi/2$ rotation of the dumb-bell axis in the manner shown in Fig. 2.23. The jump frequencies of the mixed dumb-bells and self-interstitial dumb-bells are denoted by $w_{\mathrm{I}}$ and $w_0$ respectively. A mixed dumb-bell may also rotate by $\pi/2$ on its lattice site with frequency $w_{\mathrm{R}}$.

The presence of a nearby substitutional solute atom can perturb the jump frequency of a self-interstitial. It is assumed that such pairs are bound only when they are nearest-neighbours. Since the energy of interaction depends on the orientation of the dumb-bell two distinct types of pair, denoted *a* and *b*, must be distinguished, as shown in Fig. 2.23. The following perturbed jumps of the self-interstitial are then distinguished: $w_1(w_1')$ for the jump of an A–A dumb-bell between sites which are both nearest neighbour to a B atom in a type-*a* (type-*b*) pair, $w_3(w_3')$ for A–A dumb-bell jumps leading to dissociation of a type-*a* (type-*b*) pair, and $w_4(w_4')$ for the inverse association jumps. Finally, the decomposition of a mixed dumb-bell to form a type-*a* pair is assigned a frequency $w_2$, and the inverse jump which forms a mixed dumb-bell is assigned the frequency $w_2'$.

The concentrations of paired and unpaired defects are readily related by the principle of detailed balance, or alternatively, by the methods of Chapter 3:

$$c_{P_b}/c_I' c_B' = 4w_4'/w_3' \tag{11.6.1a}$$

$$c_{P_a}/c_I' c_B' = 8w_4/w_3 \tag{11.6.1b}$$

$$c_{IB}/c_{P_a} = w_2'/4w_2. \tag{11.6.1c}$$

Here, $c_I'$ and $c_B'$ are the site fractions of unpaired self-interstitials and solute atoms, $c_{P_a}$ and $c_{P_b}$ are the concentrations of the two kinds of pair and $c_{IB}$ is the concentration of mixed dumb-bells.

The kinetics of this model include two sequences of jumps referred to as *caging* and *looping*. A sequence of mixed dumb-bell jumps, frequency $w_I$, merely moves the solute atom around a cage of sites which lie at the corners of an octahedron. Long-range migration can occur either by alternating displacement–rotation jumps of frequency $w_I$, with in-place rotation jumps of frequency $w_R$, or by decomposition of the mixed dumb-bell (jump frequency $w_2$) followed by a looping sequence of jumps of the self-interstitial, ending in re-formation of a mixed dumb-bell in a new orientation. These effects are all taken account of by a straightforward application of the kinetic theory methods of Chapter 8.

The phenomenological coefficients $L_{AA}$, $L_{AB}$ and $L_{BB}$ have been determined by Allnatt *et al.* (1983) in the one-shell approximation which assumes that a bound pair which dissociates will not associate again. We quote their results instead of those for the two-shell approximation (Okamura and Allnatt, 1986a) since the latter are rather long and complicated. Although it is more complex, the present model bears a broad similarity to the five-frequency model with dumb-bells replacing vacancies. We can thus anticipate that $L_{AB}$ and $L_{BB}$ will depend only on the concentratioon of pairs while $L_{AA}$ will also contain a term in the concentration of free interstitials. The results correct to first order in the concentrations are

$$L_{AA} \equiv L_{AA}^{f} + L_{AA}^{p} \tag{11.6.2a}$$

where

$$L_{AA}^{f} = 4ns^2c_{I}w_0/3kT \qquad (11.6.2b)$$

$$L_{AA}^{p} = \left(\frac{ns^2c_{Pa}w_3}{6kT}\right)\left(-\frac{7w_0}{w_4} + \frac{16(w_3 + 2w_1 + w_2')}{(5w_3 + 2w_1 + w_2')} + \frac{12[(w_R + w_I)(2w_3 + w_2') + 2w_3w_2]}{[(w_R + w_I)(5w_3 + w_2') + 5w_3w_2]} + \frac{6w_4'}{w_4}\frac{(w_3' + 2w_1')}{(2w_3' + w_1')}\right) \qquad (11.6.2c)$$

$$L_{AB} = \left(\frac{ns^2c_{P_a}}{6kT}\right)\left(\frac{6w_Iw_2'w_3}{A}\right) \qquad (11.6.3)$$

$$L_{BB} = \left(\frac{ns^2c_{P_a}}{6kT}\right)\left(\frac{w_2'w_I}{w_2A}\right)[w_R(5w_3 + w_2') + 5w_3w_2] \qquad (11.6.4)$$

in which

$$A = (w_R + w_I)(5w_3 + w_2') + 5w_3w_2. \qquad (11.6.5)$$

It will be noted that neither $L_{AB}$ nor $L_{BB}$ contains any of the jump frequencies pertaining to type-*b* pairs ($w_1'$, $w_3'$, $w_4'$). Such pairs play no part in the transport of B atoms, but serve merely to trap the self-interstitials for a time. Furthermore $L_{AB}$ and $L_{BB}$ will be zero if either $w_I$ or $w_2'$ is zero, i.e. if either the mixed dumb-bell cannot move ($w_I = 0$) or cannot be formed ($w_2' = 0$).

### *11.6.2 Experimental results*

From what has already been said we can anticipate two distinct areas of application of these results namely (i) to diffusion properties of systems close to thermodynamic equilibrium and (ii) to atomic transport transport in alloys subject to energetic-particle irradiation. The second subject is more complex than the first and we therefore treat it separately (§11.7). Here we return to the fast diffusing systems.

Evidence consistent with the existence of the dumb-bell complexes is provided by the reduction of the solute diffusion with increasing concentration for both Au (Warburton, 1975) and Ag (Kusunoki, Tsumuraya and Nishikawa, 1981) in Pb, i.e. in these systems $b_{B*}$ is negative. The reduction is greater at lower temperatures and cannot be understood in terms of vacancy or vacancy–interstitial pair mechanisms. On the other hand, it would be explicable if Au–Pb and Au–Au dumb-bells are present in dynamic equilibrium. If we assume that the Au–Au dumb-bells are the more strongly bound then we expect their concentration to increase as the solute concentration is increased or the temperature is decreased. Furthermore if the Au–Au dumb-bells are less mobile than the mixed dumb-bells then the observed behaviour of $D_B$ is explicable. Apparent evidence for Pb–Au and Au–Au dumb-bells in Pb has been found in internal friction measurements, but the contradictory experimental results obtained have left the interpretation uncertain, see e.g. Le Claire (1978). This is unfortunate as elastic loss experiments in principle allow a clear distinction to be made between dumb-bell interstitials and simple interstitial or substitutional solutes.

We have remarked earlier, in §1.8, that relatively fast diffusing atoms can be studied by the quasi-elastic scattering of neutrons, but for the systems of interest in this section the method is usually impracticable because of the low solute concentrations commonly encountered. An exception is provided by Co in $\beta$-Zr for which measurements with 2–5 atom% of Co have been made by Petry *et al.* (1987): some of their results were shown in Fig. 1.17. The high solubility suggests that the major part of the solute is substitutionally dissolved and possibly not typical of fast-diffusing systems. The half width $\Gamma(\mathbf{q})$ of the incoherent scattering function, (1.8.7), was interpreted through the Chudley–Elliott expression (9.3.11b). The dependence on scattering vector $\mathbf{q}$ was consistent with a nearest neighbour jump distance, but the jump frequency deduced was about 20 times that for Zr in the alloy and 8 times smaller than that deduced from Co tracer diffusion measurements. The latter comparison is based on the assumption that the correlation factor in tracer diffusion is approximately the same as that missing from (9.3.11b). This discrepancy and other aspects of the results are consistent with the existence of a second faster diffusion process which predominates in the solute tracer measurements. The slower jumps are inferred to be either with vacancies between substitutional sites or else to be associated with the release of the Co from trap sites to interstitial sites. The faster process is ascribed to migration between interstitial sites. The low isotope effect factor, Table 11.9, suggests a correlated interstitial mechanism such as the mixed dumb-bell mechanism already mentioned for Au in Pb, which has a similarly low value. A similar picture for Fe in $\beta$-Zr follows from measurements of Yoshida *et al.* (1987) of the quasi-elastic Mössbauer effect. Here, again the diffusion coefficient from the scattering experiment is smaller than the solute tracer diffusion coefficient but greater than the solvent diffusion coefficient.

## 11.7 Solute segregation induced by defect fluxes

In the previous analyses the defect concentrations have been assumed to retain their equilibrium values. Examples where this assumption is not correct are

Table 11.9. *Results of isotope effect measurements for anomalously fast diffusing solutes*

| System | | | |
|---|---|---|---|
| A | B | $T$/K | $f\Delta K$ |
| Pb | Cu | 592 | $0.23 \pm 0.08$ |
| Pb | Ag | 569–576 | $0.25 \pm 0.05$ |
| Pb | Au | 562–570 | $0.26 \pm 0.03$ |
| Pb | Hg | 498 | $0.23 \pm 0.04$ |
| Pb | Cd | 522 | $0.12 \pm 0.05$ |
| Nb | Fe | 2168–2172 | $0.025 \pm 0.066$ |
| Zr | Co | 1214–1741 | $0.23 \pm 0.03$ |

Table based on Herzig (1981) except for NbFe, which is from Ablitzer (1977), and ZrCo, which is from Herzig *et al.* (1987).

provided by various experiments and other situations in which the net fluxes of defects are considerable; for then, whether the defect populations remain close to their thermal equilibrium values depends on the magnitude of the defect fluxes relative to the rates of the reaction(s) tending to restore equilibrium. For metals and alloys containing vacancy defects this then becomes a question of the rates of flow of vacancies to dislocations, grain boundaries, surfaces and other sinks. Practical examples are offered by metals and alloys rapidly cooled from a high temperature ('quenched') and then subsequently annealed (or 'aged') at intermediate temperatures (see e.g. §1.10). In this case the task is to obtain the correct equations for the flux of vacancies to the sinks and associated movements of the solute elements. Such redistribution of solute elements can be of practical concern in various ways in the heat-treatment of commercial alloys. For example, redistribution around grain boundaries can affect mechanical properties such as toughness and ductility, while corrosion resistance may also be affected (see, e.g. Hondros and Seah, 1983).

Energetic particle irradiation can introduce similar considerations, although in these circumstances both vacancy and interstitial defects must be considered. In this section therefore we shall concentrate on the basic flux equations and their implications for solute segregation, firstly for the annealing of quenched alloys and secondly for alloys under continuous irradiation by energetic particles. We shall make use of the results described already in §§11.5 and 11.6.

### *11.7.1 Annealing of quenched alloys*

This topic brings out the limitations of the independent defect approach and we therefore begin by showing what these are. We then go on to formulate a more complete description. In the independent defect approach one writes down separate equations for the diffusion of unpaired vacancies and of solute–vacancy pairs, viz.

$$J'_V = -nD_V\nabla(c_V - c_P) \tag{11.7.1}$$

and

$$J_P = -nD_P\nabla c_P \tag{11.7.2}$$

in which $D_V$ is the vacancy diffusion coefficient in the perfect solvent lattice (presumed equal to that in the pure solvent) and $D_P$ is the diffusion coefficient of the solute–vacancy pairs. Since solute atoms can only move when a nearest neighbour position is vacant $J_P$ also gives the flux of solute atoms B. The total vacancy flux is $J_V = J'_V + J_P$. Now if the pairs and free vacancies are in local equilibrium we can use (11.5.1) and, since the fraction of solute $c_B$ will be much greater than $c_V$ and thus also $c_P$, we can take this in the form

$$\frac{c_P}{c_V - c_P} = zc_B\,e^{-(\Delta g/kT)} \equiv Kc_B \tag{11.7.3a}$$

and equivalently

$$c_P = c_V\left(\frac{Kc_B}{1 + Kc_B}\right) \tag{11.7.3b}$$

and

$$c_V - c_P = c_V\left(\frac{1}{1 + Kc_B}\right). \tag{11.7.3c}$$

Equations (11.7.1)–(11.7.3) thus give the total flux in a uniform alloy ($\nabla c_B = 0$) as

$$\begin{aligned} J_V &= J'_V + J_P \\ &= -n\left(\frac{D_V + Kc_B D_P}{1 + Kc_B}\right)\nabla c_V. \end{aligned} \tag{11.7.4}$$

This equation may be used to describe the motion of vacancies to the external surface, to grain boundaries or other sinks; and, indeed, often has been used for this purpose. The restriction that the alloy be uniform may not be serious in practice since the flux of solute atoms will frequently be much less than $J_V$, so that a large proportion of the vacancies may anneal out before the solute distribution is greatly disturbed. But, of course, the drift of the paired vacancies to the vacancy sinks implies a movement of solute atoms towards the sinks too, of amount

$$J_B = J_P = -n\left(\frac{Kc_B D_P}{1 + Kc_B}\right)\nabla c_V. \tag{11.7.5}$$

under the same conditions of negligible gradient in solute concentration. We thus have here a basis for explaining the observed segregation of solute atoms to grain boundaries and surfaces of quenched alloys. Indeed, the central ideas of this approach (eqns. (11.7.1), (11.7.2) and (11.7.3) have often been used for this purpose (e.g. Williams, Stoneham and Harries (1976); Faulkner (1981); Tingdong (1987); Karlsson (1986).

Nevertheless, this theory must be regarded as rather unsatisfactory as it stands. Firstly, it does not provide an expression for $D_P$ in terms of vacancy jump frequencies; this must be extracted somehow from results such as those given in §11.5. Secondly, and more importantly, it predicts that substitutional solute elements will always segregate towards the vacancy sinks, whereas one knows experimentally that the concentration of such solutes in many cases is depleted rather than enhanced around vacancy sinks (Anthony, 1975; Norris, 1987).

A moment's reflection shows that what is missing from the above theory is any consideration of the kinetics of the association and dissociation reactions among the solute atoms and vacancies. An attempt to remedy this deficiency was made by Johnson and Lam (1976) who added a term in $\nabla c_V$ to the equation for $J_B$ (11.7.2), together with a similar term in $J_V$. This introduced an additional parameter $\sigma_V$, a sort of cross-section, to represent the coupling between the flow of B atoms and the gradient in the vacancy concentration. Like $D_P$, this must be obtained by reference to the results provided in §11.5. It seems therefore altogether peferable to start with eqns. (4.4.9) or (4.4.14) when $L$-coefficients such as (11.5.2)–(11.5.4) are available. We shall therefore now turn to the exact description which these equations provide.

Firstly, we obtain the chemical potentials from eqns. (3.5.7) and Table 3.4 (as we did in §3.8.2) in order to define the forces to be inserted in (4.4.9) in terms of the gradients of $c_V$, $c_B$ and $c_P$. Remembering that expressions for the $L$-coefficients $L_{AA}$, $L_{AB}$ and $L_{BB}$ are correct to first-order in $c_B$, $c_V$ and $c_P$, we then obtain for the fluxes $J_V = -(J_A + J_B)$ and $J_B$ correct to zero-order in these same quantities the following expressions:

$$\begin{aligned} J_V &= -(J_A + J_B) \\ &= -kT\frac{\nabla(c_V - c_P)}{(c_V - c_P)}(L_{AA} + L_{AB} + L_{BA} + L_{BB}) \\ &\quad - kT\frac{\nabla(c_B - c_P)}{(c_B - c_P)}[(1 + z)(c_B - c_P)L_{AA}^{(f)} - L_{AB} - L_{BB}] \end{aligned} \tag{11.7.6}$$

and

$$J_B = kT\frac{\nabla(c_V - c_P)}{(c_V - c_P)}(L_{BA} + L_{BB}) - kT\frac{\nabla(c_B - c_P)}{(c_B - c_P)}L_{BB}. \qquad (11.7.7)$$

Here $z$ is the coordination number and $L_{AA}^{(f)}$ is that part of $L_{AA}$ which is proportional to the fraction of free vacancies, i.e.

$$L_{AA}^{(f)} = \frac{zns^2}{6kT}(c_V - c_P)w_0. \qquad (11.7.8)$$

We now use (11.5.1) to re-express the gradients appearing in (11.7.6) and (11.7.7) in terms of $\nabla c_V$ and $\nabla c_B$. This gives

$$J_V = \frac{-\nabla c_V}{(c_V c_B - c_P{}^2)}(c_B L_{VV} - c_P L_{VB}) + \frac{\nabla c_B}{(c_V c_B - c_P{}^2)}(c_P L_{VV} - c_V L_{VB}) \qquad (11.7.9)$$

and

$$J_B = \frac{\nabla c_V}{(c_V c_B - c_P{}^2)}(c_P L_{BB} - c_B L_{BV}) + \frac{\nabla c_B}{(c_V c_B - c_P{}^2)}(c_P L_{BV} - c_V L_{BB}) \qquad (11.7.10)$$

in which $L_{VV}$, $L_{VB}$ and $L_{BV}$ are given by (4.4.13). In writing these equations we have dropped the term $(1 + z)(c_B - c_P)L_{AA}^{(f)}$ which appears in (11.7.6) since it is not difficult to show that it will amost always be small compared to $L_{BB}$ and $L_{AB}$. For comparison with the equations of the independent defect approach, let us look first at the vacancy flux in the absence of any gradient in solute concentration. Evidently from (11.7.9) we can write

$$J_V = -D_{\text{eff}}\, n\nabla c_V \qquad (11.7.11)$$

in which the effective vacancy diffusion coefficient under these conditions is

$$D_{\text{eff}} = \left(\frac{kT}{n}\right)\frac{(c_B L_{VV} - c_P L_{VB})}{(c_V c_B - c_P{}^2)}. \qquad (11.7.12)$$

By extracting $L_{AA}^{(f)}$ and $L_{VV}$ and $L_{BB}$ from $L_{VB}$ and making the identifications

$$D_V = \frac{kTL_{AA}^{(f)}}{n(c_V - c_P)} \qquad (11.7.13)$$

for the diffusion coefficient of free vacancies and

$$D_P = \frac{kTL_{BB}}{nc_P} \qquad (11.7.14)$$

for the diffusion coefficient of solute–vacancy pairs, we can rewrite $D_{\text{eff}}$ in a form suitable for comparison with (11.7.4) namely

$$D_{\text{eff}} = \frac{[c_B(c_V - c_P)D_V + c_P(c_B - c_P)D_P + \alpha]}{(c_B c_V - c_P^2)}, \tag{11.7.15}$$

in which $\alpha$ contains all the other terms necessary to make this expression equal to (11.7.12). The identifications (11.7.13) and (11.7.14) are unambiguous and physically reasonable (e.g. Johnson and Lam, 1976) so the important question is how large is $\alpha$? To simplify the discussion we insert the condition $c_B \gg c_V$ and thus $c_P$, which will be generally true in practice. Then $D_{\text{eff}}$ becomes

$$\frac{1}{1 + Kc_B}\left\{D_V + Kc_B D_P + Kc_B\left(\frac{kT}{nc_P}\right)(2L_{AB} + L_{AA} - L_{AA}^{(f)})\right\}. \tag{11.7.16}$$

To estimate the size of the third term in braces relative to the other two we use the special expressions (11.5.2) *et seq.* or, rather, to avoid undue complexity, we shall use them in the approximation that takes $F = 1$. The third term in braces then becomes

$$\frac{10s^2}{3} w_3 K c_B \left(\frac{w_2 + 2w_1 + 2w_3}{w_2 + w_1 + 3.5w_3}\right) \tag{11.7.17}$$

to the neglect of a term $7c_B D_V$, which is certainly small compared to the first term. We can then draw the following conclusions.

(i) Since the ratio of frequencies $(w_2 + 2w_1 + 2w_3)/(w_2 + w_1 + 3.5w_3)$ lies between 4/7 and 2, the ratio of the third term to the first term in (11.7.16) is roughly

$$\frac{(\text{3rd term})}{(\text{1st term})} \sim 20c_B\left(\frac{w_4}{w_0}\right). \tag{11.7.18}$$

If we refer to the values of $(w_4/w_0)$ given in Table 11.4 we see that this ratio will generally be less than 1 for solute fractions less than 1%. However, it should be remembered that these values are for quite high temperatures, whereas we are here really considering much lower annealing temperatures for which $w_4/w_0$ will be greater than in Table 11.4 for all those cases where the activation energy for the association jump is less than that for free vacancy migration.

(ii) The ratio of the third to the second term in the numerator is

$$\frac{(\text{3rd term})}{(\text{2nd term})} = \frac{20w_3(w_2 + 2w_1 + 2w_3)}{w_2(w_1 + 3.5w_3)} \tag{11.7.19}$$

and is independent of $w_4$ and $w_0$. It is clear that this ratio can be $\ll 1$

for $w_3 \ll w_2$, $w_1$, e.g. for tightly bound but mobile pairs, although such cases seem very rare in practice. Equally it will be $\gg 1$ for $w_3 \gg w_2$. For the systems listed in Table 11.4 the ratio is considerably greater than 1. In general therefore it would clearly be unwise to neglect the third term in the way that the independent defect approach does.

(iii) The dependence of $D_{\text{eff}}$ on $c_{\text{B}}$ comes not only from the second and third terms in the numerator, but also from the term $Kc_{\text{B}}$ in the denominator. The ratio of expression (11.7.18) to this $Kc_{\text{B}}$ is $\sim w_3/w_0$, which is unlikely to be large at the same time as $w_4/w_0$ is large. Hence the denominator is relatively important.

(iv) Taking these various considerations together we conclude that the important terms in $D_{\text{eff}}$ will be $D_{\text{V}}$ in the numerator together with the denominator $(1 + Kc_{\text{B}})$. Of the two terms depending on $c_{\text{B}}$ in the numerator the second will mostly be smaller than the third. It follows that the use of (11.7.4) will be misleading if the term $Kc_{\text{B}}D_{\text{P}}$ in the numerator proves to be significant. It will be better to base any modelling calculations on (11.7.9) and (11.7.10).

We conclude this section by considering the flux of B atoms induced by the vacancy flow. To simplify the calculation let us again assume the condition $\nabla c_{\text{B}} = 0$, appropriate to early times in the annealing reaction, and take $c_{\text{B}} \gg c_{\text{V}}$ and $c_{\text{P}}$. Then (11.7.10) gives the flow of solute atoms as

$$J_{\text{B}} = -kTL_{\text{BV}}\frac{\nabla c_{\text{V}}}{c_{\text{V}}} = kT(L_{\text{BB}} + L_{\text{BA}})\frac{\nabla c_{\text{V}}}{c_{\text{V}}}. \tag{11.7.20}$$

If we now insert particular expressions for $L_{\text{BB}}$ and $L_{\text{BA}}$ such as (11.5.3)–(11.5.8) for the f.c.c. lattice we see that the flux of solute atoms $J_{\text{B}}$ can be either up or down the gradient of vacancy concentration, i.e. in either the opposite or the same sense as the vacancy flow, depending on the relative magnitudes of the relevant jump frequencies ($w_1$ and $w_3$ for f.c.c. alloys). The ratio $J_{\text{B}}/J_{\text{V}}$ can be inferred experimentally when both the enhancement, or depletion, of solute concentration around vacancy sinks and the corresponding number of vacancies annealed out can be measured. Early experiments of this kind by Anthony (1975) on dilute aluminium alloys (f.c.c) allowed estimates of $w_3/w_1$ to be obtained for various solutes, namely Zn($<0.20$), Cu(0.36) and Ge($>0.22$). These values are in line with those for other f.c.c. alloys given in Table 11.4. In the years since Anthony's experiments the use of high-resolution microprobe techniques has become quite widespread and many observations of quench-induced redistribution of solute elements have been made (e.g. Harries and Marwick, 1980).

### *11.7.2 Solute segregation under continuous particle irradiation*

Metals and alloys subjected to energetic particle irradiation (e.g. electrons, protons or neutrons of MeV energies) generally contain defect concentrations much higher than those present in thermal equilibrium. Interstitials and vacancies are produced in equal numbers as a consequence of collisions between the incoming particles and the atoms of the alloy. They migrate through the lattice in the normal way as a result of thermal activation, recombining with one another or disappearing at sinks, so that, under irradiation at constant intensity, steady-state concentrations of vacancies and interstitials are set up. These average concentrations may be calculated by rate equations when a self-consistent way of evaluating the effectiveness of the various sinks for the defects is used (e.g. Brailsford and Bullough, 1981).

We can adapt the previous theory to describe atomic migration under these conditions by taking three steps to give us what we may refer to as the quasi-thermodynamic approach. Firstly, we expect to be able to use the previously derived expressions for the $L$-coefficients by inserting the average steady-state defect concentrations appropriate to the irradiation conditions in place of those previously used. Secondly, we recognize that each $L_{ij}$ between chemical species $i$ and $j$ is now made up of two parts, $L_{ij}^{(\mathrm{V})}$ dependent on the vacancy movements and $L_{ij}^{(\mathrm{I})}$ dependent on the interstitial movements, even though only the vacancy part might have been present in the absence of the radiation. The total flux $J_i$ of chemical species $i$ is thus the sum of a vacancy contribution $J_i^{(\mathrm{V})}$ and an interstitial contribution $J_i^{(\mathrm{I})}$, each part being given by separate equations like (4.2.4). The fluxes of vacancies and interstitials can also be represented by equations like (4.2.4) with appropriate $L$-coefficients. The third step is then to take $L_{\mathrm{IV}} = L_{\mathrm{VI}} = 0$. Although the recombination of vacancies and interstitials is important in determining the steady-state concentrations of these defects, we do not expect a gradient in the concentration of vacancies to affect the flow of interstitials, nor vice versa.

As a result of taking these three steps we should now have an adequate set of equations to represent the fluxes of defects to sinks and the associated transport of the solute atoms. But, of course, these arguments are somewhat intuitive and one might suppose that it would be better to return to the kinetic theory of Chapter 8, and to enlarge that theory by including the creation of defects by irradiation and the mutual annihilation of vacancies and interstitials by recombination. Such an extension of kinetic theory was made by Murphy (1989a) who found that there were some terms additional to those provided by our intuitive arguments. However, her detailed examination of the magnitude of these terms (Murphy, 1987) indicated that the differences can be neglected unless the binding of the solute atoms to vacancies or to interstitials is very strong, say, greater than about 1 eV. In practice such values are unlikely. The intuitive adaptation of our

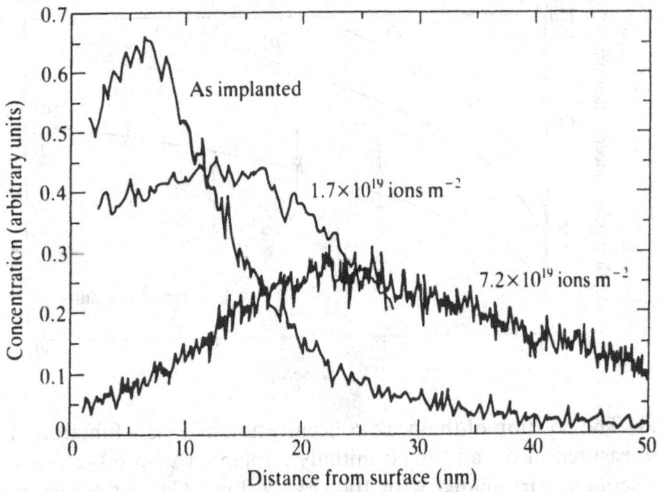


Fig. 11.5. Distribution of Mn ion-implanted into Ni, initially and after various doses of Ni ions as indicated. (After Hobbs and Marwick, 1985.)

earlier transport equations thus appears to be fully adequate for most practical situations.

This quasi-thermodynamic approach has therefore been used in a number of studies of solute transport in dilute f.c.c. alloys subject to different irradiation conditions. It would take us too far afield to describe these studies comprehensively and to discuss their practical implications in detail (but the reader may be interested by reviews of the earlier work presented by Nolfi, 1983). Instead we draw attention to several recent modelling calculations by Murphy who used the quasi-thermodynamic approach for the solute migration parts of this work. To this end she used eqns. (11.5.2)–(11.5.8) in the Lidiard–Le Claire approximation ($F = 1$) for the vacancy $L$-coefficients and eqns. (11.6.2)–(11.6.5) for the interstitial $L$-coefficients. In particular, studies were made of the re-distribution of Mn and P in Ni under subsequent bombardment by energetic Ni ions, for which corresponding experimental results were available (Murphy, 1989b; Murphy and Perks, 1989). The specimens were initially prepared by ion-implanting the solutes into Ni specimens, giving a solute distribution like that shown in Fig. 11.5. Under subsequent irradiation with energetic (100 keV) Ni ions the peak moves away from the surface into the specimen and broadens at the same time. These characteristics were modelled successfully. Of course, in order to make such calculations many parameters have to be fixed, but in those examples indications of these were available from independent experiments or could be obtained from atomistic arguments. By these arguments the Mn is believed to interact mainly with the vacancies and the P mainly with the interstitials, so that these two examples test

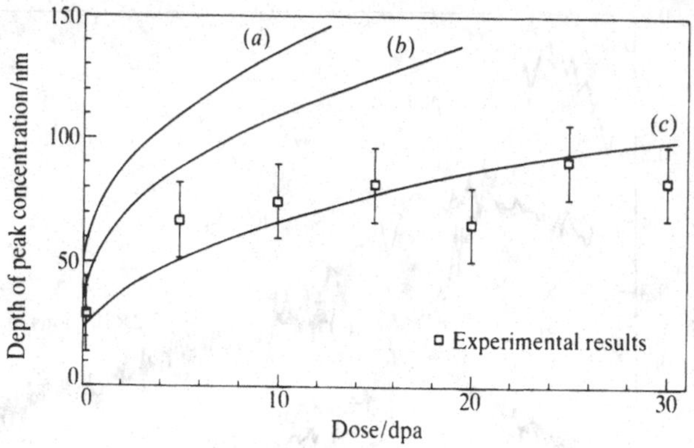


Fig. 11.6. The position of the peak concentration of P as a function of fluence ('dose' measured in d.p.a.) for Ni initially implanted with 60 keV P ions and then subsequently irradiated with 100 keV Ni ions. Curve (*a*) was calculated by employing the usual NRT damage rate, curve (*b*) by assuming a rate of 0.1 NRT and curve (*c*) by assuming a rate of 0.01 NRT. (After Murphy and Perks, 1989.)

different aspects of the theory. One major remaining uncertainty, however, is the rate at which defects are created in the Ni ion bombardment, and in order to obtain quantitative agreement with the observations (Fig. 11.6) it was necessary to reduce the rate of production of defects to 1% of that predicted by one standard way of estimating this (due to Norgett, Robinson and Torrens and widely known by their initials NRT). There is independent evidence that this is correct for ion irradiations; the reduction is believed to be the result of the recombination of vacancies and interstitials in the collision cascades. In any case, these predictions are rather insensitive to the rate at which defects are introduced (see Fig. 11.6).

Murphy (1988) also used this same approach to the transport equations to describe a quite different phenomenon which occurs under irradiation, namely the creation of spatial variations in the composition of initially uniform alloys. A general theory of such instabilities had indeed been presented earlier (Martin, Cauvin and Barbu, 1983). Murphy's calculations show that for the particular case of Ni containing low concentrations of Si it is possible to model the oscillatory characteristics of the Si redistribution observed under irradiation in related austenitic alloys (Fe–Cr–Ni) and to predict the dependence of these characteristics on variables such as Si concentration, defect production rate and the density of sinks for the defects. The occurrence of the phenomenon appears to depend on the existence of a strong interaction of the solute atoms (Si) with both vacancies and interstitials.

In summary then, although the prediction of radiation effects is a complex task, the quasi-thermodynamic approach to the movement of solute atoms in defect fluxes seems to provide a satisfactory component in the overall description. We also note that before Murphy's work various related modelling calculations were made which emphasized the transport equations proposed originally by Wiedersich, Okamoto and Lam (1979). These are also in the spirit of the quasi-thermodynamic approach but the transport coefficients are not as fundamentally based as those used by Murphy. In particular, they imply that $L_{AB} = 0$, which we have seen is not correct for defect mechanisms generally.

## 11.8 Ionic solids: the normal ionic conductors

In §1.3 we distinguished between the normal ionic conductors and the fast ion conductors; normal ionic conductors being characterized by low defect concentrations and thus low conductivities (which are sensitive to purity and the presence of structural imperfections), while fast ion conductors are highly disordered (and thus insensitive to these influences). Here we consider just the normal ionic conductors, exemplified by the alkali halides, the silver halides AgCl and AgBr, the alkaline earth fluorides (below their transition temperature to the fast-ion conducting state) and various other ionic compounds including some oxides (e.g. $Li_2O$, MgO and $ThO_2$). These materials can be made non-stoichiometric, but with difficulty. As a result transport measurements are usually made on materials which, for practical purposes, can be taken to be perfectly stoichiometric (although the use of dry, inert atmospheres may be necessary in, for example, the fluorites). The common interest in the properties of the pure crystals is centred on their electrical properties (conductivity, transport numbers, electrode effects and thermopower), tracer diffusion coefficients and nuclear magnetic relaxation times (considered in Chapter 12). But as we have already indicated in §3.3.1 and elsewhere these properties are sensitive to the presence of foreign solute ions of a valency different from the host (aliovalent ions). Hence the properties are usually studied not only in crystals of high purity but also in crystals containing deliberate additions of these aliovalent solutes, the basic idea being to control the defect concentrations according to the principles set out in §3.3.1. At the same time, one may assume, as a first approximation, that the jump frequencies of the defects remain unaffected. Analysis of measurements on this basis allows the inference of intrinsic defect concentrations and mobilities. We shall describe this approach below. However, as we have also emphasized previously, the interactions between solute ions and defects can lead to pairing reactions and other effects which demand a refinement of the approach. In addition, it may also be desirable to allow for the effects of the long-range Coulomb interactions among those defect species bearing a net charge. We shall describe the use of this refinement only rather briefly, concentrating instead on other properties which depend more exclusively on these reactions, in particular solute diffusion and solute migration in defect fluxes.

### *11.8.1 Identification of the intrinsic defect properties*

In this section we briefly consider physical properties from which information about the intrinsic defects can be extracted, principally electrical properties and tracer diffusion coefficients. The conventional approach to their analysis is that described already in general terms in §2.6.

*Ionic conductivity*

It is usual to write the electrical conductivity as a sum over all the defect species $r$, i.e. as

$$\kappa = \sum_r n_r \, |q_r| \, u_r, \tag{11.8.1}$$

in which $n_r$ is the number density, $q_r$ the effective electrical charge and $u_r$ the electrical mobility of species $r$. This is the expression of the physical idea of the defects as the charge carriers, but as we saw in §4.4 the atomic (or ionic) and the defect descriptions are formally equivalent (cf. (1.3.2)). If we can regard the defects as moving independently of one another, then we can use previous results, such as (8.2.6) and (8.2.14), to express their diffusion coefficient as

$$D_r = \tfrac{1}{6} s_r^{\,2} \Gamma_r \tag{11.8.2}$$

and their mobility

$$u_r = \frac{|q_r| D_r}{kT} \tag{11.8.3a}$$

as

$$u_r = |q_r| s_r^{\,2} \Gamma_r / 6kT, \tag{11.8.3b}$$

in which $s_r$ is the distance moved by the (effective) charge on the defect in one jump and $\Gamma_r$ is the average number of jumps made by a defect in unit time (i.e. $w_r$ times the number of distinct, but equivalent, jumps from any one site or configuration). By (11.8.1) and (11.8.3b) we have

$$\kappa = \frac{1}{6kT} \sum_r n_r q_r^{\,2} \Gamma_r s_r^{\,2}. \tag{11.8.4}$$

Evidently, the transport number for any particular defect will be $t_r = (1/6kT)(n_r q_r^{\,2} \Gamma_r s_r^{\,2} / \kappa)$; and for any one species of ion it will be the sum of $t_r$ over all defects $r$ which contribute to the movement of that species.

The ionic conductivity of AgCl was shown as an Arrhenius plot in Fig. 1.6. This shows features characteristic of many substances, namely (i) the existence of two principal, approximately straight regions which join at a curved 'knee' and (ii) the dependence of the magnitude of $\kappa$ in the low-temperature region upon the presence of solute ions of a different valency (in Fig. 1.6 $Cd^{2+}$ in place of $Ag^+$), the high-temperature region being an intrinsic property of the material. These characteristics can be understood on the basis of (11.8.4) and the dependence of the defect concentrations $n_r$ upon solute concentrations as predicted in §3.3.1. At high temperatures the defect concentrations will have their intrinsic values appropriate to the type of disorder (eqns. (3.3.24) *et seq.*), complementary defects being present in equal numbers and depending exponentially on temperature in the same way. Their relative contributions thus depend on their individual jump frequencies and unless their activation energies are closely equal one defect will dominate over the other. The slope of the intrinsic region in the Arrhenius plot is thus the sum of one-half the formation enthalpy ($h_F$ or $h_S$ as appropriate) plus the activation enthalpy of the dominant defect. On the other hand, in the low-temperature impurity-controlled region the compensating defects are present in effectively constant concentration so that $Q$ is now just the activation energy for their movement. Thus, important information about the defect properties is immediately available. In practice the techniques of analysis are more refined than indicated by this simple discussion (see, e.g. Jacobs, 1983 or Chadwick, 1991). They will nowadays almost certainly include representations of the interactions among the solute ions and defects which use the principles of §§3.4 *et seq.* In these representations the formation of pairs of oppositely charged species is paramount. The influence of the long-range Coulomb interactions among species is generally described by borrowing from the theory of liquid electrolyte solutions, in particular by using the Debye–Hückel theory in the way described in §3.9. In such an approach one should also allow for the effect of the 'ionic atmosphere' upon the mobility of any species. Equation (11.8.3) then gives us the limiting mobility at infinite dilution to which there is a (negative) correction resulting from the 'drag' due to the oppositely charged atmosphere. This correction is generally taken from the theory of Onsager (see Appendix 11.4).

A few of the conclusions and parameters obtained from analyses based on these principles were given in Tables 2.2, 3.2 and 3.3. For more extensive compilations see Corish and Jacobs (1973a) and Samara (1984).

### *Thermoelectric power*

We can also obtain an expression for the homogeneous part of the thermopower, $\theta_{hom}$, to the same approximation as used in (11.8.4). To do so we imagine the crystal to be in a temperature gradient (cf. Fig. 1.7) and find the diffusion potential corresponding to the open-circuit condition $J_e = 0$ (cf. §5.10). If we can suppose that the defects are locally in thermal equilibrium (i.e. that $n_r$ at each position has the value appropriate to the local temperature) then the result is

$$T\theta_{\text{hom}} = -\frac{\sum_r n_r q_r D_r (Q_r^* + kT^2\, \partial(\ln n_r/\partial T))}{\sum_r n_r q_r^2 D_r}, \tag{11.8.5}$$

in which $Q_r^*$ is the (effective) heat of transport of defect $r$ (in the sense of the left side of eqns. (4.4.18); for simplicity, we here drop the prime from $Q_r'^*$). Alternatively, (11.8.5) could be written in terms of mobilities $u_r$ in place of the $D_r$, but note that in any case the sign of the contribution to $\theta_{\text{hom}}$ depends on the sign of the charge $q_r$ associated with the defect. Thus in the example of AgCl shown in Fig. 1.8 we might expect the $Ag^+$ interstitials ($q_r = +e$) to yield a negative contribution and the $Ag^+$ vacancies ($q_r = -e$) to yield a positive contribution. The opposing nature of the two contributions is part of the reason for the large change in thermopower which results when the relative concentrations of $Ag^+$ vacancies and interstitials are changed by doping the crystals (cf. Fig. 1.8).

However, $\theta_{\text{hom}}$ is only one part of the total thermopower; the remaining part, $\theta_{\text{het}}$, is given by the temperature derivative of the contact potential at the electrode (see Fig. 1.7). Reproducible measurements in principle require reversible electrodes, i.e. ones which can exchange ions with the crystal in a thermodynamically reversible way. We can then extend the theory given in §3.3.1 to obtain the contact potential from the equality of the electro-chemical potential to the transferred ion on the two sides of the interface. For example, Ag electrodes on AgCl are reversible to the $Ag^+$ ions so that $\phi$, the contact potential difference between the Ag metal and AgCl, by (3.3.5c) is

$$-e\phi = \mu(\text{Ag}^+ \text{ in Ag}) - \mu(\text{Ag}^+ \text{ in AgCl}). \tag{11.8.6}$$

The chemical potential $\mu(\text{Ag}^+ \text{ in AgCl})$ is none other than the chemical potential of the interstitial $Ag^+$ ions, i.e. $\mu_3$ in (3.3.5c). The heterogeneous part of the thermopower is $\theta_{\text{het}} = \partial\phi/\partial T$. It is easily confirmed that the dominant contribution comes from the activity term ($kT \ln c_3$) in the chemical potential of the $Ag^+$ interstitials. It is also possible to devise electrodes reversible to the anion (e.g. halogens, oxygen) and these have been used to measure the thermopower of KCl (Jacobs and Knight, 1970), $SrCl_2$ (Zeqiri and Chadwick, 1983) and some oxides (Chadwick *et al.*, 1986). The theory of $\theta_{\text{het}}$ in these cases is the analogue of that for electrodes reversible to the cations.

By comparing the thermopower of pure and doped crystals, it is possible, when the disorder is predominantly either Schottky disorder or Frenkel disorder in one sublattice (as in the fluorites and in AgCl and AgBr), to infer the sum of the heats

Table 11.10. *Reduced heats of transport as inferred from measurements of the thermopower of ionic crystals*

| Substance | Defect types | Defect | $Q^*$/eV | Reference |
|---|---|---|---|---|
| AgCl | Cation Frenkel | $I^{\bullet}_{Ag}$ | $-0.18$[a] | (1) |
| | | | $-1.2$[b] | |
| AgCl | Cation Frenkel | $V'_{Ag}$ | $-0.31$ | (1) |
| NaCl | Schottky | $V'_{Na}$ | $-0.5$ to $-0.8$ | (2, 3) |
| NaCl | Schottky | $V^{\bullet}_{Cl}$ | $-1.5$ | (3) |
| KCl | Schottky | $V'_{K}$ | $-0.72$ | (4) |
| | | $V^{\bullet}_{Cl}$ | $-1.27$ | (4) |
| RbCl | Schottky | $V'_{Rb}$ | $\sim -0.7$ | (3) |
| | Schottky | $V^{\bullet}_{Cl}$ | $\sim -0.9$ | (3) |
| CsCl | Schottky | $V'_{Cs}$ | $-1.3$ to $-1.5$ | (3) |
| | Schottky | $V^{\bullet}_{Cl}$ | $-2.4$ to $-3$ | (3) |
| TlCl | Schottky | $V'_{Tl}$ | $-0.9$ | (5) |
| | | $V^{\bullet}_{Cl}$ | $-0.5$ | (5) |
| $SrCl_2$ | Anion Frenkel | $I'_{Cl}$ | $-1.8$ | (6) |
| | | $V^{\bullet}_{Cl}$ | $-0.44$ | (6) |

[a] Collinear interstitialcy.
[b] Non-collinear interstitialcy.
References:
(1) Corish and Jacobs (1973b).
(2) Rahman and Blackburn (1976).
(3) Hamman (1986).
(4) Jacobs and Knight (1970).
(5) Christy and Dobbs (1967).
(6) Zeqiri and Chadwick (1983).

of transport of the complementary defects without making further assumptions. Furthermore, by making additional, rather weak, assumptions it is possible to infer values of the separate defect heats of transport. Some values obtained in this way are given in Table 11.10. They appear to be largely independent of temperature, despite earlier indications to the contrary (Howard and Lidiard, 1964). The negative values for vacancies are perhaps to be expected, since the ion moves in the opposite sense to the vacancy and the associated (reduced) heat current (cf. 4.2.17) might be expected to follow the ion. But, by the same token, the negative values for the interstitials are surprising. However, it should be emphasized that, although an atomistic method for calculating heats of transport has been defined (Gillan, 1983), no calculations have yet been made for realistic models of ionic, or other, substances – with the result that at present we have little insight into these parameters.

*Tracer (self-) diffusion coefficients*

Measurements of self-diffusion coefficients are a valuable adjunct to conductivity measurements partly because they apply to only one chemical species at a time, and thus refer to only those defects responsible for the movement of that species. For example, in the alkali halides we would expect anion self-diffusion coefficients to tell us about the anion vacancies, whereas both anion and cation vacancies contribute to the ionic conductivity. At the same time, electrically neutral defects (e.g. anion–cation vacancy pairs) may contribute to self-diffusion but not to the ionic conductivity. Lastly, correlation effects may enter into diffusion coefficients but not into ionic conductivity, so that a comparison of the two may help in the identification of mechanisms (cf. §§5.9 and 10.9).

From the analysis in §10.3 ((10.3.6) *et seq.*) we can write $D_T$, the self-diffusion coefficient as measured by the tracer T, as

$$D_T^* = \frac{1}{6}\sum_r f_{r,T}\Gamma_{r,T}s_{r,T}^{\;2}, \tag{11.8.7}$$

the summation being over all defects $r$ which contribute to the motion of the tracer atoms. The correlation factors $f_{r,T}$ are as defined in §§10.3 *et seq.* The quantity $\Gamma_{r,T}$ is the average rate of tracer displacements, each of which is of magnitude $s_{r,T}$. The distinction between these quantities and the $\Gamma_r$ and $s_r$ appearing in (11.8.4) is important. Thus $\Gamma_{r,T}$ is the product of the probability that the tracer ion T has a defect $r$ in a neighbouring position with the average rate of tracer displacements caused by such neighbouring defects. The precise relation between $\Gamma_{r,T}$ and $\Gamma_r$ is easily established for any particular model, although the principal factor will always be the site fraction of defects. The relation between $s_{r,T}$ and $s_r$ is likewise easily established: for vacancies the two quantities are equal but for interstitialcy mechanisms (cf. Fig. 10.5) they are different. This also implies that for interstitialcy mechanisms the quantity $f$ appearing in the analysis in §5.7 differs from the corresponding Bardeen–Herring correlation factor, $f_0$.

*Analysis of transport coefficients collectively*

On account of the close relation between $D_T^*$ and $\kappa$, such measurements of $D_T^*$ and $\kappa$ taken together may be more revealing than either alone. In addition, information on ionic jump frequencies may be available from n.m.r. relaxation times (see the following chapter) and, as we saw in §10.9, measured isotope effects in $D_T^*$ may provide indications of the correlation factor (although this has not been much pursued in ionic crystals). Analyses of all this information will usually include the refinements referred to above in connection with ionic conductivity alone, e.g. the presence of solute–defect pairs in doped crystals. Direct information about these pairs may also be forthcoming from electrical and mechanical relaxation measurements (as described in Chapter 7) and from spectroscopic measurements (e.g. Fong, 1972). But we note that, since there is no net transport of matter or electrical charge during measurements of $D_{\mathrm{I}}^{*}$ and nuclear magnetic relaxation times, the Debye–Hückel atmosphere only affects these quantities through the concentration of defects. This stands in contrast to the conductivity where it also affects the electrical mobilities.

The conclusions of these studies include defect type (as in Table 2.2), defect concentrations and jump rates (and their characteristic enthalpies and other parameters), solute–defect interactions, and (sometimes) mechanisms of defect movement. The presence of minority defects can also often be inferred, e.g. anion–cation vacancy pairs in the alkali halides. In this way a consistent picture of the physical properties of strongly ionic crystals is built up. For compilations of properties and defect parameters we refer the reader to the reviews by Corish and Jacobs (1973a) and Samara (1984). The inferred properties of the defects may also be confirmed by atomistic calculations (see e.g. Harding, 1990). This knowledge can then be used to understand other systems, e.g. the diffusion of rare gas atoms in ionic solids where the trapping of gas atoms into vacancies is an important feature (Lidiard, 1980). They also point to the models appropriate for analogous compounds, such as the alkaline earth oxides, which are also always closely stoichiometric (e.g. Wuensch, 1983).

### *11.8.2 Electrical mobilities of solutes*

The previous section shows that one can understand the transport properties of ionic crystals in terms of additive contributions from different defects – vacancies, interstitials, anion–cation vacancy pairs and defect–solute pairs. All the striking features of these properties (e.g. the occurrence of distinct regions in the Arrhenius plots, the sensitivity to the presence of aliovalent ions, etc.) can be understood in terms of the dependence of defect concentrations and mobilities upon thermodynamic variables (temperature, pressure and composition) as described in §§3.3 *et seq.* In particular, the interdependence of the concentrations of the different defects is central. At the same time, the kinetic details of the reactions among the defects (e.g. vacancy–interstitial recombination, solute–vacancy pair formation, etc.) are deemed to be unimportant. In this section we consider properties where these details do enter, i.e. where we shall need results for the *L*-coefficients.

*Solute mobility*

In §5.9 we gave expressions for the electrical mobility of ions in terms of appropriate $L$-coefficients. In particular, (5.9.4) gave the ratio of mobility $u_B$ to diffusion coefficient $D_B$ for solute ions present in very low concentrations ($x_B \to 0$). This can be rewritten in the form of a simple Nernst–Einstein relation

$$\frac{u_B}{D_B} = \frac{|q_{B,\mathrm{eff}}|}{kT} \tag{11.8.8a}$$

by defining an effective charge

$$q_{B,\mathrm{eff}} = q_B + \frac{q_A L_{BA}}{L_{BB}}. \tag{11.8.8b}$$

A few direct measurements of the ratio $u_B/D_B$ have been made in the alkali halides (Chemla, 1956; Brébec and Benvenot, 1973) and in AgCl (Lur'e *et al.*, 1966). The results showed that for solute ions B having the same valency as the host, $q_{B,\mathrm{eff}}$ was close to the actual ion charge, while for aliovalent ions it was only a small fraction of it. The sub-lattices in the rocksalt structure are both f.c.c., so that we can appeal to the five-frequency model in seeking to understand these results. Thus for monovalent cations ($q_A = q_B = e$) eqns. (11.8.8), (11.5.3) and (11.5.4) give

$$(q_{\mathrm{eff}}/e) = \frac{\left[-2w_1 + w_3\left(6 + 7F + 14(1-F)\dfrac{(w_0 - w_4)}{w_4}\right)\right]}{2w_1 + 7w_3F}. \tag{11.8.9}$$

We may reasonably suppose that monovalent cations mostly exert a weak influence on vacancies so that the various jump frequencies in (11.8.9) are all much the same. Then $(q_{\mathrm{eff}}/e)$ will not differ much from the value when they are precisely the same, i.e. $(4 + 7F)/(2 + 7F) = 1/f_0$ or 1.28 for the f.c.c. lattice. For weak binding to the vacancy ($w_3/w_1 < 1$) $q_{\mathrm{eff}}$ will be slightly smaller. On the other hand, for divalent cations ($q_B = 2e$, $q_A = e$) we have

$$(q_{\mathrm{eff}}/2e) = \frac{w_3\left[3 + 7F + 7(1-F)\dfrac{(w_0 - w_4)}{w_4}\right]}{2w_1 + 7w_3F} \tag{11.8.10}$$

and, since in this case we expect there to be a strong attraction between the solute ions and the vacancies, i.e. $w_3/w_1 \ll 1$, it becomes clear why $q_{\mathrm{eff}}$ is much less than the actual ion charge of $2e$. Similar conclusions apply to anion solutes. It is clear then that the electrical mobility of solute ions depends significantly upon the off-diagonal transport coefficient $L_{AB}$.

*Ionic conductivity*

Earlier, in §5.10, we presented (5.10.13) for the ionic conductivity of a doped crystal. Here we briefly examine the relation to the methods of analysis described in §11.8.2 above. Since both assume no coupling between anion and cation fluxes, any differences will appear only in the region of extrinsic conductivity. To be specific, let us consider alkali halides (AY, Schottky disorder) doped with divalent cations ($BY_2$). Then in (5.10.13) we have $q_A = e$ and $q_B = 2e$. By appealing now to the expressions for the $L$-coefficients of the f.c.c. five-frequency model (11.5.2)–(11.5.8) we can write the conductivity as

$$\kappa = \frac{2ns^2e^2w_0}{kT}(c_V - c_P) + \kappa_c, \tag{11.8.11}$$

in which the first term is just the expression used in §11.8.2 while $\kappa_c$ is a correction to it. We can simplify the discussion of this correction by making the Lidiard–Le Claire approximation, i.e. $F = 1$. We then obtain

$$\kappa_c = \frac{2ns^2e^2w_0c_P}{kT}\left\{\frac{-7w_3}{12w_4} + \frac{20w_3(w_1 + w_2 + w_3)}{3w_0(2w_1 + 2w_2 + 7w_3)}\right\} \tag{11.8.12}$$

Whether this correction is significant is determined by the magnitude of the term in braces. For solute–vacancy binding energies of a few tenths of 1 eV we expect both $(w_3/w_4)$ and $(w_3/w_0)$ to be $\ll 1$ at reasonable temperatures. In fact, by using energies such as those calculated by Catlow *et al.* (1980) we find this term to be no more than a few parts in a hundred. It is thus considerably smaller than the reduction in conductivity which comes from the drag caused by the Debye–Hückel atmosphere. A similar conclusion follows if one uses a more elaborate model which takes account of binding at second-neighbour separations (e.g. Okamura and Allnatt, 1983b). The principal effects of solute–vacancy pairing are thus taken care of by the subtraction of $c_P$ from $c_V$ in the first term of (11.8.11), as assumed in conventional analyses.

### 11.8.3 Chemical diffusion

Chemical diffusion in ionic solids has often been discussed, although not always correctly or without unnecessary restrictions. In §5.10 we have already provided some results which apply in the presence of Schottky disorder. Here we shall use these to look at the diffusion of aliovalent solutes and to provide an explanation of the striking dependence on concentration which such diffusion may exhibit (cf. Fig. 1.2).

To be specific, we first take the example of alkali halides (AY) and divalent solute ions. The equation (5.10.11), or rather the corresponding expression of $J_B$, provides what we need to discuss the diffusion of the solute ions. Of the two factors in (5.10.11) we take the thermodynamic factor first. By using the prescription of §3.8.1 and the model of nearest-neighbour solute vacancy pairing we easily obtain to first order in $c_B$, $c_V$ and $c_P$

$$\mu_A = g_A - kT(c_B + c_{V_c} - c_P) + \cdots \qquad (11.8.13a)$$

$$\mu_B = g_B + kT\ln(c_B - c_P) + kT(zc_{V_c} - (z-1)c_P) + \cdots \qquad (11.8.13b)$$

$$\mu_{V_c} = g_{V_c} + kT\ln(c_{V_c} - c_P) + kT(zc_B - (z-1)c_P) + \cdots \qquad (11.8.13c)$$

where (as in §5.10) $c_B$ and $c_{V_c}$ are respectively the total cation site fractions of B ions and cation vacancies. The corresponding fraction of solute–vacancy pairs $c_P$ is related to $c_B$ and $c_{Vc}$ by

$$\frac{c_P}{(c_B - c_P)(c_{V_c} - c_P)} = z\,e^{-(\Delta g/kT)}, \qquad (11.8.14)$$

$z$ being the number of nearest neighbours in the cation sub-latice (= 12 in the rock-salt structure). In the present application we need only the leading terms in the gradients of the chemical potentials, whence, with $q_B = 2e$ and $q_A = e$ in (5.10.11), we easily find that the thermodynamic factor reduces simply to $-kT\nabla c_P/c_P$, i.e. the driving force for the diffusion of the B ions is essentially just the relative gradient in the fraction of B-vacancy pairs – as we might have assumed intuitively (on the grounds that only those B ions paired with vacancies are able to move).

However, to use such an argument would confuse the thermodynamic driving force with the kinetic factor, which, as (5.10.11) shows, enters quite separately. This is shown up clearly if we consider the diffusion of trivalent cations, such as $Al^{3+}$, in a divalent oxide, e.g. MgO, when we should find that the thermodynamic factor is now reduced to $(-kT)$ times the relative gradient in the fraction of triplets composed of two solute ions ($Al^{3+}$) and one cation vacancy, even though the solute–vacancy pairs are also mobile. The important conclusion is that the thermodynamic factor is governed by the gradient in the concentration of a neutral cluster, the solute–vacancy pair for divalent solutes in alkali halides, the triplet for trivalent solutes in divalent metal oxides, etc. This is consistent with the alternative form (5.10.16).

We return now to our example of divalent solutes in alkali halides. To obtain the diffusion coefficient $D_B$ we need to relate $\nabla c_P$ to $\nabla c_B$. The equations which allow us to compute $c_P$ in terms of $c_B$ are the pairing equation (11.8.14), the electroneutrality condition ($c_{V_a} = c_B + c_{V_c}$) and the Schottky product relation, which, since this is a product of the fractions of unpaired vacancies, becomes

$$(c_{V_c} - c_P)(c_{V_c} - c_B) = (c_V^{(0)})^2, \qquad (11.8.15)$$

where $c_V^{(0)}$ is the site fraction of cation (or equally anion) vacancies in the pure crystal AY. Equations (11.8.14) and (11.8.15) give us two simultaneous quadratic equations for $c_P$ and $c_{V_c}$ in terms of $c_B$. These can only be solved numerically, but we can discover the nature of the solutions without doing so. Thus we can rewrite $\nabla c_P / c_P$ as

$$\frac{\nabla c_P}{c_P} = \left( \frac{c_{V_c} - c_P}{2c_{V_c}c_B - c_P{}^2 - c_B{}^2} \right) \nabla c_B, \tag{11.8.16}$$

so that the expression for $D_B$ is

$$D_B = \frac{kT}{n} \Lambda \left( \frac{c_{V_c} - c_P}{2c_{V_c}c_B - c_P{}^2 - c_B{}^2} \right) \tag{11.8.17a}$$

where the kinetic factor $\Lambda$ in this case is

$$\Lambda = \frac{[|L_c| + L_{YY}(c_A L_{BB} - c_B L_{BA})]}{[L_{AA} + 2L_{AB} + 2L_{BA} + 4L_{BB} + L_{YY}]}. \tag{11.8.17b}$$

Two limiting cases are of special interest. Firstly, in the limit of low solute concentrations where $c_B \ll c_V{}^{(0)}$ (i.e. the intrinsic region of behaviour), we may expect $L_{AA}$ and $L_{YY}$ to be $\gg L_{AB}$ and $L_{BB}$, whence $\Lambda$ reduces to just $L_{BB}$. The thermodynamic factor in (11.8.17a) becomes simply $1/c_B$ in this limit. Hence

$$D_B = kTL_{BB}/nc_B, \quad \text{(intrinsic)}. \tag{11.8.18}$$

The opposite limit, $c_B \gg c_V{}^{(0)}$, may easily arise for moderately soluble divalent ions at normal experimental temperatures. In these extrinsic conditions $c_{V_c} \to c_B$ and the thermodynamic factor in (11.8.17a) reduces to $1/(c_B + c_P)$. At the same time $L_{YY}$ is $\ll L_{AA}$, because the fraction of anion vacancies becomes very low (cf. Fig. 3.2). Reference to eqns. (11.5.2)–(11.5.8) for the five-frequency model and equivalent results for other models shows us that as long as the fraction of bound vacancies, $c_P$, is not significantly greater than the fraction of free vacancies $(c_{V_c} - c_P)$ (i.e. the solute–vacancy binding is not too strong and $c_B$ not too large) then we may again take $L_{AA}$ to be $\gg L_{AB}$ and $L_{BB}$. The result in this limit is then

$$D_B = \frac{2kTL_{BB}}{n(c_B + c_P)} \tag{11.8.19a}$$

$$\equiv D_s \left( \frac{2p}{1 + p} \right), \quad \text{(extrinsic)}, \tag{11.8.19b}$$

in which we have introduced the degree of pairing $p = c_P/c_B$ and the limiting diffusion coefficient $D_s$ approached as $p \to 1$, i.e. at high solute concentrations. Explicitly for the rocksalt structure and nearest neighbour pairing only,

$$D_B = D_s(1 - [1 + 48c_B \exp(\Delta g/kT)]^{-1/2}) \tag{11.8.20a}$$

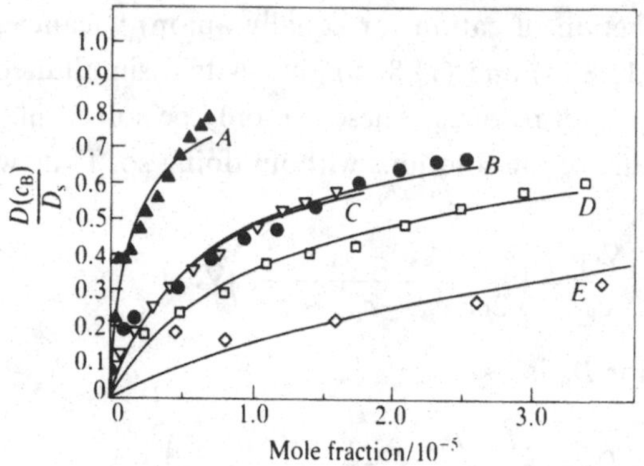


Fig. 11.7. The (intrinsic) diffusion coefficient of $Pb^{2+}$ ions in NaCl as a function of the site fraction of $Pb^{2+}$ ions at various temperatures: *A*, 348 °C; *B*, 409 °C; *C*, 451 °C; *D*, 505 °C; *E*, 553 °C. The lines represent fits to (11.8.20a). Such fits yield values of $D_s$ and $\Delta g$. The $D_s$ values so obtained have been used to obtain the experimental ordinates. (After Mannion, Allen and Fredericks, 1968.)

with

$$D_s = \tfrac{1}{6} w_2 s^2 f_B \qquad (11.8.20b)$$

The concentration dependence of $D_B$ as predicted by (11.8.20) is striking and has been verified in several systems, see e.g. Fredericks (1975). In the example shown in Fig. 11.7 the lines are fits of (11.8.20) to the data and yield values of $D_s$ and $\Delta g$. When careful experiments are made over a range of temperature, it is found that $\Delta g$ varies linearly with $T$, thus implying that $\Delta h$ and $\Delta s$ are independent of temperature, as often assumed.

### *11.8.4 Tracer diffusion of solute ions*

The experiments thus far considered are performed with an actual chemical gradient of solute B. An alternative is to follow the diffusion of a radioactive isotope B* into the crystal, possibly in the presence of an inactive carrier B. If the crystal is absolutely pure then (11.8.20) still applies. But if the crystal already contains a uniform concentration of B then a different result may apply. We can analyse this situation by beginning with the phenomenological equations for the three species A, B and B*. If the total concentration of B and B* is $c_B$ and if $c_{B0}$ is the concentration of B in the crystal before B* and B have been diffused into it then the tracer diffusion coefficient of B* is found to be

$$D_B^* = \frac{kT}{nc_B} L_{BB}\left[1 + \left(1 - \frac{c_{B0}}{c_B}\right)\frac{(c_B - c_P)^2}{2c_{V_c}c_B - c_P{}^2 - c_B{}^2}\right]. \qquad (11.8.21)$$

If the crystal is heavily doped, so $c_B \simeq c_{B0}$ throughout the experiment then

$$D_B^* = \frac{kT}{nc_B} L_{BB}. \tag{11.8.22}$$

In this case $D_B^*$ is proportional to $p$ and thus varies with $c_B$ more slowly than does $D_B$ in a 'chemical' experiment conducted on a pure crystal (cf. (11.8.19a)). On the other hand, if $c_{B0} \ll c_B$ then we recover (11.8.19), i.e. $D_B^* = D_B$, while for intermediate situations the full equation (11.8.21) must be used.

Experiments of this kind have been made for $Cd^{2+}$ in AgBr (Hanlon, 1960) and $Mn^{2+}$ in AgCl and AgBr (Süpitz and Weidmann, 1968). Of course, the intrinsic disorder in these substances is cation Frenkel disorder (Tables 2.2 and 3.3) whereas the preceding equations were obtained for crystals showing Schottky disorder. However, if we assume that the $Ag^+$ interstitials play no part in the movements of these solute ions (just as the anion vacancies play no part) then the same results will be obtained. The interpretation of these particular experiments on AgCl and AgBr was originally based (incorrectly) on (11.8.19) but was reconsidered by Friauf (1969) on the basis of (11.8.21). The difference between the two cases can be significant. Indeed, even when complete curves like those in Figs. 1.2 and 11.7 are not available the observation of a difference between $D_B$ and $D_B^*$ may still give useful information about the mechanism. For the case we have analysed the ratio $D_B/D_B^*$, by (11.8.19b) and (11.8.22), is $2/(1 + p)$, i.e. it lies between 1 and 2. The higher limit corresponds to weak association between the divalent solute ions and the cation vacancies. The same results follow for tetravalent solute ions in divalent metal oxides, e.g. $Zr^{4+}$ in MgO. On the other hand, trivalent ions in divalent metal oxides should conform to different limits, namely $\frac{1}{2} < D_B/D_B^* < \frac{3}{2}$, as long as we can still make the assumption that $L_{AA}$ is the largest of the transport coefficients entering into (5.10.11). That Osenbach, Bitler and Stubican (1981) found that $D_B/D_B^*$ was closely equal to 2 for the diffusion of chromium in MgO at three different temperatures is therefore striking. The simplest explanation would be that the chromium ions are present in $Cr^{4+}$ states under the conditions of these experiments rather than as $Cr^{3+}$ ions, but it is more likely that the methods used were not sufficiently accurate to give $D_B/D_B^*$ to the precision required for this test.

## 11.9 Transition metal oxides

While the alkaline earth oxides can be described by much the same models as the alkali halides, the open d-shells and the variable valence of transition metals give to their oxides quite different properties (see e.g. Cox, 1992). For example, some are metallic conductors while others are semiconductors; here we are principally concerned with the semiconductors. In all cases, however, the variable valency of the transition metals allows variation around the stoichiometric composition, the enthalpy of the reaction being considerably less than it would be in MgO, for example, and also less than the enthalpy of formation of Schottky or Frenkel defects in these materials. An analogous situation arises with the rare-earth and actinide oxides, although here it derives from unfilled $4f$ and $5f$ shells.

The defect picture is therefore dominated by these reactions with external oxygen phases, e.g. (3.3.30), rather than by intrinsic defect reactions. However, when the departures from stoichiometry are small the presence of aliovalent impurities can be an important influence in these materials too. They therefore offer a rich and varied field of defect behaviour.

The basic knowledge of their defect properties has been built up from measurements of the departure from stoichiometry, $\delta$, the electrical conductivity, $\kappa$ (in the semiconductors determined by the electronic defects introduced by the reaction) and the tracer diffusion coefficients of the metal and oxygen, as functions of oxygen activity. These measurements are interpreted in terms of models of independent defects in various charge states and sometimes also in terms of small defect clusters using the basic principles we have already defined (e.g. conditions governing equilibrium concentrations, expressions for tracer diffusion coefficients, etc.). Sometimes it has been supplemented with measurements of the isotope effect (allowing the inference of correlation factors, §10.9) and aided by atomistic calculations (Catlow and Mackrodt, 1982 and Harding, 1990). The theory of the various transport properties has sometimes been refined by including the Debye–Hückel representation of the effects of the long-range Coulombic interactions among the charged defects (cf. §3.9). Despite differences, the procedure has been analogous to that described in §11.8 for the ionic conductors. Reviews of the large amount of work of this kind carried out on the transition metal oxides (especially those of the $3d$-series) have been given by Monty (1990), Atkinson (1987, 1991) and Martin (1991) among others. Some of these sources contain also tables of characteristic energies and other quantities. For this reason our discussion here will be brief and selective.

One of the important groups comprises the four semiconducting transition metal monoxides NiO, CoO, FeO and MnO, all of which have the rocksalt structure (like MgO). From atomistic calculations on these oxides (Mackrodt and Stewart, 1979; Tomlinson *et al.*, 1990) we expect the *intrinsic* defects to be Schottky defects rather than either cation or anion Frenkel defects. Furthermore these oxides take up oxygen to an extent dependent upon temperature and oxygen activity, $a_{O_2}$, and they do so by the formation of metal vacancies and electron holes in the narrow $d$-like bands shown schematically in Fig. 11.8. The degrees of non-stoichiometry are effectively always much greater than the fractions of intrinsic defects, so that the extrinsic defects are predominant – rather as in a doped alkali halide crystal

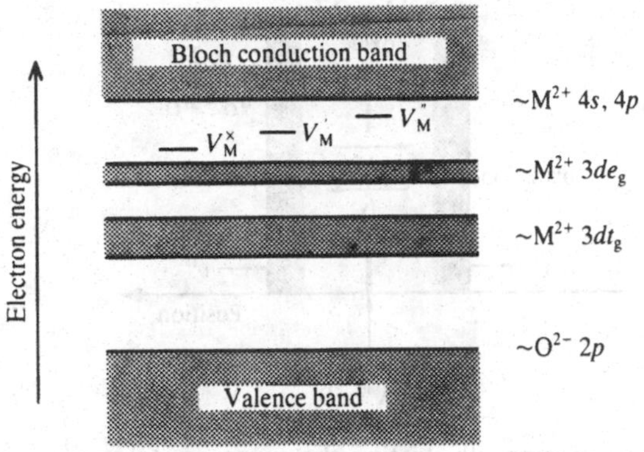


Fig. 11.8. Schematic diagram of the electronic band structure of a transition metal oxide MO having the rocksalt structure.

(§3.3). Aliovalent solute ions may be another source of extrinsic defects, depending upon the relative magnitudes of $\delta$ and the solute ion fraction. In any case, a predominance of metal vacancies will depress the concentration of oxygen vacancies by the appropriate Schottky product relation (cf. (3.3.15)). In this way we can understand why oxygen diffusion in these solids is so very slow: in practice it is more likely to occur via grain boundaries and dislocations than through the crystal lattice proper (Atkinson, 1991).

### *11.9.1 Defect flux in an activity gradient*

Since the equilibrium defect population is determined by the external oxygen activity, it follows that there will be a gradient in the defect concentration in a crystal subject to different oxygen activity on opposite faces (Fig. 11.9). As a result there will be a flux of metal vacancies and holes from the side subject to the higher oxygen activity to that subject to the lower. In chemical terms, oxygen added to the surface subject to the higher $a_{O_2}$ is matched by an equivalent flux of metal through the crystal to that same face. At the other face there is a corresponding loss of oxygen to the external phase and a flow of metal away from the surface into the bulk. In other words, the crystal grows at the high $a_{O_2}$ face and shrinks at the low $a_{O_2}$ face. Various experiments have been carried out in crystals under these conditions (Martin, 1991). The observed movements of the crystal surfaces and the associated morphological changes to the low $a_{O_2}$ surface confirm the underlying picture of the defects as metal vacancies and holes. (If the oxygen excess were accommodated instead by oxygen interstitials and holes, as in $UO_{2+x}$, there should be no displacement of the crystal faces).

To describe this process let us for simplicity suppose that the vacancies are

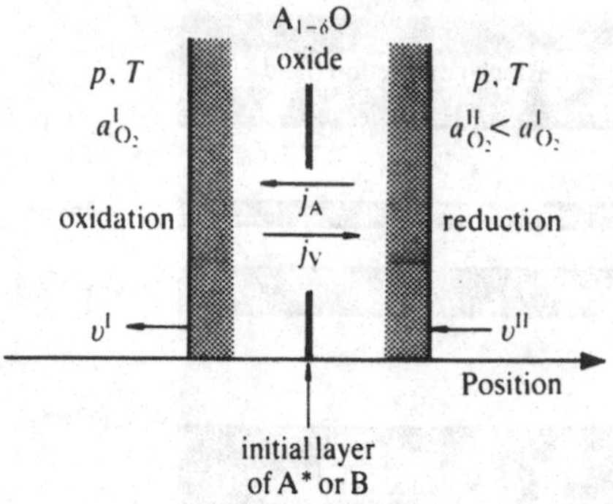


Fig. 11.9. Schematic diagram illustrating the experimental arrangement used to study the migration of tracer atoms A* and solutes B in a defect flux resulting from the different oxygen activity at the two faces of the crystal.

predominantly $V'$, as is the case in CoO, for example, over a wide range of $a_{O_2}$ (Dieckmann, 1977). The calculation can easily be extended to include vacancies $V^{\times}$ and $V''$ as well. We shall ignore the possible presence of vacancy clusters, for which there is little direct evidence save in FeO, the calculations of Tomlinson *et al.* (1990) notwithstanding. With these assumptions the flux equations in the pure oxide AO will be simply

$$J_V = L_{VV}(X_V - X_A) + L_{Vh}X_h \tag{11.9.1a}$$

$$J_h = L_{hh}X_h + L_{hV}(X_V - X_A) \tag{11.9.1b}$$

for the metal vacancies and holes respectively, $X_A$ being the thermodynamic force, $-\nabla\mu_A$, on the metal. We neglect the movement of oxygen in the lattice because it is so small for reasons given earlier. (We note in passing that the coupling coefficient $L_{Vh}$ is comparable with $L_{VV}$, as has been shown experimentally by Yoo *et al.* (1990). This can be understood in terms of the long-range Coulomb interactions between the holes and the vacancies (Janek, 1992)). Now, in the absence of applied electric fields, the flows of vacancies and holes must be such that no net electrical current flows, i.e. a Nernst electrical diffusion field will be established. Hence we can use the analysis of §5.10, and, since the mobility of the holes is much greater than that of the vacancies, (5.10.6) in particular. With $q_V = -e$ and $q_h = +e$ we get

$$J_V = L_{VV}(X_V^{(0)} - X_A^{(0)} + X_h^{(0)}). \tag{11.9.2}$$

By treating the vacancies and the holes as an ideal system (cf. §3.3) in order to obtain the chemical potentials $\mu_V$, $\mu_A$ and $\mu_h$ in terms of the corresponding defect site fractions $c_V$ and $c_h$, we get to leading order

$$J_V = kTL_{VV}\left(\frac{\nabla c_V}{c_V} + \frac{\nabla c_h}{c_h}\right). \tag{11.9.3}$$

With the vacancies moving as independent entities executing a random walk we have

$$L_{VV} = \frac{n}{6kT} \Gamma_V s_V{}^2 c_V$$

$$\equiv \frac{n}{kT} D_V c_V, \tag{11.9.4}$$

in which $D_V$ is the vacancy diffusion coefficient. From these two equations then

$$J_V = -nD_V\left(\nabla c_V + \frac{c_V}{c_h}\nabla c_h\right). \tag{11.9.5}$$

But now the condition of local electroneutrality must be brought in: in this case

$$c_V = c_h. \tag{11.9.6}$$

Hence

$$J_V = -2nD_V\nabla c_V \tag{11.9.7a}$$

or equivalently

$$J_A = -2nD_V\nabla c_A. \tag{11.9.7b}$$

The coefficient linking $-n\nabla c_A$ to $J_A$ in this connection is often called the chemical diffusion coefficient, although in the nomenclature of §5.3 it would be more properly called the intrinsic diffusion coefficient of A (cf. (5.3.5)). By (11.9.7b) this quantity

$$D_A = 2D_V, \quad (V'_A). \tag{11.9.8a}$$

If instead we had known that the vacancies were predominantly $V''_A$ with effective charge $-2e$ then by the same argument we should have obtained

$$D_A = 3D_V, \quad (V''_A). \tag{11.9.8b}$$

In cases where vacancies in all possible charge states must be considered it is usual to define their mean charge as $-\alpha e$ and to write

$$D_A = (1+\alpha)D_V. \tag{11.9.9}$$

This quantity $\alpha$ may be calculated for any specified degree of non-stoichiometry from the reaction constants for the reactions

$$V''_A + h^\bullet \rightleftharpoons V'_A \tag{11.9.10a}$$

and

$$V'_A + h^\bullet \rightleftharpoons V^\times_A. \tag{11.9.10b}$$

These constants are known for MnO, CoO and NiO.

In the steady state the two surfaces move with the same velocity $v = J_V/ac_V$, while the gradient $\nabla c_V$ is constant throughout the specimen and determined by the difference in $c_V$ at the two faces. Since these faces are individually in equilibrium with the external oxygen phases this difference will be known from independent thermogravimetric studies. Hence measurement of the velocity $v$ allows a direct determination of $D_A$ and thus of $D_V$, given that $\alpha$ has been separately determined. Since $D_A^* = fc_V D_V$ may be separately known it follows that $f$ can be inferred and the vacancy mechanism (for which $f = 0.7815$ in this lattice) confirmed or otherwise. Alternatively, if $f$ is assumed, $\alpha$ can be inferred and compared with the calculated value.

Martin and Schmalzried (1986) (see also Martin, 1991) have shown how, by studying the spread and drift of a profile of tracer A* in a crystal in the above kind of arrangement, one may simultaneously obtain $D_A^*$ and the other quantities already mentioned. In contrast to the Chemla experiment, which measures drift and diffusion in an electric field (§11.8.2), the diffusion profile here changes shape as the experiment proceeds; consequently, with an adequate theory (e.g. Martin, 1991) more information can be extracted. In this way they have confirmed the above picture in its entirety for CoO.

The substance $Fe_3O_4$, which has been investigated by the same methods, forms an interesting contrast. This material is essentially a metallic conductor at high temperatures and its electrical conductivity is largely uninfluenced by its defect state. Hence no Nernst field arises in these transport experiments and there is no enhancement factor $\alpha$. Measurements of the kind described can thus be used to infer the correlation factor for Fe* diffusion. Furthermore, since $Fe_3O_4$ can be made both hypostoichiometric (excess oxygen giving Fe vacancies) and, to a lesser extent, hyperstoichiometric (oxygen deficient giving Fe interstitials) measurements of the kind described can be particularly valuable in yielding correlation factors for both regimes. From different experiments on hypostoichiometric crystals at temperatures around 1200 °C, $f$ was inferred to be 0.54–0.56. The value calculated by the methods of Chapter 10 is 0.56, assuming migration solely within the octahedral sub-lattice. Alternatively it is 0.50 if the migration is confined to the sub-lattice of tetrahedral sites. Clearly, the measurements do not allow us to distinguish between these two possibilities, although they do exclude the possibility of movement on both sub-lattices indiscriminately, for which $f$ would be 0.73 (Becker and von Wurmb, 1986). However, both atomistic simulations (Lewis, Catlow and Cormack, 1985) and Mössbauer experiments (Becker and von Wurmb, 1986) argue that the motion is confined to the octahedral sites.

Experiments conducted on the oxygen deficient $Fe_3O_4$ in the interstitial region lead to analogous results with $f$ of about 0.4. The Mössbauer experiments lead to the conclusion that the Fe interstitials responsible for diffusion in this regime are located on tetrahedrally co-ordinated sites, but several modes of movement would each be compatible with the value of $f$.

### 11.9.2 Flux of solute ions in an activity gradient

The significant effects of aliovalent solutes upon the defect properties of these oxides make it natural to extend the experiments on the drift of A* tracers (as described in the preceding section) to measurements on the drift of solute ions, B. The basic flux equations for this situation are now

$$J_V = L_{VV}(X_V - X_A) + L_{VB}(X_B - X_A) + L_{Vh}X_h \qquad (11.9.11a)$$

$$J_B = L_{BV}(X_V - X_A) + L_{BB}(X_B - X_A) + L_{Bh}X_h \qquad (11.9.11b)$$

and

$$J_h = L_{hV}(X_V - X_A) + L_{hB}(X_B - X_A) + L_{hh}X_h \qquad (11.9.11c)$$

in place of eqns. (11.9.1). As before, $X_A$ appears only in the material forces since the requirement (4.4.8) does not involve the electron holes.

Movements in a gradient of oxygen activity in the absence of any external electric fields are subject to the condition of zero electric current, i.e. $J_e = 0$. We therefore proceed as in §5.10 and reduce these to the form of (5.10.6), i.e.

$$J_V = L_{VV}\left(X_V^{(0)} - X_A^{(0)} - \frac{q_V X_h^{(0)}}{q_h}\right) + L_{VB}\left(X_B^{(0)} - X_A^{(0)} - \frac{q_B}{q_h} X_h^{(0)}\right) \qquad (11.9.12a)$$

$$J_B = L_{BV}\left(X_V^{(0)} - X_A^{(0)} - \frac{q_V X_h^{(0)}}{q_h}\right) + L_{BB}\left(X_B^{(0)} - X_A^{(0)} - \frac{q_B}{q_h} X_h^{(0)}\right) \qquad (11.9.12b)$$

in which the $X_i^{(0)}$ are the gradients of the corresponding chemical potentials, i.e. $-\nabla\mu_i$. The effective charges on the vacancies, B ions and holes are denoted by $q_V$, $q_B$ and $q_h$ respectively. We shall allow for the formation of B–V pairs as before so that to leading order

$$X_V^{(0)} = -kT\nabla c'_V/c'_V \qquad (11.9.13a)$$

$$X_B^{(0)} = -kT\nabla c'_B/c'_B \qquad (11.9.13b)$$

in which $c'_V$ and $c'_B$ are again the site fractions of unpaired vacancies and B ions respectively. Likewise

$$X_h^{(0)} = -kT\nabla c_h/c_h \qquad (11.9.13c)$$

where $c_h$ is the site fraction of (free) holes. To the same order of approximation we can drop $X_A^{(0)}$ (cf. (3.5.8)). The B–V pairing reaction is again described by (11.5.1) and (11.8.14).

Since the solute ions may be assumed to be present in only trace concentrations in these experiments, we have $c_h$, $c'_V \gg c_B$ and the local electroneutrality condition becomes simply

$$c_h = \left(\frac{|q_V|}{e}\right)c'_V \equiv z_V c'_V. \tag{11.9.14}$$

We can use this condition to eliminate $c_h$ from eqns. (11.9.12) to get

$$J_V = -kTL_{VV}(1 - z_V)\frac{\nabla c'_V}{c'_V} - kTL_{VB}\left(\frac{\nabla c'_B}{c'_B} - z_B\frac{\nabla c'_V}{c'_V}\right) \tag{11.9.15a}$$

and

$$J_B = -kTL_{BV}(1 - z_V)\frac{\nabla c'_V}{c'_V} - kTL_{BB}\left(\frac{\nabla c'_B}{c'_B} - z_B\frac{\nabla c'_V}{c'_V}\right), \tag{11.9.15b}$$

in which $z_B$ is the effective charge number of the B ions. Now from (4.4.13)

$$L_{BV} = -(L_{AB} + L_{BB}) = L_{VB} \tag{11.9.16}$$

whence

$$J_B = kT[L_{AB}(1 - z_V) + L_{BB}(1 - z_V + z_B)]\left(\frac{\nabla c'_V}{c'_V}\right) - kTL_{BB}\left(\frac{\nabla c'_B}{c'_B}\right). \tag{11.9.17}$$

Lastly, from the pairing equation we can re-express $(\nabla c'_B/c'_B)$ in terms of the relative gradient of the total B fraction

$$\frac{\nabla c'_B}{c'_B} = \frac{\nabla c_B}{c_B} - \left(\frac{K_p c'_V}{1 + K_p c'_V}\right)\frac{\nabla c'_V}{c'_V} \equiv \frac{\nabla c_B}{c_B} - p\left(\frac{\nabla c'_V}{c'_V}\right) \tag{11.9.18}$$

where

$$p \equiv \frac{c_P}{c_B} = \frac{K_p c'_V}{1 + K_p c'_V} \tag{11.9.19}$$

is the degree of association of B ions with vacancies. On using (11.9.18) to replace $(\nabla c'_B/c'_B)$ in (11.9.17) by $\nabla c_B c_B$ we get

$$J_B = kT[L_{AB}(1 - z_V) + L_{BB}(1 - z_V + z_B + p)]\frac{\nabla c'_V}{c'_V} - kTL_{BB}\left(\frac{\nabla c_B}{c_B}\right). \tag{11.9.20}$$

We observe that this expression for $J_B$ is made up of a pure diffusion term in $\nabla c_B$ and a drift term associated with the vacancy wind. Since the tracers will have only a small effect on the vacancy concentration we can take the vacancy flux $J_V$ to be the same as in the pure oxide (cf. (11.9.3)) when the ratio of the drift component of $J_B$ to $J_V$ is

$$\frac{J_{B\text{drift}}}{J_V} = -\frac{[L_{AB}(1 - z_V) + L_{BB}(1 - z_V + z_B + p)]}{L_{AA}(1 - z_V)}. \tag{11.9.21}$$

Since $L_{AA}$ and $L_{BB}$ are both positive (cf. (4.2.6a)) the sign and magnitude of $L_{AB}$, here as in previous examples, determine the direction of the solute drift relative to the vacancy flux. When B is an isotope of A, i.e. B = A* as in §11.9.1, $L_{AB}$ is positive (cf. (5.7.7b)) and the drift of A* is necessarily opposite to $J_V$.

However, when B is chemically distinct from A this is no longer the case. Experiments with Fe tracer in CoO show that the drift may be of either sense depending on the thermodynamic conditions. Since CoO has the rocksalt structure it is reasonable to interpret these results via the five-frequency model. As Martin has shown, a full analysis of the solute profile (drift and changes in shape) taken in conjunction with other information (solute diffusion in a homogeneous specimen, isotope effects, Mössbauer measurements, etc.) allows one to test the five-frequency model and to infer the parameters in it. In this way a consistent representation is obtained. However, the picture which emerges is puzzling in one respect: not because the characteristics of Fe solute atoms depend upon the oxygen potential. That we might expect; but because, at high $a_{O_2}$, where we expect $Fe^{3+}$ ions, the Fe–V interaction is weak, while at low $a_{O_2}$ where we expect predominantly $Fe^{2+}$ ions, the Fe–V interaction is relatively strong. This is shown not only qualitatively (e.g. by the near equality of the diffusion coefficients of Fe and Co* at high $a_{O_2}$) but also by the considerable Fe–V binding energy at low $a_{O_2}$ (−0.7 eV) and the smaller ratio $w_3/w_1$ at low $a_{O_2}$ compared to that at high $a_{O_2}$ (ratio about $\frac{1}{15}$ at 1200 °C).

## 11.10 Covalent semiconductors

The principal examples in this class are the Group IV elemental semiconductors C(diamond), Si, Ge and α-Sn, the III–V compounds such as GaAs, and the II–VI compounds such as CdTe. In addition, there are wide ranges of solid solution within these groups, e.g. Si–Ge alloys, alloys of different III–V and II–VI compounds such as (Ga, Al)As and (Cd, Hg)Te respectively. An enormous amount of experimental work has been carried out on these materials, mainly because of their extensive applications as electronic devices (see, e.g. Fair, 1981; Frank *et al.*, 1984; Fahey, Griffin and Plummer, 1989; Tuck, 1988). Among these studies are spectroscopic investigations which have provided vivid insights into defect structures, interactions and movements (see, e.g. Watkins, 1975, 1976; Newman, 1982). Even so, in considering this work as a whole it seems fair to say that, although it does not violate the principles laid down previously in this and earlier chapters, there are many complications and uncertainties about basic details. These not only act as a brake on the modelling of diffusion and transport processes for practical purposes (Fahey *et al.*, 1989), they also make it difficult to find clear tests and illustrations of the basic principles in the way we were able to for metals and ionic solids.

There appear to be several reasons for this unsatisfactory situation. Firstly, in the elemental semiconductors, the open diamond lattice allows both vacancy and interstitial modes of migration. Secondly, these defects as well as the solute elements may be present in more than one charge state and the structure, symmetry and mobility of the defect may alter when the charge state changes. This means that there is a strong interplay between electronic characteristics (n-type, p-type or intrinsic, degenerate or non-degenerate) and atomic transport. Conversely, for solutes which are donors or acceptors, it is easy for the presence of the diffusing species to alter the electronic carrier density sufficiently for Nernst diffusion fields to arise. Thirdly, in the compound semiconductors there are the effects of non-stoichiometry and the additional possibility of anti-site disorder to be considered. Although the principles governing these effects are all understood it is easy to see from this complexity that the planning and analysis of incisive experiments is difficult and challenging. At the same time atomistic calculations of defect properties have been less useful for these materials than for others on account of the interplay of the strong electron–lattice coupling (Jahn–Teller effects) with the strong covalency or exchange interactions (e.g. in the re-formation of bonds at defects); see, e.g. Lannoo (1986). On the other hand there is a special phenomenon which occurs with Si, but probably more widely too, namely the injection of vacancies or Si interstitials (according to circumstances) into Si at a reaction boundary, e.g. at the $Si/SiO_2$ interface during the oxidation of Si. This allows a wide variety of drift experiments to be made in which the drift of solute elements is observed.

For these reasons the subject of atomic transport in semiconductors is complex and often uncertain. We shall therefore provide only a number of qualitative summary statements without attempting detailed mathematical illustrations of the preceding theory. For fuller accounts we refer the reader to the reviews already cited.

## *Point defects*

Detailed information about the structure and electronic states of vacancies, impurity defects and their associates has been obtained from electron spin resonance (Watkins, 1976) and infra-red spectroscopy (Newman, 1982). The vacancies were generally introduced by electron irradiation, which, of course, also introduces interstitial atoms. The work concentrated principally on Si and various II–VI compounds such as the zinc chalcogenides. It has provided details of structure, symmetry, electronic states and energy levels, etc. at low temperatures. The Arrhenius energy for the thermally activated movement of vacancies in Si is low ($\sim$0.3 eV for $V^{\times}$), while that for the re-orientation of Group V solute–vacancy pairs is higher (~1 eV). The clear-cut nature of these experiments and their unambiguous conclusions would argue that interpretations of diffusion and other transport phenomena should build upon them. In practice, however, this has not proved to be straightforward, probably because only a part of what one needs to know has been obtained with this certainty.

*Self-diffusion*

Despite the evidence that vacancies are highly mobile, self-diffusion in Si and Ge is relatively very slow, being characterized by high Arrhenius energies ($Q_{Si} \sim 4\text{–}5$ eV, $Q_{Ge} \sim 3$ eV). At the melting point the self-diffusion coefficient is $\sim 10^{-16}\,m^2\,s^{-1}$ in both cases, whereas in metals it is very much larger, typically $10^{-12}\,m^2\,s^{-1}$. Self-diffusion coefficients of both elements in the III–V compounds are similarly very small, despite the possibility of enhanced defect concentrations as a result of departures from stoichiometric proportions (Tuck, 1988). The activation energy for self-diffusion in diamond is believed from calculation also to be high (perhaps ~7 eV), although the process has never been observed experimentally.

The isotope effect for self-diffusion in Ge (and for the diffusion of Ge in Si) of about $\frac{1}{4}$ is relatively small (cf. the correlation factor $f_0 = \frac{1}{2}$ for simple vacancy migration in this lattice). Likewise the volume of activation for self-diffusion in Ge is also small, being ~0.3–0.4 of an atomic volume.

The most widely accepted view at the present time is that the principal intrinsic defects are vacancies; but that in Si the free energy of formation of interstitials is only slightly greater, so that at high temperatures they too contribute significantly to self-diffusion. (It is interesting that the vacancy and interstitial formation energies also appear to be closely equal in graphite (Thrower and Mayer, 1978), although this substance is a semi-metal rather than a semiconductor).

*Solute-enhanced self-diffusion*

The vacancy in Si is known from e.s.r. studies to exist in several charge states, depending upon the position of the Fermi level; and the same is believed to be true of the interstitial in Si and of the vacancy in Ge. Accordingly, changes in the Fermi level brought about by doping are expected to affect the self-diffusion according to the principles demonstrated in §3.3, i.e. by leading to the formation of vacancies having net charges in addition to the neutral, intrinsic vacancies $V^{\times}$. Such effects are observed. It should be noted that the theoretical representation of this example of solute-enhanced diffusion will differ from that for metals (as it also did for ionic solids).

*Solute diffusion*

Classed by their diffusion rates, solutes in Si and Ge fall into several different groups (cf. Fig. 1.1; also Fahey *et al.*, 1989). Firstly, there are metals such as Li, the noble metals and some transition metals which diffuse very rapidly with low Arrhenius energies. This is also true of these elements in the III–V compounds (Tuck, 1988). Such high mobility is believed to be because these atoms are moving interstitially by a direct interstitial mechanism through the open diamond lattice. At the same time, some of these elements may take up substitutional positions in the lattice, either by dropping into existing vacancies or by pushing Si atoms off their normal lattice sites into interstitial positions (so-called *kick-out* mechanism). Some also may precipitate on dislocations, *decorating* them and thus making them visible optically. Hence in the presence of other defects the diffusion kinetics of these elements may be quite complicated.

Secondly, there is oxygen, which from spectroscopic studies of Si is known to take up a special interstitial position in the middle of a Si–Si bond, although off-axis in one of six equivalent positions. Motion among these positions about one bond occurs relatively easily with a low activation energy, but oxygen diffusion requires movement from one bond to another and this demands a higher activation energy ($\sim 2.5$ eV, Newman and Jones, 1993).

Thirdly, there are the substitutional Group III acceptor elements (B, Al, Ga, In) and the Group V donor elements (P, As, Sb). That the diffusion coefficients of these elements (dopants) are substantially greater than those for self-diffusion in Si and Ge suggests that they bind vacancies to them, thus forming solute–vacancy pairs and enhancing the diffusion rate accordingly (cf. the examples shown in Fig. 1.1. with those presented in Figs. 2.20 and 2.21 for Al metal and for KCl respectively). For the Group V donors these binding energies are believed to be quite large, ranging from $\geq 1$ eV (P) to $\geq 1\frac{1}{2}$ eV (Sb and Bi); see e.g. Fahey *et al.* (1989). These values, however, were obtained from observed rates of annealing of irradiated samples by means of an analysis like that given in §11.7.1 and must therefore be regarded as rather uncertain. Nevertheless, these donor–vacancy pairs have been identified and studied spectroscopically and are known as E-centres. They provide an obvious starting point for the theory of diffusion of the corresponding elements, as proposed originally by Yoshida (1971). They fall into the class of models considered in §11.5. For detailed evaluations of the corresponding transport coefficients see Okamura and Allnatt (1983b): for the diffusion coefficients alone the approximate analyses of Le Claire (1970) by the random-walk method and Lidiard (1983) by the method of §8.3 are useful for the explicit forms they yield for the dependence of these coefficients upon the various jump frequencies involved. Since the open nature of the diamond lattice requires the vacancy to make a sequence of moves out at least as far as third neighbour positions if the solute is to diffuse (i.e. $f_B > 0$), these expressions enable one to avoid the pitfalls which confront oversimplified arguments. At present the component jump frequencies have not been determined in the way that they have been in metals and ionic solids. That no satisfactory reconciliation of the data on solute–vacancy pairs with solute diffusion has yet been obtained may be partly because there are some other indications of interstitial contributions to the diffusion of dopants.

*Transport in defect fluxes*

Defect fluxes into Si are observed to arise at various reaction boundaries, e.g. at $Si/SiO_2$ and $Si/Si_3N_4$ boundaries during oxidation and nitridation reactions. That such fluxes occur is shown by the observation of the growth or shrinkage of pre-existing dislocation loops of vacancy or interstitial character. Whether the injected defects are principally vacancies or interstitials depends on the reaction and the conditions. For example, oxidation of Si at temperatures beneath $\sim 1000\,°C$ injects Si interstitials, although at higher temperatures vacancies are injected.

Many experiments have shown that solute elements in the Si are moved in these fluxes and that both vacancy and interstitial fluxes are effective in moving dopant elements. The process almost certainly provides further examples of the sort we have already analysed in §11.7. Nevertheless, intriguing though the phenomena are, their quantitative description is still incomplete; first because, although there are models of the defect injection process, we do not yet have a fundamental, quantitative description of the defect fluxes comparable with those available for use in previous sections, and secondly because uncertainty about Si interstitials and the way they interact with substitutional solute elements has inhibited the calculation of the corresponding $L$-coefficients. Much remains to be done in this field.

## 11.11 Summary

In this chapter we have dealt with three important aspects of the theory of atomic transport coefficients of dilute solid solutions, namely (i) techniques of evaluation, (ii) results for particular models and (iii) applications. Firstly, in §§11.2 and 11.3, we dealt with techniques for the evaluation of the expressions for these coefficients which we arrived at in Chapters 8 and 10. (One important technique which we did not describe is the purely numerical, Monte Carlo method; this is described in Chapter 13 since it follows the same path whether it is being applied to a system of low disorder or to one of high disorder). Then we turned to the results of using these techniques on particular models. This was done first in the context of metals (§§11.5–11.6) and as a preamble to specific applications (§11.7) in two areas of widespread practical interest (solute segregation in alloys under irradiation and during the annealing of quenched alloys). Since the same models can sometimes be used in different types of substance, the later sections dealing with applications to ionic conductors (§11.8), transition metal oxides (§11.9) and covalent semiconductors (§11.10) used the previous results as required. But we pointed out where differences may arise, as e.g. with solute-enhanced self-diffusion. Despite the length of this chapter these discussions of applications have had to be very selective since so much is now known. We have therefore emphasized those applications which make use of the less well-known results (e.g. coupling effects), which may nevertheless be of the kind needed in future research.

# Appendix 11.1

## Methods of extrapolation for the computation of vacancy escape factors ($F$)

In this appendix we describe methods of extrapolation used in the calculation of the vacancy escape factor $F$ and similar quantities referred to in §11.2. One straightforward method is to make a linear extrapolation of $B_j$ versus the inverse of the number of sites in the coordination shells retained in the approximation. However, the relation between these quantities is linear only if enough coordination shells are retained, and in practice for calculations on cubic and diamond lattices (Okamura and Allnatt, 1983b) it was necessary to use matrices containing the 11th to 34th neighbour pairs for the extrapolation and to employ a computer to evaluate the determinants. Although this method gave satisfactory agreement with results previously obtained by Manning (1964), his method is apparently more efficient. Therefore we now describe this.

In order to obtain equations equivalent to those employed by Manning (1964, 1972) we first follow the procedure of Chapter 9 (§9.2 *et seq.*) and express the matrix of interest, $\mathbf{G} \equiv \mathbf{Q}^{-1}$, in terms of a matrix $\mathbf{U}$ defined by the equation

$$\mathbf{G} = (\mathbf{Q}^{(d)})^{-1}\mathbf{U}. \tag{A11.1.1}$$

Here $\mathbf{Q}$ has been separated into the sum of its diagonal and non-diagonal parts. The expression for $\mathbf{U}$ may be written

$$\mathbf{U} = \mathbf{R}^{-1}, \tag{A11.1.2}$$

where

$$\mathbf{R} = \mathbf{1} - \mathbf{A}, \tag{A11.1.3}$$

and

$$\mathbf{A} = -\mathbf{Q}^{(nd)}(\mathbf{Q}^{(d)})^{-1}. \tag{A11.1.4}$$

(Incidentally, it may be noted that we have here used a simplified notation. By the conventions of Chapter 10 the quantities $\mathbf{G}$, $\mathbf{U}$ and $\mathbf{R}$ introduced here would be $\mathbf{G}^{(3)}$, $\mathbf{U}^{(3)}$ and $\mathbf{R}''$, but we do not need these superscripts here. They should, however, be remembered if comparisons with §10.6 are made.)

The problem at hand is the calculation of $U_{11}$, from which $G_{11}$ follows trivially. It is readily shown by matrix partitioning that, because the block structure of $\mathbf{R}$ is the same as that of $\mathbf{Q}$, we may write

$$U_{11} = (R_{11} + V_{12}R_{21})^{-1}, \tag{A11.1.5}$$

where

$$V_{12} = -R_{12}U_{22}^{(1)}. \tag{A11.1.6}$$

Here we employ the notation that $\mathbf{U}^{(n)}$ denotes $(\mathbf{R}^{(n)})^{-1}$ where $\mathbf{R}^{(n)}$ is constructed from $\mathbf{R}$ by omitting the first $n$ rows and $n$ columns of its block matrix form. It follows that $U_{22}^{(1)}$ describes random walks of the vacancy that begin and end on second-shell sites and never visit first-shell sites. Neglect of such random walks (by putting $V_{12} = 0$) is equivalent to the one-shell (Lidiard–Le Claire) approximation.

To improve on this we write the analogue of (A11.1.5) for $U_{22}^{(1)}$:

$$U_{22}^{(1)} = (R_{22} + V'_{23}R_{32})^{-1} \tag{A11.1.7}$$

with

$$V'_{23} = -R_{23}U_{33}^{(2)}. \tag{A11.1.8}$$

$U_{33}^{(2)}$ describes random walks between third-shell sites which never visit first- or second-shell sites. For the five-frequency model these equations are particularly convenient. Firstly, the perturbed jump frequencies of the first two shells appear in (A11.1.7) only in $R_{22}$, while the portions of random walks involving higher-order shells are lumped into $V'_{23}$ ready to be approximated. Secondly, all the jump frequencies of the more distant shells are equal to the unperturbed frequency $w_0$, with the result that the elements of $U_{33}^{(2)}$ needed to obtain $V'_{23}$ are pure numbers.

For other models, in which the region of perturbed jump frequencies extends to the third- or even higher-order shells, further equations (beginning with that for $U_{33}^{(2)}$) would be written down, with the contributions of more distant shells in which there are only unperturbed jumps lumped into an appropriate matrix of the terminal equation.

In the case of the five-frequency and other models where we terminate the equations at (A11.1.7), it is convenient to rewrite this as an equation for the unknown $V_{12}$ of equation (A11.1.5):

$$V_{12} = A_{12} + V_{12}B_{22} = A_{12}(1 - B_{22})^{-1}, \tag{A11.1.9}$$

where $B_{22}$ is defined by the equation

$$B_{22} = A_{22} + A_{23}U_{33}^{(2)}A_{32}. \tag{A11.1.10}$$

These results are equivalent to those used by Manning (1972); in particular his equations (15) and (35)–(37) are equivalent to (A11.1.9). For the five-frequency model $V_{12}$ and $B_{22}$ are $1 \times 4$ and $4 \times 4$ matrices so the calculation of $V_{12}$ from (A11.1.9) is not difficult. The calculation of the elements of $B_{22}$ can be done in more than one way, but that employed by Manning is particularly efficient. By expanding $U_{33}^{(2)}$ the matrix $B_{22}$ can be written as a sum of contributions from random walks of increasing length:

$$B_{22} = A_{22} + A_{23}(1 + A^{(2)} + [A^{(2)}]^2 + \cdots)_{33}A_{32}. \qquad \text{(A11.1.11)}$$

By 'pencil and paper' calculations one may follow the vacancy paths for, say $N$, jumps after leaving the second shell and note the contributions to the various elements of $B_{22}$ arising from walks of length $n$ $(1 \leq n \leq N)$. These contributions were found by Manning to decrease smoothly and monotonically for larger values of $n$. Typical values of the maximum number of steps followed were $N = 6$ for the five-frequency model and $N = 14$ for a model of the diamond lattice (Manning, 1964). The contributions from the next few jumps, $n > N$, were estimated from the region of monotonic behaviour and an extrapolation to $n = \infty$ was made. These particular calculations were based on the evaluation of $\langle \cos \theta_1 \rangle$. For this route some care is needed when establishing the number of independent functions comprising $V_{12}$ (Manning, 1964; Allnatt, 1981). The values of $B_i$ for the five-frequency model as obtained by this method are compared in Table 11.2 with those obtained by the alternative (non-extrapolation) method due to Koiwa and Ishioka (Appendix 11.2). Although the latter are accurate to the number of figures shown, they required a greater amount of numerical work. The excellent agreement of the Manning results with them illustrates the remarkably efficiency of the method.

An extensive discussion of the problems arising in such calculations has been given by Manning (1972). He emphasizes that different choices of diffusion direction may lead to different numbers of distinct types of site in the second coordination shell. Generally it is best to choose the direction so that this number is a minimum. Such a choice of direction is not possible if one chooses to evaluate $\langle \cos \theta_1 \rangle$ and this may be why this route has gradually become less popular. Manning further emphasizes that in evaluating expressions such as (10.6.21) for $\langle x_1 x_2 \rangle$ and (10.6.26) for $f$ one is *not* limited to directions such that the plane through the impurity site at right angles to the diffusion direction is a plane of mirror symmetry (as it is for the $\langle 110 \rangle$ but not for the $\langle 100 \rangle$ direction is diamond, for example). When there is such a plane of symmetry it is possible to exploit the symmetry between forward and backward half-spaces used to obtain (10.6.21) and (10.6.26) in a slightly different manner (see e.g. Bakker, 1970; Mehrer, 1971). However, the use of this technique may unnecessarily restrict one to a direction in which the number of second-shell sites is not minimal.

# Appendix 11.2

## The accurate computation of vacancy escape factors (F)

In this appendix we describe the integral method of Koiwa and Ishioka (1983a,b) for the accurate numerical calculation of the vacancy escape factor $F$. We first express the matrix $\mathbf{Q}$ as the sum of $\mathbf{Q}^{(0)}$, its value for a perfect lattice in which all jump frequencies of the vacancy to any neighbour site are $w_0$, plus a perturbation $\Delta\mathbf{Q}$:

$$\mathbf{Q} = \mathbf{Q}^{(0)} + \Delta\mathbf{Q}. \tag{A11.2.1}$$

The function $\mathbf{G} = \mathbf{Q}^{-1}$ is then readily related to the unperturbed function $\mathbf{G}^{(0)} = [\mathbf{Q}^{(0)}]^{-1}$ by

$$\mathbf{G} = \mathbf{G}^{(0)} - \mathbf{G}^{(0)}\,\Delta\mathbf{Q}\mathbf{G}. \tag{A11.2.2}$$

The matrix $\Delta\mathbf{Q}$ has a relatively small number of non-zero elements and is therefore conveniently partitioned in the form

$$\Delta\mathbf{Q} = \begin{bmatrix} \Delta\mathbf{q} & 0 \\ 0 & 0 \end{bmatrix}, \tag{A11.2.3}$$

where $\Delta\mathbf{q}$ is a small, say $n \times n$, matrix. For the five-frequency model, for example, the elements $\Delta q_{ij}$ have $i$ and $j$ spanning only the first and second coordination shells and $n = 5$. The matrices $\mathbf{G}$ and $\mathbf{G}^{(0)}$ can be partitioned in the same way. If $\mathbf{g}$ and $\mathbf{g}^{(0)}$ denote the $n \times n$ matrices with elements $G_{ij}$ and $G_{ij}^{(0)}$, where $i$ and $j$ span exactly the same set of sites as for $\Delta q_{ij}$, then we find, by equating the corresponding block matrices in (A11.2.2),

$$\begin{aligned} \mathbf{g} &= \mathbf{g}^{(0)} - \mathbf{g}^{(0)}\,\Delta\mathbf{q}\,\mathbf{g} \\ &= (\mathbf{1} + \mathbf{g}^{(0)}\,\Delta\mathbf{q})^{-1}\mathbf{g}^{(0)}. \end{aligned} \tag{A11.2.4}$$

The unperturbed matrix $\mathbf{g}^{(0)}$ may be written (using the analogue of (9.2.1)) as

$(zw_0)^{-1}\mathbf{U}^{(0)}$, where the elements of $\mathbf{U}^{(0)}$ are linear combinations of the perfect lattice Green functions for unrestricted random walks ($U(\mathbf{l}, 1)$ of §9.3.2). The latter may be calculated numerically (9.3.16)). Finally, the calculation of $F$ requires the evaluation of $g_{11}$ by inversion of a $n \times n$ matrix.

Koiwa and Ishioka (1983b) have performed calculations of the vacancy escape factor $F$ for five- (or four-) vacancy jump-frequency models for b.c.c., f.c.c. and diamond lattices. The earlier results of Manning (1964) are in excellent agreement with these results (cf. Table 11.2).

# Appendix 11.3

## Solvent enhancement factors

In this appendix we outline the modifications made by Ishioka and Koiwa (1984) in the calculations of solvent enhancement factors, $b_{A*}$, for the five-frequency and corresponding b.c.c. models. Partial correlation factors are calculated for all types of jump occurring entirely inside a certain 'correlation volume' around the solute atom. For the f.c.c. model, for example, this volume comprises the solute atom site and its first two coordination shells. Consequently there are eight new jump types for $w_0$ jumps compared with the classification of Howard and Manning (1967). The expression (11.3.8) for $b_{A*}$ therefore contains an additional term $16(\chi_3/f_0 - 1)$ where $\chi_3$ is 1/8 of the sum of the additional partial correlation factors. In the modified calculation procedure type $\alpha$ jumps with positive and negative projections along the $x$-axis are treated as distinct types. The configurations just before a solute jump or a tracer jump (including those with both zero and non-zero $x$-projections) are called 'events' and the calculation of the $R_{\beta\alpha}$ is organized in terms of these events. Sets of equivalent events are combined in the usual manner. The calculation requires the evaluation of the probability that an event of kind $m$ is followed by one of kind $n$ for the various events $m, n$ that occur. Each such probability requires the solution of a random-walk problem for a vacancy moving on an imperfect lattice with two sinks (the tracer and the solute sites). This can be done numerically by Green function methods, which do not impose any restriction on the length of vacancy random walks included. This is in contrast to the earlier formulations of Howard and Manning (1967) and Jones and Le Claire (1972) in which the vacancy path is restricted to a small region, typically consisting of some 20 sites (but see the closing remarks in §11.3).

# Appendix 11.4

## *Transport coefficients in the presence of long-range Coulombic interactions*

The theoretical basis of the analysis of the transport properties of ionic crystals has much in common with that used for metals (§§11.3–11.7). In so far as we are concerned with movements on a single sub-lattice (e.g. cation and anion) we can use similar defect models. For example, in the rocksalt and zincblende structures, where the cation (anion) sub-lattice is f.c.c., we can use the five-frequency model to describe the effect of solute ions upon vacancy movements on the same sub-lattice. In this example we then simply take over the *L*-coefficients given by eqns. (11.5.2)–(11.5.8).

However, we have already emphasized in various places that these models provide descriptions of the short-range interactions. Long-range Coulombic interactions between defects bearing effective electrical charges must be treated differently. At present the only practicable approach is to ignore the discrete nature of the lattice and to adapt the theory of liquid electrolyte solutions (for which see e.g. Harned and Owen, 1958; Conway, 1970). We can do this very simply by regarding the defects as equivalent to the various ionic species in the liquid solution. This is justified because the slowly varying nature of the Coulombic forces makes the distinction between a lattice and a continuum of allowed positions of little significance. Quantitative support for this argument is provided by the calculations of Allnatt and Rowley (1970).

The principal practical points in the adaptation of electrolyte theory are as follows.

(1) The theories of Debye, Hückel, Onsager and others provide convenient expressions for the electrical mobilities and the ionic conductivity. These can be used directly if (i) we remove the electrophoretic terms (simply done by setting the solvent viscosity to infinity) and (ii) replace the limiting mobilities at infinite dilution (i.e. those in the absence of interionic Coulomb forces) by the correct

expressions for the defect models in question. This is what is commonly done. The expression for the mobility in a binary electrolyte (ion charges $q_1$ and $q_2$) as given by Onsager is

$$u_i = u_i^{(0)}\left[1 + \frac{\kappa_D q_1 q_2}{12\pi\varepsilon kT}\left(\frac{q^*}{1+\sqrt{q^*}}\right)\right], \tag{A11.4.1}$$

in which $\varepsilon$ is the electric permittivity of the solvent and $q^*$ is a dimensionless quantity defined in terms of the limiting electrical mobilities $u_i^{(0)}$

$$q^* = \frac{|q_1 q_2|}{(|q_1| + |q_2|)}\frac{(u_1^{(0)} + u_2^{(0)})}{(|q_2|u_2^{(0)} + |q_1|u_2^{(0)})}. \tag{A11.4.2}$$

Since $q_1$ and $q_2$ are necessarily of opposite sign we see that the term in $\kappa_D$ is negative. It represents the drag exerted on the movement of any one ion by its (oppositely charged) ionic atmosphere.

This expression applies when the ion size is very much less than the Debye–Hückel screening radius $\kappa_D{}^{-1}$. A slightly more elaborate expression which allows for the effect of ion size was derived by Pitts (1953) and is often used.

(2) The total ionic conductivity $\kappa$ is obtainable immediately from (A11.4.1).

$$\begin{aligned}\kappa &= n_1|q_1|u_1 + n_2|q_2|u_2\\ &= \kappa^{(0)}\left[1 - \frac{\kappa_D|q_1 q_2|}{12\pi\varepsilon kT}\left(\frac{q_*}{1+\sqrt{q^*}}\right)\right],\end{aligned} \tag{A11.4.3}$$

in which $\kappa^{(0)}$ is the conductivity at infinite dilution. Since $\kappa_D$ is proportional to the square root of the electrolyte concentration, (A11.4.3) provides the well known $\sqrt{c}$ correction to $\kappa^{(0)}$ (Kohlrausch's law).

(3) For chemical diffusion of a binary electrolyte solution the demand of electroneutrality means that the total electric current $J_e = q_1 J_1 + q_2 J_2$ must be zero (cf. §5.10). Overall electroneutrality also demands that $n_1 q_1 + n_2 q_2$ be zero. It follows immediately then that the mean velocities of the two kinds of ion must be the same. Hence the ionic atmosphere around any given ion is undistorted in the diffusion process and the atmosphere therefore exerts no drag on the diffusion. The chemical diffusion coefficient is therefore expressible in terms of the limiting mobilities $u_i^{(0)}$ alone.

(4) While the above results can be transcribed to corresponding defect systems (e.g. Lidiard, 1957; Corish and Jacobs, 1973a) it is of greater value to have the expressions for the $L$-coefficients themselves. Written in terms of the limiting mechanical mobilities $u_i^{\mathrm{M}}$ ($u_i^{(0)} = |q_i|u_i^{\mathrm{M}}$), these are

$$L_{11} = n_1 u_1^{\mathrm{M}}(1 - n_2 u_1^{\mathrm{M}}\phi'), \tag{A11.4.4a}$$

$$L_{12} = (n_1 u_1^{\mathrm{M}})(n_2 u_1^{\mathrm{M}})\phi' \tag{A11.4.4b}$$

and

$$L_{22} = n_2 u_2^{\mathrm{M}}(1 - n_1 u_2^{\mathrm{M}} \phi'), \tag{A11.4.4c}$$

where

$$\phi' = \frac{\kappa_{\mathrm{D}}}{12\pi\varepsilon kT}\left(\frac{q^*}{1+\sqrt{q^*}}\right)\frac{q_1{}^2 q_2{}^2}{(n_1 q_1{}^2 u_1^{\mathrm{M}} + n_2 q_2{}^2 u_2^{\mathrm{M}})} \tag{A11.4.5}$$

and $q^*$ is given by (A11.4.2). By appeal to the appropriate equations (cf. §§5.9 and 5.10) it is easy to verify the correctness of the statements in paragraphs (1)–(3) above.

(5) To see how to use eqns. (A11.4.4) in conjunction with a model for localized defect–solute interactions (e.g. the five-frequency model), let us take the case of an alkali halide crystal doped with divalent cations to such an extent that anion vacancies are negligible. The system can then be treated as a binary system with cation vacancies ($q_1 = -e$) and divalent cations ($q_2 = -e$) as the two species. Pairs formed from these primary species are electrically neutral and play no part in the ionic atmosphere. We can therefore use eqns. (A11.4.4) for the unpaired cation vacancies and the unpaired solute ions. But the mobility of the unpaired solute ions is zero. Hence with the densities of the unpaired species as $n'_{\mathrm{V}}$ and $n'_{\mathrm{B}}$ we see that in place of eqns. (8.5.7) we should write:

$$L_{\mathrm{VV}} = L_{\mathrm{VV}}^{(\mathrm{p})} + n'_{\mathrm{V}} u_{\mathrm{V}}^{\mathrm{M}}(1 - n'_{\mathrm{B}} u_{\mathrm{V}}^{\mathrm{M}} \phi'), \tag{A11.4.6}$$

in which

$$u_{\mathrm{V}}{}^{\mathrm{M}} = \frac{4a^2 w_0}{kT} \tag{A11.4.7}$$

and $L_{\mathrm{VV}}^{(\mathrm{p})}$ stands for the first two terms of the expression (8.5.7c) in $nc_{\mathrm{p}}$ ($\equiv n_{\mathrm{p}}$); while $L_{\mathrm{BV}} = L_{\mathrm{VB}}$ and $L_{\mathrm{BB}}$ are given by (8.5.7b) and (8.5.7a) as they stand ($u_{\mathrm{B}}^{\mathrm{M}} = 0$). The quantity $\phi'$ (A11.4.5) in these circumstances becomes

$$\phi' = e^2 \kappa_{\mathrm{D}}(1 - 1/\sqrt{2})/12\pi\varepsilon kT n_{\mathrm{V}}' u_{\mathrm{V}}^{\mathrm{M}} \tag{A11.4.8}$$

(6) The theory of liquid electrolyte solutions provides results for tracer diffusion which indicate an effect from the drag of the ionic atmosphere. However, since in solid state experiments the tracer ions move only by the action of a uniform, equilibrium distribution of charged defects, it seems unlikely that we can usefully transcribe these particular results.

(7) The preceding remarks (1)–(6) all apply to binary systems of just two types of ion. When three or more kinds of charged species are present the theory is much more complicated. It is provided in the original papers of Onsager and Fuoss (1932) and Onsager and Kim (1957). This work has been described in the context of liquid electrolyte theory by Harned and Owen (1958) and reviewed in connection with solid state transport by Janek (1992).

# 12

# The evaluation of nuclear magnetic relaxation rates

## 12.1 Introduction

In the previous chapter we considered the evaluation of the macroscopic transport coefficients for models of substances displaying small degrees of disorder. In this chapter we turn to the evaluation of the effects of atomic movements upon the relaxation processes arising in nuclear magnetic resonance, as defined in §1.7 in Chapter 1. Our reason for devoting a whole chapter to this subject is that nuclear magnetic relaxation is the most widely applicable nuclear technique for studying atomic migration in solids after the direct application of radioactive tracers. Furthermore, it is still extending in various new directions, most notably for our purposes by the use of nuclear methods of alignment and detection with unstable nuclei, especially $\beta$-emitters (see e.g. Ackermann, Heitjans and Stöckmann (1983), Heitjans (1986) and Heitjans, Faber and Schirmer (1991)).

Except for self-diffusion coefficients determined by pulsed field gradient methods, the task which this subject presents to theory is rather different from that tackled in the previous chapter and it presents different mathematical problems, some of which derive from the need to deal with a two-particle correlation function (cf. (1.7.4)), while others derive from the nature of the physical interaction between the particles. For the most part the interaction we are concerned with is the nuclear magnetic dipole–dipole interaction, since this is always present (so long as the nuclear spin $I > 0$, i.e. so long as n.m.r. is possible). We shall also give some attention to the interaction between internal electric-field gradients and nuclear electric-quadrupole moments ($I > 1/2$). The dependence of these interactions upon relative position and orientation introduces mathematical complications, even if we choose to omit correlations between the movements of the two particles in question. The accurate representation of these correlation effects presents additional difficulties. Here as elsewhere, these can be overcome by purely numerical Monte Carlo computations. We shall refer to these as appropriate, but the greater convenience of analytical solutions and the greater insight which they convey obliges us to place most emphasis on them. It is fortunate that the very simplest model (that of Bloembergen, Purcell and Pound (1948) to be referred to as the B.P.P. model) correctly describes important qualitative features in a simple mathematical form, notwithstanding the complications of interactions and correlations.* The more rigorous theories on the whole provide quantitative refinements, and these are needed by the experimenter if he (she) is to extract accurate and detailed inference from his (her) data. The experimenter can measure several distinct relaxation times and can vary several parameters, e.g. static magnetic field $B_0$ (i.e. spin precession frequency), temperature (i.e. rates of atomic jumps) and crystal orientation in the case of single crystals. Theory has therefore to provide predictions for particular models as a function of these experimental parameters.

All this adds up to a substantial task and we cannot deal with the matter comprehensively. Fortunately, it is sufficient to concentrate on the spectral functions which yield the relaxation rates (eqns. (1.7.6)). As indicated in Chapter 1 their dependence on crystal orientation can also be separated out in a way that is determined by the crystal symmetry but independent of the defect mechanism. We shall therefore concentrate on presenting the principal features of theories of the spectral functions of systems containing a single type of spin and their relation to atomic migration in pure solids and dilute solid solutions. When we have to be specific we shall suppose that the systems of concern are dilute in defects. Even with this restriction the breadth of this subject means that we can deal with only a narrower range of physical models than were analysed in Chapter 11.

The plan of this chapter is therefore as follows. We begin (§12.2) by presenting the general expression for the spectral functions $J^{(p)}(\omega)$ in terms of appropriate probability functions. The following section then presents one or two simple models and approximations, including the basic B.P.P. or single-relaxation-time model. In moving towards more precise theories it is of considerable help to know the high-frequency and low-frequency limits of the $J^{(p)}(\omega)$. General expressions for these limits can be obtained and these are the subject of §12.4. Then in §12.5 we turn to the use of the encounter model (cf. §9.5) as a precise way of evaluating $J^{(p)}(\omega)$ for systems dilute in vacancies. In practice, this evaluation is aided by the use of the high- and low-frequency limits presented in §12.4. The penultimate section §12.6 deals with corresponding results for the e.f.g./quadrupole interaction while §12.7 summarizes the chapter.

* At least, this is true of the usual three-dimensional solids containing low concentrations of defects. In fast-ion conductors, especially those of lower dimensionality, qualitative differences from B.P.P. may become obvious (Funke, 1989 and Heitjans *et al.*, 1991)

## 12.2 Spectral functions in n.m.r.

In Chapter 1 we have already given the functions $J^{(p)}(\omega)$, which are needed to calculate the various relaxation times ($T_1$, $T_{1\rho}$, $T_2$, etc.) measurable by n.m.r. techniques. These are the spectral functions, i.e. the Fourier transforms in time, of the correlation function (1.7.4) which we repeat here

$$F^{(p)}(s) = \sum_j \langle F_{ij}^{(p)*}(t+s) F_{ij}^{(p)}(t) \rangle, \tag{12.2.1}$$

and

$$J^{(p)}(\omega) = 2\,\mathrm{Re} \int_0^\infty F^{(p)}(s)\, \mathrm{e}^{-\mathrm{i}\omega s}\, \mathrm{d}s. \tag{12.2.2}$$

As in §1.7 we shall have in mind mostly nuclear spin relaxation resulting from the dipole–dipole coupling of the nuclear magnetic moments. In this case the function $F_{ij}^{(p)}$, which depends upon the relative positions of the two nuclei $i$ and $j$, is a stochastic function of time through the random, thermally activated motion of the nuclei. For atoms of the same type the ensemble average (12.2.1) will be the same for every pair $(i, j)$: we therefore need consider only a single pair. The form of $F^{(p)}$ derives from the form of the magnetic dipole–dipole interaction and, apart from a normalizing constant, is simply $(1/r^3)$ times the spherical harmonic function $Y_{2,p}(\theta, \phi)$, in which $(r, \theta, \phi)$ are the spherical polar co-ordinates giving the position of one nucleus relative to that of the other. The polar axis is the direction of the static magnetic field $\mathbf{B}_0$. (The definition and relevant properties of spherical harmonic functions $Y_{l,m}(\theta, \phi)$ are summarized in Appendix 12.1.) For convenience we shall now use $\Omega$ to denote the pair of angular co-ordinates $\theta, \phi$. The correlation function for a system of like nuclei can therefore be written as

$$F^{(p)}(t) = c d_p^{\,2} \sum_{l,m} \frac{Y_{2,p}^{*}(\Omega_m)}{r_m^{\,3}} \frac{Y_{2,p}(\Omega_l)}{r_l^{\,3}} \pi(\mathbf{r}_m, \mathbf{r}_l; t), \tag{12.2.3}$$

in which $\pi(\mathbf{r}_m, \mathbf{r}_l; t)$ is the probability that a particular pair of nuclei ($i, j$ above) initially separated by $\mathbf{r}_l$ are separated by $\mathbf{r}_m$ at time $t$ (Fig. 12.1). (Henceforth we denote these positions simply by $\mathbf{l}$ and $\mathbf{m}$.) The quantity $c$ is the site fraction of the nuclei in question while the $d_p$ are the coefficients of proportionality between the functions $F^{(p)}$ as defined in eqns. (1.7.3) and the spherical harmonics $Y_{2,p}/r^3$. They have the values

$$d_0^{\,2} = \frac{16\pi}{5}, \tag{12.2.4a}$$

$$d_1^{\,2} = \frac{8\pi}{15} \tag{12.2.4b}$$

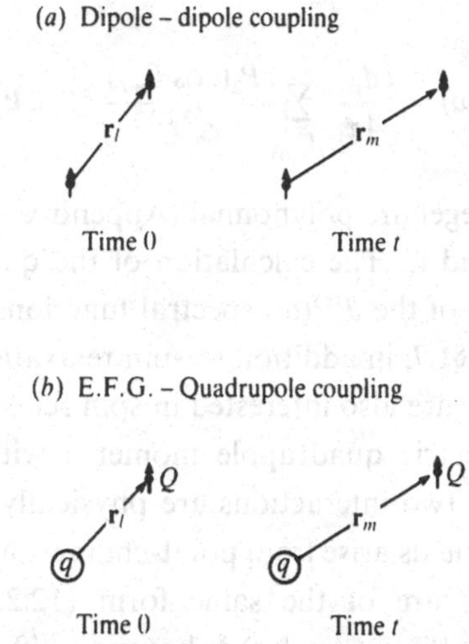

Fig. 12.1. (a) A pair of nuclei initially separated by the lattice vector $\mathbf{r}_l$ are separated by a lattice vector $\mathbf{r}_m$ at time $t$ later. (b) A nucleus initially at position $r_l$ from a point charge $q$ is at position $r_m$ at time $t$ later.

and

$$d_2{}^2 = \frac{32\pi}{15}, \tag{12.2.4c}$$

as is easily verified from the definition of the normalized spherical harmonics (Appendix 12.1).

From (12.2.2) and (12.2.3) the required spectral functions can therefore be written as

$$J^{(p)}(\omega) = cd_p{}^2 \sum_{\substack{m,l \\ (\neq 0)}} \frac{Y^*_{2,p}(\Omega_m)}{r_m{}^3} \frac{Y_{2,p}(\Omega_l)}{r_l{}^3} \tilde{\pi}(\mathbf{m}, \mathbf{l}; \omega), \tag{12.2.5}$$

in which

$$\tilde{\pi}(\mathbf{m}, \mathbf{l}; \omega) = 2 \operatorname{Re} \int_0^\infty \pi(\mathbf{m}, \mathbf{l}; s)\, \mathrm{e}^{-i\omega s}\, \mathrm{d}s \tag{12.2.6}$$

is the transform of the probability $\pi(\mathbf{m}, \mathbf{l}; t)$.

These expressions allow us to calculate the relaxation rates $T_1{}^{-1}$, $T_2{}^{-1}$, $T_{1\rho}{}^{-1}$ in single crystals as functions of orientation. In polycrystals and powders we are only interested in averages over all orientations. In practice, the measured rate of decay of the total magnetization of such materials is closely equal to the average decay rate of the contributions from the component crystals; this we obtain by inserting the average of $J^{(p)}(\omega)$ over all orientations (denoted by $J^{(p)}_{\mathrm{poly}}(\omega)$) into (1.7.6). This can be obtained by making use of the known properties of the spherical harmonics to give

$$J^{(p)}_{\text{poly}}(\omega) = \frac{cd_p{}^2}{4\pi} \sum_{\substack{m,l \\ (\neq 0)}} \frac{P_2(\cos\theta_{ml})}{r_m{}^3 r_l{}^3} \tilde{\pi}(\mathbf{m}, \mathbf{l}; \omega), \tag{12.2.7}$$

in which $P_2$ is the usual Legendre polynomial (Appendix 12.1) and $\theta_{ml}$ is the angle between the vectors $\mathbf{r}_m$ and $\mathbf{r}_l$. The calculation of the quantity $\tilde{\pi}(\mathbf{m}, \mathbf{l}; \omega)$ is thus the key to the calculation of the $J^{(p)}(\omega)$ spectral functions.

As already indicated in §1.7, in addition to spin relaxation via nuclear magnetic dipole–dipole coupling we are also interested in spin relaxation resulting from the interaction of nuclear electric quadrupole moments with internal electric field gradients. Although these two interactions are physically quite different, it turns out that when the electric fields arise from point-charge sources then the associated spectral functions $J^{(p)}(\omega)$ are of the same form (12.2.5) in both cases. The interpretation of the probability $\pi(\mathbf{m}, \mathbf{l}; t)$ is however different. For dipole–dipole coupling it is the probability that a particular pair of nuclei initially separated by the lattice vector $\mathbf{r}_l$ will be separated by $\mathbf{r}_m$ at time $t$. For the coupling of nuclear quadrupoles with internal field gradients it is the probability that a nuclear quadrupole initially at $\mathbf{r}_l$ relative to a point-charge source will be at $\mathbf{r}_m$ from the same or another such source at time $t$. The evaluation of these probabilities in terms of atomic movements therefore involves different physical considerations in the two cases. This problem has received most attention for the dipole–dipole coupling, largely because this interaction is always present in systems where nuclear magnetic resonance is possible (i.e. $I \geq 1/2$), whereas the other is only present when the nuclei possess a quadrupole moment ($I > 1/2$) and point sources of electric and/or elastic strain fields exist (as e.g. charged defects). We shall therefore also give most attention to the evaluation of the $J^{(p)}(\omega)$ in the context of the dipole–dipole coupling.

The calculation of $J^{(p)}(\omega)$ in this case is made difficult partly by the need to perform the summations indicated in (12.2.5), but mainly because the probability function $\pi(\mathbf{m}, \mathbf{l}; t)$ is a two-particle correlation function when both nuclei are equally able to move. The subject has therefore given rise to its own individual formulations. Nevertheless, much of it can be set in the context of Chapters 6 and 9, as is understandable when we observe that this quantity $\pi(\mathbf{m}, \mathbf{l}; t)$ is another example of the functions $\mathbf{G}(t)$ introduced there. In the next section we shall review several well-studied simplified theories in the light of this observation.

## 12.3 Some simple models and approximations

As just remarked, it should be clear from its definition that $\pi(\mathbf{m}, \mathbf{l}; t)$ is closely related to the propagators $\mathbf{G}(t)$ introduced in Chapter 6 and developed for random walks in Chapter 9. More precisely, we can see that $\pi(\mathbf{m}, \mathbf{l}; t)$ is given by the general propagator for the entire system averaged over the motions of all the nuclei except the two specifically considered. It is helpful to keep this in mind, although not all treatments proceed formally in that way. In fact, various widely used treatments proceed more intuitively. They are the subject of this section.

For the time being we shall not specify the nature of the system, except that for simplicity we shall require the lattice to which the nuclei are confined to be a cubic Bravais lattice, the distinct sites of which are labelled **l**, **m**, **n**, etc. It could be a lattice-gas type of system having fewer atoms than available lattice sites, e.g. interstitial H atoms in metals. Or it could be a system like those considered in the previous two chapters where the atoms can move only by the action of vacancies or self-interstitials which are present in low concentrations. Naturally, we can make only limited progress without specifying the system in detail, but certain approximate, though useful, results can be obtained.

### *12.3.1 The single-relaxation-time model*

This model is due originally to Bloembergen *et al.* (1948) and has already been introduced through (1.7.8) and (1.7.9) in Chapter 1. The approximations involved in deriving its rather simple result are several. Firstly, it is necessary to make some sort of mean-field approximation to get from the general propagator $\mathbf{G}(t)$ to the required probability function for the selected pair of nuclei $\pi(\mathbf{m}, \mathbf{l}; t)$. In other words, we would consider the motion of the two selected nuclei as taking place in a mean distribution of all the others. This motion will be described by a propagator which we denote by $G_{\mathbf{ml}}(t)$. Since only the separation of the pair is significant, and not the absolute position of either, we can label the states of the system by this separation vector. Hence $G_{\mathbf{ml}}(t)$ gives the probability that the separation is $\mathbf{r}_m$ at time $t$ given that it was $\mathbf{r}_l$ initially, and is therefore the same as the quantity $\pi(\mathbf{m}, \mathbf{l}; t)$. In previous chapters we have given attention to the Laplace transform of the propagator. In this case also we introduce the transform $\bar{G}_{\mathbf{ml}}(\lambda)$ of $G_{\mathbf{ml}}(t)$. By the equivalence of $G_{\mathbf{ml}}(t)$ and $\pi(\mathbf{m}, \mathbf{l}; t)$ and by (12.2.6) we evidently have

$$\bar{\pi}(\mathbf{m}, \mathbf{l}; \omega) = 2\,\mathrm{Re}\,\bar{G}_{\mathbf{ml}}(\mathrm{i}\omega). \tag{12.3.1}$$

Next we recall the definition (9.2.1) and subsequent equations and suppose that we can roughly approximate the corresponding $\mathbf{U}(\lambda)$ by the leading term in (9.2.6), as might seem valid for large frequency $\omega$ and small off-diagonal terms in $\mathbf{P}$ relative to the diagonal terms. Since this leading term is just the unit matrix this gives

$$\begin{aligned}\bar{\mathbf{G}}(\lambda) &= \bar{\mathbf{G}}^{(\mathrm{d})}(\lambda)\\ &= (\mathbf{P}^{(\mathrm{d})} + \lambda\mathbf{1})^{-1},\end{aligned} \tag{12.3.2}$$

whence

$$2\,\mathrm{Re}\,\bar{G}_{\mathrm{ml}}(\mathrm{i}\omega) = \frac{2P^{(\mathrm{d})}_{\mathrm{ll}}\,\delta_{\mathrm{ml}}}{(P^{(\mathrm{d})}_{\mathrm{ll}})^2 + \omega^2}. \tag{12.3.3}$$

Now in §6.3 we saw that $G^{(\mathrm{d})}_{\mathrm{ll}}(t) = \exp(-P_{\mathrm{ll}}t)$ is the probability that a system initially in state l undergoes no transition in time $t$ and so is still in that state at that time. The average lifetime of the state is thus $1/P_{\mathrm{ll}}$. If now we can further assume that all these lifetimes are equal ($\tau$), then by (12.2.5), (12.3.1) and (12.3.3) we arrive at the result of Bloembergen *et al.* namely

$$J^{(p)}(\omega) = cd_p{}^2\left(\frac{2\tau}{1+\omega^2\tau^2}\right)\sum_{l\neq 0}\frac{|Y_{2,p}(\Omega_l)|^2}{r_l{}^6}. \tag{12.3.4}$$

Since the states of the system are here defined by the separation of a *pair* of like nuclei, it follows that the average lifetime $\tau$ is just one-half the average lifetime, $\tau_{\mathrm{s}}$, of one nucleus on any particular site, since movement of either nucleus changes the state of the pair. In using (12.3.4) we shall therefore set $\tau = \tau_{\mathrm{s}}/2$.

The lattice summation in (12.3.4) has been evaluated by Barton and Sholl (1976) for each of the s.c., b.c.c. and f.c.c. lattices. For the present, however, we shall content ourselves with $J^{(p)}(\omega)$ for a polycrystalline or powdered specimen, where all possible crystal orientations relative to the field direction are represented. This is obtained by taking the spherical average of (12.3.4) over all field directions (solid angle $4\pi$) and since the spherical harmonic functions are normalized to unity the result is simply

$$J^{(p)}_{\mathrm{poly}}(\omega) = \frac{cd_p{}^2}{4\pi}\left(\sum_{l\neq 0}\frac{1}{r_l{}^6}\right)\left(\frac{2\tau}{1+\omega^2\tau^2}\right), \tag{12.3.5}$$

so that in this case the different spectral functions differ only by numerical factors. In fact, from (12.2.4) and (12.3.5) $J^{(0)}_{\mathrm{poly}}(\omega):J^{(1)}_{\mathrm{poly}}(\omega):J^{(2)}_{\mathrm{poly}}(\omega)$ are in the ratio 6:1:4, as they must be since this is always true for cubic crystals (cf. eqns. (1.7.7) *et seq.*).

We can now use (12.3.5) to obtain the various magnetic relaxation times from eqns. (1.7.6). This simple result, in fact, enables us to understand many of the principal features of the measured times of magnetic relaxation via the dipole–dipole coupling mechanism. These times are mostly obtained as a function of temperature at fixed frequency. Since we expect $\tau^{-1}$ to depend on temperature in an Arrhenius fashion, i.e. we expect

$$\tau = \tfrac{1}{2}\tau_{\mathrm{s}} = \tfrac{1}{2}\tau_{\mathrm{s},0}\exp(Q/kT), \tag{12.3.6}$$

it is clear that the factor $\tau/(1+\omega^2\tau^2)$ in (12.3.5) determines the temperature dependence of the magnetic relaxation times. For any particular frequency the two important regimes are (i) $\omega\tau \ll 1$, i.e. high temperatures, when the factor is just $\tau$, and (ii) $\omega\tau \gg 1$, i.e. low temperatures, when the factor becomes $1/\omega^2\tau$. If we plot the logarithm of the factor against the reciprocal of the temperature we shall obtain a plot which becomes a straight line of slope $-Q/k$ when $\omega\tau \ll 1$ (high temperatures) and another straight line of equal but opposite slope $+Q/k$ when $\omega\tau \gg 1$ (low temperatures). In this way we can understand precisely all the qualitative features of the results already portrayed in Fig. 1.15 (although, more generally, it should be remembered that (12.3.6) may not hold over the entire temperature range of concern in ionic crystals).

For more quantitative comparisons we consider just the spin-lattice relaxation time $T_1$. If we substitute (12.2.4) and (12.3.5) into (1.7.6a) we can write $T_1^{-1}$ as

$$\frac{1}{T_1} = AS_0 f(\xi_0), \tag{12.3.7}$$

where

$$A = \frac{2\gamma^4\hbar^2 I(I+1)c(\mu_0/4\pi)^2}{5\omega_0 a^6}, \tag{12.3.8}$$

$$S_0 = \sum_{l\neq 0} (a/r_l)^6, \tag{12.3.9}$$

$$f(\xi) = \frac{\xi}{(1+\xi^2)} + \frac{4\xi}{(1+4\xi^2)} \tag{12.3.10}$$

with $\xi_0 = \omega_0\tau = \omega_0\tau_s/2$ and $a$ the edge of the elementary cube of the lattice (so that the nearest neighbour distance is $a$ (s.c.), $a\sqrt{3}/2$ (b.c.c.) and $a/\sqrt{2}$ (f.c.c.)). The lattice sums (12.3.9) have the (dimensionless) values

$$\left.\begin{aligned} S_0 &= 8.4019 \text{ (s.c.)} \\ &= 29.046 \text{ (b.c.c.)} \\ &= 115.63 \text{ (f.c.c.)} \end{aligned}\right\}. \tag{12.3.11}$$

Figure 12.2 shows the dependence of $f(\xi)$ on $\xi$. Although $f$ is a sum of two Debye functions the overall behaviour is very close to that of a single Debye function, albeit with a shifted position and height of the maximum. This is a rather common occurrence, as we have already noted (§7.3.2). The maximum value of $f$ is 1.4252 and occurs at $\xi = 0.616$, i.e. at $\omega\tau_s = 1.23$.

The simplicity of these results makes the theory very attractive. The slopes of the straight line portions in plots such as those in Fig. 1.15 allow the activation energy governing the atomic jump frequency to be determined, while the positions of the minima in $T_1$ and $T_{1\rho}$ allow the corresponding pre-exponential factor to be

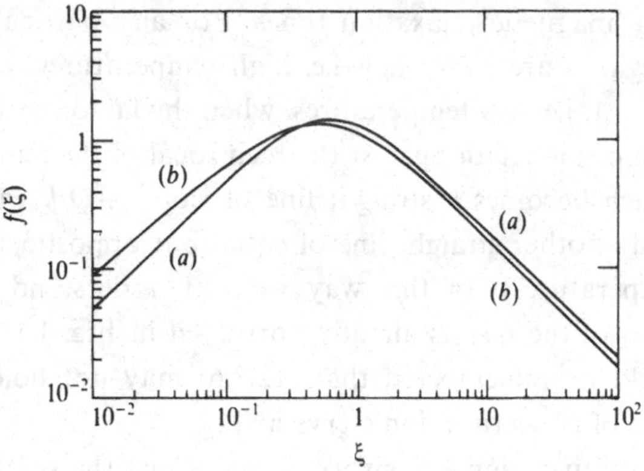


Fig. 12.2. A log–log plot of the functions $f(\xi)$ v. $\xi$, (a) according to the single-relaxation-time model and (b) according to the Resing and Torrey isotropic, random-walk model for a b.c.c. lattice.

found. These parameters by (12.3.6) are all that is needed to predict $\tau_s$ and derived transport quantities at all temperatures.

Nevertheless, the derivation of (12.3.5) must be regarded as rather unsatisfactory on account of the three unrelated approximations which were introduced, namely (i) the mean field approximation (ii) the simplification of **U** and (iii) the assumption of a single decay lifetime. We shall therefore now look at the problem from another angle.

### *12.3.2 Random-walk model*

The treatment in the previous section was largely phenomenological and independent of the details of the way the atoms move. Such information is represented by only the single parameter $\tau_s$. In the present section we describe a theory which incorporates the information in a more explicit way, even though it remains approximate. Naturally, the predictions will depend upon the details of the system considered.

We again suppose that we are dealing with a system of like spins moving on a single lattice. We shall retain the idea of the mean-field approximation, but we now incorporate specific descriptions of the motion of the individual nuclei by supposing that the two particular nuclei considered in the formula (12.2.5) move independently of one another in the average way that all nuclei do. Under these circumstances the change, $\mathbf{r}_m - \mathbf{r}_l$, in their separation in a time $t$ is made up of a series of independent movements and is thus independent of the initial separation $\mathbf{r}_l$. Furthermore, since the displacement of either nucleus changes the separation between them, the average rate at which changes in separation occur is twice the average rate at which individual nuclei jump. We can thus equate $\pi(\mathbf{m}, \mathbf{l}; t)$ to the probability function for a random walk by a single particle jumping at the average rate for twice as long, i.e.

$$\pi(\mathbf{m}, \mathbf{l}; t) = G(\mathbf{r}_m - \mathbf{r}_l; 2t). \tag{12.3.12}$$

This result can, of course, be obtained more formally if one so desires (as was first done by Torrey, 1953).

We have thus reduced the two-particle probability function $\pi(\mathbf{m}, \mathbf{l}; t)$ to a one-particle probability function – which is a substantial simplification. The model defined by this relation (12.3.12) is generally referred to as the *random-walk model.* It has been analysed by various authors in different ways, some of which we mention below. However, the assumption of independent movements underlying it can never be exact, because no two nuclei may ever occupy the same lattice site so that one nucleus will always to some extent 'get in the way' of the other.* Furthermore, when the defect concentration is low, so that both nuclei will be mostly moved by the same defect, there will be correlations between the movements of the two nuclei. These conclusions will become less important as the defect concentration increases. We conclude therefore that the random-walk model will always be better in highly defective than in less imperfect solids and better in close-packed lattices ($z = 12$) than in open lattices (e.g. diamond with $z = 4$). Also that in all cases it will be good at large separations – which means large $\mathbf{r}_m - \mathbf{r}_l$ and long times. We shall consider this long-time limit shortly, but first we pursue the consequences of (12.3.12).

For definiteness we suppose that the atoms move by a vacancy mechanism and therefore that the site fraction of nuclei $c < 1$, although we shall not limit ourselves to cases with $c_v \equiv (1 - c) \ll 1$. By the mean-field approximation the propagator $\mathbf{G}$ for a single nucleus is just like that for a particle moving on a completely empty lattice (§9.3.1), except that the jump frequency, instead of being $w$, now takes the average value $wc_v$. With this change we can now use the analysis of §9.3.1 to give us the single-particle propagator appearing in (12.3.12).

For a start we can again make the simple approximation of setting $\bar{\mathbf{G}}(\lambda)$ equal to $\bar{\mathbf{G}}^{(d)}(\lambda)$, i.e. of putting $\mathbf{U}(\lambda)$ equal to the unit matrix. Thus from (9.3.2) we see that $\bar{\mathbf{G}}^{(d)}(\lambda)$ is just $1/(2zwc_v + \lambda)$ times the unit matrix. The B.P.P. result (12.3.4) is again obtained, but with $\tau$ now given explicitly by $1/2zwc_v$. However, we can do better than this by retaining $\mathbf{U}(\lambda)$. In particular, we can use (9.3.3) and (9.3.16b) with $\lambda = \mathrm{i}\omega$ to obtain $\pi(\mathbf{m}, \mathbf{l}; \omega)$. Taking account of the symmetry of $\gamma_q$ in the

* These remarks are true when, as here, the system comprises only a single type of spin. However, in compounds, where different types of spin may be confined to different sub-lattices, the random-walk model may be exact.

Brillouin zone (cf. eqns. (9.3.9)) we obtain for cubic Bravais lattices

$$\tilde{\pi}(\mathbf{m}, \mathbf{l}; \omega) = \frac{2v}{(2\pi)^3} \int_{\mathrm{BZ}} \frac{\mathrm{e}^{-\mathrm{i}\mathbf{q}(\mathbf{r}_m - \mathbf{r}_l)} \Lambda(\mathbf{q})\, \mathrm{d}\mathbf{q}}{(\Lambda(\mathbf{q}))^2 + \omega^2}, \qquad (12.3.13)$$

where

$$\Lambda(\mathbf{q}) = 2zwc_{\mathrm{v}}(1 - \gamma_{\mathbf{q}})$$

$$\equiv \frac{2}{\tau_{\mathrm{s}}}(1 - \gamma_{\mathbf{q}}), \qquad (12.3.14)$$

and $v$ is the volume per lattice site. The quantity $\Lambda(\mathbf{q})$ is similar to the $\Gamma(\mathbf{q})$ of (9.3.11b) but contains two additional factors – namely 2, because changes in separation $\mathbf{r}_m - \mathbf{r}_l$ can arise from the movement of either nucleus, and $c_{\mathrm{v}}$ for the mean-field expression of the restriction that a jump is only possible if the site at which it aims is initially vacant. It will be seen that the integrand in (12.3.13), apart from the exponential factor, has the character of an incoherent dynamical structure factor – as is to be expected from (12.3.12) and §9.3.

Different authors have treated the task of evaluating (12.3.13) and the spectral functions for the random-walk model in different ways, so that there are various results in the literature under the same title. We shall comment separately on three of these.

*Torrey's isotropic model*

One of the earliest treatments which is still often referred to is that of Torrey (1953, 1954 with corrections in Resing and Torrey, 1963). In addition to the assumption specified by (12.3.12) this makes two further assumptions, namely (i) $\Lambda(\mathbf{q})$ is isotropic, i.e. dependent on only the magnitude of $\mathbf{q}$ and (ii) the summations over lattice points $\mathbf{m}$ and $\mathbf{l}$ in (12.2.3) can be replaced by integrations. To implement the first of these one takes the nearest-neighbour displacements to be uniformly distributed on the surface of a sphere of radius $s$. This yields

$$\gamma_q = \sin(sq)/(sq) \qquad (12.3.15)$$

in place of the forms (9.3.9). To implement the second, one must introduce a lower limit to the integrals to correspond to the fact that the smallest separations $\mathbf{r}_l$ and $r_m$ appearing in (12.2.3) are those for nearest neighbours. In practice this lower limit is treated as a parameter and chosen to ensure that $J^{(p)}(\omega)$ is correctly normalized to give the correct value of the correlation function $F^{(p)}$ at zero time, i.e. to satisfy

$$F(0) = \frac{1}{2\pi} \int_{-\infty}^{\infty} J^{(p)}(\omega)\, \mathrm{d}\omega, \qquad (12.3.16)$$

by the inverse of (12.2.2). The left side of (12.3.16) is already known in this spherically averaged form; it is just the coefficient of $2\tau/(1+\omega^2\tau^2)$ in (12.3.5).

After some mathematical manipulations the result of following this procedure is found to be

$$J^{(p)}_{\text{poly}} = \frac{d_p{}^2\tau_s c}{4\pi a^6} g(\xi), \tag{12.3.17}$$

where $\xi = \omega\tau_s/2$ and the function $g$ is

$$\begin{aligned} g(\xi) &= 3S_0 \int_0^\infty \frac{[J_{3/2}(kx)]^2\left(1 - \dfrac{1}{x}\sin x\right)}{\left(1 - \dfrac{1}{x}\sin x\right)^2 + \xi^2} \frac{dx}{x} \\ &\equiv 3S_0 G(k, \xi). \end{aligned} \tag{12.3.18}$$

The coefficient $S_0$ is the same summation as defined by (12.3.9). The equation which determines $k$, (12.3.16), is

$$k^3 = 4\pi a^6/3s^3 v S_0. \tag{12.3.19}$$

From (12.3.19) $k$ turns out to be about 3/4ths of the nearest neighbour separation (0.763 for b.c.c. lattices and 0.743 for f.c.c. lattices). Clearly $g(\xi)$ is the analogue of $S_0/(1+\xi^2)$ which appears in (12.3.5) for the B.P.P. single-relaxation-time model. The integral in (12.3.18) was evaluated by Resing and Torrey (1963) for values of $k$ appropriate to b.c.c. and f.c.c. lattices. Last of all, if we now substitute the expression (12.3.17) for $J^{(p)}_{\text{poly}}$ into (1.7.6a) we again obtain the form (12.3.7) but with the function $f$ given by

$$f(\xi) = 3\xi G(k, \xi) + 12\xi G(k, 2\xi), \tag{12.3.20}$$

where we have used the notation of Resing and Torrey for the integral. A plot of this $f(\xi)$ for the b.c.c. lattice as obtained from the tables of Resing and Torrey (1963) is also given in Fig. 12.2. The corresponding predictions for the f.c.c. lattice are much the same, differing by about only 1% or less over most of the range of $\xi$. The maximum value of $f$ is about 8% less than that in the B.P.P. model and it occurs close to $\xi = 0.53$ (both b.c.c. and f.c.c.), whereas for the B.P.P. model $\xi_{\max} = 0.62$. The most noticeable qualitative feature, however, is the slower fall-off at lower frequencies, which results in a less symmetrical curve.

We now turn to those calculations which aim to evaluate the consequences of (12.3.12) without the use of further approximations.

*Real-space summation method*

For simplicity we shall again limit the discussion to the case of a polycrystal or a powder. The starting point, as before, is (12.2.5) with the transformed (12.3.12). The average over all field directions can be taken by making use of the known transformation properties of the spherical harmonic functions. The result is again expressible in the form (12.3.17) but now $g(\xi)$ is a sum over all shells of neighbours to any particular site, i.e.

$$g(\xi) = \sum_{n=0}^{\infty} S_n \tilde{\pi}(n, \omega) \tag{12.3.21}$$

where $\tilde{\pi}(n, \omega)$ is the transform of the probability of finding the nucleus at a particular $n$th neighbour site to the origin at time $2t$, $G(n, 2t)$. It is given by (12.3.13) when we replace $\mathbf{r}_m - \mathbf{r}_l$ by $\mathbf{r}_n$. The sums are

$$S_0 = \sum_l \frac{a^6}{r_l^{\,6}}, \tag{12.3.22a}$$

and

$$S_n = \sum_l \frac{a^6}{r_l^{\,3}} \sum_m^{(n)} \frac{P_2(\cos\theta_{lm})}{r_m^{\,3}}, \tag{12.3.22b}$$

in which the sum over $m$ is over all sites which are $n$th neighbours of the site $\mathbf{r}_l$. These sums have been evaluated by Sholl (1974, 1975) for s.c., b.c.c. and f.c.c. lattices. The $\tilde{\pi}(n, \omega)$ quantities, also in (12.3.21), were obtained by performing the integrations in (12.3.13) numerically.

In this way Sholl was able to evaluate the dimensionless quantity $(AT_1)^{-1}$ (with $A$ defined by 12.3.8) for polycrystals as a function of $(\omega_0\tau_s/2)$. His tabulated results plotted on a figure of the scale of Fig. 12.2 are indistinguishable from those shown for Torrey's isotropic representation of the random-walk model, the differences being less than about 7% in all cases. Equations (12.3.17) and (12.3.18) can therefore be taken to give a faithful representation of the random-walk model applied to a polycrystal. Barton and Sholl (1976) have extended this method of calculation to make predictions for single crystals having s.c., b.c.c. and f.c.c. lattices, i.e. of the dependence of the spectral functions upon crystal orientation.

*Summation in* **q**-*space*

An alternative to the above procedure is to Fourier transform the $Y_{2,p}/r^3$ factors in (12.2.5), i.e. to find

$$T_p(\mathbf{q}) = \sum_{l\neq 0} \frac{Y_{2,p}(\Omega_l)\, e^{i\mathbf{q}\cdot\mathbf{r}_l}}{r_l^{\,3}}. \tag{12.3.23}$$

We also replace the integral (12.3.13) by a discrete summation, with the result that $J^{(p)}(\omega)$ can be written as a double summation in **q**-space over the Brillouin zone, namely

$$J^{(p)}(\omega) = cd_p{}^2 \sum_{\mathbf{q},\mathbf{q}'}^{\mathrm{BZ}} T_p(\mathbf{q})\tilde{\tilde{\pi}}(\mathbf{q},\mathbf{q}';\omega)T^*_{\bar{p}}(\mathbf{q}'), \tag{12.3.24}$$

in which

$$\tilde{\tilde{\pi}}(\mathbf{q},\mathbf{q}';\omega) = \sum_{\mathbf{l},\mathbf{m}} \pi(\mathbf{m},\mathbf{l};\omega)\,\mathrm{e}^{-\mathrm{i}\mathbf{q}\cdot\mathbf{r}_l+\mathrm{i}\mathbf{q}'\cdot\mathbf{r}_m}. \tag{12.3.25}$$

In the present case (12.3.25) is equivalent to

$$\tilde{\tilde{\pi}}(\mathbf{q},\mathbf{q}';\omega) = \delta_{\mathbf{q}\mathbf{q}'}\frac{2\Lambda(\mathbf{q})}{(\Lambda(\mathbf{q})^2+\omega^2)}. \tag{12.3.26}$$

The Kronecker $\delta$-function in this relation is a direct consequence of the assumption that the two nuclei move independently of one another (12.3.12). In more elaborate models which improve on that assumption additional terms in both **q** and **q**′ appear. The point of expressing $J^{(p)}(\omega)$ in this form is that the numerical techniques required are common to many problems in solid state physics and have been widely practiced. These have been applied in the present connection by Fedders and Sankey (1978) and by Barton and Sholl (1980). The results agree with one another and with those obtained by the real-space summation method.

We can thus conclude that, at least for the three principal cubic lattices, we now have accurate descriptions of the random-walk model as defined by the assumption (12.3.12). For polycrystals and powder specimens the isotropic approximation of Torrey (eqns. (12.3.17) and (12.3.18)) is accurate to better than 7% in the relaxation rates: correspondingly, it makes little difference to the inferred activation energies in $\tau_s$.

However, we are still left with the intrinsic limitation posed by (12.3.12), namely that when the two nuclei are close together the presence of one nucleus in the pair may hinder the motion of the other. The task of removing that limitation is difficult. Nevertheless, some assistance is at hand in the form of general expressions for the high- and low-frequency limits of the spectral functions. These can be used to 'pin down' theories in these regions with the object of improving their accuracy throughout the whole range of frequency. We therefore turn to these limits next.

## 12.4 High- and low-frequency limits of the spectral functions

From particular examples we have seen that the spectral functions $J^{(p)}(\omega)$ can be written as $\omega^{-1}a^{-6}$ times a dimensionless function of $(\omega\tau_s/2)$. It is obvious from purely dimensional considerations that this must always be so. Hence, were it not for the factor $\omega^{-1}$ the limit of small (large) frequency would be equivalent to that of high (low) temperature. Experimentally it has been much easier to span the range of $\omega\tau_s$ over several orders of magnitude by varying the temperature than by changing $\omega$ so the equivalence is often relied on. We shall also follow the corresponding usage, although one should always remember the factor $\omega^{-1}$. For example as $\omega \to 0$, $J^{(p)}(\omega)$ tends to a non-zero limit, $J^{(p)}(0)$, whereas as $T \to \infty$, i.e. $\tau \to 0$, $J^{(p)} \to 0$.

### 12.4.1 High-frequency (low-temperature) limit

The results (12.3.4), (12.3.18) and (12.3.26) all suggest that at high frequencies $\omega$ we have

$$J^{(p)}(\omega) \propto \omega^{-2}.$$

This dependence is easily seen to be general by referring back to (9.2.6) and setting $\lambda = \mathrm{i}\omega$. We recognize that $\tilde{\pi}(\mathbf{m}, \mathbf{l}; \omega)$ is equivalent to the quantity $2\,\mathrm{Re}\langle\tilde{G}_{\mathbf{ml}}(\mathrm{i}\omega)\rangle$, where we have labelled the states of the system by the separation of the pair of nuclei ($\mathbf{r}_l$ initially and $\mathbf{r}_m$ finally) and where the angular brackets denote an average over the motion of all the other atoms. In terms of the quantities introduced in (9.2.6), we see that large $\omega$ implies $\mathbf{A}$ small, so that only the early terms in the summation are significant. In fact, we easily see that the leading terms are $O(\omega^{-2})$ and, more explicitly, that

$$\tilde{\pi}(\mathbf{m}, \mathbf{l}; \omega) = 2\,\mathrm{Re}\langle\bar{G}_{\mathbf{ml}}(\mathrm{i}\omega)\rangle$$

$$= \frac{2}{\omega^2}\langle(P_{\mathbf{mm}}\delta_{\mathbf{ml}} - \Gamma_{\mathbf{ml}}P_{\mathbf{ll}})\rangle + O(\omega^{-4}),$$

which, by the definition (9.2.4),

$$= \frac{2}{\omega^2}\langle(P_{\mathbf{mm}}\delta_{\mathbf{ml}} - w_{\mathbf{ml}})\rangle.$$

$$= \frac{2}{\omega^2}\left\langle\left(\sum_{n \neq m} P_{\mathbf{nm}}\delta_{\mathbf{ml}} - w_{\mathbf{ml}}\right)\right\rangle. \tag{12.4.1}$$

The first term comes from the $G^{(\mathrm{d})}_{\mathbf{ml}}$ term (i.e. $n = 0$) in (9.2.6), while the second comes from $(G^{(\mathrm{d})}\Gamma A)_{\mathbf{ml}}$ (i.e. $n = 1$), or just one transition during the period $2\pi/\omega$. Higher terms in (9.2.6) give terms in the spectral function of $\omega^{-4}$ and higher inverse powers of $\omega$. The result (12.4.1) shows that the coefficient of $\omega^{-2}$ in the high-frequency limit of the spectral function depends upon the details of the way in which the atoms move, although the form itself does not. For simple systems the coefficient of $1/\omega^2$ in (12.4.1) may be written down by inspection, although in more complex cases the task of averaging over the movements of other atoms may be difficult.

As a simple example suppose that the system is very dilute and that there are very few nuclei relative to the number of (interstitial) lattice sites. Then we are only concerned with the movement of a single pair of nuclei and can suppose that these move on an otherwise empty lattice. Then if we evaluate the (spherically averaged) spectral function for a polycrystal by insertion of (12.4.1) into (12.2.7), we obtain for the present example

$$J^{(p)}(\omega) = \frac{c d_p^{\,2}}{\pi a^6 \tau_s \omega^2}(J_0 + J_1), \tag{12.4.2}$$

in which $\tau_s = (zw)^{-1}$ is the mean time between jumps of a single, isolated nucleus and $a$ is the cube edge length. The terms $J_0$ and $J_1$ correspond to no transition and to one transition respectively and are given by

$$J_0 = \sum_{l \neq 0} \frac{1}{(r_l/a)^6}\left(1 - \frac{1}{z}\sum_j^z \delta_{\mathbf{r}_l, \mathbf{s}_j}\right) \tag{12.4.3}$$

where the nearest-neighbour vectors, as before, are denoted by $\mathbf{s}_j$ $(j = 1, \ldots, z)$, and

$$J_1 = -\frac{a^6}{z}\sum_{l \neq 0}\sum_{m \neq 0}\sum_{k=1}^{z} \frac{\delta_{\mathbf{r}_m, \mathbf{r}_l + \mathbf{s}_k} P_2(\cos\theta_{lm})}{r_l^{\,3} r_m^{\,3}}. \tag{12.4.4}$$

These two summations can be obtained from the calculations of Barton and Sholl (1980). Values for the three cubic lattices are listed in Table 12.1.

As long as we are prepared to make the mean-field approximation we can also use this result to give us the high-frequency limit of $J^{(p)}(\omega)$ for the lattice gas model at arbitrary concentrations. In that case we merely replace $zw \equiv \tau_s^{-1}$ in (12.4.2) by $zwc_v$.

It is interesting to compare these results with the corresponding predictions of the B.P.P. and the random-walk models. The first gives a high-frequency (low-temperature) limit like (12.4.2), but without $J_1$ and with only the first part of $J_0$, i.e. in place of $J_0$ we have just the sum $\sum_{l \neq 0}(r_l/a)^6$. In the random-walk model, where the two nuclei are assumed to move independently of one another, cf. (12.3.12), we again obtain just $\sum_{l \neq 0}(r_l/a)^6$ in place of $J_0$, but in this model we do pick up $J_1$ (cf. 12.3.21). The additional term in the correct result for $J_0$, (12.4.3), comes from the way in which one nucleus of the pair hinders the movements of the other when they are nearest neighbours. (There is no similar term in $J_1$ because it would amount to the exclusion of $\mathbf{r}_m = 0$, and that is already demanded of the summation over $m$.) The magnitudes of the coefficients in these two models are compared with the exact values of $J_0 + J_1$ in Table 12.1.

Table 12.1. *Values of $J_0$ and $J_1$ appearing in the limiting form* (12.4.2) *as obtained from Table* 1 *of Barton and Sholl* (1980)

N.B. The side $a$ of the lattice cube in (12.4.3) and (12.4.4) is such that the nearest neighbour distance is $a$ (s.c.), $a\sqrt{3}/2$ (b.c.c.) and $a/\sqrt{2}$ (f.c.c.).)

The lower part of the table compares the values of $J_0 + J_1$ with the corresponding coefficients arising in the B.P.P. single-relaxation time model and in the random-walk model (r.w.m)

| Lattice | s.c. | b.c.c. | f.c.c. |
|---|---|---|---|
| $J_0$ | 7.402 | 26.68 | 107.63 |
| $J_1$ | −1.929 | −5.36 | −17.83 |
| $J_0 + J_1$ | 5.473 | 21.32 | 89.80 |
| B.P.P. | 8.402 | 29.05 | 115.63 |
| r.w.m. | 6.473 | 23.69 | 97.80 |

### 12.4.2 Low-frequency limit

The key to the determination of the limiting expression for $J^{(p)}(\omega)$ at small $\omega$ is the recognition that this is determined by $\pi(\mathbf{m}, \mathbf{l}; t)$ at long times and that this implies large differences $\mathbf{r}_m - \mathbf{r}_l$. After a long time both nuclei in the pair considered will have diffused over many lattice spacings and the discrete nature of the lattice will have become unimportant. In other words, we can describe the nuclear displacements in the continuum diffusion approximation. Furthermore, in view of the large changes $\mathbf{r}_m - \mathbf{r}_l$ we can also treat the diffusion of the two nuclei as independent of one another; for, it is only for a very few special trajectories that they will pass sufficiently close to one another that their motions interfere. But this is just the condition embodied in (12.3.12), so that to obtain the low-frequency limit it would appear that all we have to do is to make the diffusion approximation in (12.3.13), i.e. we replace $\Lambda(\mathbf{q})$ by its form for small $\mathbf{q}$, namely $2Dq^2$ with $D$ the diffusion coefficient of a *single* nucleus (cf. 9.3.12). This would then give us a limiting expression dependent on only phenomenological parameters ($D$ and the crystal structure).

However, we cannot proceed in quite so simple a way, with the result that the conclusion is only partly true. The reason is that we are really concerned with an integral of the form (9.3.16) with $\lambda = i\omega$ and, as we already remarked in §9.3, there are mathematical subleties associated with the point $\lambda = 0$. Since these have already been dealt with in our discussion of $U(\mathbf{l}, 1)$, we therefore first find $J^{(p)}(0)$ and then evaluate the difference $J^{(p)}(\omega) - J^{(p)}(0)$ by means of (12.3.13). The path we follow has already been mapped out in §9.3. From $G(\mathbf{m}-\mathbf{l}, 2t)$ we go to $G(\mathbf{m}-\mathbf{l}, i\omega/2)$ by setting $\lambda = i\omega/2$ in (9.3.3) and (9.3.16b). We can now take the limit $\omega \to 0$ and obtain

$$\tilde{\pi}(\mathbf{m}, \mathbf{l}; 0) = \frac{1}{zwc_v} U(\mathbf{m}-\mathbf{l}; 1). \tag{12.4.5}$$

The properties of the right side are known (§9.3). By (12.2.5) and (12.3.1) we then arrive at $J^{(p)}(0)$

$$J^{(p)}(0) = \frac{cd_p{}^2}{zwc_v} \sum_{m,l \neq 0} \frac{Y_{2,p}{}^*(\Omega_m)}{r_m{}^3} U(\mathbf{m}-\mathbf{l}, 1) \frac{Y_{2,p}(\Omega_l)}{r_l{}^3}. \tag{12.4.6}$$

For cubic lattices this summation is independent of the orientation of the crystal relative to the static magnetic field. This result (12.4.6) is evidently dependent on our assumption of a vacancy model and, more particularly, on the mean-field approximation (atomic jump rate $= zwc_v$).

We now return to (12.3.13), subtract the part corresponding to $\omega = 0$ and evaluate the remainder in the diffusion limit. This is then inserted into (12.2.5) to give the leading term in $J^{(p)}(\omega)$ after $J^{(p)}(0)$. Thus

$$\tilde{\pi}(\mathbf{m}, \mathbf{l}; \omega) = \tilde{\pi}(\mathbf{m}\, \mathbf{l}; 0) - \frac{\omega^2 v}{4\pi^3} \int_{\mathrm{BZ}} \frac{e^{-i\mathbf{q}\cdot(\mathbf{r}_m - \mathbf{r}_l)}\, d\mathbf{q}}{(2Dq^2)((2Dq^2)^2 + \omega^2)}, \tag{12.4.7}$$

in which

$$D = \tfrac{1}{6} z w s^2 c_v. \tag{12.4.8}$$

Since the integral is dominated by the region of small $\mathbf{q}$ we can extend the region of integration to all $\mathbf{q}$ without error. The angular integrations (those in the polar and azimuthal angles of $\mathbf{q}$ relative to the polar axis defined by $\mathbf{r}_m - \mathbf{r}_l$) can be carried out straightforwardly. We are then left with the task of evaluating the following integral

$$\begin{aligned} I &= 4\pi \int_0^\infty \frac{\sin qr}{qr} \frac{dq}{q^4 + \chi^2} \\ &= 2\pi \int_{-\infty}^\infty \frac{\sin qr}{qr} \frac{dq}{q^4 + \chi^2}. \end{aligned} \tag{12.4.9}$$

in which $\chi = \omega/2D$ and $r = |\mathbf{r}_m - \mathbf{r}_l|$. At low frequencies the integrand is dominated by the region of small $q < \chi^{1/2}$ around the origin. We can therefore impose the low-frequency condition in the form $r \ll \chi^{-1/2}$, replace $(\sin qr)/qr$ by its limiting value 1 and look up the resulting integral in tables. This gives

$$I = \pi^2 \sqrt{2}/\chi^{3/2}. \tag{12.4.10}$$

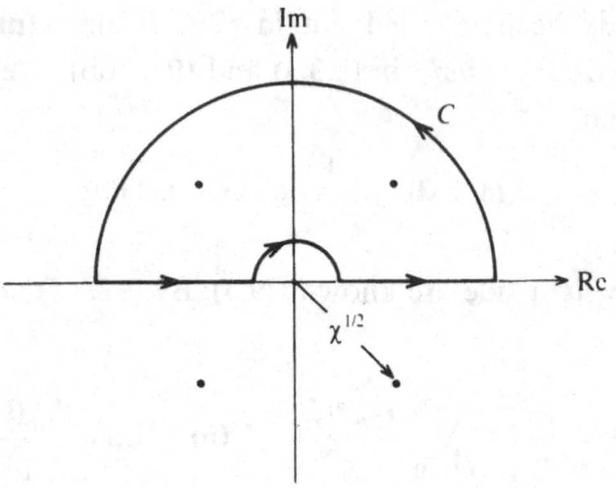


Fig. 12.3. Contour $C$ in the complex plane used to evaluate the integral $I$ defined by (12.4.9). The integrand has simple poles at the points indicated. The radius of the outer semi-circle is extended to $\infty$ and that of the inner semi-circle is shrunk to zero. In these limits there is a contribution to the contour integral from the region around the origin, but none from the outer arc.

Alternatively, by re-expressing $I$ as

$$I = 2\pi \,\mathrm{Im} \int_{-\infty}^{\infty} \frac{e^{iqr}}{qr} \frac{dq}{q^4 + \chi^2}, \tag{12.4.11}$$

we can relate it to the integral round the contour $C$ in the upper half of the complex plane (Fig. 12.3) and evaluate it by the method of residues. This gives, without approximation,

$$I = \frac{2\pi^2}{r\chi^2} \left\{ 1 - \exp\left( -r\sqrt{\frac{\chi}{2}} \right) \cos\left( r\sqrt{\frac{\chi}{2}} \right) \right\}, \tag{12.4.12}$$

which reduces to (12.4.10) when $r\chi^{1/2} \ll 1$.

Finally, therefore, if we insert the low-frequency result (12.4.10) into (12.4.6) and then into (12.2.5) we arrive at the result

$$J^{(p)}(\omega) = J^{(p)}(0) - \frac{c_v(\omega/D)^{1/2}}{8\pi D} d_p^{\,2} \sum_{m,l\neq 0} \frac{Y^*_{2,p}(\Omega_m)}{r_m^{\,3}} \frac{Y_{2,p}(\Omega_l)}{r_l^{\,3}}. \tag{12.4.13}$$

The summation is independent of the orientation of the crystal relative to the magnetic field. Furthermore, the only dependence on the index $p$ arises through the factors $d_p^{\,2}$. Hence we can write $J^{(p)}(\omega)$ in the form

$$J^{(p)}(\omega) = J^{(p)}(0) - \beta(cd_p^{\,2}/Dv)(\omega/D)^{1/2} + \cdots \tag{12.4.14}$$

The coefficient $\beta$ has been evaluated by Sholl (1981a, 1992) who found it to be 1/18 for the cubic lattices as well as for simple liquids.

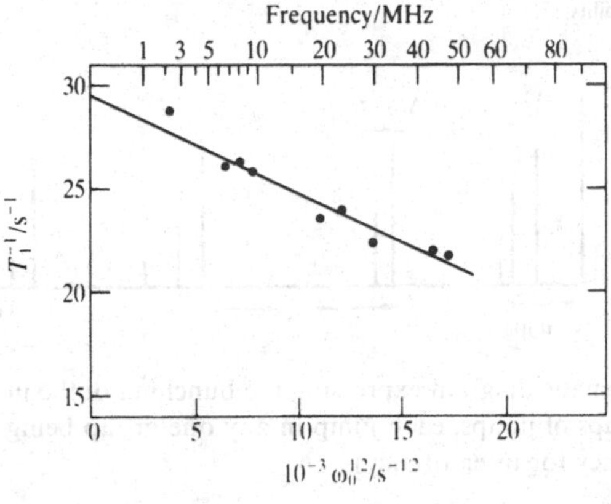


Fig. 12.4. Dependence of the spin–lattice relaxation rate $T_1^{-1}$ of protons in $TiH_{1.63}$ at 725 K as a function of $\omega_0^{1/2}$. The circles are the experimental data and the straight line is a least-squares fit to this data. (After Salibi and Cotts, 1983.)

The important qualitative result is the dependence on $\omega^{1/2}$. It is not always easy to confirm this experimentally since it is necessary to vary $\omega$ while holding the temperature constant (to maintain $J^{(p)}(0)$ and $D$ constant). However, one example of measurements which do seem to confirm the $\omega^{1/2}$ dependence is shown in Fig. 12.4. Beyond this, the form (12.4.14) indicates how particular theories should behave in this region.

Lastly, we note that the $\omega$-dependence of $J^{(p)}(\omega)$ at low-frequencies is characteristic of the dimensionality of the lattice. Thus whereas the $\omega^{1/2}$-dependence shown by (12.4.14) applies to three-dimensional lattices the corresponding variation for two- and one-dimensional structures is as $\ln \omega$ and $\omega^{-1/2}$ respectively. The experimental determination of $J^{(p)}(\omega)$ at low frequencies can therefore be very helpful in the investigation of atomic movements in complex structures (e.g. Heitjans *et al.*, 1991).

## 12.5 Systems dilute in defects – the encounter model

In the previous sections we have developed theories of the spectral functions with a minimum of detail about the mechanism of atomic movements. The use of the mean-field approximation has made it possible to do this without regard to the fraction of vacant sites and, at some points, without regard to the mechanism at all. However, we are already aware that at low defect concentrations successive movements of the atoms are correlated both in space and time and that this feature is omitted in the mean-field approximation. We shall therefore now turn to the

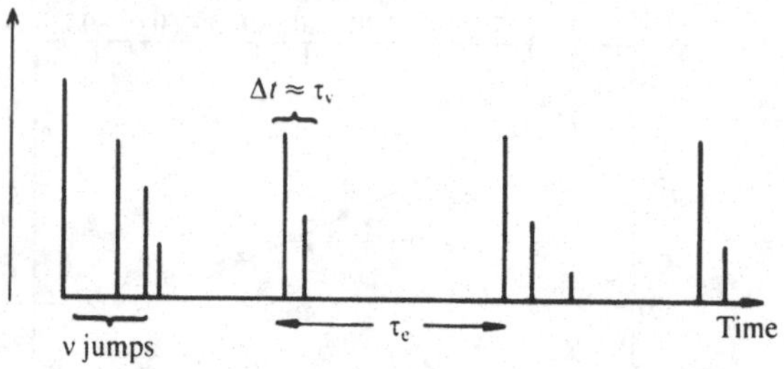


Fig. 12.5. Schematic diagram expressing the bunching of the jumps of a tracer atom into groups of jumps, each jump in any one group being occasioned by the same vacancy (or interstitialcy).

encounter theory of such systems; this enables us to include the effects of correlations among atomic movements.

At points in Chapters 8–10 we introduced the central idea of the encounter model, namely that the displacements of any particular atom are bunched in time, each bunch corresponding to a sequence of exchanges with a single vacancy, i.e. to an *encounter* between the atom and a vacancy (Fig. 12.5). (Here and henceforth we assume that the defects are vacancies, although the same broad ideas apply to movement via interstitialcy mechanisms; however such cases have received less attention.)

If now the atom in question is an atom of the host species then the average number of these exchanges is the quantity $\nu$ obtained in §9.5. As we see from the values given in Table 9.2 it is quite small for three-dimensional lattices, in all cases less than two. For solute atoms of another chemical species the number may be somewhat greater, depending on factors such as the magnitude of the solute–vacancy binding energy and the relative rates at which the vacancy exchanges position with the solute atom and the host atom. However, in all cases the vacancy soon is lost to the vicinity of the atom and the sequence of correlated displacements of the atom comes to an end. The next vacancy to interact with the atom will arrive from a direction which is random relative to the first, so that there is no correlation between the movements undergone in successive encounters. Since the vacancy fraction is small the time between encounters is necessarily long compared to the average duration of one encounter.

When we are concerned only with very slow processes (e.g. macroscopic diffusion) this division of the motion of atoms into a sequence of encounters is of no significance for the end result. On the other hand, if we are concerned with processes like nuclear magnetic relaxation where significant changes occur in times less than the interval between encounters, then the model becomes particularly useful. Indeed, it was in the context of n.m.r. that it was first defined (Eisenstadt and Redfield, 1963). The randomness in the occurrence of encounters, in fact, allows us to write down a kinetic equation for $\pi(\mathbf{m}, \mathbf{l}; t)$ valid for times longer than the duration of an encounter. That equation and its solution are the principal topic of this section. Before we can proceed we have first to see how we can adapt the central idea to represent the relative motion of a pair of nuclei, i.e. we have to enlarge the idea of an encounter to embrace the interaction of one vacancy with a *pair* of atoms. Then we shall need to calculate some additional quantities which enter into the kinetic equation for $\pi(\mathbf{m}, \mathbf{l}: t)$.

### *12.5.1 Enlargement of the encounter model*

It is evident that if we want to set down an equation for $\pi(\mathbf{m}, \mathbf{l}: t)$ we shall need two quantities, one the mean time between encounters of a vacancy with a pair of atoms and the other the probability $\varpi(\mathbf{m}, \mathbf{l})$ that as the result of an encounter with a vacancy a pair of atoms initially separated by $\mathbf{r}_l$ end up separated by $\mathbf{r}_m$. We deal with these quantities in turn, considering only the case of a pure monatomic substance in which all the atoms are the same.

*The mean time between encounters of a pair with a vacancy*

As before let the mean time between the moves of any one atom be $\tau_s$ and let the mean time between movements of a vacancy be $\tau_v$. Then

$$\frac{c_v}{\tau_v} = \frac{(1 - c_v)}{\tau_s} \simeq \frac{1}{\tau_s} \tag{12.5.1}$$

We have already seen that, in the course of an encounter with a vacancy, an atom will move $\nu$ times on average. Hence the average time, $\tau_e$, that an atom spends at rest between encounters with vacancies must be given by

$$\frac{1}{\tau_s} = \frac{\nu}{\tau_e} = \frac{U(0, 1)}{\tau_e}. \tag{12.5.2}$$

In other words, the average number of atomic jumps per unit time must equal the number of encounters experienced by an atom in unit time multiplied by the average number of movements which it undergoes per encounter.

These definitions relate to the encounter between a vacancy and a single atom. We now want the corresponding mean time $\tau_l$ between encounters of a pair of atoms (separated by $\mathbf{r}_l$) with vacancies. This mean time may be related to $\tau_s$, the mean time between jumps of a particular atom, by finding the mean number of jumps of a particular atom of the pair during an encounter and setting down the relation corresponding to (12.5.2). One particular atom of the pair (taken to be at the origin) will be affected first in the encounter with a probability of 1/2. In this case it makes a total of $\nu$ jumps during the encounter. Equally, the other atom of the pair (at $\mathbf{r}_l$) will be the first to be affected in the encounter with a probability of 1/2. In that case the atom at the origin will make $\nu_l$ jumps, where this number is given by (9.4.15). Hence the mean number of jumps of the selected atom per encounter is $(\nu + \nu_l)/2$ and the average jump frequency of this atom is thus $(\nu + \nu_l)/2\tau_l$. But this quantity, by definition, is just $1/\tau_s$. Hence

$$\frac{1}{\tau_s} = \frac{(\nu + \nu_l)}{2\tau_l}. \tag{12.5.3}$$

Finally by (9.4.15), (9.5.4) and (9.5.6) we see that

$$\tau_l = \tfrac{1}{2}(U(0, 1) + U(l, 1))\tau_s. \tag{12.5.4}$$

Values of the coefficients of $\tau_s$ are obtainable from Table 9.1. By (12.5.2) $\tau_l$ is greater than $\tau_e/2$ but never as large as $\tau_e$ (since $l > 0$).

*The probability $\varpi(\mathbf{r}_m, \mathbf{r}_l)$*

The other quantity that we need is the probability that the separation of a pair of atoms after an encounter is $\mathbf{r}_m$ given that initially it was $\mathbf{r}_l$. This quantity is more difficult to calculate than $\tau_l$. It has been computed by Wolf (1974) and by Wolf, Figueroa and Strange (1977) by purely numerical Monte Carlo methods. It has also been calculated by Sholl (1982, 1992) by methods based on the theory of random walks as described in Chapter 9. Table 12.2 contains illustrative results for the three cubic lattices when $\mathbf{r}_l$ is a nearest-neighbour vector.

### 12.5.2 The kinetic equation and its solution

Given that we now have the frequency $\tau_l^{-1}$ with which encounters occur and the probability that the separation of a pair of atoms changes from $\mathbf{r}_l$ to $\mathbf{r}_m$ in the course of an encounter, we can now set down the kinetic equation for $\pi(\mathbf{m}, \mathbf{l}; t)$. The rate of change $\partial\pi(\mathbf{m}, \mathbf{l}; t)/\partial t$ is obviously the difference between the rate of gain to $\pi(\mathbf{m}, \mathbf{l}; t)$ by those encounters of $\mathbf{r}_n$-pairs $(n \neq m)$ with vacancies which generate $\mathbf{r}_m$-pairs and the rate of loss from $\pi(\mathbf{m}, \mathbf{l}; t)$ by encounters of $\mathbf{r}_m$-pairs with vacancies which result in transitions to other separations. The equation is thus

$$\frac{\partial\pi(\mathbf{m}, \mathbf{l}; t)}{\partial t} = \sum_{n \neq m} \frac{\varpi(\mathbf{m}, \mathbf{n})\pi(\mathbf{n}, \mathbf{l}; t)}{\tau_n} - \frac{\pi(\mathbf{m}, \mathbf{l}; t)}{\tau_m} \sum_{n \neq m} \varpi(\mathbf{n}, \mathbf{m}), \tag{12.5.5}$$

Table 12.2. *Values of the probability $\varpi(\mathbf{m}, \mathbf{l})$ that a pair of nuclei initially nearest neighbours of one another are separated by the lattice vector $\mathbf{r}_m$ after an encounter with a vacancy*

The results for each crystal structure are presented in groups of sites which are equivalent by symmetry. (After Sholl, 1982.)

| | **m** | $\varpi$**(m, l)** |
|---|---|---|
| s.c. **l** = (1, 0, 0) | (1, 0, 0) | 0.176 |
| | (1, ±1, 0)(1, 0, ±1) | 0.109 |
| | (2, 0, 0) | 0.108 |
| | (0, ±1, 0)(0, 0, ±1) | 0.031 |
| | (1, −1, ±1)(1, 1, ±1) | 0.011 |
| | (2, ±1, 0)(2, 0, ±1) | 0.011 |
| | (1, ±2, 0)(1, 0, ±2) | 0.004 |
| b.c.c. **l** = (1, 1, 1) | (0, 0, 0) | 0 |
| | (1, 1, 1) | 0.128 |
| | (2, 2, 2) | 0.087 |
| | (2, 2, 0)(2, 0, 2)(0, 2, 2) | 0.087 |
| | (2, 0, 0)(0, 2, 0)(0, 0, 2) | 0.086 |
| | (−1, 1, 1)(1, −1, 1)(1, 1, −1) | 0.024 |
| | (−1, −1, 1)(−1, 1, −1)(1, −1, −1) | 0.016 |
| | (1, 1, 3)(1, 3, 1)(3, 1, 1) | 0.013 |
| f.c.c **l** = (1, 1, 0) | (0, 0, 0) | 0 |
| | (1, 1, 0) | 0.0860 |
| | (0, 1, ±1)(1, 0, ±1) | 0.0673 |
| | (1, 2, ±1)(2, 1, ±1) | 0.0662 |
| | (2, 2, 0) | 0.0661 |
| | (2, 0, 0)(0, 2, 0) | 0.0660 |
| | (−1, 1, 0)(1, −1, 0) | 0.0118 |
| | (1, 1, ±2) | 0.0073 |

with the initial condition

$$\pi(\mathbf{m}, \mathbf{l}; 0) = \delta_{\mathbf{ml}}. \tag{12.5.6}$$

If we write

$$w(\mathbf{m}, \mathbf{n}) = \frac{\varpi(\mathbf{m}, \mathbf{n})}{\tau_n}, \tag{12.5.7}$$

then (12.5.5) is clearly of the same form as the master equation (6.2.1) of Chapter 6. We can therefore pursue some of the same formal manipulations as were given there. In particular, (12.5.5) can be rewritten in matrix form like (6.2.2), i.e.

$$\frac{\mathrm{d}\boldsymbol{\pi}}{\mathrm{d}t} = -\mathbf{P}\boldsymbol{\pi}, \tag{12.5.8}$$

where $\boldsymbol{\pi}$ is the column matrix of the $\pi(\mathbf{m}, \mathbf{l}; t)$ arranged in order of $\mathbf{m}$. (The initial separation $\mathbf{r}_l$ is essentially just a parameter as far as (12.5.5) and (12.5.6) go.) The elements of the matrix $\mathbf{P}$ in this case are

$$P_{mn} = -w(\mathbf{m}, \mathbf{n}), \qquad \mathbf{m} \neq \mathbf{n}, \tag{12.5.9a}$$

$$P_{mm} = \sum_{n \neq m} w(\mathbf{n}, \mathbf{m}). \tag{12.5.9b}$$

However, in the present application the sum of the probabilities $\varpi(\mathbf{m}, \mathbf{n})$ over all final separations $\mathbf{r}_m$ (including $\mathbf{m} = \mathbf{n}$) must be unity, i.e.

$$\sum_m \varpi(\mathbf{m}, \mathbf{n}) = 1, \qquad \text{all } \mathbf{n}. \tag{12.5.10}$$

Hence

$$\sum_m w(\mathbf{m}, \mathbf{n}) = \frac{1}{\tau_n} \tag{12.5.11}$$

and $P_{nn}$ by (12.5.9b) is

$$P_{nn} = \frac{1}{\tau_n}(1 - \varpi(\mathbf{n}, \mathbf{n})). \tag{12.5.12}$$

The quantity we want is the propagator $\mathbf{G}(t)$ corresponding to $\mathbf{P}$, i.e. $\mathbf{G}(t) = \exp(-\mathbf{P}t)$, since the $(\mathbf{m}, \mathbf{l})$ element of $\mathbf{G}(t)$ gives just the probability that the pair is separated by $\mathbf{r}_m$ at $t$ given that the initial separation was $\mathbf{r}_l$, i.e. just the solution of (12.5.5) and (12.5.6). Then, as before, we obtain $J^{(p)}(\omega)$ from the Laplace transform $\bar{\mathbf{G}}(\lambda)$ of this propagator by setting $\lambda = \mathrm{i}\omega$.

Before pursuing this strategy for the encounter model proper it is instructive to take a step back and examine the consequences of omitting correlations between successive moves of the atoms in question. We do this by supposing that each encounter results in just one exchange of positions between the vacancy and one of the two atoms, after which the vacancy wanders away. This is equivalent to setting $\pi'$ in (9.5.4) equal to zero and thus to $U(0, 1) = 1$, whence by (9.3.18) to $U(\mathbf{l}, 1) = 0$ $(\mathbf{l} \neq 0)$ and by (12.5.4) to $\tau_l = \tau_s/2$. At the same time it is clear that from just one exchange the only changes in separation are by nearest neighbour vectors, $\mathbf{s}_j$. Hence

$$\varpi(\mathbf{m}, \mathbf{n}) = \frac{1}{z}, \qquad \mathbf{m} - \mathbf{n} = \mathbf{s}_j$$

$$= 0, \qquad \text{otherwise}.$$

Actually, if we also ignore the hindrance to the motion of one atom caused by the presence of the other we can use this relation for all $\mathbf{n}$. As a result of these simplifications we can use the arguments presented in §9.3 to follow the strategy defined above for dealing with (12.5.5). The consequences are then just those we obtained for the random-walk model in §12.3, e.g. (12.3.13) and (12.3.14). The random-walk model therefore provides a reference with which the encounter model can be compared when we wish to assess the influence of correlation effects (and hindrances) upon the results.

We now turn to the calculation for the encounter model proper which includes both correlations and hindrances. We can no longer follow the strategy in the same way as we did in §9.3 because now the transition rates $w(\mathbf{m}, \mathbf{n})$ are no longer simply functions of the difference $\mathbf{m} - \mathbf{n}$, but depend separately on both $\mathbf{m}$ and $\mathbf{n}$. Fourier transformation in the space variables $\mathbf{m}, \mathbf{n}$, etc. of the set of equations represented by (12.5.5) does not therefore lead to a set of uncoupled equations in the Fourier amplitudes of the $\pi(\mathbf{m}, \mathbf{l}; t)$, and we are no longer able to infer simple forms like (9.3.8). We must therefore develop the equations in other ways.

One way is to treat the task of solving (12.5.5) as a perturbation or interaction problem in the following sense. At large separations $\tau_n$ becomes independent of $\mathbf{r}_n$ while $\varpi(\mathbf{m}, \mathbf{n})$ depends only on the change in separation $\mathbf{r}_m - \mathbf{r}_n$, because then the motions of one nucleus are uncorrelated with those of the other. At close separations, by contrast, the two nuclei are moved by the same defect. As a result the diagonal terms $P_{mm}$ tend to a constant value as $\mathbf{r}_m$ increases. The limiting value of $\varpi(\mathbf{m}, \mathbf{m})$ is just $\varpi(0)$ calculated in §9.5 ($\simeq 1/z$) and the limiting value of $\tau_m$ is just $\tau_e/2$; hence $w(\mathbf{m}, \mathbf{m}) \to 2(1 - \varpi(0))/\tau_e \equiv \Gamma_e$ as $\mathbf{r}_m$ increases. We now regard the non-diagonal terms $P_{mn}$ and the departures of $P_{mm}$ from the limiting value (large $\mathbf{m}$) as the perturbation and so write

$$\mathsf{P} = \mathsf{P}^{(1)} + \mathsf{P}^{(2)}, \tag{12.5.13}$$

with

$$\mathsf{P}^{(1)} = \frac{2(1 - \varpi(0))\mathbf{1}}{\tau_e} = \Gamma_e \mathbf{1} \tag{12.5.14}$$

and where $\mathsf{P}^{(2)}$ contains all the perturbation terms. The quantity $\bar{\mathsf{G}}(\lambda)$ corresponding to $\mathsf{P}$ is now expanded in terms of $\bar{\mathsf{G}}^{(1)}(\lambda)$ and $\mathsf{P}^{(2)}$ as was done in (9.2.2), where in place of $\mathbf{P}^{(1)}$ and $\mathbf{P}^{(2)}$ we had $\mathbf{P}^{(d)}$ and $\mathbf{P}^{(nd)}$. The algebra is the same but in this case, as $\mathbf{P}^{(1)}$ is a multiple of the unit matrix, the expansion conveniently separates the factors in $\lambda$ from the matrix terms so yielding

$$\bar{\mathbf{G}}(\lambda) = \sum_{r=1}^{\infty} \frac{(-1)^{r-1}(\mathbf{P}^{(2)})^{r-1}}{(\Gamma_e + \lambda)^r}. \tag{12.5.15}$$

For the spectral function $J^{(p)}(\omega)$ we want $2\,\mathrm{Re}\,\bar{\mathbf{G}}(\mathrm{i}\omega)$. If we form the $Y_{2,p}(\Omega_m)/r_m^{\,3}$ elements in $J^{(p)}(\omega)$ into row and column matrices $\tilde{\mathbf{Y}}^*$ and $\mathbf{Y}$ respectively, then $J^{(p)}(\omega)$ is compactly written as

$$J^{(p)}(\omega) = cd_p^{\,2} \sum_{r=1}^{\infty} S_r^{(p)}\, 2\,\mathrm{Re} \frac{1}{(\Gamma_e + \mathrm{i}\omega)^r}, \tag{12.5.16}$$

with

$$S_r^{(p)} = (-1)^{r-1}\tilde{\mathbf{Y}}^*(\mathbf{P}^{(2)})^{r-1}\mathbf{Y}. \tag{12.5.17}$$

The convenience of (12.5.16) is that the sums $S_r^{(p)}$ do not depend upon the frequency $\omega$ and so can be evaluated once and for all, when $\Gamma_e$ has been chosen. Their dependence on the choice of $\Gamma_e$ must however be remembered. The real part of $(\Gamma_e + \mathrm{i}\omega)^{-r} = (\Gamma_e - \mathrm{i}\omega)^r/(\Gamma_e^{\,2} + \omega^2)^{2r}$ is straightforwardly obtained from the binomial expansion of $(\Gamma_e - \mathrm{i}\omega)^r$.

Macgillivray and Sholl (1986) have used (12.5.16) to evaluate $J^{(p)}(\omega)$ for a vacancy model of the three cubic lattices. One or two practical aspects of this evaluation should be mentioned. Firstly, the authors chose a slightly different breakdown of $\mathbf{P}$ corresponding to $\Gamma_e = 2/\tau_e$ rather than the form suggested by the perturbation argument (i.e. 12.5.14); the difference is, however, small and presumably has little effect on the convenience of this approach. Secondly, no finite number of terms in the series will give the required $\omega^{1/2}$ dependence of $J^{(p)}(\omega)$ as $\omega \to 0$. This difficulty was avoided by using (12.4.13) to extrapolate the predictions of (12.5.16) downwards from values of $\omega$ where the two predictions matched. Thirdly, there is the fact that the high-frequency limit of (12.5.16) is different from that predicted by (12.4.1) because the encounter model itself presupposes that the period $\omega^{-1}$ is greater than the duration of an encounter, i.e. $\omega\tau_e \ll 1$. On the other hand, the condition for (12.4.1) to apply is $\omega\tau_s \gg 1$. The high-frequency limit for the encounter model is obtainable from (12.5.16) in an analogous way and again shows that $J^{(p)}(\omega)$ is proportional to $\omega^{-2}$, although the coefficients differ. These small differences at high frequencies can be regarded as a measure of the effects of the motions of the spins *during* an encounter.

### *12.5.3 Some results and conclusions*

The encounter model has been evaluated for vacancies in the three cubic lattices (MacGillivray and Sholl, 1986 and references cited there), for vacancy pairs in b.c.c. and f.c.c. lattices (Wolf, 1977) and for anion interstitialcies in the fluorite lattice (Wolf, Figueroa and Strange, 1977 and Figueroa, Strange and Wolf, 1979). We call on these sources in drawing the following conclusions and in making comparisons with the simpler analytical theories. We do so from the point of view of one attempting to analyse experimental data on relaxation times for information about the mechanism and characteristics of atomic movements.

(1) Calculations of the spectral functions for the encounter model of vacancy diffusion made by different workers generally agree to within 20% or better. Those of MacGillivray and Sholl (1986) using the method described in §12.5.2 above are probably the most accurate. Their values for the spin–lattice relaxation rate (i.e. $T_1^{-1}$) are some 6–12% higher (depending on lattice and frequency) than those obtained by the Monte Carlo method by Wolf (1974) and Wolf *et al.* (1977).

(2) At low values of $\omega\tau$ the slope of $\ln T_1$ v. $\ln(\omega\tau)$ from the calculations of MacGillivray and Sholl (1986) as shown in Figs. 12.6–12.8 is slightly less than that given by the B.P.P. model, which in turn is slightly less than is indicated by the Torrey isotropic model. However, all these theories actually have asymptotes with the same slope ($-1$). In practice therefore if activation energies (in $\tau_s$) are to be obtained to better than $\sim 5\%$ a wider range of high temperature measurements is required than would correspond to Figs. 12.6–12.8 (i.e. wider than two orders of magnitude). If this is not possible, a fit to the analytic representation of the predictions of MacGillivray and Sholl proposed by Sholl (1988) may be the best way to determine the parameters in $\tau_s$.

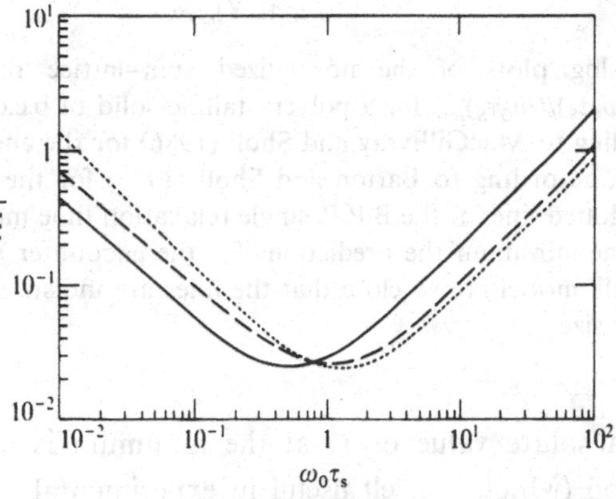


Fig. 12.6. Log–log plots of the (dimensionless) spin–lattice relaxation time ($AT_1$) v. $\omega_0\tau_s$ for a polycrystalline solid of b.c.c. structure: (*a*) full line, according to MacGillivray and Sholl (1986) for the encounter model, (*b*) dashed line, according to Barton and Sholl (1976) for the random-walk model (no correlations) and (*c*) dotted line, for the B.P.P. single-relaxation-time model.

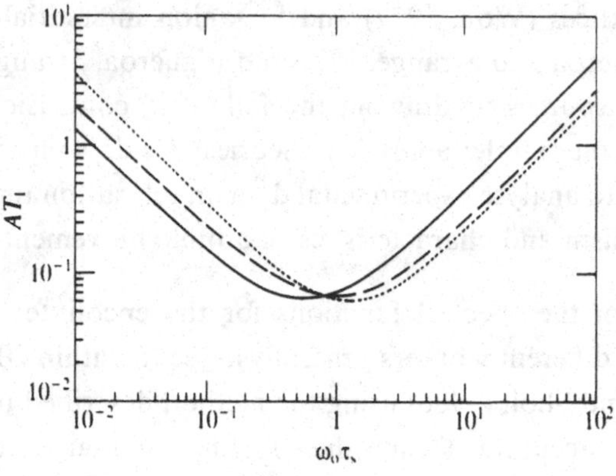


Fig. 12.7. Log–log plots of the (dimensionless) spin–lattice relaxation time ($AT_1$) v. $\omega_0\tau_s$ for a polycrystalline solid of f.c.c. structure. The significance of the different lines is as in Fig. 12.6.

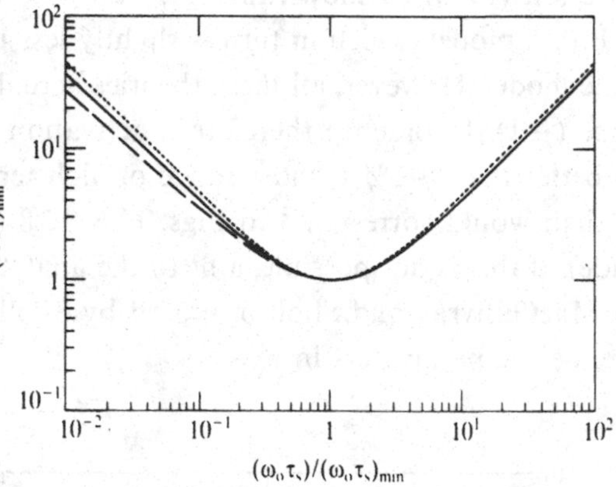


Fig. 12.8. Log–log plots of the normalized spin–lattice relaxation time ($T_1/T_{1,min}$) v. $(\omega_0\tau_S)/(\omega_0\tau_S)_{min}$ for a polycrystalline solid of b.c.c. structure (*a*) full line, according to MacGillivray and Sholl (1986) for the encounter model (*b*) dashed line, according to Barton and Sholl (1976) for the random walk model and (*c*) dotted line for the B.P.P. single relaxation time model. At values of $\omega\tau_S$ above the minimum the predictions for the encounter model and for the random-walk model are so close that the lines are indistinguishable on a diagram of this size.

(3) Although the absolute value of $T_1$ at the minimum is insensitive to these variations in method (which is itself useful in experimental work), the value of $(\omega\tau_s)$ at which the minimum occurs does vary substantially. It is considerably lower according to MacGillivray and Sholl at 0.51 (b.c.c.) and 0.57 (f.c.c.) compared to 1.23 from the B.P.P. equation and 1.0 (b.c.c) and 1.1 (f.c.c.) from the random-walk model. This means that the true value of $\tau_s$ at the temperature where

the minimum of $T_1$ is observed could be only half that which we should infer if we relied on the B.P.P. equation. Such a difference is obviously significant if the aim is to obtain precise data on $\tau_s$. We return to the significance of this value in (6) below.

(4) Despite these differences, different models lead to curves of $T_1$, $T_{1\rho}$ and $T_2$ v. $\omega_0\tau$ of very similar shape, i.e. of very similar temperature dependence. This is apparent from Figs. 12.6–12.8. In order to show how close the shapes are, in Fig. 12.8 we plot $\log_{10}(T_1/T_{1,\min})$ v. $\log_{10}(\omega\tau/\omega\tau_{\min})$ for the encounter model and the B.P.P. model. The most noticeable difference is the slightly narrower minimum of the B.P.P. curve. A detailed examination of this matter of shape (width and asymmetry of the minimum) was made by Wolf (1977). However, from the point of view of the analysis of experimental results it would appear generally difficult to discriminate among different physical models on the basis of shape – especially when it is recalled that shape as the term is used here only equates directly to temperature variation for systems where the Arrhenius equation (12.3.6) is strictly obeyed. This is borne out by the conclusions of a detailed attempt to do just that by Figueroa *et al.* (1979).

(5) The theory shows that the dependence of the relaxation times of single crystals upon orientation also provides a possible way of discriminating among models. In general, it may provide a better method than does the temperature dependence, or shape, considered in (4). In practice the discrimination can remain rather limited (see e.g. Wolf *et al.*, 1977). Since measurements of $T_1$ and $T_{1\rho}$ may not be accurate to better than 10%, even in favourable circumstances, this insensitivity to details of the mechanism of atomic movements makes the direct inference of the responsible mechanism very difficult in most cases. This is possibly one of the reasons why the B.P.P. formulae have remained in use so extensively.

(6) Lastly, we consider the sense in which we can speak of a *correlation factor* in the expressions for the relaxation times analogous to that evaluated for tracer and solute atom diffusion in Chapter 10. The comparison is between the encounter model (which includes correlation effects) and the random-walk model (which does not). To make this comparison we return to Figs. 12.6 and 12.7 and observe that the predictions *with* correlations can be obtained approximately from those *without* by a rigid shift along the $\ln(\omega\tau_s)$ axis. In other words, the effective rate of encounters ($2/\tau_s$ in the uncorrelated random-walk model) is reduced by about one-half when correlations are included. In this sense the correlation factor for the spin-lattice relaxation time is about 0.5. Two factors enter into this figure, namely (i) the frequency of encounters and (ii) the chance of a displacement during an encounter (cf. (12.5.7)). For nearest-neighbour pairs (which provide the greatest contribution to the interactions) the first factor, by (12.5.4) is $1/(U(0, 1) + U(s, 1))$,

or 0.56 (b.c.c.) and 0.59 (f.c.c.) by Table 9.1. Correlation effects result in the chance of a net displacement in an encounter being only about $(1 - z^{-1})$, actually 0.872 (b.c.c.) and 0.914 (f.c.c.) by Table 12.2. By multiplying these two factors together we get 0.49 (b.c.c.) and 0.54 (f.c.c.), more or less as Figs. 12.6 and 12.7 require. As with the correlation factors for tracer diffusion, these figures depend upon the lattice and the defect mechanism of migration. Again therefore any means of determining them separately is valuable. In the case of ionic conductors this may be achieved if both n.m.r. and conductivity data are available. Some illustrations are given in the review by Chadwick (1988).

## 12.6 Spin relaxation via the nuclear quadrupole – electric field gradient interaction

As we have already mentioned, nuclei having a spin $I > 1/2$ may possess an electric quadrupole moment resulting from a non-spherical charge distribution in the nucleus. This quadrupole moment will interact electrostatically with any electric field gradient (e.f.g.) existing at the nucleus. Depending upon the source of the electric field, this interaction can either be large or small compared with the Zeeman energy ($\hbar\omega_0$) of interaction of the nuclear magnetic dipole with the static magnetic field $B_0$. When it is large the Zeeman energy is largely irrelevant and nuclear resonance experiments in zero or small fields are possible ('pure quadrupole resonance'). Such situations commonly arise in covalently bonded molecules where the bonding p-electrons give rise to large, non-centrally-symmetric electric fields at the nuclei. A review of such low-field experiments which give information about chemical structure and bonding was given by Das and Hahn (1958). Here we shall be largely concerned with the other limit where the quadrupole–e.f.g. interaction represents a perturbation on the usual Zeeman resonance (for which a corresponding review was given by Cohen and Reif, 1957).

Even with this restriction there is much that could be said but which does not pertain to our interest in atomic migration. For example, we exclude static effects coming from e.f.g.s resulting from the low symmetry of the crystal structure. We limit ourselves to cubic crystals, for when such crystals are perfect there can be no electric field gradients and thus no effect of the quadrupoles on the nuclear magnetic resonance. Any effect in these crystals thus comes solely from the presence of imperfections (unless the crystal is distorted from cubic by externally applied stresses). Any imperfection can in principle generate effects either through the presence of electric charges (e.g. point defects bearing net charges) or through associated inhomogeneous strain fields which, by disturbance of the electron distribution around the atoms, in turn give rise to electric field gradients at the nuclei. Again there are static effects visible in (c.w.) resonance experiments (e.g. line broadening) coming from the presence of dislocations, solute atoms etc.; and these have been very useful in the study of imperfections *per se*. By contrast, our interest here is with dynamical effects, specifically with the way in which atomic displacements give rise to varying quadrupole–e.f.g. interactions and thus to effects upon the spin–lattice relaxation rates. (It may be noted that the thermal vibrations of the lattice may also lead to significant variations in these interactions and thus to associated contributions to the spin–lattice relaxation rate, but these generally increase more slowly with temperature ($\sim T^2$) than those arising from thermally activated defect motion.) In considering the theory of these effects it should be remembered that they are additional to the contribution from the magnetic dipole–dipole interaction, which, of course, is always present. We can, however, anticipate that there may be a region of temperature in which they are dominant. Thus we have seen that the maximum spin–lattice relaxation rate (minimum $T_1$) arising from the dipole–dipole coupling occurs when $\omega_0 \sim \tau_s^{-1}$, i.e. when $\omega_0$ is roughly equal to the mean jump rate of the nuclei in resonance. Likewise, the maximum contributions to the rate from the interaction of the quadrupoles on these nuclei with the e.f.g.s generated by mobile point defects would be expected to occur when $\omega_0$ is roughly equal to the mean jump rate of the defects. This jump rate is, of course, greater than $\tau_s^{-1}$ by a factor equal to $1/c_V$ (vacancies) or $1/c_I$ (interstitials). In other words, if $\omega_0$ is kept constant but the temperature is varied we would expect the quadrupole–e.f.g. contribution to be dominant at temperatures well beneath that where $T_1$ due to the dipole–dipole coupling is least. This is indeed observed (Fig. 12.9).

### *12.6.1 Theory*

The theory of these effects has a structure broadly similar to that of dipole–dipole relaxation. This structure is generally as set out in the original review by Cohen and Reif (1957). Subsequent developments of particular aspects by others will be referred to individually. We shall suppose that the density of defects giving rise to the field gradients is low. This is the case in most of the ionic crystals on which relevant measurements have been made.

(1) Firstly, there is the specification of the interaction Hamiltonian between a nucleus and the electric field gradient acting at it (cf. (1.7.2)). Matrix elements for transitions between the magnetic sub-levels of a nucleus of given total angular momentum quantum number $I$ can then be written down in terms of appropriate elements of the e.f.g. tensor.

(2) Secondly, the e.f.g. tensor is specified in terms of the defects responsible. This part presents a sizeable solid state problem, since not only must the source of the field be described, but one must determine the way the electron distribution around

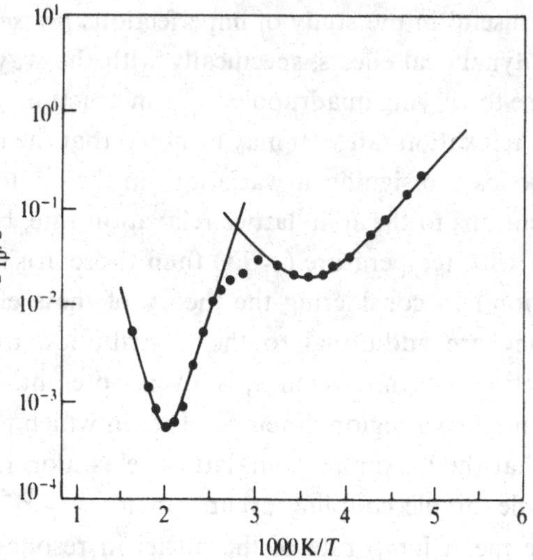


Fig. 12.9. The spin–lattice relaxation time, $T_1$, of Li ions in polycrystalline $Li_2O$ doped with LiF as a function of the inverse absolute temperature, $T^{-1}$. The high-temperature minimum (*a*) is associated with the dipole–dipole mechanism of relaxation and the low-temperature minimum (*b*) with the quadrupole–e.f.g. mechanism. (Courtesy S. M. Rageb and J. H. Strange, unpublished work.)

the nucleus in question reacts, for this reaction can lead to a considerable amplification of the bare field. This is especially so in the case of heavy atoms. For charged point defects in insulators (as in ionic crystals) the two relevant elements of the e.f.g. are

$$V^{(1)} = 3d_1\beta \sum_n \frac{q_n}{r_n^{\,3}} Y_{2,1}(\theta_n, \phi_n) \tag{12.6.1a}$$

and

$$V^{(2)} = 3d_2 \frac{\beta}{2} \sum_n \frac{q_n}{r_n^{\,3}} Y_{2,2}(\theta_n, \phi_n), \tag{12.6.1b}$$

in which the nucleus is assumed to be at the origin, $\mathbf{r}_n$ $(\equiv r_n, \theta_n, \phi_n)$ is the position vector of point defect $n$ bearing (effective) charge $q_n$. The polar axis is that defined by the static magnetic field $\mathbf{B}_0$. The factor $\beta$ represents the amplification of the bare e.f.g. by electronic polarization and distortion, while $d_1$ and $d_2$ are given by (12.2.4). The accurate calculation of $\beta$ is quite difficult, but appropriate methods have been developed notably by Schmidt, Sen, Das and Weiss (1980).

Point defects bearing no net charge can still give rise to electric field gradients by the inhomogeneous lattice strain they give rise to, and, if this strain field can be approximated by that generated by a point defect in an elastically isotropic solid, then the same forms (12.6.1) will apply, although of course, the coefficient $q_n$ is no longer an electric charge but is related to the elastic strength of the defect (i.e. to the trace of the elastic dipole strain tensor, $\lambda_{xx} + \lambda_{yy} + \lambda_{zz}$). When both sources are present the contribution of the strain to $V^{(1)}$ and $V^{(2)}$ is smaller than that of the electric charge, but may still be significant. This isotropic elastic continuum approximation has often been made (e.g. Becker, 1978), whereas atomistic calculations are much rarer.

The similarity of the summands in (12.6.1) to the $F_{ij}$ functions arising in the theory of dipole–dipole relaxation is accidental, but convenient for the subsequent analysis, as we show below.

(3) The third stage of the theory recognizes that the quadrupole–e.f.g. interaction varies stochastically with time as a result of the movement of the defects. One thus obtains average transition rates among the magnetic sub-levels (in particular for $\Delta m = \pm 1$ and $\Delta m = \pm 2$ transitions) in terms of the spectral functions of $V^{(1)}$ and $V^{(2)}$, i.e. in terms of quantities like the $J^{(1)}(\omega)$ and $J^{(2)}(\omega)$ defined by (1.7.5). Of course, to evaluate these spectral functions we have now to describe the motion of the point defects relative to the probe nuclei, not the relative motion of a pair of nuclei, as with dipole–dipole relaxation.

(4) Finally these transition rates are inserted into appropriate expressions giving the kinetics of the decay of the magnetization. These are obtained either from appropriate rate equations describing the occupation of the various magnetic sub-levels or by corresponding quantum mechanical analysis (density matrix method). Various time constants are found, according to the experimental set-up and the nature of the system (magnitude of $I$, etc.). The original analysis of Cohen and Reif (1957) has been developed subsequently by Andrew and Tunstall (1961), Hubbard (1970) and Becker (1982) among others. In general, there is a greater diversity of results here than is shown by the relations (1.7.6) for dipole–dipole relaxation. We shall not go into details since to do so would take us too far afield from our principal purpose. We confine our subsequent discussion therefore to the evaluation of the required spectral functions and the associated transition rates. The above references should be consulted for the way these transition rates enter into the expressions which describe the decay of the nuclear magnetization in various experiments. This decay is normally more complex than can be described by a single exponential function, because for $I > 1/2$ there are more than two magnetic sub-levels and the kinetic equations describing their population inevitably possess more than one eigenmode.

### 12.6.2 *Theoretical results: transition rates*

In general, if the interaction Hamiltonian $H^{(Q)}$ has matrix elements between magnetic sub-levels $m'$ and $m$, $H^{(Q)}_{m'm}$, then the average transition rate is

$$w_{m'm} = \frac{1}{\hbar^2} J^{(m'-m)}(\omega_{m'm}), \tag{12.6.2}$$

in which $J^{(p)}(\omega)$ $(p \equiv m' - m)$ is the spectral function of the time correlation function of the interaction, i.e.

$$J^{(p)}(\omega) = \int_{-\infty}^{\infty} \langle H^{(Q)}_{m'm}(t) H^{(Q)*}_{m'm}(t+s)\rangle \, \mathrm{e}^{-\mathrm{i}\omega s} \, \mathrm{d}s. \tag{12.6.3}$$

Analysis of the matrix elements of $H^{(Q)}$, as carried out by Cohen and Reif (1957, especially §6), then allows (12.6.2) to be rewritten in terms of the spectral functions of the $V^{(1)}$ and $V^{(2)}$. For the important case of $I = 3/2$ (exemplified by elements such $^7$Li, $^{23}$Na, $^{63}$Cu, $^{65}$Cu and the Cl and Br isotopes) the result is

$$w_1 \equiv w_{1/2,3/2} = \frac{1}{12}\left(\frac{eQ}{\hbar}\right)^2 J^{(1)}(\omega_0), \tag{12.6.4a}$$

$$w_2 = w_{-1/2,3/2} = \frac{1}{12}\left(\frac{eQ}{\hbar}\right)^2 J^{(2)}(2\omega_0), \tag{12.6.4b}$$

in which $eQ$ is the quadrupole moment parameter, $\hbar\omega_0$, as before, is the Zeeman splitting of the magnetic sub-levels, while $J^{(p)}(\omega)$ is the spectral function of $V^{(p)}$, i.e.

$$J^{(p)}(\omega) = \int_{-\infty}^{\infty} \langle V^{(p)}(t) V^{(p)*}(t+s)\rangle \, \mathrm{e}^{-\mathrm{i}\omega s} \, \mathrm{d}s. \tag{12.6.5}$$

When it is appropriate to use (12.6.1) for $V^{(p)}$ we can expect close similarities to the theory of dipole–dipole relaxation. In particular, when the defects contributing to $V^{(1)}$ and $V^{(2)}$ move independently of one another there will be no correlations between the contributions of different defects and, in place of (12.6.5), we get

$$J^{(1)}(\omega) = 9\beta^2 \sum_n q_n^2 \int_{-\infty}^{\infty} \langle F_n^{(1)*}(t+s) F_n^{(1)}(t)\rangle \, \mathrm{e}^{-\mathrm{i}\omega s} \, \mathrm{d}s, \tag{12.6.6a}$$

and

$$J^{(2)}(\omega) = \frac{9\beta^2}{4} \sum_n q_n^2 \int_{-\infty}^{\infty} \langle F_n^{(2)*}(t+s) F_n^{(2)}(t)\rangle \, \mathrm{e}^{-\mathrm{i}\omega s} \, \mathrm{d}s, \tag{12.6.6b}$$

where, by analogy with the notation for dipole–dipole coupling, we have introduced

$$F_n^{(p)} = \frac{d_p}{r_n^3} Y_{2,p}(\theta_n, \phi_n). \tag{12.6.7}$$

Three significant observations can now be made. Firstly, it is obvious that each defect $n$ of the same type ($i$, say) has the same charge and the same correlation function. The sum over $n$ in (12.6.6) is therefore replaced by the product of $N_i$ (the number of defects of type $i$) and a sum over $i$, whence by (12.6.4) and (12.6.6)

$$w_1 = \frac{3}{4}\left(\frac{e\beta Q}{\hbar}\right)^2 \sum_i N_i J_i^{(1)}(\omega_0), \tag{12.6.8a}$$

$$w_2 = \frac{3}{16}\left(\frac{e\beta Q}{\hbar}\right)^2 \sum_i N_i J_i^{(2)}(2\omega_0). \tag{12.6.8b}$$

Secondly, the probe nuclei can be assumed to be effectively immobile in comparison with the point defects. Thus, even when they may themselves be moved by the defects, the average frequency with which this happens is less than the frequency of defect movement by a factor roughly equal to the defect fraction ($\ll 1$). This has the important consequence that the spectral functions in (12.6.6) require the evaluation of just single-particle correlation functions, $J_i^{(p)}(\omega)$. Lastly, the equivalence of (12.6.7) to the $F_{ij}$ in previous sections means that we already have effectively all the mathematical results we need. We shall show how to use the previous results by considering two important classes of system.

### *12.6.3 Extension of encounter model to doped crystals*

Various experiments have been carried out on ionic crystals, particularly halides and bromides, in which the effects of quadrupole couplings have been detected. Some of the earliest of these were made on AgBr by Reif (1955). Related experiments on various other bromides as well as AgBr followed later (Becker and Richtering, 1974; Becker, Hamann, Kozubek and Richtering, 1975). The Br nuclei ($Br^{79}$ and $Br^{81}$) were used as the probe nuclei and the relevant defects were the defects in the cation sub-lattice (e.g. $Ag^+$ Frenkel defects in AgBr). The density of such defects was varied by doping the crystals with aliovalent cations (e.g. $Cd^{2+}$ in AgBr) in the manner analysed in Chapter 3. If we can neglect any pairing between oppositely charged defects in these systems then the independence of the complementary defects (e.g. $Ag^+$ vacancies and $Ag^+$ interstitials) follows and we can use eqns. (12.6.8). Furthermore the defects each execute an uncorrelated random-walk. But $J^{(p)}$ for an uncorrelated random walk is precisely what was evaluated in §12.3.2 (although there it was considered as an approximation to the spectral function for the two-particle correlation function required for relaxation by the dipole–dipole coupling). We can therefore take over the calculations and results described there. In particular, if variations with crystal orientation are not of concern (as for example in experiments on polycrystals) the Torrey isotropic formula (12.3.17) provides a good approximation to the exact result. Even the B.P.P. single-relaxation-time model provides a fairly accurate description (eqns. (12.3.4) and (12.3.5)) together with the advantage of analytical simplicity. In any case, these results show that the maximum in $J_i^{(p)}(\omega)$ occurs when $\omega\tau_i \sim 1$, in which $\tau_i$ is now to be interpreted as the mean time between jumps of defects of type $i$. This confirms our earlier expectation that the observation of relaxation via the quadrupolar–e.f.g. coupling can yield direct information about defect movements. In analysing such data there is one interesting contrast with dipole–dipole relaxation, which is immediately visible if we refer to the B.P.P. eqns. (12.3.4) and (12.3.5). From dipole–dipole relaxation we get information about $\tau_s$, i.e. about the quotient $(\tau_i/c_i)$ of the defect time of stay $\tau_i$ and the defect fraction $c_i$ $(=c_V$ or $c_I)$. From (12.6.8) and (12.3.5) we see that we can in principle also get information about the product $c_i\tau_i$ from the region where $\omega_0\tau_i \ll 1$. Analyses of a variety of ionic systems both doped and pure have, in fact, been made by means of the above principles and results.

That these methods have been applied to doped crystals raises one last point of principle. Much of our previous theory (e.g. Chapters 7, 8 and 11) has been concerned with the existence and consequences of the pairing of solute atoms (or ions) and defects. In doped crystals one certainly expects such pairing to occur and this must immediately raise the question of the influence of the movement of the solute ions upon the relaxation of the probe nuclei. In turn this also suggests that the theory we have described may be complicated by the occurrence of correlations in the motions of the solute ions and the vacancies. In fact, in most practical circumstances both these concerns can be met by a suitable extension of the encounter model.

In the preceding secction, §12.5, we were concerned with encounters between a defect and a particular host atom or pair of host atoms, but as we saw earlier in Chapter 8 the idea is of wider application. The essential restrictions for our present purposes are (i) that the average duration of an encounter should be much less than the time, $\tau_e$, between encounters and (ii) that the period of the radiation, $\omega_0^{-1}$, should be much greater than the duration of an encounter. Both conditions may still be satisfied in the presence of a binding between the solute atoms and vacancies, but clearly the stronger the binding the longer the lifetime of the pair and thus the more stringent do these conditions become. Nevertheless, given that these two essential conditions can be met, we can adapt the results already obtained. To do so we need the analogues of $\nu$ (the mean number of jumps made by an atom in an encounter) and $\varpi(\mathbf{r}_l)$ the probability that an atom is displaced by the lattice vector $\mathbf{r}_l$ as the result of an encounter. These quantities, of course, depend upon the way the solute atom perturbs the movements of the vacancy. To be

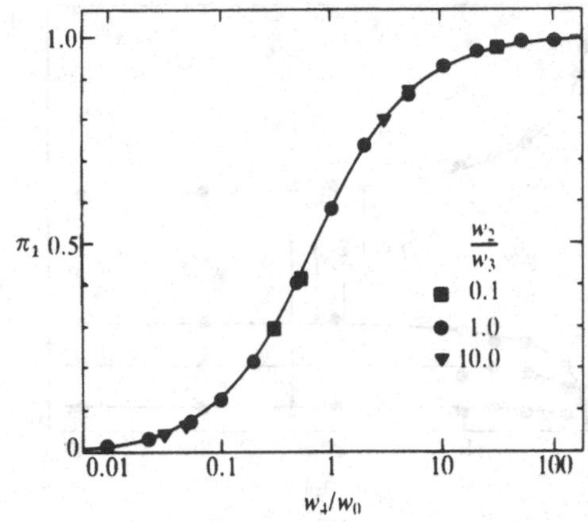


Fig. 12.10. For the five-frequency model of an f.c.c. lattice: the probability $\pi_1$ that a vacancy which leaves the first coordination shell around a solute atom will subsequently return to it, shown as a function of $w_4/w_0$ ($\pi_1$ does not depend on the other frequencies). The solid points were obtained from Monte Carlo simulations and the line was obtained from an approximate, though quite accurate, analytical expression. (After Becker, 1984.)

specific, let us take the five-frequency model of f.c.c. lattices and consider $v$ first. By adapting the argument given at the end of §9.5 we see that

$$v = 1 + \left(\frac{w_2}{7w_3}\right)\frac{1}{(1-\pi_1)}. \tag{12.6.9}$$

Here $(7w_3)^{-1}$ is the average lifetime of a solute–vacancy pair and $w_2$ as before is the average rate at which the solute atom jumps into an adjoining vacancy. The quantity $\pi_1$ is the probability that a vacancy returns to one of the first-neighbour positions (of the solute) once it has left it. It will be a function solely of $(w_4/w_0)$ in this model and may be obtained either by using the method of §9.4.1 or by purely numerical Monte Carlo computations. Fig. 12.10 shows this quantity as a function of $(w_4/w_0)$ as obtained by such means by Becker (1984). The form (12.6.9) is the result of adding up the probabilities of not returning, of returning once, twice and so on.

The second quantity $\varpi(\mathbf{r}_l)$ we require, namely the probability of displacement of the solute in an encounter, has been obtained by both Monte Carlo simulations (Becker, 1978) and by methods based on the principles established in Chapter 9 (Sholl, 1992). Becker's Monte Carlo simulations were for a dilute version of the random alloy model (i.e. all solvent jump frequencies equal, $w_1 = w_3 = w_4 = w_0$). Some of his results are shown in Fig. 12.11. Sholl's analysis is for the full five-frequency model (f.c.c.): his results agree with Becker's when the appropriate simplification is made. As we shall see shortly, the quantity $1 - \varpi(0)$, i.e. the

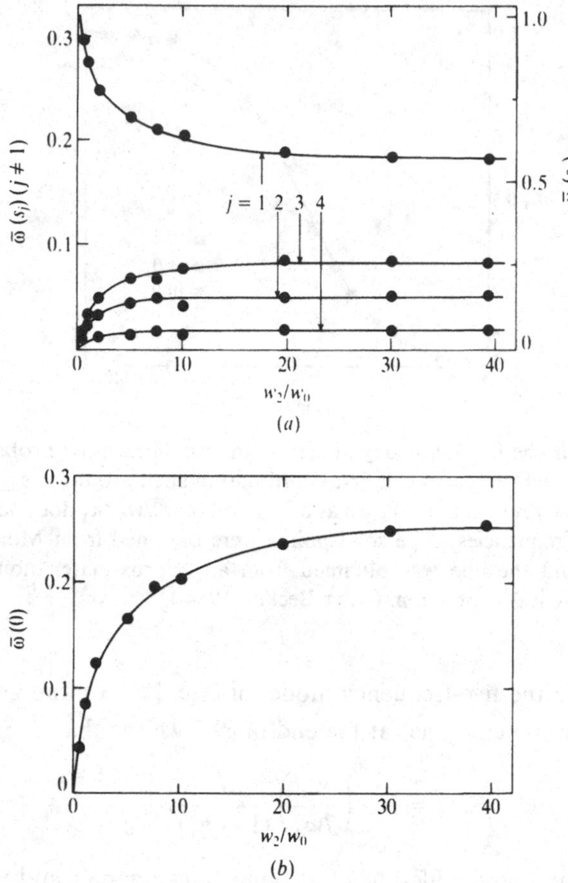


Fig. 12.11. For the two-frequency random alloy model of a f.c.c. lattice: (*a*) the probability $\varpi(s_j)$ that a solute atom, as the result of its encounter with a vacancy, is moved from its original position to one of the shells of neighbours ($j = 1, 2, 3, \ldots$) surrounding that position, shown as a function of $w_2/w_0$. Also shown (*b*) is the probability $\varpi(0)$ that no net displacement results from the encounter. These results were obtained from Monte Carlo simulations. (After Becker, 1978.)

probability that the encounter does result in a non-zero displacement, is of practical importance.

Having now obtained these quantities $\nu$ and $\varpi(\mathbf{r}_l)$, we are in a position to calculate the necessary spectral functions. The encounter idea greatly simplifies the task because it allows us to treat the motions of the vacancies and the solute atoms as independent of one another. Their movements are only correlated during the relatively short periods of the encounters. We can therefore adapt the methods described in §12.3. In particular, if we are not concerned with orientational effects (which will in any case be smaller here) we can use the approach of Torrey's isotropic model. Since the vacancies move by direct nearest-neighbour jumps we can use Torrey's results (or the more precise results of Sholl) as they stand to get the corresponding partial spectral function. However, for the solute atoms we must allow for all those displacements for which $\varpi(\mathbf{r}_l)$ is appreciably different from zero, whereas by (12.3.14) and (12.3.15) we only included direct nearest-neighbour hops. The result is that the partial spectral function contributed by the solute atoms (eqns. (12.6.8)) is

$$J^{(p)}(\omega) = \frac{3d_p{}^2\tau_e}{2\pi}\left(\sum_l \frac{1}{r_l{}^6}\right)G'(k, \omega\tau), \tag{12.6.10}$$

with

$$G'(k, \omega\tau) = \int_0^\infty \frac{[J_{3/2}(ksq)]^2[1 - A(q)]}{[(1 - A(q))^2 + \omega^2\tau_e{}^2]}\frac{\mathrm{d}q}{q}. \tag{12.6.11}$$

This integral $G'$ is like $G$ in (12.3.18) except that in place of $\sin(sq)/(sq)$, which applies when only nearest-neighbour displacements of magnitude $s$ take place, we now have $A(q)$, which is a sum over all displacements that can occur as the result of an encounter, i.e.

$$A(q) = \sum_{s_j > 0} \varpi(s_j)\frac{\sin qs_j}{qs_j}. \tag{12.6.12}$$

In these expressions (12.6.10) and (12.6.11), $\tau_e$ is the mean time between encounters leading to a non-zero displacement (which is why there is an apparent difference of a factor of 2 with (12.3.17), $\tau_s$ there being twice the time between encounters with a pair of spins). This adaptation of Torrey's integral was first suggested by Wolf (1971) and used in the way described above for solute atoms by Becker (1978).

Some of Becker's results for $J^{(1)}(\omega)$ for the dilute random alloy model are shown in Fig. 12.12. The broad conclusions that one can draw from these are like those which we drew for the dipole–dipole coupling. Firstly, it will be observed from Fig. 12.12 that the shape of the curve of $J^{(1)}$ as a function of $\omega\tau_s$ is very insensitive to the ratio of jump frequencies $w_2/w_0$. In fact, a rigid shift along the $\omega\tau_s$ axis brings them virtually into coincidence. This result can be rationalized in terms of correlation effects in the same way as the related results for relaxation via the dipole–dipole coupling. In brief, this amounts to saying that the maximum in the spectral function occurs not when $\omega = \tau_s{}^{-1}$ (i.e. not when $\omega$ equals the average frequency of solute atom jumps) but when $\omega = \tau_e{}^{-1}$ with $\tau_e$ related to $\tau_s$ as

$$\tau_e = \tau_s\nu/(1 - \varpi(0)). \tag{12.6.13}$$

The expression $(1 - \varpi(0))/\nu$ may be referred to as a correlation factor for nuclear magnetic relaxation, since in effect it reduces the average jump frequency of the

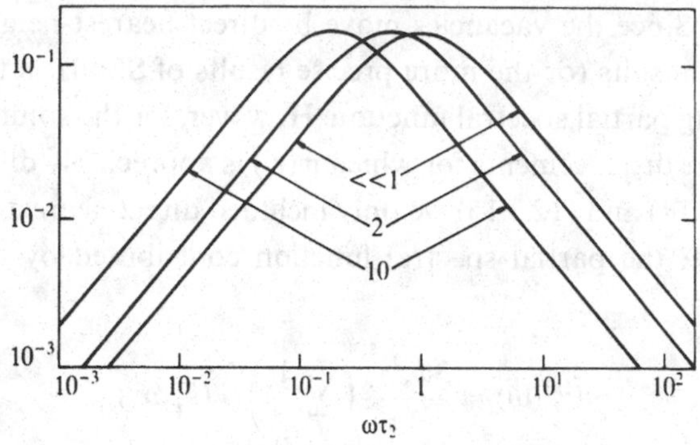


Fig. 12.12. For the two-frequency random-alloy model of an f.c.c. lattice: the spectral function $J^{(1)}(\omega)$ for the interaction between the nuclear quadrupoles and electric field gradients arising from aliovalent solute ions, as predicted by (12.6.10) and (12.6.11), shown as a function of $\omega\tau_2$ for three values of $w_2/w_0$ as indicated. Here $\tau_2$ is the overall mean time between jumps of the solute ions, i.e. the reciprocal of the product of $w_2$ with the probability that there is a vacancy in one of the nearest neighbour positions. (After Becker, 1978.)

solute atoms, $\tau_s^{-1}$, to an effective jump frequency, $\tau_e^{-1}$, which determines the frequency $\omega$ at which the spectral function is a maximum.

## 12.7 Summary

In this chapter we have been concerned with the influence of atomic migration upon nuclear magnetic relaxation rates. There are many other aspects to the theory of nuclear magnetic relaxation, but here we have concentrated on those which relate to diffusive atomic movements. To this end we have made several important simplifications, namely (i) only three-dimensional cubic lattices have been considered (ii) only a single type of spin has been assumed and (iii) little attention has been paid to the effects of crystal orientation (but see §1.7). The key to the representation of these motional effects is the calculation of appropriate spectral functions $J^{(p)}(\omega)$, i.e. the Fourier transforms of the time correlation function of the appropriate interaction. It is these spectral functions which have been the principal subject of the chapter. Two important interactions have been considered (i) the nuclear magnetic dipole–dipole interaction and (ii) the interaction of nuclear electric quadrupoles with internal electric field gradients. The first of these is always present ($I > 0$), while the second is present when $I \geq 1$.

We have considered the calculation of the spectral functions for the dipole–dipole coupling at some length in §§12.2–12.5. One of the rather surprising conclusions of this study is that for three-dimensional systems the shape of the

$J^{(p)}(\omega)$ functions is insensitive to details of the system considered (lattice structure, nature and concentration of defects involved, etc.). In fact, to a good first approximation one can regard $J^{(p)}(\omega)/J^{(p)}_{max}$ v. $\omega/\omega_{max}$ as a universal curve, which is one of the reasons why the simple form for $J^{(p)}(\omega)$ first derived by Bloembergen, Purcell and Pound (§12.3.1) continues in vogue. Furthermore, treatment of the same system in different approximations leads to shifts in $\omega_{max}$, but with almost no change in $J^{(p)}_{max}$. This fact allows one to define a correlation factor by comparing the $\omega_{max}$ predicted by a treatment including correlation effects with that predicted by neglecting them. For low defect concentrations the encounter model provides a way to calculate this factor – which turns out to be smaller than the corresponding correlation factor for tracer diffusion (Chapter 10).

The second important interaction (i.e. the quadrupole–e.f.g. coupling) was considered in §12.6. In perfect cubic crystals there can be no internal electric field gradients so that any effects seen can arise only from imperfections. This has the important consequence that motional effects relate directly to the movements of vacancies, interstitials and solute atoms, whereas relaxation via the dipole–dipole coupling relates to the average movements of all the nuclei. The maximum rates of relaxation by these two mechanisms thus occur at quite different frequencies, other things being equal. This makes it possible to separate the two contributions experimentally. A further convenience occurs with electrically charged point defects in insulators, for then the characteristics of the e.f.g. tensor are such that the form of the spectral function $J^{(p)}(\omega)$ is the same as for the dipole–dipole coupling. This is also approximately true of point defects in other circumstances too. Where the defects concerned are foreign ions it is possible to extend the encounter model and to use this to obtain corresponding n.m.r. correlation factors.

In view of the need to supply much of the context in which these spectral functions are used we have had to curtail the discussions in several important ways. Some of the restrictions are easily removed, others less so. Perhaps the most obvious is the limitation to systems with a single type of spin. Nuclear magnetic relaxation in a compound often involves at least two types of spin. If these lie on different sub-lattices (e.g. cation and anion sub-lattices) the motions of a pair of unlike spins may not be correlated as those of a pair of like spins are: this simplifies the task of calculating the corresponding spectral function. On the other hand, to remove the limitation to a low concentration of defects is more difficult. Important examples of this type are provided by the fast-ion conductors where qualitative differences from B.P.P. behaviour have been observed (Funke, 1989 and Heitjans *et al.*, 1991). Other examples are provided by the (interstitial) solid solutions of hydrogen in metals, in which although the H atoms are present in large numbers they by no means fill all the available interstitial sites. To go beyond the mean-field approximation in these cases demands new methods. (The description of such highly defective systems is the subject of the next chapter.) Lastly, we have not had space to review the findings on particular systems. For this we must refer the reader to individual reviews on particular classes of solid, e.g. (i) Stokes (1984) – metals, (ii) Cotts (1978) – metal–hydrogen systems, (iii) Chezeau and Strange (1979) – molecular solids, (iv) Chadwick (1988) – ionic and molecular solids and (v) Heitjans *et al.* (1991) – glassy and layered compounds.

# Appendix 12.1

## The spherical harmonic functions

The spherical harmonic functions $Y_{l,m}(\theta, \phi)$ are important in atomic theory and in other centro-symmetric problems and possess a number of convenient and well-understood mathematical properties. They are defined as follows

$$Y_{l,m}(\theta, \phi) = \frac{(-1)^{\frac{1}{2}(m+|m|)}}{\sqrt{2\pi}} \left\{ \frac{(2l+1)(l-|m|)!}{2(l+|m|)!} \right\}^{1/2} P_l^{|m|}(\cos\theta)\, e^{im\phi} \quad \text{(A12.1.1)}$$

in which $l$ is a positive integer or zero and $m$ is an integer in the range $-l$ to $l$. (These symbols are those conventionally used in atomic theory and, of course, are quite distinct from our use of **l**, **m**, etc. to denote lattice points.) The factor $P_l^{|m|}(\cos\theta)$ is the associated Legendre function obtained from the Legendre polynomial $P_l(\cos\theta)$ according to the definition

$$P_l^{|m|}(z) = (1 - z^2)^{|m|/2} \frac{d^{|m|}}{dz^{|m|}} P_l(z). \quad \text{(A12.1.2)}$$

The Legendre polynomials are defined in the usual way from the generating function $(1 - 2tz + z^2)^{-1/2}$, i.e.

$$\sum_{l=0}^{\infty} P_l(z) t^l = (1 - 2tz + z^2)^{-1/2}. \quad \text{(A12.1.3)}$$

The first few Legendre polynomials are thus easily seen to be

$$P_0(z) = 1, \quad \text{(A12.1.4a)}$$

$$P_1(z) = z, \quad \text{(A12.1.4b)}$$

$$P_2(z) = \tfrac{1}{2}(3z^2 - 1), \quad \text{(A12.1.4c)}$$

$$P_3(z) = \tfrac{1}{2}(5z^3 - 3z). \quad \text{(A12.1.4d)}$$

Then from (A12.1.2) we see that

$$P_0{}^0(z) = 1, \tag{A12.1.5}$$

$$P_1{}^0(z) = z, \tag{A12.1.6a}$$

$$P_1{}^1(z) = (1 - z^2)^{1/2}, \tag{A12.1.6b}$$

$$P_2{}^0(z) = \tfrac{1}{2}(3z^2 - 1), \tag{A12.1.7a}$$

$$P_2{}^1(z) = 3(1 - z^2)^{1/2}z, \tag{A12.1.7b}$$

and

$$P_2{}^2(z) = 3(1 - z^2). \tag{A12.1.7c}$$

The spherical harmonics as defined by (A12.1.1) form an orthonormal set of functions, i.e.

$$\int_0^\pi \int_0^{2\pi} Y^*_{l,m}(\theta, \phi) Y_{l',m'}(\theta, \phi) \sin\theta \, d\theta \, d\phi = \delta_{ll'}\delta_{mm'}. \tag{A12.1.8}$$

Lastly, this set is complete in the sense that any function of $\theta$, $\phi$ can be expanded linearly in terms of them, i.e.

$$\psi(\theta, \phi) = \sum_{l=0}^{\infty} \sum_{m=-l}^{l} c_{l,m} Y_{l,m}(\theta, \phi) \tag{A12.1.9}$$

where the $c_{l,m}$ are constant coefficients.

# 13

# *Theories of concentrated and highly defective systems*

## 13.1 Introduction

In the preceding Chapters 6–12 we have dealt with the general structure of the various statistical theories of atomic transport in solids and with the relations between them. However, in obtaining analytical results for specific classes of model the emphasis has been mostly on dilute alloys and solid solutions containing only low concentrations of defects. The theory of such systems is made easier because we need to retain only the lower-order terms in defect and solute concentrations. In this chapter we turn to theories of so-called *lattice-gas* models which give physically simplified representations of both concentrated alloys and systems containing high concentrations of defects. In such systems the concentrations of the components are not useful expansion parameters. As a result, and as is usual in statistical mechanics, an accurate theory is much more difficult to formulate in these circumstances.

The rather broad range of systems of current interest includes (i) ideal or random alloys and solid solutions (such as were discussed in §§5.6 and 10.8), (ii) alloys displaying short-range or long-range order and (iii) various systems where the number of mobile atoms is significantly less than the number of sites available to them. In this third group these sites are commonly interstitial sites within a structure formed by relatively immobile atoms. Examples are provided by the $\beta$-aluminas and other fast ion conductors (Laskar and Chandra, 1989), metals containing interstitial hydrogen or deuterium atoms and various other interstitial solid solutions (e.g. the tungsten 'bronzes'; Cox, 1992). We shall refer to the sublattice of interstitial sites in these systems simply as the lattice.

The models of these systems which have been studied theoretically are almost all members of the class of lattice-gas models defined in Chapter 3, i.e. ones for which the (Gibbs) energy of a given configuration of atoms on the available lattice sites is determined by the numbers of the chemically distinct pairs of atoms in that configuration (e.g. by the numbers of AA, AB, BB pairs in a solid solution of A and B). Mostly, the energy is assumed to be defined by the numbers of nearest-neighbour pairs alone, but sometimes models with interactions of longer range are considered. Additional assumptions must be made to define the atomic jump frequencies.

The simplest model from this class which remains of practical interest is the random lattice-gas model of a multicomponent system with an arbitrary vacancy concentration. The model is defined by assuming (i) that the vacancies and atoms mix randomly, i.e. that there are no pair terms in the energy, and (ii) that the rates of atom–vacancy exchange are determined solely by the nature of the jumping atom and are independent of the detailed configuration adopted by nearby atoms. In the limit of a very small vacancy content this is the random alloy model for which the theoretical results of Manning (1968, 1970, 1971) have already been discussed in §§5.6 and 10.8. A number of successful studies of the random lattice-gas model have been made by different theoretical methods and the insights gained have been useful in guiding the formulation of theories of more complex models in which the atomic and defect distributions are not random and where the jump frequencies depend on the local environment of the jumping atoms. At present, however, the theories for these more complex models sometimes contain *ad hoc* elements and no widely accepted methods are available other than numerical simulations by Monte Carlo methods.

With these considerations in mind we begin (§13.2) with the theory of the random lattice-gas model. We show that several useful approximate results, including those of Manning for the random alloy, follow from the general linear response expressions for the phenomenological coefficients when we use the particular statistical mechanical techniques introduced into this field by Tahir-Kheli and Elliott (1983) and Holdsworth and Elliott (1986). We then turn in §13.3 to non-random systems. We first discuss expressions for the jump frequency which depend upon the local environment of the jumping atoms and then illustrate some characteristic features of the phenomenological coefficients in alloys with short-range or long-range order as found by Monte Carlo simulations.

The extension of the analytical methods introduced in §13.2 to these systems is still a matter of current research and it would therefore be premature to address this topic here. The path probability method of Kikuchi (e.g. Sato, 1984) has been widely applied to the models of interest in this chapter, but it has not yet been related to the theoretical structure developed here.

Monte Carlo simulations thus play a central role in understanding the transport properties of the systems of interest in this chapter. This is true not only for those in which there are significant interatomic interactions, but also for random alloys and solid solutions where they also provide important tests of the analytical theories. There are a number of ways in which Monte Carlo simulations can be used for these transport problems and we therefore provide a review of the techniques in Appendix 13.1. To appreciate the nature of these simulations and the information they provide, several general features should be recognized. Firstly, they set out to calculate transport coefficients either by evaluating the expression (6.5.14) in thermal equilibrium or by evaluating flows in the presence of a force. In both cases, by the nature of the underlying equations, one has to obtain an average quantity per unit of real time $t$ (cf. (6.5.14), (10.2.4) and (10.2.5)). Yet in the computer simulation one is necessarily confined to sequences of atomic jumps, and these sequences have to be related separately to the real time in which they are supposed to occur. The link is, of course, the overall jump frequency. As is already clear from previous considerations, this is made up, broadly speaking, of two factors, the defect concentration and the conditional atomic jump frequency (i.e. the jump frequency given that there is a defect in place to allow or effect it). Both have to be independently specified. The conditional jump frequency will either be given as a parameter (as in the random alloy model where it depends only on the nature of the atom which jumps) or it will be prescribed in terms of the local configuration of atoms (as in the dilute alloy models analysed in Chapter 11 and as in §13.3 below). As to the defect concentration, it is impracticable to determine this in these simulations in any case where it is low, simply because the simulation contains too few atoms. As in the analytic theories, defect concentrations must be independently determined by the methods of Chapter 3.

It follows immediately therefore that on account of these limitations Monte Carlo simulations cannot alone provide absolute transport coefficients, but only reduced or relative quantities. Correlation factors and correlation functions are, however, obtained absolutely; they therefore figure prominently among the results.

## 13.2 The random lattice-gas model

We have already introduced this model in §§8.6 and 10.8. Here we shall first (§13.2.1) show how to evaluate the linear-response expressions for the phenomenological coefficients, $L$, of this model by exploiting the simplifications that arise because each jump frequency is independent of the environment of the jumping atom. We do so by starting from (6.5.10)–(6.5.12) and then developing the kinetic equation for a certain time-correlation function related to $\Lambda_{AB}(t)$. This kinetic equation then appears as a member of a coupled hierarchy of kinetic equations. In §13.2.2 we employ a physical argument to suggest a decoupling approximation, which is subsequently shown to lead to Manning's theory of the random alloy. We then describe in §§13.2.3–13.2.5 variants of this procedure which lead to useful approximate results for the random lattice gas with an arbitrary vacancy concentration.

### 13.2.1 *Kinetic equations for the (cubic) random lattice-gas model*

To evaluate the phenomenological coefficients $L_{ij}$ we need to interpret the general many-body expressions (6.5.10)–(6.5.12) in terms of the particular features of the random lattice gas. These are several.

(i) The lattice in question is assumed to be cubic so that all lattice sites and all nearest-neighbour displacement vectors, **s**, are crystallographically equivalent. This feature has the necessary consequence that the tensor $L_{ij}$ is diagonal in the Cartesian components with $L_{ijxx} = L_{ijyy} = L_{ijzz}$ and can thus be treated as a scalar. We shall use this fact in evaluating the correlated part of $L_{ij}$.

(ii) Since the mixture of atoms and vacancies is thermodynamically *ideal*, all configurations $\alpha, \beta, \ldots$ have the same energy and thus $p_{\alpha}^{(0)}$ is the same for all $\alpha$.

(iii) The transition $w_{\beta\alpha}^{(0)}$ is non-zero only for transitions $\alpha \to \beta$ in which just one atom exchanges places with a vacancy in a nearest-neighbour position: no other transitions occur.

(iv) Such transitions occur at a rate which depends only on the chemical nature of the atom which moves, and which is independent of the arrangement of atoms in the vicinity.

These features allow us to carry out the summations in (6.5.11) and (6.5.12) over atoms $m$ and $n$ and over two of the state variables. Firstly, it is easily seen that the uncorrelated part of $L_{ij}$, as given by (6.5.11), is simply

$$L_{ij}^{(0)} = \delta_{ij} w_i s^2 \left( \sum_{\mathbf{l},\mathbf{s}} \sum_{\alpha} \rho_{\alpha}^{(\mathrm{v})}(\mathbf{l}) \rho_{\alpha}^{(i)}(\mathbf{l} - \mathbf{s}) p_{\alpha}^{(0)} \right) \Big/ 6VkT, \tag{13.2.1}$$

in which $w_i$ is the jump frequency for atoms of species $i$, $s$ the jump distance and **s** a nearest-neighbour vector. We have employed occupancy variables for vacancies and atoms of species $i$ such that that, for example, $\rho_{\alpha}^{(i)}(\mathbf{l})$ is unity when there is an $i$-atom at site **l** and is zero otherwise. The quantity in brackets is evidently just the number of lattice sites multiplied by $zc_{\mathrm{V}}c_i$ where $z$ is the coordination number of the lattice and $c_{\mathrm{V}}$ and $c_i$ the fractions of sites occupied by vacancies and by atoms of species $i$. The uncorrelated part of $L_{ij}$ is therefore

$$L_{ij}^{(0)} = \delta_{ij} n z c_{\mathrm{V}} c_i w_i s^2 / 6kT, \tag{13.2.2}$$

in which $n$ is the number of lattice sites per unit volume.

The general expression (6.5.12) for the correlated part, $L_{ij}^{(1)}$, may also be simplified by making use of these features. We first employ the principle of detailed balance (6.2.7) to write it in the form

$$L_{ij}^{(1)} = -\frac{1}{3VkT}\sum_{\alpha,\beta,\gamma,\delta}\sum_{m,n}(\mathbf{r}_{\delta\gamma}(i_m)\cdot w_{\delta\gamma})\bar{G}_{\gamma\beta}(0)\,p_{\beta}^{(0)}(\mathbf{r}_{\alpha\beta}(j_n)w_{\alpha\beta}). \qquad (13.2.3)$$

Here we have again employed the notation

$$\bar{f}(\lambda) = \int_0^{\infty} \mathrm{d}t\, \exp(-\lambda t) f(t) \qquad (13.2.4)$$

for the Laplace transform of a function $f(t)$ of the time $t$ (in this case the propagator $\mathbf{G}(t)$). It is also to be understood that the product of $\mathbf{r}_{\delta\gamma}$ with $\mathbf{r}_{\alpha\beta}$ in (13.2.3) is to be the scalar product as indicated by the dot following $\mathbf{r}_{\delta\gamma}$. We next interpret the sequence of transitions $\alpha \ldots \delta$ as sequences which begin ($\alpha \to \beta$) with the exchange of a vacancy at site $\mathbf{m}_0$ with a $j$ atom at site $\mathbf{l}_0$ and end ($\gamma \to \delta$) at time $t$ with the exchange of a vacancy at site $\mathbf{l}$ with an $i$ atom at site $\mathbf{m}$. Then to evaluate the summations in (13.2.3) it is convenient to introduce the joint probability of observing the states *after* the first jump and *before* the final jump. This, in the many-body notation of Chapter 6, is

$$\psi_{\mathrm{V}i:\mathrm{V}j}(\mathbf{l}, \mathbf{m}; t{:}\, \mathbf{l}_0, \mathbf{m}_0) = \sum_{\gamma,\beta} \rho_{\gamma}^{(\mathrm{V})}(\mathbf{l})\,\rho_{\gamma}^{(i)}(\mathbf{m})G_{\gamma\beta}(t)\,\rho_{\beta}^{(\mathrm{V})}(\mathbf{l}_0)\,\rho_{\beta}^{(j)}(\mathbf{m}_0)\,p_{\beta}^{(0)}. \qquad (13.2.5)$$

The contribution of the sequence of states considered to $L_{ij}^{(1)}$, from (13.2.3) is just this probability multiplied by the product of $w_i w_j/3VkT$ with the scalar product of the jump vectors of the initial and final jumps (cf. the assumption that the lattice is cubic). When this expression is summed over all sites $\mathbf{l}$ and $\mathbf{m}$ which are nearest neighbours and over all sites $\mathbf{l}_0$ and $\mathbf{m}_0$ which are nearest neighbours and finally integrated over all time $t \geq 0$ then one obtains $L_{ij}^{(1)}$. By invoking the translational invariance of a system at equilibrium we finally obtain

$$L_{ij}^{(1)} = -(nzw_i w_j s^2/3kT)\left[\sum_{\mathbf{s}} \mathbf{s}\cdot\mathbf{s}_0/s^2 \sum_{l}\int_0^{\infty} \mathrm{d}t\psi_{\mathrm{V}i:\mathrm{V}j}(\mathbf{l}, \mathbf{l}-\mathbf{s}; t{:}\, \mathbf{l}_0, \mathbf{l}_0-\mathbf{s}_0)\right]. \qquad (13.2.6)$$

Here $-\mathbf{s}_0$ and $\mathbf{s}$ are the jump vectors of the initial $i$-atom jump and of the final $j$-atom jump respectively.

Throughout this chapter it will be convenient to express the $L$-coefficients as ratios of their uncorrelated parts $L^{(0)}$, which we shall call *correlation functions*, defined as follows:

$$f_{ii} = L_{ii}/L_{ii}^{(0)}, \tag{13.2.7a}$$

$$f_{ij}^{(i)} = L_{ij}/L_{ii}^{(0)} \qquad (i \neq j) \tag{13.2.7b}$$

and

$$f_{ij}^{(j)} = L_{ij}/L_{jj}^{(0)} \qquad (i \neq j). \tag{13.2.7c}$$

We note that, although $L_{ij} = L_{ji}$, $f_{ij}^{(i)} \neq f_{ij}^{(j)}$. These definitions are obviously of the same general type as that of the Bardeen–Herring tracer correlation factor (cf. (10.3.3)). This aspect is sometimes over-emphasized in the literature where, for example, $f_{AA}$ for a pure substance A may be called the 'physical', or 'electrical conductivity', or 'charge' correlation factor. The two kinds of quantity are, however, physically quite distinct, since the tracer correlation factor relates only to self-correlations in the motion of individual atoms whereas the correlation functions contain additional correlations between pairs of atoms (cf. §6.5).

From the preceding equations we then readily find for the random lattice gas

$$f_{ij}^{(j)} = \delta_{ij} - 2w_i h_{Vi:Vj}/c_V c_j, \tag{13.2.8}$$

where

$$h_{Vi:Vj} = \sum_{\mathbf{s}} \mathbf{s}\cdot\mathbf{s}_0/s^2 \sum_{\mathbf{l}} \int_0^\infty dt \psi_{Vi:Vj}(\mathbf{l}, \mathbf{l}-\mathbf{s}; t: \mathbf{l}_0, \mathbf{l}_0 - \mathbf{s}_0). \tag{13.2.9}$$

Eqn. (13.2.8) applies both when $i = j$ and when $i \neq j$ (but when $i = j$ the superscript is superfluous and may be dropped in accordance with (13.2.7a)).

After these preliminaries our next step is to obtain an equation of motion for the function $\psi$ defined by (13.2.5). The solutions to this equation can then be used to obtain the correlation functions via (13.2.8) and (13.2.9). This seemingly roundabout procedure proves, in fact, to be very fertile, essentially because equations of motion may allow one to see workable approximations which would not be apparent otherwise. At the same time the development of such an approach for the random lattice gas may point the way to the generation of kinetic theories for interacting systems.

We could obtain the kinetic equation for $\psi$ formally by differentiating (13.2.5) with respect to time and using the definition (6.3.1) of $\mathbf{G}(t)$, but the same result is also obtained by simply enumerating the processes which can change the occupancy of sites $\mathbf{l}$ and $\mathbf{m}$ at time $t$ by atom–vacancy exchanges. In writing the equation of motion we can, for the time being, suppress all the labels on the correlation functions $\psi$ which refer to the initial state – because they will be the same everywhere: so that, for example, we write $\psi_{Vi}(\mathbf{l}, \mathbf{m})$ for $\psi_{Vi:Vj}(\mathbf{l}, \mathbf{m}; t: \mathbf{l}_0, \mathbf{m}_0)$.

The equation of motion is then

$$\frac{d\psi_{Vi}(\mathbf{l},\mathbf{m})}{dt} = w_i\theta(\mathbf{l}-\mathbf{m})[\psi_{Vi}(\mathbf{m},\mathbf{l}) - \psi_{Vi}(\mathbf{l},\mathbf{m})]$$

$$+ \sum_k \sum_{\mathbf{n}\neq\mathbf{m}} w_k\theta(\mathbf{l}-\mathbf{n})[\psi_{Vik}(\mathbf{n},\mathbf{m},\mathbf{l}) - \psi_{Vik}(\mathbf{l},\mathbf{m},\mathbf{n})]$$

$$+ \sum_{\mathbf{n}\neq\mathbf{l}} w_i\theta(\mathbf{m}-\mathbf{n})[\psi_{ViV}(\mathbf{l},\mathbf{n},\mathbf{m}) - \psi_{ViV}(\mathbf{l},\mathbf{m},\mathbf{n})], \qquad (13.2.10)$$

where $\theta(\mathbf{l}-\mathbf{m})$ is unity if $\mathbf{l}$ and $\mathbf{m}$ are nearest-neighbour sites and is zero otherwise. This equation introduces additional time correlation functions such as $\psi_{Vik}(\mathbf{l},\mathbf{m},\mathbf{n})$ which specifies the probability of observing a vacancy at site $\mathbf{l}$, an $i$-atom at site $\mathbf{m}$ and a $k$-atom at site $\mathbf{n}$ at time $t$, given the same initial state as before, i.e. it is

$$\psi_{Vik:Vj}(\mathbf{l},\mathbf{m},\mathbf{n};t:\mathbf{l}_0,\mathbf{m}_0)$$

$$= \sum_{\beta,\alpha} \rho_\beta^{(V)}(\mathbf{l})\rho_\beta^{(i)}(\mathbf{m})\rho_\beta^{(k)}(\mathbf{n})G_{\beta\alpha}(t)\rho_\alpha^{(V)}(\mathbf{l}_0)\rho_\alpha^{(j)}(\mathbf{m}_0)p_\alpha^{(0)}. \qquad (13.2.11)$$

We refer to functions like $\psi_{Vi}$, $\psi_{Vik}$ ... as two-, three- ... site functions. A diagram showing the significance of the various terms on the right side of (13.2.10) is provided by Fig. 13.1.

The equation of motion (13.2.10) is clearly a member of a coupled hierarchy of equations of the same character as other hierarchies of kinetic equations encountered in statistical physics, e.g., the BBGKY hierarchy in the theory of classical fluids (Kreuzer, 1981; McQuarrie, 1976). The kinetic equation (13.2.10) for the two-site functions contains three-site functions, the kinetic equations for the three-site functions in turn contain four-site functions, and so on. Some approximation is required which expresses the three-site functions in terms of two-site functions and so *decouples* the equations.

Before proceeding to the discussion of such decoupling approximations we must mention one exact result governing the $L$-coefficients of a random lattice gas which all such approximations should satisfy. It was discovered by Moleko and Allnatt (1988) and, although simple, does not appear to be physically obvious. It is

$$\sum_i \frac{L_{ij}}{w_i} = \frac{L_{jj}^{(0)}}{w_j}, \qquad \text{(all } j\text{)}. \qquad (13.2.12a)$$

Written in terms of the correlation functions defined by (13.2.7), it takes the form of a *sum rule*, namely

$$\sum_i f_{ij}^{(j)} w_j / w_i = 1. \qquad (13.2.12b)$$

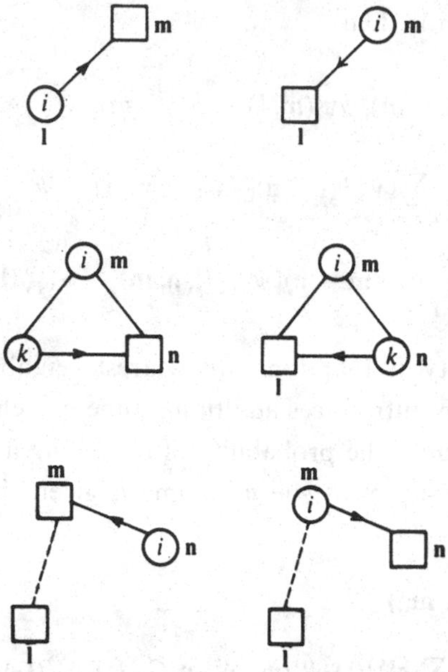


Fig. 13.1. Schematic diagrams showing how the right side of (13.2.10) is made up. The layout of the diagrams is the same as that of the corresponding terms in the equation.

It follows from this relation, which is derived in Appendix 13.2, that in a random alloy of $n$ atomic components there are $n$ linear relations among the $n^2$ phenomenological coefficients in addition to the $n(n-1)/2$ Onsager relations. The number of independent phenomenological coefficients in a random lattice gas is therefore only $n(n-1)/2$, i.e. there is only one independent coefficient in a two-component system.

In the particular case of a mixture of two isotopes A and A* (for which both jump frequencies will be the same) eqn. (13.2.12a) gives

$$L_{AA^*} + L_{A^*A^*} = L^{(0)}_{A^*A^*}$$

and

$$L_{AA} + L_{A^*A} = L^{(0)}_{AA},$$

which by (13.2.2) are seen to be the same as eqns. (5.7.5). The general relations (13.2.12) have been confirmed by recent computer simulations of the binary random lattice gas by Kehr, Binder and Reulein (1989). All the approximate theories of the random lattice-gas model which we shall encounter in this chapter also satisfy the relations.

### *13.2.2 Decoupling leading to the Manning theory for the random alloy*

We shall assume in this section, but not in those following, that the vacancy concentration is very small so that the model becomes identical with Manning's random alloy model. The contributions of the terms in $\psi_{ViV}$ to the kinetic equation (13.2.10) can then be neglected since their contributions are of higher order in $c_V$ than the others. Even with this simplification the analysis is rather complicated. It is nevertheless important. The main steps are several: (i) the expression of the occupancy variables, $\rho$, as their mean values plus a fluctuation $\Delta\rho$; (ii) the derivation of kinetic equations for $\psi$ functions defined in terms of the fluctuation variables; (iii) decoupling the hierarchy of equations for these $\psi$ functions by appeal to the macroscopic limit (in other connections often called the 'hydrodynamic limit'); and (iv) manipulation of the decoupled equations to give the correlation functions via (13.2.8) and (13.2.9). The results then emerge as (13.2.35)–(13.2.37). One further appeal to the macroscopic equations then discloses that the quantity $\chi$ in these results is just unity, whence they yield all the correlation factors, $f_i$, of Manning's theory as presented in §10.8 and all the correlation functions $f_{ij}$ in accordance with §5.8. The importance of this section derives from the rigorous basis of the calculation and the systematic nature of the development. This not only gives us better insight into the basis of the Manning theory, it also points the way to improved theories (see §13.2.4) and new applications. We turn now to the detailed calculation.

Before any approximation is made towards solving coupled sets of kinetic equations it is usually convenient to introduce a method of displaying and classifying the correlation between particles, for example by means of a cluster expansion of the probability functions (see e.g. Kreuzer, 1981). We shall do this here by a procedure introduced into solid state diffusion theory by Tahir-Kheli and Elliott (1983). They employ the device of writing each occupancy variable, e.g. $\rho_\beta^{(i)}(\mathbf{l})$, appearing in a probability function as the sum of its mean value, which is just a fractional concentration, plus a fluctuation from this mean value, viz.

$$\rho_\beta^{(i)}(\mathbf{l}) = c_i + \Delta\rho_\beta^{(i)}(\mathbf{l}). \tag{13.2.13}$$

If now we insert this identify into the definition of the two-site function $\psi_{Vi}$, we obtain

$$\psi_{Vi}(\mathbf{l}, \mathbf{m}) = c_i\psi_V(\mathbf{l}) + \psi_{V\Delta i}(\mathbf{l}, \mathbf{m}) \tag{13.2.14}$$

where

$$\psi_V(\mathbf{l}) = \sum_{\beta,\alpha} \rho_\beta^{(V)}(\mathbf{l}) G_{\beta\alpha}(t) \rho_\alpha^{(V)}(\mathbf{l}_0) \rho_\alpha^{(j)}(\mathbf{m}_0) p_\alpha^{(0)} \tag{13.2.15}$$

and

$$\psi_{\mathrm{V}\Delta i}(\mathbf{l}, \mathbf{m}) = \sum_{\beta,\alpha} \rho_\beta^{(\mathrm{V})}(\mathbf{l})\, \Delta\rho_\beta^{(i)}(\mathbf{m}) G_{\beta\alpha}(t) \rho_\alpha^{(\mathrm{V})}(\mathbf{l}_0) \rho_\alpha^{(j)}(\mathbf{m}_0) p_\alpha^{(0)}. \quad (13.2.16)$$

To approximate $\psi_{\mathrm{V}i}$ by just the first term in (13.2.14) in effect would replace the alloy around the vacancy by an average uniform alloy in which the probability of finding an $i$ atom at the specified site is $c_i$ at all times.

A similar expansion may be made for three-site functions, e.g.

$$\psi_{\mathrm{V}ik}(\mathbf{l}, \mathbf{m}, \mathbf{n}) = c_i c_k \psi_{\mathrm{V}}(\mathbf{l}) + c_i \psi_{\mathrm{V}\Delta k}(\mathbf{l}, \mathbf{n}) + c_k \psi_{\mathrm{V}\Delta i}(\mathbf{l}, \mathbf{m}) + \psi_{\mathrm{V}\Delta i \Delta k}(\mathbf{l}, \mathbf{m}, \mathbf{n}). \quad (13.2.17)$$

By retaining terms up to zero, first and second order respectively in the fluctuations we obtain a successively more detailed description of the correlations between the three particles.

Let us make these substitutions for the occupancy variables of species $i$ and species $k$ at time $t$ in all the correlation functions appearing in the kinetic equation (13.2.10). The result is

$$\begin{aligned}\frac{\mathrm{d}\psi_{\mathrm{V}i}(\mathbf{l}, \mathbf{m})}{\mathrm{d}t} &= w_i \theta(\mathbf{l} - \mathbf{m}) \{\psi_{\mathrm{V}\Delta i}(\mathbf{m}, \mathbf{l}) - \psi_{\mathrm{V}\Delta i}(\mathbf{l}, \mathbf{m}) + c_i[\psi_{\mathrm{V}}(\mathbf{m}) - \psi_{\mathrm{V}}(\mathbf{l})]\} \\ &\quad + \sum_k \sum_{n \neq m} w_k \theta(\mathbf{l} - \mathbf{n}) \{\psi_{\mathrm{V}\Delta i \Delta k}(\mathbf{n}, \mathbf{m}, \mathbf{l}) - \psi_{\mathrm{V}\Delta i \Delta k}(\mathbf{l}, \mathbf{m}, \mathbf{n}) \\ &\qquad + c_i[\psi_{\mathrm{V}\Delta k}(\mathbf{n}, \mathbf{l}) - \psi_{\mathrm{V}\Delta k}(\mathbf{l}, \mathbf{n})] \\ &\qquad + c_k[\psi_{\mathrm{V}\Delta i}(\mathbf{n}, \mathbf{m}) - \psi_{\mathrm{V}\Delta i}(\mathbf{l}, \mathbf{m})] \\ &\qquad + c_i c_k[\psi_{\mathrm{V}}(\mathbf{n}) - \psi_{\mathrm{V}}(\mathbf{l})]\}. \end{aligned} \quad (13.2.18)$$

If we now neglect all the terms $\psi_{\mathrm{V}\Delta i \Delta k}$ of second order in the fluctuations then we obtain an equation containing only one- and two-site functions from which the $L$-coefficients can be straightforwardly determined. As we shall see below, these results, although interesting, are not really adequate for concentrated alloys when there is more than a small disparity between the jump frequencies of the various species of atom. We shall therefore first obtain a more useful approximation for the second-order terms by a method modelled on that used by Holdsworth and Elliott (1986) in a similar context. The $L$-coefficients will then be calculated both with and without the approximated three-site terms.

*The method of Holdsworth and Elliott*

The essence of this approach is contained in two principal steps. In the first, one examines the spatial Fourier transform of the kinetic equation for a one-site time-correlation function. The exact equation contains two-site functions, but for times and wavelengths long on microscopic scales it must nevertheless be consistent with the form of the kinetic equation for one-site functions which is provided by macroscopic transport theory. This implies a relation between terms containing two-site functions and certain terms containing one-site functions in the two equations. The result is not used directly (it would lead to a mean-field description), but rather to point to the second step which is the assumption of an analogous relation between the three-site contributions in the equation of interest and the two-site functions. By assuming that this approximation can be employed under all circumstances and not just in the macroscopic limit of long times and wavelengths, a closed set of equations for two-site functions is obtained from which the transport coefficients can be calculated.

In order to carry out this procedure in the present context we begin with the equation of motion for $\psi_{\mathrm{V}}(\mathbf{l})$, i.e. for the probability of observing a vacancy at site $\mathbf{l}$ at time $t$ after starting from the same initial state as before (equation (13.2.15)),

$$\frac{\mathrm{d}\psi_{\mathrm{V}}(\mathbf{l})}{\mathrm{d}t} = \sum_{k}\sum_{\mathbf{n}} w_k \theta(\mathbf{l}-\mathbf{n})[\psi_{\mathrm{V}k}(\mathbf{n},\mathbf{l}) - \psi_{\mathrm{V}k}(\mathbf{l},\mathbf{n})]. \tag{13.2.19}$$

After employing (13.2.13) for atoms of species $k$ and making a Fourier transformation with respect to space coordinates (i.e. multiplying by $\exp(\mathrm{i}\mathbf{q}\cdot\mathbf{l})$ throughout and then summing over all lattice sites $\mathbf{l}$) we obtain

$$\frac{\mathrm{d}\tilde{\psi}_{\mathrm{V}}(\mathbf{q})}{\mathrm{d}t} = -zw(1-\tilde{\gamma}_{\mathbf{q}})\tilde{\psi}_{\mathrm{V}}(\mathbf{q}) + \tilde{F}_{\mathrm{V}}(\mathbf{q}) \tag{13.2.20}$$

where

$$w = \sum_{k} c_k w_k. \tag{13.2.21}$$

This quantity $w$ is the average jump frequency of the vacancies: for, the total number of atom jumps per unit volume and per unit time is $\sum_j n_j z c_{\mathrm{v}} w_j$ and this, by definition, must equal the total number of vacancy jumps per unit volume per unit time, i.e. $n_{\mathrm{V}} z w_{\mathrm{v}}$. Hence $w_{\mathrm{v}} = w$ as given by (13.2.21). The other quantities appearing in (13.2.20), $\tilde{\gamma}_{\mathbf{q}}$ and $\tilde{F}_{\mathrm{V}}(\mathbf{q})$, are, respectively, the Fourier transforms of $\theta(\mathbf{l})/z$ and of

$$\sum_{k}\sum_{\mathbf{n}} w_k \theta(\mathbf{l}-\mathbf{n})[\psi_{\mathrm{V}\Delta k}(\mathbf{n},\mathbf{l}) - \psi_{\mathrm{V}\Delta k}(\mathbf{l},\mathbf{n})]. \tag{13.2.22}$$

In the limit of long wavelengths ($\mathbf{q} \to 0$) the factor $(1-\tilde{\gamma}_{\mathbf{q}})$ is proportional to $q^2$ (cf. §9.3.1) and (13.2.20) takes on the form

$$\frac{\mathrm{d}\tilde{\psi}_{\mathrm{V}}(\mathbf{q})}{\mathrm{d}t} = -D_{\mathrm{V}}^{(0)} q^2 \tilde{\psi}_{\mathrm{V}}(\mathbf{q}) + \tilde{F}_{\mathrm{V}}(\mathbf{q}), \tag{13.2.23}$$

in which

$$D_V^{(0)} = \tfrac{1}{6}zws^2. \tag{13.2.24}$$

By results such as (8.2.6) and those presented in §10.3 we see that $D_V^{(0)}$ may be identified with the vacancy diffusion coefficient in the absence of correlation effects, since as noted above $w$ is the average jump frequency.

Now in the limit of long times and long wavelengths the decay of a fluctuation in the vacancy concentration in a system at equilibrium may be assumed to obey Fick's equation. We therefore assume that $\psi_V(\mathbf{q})$ will satisfy the Fourier transformed version of Fick's equation in this limit, i.e. that

$$\frac{d\tilde{\psi}_V(\mathbf{q})}{dt} = -D_V q^2 \tilde{\psi}_V(\mathbf{q}). \tag{13.2.25}$$

Comparison of (13.2.25) with (13.2.23) therefore leads us to expect that, in the same limit,

$$\tilde{F}_V(\mathbf{q}) = -(D_V - D_V^{(0)})q^2\tilde{\psi}_V(\mathbf{q}). \tag{13.2.26}$$

Since the factor $q^2$ in (13.2.23) arose from the form of $(1 - \gamma_\mathbf{q})$ at small $\mathbf{q}$, this suggests that when $\mathbf{q}$ is *not* small we may be able to replace $D_V^{(0)}q^2$ by $zw(1 - \gamma_\mathbf{q})$. In addition we define a vacancy correlation factor $f_V \equiv D_V/D_V^{(0)}$. Then with these two changes we are able to rewrite this as

$$\tilde{F}_V(\mathbf{q}) = -zw(f_V - 1)(1 - \tilde{\gamma}_\mathbf{q})\tilde{\psi}_V(\mathbf{q}). \tag{13.2.27}$$

If we were to adopt this as an approximation for all values of $\mathbf{q}$ and $t$ and invert the Fourier transform we would obtain the term (13.2.22) in the form

$$\sum_k \sum_\mathbf{n} w_k \theta(\mathbf{l} - \mathbf{n})[\psi_{V\Delta k}(\mathbf{n}, \mathbf{l}) - \psi_{V\Delta k}(\mathbf{l}, \mathbf{n})]$$
$$= w(f_V - 1)\sum_\mathbf{n} \theta(\mathbf{l} - \mathbf{n})[\psi_V(\mathbf{n}) - \psi_V(\mathbf{l})] \tag{13.2.28}$$

However, we do not use this equation directly, since, taken with (13.2.19), it would yield merely a mean-field theory of $\psi_V$. As previously indicated, the real purpose of (13.2.28) is to suggest a relation between the three-site terms in (13.2.18) and the two-site terms, thereby decoupling the hierarchy of equations. We therefore assume that an analogous equation can be used in which $\psi_{V\Delta k}$ and $\psi_V$ in (13.2.28) are replaced by $\psi_{V\Delta i\Delta k}$ and $\psi_{V\Delta i}$ respectively, where the species $i$ refers to the occupancy of the same site, $\mathbf{m}$, in every case, i.e.

$$\sum_k \sum_{\mathbf{n}\neq\mathbf{m}} w_k \theta(\mathbf{l} - \mathbf{n})[\psi_{V\Delta i\Delta k}(\mathbf{n}, \mathbf{m}, \mathbf{l}) - \psi_{V\Delta i\Delta k}(\mathbf{l}, \mathbf{m}, \mathbf{n})]$$
$$= w(f_V - 1)\sum_{\mathbf{n}\neq\mathbf{m}} \theta(\mathbf{l} - \mathbf{n})[\psi_{V\Delta i}(\mathbf{n}, \mathbf{m}) - \psi_{V\Delta i}(\mathbf{l}, \mathbf{m})]. \tag{13.2.29}$$

The left side of this equation gives the three-site contributions to the kinetic equation (13.2.18) and the right side provides our approximation for them.

*Results*

The approximate kinetic equation obtained by insertion of (13.2.29) into (13.2.18) provides a set of linear relations between the two-site functions $\psi_{\mathrm{V}\Delta i:\mathrm{V}j}$ of the various atom species. (N.B. We have now restored the full suffix notation.) The functions $h_{\mathrm{V}i:\mathrm{V}j}$ defined in (13.2.9), which are related through (13.2.8) to the correlation functions of interest, can be calculated from these two-site functions by appropriate summation and time integration:

$$h_{\mathrm{V}i:\mathrm{V}j} = \sum_{\mathbf{s}} \mathbf{s}\cdot\mathbf{s}_0/s^2 \sum_{\mathbf{l}} \int_0^\infty dt \psi_{\mathrm{V}\Delta i:\mathrm{V}j}(\mathbf{l}, \mathbf{l}-\mathbf{s}; t: \mathbf{l}_0, \mathbf{l}_0 - \mathbf{s}_0). \qquad (13.2.30)$$

(Although this equation contains $\psi_{\mathrm{V}\Delta i:\mathrm{V}j}$ rather than $\psi_{\mathrm{V}i:\mathrm{V}j}$ it is equivalent to (13.2.9) by (13.2.14) because the summation over $\mathbf{s}$ annihilates the $c_i\psi_{\mathrm{V}}(\mathbf{l})$ term in (13.2.14).) One finds that, rather than solving the kinetic equation explicitly, it is simpler to convert it directly to a linear relation between the functions $h_{\mathrm{V}i:\mathrm{V}j}$; this operation also eliminates the one-site functions of the kinetic equation. The final result of this rather tedious operation, details of which are outlined in Appendix 13.3, is the equation

$$h_{\mathrm{V}i:\mathrm{V}j} = (f_{\mathrm{V}}H_0 + 2w_i)^{-1}[c_{\mathrm{V}}c_j(\delta_{ij} - c_i) + 2c_iS_j] \qquad (13.2.31)$$

where

$$S_j = \sum_k w_k h_{\mathrm{V}k:\mathrm{V}j}, \qquad (13.2.32)$$

$$H_0 = M_0 w, \qquad (13.2.33)$$

and

$$M_0 = 2f_0/(1 - f_0). \qquad (13.2.34)$$

Here $f_0$ is the isotope correlation factor for the pure lattice (Table 10.1).

The linear equation (13.2.31) is simply solved by multiplying it by $w_i$ and summing over all species $i$ to obtain an expression for $S_j$. The final result for the correlation function is then

$$f_{ij}^{(j)} = f_i\delta_{ij} + \frac{2c_i w_i f_i f_j \chi}{M_0(\sum_k c_k w_k f_k)} \qquad (13.2.35)$$

where

$$f_i = f_{\mathrm{V}}H_0/(f_{\mathrm{V}}H_0 + 2w_i) \qquad (13.2.36)$$

and

$$\chi = \left(\sum_k c_k w_k f_k\right) \Big/ \left(w f_V - 2M_0^{-1} \sum_k c_k w_k f_k\right). \tag{13.2.37}$$

The notation $f_i$ for the quantity defined by (13.2.36) is allowed because a comparison of the structure of (13.2.35) with that of the general linear response theory expression for $L_{ij}$, (6.5.10)–(6.5.12), allows us to identify it with the tracer correlation factor for species $i$. The general linear response expression for $L$ is the sum of two terms (cf. (6.5.18)). The first, which contributes only for $i = j$, is an average over the motion of individual atoms (self-correlation term) while the second is an average over the motion of pairs of different atoms (pair-correlation term). When the general equation is reduced to an expression for $f_{ij}^{(j)}$ then the first term contributes $\delta_{ij} f_i$, in which $f_i$ is the tracer correlation factor for $i$. Equation (13.2.35) evidently has the same structure as the general equation in terms of the division into self-correlation and pair-correlation contributions and $f_i$ in (13.2.35) can therefore be identified with the correlation factor.

An explicit expression for the vacancy correlation factor $f_V$ which appears in these equations can then be found self-consistently by using (13.2.35) to calculate it from the sum of the phenomenological coefficients, which by (4.4.13) gives $L_{VV} = nc_V D_V/kT = nc_V f_V D_V^{(0)}/kT$. The result is then Manning's approximation for the vacancy correlation factor (10.8.5) when, as here, $c_V \ll 1$

$$f_V = \left(\sum_k c_k w_k f_k\right) \Big/ w f_0. \tag{13.2.38}$$

From this result and the definition (13.2.37) it follows that $\chi = 1$ and it is then readily recognized that the results (13.2.35) and (13.2.36) are equivalent to Manning's expressions for the phenomenological coefficients and tracer correlation factors quoted earlier in §§5.8 and 10.8. In particular, (13.2.36) is seen to be the equation for the tracer correlation factors presented earlier as (10.8.1) and (10.8.4). Furthermore, by multiplying (13.2.35) through by $L_{jj}^{(0)}$ (13.2.2) we obtain the equation for $L_{ij}$ in terms of the $D_i^*$ (i.e. (5.8.7)).

As already mentioned, the decoupling approximation used here to derive Manning's results is modelled on a method due to Holdsworth and Elliott (1986). Their calculation is described in Appendix 13.4, while their own results are related to those of others in §13.2.4 below.

### *13.2.3 Further theoretical aspects of the Manning and related theories*

We have already seen in §10.8 that the Manning theory is very accurate. In this section we shall examine the importance of the terms of second order in

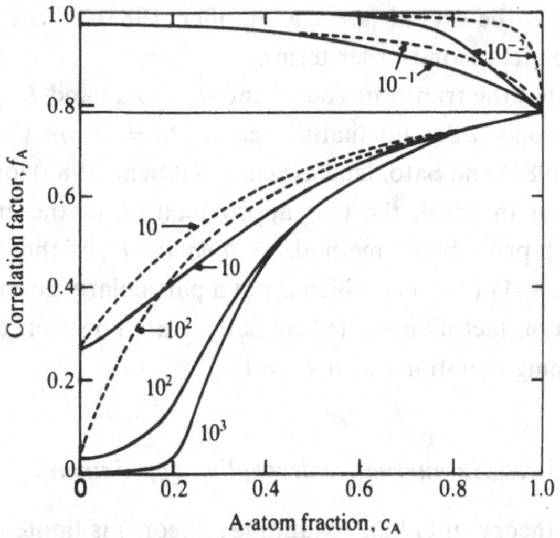


Fig. 13.2. The correlation factor $f_A$ for the tracer diffusion coefficient versus the concentration of A atoms for a f.c.c. binary random alloy for various values of the ratio $w_A/w_B$ as indicated. The full lines refer to Manning's theory and the dashed lines to that approximation to his theory that neglects terms of second order in fluctuations in the kinetic equations.

fluctuations to this accuracy. The contribution of these terms can be assessed by putting $f_V = 1$ in the final expressions (13.2.35) and (13.2.36), since by (13.2.29) this removes all such terms from (13.2.18). The correlation factor $f_i$ is then equal to that of a tracer of $i$ in a uniform system in which all non-tracer jumps are made with the average jump frequency $w$ of the actual system.

A comparison of the full Manning theory with this approximation to it which neglects second-order fluctuation terms is given in Fig. 13.2, which shows the tracer correlation factor $f_A$ in a binary random alloy AB. The discrepancies between the two theories increase as the ratio $w_A/w_B$ is increased or decreased from unity. At some concentrations the approximation is already poor for the ratios $10^{-1}$ and 10 and is of little use at more extreme ratios. The two calculations of the intrinsic (chemical) diffusion coefficient $D_A$ (5.3.6) show corresponding discrepancies when $w_A/w_B$ is either very large or very small.

An important feature of Manning's theory is that, in the limit in which B is immobile, $f_A$ is predicted to be zero from $c_A = 0$ up to a concentration $c_A = (1 - f_0)$, in excellent agreement with simulation calculations (Murch and Rothman, 1981). The approach to this limiting behaviour is apparent in the curve for $w_A/w_B = 10^3$ in Fig. 13.2. This is a percolation effect which arises because an infinite nearest-neighbour-connected cluster of A atoms is required to sustain bulk transport of A across the crystal (see, e.g. Stauffer, 1985). This effect is absent in the theory which omits second-order terms.

Lastly, we note that the transport coefficients $L_{AA}$, $L_{AB}$ and $L_{BB}$ calculated with omission of the second-order fluctuation terms ($f_V = 1$) are the same as those obtained by Sato (1984) and Sato, Ishikawa and Kikuchi (1985) for a body-centred cubic lattice in what they call the pair approximation of the ensemble-average version of the path-probability method, *as long as* $f_0$ in the present result is approximated by $(z - 1)/(z + 1)$ – which is not a particularly good approximation. The tracer correlation factors obtained by Sato *et al.*, however, differ from those given by the Manning equations with $f_V = 1$.

### *13.2.4 An alternative decoupling approximation*

We recall that the theory of §13.2.2 (Manning's theory) is limited to low vacancy concentrations. Various calculations have been made in an attempt to avoid this limitation. In this section we shall therefore describe an alternative decoupling approximation which enables us to obtain accurate predictions at arbitrary $c_V$ and so to set various other such calculations in context. As before, the calculations are designed to give all the correlation factors and functions of interest (cf. (13.2.8)). We retain the same formal structure for this purpose, in particular we start from (13.2.10) and again use the device of expanding the correlation functions by introducing fluctuation variables (cf. (13.2.13)). The difference is that in this section we choose to deal with atoms and vacancies on the same footing and to expand all occupancy variables in the manner of (13.2.13) instead of expanding only atom occupancy variables as was done earlier. For example, in place of (13.2.17) we then have

$$\begin{aligned}\psi_{Vik}(\mathbf{l}, \mathbf{m}, \mathbf{n}) = {} & c_V c_i c_k \psi + c_V c_i \psi_{\Delta k}(\mathbf{n}) + c_V c_k \psi_{\Delta i}(\mathbf{m}) + c_i c_k \psi_{\Delta V}(\mathbf{l}) \\ & + c_V \psi_{\Delta i \Delta k}(\mathbf{m}, \mathbf{n}) + c_i \psi_{\Delta V \Delta k}(\mathbf{l}, \mathbf{n}) \\ & + c_k \psi_{\Delta V \Delta i}(\mathbf{l}, \mathbf{m}) + \psi_{\Delta V \Delta i \Delta k}(\mathbf{l}, \mathbf{m}, \mathbf{n}). \end{aligned} \tag{13.2.39}$$

in which $\psi$ is the probability that there is a vacancy at $\mathbf{l}_0$ and a $j$-atom at $\mathbf{m}_0$ initially (obtained from (13.2.5) by replacing both $\rho_\gamma^{(V)}(\mathbf{l})$ and $\rho_\gamma^{(i)}(\mathbf{m})$ by unity). Approximations generated by employing this alternative expansion in all the terms of the kinetic equation (13.2.10) are the subject of this section.

The obvious first approximation is to omit all three-site terms (i.e., all terms of third order in the fluctuation variables). We may refer to this as the *simple* decoupling approximation. It differs from that obtained by neglecting all three-site terms in the earlier formulation and leads to useful results for the phenomenological coefficients, some of which have been derived by other methods in the literature. The results are not however satisfactory when the various jump frequencies differ greatly from one another.

A much more satisfactory theory may be obtained by again seeking an approximation to the third-order terms in the spirit of Holdsworth and Elliott. Such a theory was devised by Moleko, Allnatt and Allnatt (1989). The decoupling approximation is based on an analogy with the decoupling of the equation of motion for $\psi_A(\mathbf{l})$ rather than, as in §13.2.2, the equation of motion for $\psi_V(\mathbf{l})$. This approximation, has the practical advantage that it enables all three-site terms to be treated consistently (whereas that used in §13.2.2 applied only to one set of three-site terms, i.e. those arising from $\psi_{V\Delta i\Delta k}$ but not those from $\psi_{V\Delta i\Delta V}$ which is what inhibits the extension of that calculation to arbitrary $c_V$). The resulting theory is algebraically more complex than that of §13.2.2, but the steps and techniques are similar. We therefore pass directly to the results obtained.

*Simple decoupling approximation*

In this approximation (which neglects all terms of third order in fluctuations) the results are relatively simple in form. The tracer correlation factor is

$$f_i = M_0 Z_i/(2w_i + (M_0 - 2)Z_i), \qquad (13.2.40)$$

in which

$$Z_i = c_V w_i + \sum_k c_k w_k w_i (w_i + w_k)^{-1} \qquad (13.2.41)$$

and $M_0$ is given, as before, by (13.2.34).

The correlation functions for $L_{AA}$ and $L_{AB}$ for a binary lattice gas AB are

$$\left.\begin{aligned} f_{AA} &= 1 - 2c_B w_A \Omega^{-1}, \\ f_{AB}^{(B)} &= 2c_A w_A \Omega^{-1}, \end{aligned}\right\} \qquad (13.2.42)$$

where

$$\Omega = (M_0 - 2)[c_A w_A + c_B w_B + c_V(w_A + w_B)] + 2(w_A + w_B). \qquad (13.2.43)$$

These results, or particular cases of them, have also been obtained by other methods. They are identical to those obtained by Moleko, Okamura and Allnatt (1988) by a method closely related to the kinetic theory of dilute alloys described in Chapter 8. The kinetic equations were here decoupled by using the Kirkwood superposition approximation to express the three-particle correlation functions in terms of pair correlation functions in the steady state. The results for a binary alloy in the limit $c_V \to 0$ reduce to those obtained in the pair approximation of the ensemble average version of the path probability method when $f_0$ in the present theory is approximated by $(z - 1)/(z + 1)$. In this limit the correlation function $f_{AA}$ and $f_{AB}^{(B)}$, but not the tracer correlation factors, are the same as in the approximation to the Manning theory which neglects second-order fluctuation terms. As we have already noted, such results are useful only when the ratio of jump frequencies of fast and slow diffusing components is not too large ($\lesssim 10$) for they fail to show the percolation effects characteristic of very large ratios.

When there is just one type of atom the expression for the tracer correlation factor at arbitrary vacancy content reduces to an expression derived by Nakazato and Kitahara (1980) by a projection operator method and by Tahir-Kheli and Elliott (1983) by a time-dependent Green function method already outlined in §8.6, namely

$$f = \left(1 + \frac{(1 - c_V)(1 - f_0)}{(1 + c_V)f_0}\right)^{-1}. \qquad (13.2.44)$$

This expression agrees with the Monte Carlo simulation results of Murch (1984b) for the simple cubic lattice to better than 1/2%.

*Self-consistent theory*

The results obtained in the fully self-consistent theory of this section are rather lengthy (Moleko *et al.*, 1989). The essential new feature is that the various quantities calculated are now functions of the correlation functions ($f_{AA}, f_{AB}^{(B)}$, etc.) and, in the case of the tracer correlation factor, they depend on the correlation factors as well. The theory lacks the simplicity of the Manning theory in which all quantities are functions only of the tracer correlation factors and not of the correlation functions, $f_{AA}$, etc.

The self-consistent theory of Holdsworth and Elliott (1986) previously referred to, is limited to the tracer correlation factor. This theory and the present one both agree well with the Monte Carlo results of El-Meshad and Tahir-Kheli (1985) for $w_B/w_A = 0.1$, $c_A/c_B = 1/2, 1, 2$ in the simple cubic lattice. An example is shown in Fig. 13.3. In the case of the correlation functions of $L_{AA}$, $L_{AB}$ and $L_{BB}$, the self-consistent theory of Moleko *et al.* (1989) is in good agreement with Monte Carlo simulations on the same system.

### *13.2.5 Experimental tests*

We have already described some experimental tests of Manning's theory, firstly for chemical diffusion in §5.8 and secondly for tracer diffusion in §§10.8 and 10.9. Given the limitations of the random alloy model, it will be apparent that we should hardly expect experimental observations to distinguish sharply between Manning's theory (§§10.8 and 13.2.2) and those improvements of it described in §13.2.4. However, there is one difference between their predictions which might be

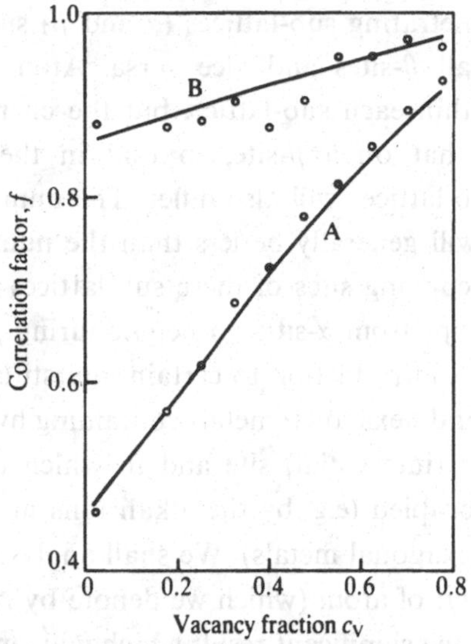


Fig. 13.3. The tracer correlation factors $f_A$ and $f_B$ as indicated for a binary random lattice gas on a simple cubic lattice as a function of vacancy fraction $c_V$ for $c_A = 2c_B$ and $w_B/w_A = 0.1$. The circles are Monte Carlo estimates (El-Meshad and Tahir-Kheli, 1985). The full lines are the predictions of the self-consistent theory of Moleko, Allnatt and Allnatt (1989) described in §13.2.4.

significant in the analysis of experimental observations and that concerns the isotope effect.

In §10.9 we saw that the analysis of isotope effects depended upon (10.9.1) and, in turn, that this depended upon the correlation factor taking the form (10.6.28), in which $H$ is independent of the jump frequency $w_s$ of the tracer considered. While this is certainly true for the Manning theory (cf. (10.8.4)) it does not follow from the self-consistent theories of Moleko *et al.* (1989) and of Holdsworth and Elliott (1986) described in §13.2.4. The self-consistent theories predict that the quantity $(\Delta D/D)/(\Delta w_s/w_s)$ differs from the correlation factor $f$, especially when there is a large difference between the jump frequencies of the different atomic species. For the rapidly diffusing component (A, say) in a binary, simple cubic alloy AB it is up to 8% larger than $f_A$ when $w_A/w_B = 5$ and as much as 20% larger when $w_A/w_B = 10$ according to the theory of Holdsworth and Elliott (1986). Such predictions could be confirmed by suitable Monte Carlo simulations; but this has not yet been done for the random alloy.

### 13.2.6 *A two-sublattice model*

We conclude our discussion of random lattice-gas models by considering a two-sub-lattice version. We shall suppose that the overall lattice structure defines two inequivalent interpenetrating sub-lattices ($\alpha$ and $\beta$) such that the ($z$) neighbours of an $\alpha$-site are all $\beta$-sites and vice versa. Atoms (and vacancies) are distributed randomly within each sub-lattice, but the energy of an atom on an $\alpha$-site is different from that on a $\beta$-site, so that in thermal equilibrium the occupancy of the two sub-lattices will also differ. The number of atoms disposed on the two sub-lattices will generally be less than the number of available sites. Movements of atoms occupying sites of these sub-lattices are assumed to occur by nearest-neighbour jumps from $\alpha$-sites to neighbouring $\beta$-sites and vice versa.

This model serves as an introduction to certain interstitial solid solutions such as the alkali $\beta$-aluminas and hexagonal metals containing hydrogen in which there are inequivalent types of (interstitial) site and in which only a fraction of the available sites may be occupied (e.g. by the alkali ions in the $\beta$-aluminas or the hydrogen atoms in the hexagonal metals). We shall analyse this simplified model when there is just one kind of atom (which we denote by A) distributed over the available $\alpha$- and $\beta$-sites. One significant result which will emerge is the occurrence of concentration-dependent correlation effects, not only in tracer diffusion ($f_A$) but also in $f_{AA}$, i.e. in the electrical conductivity, $\kappa$, of an ionic compound. This is an interesting contrast to the behaviour of the one-sub-lattice model where the corresponding quantities ($f_{AA}$ and $\kappa$) are independent of concentration (§8.6).

Let the sub-lattice of lower energy be the $\alpha$-sub-lattice and the one of higher energy the $\beta$-sub-lattice. The fraction of sites in the $\alpha$-sub-lattice occupied by atoms of species A is denoted by $c_A^\alpha$ and the exchange frequency of an A atom with a vacancy is $w_A^\alpha$ or $w_A^\beta$ according as the atom jumps from the $\alpha$- or the $\beta$-sub-lattice respectively. According to the principle of detailed balance, at equilibrium we have

$$c_A^\alpha c_V^\beta w_A^\alpha = c_A^\beta c_V^\alpha w_A^\beta. \tag{13.2.45}$$

or

$$\frac{c_V^\alpha c_A^\beta}{c_V^\beta c_A^\alpha} = \frac{w_A^\alpha}{w_A^\beta} \equiv w. \tag{13.2.46}$$

In addition

$$c_A^\alpha + c_V^\alpha = 1 \qquad (\alpha\text{-sites}) \tag{13.2.47a}$$

and

$$c_A^\beta + c_V^\beta = 1 \qquad (\beta\text{-sites}), \tag{13.2.47b}$$

while constancy in the total number of A-atoms requires

$$c_A^\alpha + c_A^\beta = 2c_A, \tag{13.2.48}$$

in which $c_A$ is the number of A atoms as a fraction of all sites ($\alpha$ and $\beta$). If we write

$$c_A^\alpha = c_A + \tfrac{1}{2}\xi, \tag{13.2.49a}$$

$$c_A^\beta = c_A - \tfrac{1}{2}\xi \tag{13.2.49b}$$

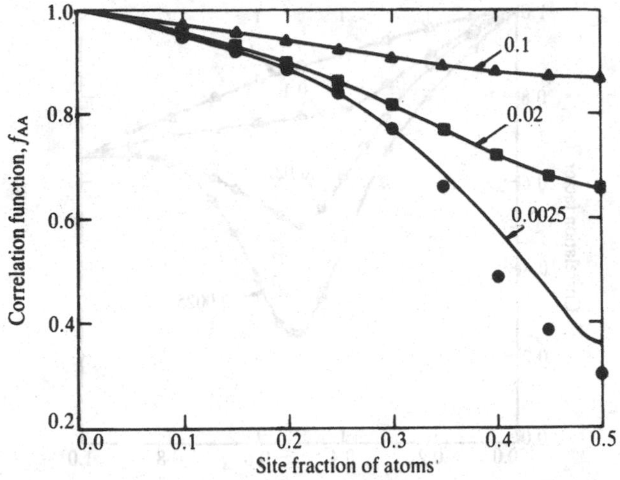


Fig. 13.4. Monte Carlo simulation results (Murch, 1982e) for $f_{AA}$ for a simple cubic lattice gas of two interpenetrating f.c.c. lattices. Results are shown as a function of the site fraction of atoms $c_A$ for various values of the ratio $w_A^\alpha/w_A^\beta$ of jump frequencies. The solid lines represent Richards' equation (13.2.51).

and use (13.2.47) and (13.2.48), then (13.2.46) becomes

$$\frac{[(1-c_A)-\frac{1}{2}\xi][c_A-\frac{1}{2}\xi]}{[(1-c_A)+\frac{1}{2}\xi][c_A+\frac{1}{2}\xi]}=w. \tag{13.2.50}$$

We see immediately that the solution of this equation is unchanged if we substitute $(1-c_A)$ for $c_A$, i.e. it is symmetrical about $c_A = 1/2$.

We turn now to the transport coefficients of this model, specifically $f_{AA}$, $f_A$ and the Haven ratio $f_A/f_{AA}$. An analytic expression for the correlation function $f_{AA}$ was first obtained by Richards (1978):

$$f_{AA}=1-\frac{x(c_A^\alpha-c_A^\beta)}{z-1-x(c_A^\alpha-c_A^\beta)(z-2)}, \tag{13.2.51}$$

in which

$$x=(1-w)/(w+1). \tag{13.2.52}$$

Although the method by which Richards arrived at (13.2.51) was special, the same result may also be derived by way of the simple decoupling approximation presented in §13.2.4. That $f_{AA}$ should be symmetrical about $c_A = 1/2$ may or may not be physically obvious, but this expression, because it is a function of just the difference $c_A^\alpha - c_A^\beta \equiv \xi$, predicts that it is so (cf. (13.2.50)). The dependence of $f_{AA}$ on $c_A$ is shown in Fig. 13.4 for three values of the ratio $w$ of the jump frequencies. This figure also shows the results of Monte Carlo simulations made by Murch

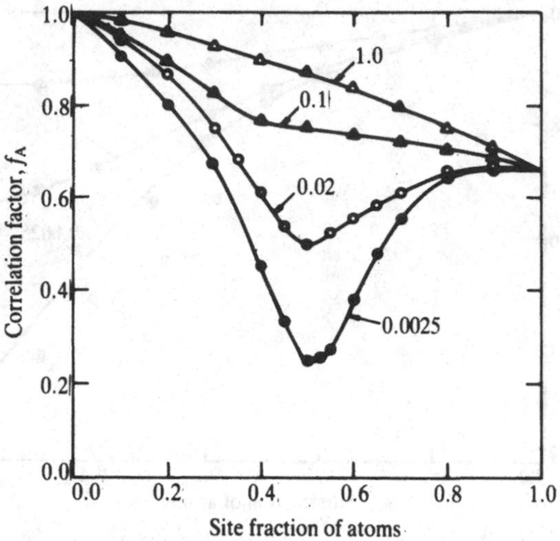


Fig. 13.5. Monte Carlo simulation results (Murch, 1982e) for $f_A$ for a simple-cubic lattice gas of two interpenetrating f.c.c. lattices. Results are shown as a function of the site fraction of atoms, $c_A$, for various values of the ratio $w_A^\alpha/w_A^\beta$ as indicated. The solid lines are merely a guide to the eye.

(1982e) for the case where the two interpenetrating sub-lattices are each face-centred cubic. They agree well with the predictions of (13.2.50) and (13.2.51).

The tracer correlation factor $f_A$ has been investigated both analytically (by Holdsworth, Elliott and Tahir-Kheli, 1986) and by Monte Carlo simulations (by Murch, 1982e and Tahir-Kheli, Elliott and Holdsworth, 1986). However, no simple analytic formula for $f_A$ is yet available. From the Monte Carlo results shown in Fig. 13.5 we see that there is a characteristic minimum of $f_A$ (just as there is for $f_{AA}$) in the vicinity of $c_A = 1/2$. According to Murch (1982e), at low temperatures ($w \ll 1$) the sequences of configurations computed in these simulations show that at these concentrations the migration process has some interstitialcy character, in the sense that two atoms may follow one another – although not simultaneously. Thus if an atom jumps from an $\alpha$- to a $\beta$-site then another atom nearby may make a similar jump and eventually jump into the vacancy left by the first atom. At higher temperatures the system is sufficiently disordered that such processes are infrequent.

The Haven ratio (cf. §5.9) for this model $H_R = f_A/f_{AA}$. The simulations show that at low temperatures $H_R$ varies rapidly with concentration $c_A$ in the region around $c_A = 1/2$, where these interstitialcy-like movements occur (Fig. 13.6). Although the model is too simple to give a precise representation of $\beta$-alumina, nevertheless somewhat similar behaviour is seen experimentally in this substance.

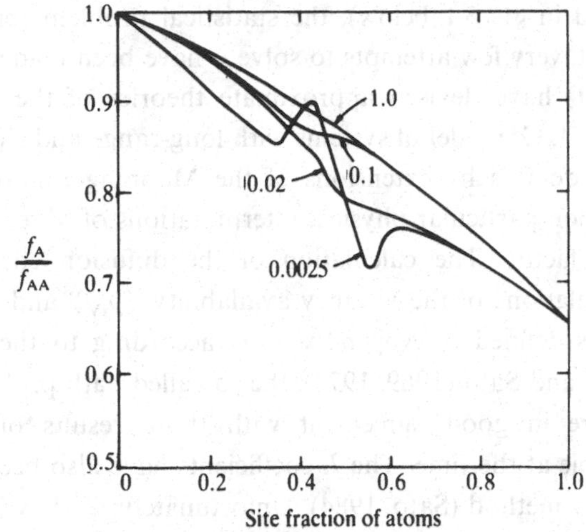


Fig. 13.6. Values of the Haven ratio, $f_A/f_{AA}$, for the simple-cubic lattice gas of two interpenetrating f.c.c. lattices as derived from the results shown in Figs. 13.4 and 13.5 (Murch, 1982e).

The simulation therefore appears to disclose some of the characteristics of real substances containing inequivalent sub-lattices.

We conclude this section by noting that no extensive analytical or simulation results for the multicomponent version of this model are available.

## 13.3 Models of non-random systems

The thermodynamic properties of concentrated alloys and lattice gases in which the equilibrium distribution of atomic species and defects on the lattice sites is non-random were referred to at an elementary level in Chapter 3. There we introduced the interacting lattice-gas (I.L.G.) model having nearest-neighbour interactions between the atoms. This model and variants of it, although physically simple as representations of real systems, have provided much insight into the equilibrium properties of such systems, including the occurrence of phase separation and of order–disorder transitions. Although no exact analytic theories of these thermodynamic properties have been obtained for three-dimensional systems, useful approximate theories exist and have been widely used (see e.g. the reviews by Sato, 1970 and de Fontaine, 1979). The approximations involved have long been understood in terms of systematic expansions of the exact thermodynamic functions. Unfortunately, when it comes to the atomic transport properties of these systems we are still at an early stage. Even when we have decided how the atomic jump frequencies are to be determined (one widely used

Ansatz is described in §13.3.1 below), the statistical problem remaining is very difficult. As a result, very few attempts to solve it have been made. Bakker (1979) and Stolwijk (1981) have devised approximate theories of the tracer diffusion coefficients for the I.L.G. model of systems with long-range and short-range order respectively. They do this by extensions of the Manning random-alloy theory which are based upon particular physical interpretations of his expression for the tracer correlation factor. The calculation of the diffusion coefficient is then completed by calculations of the vacancy availability $\langle p_{iV} \rangle$ and the mean jump frequency $\langle w_i \rangle$ (as defined in Appendix 13.1) according to the (approximate) method of Kikuchi and Sato (1969, 1972), the so-called path probability method. Both theories were in good agreement with those results of Monte Carlo simulations available at the time. The $L$-coefficients have also been calculated by the path probability method (Sato, 1984). Unfortunately, we have been unable to place any of these calculations within an appropriate generalization of the structure set up in §13.2; nor have we been able to set them in any other rigorous scheme of development. That remains a task for the future; so that we shall not discuss these calculations further. Instead we turn to consider the insights obtained by means of Monte Carlo computations.

In a system in which there are interactions among the atoms it is clear that the jump frequency of an atom will in general depend upon the configuration of nearby atoms and defects. We therefore begin our consideration of transport in non-random systems by a discussion in §13.3.1 of the expressions which have been adopted for atomic jump frequencies in them. We then discuss some of the characteristics of the atomic transport coefficients in such systems, especially the results obtained for binary order–disorder alloys ($E > 0$). We shall rely particularly on the results of Murch and co-workers for the simple-cubic lattice. For comparison, we shall also refer to results obtained for a strictly regular solution ($E < 0$) on a f.c.c. lattice. We shall, however, resist the temptation to describe all the results of this considerable body of work, partly because it would demand a lot of space and partly because any general statements about numerical results are in the nature of inferences and not deductions. Rather we shall use the results to throw light on two particular matters. The first (the subject of §13.3.2) concerns certain characteristic features of tracer diffusion ($D^*$) in ordered alloys. The second (dealt with in §13.3.3) is the accuracy of the Darken and Manning relations between the $D^*$ and the $L$-coefficients considered earlier in Chapter 5.

### *13.3.1 The jump frequency*

As pointed out in the introduction, Monte Carlo simulations require a specification of the atomic jump frequencies. In this section we therefore consider expressions for jump frequencies adopted for concentrated alloy and lattice-gas models in the light of the classical rate theory expression (3.11.11). We do this with the vacancy mechanism and the I.L.G. and similar models in mind.

The frequency factor $\nu$ and the Gibbs energy of activation $g_m$ appearing in §3.11 both depend, in general, not only on the species of the jumping atom but also on the configuration of surrounding atoms. However, the dominant dependence on configuration is contained in $g_m$, so that $\nu$ is usually taken as constant for each atomic species. Furthermore, it is usually assumed that $g_m$ can be approximated as a temperature-independent energy $\Delta g = g^+ - g^0$, which is the difference in energy of the atom in the saddle-point configuration and in the configuration preceding the jump.

The most widely used model makes two further simplifications. The first is to assume that $g^+$ is the same for all jumps, so that the jump frequency from configuration $\alpha$ to configuration $\beta$ takes the form

$$w_{\beta\alpha} = \nu \exp(-g^+/kT) w_\alpha, \tag{13.3.1}$$

where the only configuration-dependent term on the right side is

$$w_\alpha = \exp(g_\alpha^0/kT). \tag{13.3.2}$$

The second simplification is to express $g_\alpha^0$ in terms of the pairwise energies which appear in the calculation of equilibrium thermodynamic properties. For the Bragg–Williams model of a binary alloy for example, if the jumping atom is A then

$$g_\alpha^0 = (z_A^\alpha g_{AA} + z_B^\alpha g_{AB}), \tag{13.3.3a}$$

in which $z_A^\alpha$ and $z_B^\alpha$ are the numbers of A and B atoms which are nearest-neighbour to the jumping A atom before its jump, and $g_{AA}$ etc. are the nearest-neighbour interaction energies introduced in §3.10. Likewise, for a jump by a B atom

$$g_\alpha^0 = z_B^\alpha g_{BB} + z_A^\alpha g_{AB}. \tag{13.3.3b}$$

Mostly, these simulations have been made for the symmetric model for which $g_{AA} = g_{BB}$. Any property of A at concentration $c_A = x$ is then identical to the same property of B at concentration $c_A = 1 - x$.

This model was introduced by Kikuchi and Sato (1969, 1970) in analytical studies of the I.L.G. model and was subsequently adopted in many Monte Carlo simulations of both one- and two-component systems (see e.g. Murch, 1984a). There is very little theoretical guidance on its correctness, although we can compare it with the models which have been used successfully for dilute alloys. Such a comparison indicates caution. Thus, if we take the lattice to be f.c.c. and adopt the commonly made simplification that $g_{AA} = g_{BB}$, we find (in the notation of the five-frequency model) that $w_2/w_1 = \exp(-5E/kT)$, $w_3/w_1 = \exp(E/2kT)$ and $w_4/w_0 = \exp(-E/2kT)$, in which $E$ is positive for an order–disorder alloy and negative for a regular solution. Hardly any of the alloys listed in Tables 11.1 and 11.6 have values of these ratios even qualitatively in agreement with these predictions. In particular, CuZn, which we should expect to conform to the order–disorder model (i.e. $E > 0$), by these frequency ratios demands $E < 0$! Refinements of it have therefore been proposed by others; see e.g. Radelaar (1970), Kinoshita, Tomokiyo and Eguchi (1978) and Dietrich, Fulde and Peschel (1980). For example, $g^+$ may also be allowed to depend on configuration and expressed as a sum of interaction energies. However, such elaborations introduce further parameters as well as making the calculations more complex and the simulations more expensive. In consequence such refinements have not yet been much taken up. For the time being therefore we have little choice but to examine the results of simulations made with the assumptions (13.3.1)–(13.3.3) and to see whether these may yield a qualitative understanding of correlation effects in non-ideal alloys.

### *13.3.2 The tracer diffusion coefficients*

As already mentioned in §13.1, the vacancy concentration is a given (usually fixed) quantity in these simulations. Since it actually depends on both composition and temperature, only kinetic effects can be disclosed by the results. $D_A^*$ itself is therefore not given, but only a reduced or relative quantity $\bar{D}_A^*$ equal to $\bar{p}_{AV}\bar{w}_A f_A$ in which, in terms of the quantities defined in Appendix 13.1, $\bar{p}_{AV} = \langle p_{AV}\rangle/c_V$ and $\bar{w}_A = \langle w_A\rangle/w_0$, $w_0$ being the jump frequency in pure A. Although $\bar{D}_A^*$ will vary with temperature and composition it can thus provide only a part of the variation of $D_A^*$ with these quantities.

Some results are shown in Figs. 13.7 and 13.8 which show $\bar{D}_A^*$ as a function of $c_A$ at various temperatures below $T_c$ and as a function of temperature at various compositions. From the first of these it will be seen that $\bar{D}_A^*$ displays a maximum, whose locus lies on the order–disorder boundary, and a minimum in the vicinity of the stoichiometric composition. Above $T_c$, which is the highest temperature represented in Fig. 13.7, $\bar{D}_A^*$ simply increases monotonically with $c_A$ from a low value when $c_A$ is small up to $f_0$ at $c_A = 1$. From the second figure, which is an Arrhenius plot, it will be seen that for $T < T_c$ the Arrhenius energy (apparent activation energy) associated with $\bar{D}_A^*$ is a maximum at the stoichiometric composition. Correspondingly, at a given composition the Arrhenius energy increases sharply as the temperature is lowered into the region where the alloy is ordered.

Each of these three features is shown by some measured tracer diffusion coefficients of ordered alloys (reviewed by Adda and Philibert, 1966 and Bakker,

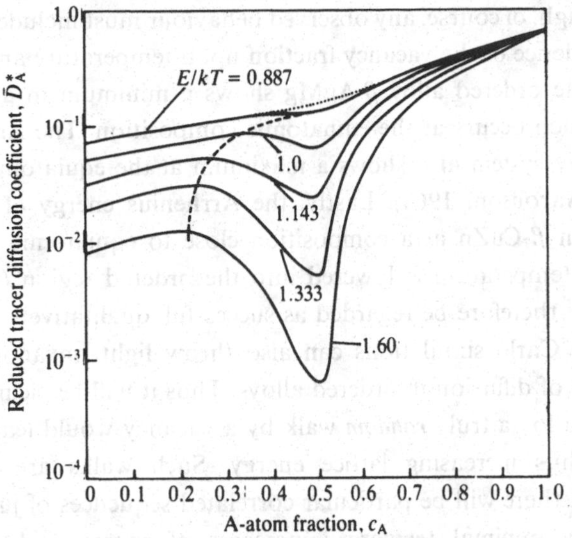


Fig. 13.7. Monte Carlo simulation results (Murch, 1982c) for the reduced tracer diffusion coefficient $\bar{D}_A^*$ as a function of concentration for a simple-cubic order–disorder alloy. The value of $E/kT_c$ in this structure is 0.887.

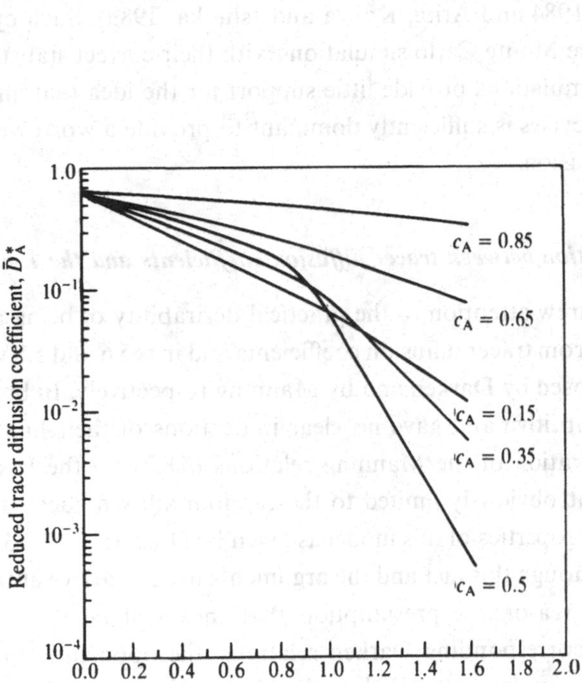


Fig. 13.8. An Arrhenius plot of the Monte Carlo simulation results (Murch, 1982c) for the reduced tracer diffusion coefficient $\bar{D}_A^*$ of a simple-cubic, order–disorder alloy as in Fig. 13.7.

1984), even though, of course, any observed behaviour must include a contribution from the dependence of the vacancy fraction upon temperature and composition. For example, the ordered alloy β-AgMg shows a minimum in a plot of $D^*_{Ag}$ v. composition which occurs at the equiatomic composition. The Arrhenius energy of the same alloy system also shows a maximum at the equiatomic composition (Domian and Aaronson, 1964). Lastly, the Arrhenius energy of both $D^*_{Cu}$ and $D^*_{Zn}$ measured in β-CuZn at a composition close to equiatomic shows a sharp increase as the temperature is lowered into the ordered region (Fig. 2.19). The simulations may therefore be regarded as successful, qualitatively at least.

These Monte Carlo simulations can also throw light on another frequently discussed aspect of diffusion in ordered alloys. Thus it will be clear that in a fully ordered binary alloy a truly *random* walk by a vacancy would lead to increasing disorder and thus increasing lattice energy. Such walks are therefore very improbable. But there will be particular correlated sequences of jumps which are accompanied by minimal *temporary* increases of energy and which lead to long-range matter transport without finally increasing the lattice disorder. The smallest such sequence in the CsCl structure is a six-jump cycle. Various discussions of diffusion in ordered alloys in terms of these cycles have been given (see, e.g. Bakker, 1984 and Arita, Koiwa and Ishioka, 1989). Such cycles will, of course, occur in the Monte Carlo simulations with their correct statistical weights. However, these simulations provide little support for the idea that any particular small set of jump cycles is sufficiently dominant to provide a worthwhile basis for an analysis of diffusion.

### *13.3.3 The relation between tracer diffusion coefficients and the L-coefficients*

In Chapter 5 we drew attention to the practical desirability of being able to infer the *L*-coefficients from tracer diffusion coefficients and in §§5.6 and 5.8 we discussed the relations proposed by Darken and by Manning respectively. In both cases our arguments were intuitive and gave no clear indications of their limitations. For example, our derivation of the Manning relations (5.8.7) on the basis of (5.8.4) and (5.8.6) was not obviously limited to the random alloy model – although, as we have seen, the properties of this model as given by (13.2.35)–(13.2.37) do satisfy (5.8.7) exactly. Although this fact and the arguments used to arrive at the Manning relations create a reasonable presumption that they will be more accurate in practice than the corresponding Darken relations, the opportunity to test them with the results of simulations of order–disorder and regular solution models is clearly important. That is the subject of this section.

In order to test the Darken and Manning relations it is convenient to write them as expressions for the correlation functions defined by eqns. (13.2.7) rather

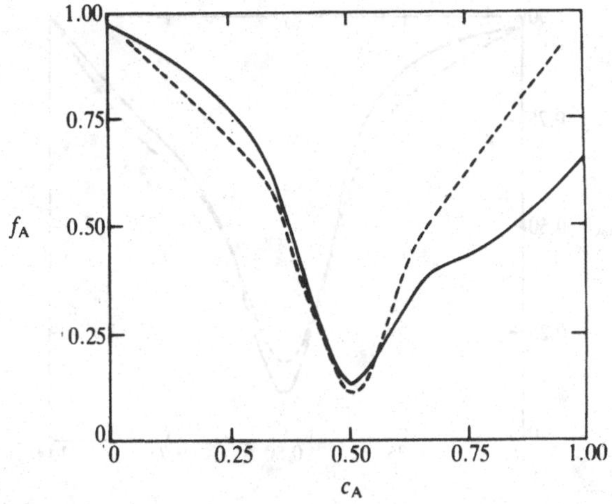


Fig. 13.9. Dependence of the tracer correlation factor, $f_A$, on the fraction of A atoms for a simple-cubic, order–disorder alloy at $E/kT = 1.143$ (i.e. beneath the critical temperature of the stoichiometric AB alloy). The full line is from the direct calculation of Zhang *et al.* (1989); the dashed line is calculated from the Darken eqn. (13.3.4), again using the results of Zhang *et al.* on the right side.

than for the phenomenological coefficients. For the Darken relation (5.6.4) in a binary alloy AB we obtain

$$f_A = f_{AA} - c_A f_{AB}^{(A)}/c_B, \tag{13.3.4}$$

and for the Manning relations (5.8.7)

$$f_{AA} = f_A \left[ 1 + \frac{1 - f_0}{f_0} \frac{c_A D_A^*}{c_A D_A^* + c_B D_B^*} \right], \tag{13.3.5a}$$

$$f_{AB}{}^{(A)} = f_A \left( \frac{1 - f_0}{f_0} \right) \frac{c_B D_B^*}{c_A D_A^* + c_B D_B^*}, \tag{13.3.5b}$$

in which we can replace the $D_A^*$ and $D_B^*$ by the corresponding reduced coefficients $\bar{D}_A^*$ and $\bar{D}_B^*$.

The simulations of Zhang, Oates and Murch (1989) can be used to test these relations. We take the Darken relation first. Figure 13.9 shows $f_A$ for a temperature such that $E/kT = 1.143$: (a) as directly computed and (b) as calculated from (13.3.4) by using the computed values of $f_{AA}$ and $f_{AB}^{(A)}$. Qualitatively the two agree, but quantitatively the relation is poor when A is the major component of the alloy. A similar result holds at higher temperatures in the wholly disordered region.

A similar test can be made for the Manning relations (Zhang *et al.*, 1989). The results in Fig. 13.10 (again for $E/kT = 1.143$) show good agreement between the two sides of (13.3.5a) for $f_{AA}$ and a similar agreement for $f_{AB}^{(A)}$ except at low

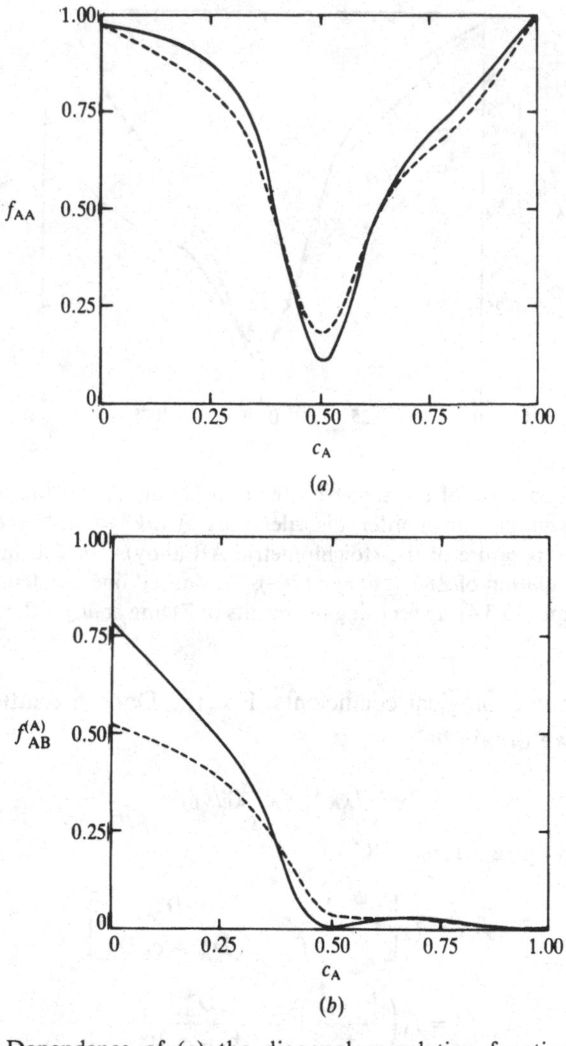


Fig. 13.10. Dependence of (*a*) the diagonal correlation function, $f_{AA}$, and (*b*) the off-diagonal function, $f^{(A)}_{AB}$, on the fraction of A atoms for a simple-cubic, order–disorder alloy at $E/kT = 1.143$. As in Fig. 13.9 the full line is from the direct calculation of Zhang *et al.* (1989), while the dashed line is as calculated from the Manning equations (13.3.5).

concentrations where $c_A \to 0$. The agreement is slightly better above the critical temperature. One may conclude that, for alloys with a tendency to order, the Manning relations are more closely satisfied than the Darken relation. Whether a similar conclusion holds for alloys with a tendency to separate into two phases at low temperatures (i.e. $E < 0$) is less clear. Simulations of a f.c.c. regular solution (Allnatt and Allnatt, 1991) show that here too both relations (13.3.4) and (13.3.5a) are satisfied rather well; but not (13.5.5b), although the form of the variation of $f^{(A)}_{AB}$ with $c_A$ is correct.

In making these comparisons it is helpful to remember that we know the behaviour in the dilute limits $c_A \to 0$ and $c_B \to 0$ from the previous calculations on dilute solid solutions. In particular, as $c_B \to 0$, $f_A \to f_0$ and $f_{AA} \to 1$. At the other extreme $c_A \to 0$ and $f_{AA} \to f_A$. Clearly, (13.3.5a) satisfies these requirements. On the other hand, the Darken equation (13.3.4) satisfies the second, but not necessarily the first; because, although the term $f^{(A)}_{AB}/c_B$ becomes independent of $c_B$ as $c_B \to 0$, it remains dependent upon the *properties* of B even though there may be none present! How good (13.3.4) is in this limit thus depends on details of the way a B atom may influence the vacancy motion in its vicinity. Likewise the reliability of the Manning equation for $f^{(A)}_{AB}$ in the extremes of small $c_A$ and small $c_B$ also depends on these details. Quantitative estimates can always be made by appealing to the results presented in Chapter 11.

A further consideration in these comparisons is that in practice we are generally concerned with some combination of the $L_{ij}$, for example the intrinsic (chemical) diffusion coefficients, rather than the $L_{ij}$ separately. From (5.3.6) and (5.8.7) the intrinsic diffusion coefficient, $D_A$, in the Manning approximation is given by

$$D_A = \alpha r_A D^*_A, \tag{13.3.6}$$

in which $\alpha$ is the thermodynamic factor

$$\alpha = 1 + \frac{\partial \ln \gamma}{\partial \ln c} \tag{13.3.7}$$

and $r_A$ – which would be unity in Darken's approximation – is

$$r_A = \frac{f_0 D^*_B + c_A(D^*_A - D^*_B)}{f_0(c_A D^*_A + c_B D^*_B)}. \tag{13.3.8}$$

The simulation results for an order–disorder model presented in Fig. 13.11 show that this expression for $r_A$ is accurate both above and below the order–disorder transition temperature. However, the particular circumstances of this test may be favourable to (13.3.8). Thus, corresponding results for a regular solution ($E < 0$) indicate that, at temperatures near (but above) the critical temperature for separation into two phases, Darken's equation (i.e. $r_A = 1$) may predict $D_A$ better than (13.3.6) and (13.3.8) do, especially when A is the major constituent.

On the whole though, the indications are that Manning's equations (13.3.5) give reliable predictions in the presence of significant interatomic interactions (of whatever sign) and that this is true over a wide range of composition. In view of this it is reasonable also to examine the two intuitive assumptions from which the Manning relations were derived in §5.8. The first of these, (5.8.4), was obtained by

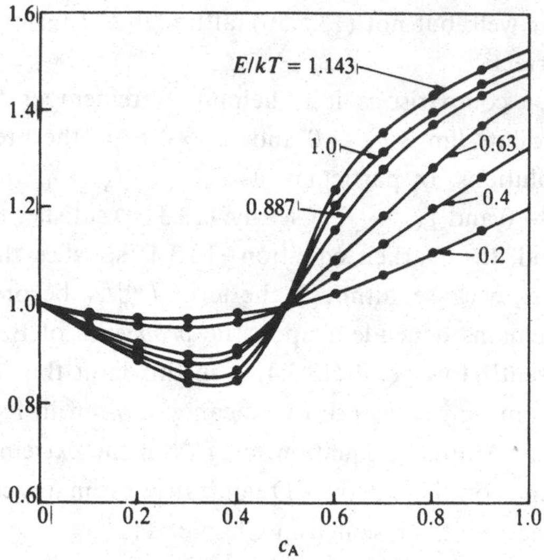


Fig. 13.11. The factor $r_A$ (13.3.6) as a function of the fraction of A atoms in a simple-cubic, order–disorder alloy at various values of $E/kT$ as indicated. The ● symbols are from direct Monte Carlo calculations, while the full lines are calculated from Manning's expression (13.3.8) using values of $\bar{D}_A^*$ and $\bar{D}_B^*$ from the simulation. (After Murch, 1982d.)

modifying the argument which led to Darken's equation so as to correct it for the vacancy wind. The second, (5.8.6), was obtained by assuming that a certain mobility bore the same relation to the tracer diffusion coefficient as in the pure material. Expressed in terms of the correlation functions, the two expressions become

$$f_A = f_{AA} - \frac{f_{AB}^{(B)}(f_{AA} + f_{AB}^{(A)})}{(f_{BB} + f_{AB}^{(B)})}, \tag{13.3.9}$$

with an equivalent expression for $f_B$, and

$$\frac{f_A}{f_0} = f_{AA} + f_{AB}^{(A)}, \tag{13.3.10}$$

again with an equivalent expression for $f_B$.

Both sides of these expressions can be found from the simulation data. The results for an order–disorder alloy in Fig. 13.12 indicate that both the expressions (13.3.9) and (13.3.10) are quite accurate, except in the region of small $c_A$. Corresponding results for a regular solution indicate that (13.3.10) is again quite accurate but that (13.3.9) steadily diverges from the directly obtained values of $f_A$ as $c_A$ increases. This weakness of (13.3.9), like the corresponding feature of the Darken relation (13.3.4), is understandable since the second term on the right side

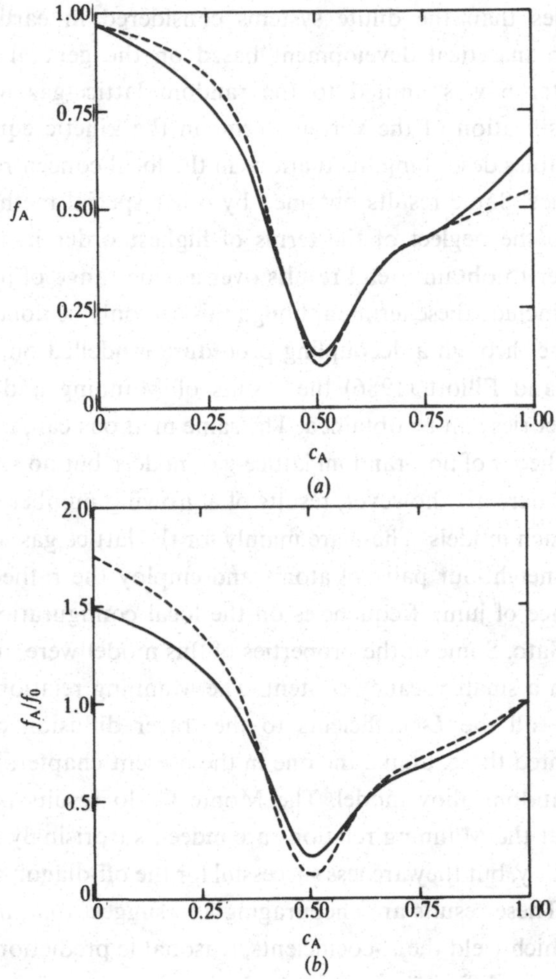


Fig. 13.12. (*a*) Dependence of the correlation factor $f_A$ on the fraction of A atoms for a simple-cubic, order–disorder alloy at $E/kT = 1.143$: full line as calculated directly by a Monte Carlo simulation (Zhang *et al.*, 1989), dashed line as obtained from (13.3.9). (*b*) Ditto but with the dashed line obtained from (13.3.10).

depends on the properties of B even as $c_B \to 0$. The extent of the failure at this extreme can again be obtained by using the dilute solution results of Chapter 11.

## 13.4 Summary

In this Chapter we have described some of the lattice gas models which have been used as representations of both concentrated alloys and systems with high concentrations of vacancies. Such models pose much more difficult problems for analytical theories than the dilute systems considered in earlier chapters. In consequence our analytical development based on the general linear response results of Chapter 6 was limited to the random lattice-gas model. Here we employed a classification of the various terms in the kinetic equations by their order in the variables describing fluctuations in the local concentrations of defects and atomic species. Some results obtained by other special methods were found to correspond to the neglect of the terms of highest order in the fluctuations. However, in order to obtain useful results over a wide range of jump frequencies it is necessary to include these terms, although this can only be done approximately. When this is done through a decoupling procedure modelled on that introduced by Holdsworth and Elliott (1986) the results of Manning and other accurate self-consistent theories can be obtained. The same methods can, in principle, form the basis for the theory of non-random lattice-gas models, but no such development is yet available. There are, however, results of a growing number of Monte Carlo simulations for such models. These are mainly for the lattice gas with interactions between nearest-neighbour pairs of atoms and employ the rather simple Ansatz for the dependence of jump frequencies on the local configuration of atoms due to Kikuchi and Sato. Some of the properties of this model were illustrated for the binary alloy with a small vacancy content. The Manning relations introduced in Chapter 5 relate all the $L$-coefficients to the tracer diffusion coefficients. The derivation presented there, unlike the one in the present chapter, is not obviously limited to the random alloy model. The Monte Carlo results presented in this chapter show that the Manning relations are indeed surprisingly accurate for the ordered binary alloy, but they are less successful for the off-diagonal $L_{ij}$ in a regular solution model. These results are encouraging and suggest that, in the absence of measurements which yield the $L$-coefficients, reasonable predictions of them could be made from tracer diffusion coefficients alone.

The extension of these analyses and tests to models of interstitial defects (as required for the theory of the effects of irradiation on concentrated alloys) is not far advanced. The results of Bocquet (1987) for random alloys show that the Manning relations (5.8.7) may be less reliable than they are for vacancy mechanisms, although they do suggest that it will usually be better to employ these relations than to neglect the effect of correlations entirely. A kinetic equation analysis along the lines of §13.2.2 (Chaturvedi and Allnatt, to be published) leads to disappointingly unwieldy results even when second-order fluctuations are neglected. Much further work will be needed to obtain a self-consistent theory.

Another topic on which much theoretical work remains to be done is that of the fast ion conductors – despite the existing extensive literature on these materials. The present position is typified by analyses (e.g. of the $\beta$- and $\beta''$-aluminas) based on the I.L.G. model (with nearest-neighbour interactions and the Kikuchi–Sato

Ansatz for the jump frequency, §13.3.1). In practice, both Monte Carlo simulations (Murch, 1982f) and analyses made with the path probability method (Sato, 1984) have been only partially successful in describing the transport properties of these materials via this model. More recent studies appear to be developing in two broad directions. On the one hand, both analytical studies (e.g. Funke, 1989, 1991) and Monte Carlo simulations of Markovian lattice-gas models (e.g. Maass *et al.*, 1991) are exploring the extent to which the omission from the earlier models of the long-range Coulomb interactions among the mobile ions is responsible for their failure to give an adequate account of properties such as the frequency dependence of the dielectric loss, quasi-elastic neutron scattering and nuclear magnetic relaxation. On the other hand, full molecular dynamics simulations of models (specified by detailed intermolecular forces) are being made more widely (see e.g. Smith and Gillan, 1992 for $\beta''$-alumina). Such simulations should eventually clarify the degree to which Markovian models in general are applicable to these substances as well as defining the specific models appropriate for future extensions of the kinetic theory presented in this chapter.

# Appendix 13.1

## Monte Carlo simulations of matter transport

As this chapter shows it is difficult to devise a satisfactory analytical theory of concentrated alloys and solid solutions, even when these are thermodynamically ideal. In the presence of interatomic interactions the task is seemingly yet more difficult. The alternative approach of using fully numerical Monte Carlo simulations of the system has therefore received considerable attention in recent years and is the subject of this appendix. In the present application these methods are naturally restricted to models whose time evolution is governed by the Markovian master equation introduced in Chapter 6. They should not be confused with the standard Metropolis Monte Carlo method for the calculation of equilibrium thermodynamic properties by the method of importance sampling (see, e.g. Allen and Tildesley, 1987). The Metropolis method is also based on a master equation, but the sequence of states generated in the simulation is not a realization of the time evolution of the system and in consequence the simulation cannot be used to calculate transport properties.

The methods to be discussed are, in principle, applicable to any model, irrespective of the mechanism of atomic migration, defect concentration, etc.; but, in practice, limitations of computer speed and memory restrict both the complexity of the model and the properties which can be successfully studied in any particular case. Nevertheless, such simulations play an important role in providing results for models for which analytical methods are not yet available and in providing checks on the accuracy of current analytical results. Our discussion of these topics will be brief since several detailed reviews are available (see, e.g. Murch, 1984a, Murch and Zhang, 1990, Kehr and Binder, 1984).

Transport properties in the linear response region may be calculated in two ways, i.e. either from the simulation of a system at thermodynamic equilibrium or from the simulation of a system in the presence of some thermodynamic force. In the first, the transport coefficients are calculated from the linear response expressions of Chapter 6, as we describe below. In the second several different 'computer experiments' are possible (see, e.g. Murch, 1984a) of which we describe only one, which is perhaps the simplest and most widely used. In this, the transport coefficients are calculated from the matter fluxes induced by a known external field applied to a chemically homogeneous system. We describe both methods as they apply to a simple-cubic binary alloy containing vacancies, but the modifications required for other models are usually minor.

In a typical simulation one employs a lattice of sites with periodic boundary conditions. For the alloy model a system of $20 \times 20 \times 20$ or more sites is needed. An array in the computer memory stores the current occupancy of each lattice site (e.g. A, B or vacancy). For systems without long-range order the atoms and vacancies are initially assigned at random to the sites and the system is then taken to equilibrium by performing a Metropolis grand canonical ensemble simulation (see, e.g. Murch, 1982a). If the system of interest has long-range order the initial configuration is taken as a perfect, ordered structure in order to avoid the occurrence of metastable, but long-lived antiphase boundaries. The use of the grand canonical ensemble in this initial simulation implies that the numbers of atoms of the various species present fluctuate and this provides a more rapid and surer approach to thermal equilibrium than the use of the canonical ensemble where these numbers are held constant. Once equilibrium is attained the main simulation is carried out on a closed system in the following way.

The first step is to select a vacancy at random and then to select, also at random, a site next to the chosen vacancy. If the chosen neighbour site is vacant one returns to the start and again selects a vacancy at random, but when it is occupied by an atom the jump frequency $w$ for exchange of the vacancy with that atom is calculated. For a random alloy model this frequency, by definition (§10.8), depends just on the species of the atom, but for non-random alloy models the frequency will also depend on the nature and configuration of other nearby atoms (see §13.3.1). If an external field is applied then the jump frequency contains an additional factor. For example, when an electric field of strength $E$ is applied along the $x$-axis, this factor for a jump along the $x$-axis will be $\exp(\pm\alpha/2)$, where $\alpha = qEa/kT$. Here $q$ is the charge on the jumping atom, $a$ is the nearest-neighbour spacing along the $x$-axis and the sign depends on the relative directions of the jump and of the field. Throughout the simulation the jump frequencies are normalized so that the highest possible jump frequency is unity. The calculated value of $w$ is compared with a random number, $R$, uniform on the interval $(0, 1)$. If $R < w$ the jump is accepted and the atom and vacancy are interchanged, thus creating the next configuration. If $R \geq w$ the jump attempt is rejected and the configuration remains unchanged. The process is now repeated by again selecting a vacancy at random and following the same sequence of operations.

It follows that over a run of very many vacancy jumps a given type of jump is made with a probability equal to its normalized jump frequency. A typical run may contain of the order of $10^6$–$10^7$ vacancy moves, depending on the problem to hand and the precision sought. The manner in which the run is monitored and the data used depends on which of the two types of simulation 'experiment' is employed. We now describe the theory of these in greater detail, dealing first with the system in thermodynamic equilibrium and then with the system subject to a force field.

*Use of generalized Einstein expressions for equilibrium simulations*

The generalized Einstein relation (6.5.14) may be written for cubic lattices as

$$L_{ij} = \langle \Delta \mathbf{R}_i \cdot \Delta \mathbf{R}_j \rangle / (6VkTt), \qquad \text{(A13.1.1)}$$

in which $\Delta \mathbf{R}_i$ is the sum of the vector displacements of all atoms of species $i$ in time $t$. We shall follow the procedure of the rest of this chapter in separating the calculation of the $L$-coefficients into two parts, the calculation of the uncorrelated parts, e.g. $L_{ii}^{(0)}$, and of the correlation functions defined, as in (13.2.7), by

$$f_{ii} = L_{ii}/L_{ii}^{(0)}, \qquad \text{(A13.1.2a)}$$

$$f_{ij}^{(i)} = L_{ij}/L_{ii}^{(0)} = L_{ji}/L_{ii}^{(0)} \qquad \text{(A13.1.2b)}$$

and

$$f_{ij}^{(j)} = L_{ij}/L_{jj}^{(0)} = L_{ji}/L_{jj}^{(0)}. \qquad \text{(A13.1.2b)}$$

For the vacancy mechanism, non-diagonal phenomenological coefficients are zero when correlations between successive jumps are neglected, while the diagonal coefficients in the same approximation may be written in the form

$$L_{ii}^{(0)} = \nu_i s^2/6VkTt. \qquad \text{(A13.1.3)}$$

Here $\nu_i$ is the total number of jumps made by all atoms of species $i$ in time $t$ and $s$ is the magnitude of the jump distance. From (A13.1.1)–(A13.1.3) we obtain the following expression for the correlation function:

$$f_{ii} = \langle \Delta \mathbf{R}_i \cdot \Delta \mathbf{R}_i \rangle / \nu_i s^2 \qquad \text{(A13.1.4a)}$$

$$f_{ij}^{(j)} = \langle \Delta \mathbf{R}_i \cdot \Delta \mathbf{R}_j \rangle / \nu_j s^2, \qquad (i, j = \text{A, B}). \qquad \text{(A13.1.4b)}$$

These expressions (A13.1.2) and (A13.1.4) separate the calculation of the $L$-coefficients into two parts, the calculation of the uncorrelated factors $L_{ii}^{(0)}$ and the correlation functions. They are generalizations of the expression of the tracer diffusion coefficient $D_i^*$ of species $i$ through the Einstein relation (10.2.4) as the product of its value $D_i^{(0)}$, neglecting correlations, with the tracer correlation factor $f_i$, (cf. 10.3.3),

$$D_i^* = D_i^{(0)} f_i, \tag{A13.1.5}$$

where

$$\begin{aligned} D_i^{(0)} &= kTL_{ii}^{(0)}/nc_i \\ &= \nu_i s^2/(6Vnc_i t) \equiv m_i s^2/6t, \end{aligned} \tag{A13.1.6}$$

and

$$f_i = \langle(\Delta \mathbf{r}_i)^2\rangle/(m_i s^2), \tag{A13.1.7}$$

in which $\langle(\Delta \mathbf{r}_i)^2\rangle$ is the mean-square displacement of *one* atom of species $i$ and $m_i$ is the mean number of jumps made by such an atom in time $t$.

The calculation of the correlation functions and the tracer correlation factors from a simulation by means of eqn. (A13.1.4) is fundamentally straightforward, since only the various vectors $\Delta\mathbf{R}_i$, $\Delta\mathbf{r}_i$ and the numbers of jumps $\nu_i$ and $m_i$ are required. The simulation may be divided into consecutive blocks according to some criterion (e.g. that each block contains that number of vacancy jumps which will give on average a certain minimum number of jumps of the slowest moving atom in each block), and the correlation functions calculated for each block. The mean values, their standard errors and possibily other statistical data are then calculated from these results.

The size of the system, the number of jumps made in a block and the number of blocks have all to be determined by trial and error, but there are two further considerations (see, e.g. Murch, 1984b). In the first place, it is clear from the nature of the analytical calculations of the $L$-coefficients in Chapter 11 that each atom must undergo enough jumps in each block to yield a proper measure of its correlations. In addition, it is important to average over the histories of a large enough number of atoms in order to reduce statistical fluctuations. However, rather than use a very big system of, say, $10^6$ sites it is usually more economical with current computers to use a smaller system and to do the same run many times, to take many blocks and average the results. It is, of course, also important to determine the lower limit of system size below which spurious correlations arise solely from the smallness of the system.

The uncorrelated parts $L_{ii}^{(0)}$ and $D_i^{(0)}$ can be expressed as equilibrium ensemble averages as in (6.5.11), and they are usually calculated as such. A convenient form is

$$D_i^{(0)} = z\langle p_{iv}\rangle\langle w_i\rangle s^2/6, \tag{A13.1.8}$$

in which $\langle p_{iv}\rangle$ is the mean site-fraction of vacancies at a particular nearest-neighbour site of a particular atom of species $i$ and $\langle w_i \rangle$ is the mean jump rate for an $i$ atom which is nearest neighbour to a vacancy. For the random alloy model these quantities are just $c_v$ and $w_i$ respectively, but for other models they have to be calculated in the simulation. In a system with only a small concentration of vacancies the quantity $\langle p_{vi} \rangle$, i.e. the mean site-fraction of atoms of species $i$ at a particular nearest-neighbour site to a particular vacancy, is more easily calculated in a simulation, whence $\langle p_{iv} \rangle$ can be found since $c_v \langle p_{vi} \rangle = c_i \langle p_{iv} \rangle$. In practice, the functions $\langle p_{vi} \rangle$ and $\langle w_i \rangle$ are conveniently calculated in the same simulation as the tracer correlation factors $f_i$. For a typical nearest-neighbour model of a binary alloy such a simulation run requires only a few minutes of mainframe computer central processing time for results of high precision (Murch, 1982a), whereas a simulation run for the correlation functions may require several hours to attain even a lower precision (see e.g. Zhang, Oates and Murch, 1988).

The use of the Einstein expression has usually provided the preferred route to the tracer diffusion coefficients. However, for other transport coefficients a second approach has often been employed, which we now consider.

*Use of simulations with an external field*

This method of computing the transport coefficients, $L_{ij}$, is based on the recognition that they are independent of the forces applied. Electrical mobilities obtained by attaching notional charges to the atoms and computing the fluxes of atoms in an electrical field are then equated to those obtained from the phenomenological equations of §5.9. But these charges may be varied at our convenience, with the result that all the transport coefficients may be separately obtained.

We therefore suppose that an electric field of strength $E$ is applied along the $x$-axis, defined as the $\langle 100 \rangle$ direction in the model simple-cubic binary alloy used as an example in this section. The phenomenological equations are then (cf. §5.9)

$$\left.\begin{aligned} J_A &= (L_{AA} q_A + L_{AB} q_B) E, \\ J_B &= (L_{BA} q_A + L_{BB} q_B) E. \end{aligned}\right\} \qquad \text{(A13.1.9)}$$

The electrical mobilities which follow from these equations have already been given in §5.9 (see eqn. (5.9.3), etc.).

Now in the simulation, one can determine the number $\Delta \nu_i^x$ of $i$-atom jumps in the $+x$-direction minus the number in the $-x$-direction as well as the total numbers $\nu_i^{(\alpha)}$ ($\alpha = x, y, z$) of $i$ atom jumps along the $\alpha$-axis (in both directions) for each species of atom. The mobility $u_i$ of atoms of species $i$ is then

$$u_i = a \Delta \nu_i^x / (V n c_i E t). \qquad \text{(A13.1.10)}$$

Since all the jumps in the $y$- and $z$-directions occur at the equilibrium rate, the expression for the uncorrelated part of $L_{ii}$, (A13.1.3), becomes, for the present 'computer experiment',

$$L_{ii}^{(0)} = \tfrac{3}{2}(v_i^y + v_i^z)a^2/(6VkTt) \qquad \text{(A13.1.11)}$$

and the mobility may therefore be expressed as

$$u_i = 4\Delta v_i^x L_{ii}^{(0)} kT/(nc_i Ea(v_i^y + v_i^z)). \qquad \text{(A13.1.12)}$$

(We note that, for the simple-cubic lattice for which nearest-neighbour jumps occur only along the $x$, $y$ and $z$ crystal axes, $s = a$.)

Three different choices of the atom charges are particularly useful in the simulation experiments. Firstly, if one chooses $q_A = q$, $q_B = 0$ and, for each species, equates the mobility calculated from the flux equation (A13.1.9) with that in (A13.1.12), one finds the following expressions for the correlation functions

$$\left.\begin{aligned} f_{AA} &= 4\Delta v_A^x/\alpha(v_A^y + v_A^z), \\ f_{AB}^{(B)} &= 4\Delta v_B^x/\alpha(v_B^y + v_B^z), \end{aligned}\right\} \qquad \text{(A13.1.13)}$$

where $\alpha = qEa/kT$ is, as noted earlier, precisely that quantity which describes the effect of the external field upon the jump frequencies. In a similar manner the choice $q_A = 0$, $q_B = q$ gives similar expressions for $f_{BB}$ and $f_{AB}^{(A)}$, while the choice $q_A = q_B = q$ leads to expressions for $(f_{AA} + f_{AB}^{(A)})$ and $(f_{BB} + f_{AB}^{(B)})$ in terms of the evaluated quantities.

The disadvantage of this method compared to that based on the Einstein expression is that, in order to avoid excessively long computer runs, it may be necessary to use values of $\alpha$ so large that the linear response approximation is not valid. Some extrapolation of the results to small values of $\alpha$ is then necessary to obtain precise correlation functions (see e.g. Moleko *et al.*, 1989). Nevertheless in simulations of $f_{ii}$ and $f_{ij}^{(j)}$ Zhang *et al.* (1988) have compared the two methods and shown that they make closely similar predictions.

# Appendix 13.2

## Derivation of Moleko–Allnatt sum rule

In this appendix we show how to derive (13.2.12). For this purpose it is convenient to let the suffixes $i, j, k$, etc. denote all the types of structural element, i.e. vacancies (V) as well as species of atom. With this notation we then have the relation

$$\sum_i \rho_\beta^{(i)}(\mathbf{l}) = 1, \qquad \text{(A13.2.1)}$$

which expresses the fact that in any state $\beta$ of the system a lattice site $\mathbf{l}$ must be occupied by one of the species $i$, i.e. either by an atomic component or by a vacancy.

We next define a set of functions $h_{ij:i_o j_o}$, which includes the particular functionsy $h_{\mathrm{V}i:\mathrm{V}j}$ required to calculate the correlation functions $f_{ij}^{(j)}$:

$$h_{ij:i_o j_o} = \sum_{\mathbf{s}} \mathbf{s}\cdot\mathbf{s}_0/s^2 \sum_{\mathbf{l}} \bar{\psi}_{ij:i_o j_o}(\mathbf{l}, \mathbf{l}-\mathbf{s}; 0{:}\, \mathbf{l}_o, \mathbf{l}_o - \mathbf{s}_0). \qquad \text{(A13.2.2)}$$

By summing over all species $j$ and applying (A13.2.1) one obtains

$$\sum_j h_{ij:i_o j_o} = 0, \qquad \text{(A13.2.3)}$$

since the summation over $\mathbf{s}$ contained in this expression leads to zero (Bravais lattice). It is also straightforward to obtain from the definition (A13.1.2) the relation

$$h_{ij:i_o j_o} = -h_{ji:i_o j_o}. \qquad \text{(A13.2.4)}$$

Since this is valid when $i = j$, which shows that $h_{ii:i_o j_o} = 0$, we may rewrite (A13.2.3) as

$$\sum_{j \neq i} h_{ij:i_o j_o} = 0. \qquad \text{(A13.2.5)}$$

By choosing $i = i_0 = \mathrm{V}$ and replacing $j$ by $i$ and $j_0$ by $j$, we therefore have

$$\sum_{i \neq \mathrm{V}} h_{\mathrm{V}i:\mathrm{V}j} = 0 \tag{A13.2.6}$$

and the sum rule (13.2.12) follows at once by expressing $h_{\mathrm{V}i:\mathrm{V}j}$ in terms of $f_{ij}^{(j)}$ by means of (13.2.8).

# Appendix 13.3

## Analysis for correlation functions of a random lattice-gas

In this appendix we sketch how the kinetic equation obtained by inserting the approximate expression (13.2.29) for the three-site terms into (13.2.18) is transformed into the linear relation (13.2.31) between the correlation functions $h_{Vi:Vj}$. We first set $\mathbf{m} = \mathbf{l} - \mathbf{r}$ and $\mathbf{m}_0 = \mathbf{l}_0 - \mathbf{r}_0$, and define

$$\phi_{V\Delta i:Vj}(\mathbf{r}; t:\mathbf{r}_0) = \sum_{\mathbf{l}} \psi_{V\Delta i:Vj}(\mathbf{l}, \mathbf{l} - \mathbf{r}; t:\mathbf{l}_0, \mathbf{l}_0 - \mathbf{r}_0). \tag{A13.3.1}$$

By (13.2.9) and (13.2.14) we can express the required correlation function $h_{Vi:Vj}$ in terms of this function, thus

$$h_{Vi:Vj} = \sum_{\mathbf{s}} (\mathbf{s}\cdot\mathbf{s}_0/s^2) \int_0^\infty \mathrm{d}t \phi_{V\Delta i:Vj}(\mathbf{s}; t:\mathbf{s}_0). \tag{A13.3.2}$$

The next step is to obtain the kinetic equation for $\phi$. This we do by replacing the three-site terms in (13.2.18) by the approximate expression (13.2.29) and then summing both sides of the resulting equation over all $\mathbf{l}$. We again contract the notation as in §§13.2.1 *et seq.* and obtain the result

$$-zA(\mathbf{r}) + \sum_{\mathbf{r}'} \theta(\mathbf{r}')A(\mathbf{r} - \mathbf{r}') = C(\mathbf{r}) + \theta(\mathbf{r})D(\mathbf{r}), \tag{A13.3.3}$$

in which

$$A(\mathbf{r}) = f_V w \phi_{V\Delta i}(\mathbf{r}), \tag{A13.3.4a}$$

$$C(\mathbf{r}) = \frac{\mathrm{d}\phi_{Vi}(\mathbf{r})}{\mathrm{d}t} \tag{A13.3.4b}$$

and

$$D(\mathbf{r}) = -w_i\phi_{V\Delta i}(-\mathbf{r}) + (w_i - wf_V)\phi_{V\Delta i}(\mathbf{r}) + c_i \sum_{k'} w_k(\phi_{V\Delta k}(-\mathbf{r}) - \phi_{V\Delta k}(\mathbf{r})) + f_V w\phi_{V\Delta i}(0). \qquad \text{(A13.3.4c)}$$

In the course of the algebraic manipulations leading to (A13.3.3) we have set the dummy index $\mathbf{n}$ equal to $\mathbf{l} - \mathbf{r}'$ and we have also used the fact that the sum $\sum_{\mathbf{l}} \psi_V(\mathbf{l} - \mathbf{r})$, by its definition, is independent of $\mathbf{r}$: this has the consequence that all these one-site terms cancel out. Lastly, the assumed Bravais nature of the lattice ensures the cancellation of some remaining terms in $f_V$.

Since $\mathbf{r}$ and $\mathbf{r}'$ refer to discrete lattice points we can regard (A13.3.3) as a matrix equation of the form

$$-\mathbf{PA} = \mathbf{E} \qquad \text{(A13.3.5)}$$

in which $\mathbf{A}$ and $\mathbf{E}$ are column matrices formed from the elements $A(\mathbf{r})$ and $C(\mathbf{r}) + \theta(\mathbf{r})D(\mathbf{r})$ respectively. Most importantly, the matrix $\mathbf{P}$ is just that defined by (9.3.1a) with $w = 1$. We can therefore use the mathematics already set out in §9.3 to obtain the required solution $A$. Evidently then

$$A(\mathbf{r}) = -\frac{1}{z}\sum_{\mathbf{r}'} U(\mathbf{r} - \mathbf{r}', 1)(C(\mathbf{r}') + \theta(\mathbf{r}')D(\mathbf{r}')), \qquad \text{(A13.3.6)}$$

in which $U(\mathbf{r}, 1)$ is the generating function for vacancy random walks from the origin to $\mathbf{r}$ in a perfect lattice (cf. §9.3.2).

Equation (A13.3.6) may be transformed into a linear relation between the correlation functions $h_{Vi:Vj}$, after setting $\mathbf{r} = \mathbf{s}$ and $\mathbf{r}' = \mathbf{s}_0$, by multiplication by $\mathbf{s}\cdot\mathbf{s}_0/s^2$ followed by summation over $\mathbf{s}$ and integration with respect to time. In carrying out these operations we need to recognize three relations:

(i) The average cosine of the angle between successive jumps of a tracer, $T_0$, may be written as

$$T_0 = -\frac{1}{z}\sum_{\mathbf{s}} (\mathbf{s}\cdot\mathbf{s}_0/s^2)U(|\mathbf{s} - \mathbf{s}_0|), \qquad \text{(A13.3.7)}$$

in which the first tracer jump is from $\mathbf{s}_0$ to the origin and the second from the origin to $\mathbf{s}$. The average cosine $T_0$ may be re-expressed in terms of the tracer correlation factor $f_0$ in the pure lattice by (10.4.3).

(ii) For cubic lattices we have the identity

$$-\frac{1}{z}\sum_{\mathbf{s}} (\mathbf{s}\cdot\mathbf{s}_0/s^2)\sum_{\mathbf{s}'} U(|\mathbf{s} - \mathbf{s}'|)D(\mathbf{s}':\mathbf{s}_0) = T_0 \sum_{\mathbf{s}} (\mathbf{s}\cdot\mathbf{s}_0/s^2)D(\mathbf{s}:\mathbf{s}_0). \qquad \text{(A13.3.8)}$$

(iii) The contribution of the term in $C(\mathbf{r})$ on the right side of (A13.3.6) becomes

$$\frac{1}{z}\sum_{\mathbf{s}}(\mathbf{s}\cdot\mathbf{s}_0/s^2)\sum_{\mathbf{r}}U(|\mathbf{s}-\mathbf{r}|)\phi_{\mathrm{V}i:\mathrm{V}j}(\mathbf{r};0:\mathbf{s}_0) = -T_0c_\mathrm{V}c_j[(1-c_\mathrm{V})\delta_{ij}-c_i], \tag{A13.3.9}$$

where the last form follows by evaluation of the equilibrium average $\phi_{\mathrm{V}i:\mathrm{V}j}(\mathbf{r};0:\mathbf{s}_0)$.

The result of these manipulations is (13.2.31), in which terms in $c_\mathrm{V}{}^2$ have, of course, been neglected.

# Appendix 13.4

## The calculation of Holdsworth and Elliott

Frequent reference has been made in §§13.2.2–13.2.4 to the calculations of Holdsworth and Elliott (1986). The following brief account is included in order to relate the structure of their calculations to the development in §13.2.

Holdsworth and Elliott (1986) base their calculation of the tracer correlation factor in a random lattice gas on an analysis of the spatial Fourier transform of the equation of motion for the time correlation function

$$\chi_{T:T}(\mathbf{l}; t:\mathbf{l}_0) = \sum_{\beta,\alpha} \rho_{\beta}^{(T)}(\mathbf{l}) G_{\beta\alpha}(t) \rho_{\alpha}^{(T)}(\mathbf{l}_0) p_{\alpha}^{(0)}, \tag{A13.4.1}$$

in which T denotes a tracer component. The tracer diffusion coefficient can be identified by taking the long-wavelength and long-time limits of the equation of motion for $\chi_{T:T}$. The advantage of this procedure is that the self-correlation function $S_s(\mathbf{q}, \omega)$ appearing in incoherent quasi-elastic scattering (cf. §1.8) can be simultaneously studied.

The calculation requires approximations for two-site correlation functions of the type $\chi_{T\Delta V:T}(\mathbf{l}, \mathbf{m}; t:\mathbf{l}_0)$ and $\chi_{T\Delta i:T}(\mathbf{l}, \mathbf{m}; t:\mathbf{l}_0)$ whose equations of motion contain three-site functions. The self-consistent decoupling approximation employed, is

$$\theta(\mathbf{l} - \mathbf{n}) \chi_{T\Delta V\Delta i}(\mathbf{l}, \mathbf{n}, \mathbf{m}) = -\theta(\mathbf{l} - \mathbf{n}) c_V (1 - f_T) \chi_{T\Delta i}(\mathbf{l}, \mathbf{m}), \tag{A13.4.2}$$

and a similar equation with $i$ replaced by V. This approximation is suggested by the observation that the rather similar relation

$$\theta(\mathbf{l} - \mathbf{n}) \chi_{T\Delta V}(\mathbf{l}, \mathbf{n}) = -\theta(\mathbf{l} - \mathbf{n}) c_V (1 - f_T) \chi_T(\mathbf{l}) \tag{A13.4.3}$$

is sufficient to regain the expected Fick's law form for the equation of motion of $\chi_T(\mathbf{l})$ in the long-wavelength and long-time limit. The derivation follows the lines of that given for the decoupling approximation (13.2.29) which led to the Manning theory.

When terms of second order in the fluctuation variables are neglected ($f_T = 1$ in (A13.4.2)) the earlier approximate theory of Tahir-Kheli and Elliott (1983) is regained. The result for $f_T$ in a pure substance as a function of $c_V$ obtained from the latter theory and given as eqn. (13.2.44) is remarkably accurate.

# *References*

Abbink, H. C., and Martin, D. S. (1966) *J. Phys. Chem. Solids* **27**, 205.
Ablitzer, D. (1977) *Phil. Mag.* **36**, 391.
Ablitzer, D., Haeussler, J. P., and Sathyraj, K. V. (1983) *Phil. Mag.* **A47**, 515.
Aboagye, J. K., and Friauf, R. J. (1975) *Phys. Rev.* **B11**, 1654.
Abragam, A. (1961) *The principles of nuclear magnetism* (Clarendon Press: Oxford).
Ackermann, H., Heitjans, P., and Stöckmann, H.-J. (1983) in *Hyperfine interactions of radioactive nuclei*, Christiansen, J. (ed.) (Springer-Verlag: Berlin) p. 291.
Adda, Y., and Philibert, J. (1966) *La diffusion dans les solides* (Presses Universitaires de France: Paris).
Agullo-Lopez, F., Catlow, C. R. A., and Townsend, P. (1988) *Point defects in materials* (Academic: London).
Allen, M. P., and Tildesley, D. J. (1987) *Computer simulation of liquids* (Clarendon Press: Oxford).
Allnatt, A. R. (1965) *J. Chem. Phys.* **43**, 1855.
(1981) *J. Phys.* **C14**, 5453, 5467.
(1982) *J. Phys.* **C15**, 5605.
Allnatt, A. R., and Allnatt, E. L. (1984) *Phil. Mag.* **A49**, 625.
(1991) *Phil. Mag.* **A64**, 341.
Allnatt, A. R., Barbu, A., Franklin, A. D., and Lidiard, A. B. (1983) *Acta Metall.* **31**, 1307.
Allnatt, A. R., and Lidiard, A. B. (1987a) *Acta Metall.* **35**, 1555.
(1987b) *Rep. Prog. Phys.* **50**, 373.
(1988) *Proc. Roy. Soc.* **A420**, 417.
Allnatt, A. R., and Loftus, E. (1973) *J. Chem. Phys.* **59**, 2250 (errata 1979, *J. Chem. Phys.* **71**, 5388).
Allnatt, A. R., and Okamura, Y. (1986) in *Solute–defect interaction – theory and experiment*, Saimoto, S., Purdy, G. R., and Kidson, G. V. (eds.) (Pergamon: Toronto) p. 205.
Allnatt, A. R., and Rowley, L. A. (1970) *J. Chem. Phys.* **53**, 3217, 3232.
Alonso, J. A., and March, N. H. (1989) *Electrons in metals and alloys* (Academic: London and San Diego).
Andrew, E. R., and Tunstall, D. P. (1961) *Proc. Phys. Soc.* **78**, 1.
Anthony, T. R. (1975) in *Diffusion in solids – recent developments*, Nowick, A. S., and Burton, J. J. (eds.) (Academic: New York) p. 353.

Arita, M., Koiwa, M., and Ishioka, S. (1989) *Acta Metall.* **37**, 1363.
Atkinson, A. (1987) in *Adv. in Ceramics*, Vol. 23: Non-Stoichiometric Compounds, Catlow, C. R. A., and Mackrodt, W. C. (eds.) (Amer. Ceram. Soc.: Ohio) p. 3.
(1991) *UKAEA Report* AEA-InTech-0609 to be published in *Materials Science and Technology – a comprehensive treatment*, Cahn, R. W., Haasen, P., and Kramer, E. J. (eds.).
Bacon, D. J. (1988) *J. Nucl. Mater.* **159**, 176.
(ed.) (1991) *Phil. Mag.* **A63**, No. 5.
Bakker, H. (1970) *Phys. Stat. Solidi* **38**, 167.
(1979) *Phil. Mag.* **A40**, 525.
(1984) in *Diffusion in crystalline solids*, Murch, G. E., and Nowick, A. S. (eds.) (Academic: New York) p. 189.
Bakker, H., and van Winkel, A. (1980) *Phys. Stat. Sol. (a)* **61**, 543.
Bakker, H., van Winkel, A., Waegemaekers, A. A. H. J., van Ommen, A. H., Stolwijk, N. A., and Hatcher, R. D. (1985) in *Diffusion in solids: recent developments*, Dayananda, M. A. and Murch, G. E. (eds.) (Met. Soc. AIME: Warrendale, Pennsylvania) p. 39.
Barber, M. N., and Ninham, B. W. (1970) *Random and restricted walks* (Gordon and Breach: New York).
Barbu, A. (1980) *Acta Metall.* **28**, 499.
Bardeen, J., and Herring, C. (1952) in *Imperfections in nearly perfect crystals*, Shockley, W., Hollomon, J. H., Maurer, R., and Seitz, F. (eds.) (Wiley: New York) p. 261. (N.B. Eqns. (A.2) and (A.5) of this paper are incorrect.)
Bartels, A. (1987) *Mater. Sci. Forum* **15–18**, 1183.
Bartels, A., Bartusel, D., Kemkes-Sieben, C., and Lücke, K. (1987) *Mater. Sci. Forum* **15–18**, 1237.
Barton, W. A., and Sholl, C. A. (1976) *J. Phys.* **C9**, 4315.
(1980) *J. Phys.* **C13**, 2579.
Bauerle, J. E., and Koehler, J. S. (1957) *Phys. Rev.* **107**, 1493.
Becker, K. D. (1978) *Phys. Stat. Solidi (b)* **87**, 589.
(1982) *Zeits. für Naturforschung* **37A**, 697.
(1984) *Phys. Stat. Solidi (b)* **121**, 91.
Becker, K. D., Hamann, H., Kozubek, N., and Richtering, H. (1975) *Ber. Bunsenges. Phys. Chem.* **78**, 461.
Becker, K. D., and Richtering, H. (1974) *Ber. Bunsenges. Phys. Chem.* **78**, 461.
Becker, K. D., and von Wurmb, V. (1986) *Z. Phys. Chem. Neue Folge* **149**, 77.
Bée, M. (1988) *Quasi-elastic neutron scattering* (Adam Hilger: Bristol).
Bénière, F. A. (1983) in *Mass transport in solids*, Bénière, F. A., and Catlow, C. R. A. (eds.) (Plenum Press: New York) p. 21.
Bénière, F., Kostopoulos, D., and Bénière, M. (1980) *J. Phys. Chem. Solids* **41**, 727.
Bennett, C. H. (1975) in *Diffusion in solids – recent developments*, Nowick, A. S., and Burton, J. J. (eds.) (Academic Press: New York) p. 74.
Benoist, P., Bocquet, J.-L., and Lafore, P. (1977) *Acta Metall.* **25**, 265.
Bérces, G., and Kovács, I. (1983) *Phil. Mag.* **A48**, 883.
Beshers, D. N. (1973) in *Diffusion*, Aaronson, H. I. (ed.) (American Society for Metals: Ohio) p. 209.
Bloembergen, N., Purcell, E. M., and Pound, R. V. (1948) *Phys. Rev.* **73**, 679.
Bocquet, J.-L. (1974a) *CEA Rep.* R-4565.
(1974b) *Acta Metall.* **22**, 1.
(1986) *Acta Metall.* **34**, 571.
(1987) *Res. Mechanica* **22**, 1.

Bocquet, J.-L., Brébec, G., and Limoge, Y. (1983) in *Physical metallurgy*, 3rd edn., Vol. 1, Cahn, R. W., and Haasen, P. (eds.) (North-Holland: Amsterdam) p. 385.
Boswarva, I. M., and Franklin, A. D. (1965) *Phil. Mag.* **11**, 335.
Bowker, M., and King, D. A. (1978) *Surface Sci.* **71**, 583.
Brailsford, A. D., and Bullough, R. (1981) *Phil. Trans. Roy. Soc.* **302**, 87.
Brébec, G. (1973) *J. Physique* **34**, C9–421.
Brébec, G., and Benvenot, J. (1973) *Acta Metall.* **21**, 585.
Brown, F. C. (1972) in *Point defects in solids*, Vol. 1, Crawford Jr., J. H., and Slifkin, L. M. (eds.) (Plenum: New York and London) p. 491.
de Bruin, H. J., Bakker, H., and van der Mey, L. P. (1977) *Phys. Stat. Solidi* (*b*) **82**, 581.
de Bruin, H. J., Murch, G. E., Bakker, H., and van der Mey, L. P. (1975) *Thin Solid Films* **25**, 47.
Bucci, C., Fieschi, R., and Guidi, G. (1966) *Phys. Rev.* **148**, 816.
Bullough, R., and Newman, R. C. (1970) *Rep. Prog. Phys.* **33**, 101.
Burton, C. H., and Dryden, J. S. (1970) *J. Phys.* **C3**, 523.
Butrymowicz, D. B., and Manning, J. R. (1978) *Met. Trans.* **A9**, 947.
Cahn, J. W., and Larché, F. C. (1983) *Scr. Metall.* **17**, 927.
Capelletti, R. (1986) in *Defects in solids – modern techniques*, Chadwick, A. V., and Terenzi, M. (eds.) (Plenum: New York) p. 407.
Carlson, P. T. (1976) *Met. Trans. AIME* **7A**, 199.
Carslaw, H. W., and Jaeger, J. C. (1959) *Conduction of heat in solids*, 2nd edn. (Clarendon Press: Oxford).
Casimir, H. B. G. (1945) *Rev. Mod. Phys.* **17**, 343.
Catlow, C. R. A., Corish, J., Jacobs, P. W. M., and Lidiard, A. B. (1981) *J. Phys.* **C14**, L121.
Catlow, C. R. A., Corish, J., Quigley, J. M., and Jacobs, P. W. M. (1980) *J. Phys. Chem. Solids* **41**, 237.
Catlow, C. R. A., and Mackrodt, W. C. (eds.) (1982) *Computer simulation of solids* (Springer-Verlag: Berlin).
Cauvin, R. (1981) *CEA Rep.* R-5105.
Cauvin, R., and Martin, G. (1981) *Phys. Rev.* **B23**, 3333.
Chadwick, A. V. (1988) *Int. Rev. Phys. Chem.* **7**, 251.
(1991) *Phil. Mag.* **A64**, 983.
Chadwick, A. V., and Glyde, H. R. (1977) in *Rare gas solids*, Vol. 2, Klein, M. L., and Venables, J. A. (eds.) (Academic Press: New York and London) p. 1151.
Chadwick, A. V., Hammam, E.-S., Zeqiri, B., and Beech, F. (1986) *Mater. Sci. Forum* **7**, 317.
Chandra, S. (1981) *Superionic solids* (North Holland: Amsterdam).
Chandra, S., and Rolfe, J. (1971) *Can. J. Phys.* **49**, 2098.
Chaturvedi, D. K., and Allnatt, A. R. (1992) *Phil. Mag.* **A65**, 1169.
Chang, Y. A., and Neumann, J. P. (1982) *Prog. Solid State Chem.* **14**, 221.
Chemla, M. (1956) *Ann. Phys. Paris* **13**, 1959.
Cheng, C. Y., Wynblatt, P. P., and Dorn, J. E. (1967) *Acta Metall.* **15**, 1035, 1045.
Chezeau, J. M., and Strange, J. H. (1979) *Phys. Rep.* **53**, 1.
Christy, R. W. (1961) *J. Chem. Phys.* **34**, 1148.
Christy, R. W., and Dobbs, H. S. (1967) *J. Chem. Phys.* **46**, 722.
Chudley, C. T., and Elliott, R. J. (1961) *Proc. Phys. Soc.* **77**, 353.
Cohen, E. R., and Giacomo, P. (1987) *Symbols, units, nomenclature and fundamental constants in physics* (I.U.P.A.P.), reprinted from *Physica* **146A**, 1 (1987).
Cohen, M. H. and Reif, F. (1957) *Solid State Physics*, Vol. 5, Seitz, F., and Turnbull, D., (eds.) (Academic: New York) p. 322.

Collins, A. T. (1980) *J. Phys.* **C13**, 2641.
Compaan, K., and Haven, Y. (1956) *Trans. Faraday Soc.* **52**, 786.
(1958) *Trans. Faraday Soc.* **54**, 1498.
Conway, B. E. (1970) in *Physical Chemistry – an advanced treatise*, Vol. 9A, Eyring, H. (ed.) (Academic: New York) p. 1.
Corish, J., and Jacobs, P. W. M. (1973a) *Specialist periodical reports – surface and defect properties of solids*, Vol. 2 (Chemical Society: London) p. 160.
(1973b) *J. Phys.* **C6**, 57.
Corish, J., Jacobs, P. W. M., and Radhakrishna, S. (1977) *Specialist periodical reports – surface and defect properties of solids*, Vol. 6 (Chemical Society: London) p. 218.
Corish, J., and Mulcahy, D. C. A. (1980) *J. Phys.* **C13**, 6459.
Cotton, F. A. (1971) *Chemical applications of group theory*, 2nd edn. (Wiley-Interscience: New York).
Cottrell, A. H. (1975) *An introduction to metallurgy*, 2nd edn. (Edward Arnold: London).
Cotts, R. M. (1978) in *Topics in Applied Physics No. 28* (*Hydrogen in Metals. I. Basic Properties*), Alefeld, G., and Völkl, J. (eds.) (Springer-Verlag: Berlin) p. 227.
Cox, P. A. (1992) *Transition metal oxides – an introduction to their electronic structure and properties* (Clarendon Press: Oxford).
Crank, J. (1975) *The mathematics of diffusion*, 2nd edn. (Clarendon Press: Oxford).
Crawford, J. H., and Slifkin, L. M. (eds.) (1972) *Point defects in solids*, Vol. 1 (Plenum: New York).
(eds.) (1975) *Point defects in solids*, Vol. 2 (Plenum: New York).
Crolet, J. L., and Lazarus, D. (1971) *Solid State Commun.* **9**, 347.
Curtis, A. R., and Sweetenham, W. P. (1985) *FASCIMILE Release-H User's Manual Harwell Report* AERE-R 11771.
Dallwitz, M. J. (1972) *Acta Metall.* **20**, 1229.
Damköhler, R., and Heumann, T. (1982) *Phys. Stat. Sol. (a)* **73**, 117.
Darken, L. S. (1948) *Trans. Am. Inst. Min. (Metall.) Engrs.* **175**, 184.
Das, T. P., and Hahn, E. L. (1958) *Solid State Physics, Suppl.* **1**, *Nuclear Quadrupole Resonance Spectroscopy*.
Dattagupta, S., and Schroeder, K. (1987) *Phys. Rep.* **150**, 263.
Davies, G. (1977) *Chemistry and Physics of Carbon* **13**, 1.
Dederichs, P. H., Lehman, C., Schober, H. R., Scholz, A., and Zeller, R. (1978) *J. Nucl. Mater.* **69/70**, 176.
Denbigh, K. G. (1951) *The thermodynamics of the steady state* (Methuen: London).
Dieckmann, R. (1977) *Z. Phys. Chem. Neue Folge* **107**, 189.
Dieckmann, R., and Schmalzried, H. (1975) *Ber. Bunsenges. Phys. Chem.* **79**, 1108.
Dietrich, W., Fulde, P., and Peschel, I. (1980) *Adv. Phys.* **29**, 527.
Differt, K., Seeger, A., and Trost, W. (1987) *Mater. Sci. Forum* **15–18**, 99.
Doan, N. V. (1971) *J. Phys. Chem. Solids* **32**, 2135.
(1972) *J. Phys. Chem. Solids* **33**, 2161.
Domian, H. A., and Aaronson, H. I. (1964) *Trans. AIME* **230**, 44.
Dorn, J. E., and Mitchell, J. B. (1966) *Acta Metall.* **14**, 71.
Dreyfus, R. W. (1961) *Phys. Rev.* **121**, 1675.
Dryden, J. S., and Heydon, R. G. (1978) *J. Phys.* **C11**, 393.
Durham, W. B., and Schmalzried, H. (1987) *Ber. Bunsenges, Phys. Chem.* **91**, 556.
Dworschak, F., and Koehler, J. S. (1965) *Phys. Rev.* **140**, A941.
Eisenstadt, M., and Redfield, A. G. (1963) *Phys. Rev.* **132**, 635.
El-Meshad, N., and Tahir-Kheli, R. A. (1985) *Phys. Rev.* **B32**, 6176.
Emsley, J. (1989) *The elements* (Clarendon Press: Oxford).

Eshelby, J. D. (1956) *Solid State Physics*, Vol. 3, Seitz, F., and Turnbull, D. (eds.) (Academic: New York) p. 79.
Evans, T., and Qi, Z. (1982) *Proc. Roy. Soc.* **A381**, 159.
Fahey, P. M., Griffin, P. B., and Plummer, J. D. (1989) *Rev. Mod. Phys.* **61**, 289.
Fair, R. B. (1981) *Applied Solid State Science Suppl. 2 Part B (Silicon integrated circuits)*, Kahng, D. (ed.) (Academic: New York) p. 1.
Faulkner, R. G. (1981) *J. Mater. Sci.* **16**, 373.
Faupel, F., and Hehenkamp, T. (1986) *Phys. Rev.* **B34**, 2116.
Fedders, P. A., and Sankey, O. F. (1978) *Phys. Rev.* **B18**, 5938.
Feder, R., and Nowick, A. S. (1958) *Phys. Rev.* **109**, 1959.
Field, J. E. (ed.) (1979) *The properties of diamond* (Academic: New York).
Figueroa, D. R., Strange, J. H., and Wolf, D. (1979) *Phys. Rev.* **B19**, 148.
Fitts, D. D. (1962) *Non-equilibrium thermodynamics* (McGraw-Hill: New York).
Fletcher, N. H. (1970) *The chemical physics of ice* (Cambridge University Press: London).
Flynn, C. P. (1972) *Point defects and diffusion* (Clarendon Press: Oxford).
(1987) *Mater. Sci. Forum* **15–18**, 281.
Flynn, C. P., and Stoneham, A. M. (1970) *Phys. Rev.* **B1**, 3966.
Fong, E. K. (1972) in *Physics of electrolytes*, Vol. 1, Hladik, J. (ed.) (Academic: London and New York) p. 79.
de Fontaine, D. (1979) *Solid State Physics*, Vol. 34, Ehrenreich, H., Seitz, F., and Turnbull, D. (eds.) (Academic: New York) p. 73.
Forster, D. (1975) *Hydrodynamic fluctuations, broken symmetry and correlation functions* (Benjamin: Reading, MA).
Fox, L. (1974) *Moving boundary problems in heat flow and diffusion* (Clarendon Press: Oxford).
Frank, W., Gösele, U., Mehrer, H., and Seeger, A. (1984) in *Diffusion in crystalline solids*, Murch, G. E., and Nowick, A. S. (eds.) (Academic: Orlando) p. 63.
Franklin, A. D. (1965) *J. Res. N.B.S.* **69A**, 301.
(1972) in *Point defects in solids*, Vol. 1, Crawford, J. H., and Slifkin, L. M. (eds.) (Plenum: New York) p. 1.
Franklin, A. D., Crissman, J. M., and Young, K. F. (1975) *J. Phys.* **C8**, 1244.
Franklin, A. D., and Lidiard, A. B. (1983) *Proc. Roy. Soc.* **A389**, 405.
(1984) *Proc. Roy. Soc.* **A392**, 457.
Franklin, A. D., Shorb, A., and Wachtman, J. B. (1964) *J. Res. N.B.S.* **68A**, 425.
Franklin, A. D., and Young, K. F. (1982) *J. Phys. Chem. Solids* **43**, 357.
Franklin, W. M. (1975) in *Diffusion in solids – recent developments*, Nowick, A. S., and Burton, J. J. (eds.) (Academic: New York) p. 1.
Fredericks, W. J. (1975) in *Diffusion in solids – recent developments*, Nowick, A. S., and Burton, J. J. (eds.) (Academic: New York) p. 381.
Friauf, R. J. (1969) *J. Phys. Chem. Solids* **30**, 429.
Fröhlich, H. (1958) *Theory of dielectrics*, 2nd edn. (Clarendon Press: Oxford).
Fuller, R. G. (1972) in *Point defects in solids*, Vol. 1, Crawford, J. H., and Slifkin, L. M. (eds.) (Plenum: New York) p. 103.
Funke, K. (1989) in *Superionic solids and solid electrolytes*, Lasker, A. L., and Chandra, S. (eds.) (Academic: San Diego) p. 569.
(1991) *Phil. Mag.* **A64**, 1025.
Gerl, M. (1967) *J. Phys. Chem. Solids* **28**, 725.
Gerstein, B. C., and Dybowski, C. R. (1985) *Transient techniques in N.M.R. of solids* (Academic: New York).

Gillan, M. J. (1983) in *Mass transport in solids*, Benière, F. A., and Catlow, C. R. A. (eds.) (Plenum: New York) p. 227.
(1985) *Physica* **131 B&C**, 157.
(1988) *Phil. Mag.* **A58**, 257.
(1989) in *Ionic solids at high temperatures*, Stoneham, A. M. (ed.) (World Scientific: Singapore) p. 170.
Gorsky, V. S. (1935) *Phys. Z. Sowjetunion* **8**, 457.
Greene, M. H., Batra, A. P., Lowell, R. C., Meyer, R. O., and Slifkin, L. M. (1971) *Phys. Stat. Solidi (a)* **5**, 365.
de Groot, S. R. (1952) *Thermodynamics of irreversible processes* (North-Holland: Amsterdam).
de Groot, S. R., and Mazur, P. (1962) *Non-equilibrium thermodynamics* (North-Holland: Amsterdam).
Guggenheim, E. A. (1952) *Mixtures* (Clarendon Press: Oxford).
(1967) *Thermodynamics*, 5th edn. (North-Holland: Amsterdam).
Gunther, L. (1976) *J. Physique* **37**, C6–15.
(1986) in *Solute-defect Interaction – theory and experiment*, Saimoto, S., Purdy, G. R., and Kidson, G. V. (eds.) (Pergamon: Toronto) p. 175.
Gunther, L., and Gralla, B. (1983) *J. Phys.* **C16**, 1863.
Haase, R. (1969) *Thermodynamics of irreversible processes* (Addison Wesley: Reading MA).
Hagenschulte, H., and Heumann, T. (1989a) *J. Phys.: Condens. Matter* **1**, 3601.
(1989b) *Phys. Stat. Solidi (b)* **154**, 71.
Hamman, E.-S. (1986) *Studies of the electrical properties of solid ionic halides and oxides*, Ph.D. thesis, University of Kent.
Hanlon, J. E. (1960) *J. Chem. Phys.* **32**, 1492.
Harding, J. H. (1985) *Phys. Rev.* **B32**, 6861.
(1990) *Rep. Prog. Phys.* **53**, 1403.
Harding, J. H., and Tarento, R. J. (1986) *Materials Researach Society Symposium Proceedings* **60**, 299.
Harned, H. S., and Owen, B. B. (1958) *The physical chemistry of electrolyte solutions*, 3rd edn. (Reinhold: New York).
Harper, P. G., Hodby, J. W., and Stradling, R. A. (1973) *Rep. Prog. Phys.* **36**, 1.
Harries, D. R., and Marwick, A. D. (1980) *Phil. Trans. Roy. Soc.* **A295**, 197.
Haus, J. W., and Kehr, K. W. (1987) *Phys. Rep.* **150**, 263.
Haven, Y., and van Santen, J. H. (1958) *Nuovo Cimento* **6** Suppl. 2, 605.
Havlin, S., and Ben-Avraham, D. (1987) *Adv. Phys.* **36**, 695.
Hayes, W. (ed.) (1974) *Crystals with the fluorite structure* (Clarendon Press: Oxford).
Hänngi, P., Talkner, P., and Borkovec, M. (1990) *Rev. Mod. Phys.* **62**, 251.
Hehenkamp, T. (1981) *Z. Metallkunde* **72**, 623.
(1986) in *Solute–defect interaction – theory and experiment*, Saimoto, S., Purdy, G. R., and Kidson, G. V. (Pergamon: Toronto) p. 241.
Heitjans, P. (1986) *Solid State Ionics* **18** & **19**, 50.
Heitjans, P., Faber, W., and Schirmer, A. (1991) *J. Non-Crystalline Solids* (in press).
Henderson, B., and Hughes, A. E. (eds.) (1976) *Defects and their structure in non-metallic solids* (Plenum: New York).
Hermeling, J., and Schmalzried, H. (1984) *Phys. Chem. Minerals* **11**, 161.
Herzig, C. (1981) *Z. Metallkunde* **72**, 601.
(1990) in *Diffusion in materials*, Laskar, A. L., Bocquet, J.-L., Brébec, G., and Monty, C. (eds.) (Kluwer: Dordrecht) p. 287.
Herzig, C., and Köhler, U. (1987) *Mater. Sci. Forum* **15–18**, 301.

Herzig, C., Neuhaus, J., Vieregge, K., and Manke, L. (1987) *Mater. Sci. Forum* **15–18**, 481.
Heumann, T. (1979) *J. Phys.* **F9**, 1997.
Heumann, T., and Rottwinkel, T. (1978) *J. Nucl. Mater.* **69/70**, 567.
Hill, T. L. (1956) *Statistical mechanics* (McGraw-Hill: New York).
(1960) *An introduction to statistical thermodynamics* (Addison-Wesley: Reading, MA).
Hirth, J. P., and Lothe, J. (1982) *Theory of dislocations*, 2nd edn. (Wiley Interscience: New York).
Ho, P., and Kwok, T. (1989) *Rep. Prog. Phys.* **52**, 301.
Hobbs, J. E., and Marwick, A. D. (1985) *Nucl. Instr. and Methods* **B9**, 169.
Holdsworth, P. C. W. (1985) D. Phil. thesis, University of Oxford.
Holdsworth, P. C. W., and Elliott, R. J. (1986) *Phil. Mag.* **A54**, 601.
Holdsworth, P. C. W., Elliott, R. J., and Tahir-Kheli, R. A. (1986) *Phys. Rev.* **B34**, 3221.
Hondros, E. D., and Seah, M. P. (1983) in *Physical metallurgy*, 3rd edn., Cahn, R. W., and Haasen, P. (eds.) (North-Holland: Amsterdam) p. 855.
Howard, R. E. (1966) *Phys. Rev.* **144**, 650.
Howard, R. E., and Lidiard, A. B. (1963) *J. Phys. Soc. Japan* **18** Suppl. 2, 197.
(1964) *Rep. Prog. Phys.* **27**, 161.
Howard, R. E., and Manning, J. R. (1967) *Phys. Rev.* **154**, 561.
Hubbard, P. S. (1970) *J. Chem. Phys.* **53**, 985.
Hughes, A. E. (1986) *Radiation Effects* **97**, 161.
Hull, D., and Bacon, D. J. (1984) *Introduction to dislocations*, 3rd edn. (Pergamon: Oxford).
Huntington, H. B. (1975) *Diffusion in solids – recent developments*, Nowick, A. S., and Burton, J. J. (eds.) (Academic: New York) p. 303.
Huntington, H. B., and Seitz, F. (1942) *Phys. Rev.* **61**, 315.
Huntley, F. A. (1974) *Phil. Mag.* **30**, 1075.
Hutchings, M. T. (1989) *Crystal Lattice Defects* **18**, 205.
Iorio, N. R., Dayananda, M. A., and Grace, R. E. (1973) *Met. Trans.* **4**, 1339.
Ishioka, S., and Koiwa, M. (1978) *Phil. Mag.* **37**, 517.
(1984) *Phil. Mag.* **A50**, 503.
Jackson, R. A., Murray, A. D., Harding, J. H., and Catlow, C. R. A. (1986) *Phil. Mag.* **A53**, 27.
Jacobs, P. W. M. (1983) in *Mass transport in solids*, Bénière, F. A., and Catlow, C. R. A. (eds.) (Plenum: New York) p. 81.
Jacobs, P. W. M., and Knight, P. C. (1970) *Trans. Faraday Soc.* **66**, 1227.
Jacobs, P. W. M., Rycerz, Z. A., and Mościński, J. (1991) *Adv. Solid-State Chem.* **2**, 113.
Jacucci, G. (1984) in *Diffusion in crystalline solids*, Murch, G. E., and Nowick, A. S. (eds.) (Academic: New York) p. 429.
Janek, J. (1992) *Dynamische Wechselwirkungen bei Transportvorgängen in Übergangsmetalloxyden* Dr. rer. Nat. thesis, University of Hanover.
Johnson, H. B., Tolar, N. J., Miller, G. R., and Cutler, I. (1969) *J. Phys. Chem. Solids* **30**, 31.
Johnson, R. A., and Lam, N. Q. (1976) *Phys. Rev.* **B13**, 4364.
Jones, M. J., and Le Claire, A. D. (1972) *Phil. Mag.* **26**, 1191.
Joyce, G. S. (1972) in *Phase transformations and critical phenomena*, Vol. 2, Domb, C., and Green, M. S. (eds.) (Academic: London) p. 375.
van Kampen, N. G. (1981) *Stochastic processes in physics and chemistry* (North-Holland: Amsterdam).

Karlsson, K. (1986) *Segregation and precipitation in austenitic stainless steels*, Ph.D. thesis, Chalmers University, Gothenburg.

Kaur, I., and Gust, W. (1989) *Fundamentals of grain and interphase boundary diffusion*, 2nd edn. (University of Stuttgart).

Kehr, K. W., and Binder, K. (1984) *Applications of the Monte Carlo method in statistical physics*, Binder, K. (ed.) (Springer-Verlag: Berlin) p. 181.

Kehr, K. W., Binder, K., and Reulein, S. M. (1989) *Phys. Rev.* **B39**, 4891.

Kehr, K. W., Kutner, R., and Binder, K. (1981) *Phys. Rev.* **B23**, 4931.

Kelly, S. W., and Sholl, C. A. (1987) *J. Phys.* **C20**, 5293.

Keneshea, F. J., and Fredericks, W. J. (1963) *J. Chem. Phys.* **38**, 1952.

(1965) *J. Phys. Chem. Solids* **26**, 1787.

Kidson, G. V. (1978) *Phil. Mag.* **A37**, 305.

(1985) *On the statistical thermodynamics of point defects in the dilute alloy model, Atomic Energy of Canada Ltd Report AECL*-8685.

Kikuchi, R. (1951) *Phys. Rev.* **81**, 988.

Kikuchi, R., and Sato, H. (1969) *J. Chem. Phys.* **51**, 161.

(1970) *J. Chem. Phys.* **53**, 2702.

(1972) *J. Chem. Phys.* **57**, 4962.

Kinoshita, C., Tomokiyo, Y., and Eguchi, T. (1978) *Phil. Mag.* **B38**, 221.

Kittel, C. (1986) *Solid state physics*, 6th edn. (Wiley: New York).

Kluin, J.-E. (1992) *Phil. Mag.* (in press).

Kohl, W., Mais, B., and Lücke, K. (1987) *Mater. Sci. Forum* **15–18**, 1219.

Köhler, U., Neuhaus, P., and Herzig, C. (1985) *Z. Metallkunde* **76**, 170.

Koiwa, M. (1978) *J. Phys. Soc. Japan* **45**, 781.

Koiwa, M., and Ishioka, S. (1979) *Phil. Mag.* **A40**, 625.

(1983a) *J. Stat. Phys.* **30**, 477.

(1983b) *Phil. Mag.* **A47**, 927.

Kreuzer, J. H. (1981) *Non-equilibrium thermodynamics and its statistical foundations* (Clarendon Press: Oxford).

Krivoglaz, M. A. (1969) *Theory of x-ray and thermal neutron scattering by real crystals* (Plenum: New York).

Kröger, F. A. (1974) *The chemistry of imperfect crystals*, 2nd edn., Vols. 2 and 3 (North-Holland: Amsterdam).

Kumagai, A., Iijima, Y., and Hirano, K. (1983) in *DIMETA-82 Diffusion in metals and alloys*, Kedves, F. J., and Beke, D. L. (eds.) (Trans. Tech. Publications: Switzerland) p. 389

Kuper, A. B., Lazarus, D., Manning, J. R., and Tomizuka, C. T. (1956) *Phys. Rev.* **104**, 1536.

Kusunoki, K., Tsumuraya, K., and Nishikawa, S. (1981) *Trans. Jap. Inst. Met.* **22**, 501.

Kutner, R. (1981) *Phys. Lett.* **A81**, 239.

Landau, D. P. (1979) in *Monte Carlo methods in statistical physics*, Binder, K. (ed.) (Springer-Verlag: Berlin) p. 121.

(1984) in *Applications of the Monte Carlo method in statistical physics*, Binder, K. (ed.) (Springer-Verlag: Berlin) p. 93.

Lannoo, M. (1986) in *Current issues in semiconductor physics*, Stoneham, A. M. (ed.) (Adam Hilger: Bristol and Boston) p. 27.

Larché, F. C., and Cahn, J. W. (1973) *Acta Metall.* **21**, 1051.

(1978a) *Acta Metall.* **26**, 53.

(1978b) *Acta Metall.* **26**, 1579.

(1982) *Acta Metall.* **30**, 1835.

Laskar, A. L., and Chandra, S. (eds.) (1989) *Superionic solids and solid electrolytes – recent trends* (Academic: San Diego).
Lay, K. W., and Whitmore, D. H. (1971) *Phys. Stat. Sol. (b)* **43**, 175.
Le Claire, A. D. (1953) *Prog. Met. Phys.* **4**, 265.
(1966) *Phil. Mag.* **14**, 1271.
(1970) in *Physical chemistry – an advanced treatise*, Vol. 10, Eyring, H., Henderson, D., and Jost, W. (eds.) (Academic: New York) p. 261.
(1978) *J. Nucl. Mater.* **69/70**, 70.
(1983) in *DIMETA-82 Diffusion in metals and alloys*, Kedves, F. J., and Beke, D. L. (eds.) (Trans. Tech. Publications: Switzerland) p. 82.
Le Claire, A. D., and Rabinovitch, A. (1984) in *Diffusion in crystalline solids*, Murch, G. E., and Nowick, A. S. (eds.) (Academic: Orlando) p. 257.
Lewis, G. V., Catlow, C. R. A., and Cormack, A. N. (1985) *J. Phys. Chem. Solids* **46**, 1227.
Lidiard, A. B. (1954) *Phys. Rev.* **94**, 29.
(1955) *Phil. Mag.* **46**, 1218.
(1957) *Handbuch der Physik* (*Springer-Verlag, Berlin*) **20**, 246.
(1960) *Phil. Mag.* **5**, 1171.
(1980) *Radiation Effects* **53**, 133.
(1983) in *Mass transport in solids*, Bénière, F. A., and Catlow, C. R. A. (eds.) (Plenum: New York) p. 43.
(1984) *UKAEA Harwell Laboratory Report* AERE-R 11367.
(1985) *Proc. Roy. Soc.* **A398**, 203.
(1986a) *Acta Metall.* **34**, 1487.
(1986b) *Proc. Roy. Soc.* **A406**, 107.
(1987) *Proc. Roy. Soc.* **A413**, 429.
Lidiard, A. B., and McKee, R. A. (1980) *J. Physique* **41**, C6–90.
Lindström, R. (1973a) *J. Phys.* **C6**, L197.
(1973b) *Chemical diffusion, self diffusion and ionic conductivity in KCl–RbCl solid solutions, Rep. No. 1*, Institute of Physics, University of Turku, Finland.
Lomer, W. M. (1958) in *Vacancies and other point defects in metals and alloys* (Institute of Metals: London) p. 79.
Lovesey, S. W. (1986) *Theory of neutron scattering from condensed matter*, 2 vols (Clarendon Press: Oxford).
Lur'e, B. G., Murin, A. N., and Murin, I. V. (1966) *Soviet Physics – Solid State* **9**, 1337.
Maass, P., Petersen, J., Bunde, A., Dieterich, W., and Roman, H. E. (1991) *Phys. Rev. Letts.* **66**, 52.
McCombie, C. W. (1962) in *Fluctuations, relaxation and resonance in magnetic systems*, ter Haar, D. (ed.) (Oliver and Boyd: Edinburgh) p. 183.
McCombie, C. W., and Lidiard, A. B. (1956) *Phys. Rev.* **101**, 1210.
McCombie, C. W., and Sachdev, M. (1975) *J. Phys.* **C8**, L413.
Macdonald, J. R. (1987) *Impedance spectroscopy* (Wiley: New York).
MacGillivray, I. R., and Sholl, C. A. (1986) *J. Phys.* **C19**, 4771.
McKee, R. A. (1977) *Phys. Rev.* **B15**, 5612.
Mackrodt, W. C., and Stewart, R. F. (1979) *J. Phys.* **C12**, 431.
McQuarrie, D. A. (1976) *Statistical mechanics* (Harper and Row: New York).
Maier, J. (1989) in *Superionic solids and solid electrolytes – recent trends*, Laskar, A. L., and Chandra, S. (eds.) (Academic: San Diego) p. 137.
Manning, J. R. (1964) *Phys. Rev.* **136**, A1758.
(1968) *Diffusion kinetics for atoms in crystals* (Van Nostrand: Princeton NJ).
(1970) *Met. Trans.* **1**, 499.

(1971) *Phys. Rev.* **B4**, 1111.
(1972) *Phys. Rev.* **B6**, 1344.
Manning, J. R., and Stark, J. P. (1975) *Phys. Rev.* **B12**, 549.
Mannion, W. A., Allen, C. A., and Fredericks, W. J. (1968) *J. Chem. Phys.* **48**, 1537.
March, N. H. (1978) *J. Nucl. Mater.* **69/70**, 490.
Martin, G., Cauvin, R., and Barbu, A. (1983) in *Phase transformations during irradiation*, Nolfi Jr., F. V. (ed.) (Applied Science Publishers: New York) p. 47.
Martin, M. (1991) *Mater. Sci. Rep.* **7**, 1.
Martin, M., and Schmalzried, H. (1985) *Ber. Bunsenges. Phys. Chem.* **89**, 124.
(1986) *Solid State Ionics* **20**, 75.
Matzke, Hj. (1987) *J. Chem. Soc. Faraday Trans.* **83**, 1124.
(1990) *J. Chem. Soc. Faraday Trans.* **86**, 1243.
Mayer, J. E., and Mayer, M. G. (1940) *Statistical mechanics* (Wiley: New York).
Mehrer, H. (1971) in *Atomic transport in solids and liquids*, Lodding, A. and Lagerwall, T. (eds.) (Verlag der Zeitschrift für Naturforschung: Tübingen) p. 221.
(ed.) (1990) Landolt–Börnstein *Numerical data and functional relationships in science and technology, new series, Group III, Vol. 26 Diffusion in solid metals and alloys* (Springer-Verlag: Berlin).
Miller, A. R. (1948) *The theory of solutions of high polymers* (Clarendon Press: Oxford).
Miller, J. W. (1969) *Phys. Rev.* **181**, 1095.
Mills, I. M. (Chairman) *et al.* (1988) *Quantities, units and symbols in physical chemistry* (Blackwell Scientific Publications: Oxford).
Moleko, L. K., and Allnatt, A. R. (1988) *Phil. Mag.* **A58**, 677.
Moleko, L. K., Allnatt, A. R., and Allnatt, E. L. (1989) *Phil. Mag.* **A59**, 141.
Moleko, L. K., Okamura, Y., and Allnatt, A. R. (1988) *J. Chem. Phys.* **88**, 2706.
Montet, G. (1973) *Phys. Rev.* **B7**, 650.
Montroll, E. G. (1964) *Symp. Appl. Math.* **16**, 193.
Monty, C. (1990) in *Diffusion in materials*, Laskar, A. L., Bocquet, J. L., Brébec, G., and Monty, C. (eds.) (Kluwer: Dordrecht) p. 359.
Mott, N. F., and Littleton, M. J. (1938) *Trans. Faraday Soc.* **34**, 485.
Mullen, J. G. (1961a) *Phys. Rev.* **121**, 1649.
(1961b) *Phys. Rev.* **124**, 1723.
(1982) in *Proc. Int. Conf. on the Applications of the Mössbauer Effect* (Indian National Academy of Sciences: New Delhi) p. 29.
(1984) in *Nontraditional methods in diffusion*, Murch, G. E., Birnbaum, H. K., and Cost, J. R. (eds.) (Met. Soc. AIME: New York) p. 59.
Müller, E. W. (1963) *J. Phys. Soc. Japan* **18**, Suppl. II, 1.
Murch, G. E. (1980) *Phil. Mag.* **A41**, 157.
(1982a) *Phil. Mag.* **A45**, 941.
(1982b) *Phil. Mag.* **A46**, 151.
(1982c) *Phil. Mag.* **A46**, 565.
(1982d) *Phil. Mag.* **A46**, 575.
(1982e) *J. Phys. Chem. Solids* **43**, 243.
(1982f) *Solid State Ionics* **7**, 177.
(1984a) in *Diffusion in crystalline solids*, Murch, G. E., and Nowick, A. S. (eds.) (Academic: New York) p. 379.
(1984b) *Phil. Mag.* **A49**, 21.
(1984c) *J. Phys. Chem. Solids* **45**, 451.
Murch, G. E., and Rothman, S. J. (1981) *Phil. Mag.* **A43**, 229.
Murch, G. E., and Zhang, L. (1990) in *Diffusion in materials*, Laskar, A. L., Bocquet, J. L., Brébec, G., and Monty, C. (eds.) (Kluwer: Dordrecht) p. 251.

Murphy, S. M. (1987) *Theory of compositional effects in radiation damage*, D.Phil. thesis, University of Oxford; also *UKAEA Harwell Laboratory Report* TP 1228.
(1988) *Fluctuations in composition in dilute alloys under irradiation, UKAEA Harwell Laboratory Report* AERE-R 13282.
(1989a) *Phil. Mag.* **A59**, 1163.
(1989b) *J. Nucl. Mater.* **168**, 31.
Murphy, S. M., and Perks, J. M. (1989) *Analysis of phosphorus segregation in ion-irradiated nickel, UKAEA Harwell Laboratory Report* AERE-R 13495.
Nabarro, F. R. N. (1967) *Theory of crystal dislocations* (Clarendon Press: Oxford).
(1979–) *Dislocations in solids* (North Holland: Amsterdam) 7 vols.
Nakazato, K., and Kitahara, K. (1980) *Prog. Theor. Phys.* **64**, 2261.
Newman, R. C. (1982) *Rep. Prog. Phys.* **45**, 1163.
Newman, R. C., and Jones, R. (1993) in *Oxygen in silicon*, Shimura, F. (ed.) (to be published as a volume of Semiconductors and Semimetals) Chap. 8.
Nolfi, F. V. (ed.) (1983) *Phase transformations during irradiation* (Applied Science Publishers: New York).
Norris, D. I. R. (ed.) (1987) *Radiation-induced sensitisation of stainless steels* (Central Electricity Generating Board: London).
Norvell, J. C., and Als-Nielson, J. (1970) *Phys. Rev.* **2**, 277.
Nowick, A. S. (1967) *Adv. Phys.* **16**, 1.
(1970a) *J. Chem. Phys.* **53**, 2066.
(1970b) *J. Phys. Chem. Solids* **31**, 1819.
Nowick, A. S., and Berry, B. S. (1972) *Anelastic relaxation in crystalline solids* (Academic: New York).
Nowick, A. S., and Heller, W. R. (1963) *Adv. Phys.* **12**, 251.
(1965) *Adv. Phys.* **14**, 101.
Nye, J. F. (1985) *Physical properties of crystals* (Clarendon Press: Oxford).
Okamura, Y., and Allnatt, A. R. (1983a) *Phil. Mag.* **A48**, 387.
(1983b) *J. Phys.* **C16**, 1841.
(1984) *Phil. Mag.* **A50**, 603.
(1985) *J. Phys.* **C18**, 4831.
(1986a) *Acta Metall.* **34**, 1189.
(1986b) *Phil. Mag.* **A54**, 773.
Onsager, L., and Fuoss, R. M. (1932) *J. Phys. Chem.* **36**, 2689.
Onsager, L., and Kim, S. K. (1957) *J. Phys. Chem.* **61**, 215.
Oppenheim, I., Shuler, K. E., and Weiss, G. H. (1977) *Stochastic processes in chemical physics: the master equation* (MIT Press: Cambridge, MA).
Osenbach, J. W., Bitler, W. R., and Stubican, V. S. (1981) *J. Phys. Chem. Solids* **42**, 599.
Peterson, N. L. (1975) in *Diffusion in solids – recent developments*, Nowick, A. S., and Burton, J. J. (eds.) (Academic: New York) p. 115.
(1978) *J. Nucl. Mater.* **69/70**, 3.
Peterson, N. L., Barr, L. W., and Le Claire, A. D. (1973) *J. Phys.* **C6**, 2020.
Peterson, N. L., and Rothman, S. J. (1970a) *Phys. Rev.* **B1**, 3264.
(1970b) *Phys. Rev.* **B2**, 1540.
Petry, W., Vogl, G., Heidemann, A., and Steinmetz, K.-H. (1987) *Phil. Mag.* **A55**, 183.
Philibert, J. (1985) *Diffusion et transport de matière dans les solides* (Editions de Physique: Les Ulis). English translation (1991) *Atom movements, diffusion and mass transport in solids.*
Pitts, E. (1953) *Proc. Roy. Soc.* **A217**, 43.
Pratt, J. N., and Sellors, R. G. R. (1973) *Electrotransport in metals and alloys* (Trans. Tech. SA: Riehen, Switzerland).

Pruthi, D. D., and Agarawala, R. P. (1982) *Phil. Mag.* **A46**, 841.
Radelaar, S. (1970) *J. Phys. Chem. Solids* **31**, 219.
Rahman, A., and Blackburn, D. A. (1976) *J. Physique* **37**, C7-359.
Rebonato, R. (1989) *Phil. Mag.* **B60**, 325.
Reif, F. (1955) *Phys. Rev.* **100**, 1597.
Resing, H. A., and Torrey, H. C. (1963) *Phys. Rev.* **131**, 1102.
Reynolds, J. E., Averbach, B. L., and Cohen, M. (1957) *Acta Metall.* **5**, 29.
Rice, S. A., and Gray, P. (1965) *The statistical mechanics of simple liquids* (Interscience: New York).
Richards, P. M. (1978) *J. Chem. Phys.* **68**, 2125.
Rothman, S. J. (1984) in *Diffusion in crystalline solids*, Murch, G. E., and Nowick, A. S. (eds.) (Academic: New York) p. 1.
Rothman, S. J., Peterson, N. L., Laskar, A. L., and Robinson, L. C. (1972) *J. Phys. Chem. Solids* **33**, 1061.
Rowe, J. M., Rush, J. J., de Graaf, L. A., and Ferguson, G. A. (1972) *Phys. Rev. Lett.* **29**, 1250.
Salibi, N., and Cotts, R. M. (1983) *Phys. Rev.* **B27**, 2625.
Salter, L. S. (1963) *Trans. Faraday Soc.* **59**, 657.
Samara, G. A. (1984) *Solid State Physics* **38**, 1.
Sato, H. (1970) in *Physical chemistry – an advanced treatise*, Vol. 10, Eyring, H., Henderson, D., and Jost, W. (eds.) (Academic: New York) p. 579.
(1984) in *Nontraditional methods in diffusion*, Murch, G. E., Birnbaum, H. K., and Cost, J. R. (eds.) (Met. Soc. AIME: New York) p. 203.
Sato, H., Ishikawa, T., and Kikuchi, R. (1985) *J. Phys. Chem. Solids* **46**, 1361.
Sato, H., and Kikuchi, R. (1983) *Phys. Rev.* **B28**, 648.
Scaife, B. K. P. (1989) *Principles of dielectrics* (Clarendon Press: Oxford).
Schapink, F. W. (1965) *Phil. Mag.* **12**, 1035.
Schmalzried, H. (1981) *Solid state reactions*, 2nd edn. (Verlag. Chemie: Weinheim).
(1986) *Reactivity of Solids* **1**, 117.
Schmalzried, H., and Laqua, W. (1981) *Oxidation of Metals* **15**, 339.
Schmalzried, H., Laqua, W., and Lin, P. L. (1979) *Zeits. für Naturforschung* **34A**, 192.
Schmalzried, H., and Pfeiffer, Th. (1986) *Z. Phys. Chem. Neue Folge* **148**, 21.
Schmidt, P. C., Sen, K. D., Das, T. P., and Weiss, A. I. (1980) *Phys. Rev.* **B22**, 4167.
Schnehage, M., Dieckmann, R., and Schmalzried, H. (1982) *Ber. Bunsenges. Phys. Chem.* **86**, 1061.
Schroeder, K. (1980) in *Point defects in metals II*, Dederichs, P. H., Schroeder, K., and Zeller, R. (eds.) (Springer-Verlag: Berlin) p. 171.
Schulze, H. A., and Lücke, K. (1968) *J. Appl. Phys.* **39**, 4861.
(1972) *Acta Metall.* **20**, 529.
Schulz, M. (ed.) (1989) Landolt–Börnstein *Numerical data and functional relationships in science and technology, new series, Group III, Vol. 22b.* Semiconductors: impurities and defects in Group IV elements and III–V compounds (Springer-Verlag: Berlin and Heidelberg).
Seidman, D. N. (1973) *J. Phys.* **F3**, 393.
Seitz, F. (1952) in *Imperfections in nearly perfect crystals*, Shockley, W., Holloman, J. H., Maurer, R., and Seitz, F. (eds.) (Wiley: New York) p. 3.
Sekerka, R. F., and Mullins, W. W. (1980) *J. Chem. Phys.* **73**, 1413 (1980).
Serruys, Y., and Brébec, G. (1982a) *Phil. Mag.* **A45**, 563.
(1982b) *Phil. Mag.* **A46**, 661.

Shewmon, P. G. (1960) *Acta Metall.* **8**, 605.
(1989) *Diffusion in solids*, 2nd edn. (Minerals, Metals and Materials Society: Warrendale, Pennsylvania).
Shockley, W., Holloman, J. H., Maurer, R., and Seitz, F. (eds.) (1952) *Imperfections in nearly perfect crystals* (Wiley: New York).
Sholl, C. A. (1974) *J. Phys.* **C7**, 3378.
(1975) *J. Phys.* **C8**, 1737.
(1981a) *J. Phys.* **C14**, 447.
(1981b) *J. Phys.* **C14**, 2723.
(1982) *J. Phys.* **C15**, 1177.
(1986) *J. Phys.* **C19**, 2547.
(1988) *J. Phys.* **C21**, 319.
(1992) *Phil. Mag.* **A65**, 749.
Siegel, R. W. (1982) in *Point defects and defect interactions in metals*, Takamura, J. I., Doyama, M., and Kiritani, M. (eds.) (University of Tokyo Press: Tokyo) p. 533.
Simmons, R. O., and Balluffi, R. W. (1960) *Phys. Rev.* **117**, 52.
(1962) *Phys. Rev.* **125**, 862.
(1963) *Phys. Rev.* **129**, 1523.
Singwi, K. S., and Sjölander, A. (1960) *Phys. Rev.* **120**, 1093.
Slichter, C. P. (1978) *Principles of magnetic resonance*, 2nd edn. (Springer-Verlag: Berlin).
Slifkin, L. (1989) in *Superionic solids and solid electrolytes – recent trends*, Laskar, A. L. and Chandra, S. (eds.) (Academic: San Diego) p. 407.
Smigelskas, A. D., and Kirkendall, E. O. (1947) *Trans. AIME* **171**, 130.
Smith, W., and Gillan, M. J. (1992) *J. Phys. Condensed Matter* **4**, 3215.
Squires, G. L. (1978) *Introduction to the theory of thermal neutron scattering* (Cambridge University Press).
Stark, J. P. (1974) *Acta Metall.* **22**, 533.
Stark, J. P., and Manning, J. R. (1974) *Phys. Rev.* **B9**, 425.
Stauffer, D. (1979) *Phys. Rep.* **54**, 1.
(1985) *Introduction to percolation theory* (Taylor and Francis: London).
Stokes, H. T. (1984) in *Nontraditional methods in diffusion*, Murch, G. E., Birnbaum, H. K., and Cost, J. R. (Met. Soc. AIME: Warrendale, Pennsylvania) p. 39.
Stolwijk, N. A. (1981) *Phys. Stat. Solidi (b)* **105**, 223.
Stoneham, A. M. (1975) *Theory of defects in solids* (Clarendon Press: Oxford).
Süptitz, P., and Weidmann, N. (1968) *Phys. Stat. Solidi* **27**, 631.
Symmons, H. F. (1970) *J. Phys.* **C3**, 1846.
(1971) *J. Phys.* **C4**, 1945.
Tahir-Kheli, R. A. (1983) *Phys. Rev.* **B28**, 3049.
Tahir-Kheli, R. A., and Elliott, R. J. (1983) *Phys. Rev.* **B27**, 844.
Tahir-Kheli, R. A., Elliott, R. J., and Holdsworth, P. C. W. (1986) *Phys. Rev.* **B34**, 3233.
Talbot, A. (1979) *J. Inst. Maths. Applications* **23**, 97.
Teltow, J. (1949) *Ann. Phys. Lpz.*, **5**, 63, 71.
Tewary, V. K. (1973) *Adv. Phys.* **22**, 757.
Thompson, M. W. (1969) *Defects and radiation damage in metals* (Cambridge University Press: London).
Thrower, P. A., and Mayer, R. M. (1978) *Phys. Stat. Sol. (b)* **47**, 11.
Tilley, R. J. D. (1987) *Defect crystal chemistry* (Blackie: Glasgow).
Tingdong, X. (1987) *J. Mater. Sci.* **22**, 337.
Tomlinson, S. M., Catlow, C. R. A., and Harding, J. H. (1990) *J. Phys. Chem. Solids* **51**, 477.

Torrey, H. C. (1953) *Phys. Rev.* **92**, 962.
(1954) *Phys. Rev.* **96**, 690.
Tsai, D., Bullough, R., and Perrin, J. C. (1970) *J. Phys.* **C3**, 2022.
Tuck, B. (1988) *Atomic diffusion in III–V semiconductors* (Adam Hilger: Bristol and Philadelphia).
Ullmaier, H. (ed.) (1991) Landolt–Börnstein. *Numerical data and functional relationships in science and technology, new series, Group III, Vol. 25.* Atomic defects in metals (Springer-Verlag: Berlin and Heidelberg).
Vineyard, G. H. (1957) *J. Phys. Chem. Solids* **3**, 121.
Vitek, V. (1985) in *Dislocations and properties of real materials* (Book 323, Institute of Metals: London) p. 30.
Völkl, J., and Alefeld, G. (1975) in *Diffusion in solids: recent developments*, Nowick, A. S., and Burton, J. J. (eds.) (Academic: New York) p. 231.
Wachtman, J. B. (1963) *Phys. Rev.* **131**, 517.
Walter, C. M., and Peterson, N. L. (1969) *Phys. Rev.* **178**, 922.
Warburton, W. K. (1973) *Phys. Rev.* **B7**, 1330.
Warburton, W. K. (1975) *Phys. Rev.* **B11**, 4945.
Warburton, W. K., and Turnbull, D. (1975) in *Diffusion in solids – recent developments*, Nowick, A. S., and Burton, J. J. (eds.) (Academic: New York) p. 171.
Watkins, G. D. (1959) *Phys. Rev.* **113**, 79, 91.
(1975) in *Point defects in solids*, Vol. 2, Crawford, J. H. and Slifkin, L. M. (eds.) (Plenum: New York and London) p. 333.
(1976) in *Defects and their structure in non-metallic solids*, Henderson, B., and Hughes, A. E. (eds.) (Plenum: New York and London) p. 203.
Weber, M. D., and Friauf, R. J. (1969) *J. Phys. Chem. Solids* **30**, 407.
Wever, H. (1973) *Elektro- und thermotransport in metallen* (Barth: Leipzig).
Whelan, M. J. (1975) in *The physics of metals*, Vol. 2, Hirsch, P. B. (ed.) (Cambridge University Press: London) p. 98.
Whitworth, R. W. (1975) *Adv. Phys.* **24**, 203.
Wiedersich, H., and Lam, N. Q. (1983) in *Phase transformations during irradiation*, Nolfi Jr., F. V. (ed.) (Applied Science Publishers: New York) p. 1.
Wiedersich, H., Okamoto, P. R., and Lam, N. Q. (1979) *J. Nucl. Mater.* **83**, 98.
Williams, T. M., Stoneham, A. M., and Harries, D. R. (1976) *Metal Sci.* **10**, 14.
Wilson, A. H. (1957) *Thermodynamics and statistical mechanics* (Cambridge University Press: Cambridge).
Wolf, D. (1971) *Zeits. für Naturforschung* **26a**, 1816.
(1974) *Phys. Rev.* **B10**, 2710.
(1977) *Phys. Rev.* **B15**, 37.
(1979) *Spin temperature and nuclear-spin relaxation in matter* (Clarendon Press: Oxford).
Wolf, D., Figueroa, D. R., and Strange, J. H. (1977) *Phys. Rev.* **B15**, 2545.
Wolfer, W. G. (1983) *J. Nucl. Mater.* **114**, 292.
Wuensch, B. J. (1983) in *Mass transport in solids*, Bénière, F. A., and Catlow, C. R. A. (eds.) (Plenum Press: New York) p. 353.
Yoo, H.-I., Schmalzried, H., Martin, M., and Janek, J. (1990) *Z. Phys. Chem. Neue Folge* **168**, 129.
Yoshida, M. (1971) *Japan J. Appl. Phys.* **10**, 702.
Yoshida, Y., Miekeley, W., Petry, W., Stehr, R., Steinmetz, K. H., and Vogl, G. (1987) *Mater. Sci. Forum* **15–18**, 487.
Yurek, G. J., and Schmalzried, H. (1975) *Ber. Bunsenges. Phys. Chem.* **79**, 255.
Zeqiri, B., and Chadwick, A. V. (1983) *Radiation Effects* **75**, 129.
Zhang, L., Oates, W. A., and Murch, G. E. (1988) *Phil. Mag.* **A58**, 937.
(1989) *Phil. Mag.* **B60**, 277.

# Index

Printed in the United States
By Bookmasters